HOLT
SOCIAL
STUDIES

Eastern World

Christopher L. Salter

HOLT, RINEHART AND WINSTON

A Harcourt Education Company

Orlando • **Austin** • New York • San Diego • London

Author

Dr. Christopher L. Salter

Dr. Christopher L. "Kit" Salter is Professor Emeritus of geography and former Chair of the Department of Geography at the University of Missouri. He did his undergraduate work at Oberlin College and received both his M.A. and Ph.D. degrees in geography from the University of California at Berkeley.

Dr. Salter is one of the country's leading figures in geography education. In the 1980s he helped found the national Geographic Alliance network to promote geography education in all 50 states. In the 1990s Dr. Salter was Co-Chair of the National Geography Standards Project, a group of distinguished geographers who created *Geography for Life* in 1994, the document outlining national standards in geography. In 1990 Dr. Salter received the National Geographic Society's first-ever Distinguished Geography Educator Award. In 1992 he received the George Miller Award for distinguished service in geography education from the National Council for Geographic Education.

Over the years, Dr. Salter has written or edited more than 150 articles and books on cultural geography, China, field work, and geography education. His primary interests lie in the study of the human and physical forces that create the cultural landscape, both nationally and globally.

Reviewers

Academic Reviewers

Elizabeth Chacko, Ph.D.
Department of Geography
The George Washington University

Altha J. Cravey, Ph.D.
Department of Geography
University of North Carolina

Eugene Cruz-Uribe, Ph.D.
Department of History
Northern Arizona University

Toyin Falola, Ph.D.
Department of History
University of Texas

Sandy Freitag, Ph.D.
Director, Monterey Bay History and
    Cultures Project
Division of Social Sciences
University of California,
    Santa Cruz

Oliver Froehling, Ph.D.
Department of Geography
University of Kentucky

Reuel Hanks, Ph.D.
Department of Geography
Oklahoma State University

Phil Klein, Ph.D.
Department of Geography
University of Northern Colorado

B. Ikubolajeh Logan, Ph.D.
Department of Geography
Pennsylvania State University

Marc Van De Mieroop, Ph.D.
Department of History
Columbia University
New York, New York

Christopher Merrett, Ph.D.
Department of History
Western Illinois University

Thomas R. Paradise, Ph.D.
Department of Geosciences
University of Arkansas

Jesse P.H. Poon, Ph.D.
Department of Geography
University at Buffalo–SUNY

Robert Schoch, Ph.D.
CGS Division of Natural Science
Boston University

Derek Shanahan, Ph.D.
Department of Geography
Millersville University
Millersville, Pennsylvania

David Shoenbrun, Ph.D.
Department of History
Northwestern University
Evanston, Illinois

Sean Terry, Ph.D.
Department of Interdisciplinary
    Studies, Geography and
    Environmental Studies
Drury University
Springfield, Missouri

Educational Reviewers

Dennis Neel Durbin
Dyersburg High School
Dyersburg, Tennessee

Carla Freel
Hoover Middle School
Merced, California

Tina Nelson
Deer Park Middle School
Randallstown, Maryland

Don Polston
Lebanon Middle School
Lebanon, Indiana

Robert Valdez
Pioneer Middle School
Tustin, California

Teacher Review Panel

Heather Green
LaVergne Middle School
LaVergne, Tennessee

John Griffin
Wilbur Middle School
Wichita, Kansas

Rosemary Hall
Derby Middle School
Birmingham, Michigan

Rose King
Yeatman-Liddell School
St. Louis, Missouri

Mary Liebl
Wichita Public Schools USD 259
Wichita, Kansas

Jennifer Smith
Lake Wood Middle School
Overland Park, Kansas

Melinda Stephani
Wake County Schools
Raleigh, North Carolina

Contents

Making This Book Work for You.................................xxviii
Scavenger Hunt...xxxii
Geography and Map Skills...H1
Reading Social Studies..H14
Social Studies Words..H18
Academic Words..H19
Standardized Test-Taking Strategies.............................H20

UNIT 1 Introduction to Geography

.... 1

CHAPTER 1 A Geographer's World

.... 2

Geography's Impact Video Series
Impact of Studying Geography

Section 1 Studying Geography...4
Section 2 Geography Themes and Essential Elements.........10
Social Studies Skills Analyzing Satellite Images...........................15
Section 3 The Branches of Geography................................16
Chapter Review.. 21
Standardized Test Practice... 23

CHAPTER 2 Planet Earth

............24

Geography's Impact Video Series
Impact of Water on Earth

Section 1 Earth and the Sun's Energy......................... 26
Section 2 Water on Earth....................................... 30
Section 3 The Land... 35
Case Study The Ring of Fire.. 42
Social Studies Skills Using a Physical Map......................... 44
Chapter Review 45
Standardized Test Practice.. 47

 CHAPTER 3 **Climate, Environment, and Resources** 48

Geography's Impact Video Series
Impact of Weather

Section 1 Weather and Climate 50
Section 2 World Climates 55
Section 3 Natural Environments 62
Geography and History Earth's Changing Environments 66
Section 4 Natural Resources 68
Social Studies Skills Analyzing a Bar Graph 74
Chapter Review 75
Standardized Test Practice 77

 CHAPTER 4 **The World's People** 78

Geography's Impact Video Series
Impact of Culture

Section 1 Culture 80
Section 2 Population 86
Section 3 Government and Economy 91
Social Studies Skills Organizing Information 96
Section 4 Global Connections 97
Chapter Review 101
Standardized Test Practice 103

Unit 1 Writing Workshop Explaining a Process 104

UNIT 2 Southwest and Central Asia 105

Regional Atlas 106
Facts about Countries 112

CHAPTER 5 **History of the Fertile Crescent, 7000–500 BC** 114

 Geography's Impact Video Series
Impact of a System of Laws

Section 1 Geography of the Fertile Crescent 116
Geography and History River Valley Civilizations 120
Section 2 The Rise of Sumer 122
Section 3 Sumerian Achievements 127
Section 4 Later Peoples of the Fertile Crescent 132
Social Studies Skills Sequencing and Using Time Lines 138
Chapter Review 139
Standardized Test Practice 141

CHAPTER 6 **Judaism and Christianity, 2000 BC–AD 1453** 142

 Geography's Impact Video Series
Impact of Location and Religion on Istanbul

Section 1 Origins of Judaism 144
Social Studies Skills Interpreting a Route Map 151
Section 2 Origins of Christianity 152
Section 3 The Byzantine Empire 160
Chapter Review 165
Standardized Test Practice 167

CHAPTER 7 **History of the Islamic World,** AD 550–1650 168

Geography's Impact Video Series
Impact of Mecca on Islam

Section 1 Origins of Islam 170

Section 2 Islamic Beliefs and Practices 174

Geography and History The Hajj 178

Section 3 Muslim Empires 180

Section 4 Cultural Achievements 186

Social Studies Skills Outlining 190

Chapter Review 191

Standardized Test Practice 193

CHAPTER 8 **The Eastern Mediterranean** 194

 Geography's Impact Video Series
Impact of Cooperation and Conflict in Jerusalem

Section 1 Physical Geography 196

Section 2 Turkey 200

Section 3 Israel 204

Social Studies Skills Analyzing a Cartogram 209

Section 4 Syria, Lebanon, and Jordan 210

Chapter Review 215

Standardized Test Practice 217

CHAPTER 9 **The Arabian Peninsula, Iraq, and Iran** 218

Geography's Impact Video Series
Impact of Oil

Section 1 Physical Geography 220

Section 2 The Arabian Peninsula 224

Case Study Oil in Saudi Arabia 228

Section 3 Iraq 230

Section 4 Iran 234

Social Studies Skills Analyzing Tables and Statistics 238

Chapter Review 239

Standardized Test Practice 241

CHAPTER 10 **Central Asia** 242

Geography's Impact Video Series
Impact of Progress in Afghanistan

Section 1 Physical Geography 244

Section 2 History and Culture 248

Section 3 Central Asia Today 253

Geography and History The Aral Sea 258

Social Studies Skills Using Scale 260

Chapter Review 261

Standardized Test Practice 263

Unit 2 Writing Workshop Compare and Contrast 264

ix

 **UNIT 3** **Africa** 265

Regional Atlas 266
Facts about Countries 272

CHAPTER 11 **History of Ancient Egypt,**
4500–500 BC 276

 **Geography's Impact Video Series**
Impact of the Egyptian Pyramids

Section 1 Geography and Early Egypt 278
Section 2 The Old Kingdom 283
Section 3 The Middle and New Kingdoms 291
Section 4 Egyptian Achievements 298
Social Studies Skills Analyzing Primary and Secondary Sources 304
Chapter Review...305
Standardized Test Practice 307

CHAPTER 12 **History of Ancient Kush,**
2300 BC–AD 350 308

 **Geography's Impact Video Series**
Impact of Iron

Section 1 Kush and Egypt 310
Section 2 Later Kush 315
Social Studies Skills Identifying Bias 320
Chapter Review...321
Standardized Test Practice 323

CHAPTER 13 # History of West Africa,
500 BC–AD 1650 324

Geography's Impact Video Series
Impact of the Salt Trade

Section 1 Empire of Ghana 326

Geography and History Crossing the Sahara. 332

Section 2 Mali and Songhai. 334

Section 3 Historical and Artistic Traditions. 340

Social Studies Skills Making Decisions 344

Chapter Review. 345

Standardized Test Practice 347

CHAPTER 14 # North Africa 348

Geography's Impact Video Series
Impact of the Nile River

Section 1 Physical Geography 350

Section 2 History and Culture. 354

Social Studies Skills Analyzing a Diagram 360

Section 3 North Africa Today. 361

Chapter Review. 367

Standardized Test Practice 369

CHAPTER 15 **West Africa** 370

Geography's Impact Video Series
Impact of Desertification

Section 1 Physical Geography 372

Section 2 History and Culture 376

Geography and History The Atlantic Slave Trade 380

Section 3 West Africa Today 382

Social Studies Skills Analyzing a Precipitation Map 388

Chapter Review 389

Standardized Test Practice 391

CHAPTER 16 **East Africa** 392

Geography's Impact Video Series
Impact of Climate Change on Mount Kilimanjaro

Section 1 Physical Geography 394
Section 2 History and Culture 398
Section 3 East Africa Today 402
Social Studies Skills Doing Fieldwork and Using Questionnaires 408
Chapter Review 409
Standardized Test Practice 411

CHAPTER 17 **Central Africa** 412

Geography's Impact Video Series
Impact of Preserving Central Africa's Forests

Section 1 Physical Geography 414
Case Study Mapping Central Africa's Forests 418
Section 2 History and Culture 420
Section 3 Central Africa Today 424
Social Studies Skills Interpreting a Population Pyramid 430
Chapter Review 431
Standardized Test Practice 433

CHAPTER 18 **Southern Africa** 434

Geography's Impact Video Series
Impact of Apartheid

Section 1 Physical Geography 436
Section 2 History and Culture 440
Section 3 Southern Africa Today 446
Social Studies Skills Evaluating a Website 452
Chapter Review 453
Standardized Test Practice 455

Unit 3 Writing Workshop Explaining Cause or Effect 456

 **UNIT 4**

South and East Asia and the Pacific 457

Regional Atlas 458
Facts about Countries 466

CHAPTER 19 **History of Ancient India,**
2300 BC–AD 500 470

Geography's Impact Video Series
Impact of Buddhism as a World Religion

Section 1 Early Indian Civilizations 472
Section 2 Origins of Hinduism 478
Section 3 Origins of Buddhism 484
Section 4 Indian Empires 490
Section 5 Indian Achievements 495
Social Studies Skills Comparing Maps 500
Chapter Review 501
Standardized Test Practice 503

CHAPTER 20 **History of Ancient China,**
1600 BC–AD 1450 504

Geography's Impact Video Series
Impact of Confucius on China Today

Section 1 Early China 506

Section 2 The Han Dynasty 510

Geography and History The Silk Road 516

Section 3 The Sui, Tang, and Song Dynasties 518

Section 4 Confucianism and Government 524

Section 5 The Yuan and Ming Dynasties 528

Social Studies Skills Making Economic Choices 536

Chapter Review 537

Standardized Test Practice 539

CHAPTER 21 **The Indian Subcontinent** 540

Geography's Impact Video Series
Impact of Population Density

Section 1 Physical Geography 542

Section 2 History and Culture of India 546

Social Studies Skills Analyzing a Line Graph 551

Section 3 India Today 552

Section 4 India's Neighbors 556

Chapter Review 561

Standardized Test Practice 563

CHAPTER 22 China, Mongolia, and Taiwan 564

Geography's Impact Video Series
Impact of the Three Gorges Dam

Section 1 Physical Geography 566
Section 2 History and Culture of China 570
Section 3 China Today 577
Section 4 Mongolia and Taiwan 582
Social Studies Skills Analyzing Points of View 586
Chapter Review 587
Standardized Test Practice 589

CHAPTER 23 Japan and the Koreas 590

Geography's Impact Video Series
Impact of Natural Hazards

Section 1 Physical Geography 592
Social Studies Skills Using a Topographic Map 596
Section 2 History and Culture 597
Section 3 Japan Today 602
Section 4 The Koreas Today 608
Chapter Review 613
Standardized Test Practice 615

CHAPTER 24 **Southeast Asia** 616

Geography's Impact Video Series
Impact of Biodiversity

Section 1 Physical Geography 618

Case Study Tsunami!.. 622

Section 2 History and Culture 624

Social Studies Skills Analyzing Visuals........................ 628

Section 3 Mainland Southeast Asia Today 629

Section 4 Island Southeast Asia Today 634

Chapter Review... 639

Standardized Test Practice 641

CHAPTER 25 **The Pacific World** 642

Geography's Impact Video Series
Impact of Nonnative Wildlife

Section 1 Australia and New Zealand 644

Geography and History Settling the Pacific 650

Social Studies Skills Locating Information 652

Section 2 The Pacific Islands 653

Section 3 Antarctica 658

Chapter Review... 663

Standardized Test Practice 665

Unit 4 Writing Workshop Persuasion...................... 666

Reference

Reading Social Studies 668

Economics Handbook 693

Facts about the World 698

Atlas 702

Gazetteer 722

Biographical Dictionary 730

English and Spanish Glossary 733

Index 745

Features

Case Study

Take a detailed look at important topics in geography.

The Ring of Fire 42
Oil in Saudi Arabia 228
Mapping Central Africa's Forests. 418
Tsunami! 622

Geography and History

Explore the connections between the world's places and the past.

Earth's Changing Environments 66
River Valley Civilizations. .120
The Hajj. 178
The Aral Sea 258
Crossing the Sahara 332
The Atlantic Slave Trade. .380
The Silk Road 516
Settling the Pacific. 650

FOCUS ON CULTURE

Learn about some of the world's fascinating cultures.

The Midnight Sun 29
The Tuareg of the Sahara. .58
Christian Holidays 154
Israeli Teens for Peace 208
Turkmen Carpets. 255
The Berbers 358
The Swahili. 400
Music of South Africa 443
The Sacred Ganges. 482
Chinese Martial Arts 576
Thai Teenage Buddhist Monks 627
Australian Sports. 648

Close-up

See how people live and what places look like by taking a close-up view of geography.

★ **Interactive** The Five Themes
of Geography 11
★ **Interactive** The Water Cycle 32
A Forest Ecosystem. 63
A Global Economy. 98
★ **Interactive** The City-State of Ur. 124
The Glory of Constantinople 162
Life in Arabia 171
The Blue Mosque 188
Early Farming Village. 200
Inside a Yurt. 250
Building the Pyramids 288
The Temple of Karnak. 300
★ **Interactive** Kush's Trade Network. 316
Rulers of Kush 318
Timbuktu. 336
A Sahara Oasis. 352
Serengeti National Park 402
Cape Town 448
Life in Mohenjo Daro 474
The Forbidden City 532
Diwali: The Festival of Lights 553
Beijing's National Day. 574
Life in Tokyo. 604
A Bangkok Canal. 631
Maori Culture. 646

Satellite View

See the world through satellite images and explore what these images reveal.

The World 1
True Color Satellite Image of Italy 15
Infrared Satellite Image of Italy 15
Southwest and Central Asia 105
Istanbul and the Bosporus 198
Pivot-Irrigated Fields 223
The Aral Sea 259
Africa 265
The Nile River 363
Great Rift Valley 396
Namib Desert 438
South and East Asia and the Pacific 457
Flooding in China 569
Antarctica's Ice Shelves 661

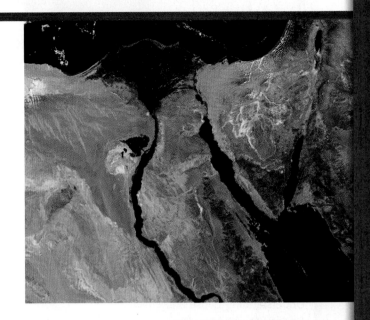

CONNECTING TO . . .

Explore the connections between geography and other subjects.

TECHNOLOGY
Computer Mapping 19
The Wheel 129
Building Small 603

SCIENCE
Soil Factory 64
The Formation of an Atoll 655

MATH
Calculating Population Density 88
Muslim Contributions to Math 225

THE ARTS
Music from Mali to Memphis 342
Masks 378

HISTORY
Bantu Languages 423

ECONOMICS
Tourism in Southern Africa 450
The Paper Trail 523
Bollywood 555

Social Studies Skills

Learn, practice, and apply the skills you need to study and analyze geography.

Analyzing Satellite Images 15
Using a Physical Map 44
Analyzing a Bar Graph 74
Organizing Information 96
Sequencing and Using Time Lines 138
Interpreting a Route Map 151
Outlining 190
Analyzing a Cartogram 209
Analyzing Tables and Statistics 238
Using Scale 260
Analyzing Primary and Secondary Sources 304
Identifying Bias 320
Making Decisions 344
Analyzing a Diagram 360
Analyzing a Precipitation Map 388
Doing Fieldwork and Using Questionnaires 408
Interpreting a Population Pyramid 430
Evaluating a Web Site 452
Comparing Maps 500
Making Economic Choices 536
Analyzing a Line Graph 551
Analyzing Points of View 586
Using a Topographic Map 596
Analyzing Visuals 628
Locating Information 652

Literature

Learn about the world's geography through literature.

The River 73
Red Brocade 214
Aké: The Years of Childhood 387
Shabanu: Daughter of the Wind 560
Antarctic Journal: Four Months at the Bottom
   of the World 662

Writing Workshop

Learn to write about geography.

Explaining a Process 104
Compare and Contrast 264
Explaining Cause or Effect .456
Persuasion 666

FOCUS ON READING

Learn and practice skills that will help you read your social studies lessons.

Using Prior Knowledge .668
Using Word Parts 669
Understanding Cause and Effect 670
Understanding Main Ideas 671
Paraphrasing 672
Understanding Implied Main Ideas 673
Sequencing 674
Setting a Purpose 675
Re-Reading 676
Using Context Clues 677
Categorizing 678
Asking Questions 679
Understanding Cause and Effect 680
Summarizing 681
Understanding Comparison-Contrast 682
Identifying Supporting Details 683
Using Word Parts 684
Making Generalizations 685
Sequencing 686
Understanding Chronological Order 687
Visualizing 688
Identifying Implied Main Ideas 689
Understanding Fact and Opinion 690
Using Context Clues–Definitions 691
Drawing Conclusions 692

FOCUS ON WRITING, SPEAKING, VIEWING

Use writing, viewing, and speaking skills to reflect on the world and its people.

Writing a Job Description 2
Writing a Haiku 24
Presenting and Viewing a Weather Report 48
Creating a Poster 78
Creating a Poster 114
Writing a Letter 142
Designing a Web Site 168
Writing a Description 194
Creating a Geographer's Log 218
Giving a Travel Presentation 242
Writing a Riddle 276
Writing a Fictional Narrative 308
A Journal Entry 324
Writing a Myth 348
Giving an Oral Description 370
Writing a Letter Home .392
Writing an Acrostic 412
Viewing a TV News Report 434
Creating a Poster 470
Writing a Magazine Article 504
Presenting and Viewing a Travelogue 540
Writing a Legend 564
Composing a Five-Line Poem 590
Presenting an Interview .616
Creating a Brochure 642

Primary Sources

Learn about the world through important documents and personal accounts.

Geography for Life 14

Robert Heinlein, on climate,
from *Time Enough for Love* 50

The Charter of the United Nations 100

Enheduanna, on the goddess Inanna,
from *Adoration of Inanna of Ur* 126

Sumerian Essay, on the importance of school,
quoted in *History Begins at Sumer* 128

Hammurabi's Code 133

The Sermon on the Mount 155

Exodus 20:12–14, on The Ten Commandments,
from *The Living Torah* 167

From *The Koran*, translated by N. J. Dawood 172

Ibn Battutah, on a caravan, from *The Travels* 193

The Dead Sea Scrolls 205

Pyramid Text, Utterance 217, on Re,
quoted in *Ancient Egypt*
by Lorna Oaks and Lucia Gahlin 290

Pen-ta-ur, on Ramses the Great, from *The Victory
of Ramses over the Khita*, in *The World's Story*,
edited by Eva March Tappan 297

On an Egyptian soldier, from *Wings of the Falcon:
Life and Thought of Ancient Egypt*,
translated by Joseph Kaster 304

Strabo, on Kush's unique culture,
from *Geography* 317

Strabo, on how Kushites live, from *Geography* 320

Strabo, on Nubia's location, from *Geography* 323

Al-Bakri, on the splendor of Ghana,
from *The Book of Routes and Kingdoms* 330

Basil Davidson, on Timbuktu,
from *A History of West Africa* 347

Kwame Nkrumah, from *I Speak of Freedom:
A Statement of African Ideology* 377

Biography of Mahommah G. Baquaqua 381

Ernest Hemingway, "The Snows of Kilimanjaro" . . . 411

Vedic hymn, on praising Indra,
in *Reading about the World, Volume I*,
edited by Paul Brians et al. 480

The Buddha, on morality, quoted in
The History of Nations: India 486

On warning listeners to think before they act,
from the *Panchatantra*,
translated by Arthur William Ryder 497

On avoiding anger, from the *Bhagavad Gita*,
translated by Barbara Stoler Miller 503

Li Bo, on being homesick,
from "Quiet Night Thoughts" 521

On Mongol destruction,
from "The Tale of the Destruction of Riazan,"
in *Medieval Russia's Epics, Chronicles,
and Tales*, edited by Serge Zenkovsky 529

Marco Polo, on a Chinese city,
from *Description of the World* 530

Lee Kuan Yew on Singapore 636

Will Steger, *Crossing Antarctica* 659

BIOGRAPHIES

Meet the people who have influenced the world and learn about their lives.

Eratosthenes 18

Alfred Wegener 37

Wangari Maathai 69

Sargon 123

Mehmed II 183

Kemal Atatürk 201

Shirin Ebadi 237

Queen Hatshepsut 292

Ramses the Great 297

Piankhi 313

Queen Shanakhdakheto 317

Tunka Manin 330

Askia the Great 337

Mansa Musa 339

Cleopatra 355

Nelson Mandela 447

Asoka 494

Emperor Shi Huangdi 508

Kublai Khan 535

Mohandas Gandhi 548

Hirohito 598

Aung San Suu Kyi 630

Sir Ernest Shackleton 660

Charts and Graphs

Quick Facts and Infographics

Geographic Dictionary H10
What Is Geography? 5
Looking at the World 6
The Geographer's Tools 8
The Five Themes of Geography 11
Geography 17
A Geographer's World 21
Solar Energy 27
The Seasons: Northern Hemisphere 28
Earth's Distribution of Water 30
The Water Cycle 32
Plate Movement 38
Planet Earth 45
Global Wind Systems 51
Rain Shadow Effect 54
Highland Climates 60
A Forest Ecosystem 63
Soil Layers 65
Climate, Environment, and Resources 75
A Global Economy 98
The World's People 101
Irrigation and Civilization 118
The City-State of Ur 124
Sumerian Achievements 130
History of the Fertile Crescent 139
Jewish Texts 148
The Glory of Constantinople 162
Judaism and Christianity 165
Life in Arabia 171
The Five Pillars of Islam 176
On the Road to Mecca 178
Islamic Achievements 186
The Blue Mosque 188
History of the Islamic World 191
Early Farming Village 200
People of Syria, Lebanon, and Jordan 212
The Eastern Mediterranean 215
The Arabian Peninsula, Iraq, and Iran 239
Influences on Central Asia 249
Inside a Yurt 250
Central Asia 261
Egyptian Society 284
Mummies and the Afterlife 286
Building the Pyramids 288
Egyptian Writing 299
The Temple of Karnak 300
History of Ancient Egypt 305

Kush's Trade Network 316
Rulers of Kush 318
History of Ancient Kush 321
Overgrazing 331
Timbuktu 336
History of West Africa 345
A Sahara Oasis 352
An Egyptian Pyramid 360
North Africa 367
A West African Village 379
West Africa 389
Lalibela, Ethiopia 399
Serengeti National Park 402
East Africa 409
Central Africa 431
Cape Town 448
Evaluating a Web Site 452
Southern Africa 453
Life in Mohenjo Daro 474
The Varnas 479
The Eightfold Path 487
Indian Science 498
History of Ancient India 501
Han Achievements 514
Chinese Inventions 522
Difficult Exams 526
The Voyages of Zheng He 531
The Forbidden City 532
History of Ancient China 537
India's History 546
Diwali: The Festival of Lights 553
The Indian Subcontinent 561
China's Early Dynasties 570
Beijing's National Day 574
Tensions between China and Taiwan 584
China, Mongolia, and Taiwan 587
Life in Tokyo 604
Japan and the Koreas 613
Tsunami! 622
Angkor Wat 625
City Life in Southeast Asia 628
A Bangkok Canal 631
Southeast Asia 639
Maori Culture 646
Settling the Pacific 650
The Pacific World 663

Charts and Graphs

Use charts and graphs to analyze geographic information.

Percentage of Students on High School
    Soccer Teams by Region 9
The Essential Elements
    and Geography Standards 13
Major Eruptions in the Ring of Fire 42
World Climate Regions 56
Climate Graph for Nice, France 59
World Energy Production Today 70
Average Annual Precipitation by Climate Region . . . 74
Top Five Aluminum Producers, 2000 77
Irish Migration to the United States, 1845–1855 . . . 89
World Population Growth, 1500–2000 90
Economic Activity 93
A Developed and a Developing Country 94
Geographical Extremes:
    Southwest and Central Asia 107
Southwest and Central Asia 112
World Oil Reserves 113
Largest Oil Reserves by Country 113
Development of Writing 128
Time Line: Major Events in the Fertile Crescent . . . 138
Basic Jewish Beliefs 147
The Western Roman and Byzantine Empires 164
Time Line: Beginnings of Islam 172
Sources of Islamic Beliefs 177
Origin of Israel's Jewish Population 206
Cartogram: Southwest and Central Asia 209
Saudi Arabia's Oil Production 228
Saudi Arabia's Exports 229
Life in Iran and the United States 236
Literacy Rates in Southwest Asia 238
Major Oil Producers 241
Reforms in Afghanistan 254
Standard of Living in Central Asia 256
Geographical Extremes: Africa 267
Africa 272
Africa's Growing Population 275
Africa and the World 275
Time Line: Periods of Egyptian History 291
Egypt's Population, 2003 362
Africa's Largest Cities 383
Population Density in East Africa 404
Major Religions of Central Africa 422
Kinshasa's Growing Population 425
Population Pyramid: Angola, 2000 430

Population Pyramid: Central African
    Republic, 2000 433
Tourism in Southern Africa 450
Geographical Extremes: South and East Asia 459
Geographical Extremes: The Pacific World 465
South and East Asia and the Pacific 466
World's Largest Populations 469
Percent of World Population 469
Economic Powers: Japan and China 469
Major Beliefs of Hinduism 480
Time Line: The Han Dynasty 510
Average Monthly Precipitation,
    Dhaka, Bangladesh 551
Religions of the Indian Subcontinent 557
Indian Film Production, 1960–2000 562
Population Growth in Indian Subcontinent 563
China's Projected Urban Population 579
Population Growth in Japan 606
Per Capita GDP in Island Southeast Asia
    and the United States 638
Ethnic Groups in Indonesia, 2005 641
Ethnic Groups in Australia and New Zealand 649
Earth Facts 698
World Population 700
Developed and Less Developed Countries 700
World Religions 701
World Languages 701

World Religions

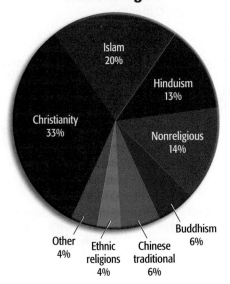

Christianity 33%
Islam 20%
Hinduism 13%
Nonreligious 14%
Buddhism 6%
Chinese traditional 6%
Ethnic religions 4%
Other 4%

The **World Almanac Education Group** is America's largest-selling reference book of all time, with more than 80 million copies sold since 1868.

FACTS ABOUT THE WORLD

Study the latest facts and figures about the world.

Eruptions in the Ring of Fire 42
World Energy Production Today 70
Geographical Extremes:
    Southwest and Central Asia 107
World Oil Reserves 113
Geographical Extremes: Africa 267
Africa's Growing Population 275
Africa and the World 275
Africa's Largest Cities 383
Major Religions of Central Africa 422
Geographical Extremes:
    South and East Asia 459
Geographical Extremes:
    The Pacific World 465
Religions of the Indian Subcontinent 557

FACTS ABOUT COUNTRIES

Study the latest facts and figures about countries.

A Developed and a Developing Country 94
Southwest and Central Asia 112
Largest Oil Reserves by Country 113
Origin of Israel's Jewish Population 206
Saudi Arabia's Oil Production 228
Saudi Arabia's Exports 229
Standard of Living in Central Asia 256
Africa 272
Egypt's Population, 2003 362
Population Density in East Africa 404
Kinshasa's Growing Population 425
Tourism in Southern Africa 450
South and East Asia and the Pacific 466
World's Largest Populations 469
Percent of World Population 469
Economic Powers: Japan and China 469
China's Projected Urban Population 579
Population Growth in Japan 606
Per Capita GDP in Island Southeast Asia
    and the United States 638
Ethnic Groups in Australia and New Zealand . . 649

Economic Powers

Japan

- World's third-largest economy
- $538.8 billion in exports
- Per capita GDP of $29,400
- Major exports: transportation equipment, cars, semiconductors, electronics

Japan is one of the most technologically advanced countries and is a leading producer of hi-tech goods.

China

- World's second-largest economy
- $583.1 billion in exports
- GDP growth rate of 9.1%
- Major exports: machinery and electronics, clothing, plastics, furniture, toys

China is an emerging economic powerhouse with a huge population and a fast growing economy.

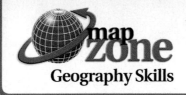

map Zone
Geography Skills

Maps

Northern Hemisphere. H3
Southern Hemisphere H3
Western Hemisphere H3
Eastern Hemisphere. H3
Mercator Projection. H4
Conic Projection H5
Flat-plane Projection. .H5
The First Crusade, 1096 . . . . .H6
Caribbean South America: Political H8
The Indian Subcontinent: Physical. H9
West Africa: Climate H9
High School Soccer Participation 8
Sketch Map. 22
The United States 23
Earth's Plates 36
Ring of Fire. 42
India: Physical 44
The World's Major Ocean Currents 52
Mediterranean Climate. 57
Arab Culture Region 82
Cultural Diffusion of Baseball 84
World Governments 92
Southwest and Central Asia: Physical 106
Size Comparison: The United States
    and Southwest and Central Asia. 107
Southwest and Central Asia: Political 108
Southwest and Central Asia: Resources 109
Southwest and Central Asia: Population. 110
Southwest and Central Asia: Climate 111
The Fertile Crescent, 7000–500 BC 115
The Fertile Crescent 117
River Valley Civilizations. 120
Sargon's Empire, c. 2330 BC. 123
Babylonian and Assyrian Empires 134
Phoenicia, c. 800 BC. .136
Mesopotamia. 141
The Jewish and Christian Worlds,
    2000 BC–AD 1453. 143
Possible Routes of the Exodus 151
Judea. 153
Paul's Journeys. 157
The Spread of Christianity, 300–400 158
The Byzantine Empire, 1025 161
Constantinople. 161
The Islamic World, AD 550–1650. 169
The Hajj. 179

The Ottoman Empire .183
The Mughal Empire. 185
Turkey: Population. 202
Israel and the Palestinian Territories 207
Southwest and Central Asia 209
Syria, Lebanon, and Jordan 212
Turkey: Physical Geography. 217
Saudi Arabia's Oil Fields. 228
Mesopotamia and Sumer 231
The Aral Sea 259
Kyrgyzstan. 260
Bishkek 260
Farmland in Central Asia 263
Africa: Physical 266
Size Comparison: The United States and Africa . . .267
Africa: Political 268
Africa: Resources 269
Africa: Population 270
Africa: Climate 271
Ancient Egypt, 4500–500 BC 277
Ancient Egypt 279
Egyptian Trade, c. 1400 BC 293
Ancient Kush, 2300 BC–AD 350. 309
Ancient Kush 311
West Africa, 500 BC–AD 1650 325
Egypt: Population 362
North Africa 369
The Atlantic Slave Trade. 380
West Africa: Precipitation. 388
West Africa: Population 391
Central Africa's National Parks 416
Michael Fay's Route 418
Madagascar: Climate .455
South and East Asia: Physical 458
Size Comparison:
    The United States and South and East Asia 459
South and East Asia: Political 460
South and East Asia: Population 461
South and East Asia: Climate 462
South and East Asia: Land Use and Resources 463
The Pacific World: Physical 464
Antarctica 465
Ancient India, 2300 BC–AD 500. 471
Harappan Civilization, c. 2600–1900 BC 473
Aryan Migrations 477
Mauryan Empire, c. 320–185 BC 491

Maps (continued)

Gupta Empire, c. 400 492
India: Physical .500
Ancient China, 1600 BC–AD 1450 505
The Silk Road 516
The Grand Canal 520
Mongol Empire, 1294 529
India: Population 554
China: Population 572
China's Environmental Challenges 580
China, Mongolia, and Taiwan: Precipitation 589
Awaji Island: Topographic Map 596
Japan: Population 606
The Demilitarized Zone 609
Japan and the Koreas 615
Indian Ocean Tsunami 623
Settling the Pacific 651
The Pacific Islands: Political 656
Australia and New Zealand: Climate 665

United States: Physical 702
United States: Political 704
World: Physical 706
World: Political 708
North America: Physical 710
North America: Political 711
South America: Physical 712
South America: Political 713
Europe: Physical 714
Europe: Political 715
Asia: Physical 716
Asia: Political 717
Africa: Physical 718
Africa: Political 719
The Pacific: Political 720
The North Pole 721
The South Pole 721

✳ Interactive Maps

Geography Skills
With map zone geography skills, you can go online to find interactive versions of the key maps in this book. Explore these interactive maps to learn and practice important map skills and bring geography to life.

To use map zone interactive maps online:
1. Go to go.hrw.com.
2. Enter the KEYWORD shown on the interactive map.
3. Press return!

Interactive Maps

Map Activity: Using a Physical Map 46
World Climate Regions 56
Map Activity: Prevailing Winds 76
World Population Density 87
Map Activity: Population Density 102
Map Activity: The Fertile Crescent 140
Jewish Migration after AD 70 146
Map Activity: The Jewish and Christian Worlds 166
The Safavid Empire 184
Map Activity: The Islamic World 192
The Eastern Mediterranean: Political 195
The Eastern Mediterranean: Physical 197
The Eastern Mediterranean: Climate 199
Map Activity: The Eastern Mediterranean 216
The Arabian Peninsula, Iraq, and Iran: Political . . . 219
The Arabian Peninsula, Iraq, and Iran: Physical . . . 221
The Arabian Peninsula, Iraq, and Iran: Climate . . . 222
Map Activity: The Arabian Peninsula,
    Iraq, and Iran 240
Central Asia: Political 243
Central Asia: Physical 245
Central Asia: Land Use and Resources 246
Languages of Central Asia 251
Map Activity: Central Asia 262
Map Activity: Ancient Egypt 306
Map Activity: Ancient Kush 322
Ghana Empire, c. 1050 327

Mali and Songhai 335
Map Activity: West Africa 346
North Africa: Political 349
North Africa: Physical 351
North Africa: Agriculture 364
Map Activity: North Africa 368
West Africa: Political 371
West Africa: Physical 373
West Africa: Climate 374
West Africa: Land Use and Resources 384
Map Activity: West Africa 390
East Africa: Political 393
East Africa: Physical 395
Official Languages of East Africa 401
East Africa: Population 404
Map Activity: East Africa 410
Central Africa: Political 413
Central Africa: Physical 415
Malaria in Central Africa 428
Map Activity: Central Africa 432
Southern Africa: Political 435
Southern Africa: Physical 437
Southern Africa: Vegetation 439
Map Activity: Southern Africa 454
Early Spread of Buddhism 488
Map Activity: Ancient India 502
Early Dynasties of China 507
Han Dynasty, c. 206 BC–AD 220 511
Chinese Dynasties, 589–1279 519
Map Activity: Ancient China 538
Indian Subcontinent: Political 541
Indian Subcontinent: Physical 543
Indian Subcontinent: Precipitation 544
Religions of the Indian Subcontinent 557
Map Activity: The Indian Subcontinent 562
China, Mongolia, and Taiwan: Political 565
China, Mongolia, and Taiwan: Physical 567
China, Mongolia, and Taiwan: Precipitation 568
Map Activity: China, Mongolia, and Taiwan 588
Japan and the Koreas: Political 591
Japan and the Koreas: Physical 593
Japan and the Koreas:
    Volcanoes and Earthquakes 594
Map Activity: Japan and the Koreas 614
Southeast Asia: Political 617
Southeast Asia: Physical 619
Southeast Asia: Climate 620
Southeast Asia: Colonial Possessions, 1914 626
Southeast Asia: Land Use and Resources 632
Map Activity: Southeast Asia 640
The Pacific World: Political 643
Australia and New Zealand: Physical 645
Map Activity: The Pacific World 664

Making This Book Work for You

Studying geography will be easy for you with this textbook. Take a few minutes now to become familiar with the easy-to-use structure and special features of your book. See how it will make geography come alive for you!

Unit

Each unit begins with a satellite image, a regional atlas, and a table with facts about each country. Use these pages to get an overview of the region you will study.

Regional Atlas

The maps in the regional atlas show some of the key physical and human features of the region.

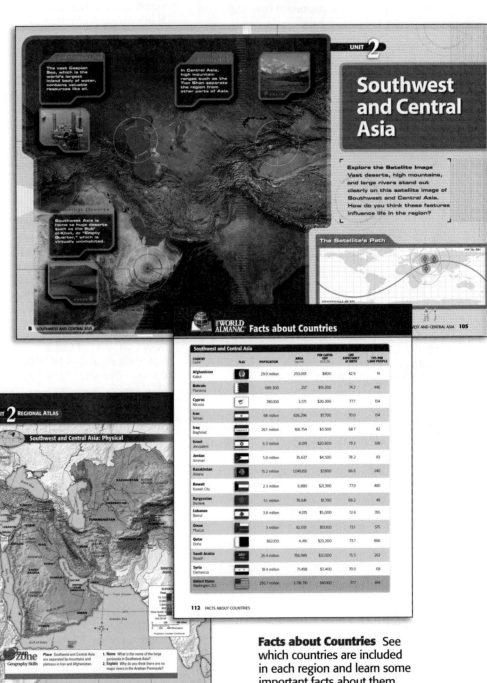

Facts about Countries See which countries are included in each region and learn some important facts about them with these helpful tables.

Chapter

Each regional chapter begins with a preview of what you will learn and a map of the region. Special instruction is also given in reading and skills.

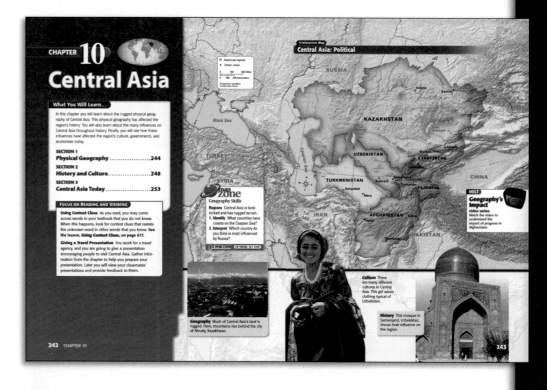

Reading Social Studies

Chapter reading lessons give you skills and practice to help you read the textbook. More help with each lesson can be found in the back of the book. Margin notes and questions in the chapter make sure you understand the reading skill.

Social Studies Skills The Social Studies Skills lessons give you an opportunity to learn, practice, and apply an important skill. Chapter Review questions then follow up on what you learned.

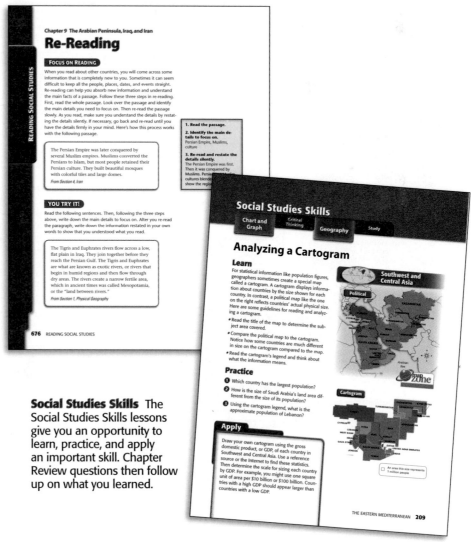

Section

The section opener pages include Main Ideas, an overarching Big Idea, and Key Terms and Places. In addition, each section includes these special features:

If YOU Lived There . . . Each section begins with a situation for you to respond to, placing you in a place that relates to the content you will be studying in the section.

Building Background Building Background connects what will be covered in each section with what you already know.

Short Sections of Content The information in each section is organized into small chunks of text that you can easily understand.

Taking Notes Suggested graphic organizers help you read and take notes on the important ideas in the section.

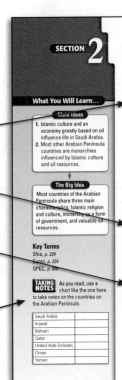

SECTION 2

The Arabian Peninsula

What You Will Learn...

Main Ideas

1. Islamic culture and an economy greatly based on oil influence life in Saudi Arabia.
2. Most other Arabian Peninsula countries are monarchies influenced by Islamic culture and oil resources.

The Big Idea

Most countries of the Arabian Peninsula share three main characteristics: Islamic religion and culture, monarchy as a form of government, and valuable oil resources.

Key Terms

Shia, p. 224
Sunni, p. 224
OPEC, p. 229

TAKING NOTES As you read, use a chart like the one here to take notes on the countries on the Arabian Peninsula.

Saudi Arabia	
Kuwait	
Bahrain	
Qatar	
United Arab Emirates	
Oman	
Yemen	

If YOU lived there...

You are a financial adviser to the ruler of Oman. Your country has been making quite a bit of money from oil exports. However, you worry that your economy is too dependent on oil. You think Oman's leaders should consider expanding the economy. Oman is a small country, but it has beautiful beaches, historic palaces and mosques, and colorful markets.

How would you suggest expanding the economy?

BUILDING BACKGROUND Oman and all the countries of the Arabian Peninsula have valuable oil resources. In addition to oil, these countries share two basic characteristics: Islamic religion and monarchy as a form of government. The largest country, and the one with the most influence in the region, is Saudi Arabia.

Saudi Arabia

Saudi Arabia is by far the largest of the countries of the Arabian Peninsula. It is also a major religious and cultural center and has one of the region's strongest economies.

People and Customs

Nearly all Saudis are Arabs and speak Arabic. Their culture is strongly influenced by Islam, a religion founded in Saudi Arabia by Muhammad. Islam is based on submitting to God and on messages Muslims believe God gave to Muhammad. These messages are written in the Qur'an, the holy book of Islam.

Nearly all Saudis follow one of two main branches of Islam. **Shia** Muslims believe that true interpretation of Islamic teaching can only come from certain religious and political leaders called imams. **Sunni** Muslims believe in the ability of the majority of the community to interpret Islamic teachings. About 85 percent of Saudi Muslims are Sunni.

With about 9 million people, Istanbul, shown here, is Turkey's largest city.

electronics. About 40 percent of Turkey's labor force works in agriculture. Grains, cotton, sugar beets, and hazelnuts are major crops.

Turkey is rich in natural resources, which include oil, coal, and iron ore. Water is also a valuable resource in the region. Turkey has spent billions of dollars building dams to increase its water supply. On one hand, these dams provide hydroelectricity. On the other hand, some of these dams have restricted the flow of river water into neighboring countries.

READING CHECK Finding Main Ideas What kind of government does Turkey have?

SUMMARY AND PREVIEW In this section you learned about Turkey's history, people, government, and economy. Next, you will learn about Israel.

Turkey Today

Turkey's government meets in the capital of Ankara, but Istanbul is Turkey's largest city. Istanbul's location will serve as an economic bridge to Europe as Turkey plans to join the European Union.

Government

Turkey's legislature is called the National Assembly. A president and a prime minister share executive power.

Although most of its people are Muslim, Turkey is a secular state. Secular means that religion is kept separate from government. For example, the religion of Islam allows a man to have up to four wives. However, by Turkish law a man is permitted to have just one wife. In recent years Islamic political parties have attempted to increase Islam's role in Turkish society.

Economy and Resources

As a member of the European Union, Turkey's economy and people would benefit by increased trade with Europe. Turkey's economy includes modern factories as well as village farming and craft making.

Among the most important industries are textiles and clothing, cement, and

Section 2 Assessment

go.hrw.com
Online Quiz
KEYWORD: SK7 HP8

Reviewing Ideas, Terms, and Places

1. a. **Recall** What city did both the Romans and Ottoman Turks capture?
   b. **Explain** In what ways did Atatürk try to modernize Turkey?
2. a. **Recall** What ethnic group makes up 20 percent of Turkey's population?
   b. **Draw Conclusions** What makes Turkey **secular**?
   c. **Elaborate** Why do you think Turkey wants to be a member of the European Union?

Critical Thinking

3. **Summarizing** Using the information in your notes, summarize Turkey's history and Turkey today.

Turkey's History	Turkey Today

FOCUS ON WRITING

4. **Describing Turkey** A description of Turkey might include details about its people, culture, government, and economy. Take notes on the details you think are important and interesting.

Reading Check Questions end each section of content so you can check to make sure you understand what you just studied.

Summary and Preview The Summary and Preview connects what you studied in the section to what you will study in the next section.

Section Assessment Finally, the section assessment boxes make sure that you understand the main ideas of the section. We also provide assessment practice online!

Features

Your book includes many features that will help you learn about geography, such as Close-up and Satellite View.

Satellite View
See and explore the world through satellite images.

Serengeti National Park

The Serengeti Plain is home to one of the world's greatest concentrations of wildlife. In Tanzania, part of the plain is a national park. About 100,000 tourists visit the Serengeti each year to view its diverse wildlife.

Tanzanian guides take visitors on a safari to view Serengeti's wildlife.

Huge herds of wildebeest migrate across the Serengeti each year.

ANALYSIS SKILL ANALYZING VISUALS
How would you describe the Serengeti landscape?

Watering holes attract wildlife, which includes flamingos, hippos, and giraffes.

Close-up These features help you see how people live and what places look like around the world.

The Nile River

From space, the Nile looks like a river of green. Actually, the areas that appear green in this satellite image are thousands of irrigated fields that line the banks of the river. The river deposits silt along its banks, which makes the land extremely fertile. Farmers also depend on the Nile's waters to irrigate their crops. Without water, they could not farm in the desert.

Notice how the river appears smaller at the bottom of this image. The Aswan High Dam controls the river's flow here, which prevents flooding and provides electricity.

Drawing Conclusions How is the Nile important to Egypt's people?

Chapter Review

At the end of each chapter, the Chapter Review will help you review key concepts, analyze information critically, complete activities, and show what you have learned.

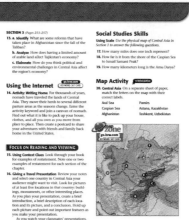

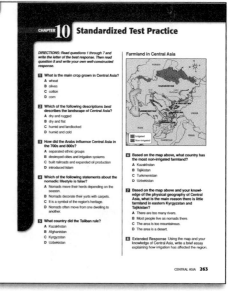

Standardized Test Practice Practice for standardized tests with the last page of each chapter before moving on to another region of the world!

Scavenger Hunt

Are you ready to explore the world of geography? *Holt Social Studies: Eastern World* is your ticket to this exciting world. Before you begin your journey, complete this scavenger hunt to get to know your book and discover what's inside.

On a separate sheet of paper, fill in the blanks to complete each sentence below. In each answer, one letter will be in a yellow box. When you have answered every question, copy these letters in order to reveal the answer to the question at the bottom of the page.

1 According to the Table of Contents, the title of Chapter 6 is ☐☐☐☐☐☐☐ and Christianity. What else does the Table of Contents show?

2 The two main ideas listed on page 552 explain what you will learn in that section. The last word of the first of these main ideas is ☐☐☐☐☐☐☐.

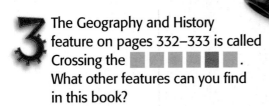

3 The Geography and History feature on pages 332–333 is called Crossing the ☐☐☐☐☐☐. What other features can you find in this book?

4 The Case Study feature on pages 42–43 is called The Ring of ☐☐☐☐. What is this feature about?

5 The second term in the English and Spanish Glossary is absolute ☐☐☐☐☐☐☐☐. How will use this glossary?

6 Look up Kabul in the Gazetteer. According to the entry, it is located in the country of ☐☐☐☐☐☐☐☐☐☐☐.

7 Page 693 is the beginning of the ☐☐☐☐☐☐☐☐ Handbook. What will you find in this section of the book?

Fact!

The oldest known permanent settlement in the world is in Southwest Asia. What is it called?

☐☐☐☐☐☐☐

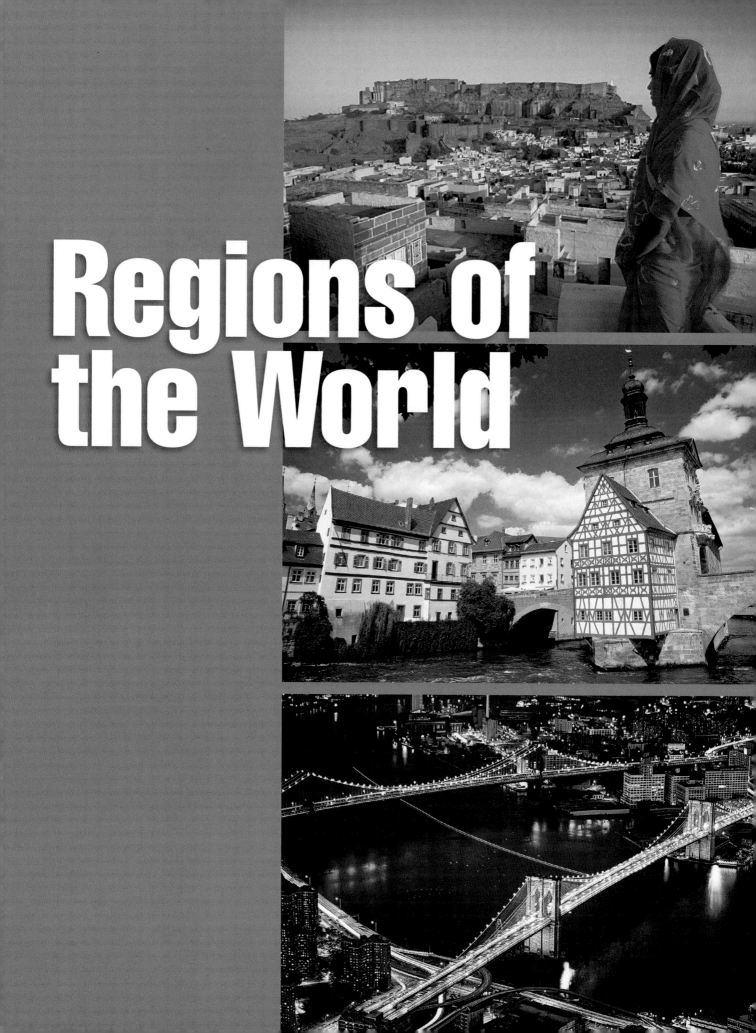

Regions of the World

Regions of the World

Geographers divide the world into regions for study. Each region has something about it that makes it unique and different from other regions. The map on the next page shows the major regions of the world. Explore this map to begin your study of geography.

How to Use the Map

The map on the next page is a special kind of map. It has transparent overlays that show different features and regions of the world. You can look at each overlay separately, or you can look at them together to see how they are connected. Just follow the steps below.

❶ **The Base Map** Start by lifting up all the transparent overlays and looking at only the base map. It shows the world's major oceans and seven continents, or large landmasses. What are the names of these continents? Where is each one located? Which oceans border each continent?

❷ **The First Overlay** Cover the base map with the first transparent overlay. It shows some of the world's major physical features, like rivers and mountains. First, study the rivers. Which rivers are shown? On which continents are they located? Next, look at the mountains. What mountain ranges can you see, and where are they?

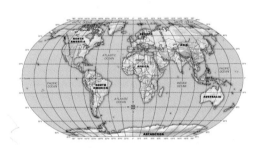

❸ **The Second Overlay** Now cover the base map and first overlay with the second overlay. It shows the major regions of the world. The name of each region is listed at the bottom. Now, put it all together. What are the five regions shown? Which continents do they include, and where are they located? What are some major mountains and rivers found in each region? What oceans surround them? Finally, where are these major world regions located in relation to one another?

Regions of the World

Regions of the World

Geographers divide the world into regions for study. Each region has something about it that makes it unique and different from other regions. The map on the next page shows the major regions of the world. Explore this map to begin your study of geography.

How to Use the Map

The map on the next page is a special kind of map. It has transparent overlays that show different features and regions of the world. You can look at each overlay separately, or you can look at them together to see how they are connected. Just follow the steps below.

❶ **The Base Map** Start by lifting up all the transparent overlays and looking at only the base map. It shows the world's major oceans and seven continents, or large landmasses. What are the names of these continents? Where is each one located? Which oceans border each continent?

❷ **The First Overlay** Cover the base map with the first transparent overlay. It shows some of the world's major physical features, like rivers and mountains. First, study the rivers. Which rivers are shown? On which continents are they located? Next, look at the mountains. What mountain ranges can you see, and where are they?

❸ **The Second Overlay** Now cover the base map and first overlay with the second overlay. It shows the major regions of the world. The name of each region is listed at the bottom. Now, put it all together. What are the five regions shown? Which continents do they include, and where are they located? What are some major mountains and rivers found in each region? What oceans surround them? Finally, where are these major world regions located in relation to one another?

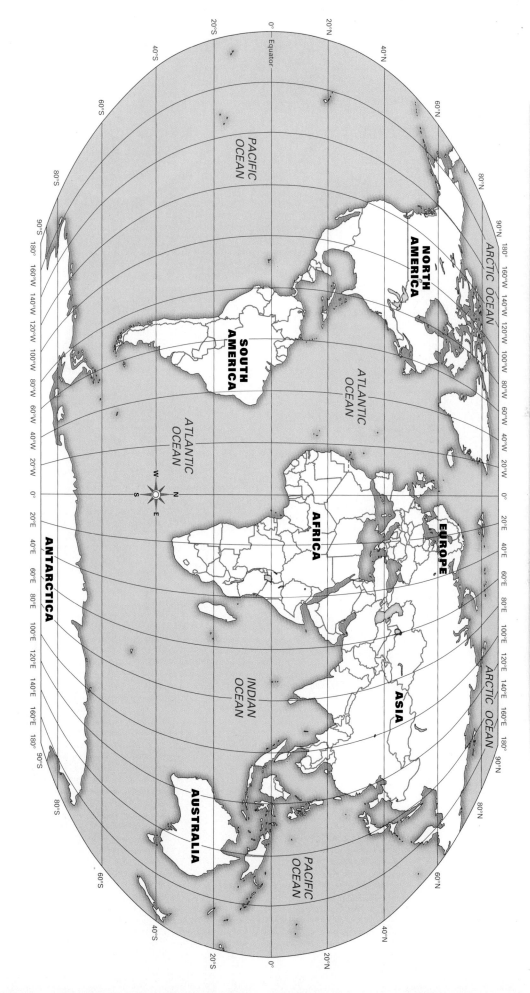

Regions of the World

PACIFIC OCEAN

NORTH AMERICA

SOUTH AMERICA

ATLANTIC OCEAN

ATLANTIC OCEAN

AFRICA

EUROPE

ASIA

ANTARCTICA

INDIAN OCEAN

AUSTRALIA

PACIFIC OCEAN

ARCTIC OCEAN

ARCTIC OCEAN

Equator

0   1,000   2,000 Miles

0   1,000   2,000 Kilometers

Projection: Azimuthal Equal Area

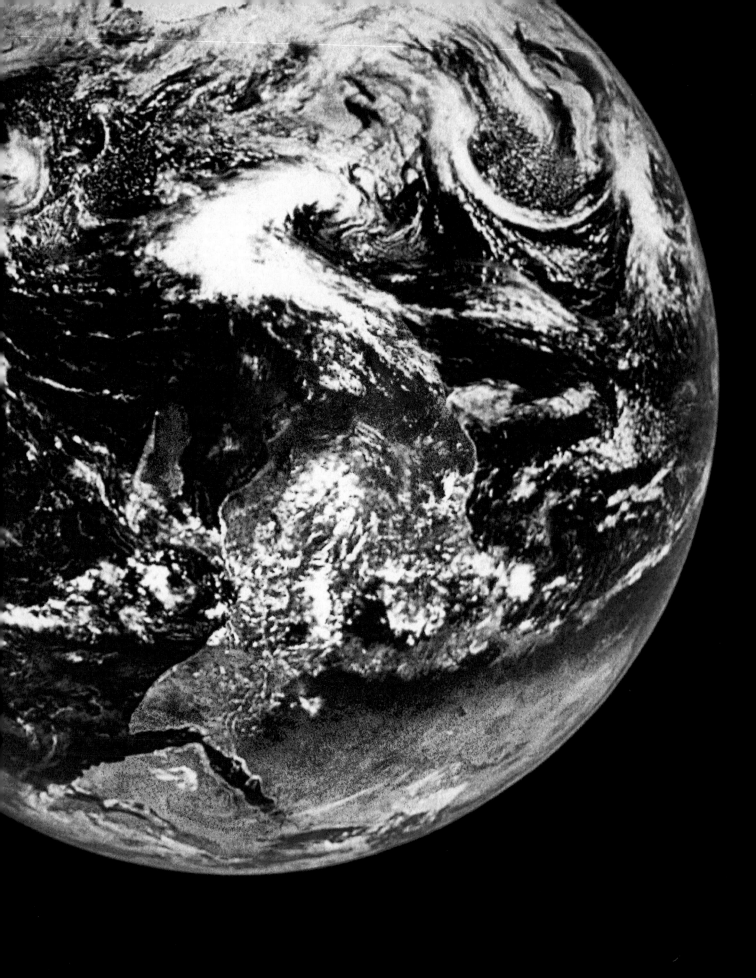

Geography and Map Skills Handbook

Contents

Mapping the Earth H2

Mapmaking H4

Map Essentials H6

Working with Maps H8

Geographic Dictionary H10

Themes and Essential Elements of Geography H12

Throughout this textbook, you will be studying the world's people, places, and landscapes. One of the main tools you will use is the map—the primary tool of geographers. To help you begin your studies, this Geography and Map Skills Handbook explains some of the basic features of maps. For example, it explains how maps are made, how to read them, and how they can show the round surface of Earth on a flat piece of paper. This handbook will also introduce you to some of the types of maps you will study later in this book. In addition, you will learn about the different kinds of features on Earth and about how geographers use themes and elements to study the world.

✳Interactive Maps

Geography Skills With map zone geography skills, you can go online to find interactive versions of the key maps in this book. Explore these interactive maps to learn and practice important map skills and bring geography to life.

To use map zone interactive maps online:

1. Go to go.hrw.com.
2. Enter the KEYWORD shown on the interactive map.
3. Press return!

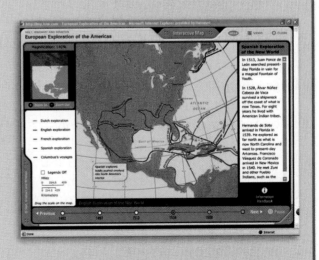

Mapping the Earth
Using Latitude and Longitude

A **globe** is a scale model of the Earth. It is useful for showing the entire Earth or studying large areas of Earth's surface.

To study the world, geographers use a pattern of imaginary lines that circles the globe in east-west and north-south directions. It is called a **grid**. The intersection of these imaginary lines helps us find places on Earth.

The east-west lines in the grid are lines of **latitude**, which you can see on the diagram. Lines of latitude are called **parallels** because they are always parallel to each other. These imaginary lines measure distance north and south of the **equator**. The equator is an imaginary line that circles the globe halfway between the North and South Poles. Parallels measure distance from the equator in **degrees**. The symbol for degrees is °. Degrees are further divided into **minutes**. The symbol for minutes is ´. There are 60 minutes in a degree. Parallels north of the equator are labeled with an N. Those south of the equator are labeled with an S.

The north-south imaginary lines are lines of **longitude**. Lines of longitude are called **meridians**. These imaginary lines pass through the poles. They measure distance east and west of the **prime meridian**. The prime meridian is an imaginary line that runs through Greenwich, England. It represents 0° longitude.

Lines of latitude range from 0°, for locations on the equator, to 90°N or 90°S, for locations at the poles. Lines of longitude range from 0° on the prime meridian to 180° on a meridian in the mid-Pacific Ocean. Meridians west of the prime meridian to 180° are labeled with a W. Those east of the prime meridian to 180° are labeled with an E. Using latitude and longitude, geographers can identify the exact location of any place on Earth.

Lines of Latitude

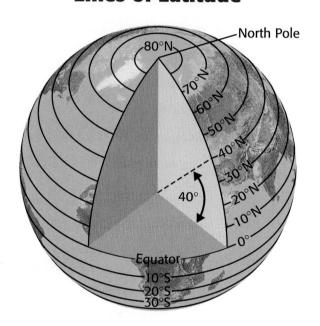

Lines of Longitude

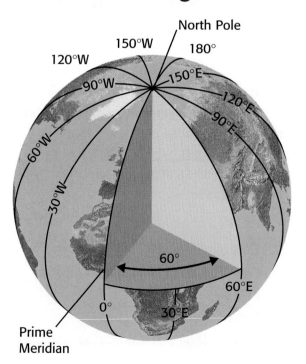

The equator divides the globe into two halves, called **hemispheres**. The half north of the equator is the Northern Hemisphere. The southern half is the Southern Hemisphere. The prime meridian and the 180° meridian divide the world into the Eastern Hemisphere and the Western Hemisphere. Look at the diagrams on this page. They show each of these four hemispheres.

Earth's land surface is divided into seven large landmasses, called **continents**. These continents are also shown on the diagrams on this page. Landmasses smaller than continents and completely surrounded by water are called **islands**.

Geographers organize Earth's water surface into major regions too. The largest is the world ocean. Geographers divide the world ocean into the Pacific Ocean, the Atlantic Ocean, the Indian Ocean, and the Arctic Ocean. Lakes and seas are smaller bodies of water.

Northern Hemisphere

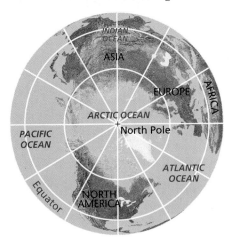

Southern Hemisphere

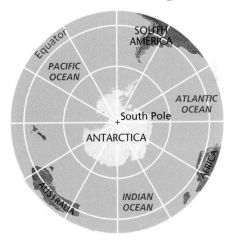

Western Hemisphere

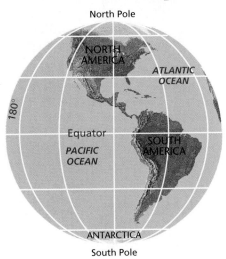

Eastern Hemisphere

Mapmaking
Understanding Map Projections

A **map** is a flat diagram of all or part of Earth's surface. Mapmakers have created different ways of showing our round planet on flat maps. These different ways are called **map projections**. Because Earth is round, there is no way to show it accurately on a flat map. All flat maps are distorted in some way. Mapmakers must choose the type of map projection that is best for their purposes. Many map projections are one of three kinds: cylindrical, conic, or flat-plane.

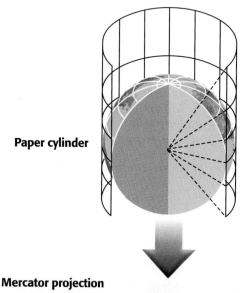

Paper cylinder

Cylindrical Projections

Cylindrical projections are based on a cylinder wrapped around the globe. The cylinder touches the globe only at the equator. The meridians are pulled apart and are parallel to each other instead of meeting at the poles. This causes landmasses near the poles to appear larger than they really are. The map below is a Mercator projection, one type of cylindrical projection. The Mercator projection is useful for navigators because it shows true direction and shape. However, it distorts the size of land areas near the poles.

Mercator projection

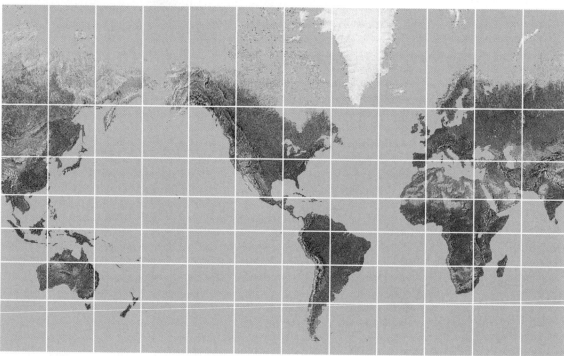

Conic Projections

Conic projections are based on a cone placed over the globe. A conic projection is most accurate along the lines of latitude where it touches the globe. It retains almost true shape and size. Conic projections are most useful for showing areas that have long east-west dimensions, such as the United States.

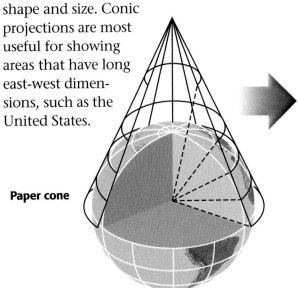

Paper cone

Conic projection

Flat-plane Projections

Flat-plane projections are based on a plane touching the globe at one point, such as at the North Pole or South Pole. A flat-plane projection is useful for showing true direction for airplane pilots and ship navigators. It also shows true area. However, it distorts the true shapes of landmasses.

Flat plane

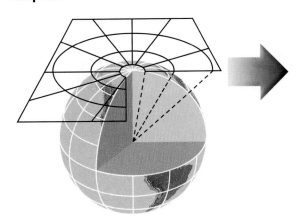

Flat-plane projection

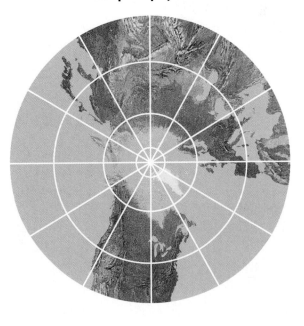

Map Essentials
How to Read a Map

Maps are like messages sent out in code. To help us translate the code, mapmakers provide certain features. These features help us understand the message they are presenting about a particular part of the world. Of these features, almost all maps have a title, a compass rose, a scale, and a legend. The map below has these four features, plus a fifth—a locator map.

❶ Title

A map's **title** shows what the subject of the map is. The map title is usually the first thing you should look at when studying a map, because it tells you what the map is trying to show.

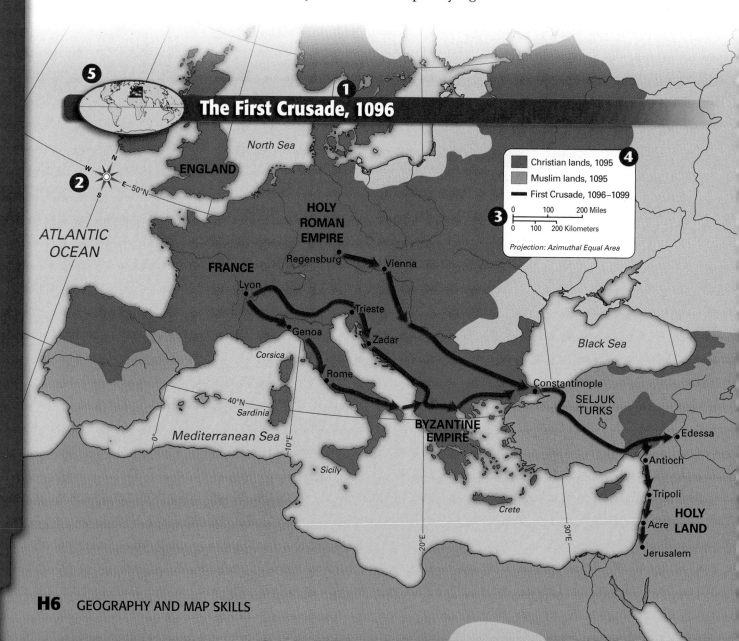

The First Crusade, 1096

Legend:
- Christian lands, 1095
- Muslim lands, 1095
- First Crusade, 1096–1099

0   100   200 Miles
0   100   200 Kilometers

Projection: Azimuthal Equal Area

❷ Compass Rose

A directional indicator shows which way north, south, east, and west lie on the map. Some mapmakers use a "north arrow," which points toward the North Pole. Remember, "north" is not always at the top of a map. The way a map is drawn and the location of directions on that map depend on the perspective of the mapmaker. Most maps in this textbook indicate direction by using a compass rose. A **compass rose** has arrows that point to all four principal directions.

❸ Scale

Mapmakers use scales to represent the distances between points on a map. Scales may appear on maps in several different forms. The maps in this textbook provide a **bar scale**. Scales give distances in miles and kilometers.

To find the distance between two points on the map, place a piece of paper so that the edge connects the two points. Mark the location of each point on the paper with a line or dot. Then, compare the distance between the two dots with the map's bar scale. The number on the top of the scale gives the distance in miles. The number on the bottom gives the distance in kilometers. Because the distances are given in large intervals, you may have to approximate the actual distance on the scale.

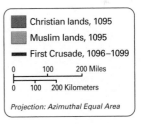

❹ Legend

The **legend**, or key, explains what the symbols on the map represent. Point symbols are used to specify the location of things, such as cities, that do not take up much space on the map. Some legends show colors that represent certain features like empires or other regions. Other maps might have legends with symbols or colors that represent features such as roads. Legends can also show economic resources, land use, population density, and climate.

Christian lands, 1095
Muslim lands, 1095
First Crusade, 1096–1099

0   100   200 Miles
0   100   200 Kilometers

Projection: Azimuthal Equal Area

❺ Locator Map

A **locator map** shows where in the world the area on the map is located. The area shown on the main map is shown in red on the locator map. The locator map also shows surrounding areas so the map reader can see how the information on the map relates to neighboring lands.

Working with Maps
Using Different Kinds of Maps

As you study the world's regions and countries, you will use a variety of maps. Political maps and physical maps are two of the most common types of maps you will study. In addition, you will use special-purpose maps. These maps might show climate, population, resources, ancient empires, or other topics.

Political Maps

Political maps show the major political features of a region. These features include country borders, capital cities, and other places. Political maps use different colors to represent countries, and capital cities are often shown with a special star symbol.

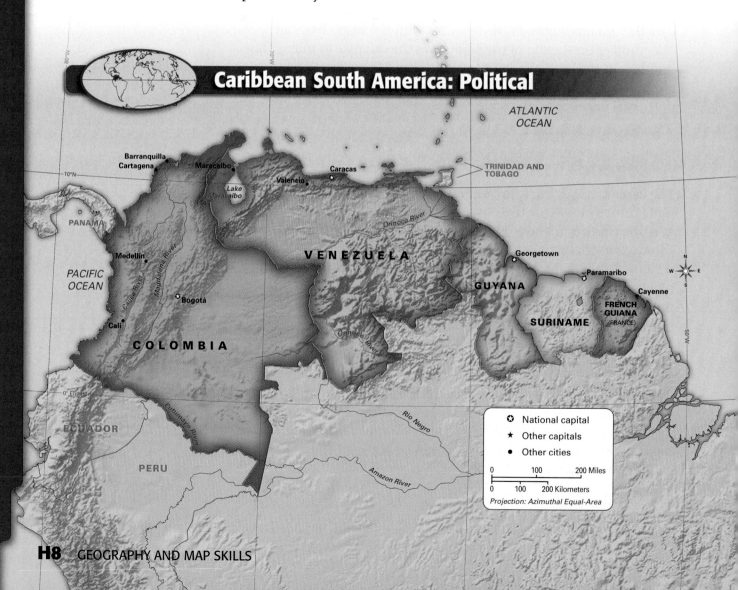

Caribbean South America: Political

- ✪ National capital
- ★ Other capitals
- ● Other cities

Projection: Azimuthal Equal-Area

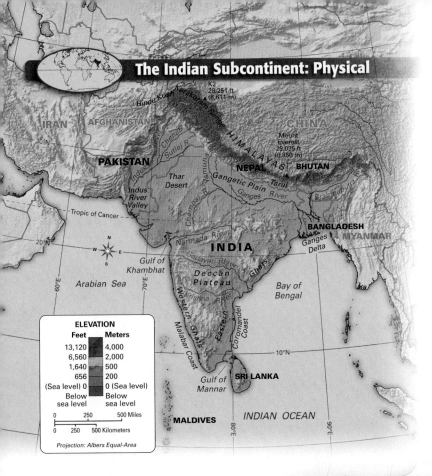

The Indian Subcontinent: Physical

Physical Maps

Physical maps show the major physical features of a region. These features may include mountain ranges, rivers, oceans, islands, deserts, and plains. Often, these maps use different colors to represent different elevations of land. As a result, the map reader can easily see which areas are high elevations, like mountains, and which areas are lower.

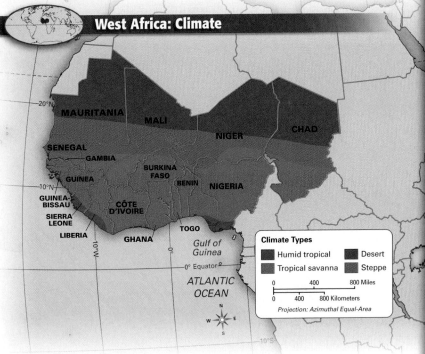

West Africa: Climate

Special-Purpose Maps

Special-purpose maps focus on one special topic, such as climate, resources, or population. These maps present information on the topic that is particularly important in the region. Depending on the type of special-purpose map, the information may be shown with different colors, arrows, dots, or other symbols.

Using Maps in Geography The different kinds of maps in this textbook will help you study and understand geography. By working with these maps, you will see what the physical geography of places is like, where people live, and how the world has changed over time.

Geographic Dictionary

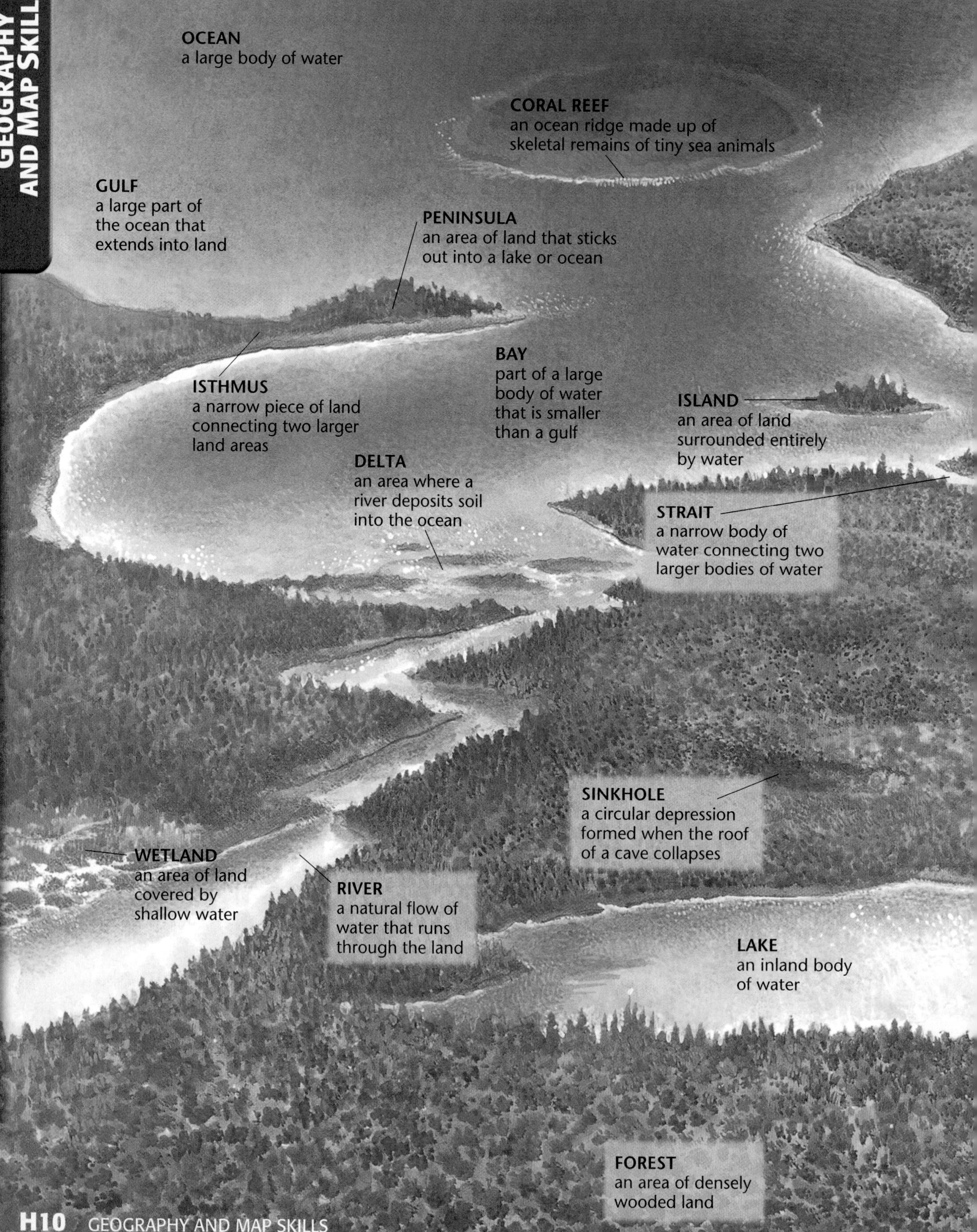

OCEAN
a large body of water

CORAL REEF
an ocean ridge made up of skeletal remains of tiny sea animals

GULF
a large part of the ocean that extends into land

PENINSULA
an area of land that sticks out into a lake or ocean

BAY
part of a large body of water that is smaller than a gulf

ISLAND
an area of land surrounded entirely by water

ISTHMUS
a narrow piece of land connecting two larger land areas

DELTA
an area where a river deposits soil into the ocean

STRAIT
a narrow body of water connecting two larger bodies of water

SINKHOLE
a circular depression formed when the roof of a cave collapses

WETLAND
an area of land covered by shallow water

RIVER
a natural flow of water that runs through the land

LAKE
an inland body of water

FOREST
an area of densely wooded land

COAST
an area of land near the ocean

MOUNTAIN
an area of rugged land that generally rises higher than 2,000 feet

VALLEY
an area of low land between hills or mountains

GLACIER
a large area of slow-moving ice

VOLCANO
an opening in Earth's crust where lava, ash, and gases erupt

CANYON
a deep, narrow valley with steep walls

HILL
a rounded, elevated area of land smaller than a mountain

PLAIN
a nearly flat area

DUNE
a hill of sand shaped by wind

OASIS
an area in the desert with a water source

DESERT
an extremely dry area with little water and few plants

PLATEAU
a large, flat, elevated area of land

Themes and Essential Elements of Geography

by Dr. Christopher L. Salter

To study the world, geographers have identified 5 key themes, 6 essential elements, and 18 geography standards.

"How should we teach and learn about geography?" Professional geographers have worked hard over the years to answer this important question.

In 1984 a group of geographers identified the 5 Themes of Geography. These themes did a wonderful job of laying the groundwork for good classroom geography. Teachers used the 5 Themes in class, and geographers taught workshops on how to apply them in the world.

By the early 1990s, however, some geographers felt the 5 Themes were too broad. They created the 18 Geography Standards and the 6 Essential Elements. The 18 Geography Standards include more detailed information about what geography is, and the 6 Essential Elements are like a bridge between the 5 Themes and 18 Standards.

Look at the chart to the right. It shows how each of the 5 Themes connects to the Essential Elements and Standards. For example, the theme of Location is related to The World in Spatial Terms and the first three Standards. Study the chart carefully to see how the other themes, elements, and Standards are related.

The last Essential Element and the last two Standards cover The Uses of Geography. These key parts of geography were not covered by the 5 Themes. They will help you see how geography has influenced the past, present, and future.

5 Themes of Geography

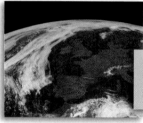

Location The theme of location describes where something is.

Place Place describes the features that make a site unique.

Regions Regions are areas that share common characteristics.

Movement This theme looks at how and why people and things move.

Human-Environment Interaction People interact with their environment in many ways.

6 Essential Elements

18 Geography Standards

1. How to use maps and other tools
2. How to use mental maps to organize information
3. How to analyze the spatial organization of people, places, and environments

I. The World in Spatial Terms

4. The physical and human characteristics of places
5. How people create regions to interpret Earth
6. How culture and experience influence people's perceptions of places and regions

II. Places and Regions

7. The physical processes that shape Earth's surface
8. The distribution of ecosystems on Earth

9. The characteristics, distribution, and migration of human populations
10. The complexity of Earth's cultural mosaics
11. The patterns and networks of economic interdependence on Earth
12. The patterns of human settlement
13. The forces of cooperation and conflict

III. Physical Systems

IV. Human Systems

14. How human actions modify the physical environment
15. How physical systems affect human systems
16. The distribution and meaning of resources

V. Environment and Society

17. How to apply geography to interpret the past
18. How to apply geography to interpret the present and plan for the future

VI. The Uses of Geography

Become an Active Reader

by Dr. Kylene Beers

Did you ever think you would begin reading your social studies book by reading about *reading*? Actually, it makes better sense than you might think. You would probably make sure you knew some soccer skills and strategies before playing in a game. Similarly, you need to know something about reading skills and strategies before reading your social studies book. In other words, you need to make sure you know whatever you need to know in order to read this book successfully.

Tip #1

Read Everything on the Page!

You can't follow the directions on the cake-mix box if you don't know where the directions are! Cake-mix boxes always have directions on them telling you how many eggs to add or how long to bake the cake. But, if you can't find that information, it doesn't matter that it is there.

Likewise, this book is filled with information that will help you understand what you are reading. If you don't study that information, however, it might as well not be there. Let's take a look at some of the places where you'll find important information in this book.

The Chapter Opener
The chapter opener gives you a brief overview of what you will learn in the chapter. You can use this information to prepare to read the chapter.

The Section Openers
Before you begin to read each section, preview the information under What You Will Learn. There you'll find the main ideas of the section and key terms that are important in it. Knowing what you are looking for before you start reading can improve your understanding.

Boldfaced Words
Those words are important and are defined somewhere on the page where they appear— either right there in the sentence or over in the side margin.

Maps, Charts, and Artwork
These things are not there just to take up space or look good! Study them and read the information beside them. It will help you understand the information in the chapter.

Questions at the End of Sections
At the end of each section, you will find questions that will help you decide whether you need to go back and re-read any parts before moving on. If you can't answer a question, that is your cue to go back and re-read.

Questions at the End of the Chapter
Answer the questions at the end of each chapter, even if your teacher doesn't ask you to. These questions are there to help you figure out what you need to review.

Tip #2

Use the Reading Skills and Strategies in Your Textbook

Good readers use a number of skills and strategies to make sure they understand what they are reading. In this textbook you will find help with important reading skills and strategies such as "Using Prior Knowledge," and "Understanding Main Ideas."

We teach the reading skills and strategies in several ways. Use these activities and lessons and you will become a better reader.

- First, on the opening page of every chapter we identify and explain the reading skill or strategy you will focus on as you work through the chapter. In fact, these activities are called "Focus on Reading."

- Second, as you can see in the example at right, we tell you where to go for more help. The back of the book has a reading handbook with a full-page practice lesson to match the reading skill or strategy in every chapter.

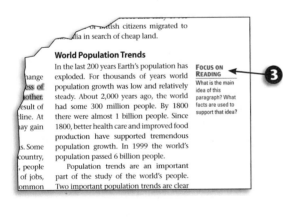

World Population Trends

In the last 200 years Earth's population has exploded. For thousands of years world population growth was low and relatively steady. About 2,000 years ago, the world had some 300 million people. By 1800 there were almost 1 billion people. Since 1800, better health care and improved food production have supported tremendous population growth. In 1999 the world's population passed 6 billion people.

Population trends are an important part of the study of the world's people. Two important population trends are clear

FOCUS ON READING
What is the main idea of this paragraph? What facts are used to support that idea?

- Third, we give you short practice activities and examples as you read the chapter. These activities and examples show up in the margin of your book. Again, look for the words, "Focus on Reading."

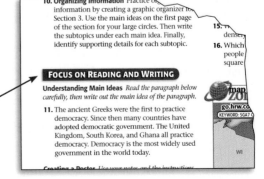

10. Organizing Information Practice or information by creating a graphic organizer for Section 3. Use the main ideas on the first page of the section for your large circles. Then write the subtopics under each main idea. Finally, identify supporting details for each subtopic.

FOCUS ON READING AND WRITING

Understanding Main Ideas *Read the paragraph below carefully, then write out the main idea of the paragraph.*

11. The ancient Greeks were the first to practice democracy. Since then many countries have adopted democratic government. The United Kingdom, South Korea, and Ghana all practice democracy. Democracy is the most widely used government in the world today.

- Finally, we provide another practice activity in the Chapter Review at the end of every chapter. That activity gives you one more chance to make sure you know how to use the reading skill or strategy.

Tip #3

Pay Attention to Vocabulary

It is no fun to read something when you don't know what the words mean, but you can't learn new words if you only use or read the words you already know. In this book, we know we have probably used some words you don't know. But, we have followed a pattern as we have used more difficult words.

- First, at the beginning of each section you will find a list of key terms that you will need to know. Be on the lookout for those words as you read through the section. You will find that we have defined those words right there in the paragraph where they are used. Look for a word that is in boldface with its definition highlighted in yellow.

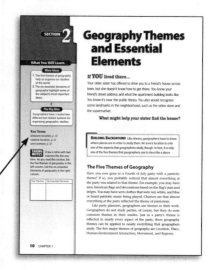

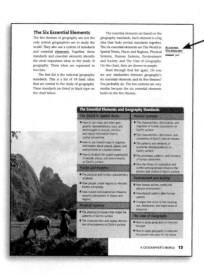

- Second, when we use a word that is important in all classes, not just social studies, we define it in the margin under the heading Academic Vocabulary. You will run into these academic words in other textbooks, so you should learn what they mean while reading this book.

Tip #4

Read Like a Skilled Reader

You won't be able to climb to the top of Mount Everest if you do not train! If you want to make it to the top of Mount Everest then you must start training to climb that huge mountain.

Training is also necessary to become a good reader. You will never get better at reading your social studies book—or any book for that matter—unless you spend some time thinking about how to be a better reader.

Skilled readers do the following:

1. They preview what they are supposed to read before they actually begin reading. When previewing, they look for vocabulary words, titles of sections, information in the margin, or maps or charts they should study.

2. They get ready to take some notes while reading by dividing their notebook paper into two parts. They title one side "Notes from the Chapter" and the other side "Questions or Comments I Have."

3. As they read, they complete their notes.

4. They read like **active readers**. The Active Reading list below shows you what that means.

5. Finally, they use clues in the text to help them figure out where the text is going. The best clues are called signal words. These are words that help you identify chronological order, causes and effects, or comparisons and contrasts.

Chronological Order Signal Words: *first, second, third, before, after, later, next, following that, earlier, subsequently, finally*

Cause and Effect Signal Words: *because of, due to, as a result of, the reason for, therefore, consequently, so, basis for*

Comparison/Contrast Signal Words: *likewise, also, as well as, similarly, on the other hand*

Active Reading

There are three ways to read a book: You can be a turn–the–pages–no–matter–what type of reader. These readers just keep on turning pages whether or not they understand what they are reading. Or, you can be a stop–watch–and–listen kind of reader. These readers know that if they wait long enough, someone will tell them what they need to know. Or, you can be an active reader. These readers know that it is up to them to figure out what the text means. Active readers do the following as they read:

Predict what will happen next based on what has already happened. When your predictions don't match what happens in the text, re-read the confusing parts.

Question what is happening as you read. Constantly ask yourself why things have happened, what things mean, and what caused certain events. Jot down notes about the questions you can't answer.

Summarize what you are reading frequently. Do not try to summarize the entire chapter! Read a bit and then summarize it. Then read on.

Connect what is happening in the section you're reading to what you have already read.

Clarify your understanding. Be sure that you understand what you are reading by stopping occasionally to ask yourself whether you are confused by anything. Sometimes you might need to re-read to clarify. Other times you might need to read further and collect more information before you can understand. Still other times you might need to ask the teacher to help you with what is confusing you.

Visualize what is happening in the text. In other words, try to see the events or places in your mind. It might help you to draw maps, make charts, or jot down notes about what you are reading as you try to visualize the action in the text.

Social Studies Words

As you read this textbook, you will be more successful if you learn the meanings of the words on this page. You will come across these words many times in your social studies classes, like geography and history. Read through these words now to become familiar with them before you begin your studies.

Social Studies Words

WORDS ABOUT TIME

AD	refers to dates after the birth of Jesus
BC	refers to dates before Jesus's birth
BCE	refers to dates before Jesus's birth, stands for "before the common era"
CE	refers to dates after Jesus's birth, stands for "common era"
century	a period of 100 years
decade	a period of 10 years
era	a period of time
millennium	a period of 1,000 years

WORDS ABOUT THE WORLD

climate	the weather conditions in a certain area over a long period of time
geography	the study of the world's people, places, and landscapes
physical features	features on Earth's surface, such as mountains and rivers
region	an area with one or more features that make it different from surrounding areas
resources	materials found on Earth that people need and value

WORDS ABOUT PEOPLE

anthropology	the study of people and cultures
archaeology	the study of the past based on what people left behind
citizen	a person who lives under the control of a government
civilization	the way of life of people in a particular place or time
culture	the knowledge, beliefs, customs, and values of a group of people
custom	a repeated practice or tradition
economics	the study of the production and use of goods and services
economy	any system in which people make and exchange goods and services
government	the body of officials and groups that run an area
history	the study of the past
politics	the process of running a government
religion	a system of beliefs in one or more gods or spirits
society	a group of people who share common traditions
trade	the exchange of goods or services

Academic Words

What are academic words? They are important words used in all of your classes, not just social studies. You will see these words in other textbooks, so you should learn what they mean while reading this book. Review this list now. You will use these words again in the chapters of this book.

Academic Words

abstract	expressing a quality or idea without reference to an actual thing
acquire	to get
affect	to change or influence
authority	power or influence
circumstances	conditions that influence an event or activity
classical	referring to the cultures of ancient Greece or Rome
complex	difficult, not simple
concrete	specific, real
consequences	the effects of a particular event or events
contracts	binding legal agreements
criteria	rules for defining
distinct	clearly different and separate
distribute	to divide among a group of people
effect	the results of an action or decision
efficient	productive and not wasteful
element	part
establish	to set up or create
execute	to perform, carry out
explicit	fully revealed without vagueness
factor	cause
features	characteristics

function	work or perform
ideals	ideas or goals that people try to live up to
impact	effect, result
implement	to put in place
implicit	understood though not clearly put into words
incentive	something that leads people to follow a certain course of action
influence	change, or have an effect on
innovation	a new idea or way of doing something
method	a way of doing something
motive	a reason for doing something
policy	rule, course of action
principle	basic belief, rule, or law
procedure	a series of steps taken to accomplish a task
process	a series of steps by which a task is accomplished
role	a part or function
structure	the way something is set up or organized
traditional	customary, time-honored
values	ideas that people hold dear and try to live by

Multiple Choice

A multiple-choice test item is a question or an incomplete statement with several answer choices. To answer a multiple-choice test item, select the choice that best answers the question or that best completes the statement.

Learn

Use these strategies to answer multiple-choice test items:

❶ Carefully read the question or incomplete statement.

❷ Look for words that affect the meaning, such as *all, always, best, every, most, never, not,* or *only.* For example, in Item 1 to the right, the word *all* tells you to look for the answer in which all three choices are correct.

❸ Read *all* the choices before selecting an answer—even if the first choice seems right.

❹ In your mind, cross off any of the answer choices that you know for certain are wrong.

❺ Consider the choices that are left and select the *best* answer. If you are not sure, select the choice that makes the most sense.

Read the questions and write the letter of the best response.

❶ ➔ ❷

1 **Which of the following are *all* physical features of geography?**

   A landforms, climates, people

   B landforms, climates, soils

   C landscapes, climates, plants

   D landscapes, communities, soils ❸

2 **A region is an area that has**

   Ⓐ one or more common features.

❺ **B** ~~no people living in it.~~ ◀ ❹

   C few physical features.

   D set physical boundaries.

Practice

Read the questions and write the letter of the best response.

1 **Which of the following is part of the study of human geography?**

   A bodies of water

   B communities

   C landforms

   D plants

2 **The economy of North Korea is *best* described as a**

   A command economy.

   B developed economy.

   C market economy.

   D traditional economy.

Primary Sources

Primary sources are materials, often called documents, created by people who lived during the times you are reading about. Examples of primary sources include text documents, such as letters and diaries, and visual documents, such as photographs.

Learn

Use these strategies to answer test questions about primary sources:

❶ Note the document's title and source line. This information can tell you the document's author, date, and purpose.

❷ Skim the document. Get an idea of its main focus.

❸ Read the question about the document. Note what information you are being asked to find.

❹ Read or examine the document carefully. As you do, identify the main idea and key details.

❺ Compare the question and answer choices to the document. Look for similar words. Then read between the lines. Use your critical-thinking skills to draw conclusions.

❻ Review the question and select the best answer.

Read the passage below. Use the passage to select the best answer to the question that follows.

❶

Geography for Life

"Geography is a field of study that enables us to find answers to questions about the world around us—about where things are and how and why they got there…With a strong grasp of geography, people are better equipped to solve issues at not only the local level but also the global level."

— from *Geography for Life*, by the Geography Education Standards Project

1 Which statement below *best* summarizes the main idea of the above passage?

A Geography helps people to read and make maps.

B Geography helps people to get where they are going.

C Geography helps people to understand the world better and to solve problems.

D Geography helps people explore Earth.

Practice

Read the passage from the primary source, "The Charter of the United Nations," in Chapter 4. Use the passage to answer the following questions.

1 What does the passage state is a central goal of the United Nations?

A democracy

B peace

C power

D war

2 Based on the passage, why might the UN send troops into a region?

A to punish a country

B to conquer a country

C to obtain valuable resources

D to protect a member country

Charts and Graphs

Charts and graphs are tables or drawings that present and organize information or data. Some standardized tests include questions about charts and graphs. These questions require you to interpret the information or data in the chart or graph to answer the question.

Learn

Use these strategies to answer test questions about charts and graphs:

❶ Read the title of the chart or graph. Identify the subject and purpose of the information shown.

❷ Read all the other labels. Note the types of information the chart or graph is showing and how the information is organized.

❸ Analyze the information or data. Look for patterns, changes over time, and similarities or differences. For example, in the graph at right, world population growth rises dramatically after 1900.

❹ Read the question carefully. Note key words in the question.

❺ Review the chart or graph to find the correct answer.

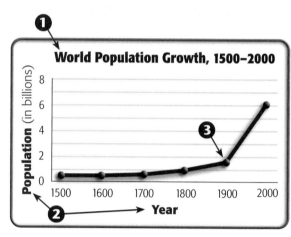

Source: *Atlas of World Population History*

❶ Based on the graph above, during which period did the world population <u>increase</u> the <u>most</u>?

   A 1600 to 1700
   B 1700 to 1800
   C 1800 to 1900
   D 1900 to 2000

Practice

Examine the chart at right. Use the chart to answer the question below.

❶ Which country in the Ring of Fire had two major volcanic eruptions?

   A Colombia
   B Indonesia
   C Philippines
   D United States

THE WORLD ALMANAC Facts about the World
Major Eruptions in the Ring of Fire

Volcano	Year
Tambora, Indonesia	1815
Krakatau, Indonesia	1883
Mount Saint Helens, United States	1980
Nevado del Ruiz, Colombia ·	1985
Mount Pinatubo, Philippines	1991

Maps

Standardized tests may include questions that refer to information in maps. These maps might show political features such as cities and states, physical features such as mountains and plains, or information such as climate, land use, or settlement patterns.

Learn

Use these strategies to answer questions about maps.

❶ Read the map title to identify the map's subject and purpose. The map below shows levels of government freedom around the world.

❷ Study the legend. It explains information in the map, such as what different colors or symbols mean.

❸ Note the map's direction and scale. The scale shows the distance between points in the map.

❹ Examine the map closely. Read all the labels and study the other information, such as colors, borders, or symbols.

❺ Read the question about the map.

❻ Review the map to find the answer.

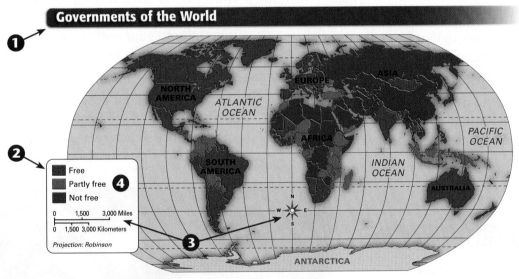

Governments of the World

Source: *Freedom House*

Practice

Examine the map above. Use the information in the map to answer the following questions.

❶ **Which two continents have the least government freedom?**

A Africa and Asia

B Africa and Europe

C Australia and Europe

D Europe and Asia

❷ **The continents with the most freedom are Australia and**

A Africa.

B Europe.

C North America.

D South America.

H23

Extended Response

Extended response questions usually require you to examine a document, such as a letter, chart, or map. You then use the information in the document to write an extended answer, often a paragraph or more in length.

Learn

Use these strategies to answer extended response questions:

1 Read the directions and question carefully to determine the purpose of your answer. For example, are you to explain, identify causes, summarize, or compare? To help determine the purpose, look for key words such as *compare, contrast, describe, discuss, explain, interpret, predict,* or *summarize.*

2 Read the title of the document. Identify its subject and purpose.

3 Study the document carefully. Read all the text. Identify the main idea or focus.

4 If allowed, make notes on another sheet of paper to organize your thoughts. Jot down information from the document that you want to include in your answer.

5 Use the question to create a topic sentence. For example, for the practice question below, a topic sentence might be, "Earth's tilt and revolution cause the seasons to change at about the same time each year in the Northern Hemisphere."

6 Create an outline or graphic organizer to help organize your main points. Review the document to find details or examples to support each point.

7 Write your answer in complete sentences. Start with your topic sentence. Then refer to your outline or organizer as you write. Be sure to include details or examples from the document.

8 Last, proofread your answer. Check for correct grammar, spelling, punctuation, and sentence structure.

Practice

Read question 1 and write your own well-constructed extended response.

1 **Extended Response** Use the diagram to explain the change of the seasons in the Northern Hemisphere. Provide an explanation for each season—winter, spring, summer, and fall.

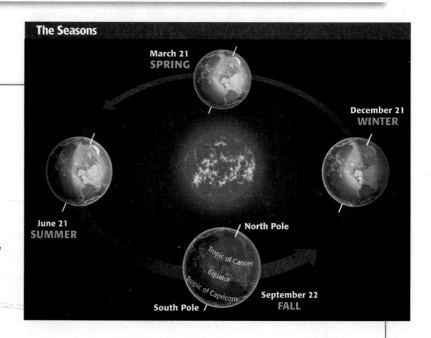

The Seasons

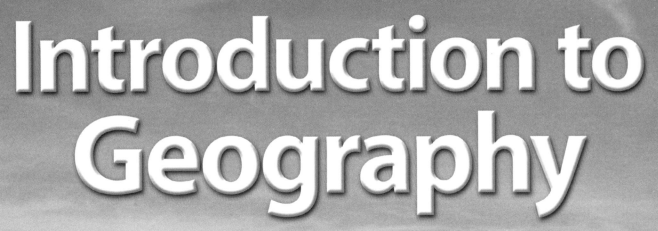

Introduction to Geography

Deserts

Huge deserts, such as the Sahara in North Africa, are visible from space and appear yellow and brown.

Oceans

About 71 percent of Earth's surface is covered by vast amounts of salt water, which form the world's oceans.

UNIT 1
Introduction to Geography

Explore the Satellite Image
Human-made machines that orbit Earth, called satellites, send back images of our planet like this one. What can you learn about Earth from studying this satellite image?

Frozen Lands

Earth's icy poles are frozen year-round and appear a brilliant white from space. These frozen lands contain much of Earth's freshwater.

CHAPTER 1

A Geographer's World

What You Will Learn...

In this chapter you will learn about the field of geography, the study of the world's people and places. You will also learn why people study geography and how they organize their studies.

SECTION 1
Studying Geography 4

SECTION 2
Geography Themes and Essential Elements 10

SECTION 3
The Branches of Geography 16

FOCUS ON READING AND WRITING

Using Prior Knowledge Prior knowledge is what you already know about a subject. Before you read a chapter, review the chapter and section titles. Then make a list of what you already know. Later, you can compare your prior knowledge with what you learned from the chapter. **See the lesson, Using Prior Knowledge, on page 668.**

Writing a Job Description Geographers are people who study geography, but what is it exactly that they do? As you read this chapter you will learn about the work that geographers do. Then you will write a job description that could be included in a career-planning guide.

Studying the World Exploring the world takes people to exciting and interesting places.

This village is in the country of Nepal. It rests high in the Himalayas, the highest mountains in the world.

What is the land around the village like? How can you tell that people live in this area?

HOLT

Geography's Impact
video series
Watch the video to understand the impact of studying geography.

Human Geography Geography is also the study of people. It asks where people live, what they eat, what they wear, and even what kinds of animals they keep.

Physical Geography Geography is the study of the world's land features, such as this windswept rock formation in Arizona.

Studying Geography

What You Will Learn...

Main Ideas

1. Geography is the study of the world, its people, and the landscapes they create.
2. Geographers look at the world in many different ways.
3. Maps and other tools help geographers study the planet.

The Big Idea

The study of geography and the use of geographic tools helps us view the world in new ways.

Key Terms

geography, *p. 4*
landscape, *p. 4*
social science, *p. 5*
region, *p. 6*
map, *p. 8*
globe, *p. 8*

TAKING NOTES Draw a large circle like the one below in your notebook. As you read this section, write a definition of geography at the top of the circle. Below that, list details about what geographers do.

Geography is ___

If YOU lived there...

You have just moved to Miami, Florida, from your old home in Pennsylvania. Everything seems very different—from the weather and the trees to the way people dress and talk. Even the streets and buildings look different. One day you get an e-mail from a friend at your old school. "What's it like living there?" he asks.

How will you describe your new home?

BUILDING BACKGROUND Often, when you are telling someone about a place they have never been, what you are describing is the place's geography. What the place looks like, what kind of weather it has, and how people live there are all parts of its geography.

What Is Geography?

Think about the place where you live. What does the land look like? Are there tall mountains nearby, or is the land so flat that you can see for miles? Is the ground covered with bright green grass and trees, or is the area part of a sandy desert?

Now think about the weather in your area. What is it like? Does it get really hot in the summer? Do you see snow every winter? How much does it rain? Do tornadoes ever strike?

Finally, think about the people who live in your town or city. Do they live mostly in apartments or houses? Do most people own cars, or do they get around town on buses or trains? What kinds of jobs do adults in your town have? Were most of the people you know born in your town, or did they move there?

The things that you have been thinking about are part of your area's geography. **Geography** is the study of the world, its people, and the landscapes they create. To a geographer, a place's **landscape** is all the human and physical features that make it unique. When they study the world's landscapes, geographers ask questions much like the ones you just asked yourself.

Geography as a Science

Many of the questions that geographers ask deal with how the world works. They want to know what causes mountains to form and what creates tornadoes. To answer questions like these, geographers have to think and act like scientists.

As scientists, geographers look at data, or information, that they gather about places. Gathering data can sometimes lead geographers to fascinating places. They might have to crawl deep into caves or climb tall mountains to make observations and take measurements. At other times, geographers study sets of images collected by satellites orbiting high above Earth.

However geographers gather their data, they have to study it carefully. Like other scientists, geographers must examine their findings in great detail before they can learn what all the information means.

Geography as a Social Science

Not everything that geographers study can be measured in numbers, however. Some geographers study people and their lives. For example, they may ask why countries change their governments or why people in a place speak a certain language. This kind of information cannot be measured.

Because it deals with people and how they live, geography is sometimes called a social science. A **social science** is a field that studies people and the relationships among them.

The geographers who study people do not dig in caves or climb mountains. Instead, they visit places and talk to the people who live there. They want to learn about people's lives and communities.

READING CHECK Analyzing In what ways is geography both a science and a social science?

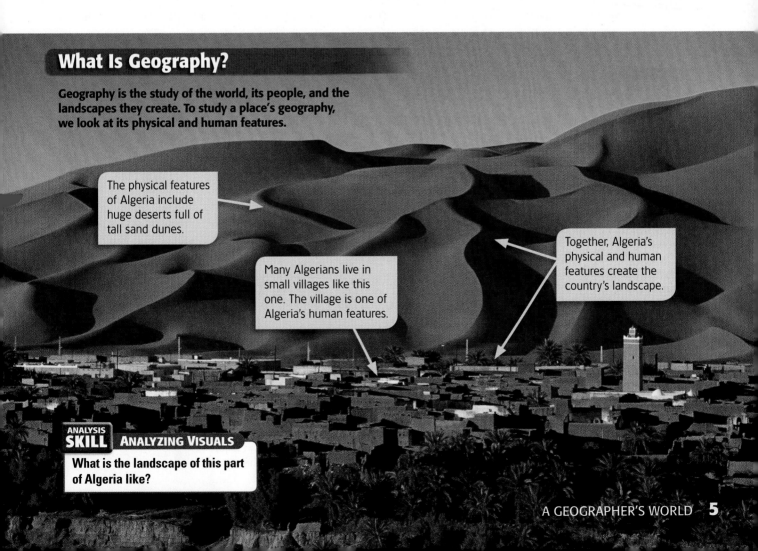

What Is Geography?

Geography is the study of the world, its people, and the landscapes they create. To study a place's geography, we look at its physical and human features.

The physical features of Algeria include huge deserts full of tall sand dunes.

Many Algerians live in small villages like this one. The village is one of Algeria's human features.

Together, Algeria's physical and human features create the country's landscape.

ANALYSIS SKILL ANALYZING VISUALS

What is the landscape of this part of Algeria like?

Looking at the World

Whether they study volcanoes and storms or people and cities, geographers have to look carefully at the world around them. To fully understand how the world works, geographers often look at places at three different levels.

Local Level

Some geographers study issues at a local level. They ask the same types of questions we asked at the beginning of this chapter: How do people in a town or community live? What is the local government like? How do the people who live there get around? What do they eat?

By asking these questions, geographers can figure out why people live and work the way they do. They can also help people improve their lives. For example, they can help town leaders figure out the best place to build new schools, shopping centers, or sports complexes. They can also help the people who live in the city or town plan for future changes.

Regional Level

Sometimes, though, geographers want to study a bigger chunk of the world. To do this, they divide the world into regions. A region is a part of the world that has one or more common features that distinguish it from surrounding areas.

Some regions are defined by physical characteristics such as mountain ranges, climates, or plants native to the area. As a result, these types of regions are often easy to identify. The Rocky Mountains of the western United States, for example, make up a physical region. Another example of this kind of region is the Sahara, a huge desert in northern Africa.

Other regions may not be so easy to define, however. These regions are based on the human characteristics of a place, such as language, religion, or history. A place in which most people share these kinds of characteristics can also be seen as a region. For example, most people in Scandinavia, a region in northern Europe, speak similar languages and practice the same religion.

Looking at the World

Geographers look at the world at many levels. At each level, they ask different questions and discover different types of information. By putting information gathered at different levels together, geographers can better understand a place and its role in the world.

ANALYZING VISUALS Based on these photos, what are some questions a geographer might ask about London?

Local Level This busy neighborhood in London, England, is a local area. A geographer here might study local foods, housing, or clothing.

Regions come in all shapes and sizes. Some are small, like the neighborhood called Chinatown in San Francisco. Other regions are huge, like the Americas. This huge region includes two continents, North America and South America. The size of the area does not matter, as long as the area shares some characteristics. These shared characteristics define the region.

Geographers divide the world into regions for many reasons. The world is a huge place and home to billions of people. Studying so large an area can be extremely difficult. Dividing the world into regions makes it easier to study. A small area is much easier to examine than a large area.

Other geographers study regions to see how people interact with one another. For example, they may study a city such as London, England, to learn how the city's people govern themselves. Then they can compare what they learn about one region to what they learn about another region. In this way, they can learn more about life and landscapes in both places.

Global Level

Sometimes geographers do not want to study the world just at a regional level. Instead they want to learn how people interact globally, or around the world. To do so, geographers ask how events and ideas from one region of the world affect people in other regions. In other words, they study the world on a global level.

Geographers who study the world on a global level try to find relationships among people who live far apart. They may, for example, examine the products that a country exports to see how those products are used in other countries.

In recent decades, worldwide trade and communication have increased. As a result, we need to understand how our actions affect people around the world. Through their studies, geographers provide us with information that helps us figure out how to live in a rapidly changing world.

READING CHECK **Finding Main Ideas** At what levels do geographers study the world?

Regional Level As a major city, London is also a region. At this level, a geographer might study the city's population or transportation systems.

Global Level London is one of the world's main financial centers. Here a geographer might study how London's economy affects the world.

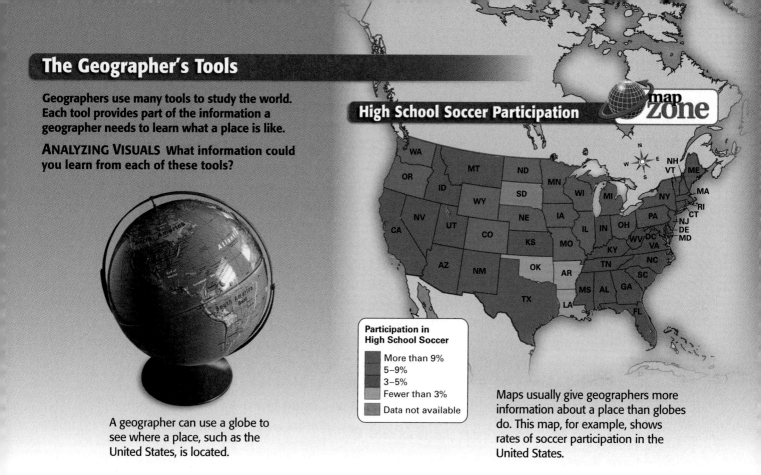

Geographers use many tools to study the world. Each tool provides part of the information a geographer needs to learn what a place is like.

ANALYZING VISUALS What information could you learn from each of these tools?

High School Soccer Participation

Participation in High School Soccer

- More than 9%
- 5–9%
- 3–5%
- Fewer than 3%
- Data not available

A geographer can use a globe to see where a place, such as the United States, is located.

Maps usually give geographers more information about a place than globes do. This map, for example, shows rates of soccer participation in the United States.

The Geographer's Tools

Have you ever seen a carpenter building or repairing a house? If so, you know that builders need many tools to do their jobs correctly. In the same way, geographers need many tools to study the world.

Maps and Globes

FOCUS ON READING

What do you already know about maps and globes?

The tools that geographers use most often in their work are maps and globes. A **map** is a flat drawing that shows all or part of Earth's surface. A **globe** is a spherical, or ball-shaped, model of the entire planet.

Both maps and globes show what the world looks like. They can show where mountains, deserts, and oceans are. They can also identify and describe the world's countries and major cities.

There are, however, major differences between maps and globes. Because a globe is spherical like Earth, it can show the world as it really is.

A map, though, is flat. It is not possible to show a spherical area perfectly on a flat surface. To understand what this means, think about an orange. If you took the peel off of an orange, could you make it lie completely flat? No, you could not, unless you stretched or tore the peel first.

The same principle is true with maps. To draw Earth on a flat surface, people have to distort, or alter, some details. For example, places on a map might look to be farther apart than they really are, or their shapes or sizes might be changed slightly.

Still, maps have many advantages over globes. Flat maps are easier to work with than globes. Also, it is easier to show small areas like cities on maps than on globes.

In addition, maps usually show more information than globes. Because globes are more expensive to make, they do not usually show anything more than where places are and what features they have.

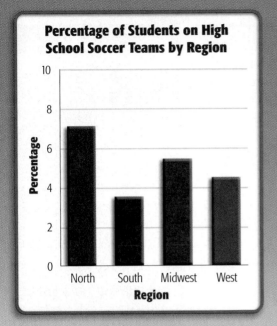

Percentage of Students on High School Soccer Teams by Region

Charts and graphs are also tools geographers can use to study information. They are often used when geographers want to compare numbers, such as the number of students who play soccer in each region of the country.

Maps, on the other hand, can show all sorts of information. Besides showing land use and cities, maps can include a great deal of information about a place. A map might show what languages people speak or where their ancestors came from. Maps like the one on the opposite page can even show how many students in an area play soccer.

Satellite Images

Maps and globes are not the only tools that geographers use in their work. As you have already read, many geographers study information gathered by satellites.

Much of the information gathered by these satellites is in the form of images. Geographers can study these images to see what an area looks like from above Earth. Satellites also collect information that we cannot see from the planet's surface. The information gathered by satellites helps geographers make accurate maps.

Other Tools

Geographers also use many other tools. For example, they use computer programs to create, update, and compare maps. They also use measuring devices to record data. In some cases, the best tools a geographer can use are a notebook and tape recorder to take notes while talking to people. Armed with the proper tools, geographers learn about the world's people and places.

READING CHECK **Summarizing** What are some of the geographer's basic tools?

SUMMARY AND PREVIEW Geography is the study of the world, its people, and its landscapes. In the next section, you will learn about two systems geographers use to organize their studies.

Section 1 Assessment

Reviewing Ideas, Terms, and Places
1. a. **Define** What is **geography**?
   b. **Explain** Why is geography considered a science?
2. a. **Identify** What is a **region**? Give two examples.
   b. **Elaborate** What global issues do geographers study?
3. a. **Describe** How do geographers use satellite images?
   b. **Compare and Contrast** How are maps and globes similar? How are they different?

Critical Thinking
4. **Summarizing** Draw three ovals like the ones shown here. Use your notes to fill the ovals with information about geography, geographers, and their tools.

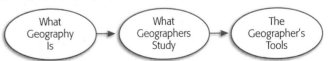

What Geography Is → What Geographers Study → The Geographer's Tools

FOCUS ON WRITING

5. **Describing the Subject** Based on what you have learned, what might attract people to work in geography? In your notebook, list some details about geography that might make people interested in working in the subject.

Geography Themes and Essential Elements

What You Will Learn...

Main Ideas

1. The five themes of geography help us organize our studies of the world.
2. The six essential elements of geography highlight some of the subject's most important ideas.

The Big Idea

Geographers have created two different but related systems for organizing geographic studies.

Key Terms

absolute location, *p. 12*
relative location, *p. 12*
environment, *p. 12*

TAKING NOTES Draw a table with two columns like the one here. As you read this section, list the five themes of geography in the left column. List the six essential elements of geography in the right column.

Five Themes	Six Essential Elements

If **YOU** lived there...

Your older sister has offered to drive you to a friend's house across town, but she doesn't know how to get there. You know your friend's street address and what the apartment building looks like. You know it's near the public library. You also would recognize some landmarks in the neighborhood, such as the video store and the supermarket.

What might help your sister find the house?

BUILDING BACKGROUND Like drivers, geographers have to know where places are in order to study them. An area's location is only one of the aspects that geographers study, though. In fact, it is only one of the five themes that geographers use to describe a place.

The Five Themes of Geography

Have you ever gone to a Fourth of July party with a patriotic theme? If so, you probably noticed that almost everything at the party was related to that theme. For example, you may have seen American flags and decorations based on the flag's stars and stripes. You may have seen clothes that were red, white, and blue or heard patriotic music being played. Chances are that almost everything at the party reflected the theme of patriotism.

Like party planners, geographers use themes in their work. Geographers do not study parties, of course, but they do note common themes in their studies. Just as a party's theme is reflected in nearly every aspect of the party, these geography themes can be applied to nearly everything that geographers study. The five major themes of geography are Location, Place, Human-Environment Interaction, Movement, and Regions.

The Five Themes of Geography

Geographers use five major themes, or ideas, to
organize and guide their studies.

go.hrw.com KEYWORD: SK7 CH1

Location The theme of location describes where
something is. The mountain shown above, Mount
Rainier, is in west-central Washington.

Place Place describes the features that make a
site unique. For example, Washington, D.C., is our
nation's capital and has many great monuments.

UNITED STATES

Regions Regions are areas that
share common characteristics.
The Mojave Desert, shown here,
is defined by its distinctive climate
and plant life.

Movement This theme looks at how
and why people and things move.
Airports like this one in Dallas, Texas,
help people move around the world.

Human-Environment Interaction
People interact with their environments
in many ways. Some, like this man in
Florida, use the land to grow crops.

ANALYSIS SKILL ANALYZING VISUALS

Which of the five themes deals with
the relationships between people
and their surroundings?

Location

Every point on Earth has a location, a description of where it is. This location can be expressed in many ways. Sometimes a site's location is expressed in specific, or absolute, terms, such as an address. For example, the White House is located at 1600 Pennsylvania Avenue in the city of Washington, D.C. A specific description like this one is called an **absolute location**. Other times, the site's location is expressed in general terms. For example, Canada is north of the United States. This general description of where a place lies is called its **relative location**.

Place

Another theme, Place, is closely related to Location. However, Place does not refer simply to where an area is. It refers to the area's landscape, the features that define the area and make it different from other places. Such features could include land, climate, and people. Together, they give a place its own character.

Human-Environment Interaction

FOCUS ON READING
What do you know about environments?

In addition to looking at the features of places, geographers examine how those features interact. In particular, they want to understand how people interact with their environment—how people and their physical environment affect each other. An area's **environment** includes its land, water, climate, plants, and animals.

People interact with their environment every day in all sorts of ways. They clear forests to plant crops, level fields to build cities, and dam rivers to prevent floods. At the same time, physical environments affect how people live. People in cold areas, for example, build houses with thick walls and wear heavy clothing to keep warm. People who live near oceans look for ways to protect themselves from storms.

Movement

People are constantly moving. They move within cities, between cities, and between countries. Geographers want to know why and how people move. For example, they ask if people are moving to find work or to live in a more pleasant area. Geographers also study the roads and routes that make movement so common.

Regions

You have already learned how geographers divide the world into many regions to help the study of geography. Creating regions also makes it easier to compare places. Comparisons help geographers learn why each place has developed the way it has.

READING CHECK Finding Main Ideas What are the five themes of geography?

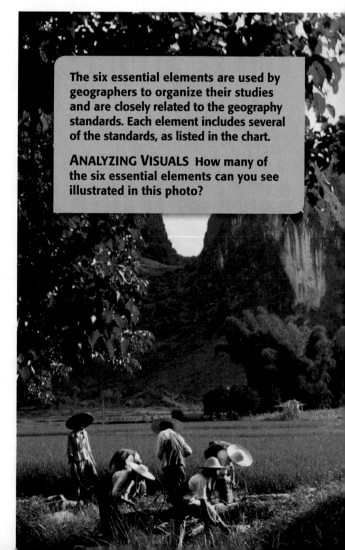

The six essential elements are used by geographers to organize their studies and are closely related to the geography standards. Each element includes several of the standards, as listed in the chart.

ANALYZING VISUALS How many of the six essential elements can you see illustrated in this photo?

The Six Essential Elements

The five themes of geography are not the only system geographers use to study the world. They also use a system of standards and essential **elements**. Together, these standards and essential elements identify the most important ideas in the study of geography. These ideas are expressed in two lists.

The first list is the national geography standards. This is a list of 18 basic ideas that are central to the study of geography. These standards are listed in black type on the chart below.

The essential elements are based on the geography standards. Each element is a big idea that links several standards together. The six essential elements are The World in Spatial Terms, Places and Regions, Physical Systems, Human Systems, Environment and Society, and The Uses of Geography. On the chart, they are shown in purple.

Read through that list again. Do you see any similarities between geography's six essential elements and its five themes? You probably do. The two systems are very similar because the six essential elements build on the five themes.

The Essential Elements and Geography Standards

The World in Spatial Terms

- How to use maps and other geographic representations, tools, and technologies to acquire, process, and report information from a spatial perspective
- How to use mental maps to organize information about people, places, and environments in a spatial context
- How to analyze the spatial organization of people, places, and environments on Earth's surface

Places and Regions

- The physical and human characteristics of places
- How people create regions to interpret Earth's complexity
- How culture and experience influence people's perceptions of places and regions

Physical Systems

- The physical processes that shape the patterns of Earth's surface
- The characteristics and spatial distribution of ecosystems on Earth's surface

Human Systems

- The characteristics, distributions, and migration of human populations on Earth's surface
- The characteristics, distribution, and complexity of Earth's cultural mosaics
- The patterns and networks of economic interdependence on Earth's surface
- The processes, patterns, and functions of human settlement
- How the forces of cooperation and conflict among people influence the division and control of Earth's surface

Environment and Society

- How human actions modify the physical environment
- How physical systems affect human systems
- Changes that occur in the meaning, use, distribution, and importance of resources

The Uses of Geography

- How to apply geography to interpret the past
- How to apply geography to interpret the present and plan for the future

BOOK
Geography for Life

The six essential elements were first outlined in a book called Geography for Life. In that book, the authors—a diverse group of geographers and teachers from around the United States— explained why the study of geography is important.

❝Geography *is* for life in every sense of that expression: lifelong, life-sustaining, and life-enhancing. Geography is a field of study that enables us to find answers to questions about the world around us—about where things are and how and why they got there.❞

❝Geography focuses attention on exciting and interesting things, on fascinating people and places, on things worth knowing because they are absorbing and because knowing about them lets humans make better-informed and, therefore, wiser decisions.❞

❝With a strong grasp of geography, people are better equipped to solve issues at not only the local level but also the global level.❞

–from *Geography for Life,*
by the Geography Education Standards Project

ANALYSIS SKILL ANALYZING PRIMARY SOURCES

Why do the authors of these passages think that people should study geography?

For example, the element Places and Regions combines two of the five themes of geography—Place and Regions. Also, the element called Environment and Society deals with many of the same issues as the theme Human-Environment Interaction.

There are also some basic differences between the essential elements and the themes. For example, the last element, The Uses of Geography, deals with issues not covered in the five themes. This element examines how people can use geography to plan the landscapes in which they live.

Throughout this book, you will notice references to both the themes and the essential elements. As you read, use these themes and elements to help you organize your own study of geography.

READING CHECK **Summarizing** What are the six essential elements of geography?

SUMMARY AND PREVIEW You have just learned about the themes and elements of geography. Next, you will explore the branches into which the field is divided.

Section 2 Assessment

go.hrw.com
Online Quiz
KEYWORD: SK7 HP1

Reviewing Ideas, Terms, and Places

1. **a. Define** What is the difference between a place's **absolute location** and its **relative location**? Give one example of each type of location.
 b. Contrast How are the themes of Location and Place different?
 c. Elaborate How does using the five themes help geographers understand the places they study?
2. **a. Identify** Which of the five themes of geography is associated with airports, highways, and the migration of people from one place to another?
 b. Explain How are the geography standards and the six essential elements related?
 c. Compare How are the six essential elements similar to the five themes of geography?

Critical Thinking

3. **Categorizing** Draw a chart like the one below. Use your notes to list the five themes of geography, explain each of the themes, and list one feature of your city or town that relates to each.

Theme					
Explanation					
Feature					

FOCUS ON WRITING

4. **Including Themes and Essential Elements** The five themes and six essential elements are central to a geographer's job. How will you mention them in your job description? Write down some ideas.

Analyzing Satellite Images

Learn

In addition to maps and globes, satellite images are among the geographer's most valuable tools. Geographers use two basic types of these images. The first type is called true color. These images are like photographs taken from high above Earth's surface. The colors in these images are similar to what you would see from the ground. Vegetation, for example, appears green.

The other type of satellite image is called an infrared image. Infrared images are taken using a special type of light. These images are based on heat patterns, and so the colors on them are not what we might expect. Bodies of water appear black, for example, since they give off little heat.

Practice

Use the satellite images on this page to answer the following questions.

1 On which image is vegetation red?

2 Which image do you think probably looks more like Italy does from the ground?

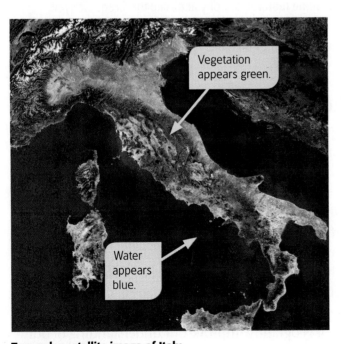

Vegetation appears green.

Water appears blue.

True color satellite image of Italy

Vegetation appears red.

Water appears black.

Infrared satellite image of Italy

Apply

Search the Internet to find a satellite image of your state or region. Determine whether the image is true color or infrared. Then write three statements that describe what you see on the image.

The Branches of Geography

What You Will Learn...

Main Ideas

1. Physical geography is the study of landforms, water bodies, and other physical features.
2. Human geography focuses on people, their cultures, and the landscapes they create.
3. Other branches of geography examine specific aspects of the physical or human world.

The Big Idea

Geography is divided into two main branches—physical and human geography.

Key Terms

physical geography, *p. 16*
human geography, *p. 18*
cartography, *p. 19*
meteorology, *p. 20*

 TAKING NOTES Draw two large circles like the ones below in your notebook. As you read this section, take notes about one of the main branches of geography in each circle.

If YOU lived there...

You are talking to two friends about the vacations their families will take this summer. One friend says that his family is going to the Grand Canyon. He is very excited about seeing the spectacular landscapes in and around the canyon. Your other friend's family is going to visit Nashville, Tennessee. She is looking forward to trying new foods at the city's restaurants and touring its museums.

Which vacation sounds more interesting? Why?

BUILDING BACKGROUND Geography is the study of the world and its features. Some features are physical, like the Grand Canyon. Others are human, like restaurants and museums. The main branches of geography focus on these types of features.

Physical Geography

Think about a jigsaw puzzle. Seen as a whole, the puzzle shows a pretty or interesting picture. To see that picture, though, you have to put all the puzzle pieces together. Before you assemble them, the pieces do not give you a clear idea of what the puzzle will look like when it is assembled. After all, each piece contains only a tiny portion of the overall image.

In many ways, geography is like a huge puzzle. It is made up of many branches, or divisions. Each of these branches focuses on a single part of the world. Viewed separately, none of these branches shows us the whole world. Together, however, the many branches of geography improve our understanding of our planet and its people.

Geography's two main branches are physical geography and human geography. The first branch, **physical geography**, is the study of the world's physical features—its landforms, bodies of water, climates, soils, and plants. Every place in the world has its own unique combination of these features.

Physical Geography

The study of Earth's physical features, including rivers, mountains, oceans, weather, and other features, such as Victoria Falls in southern Africa

Human Geography

The study of Earth's people, including their ways of life, homes, cities, beliefs, and customs, like those of these children in Malawi, a country in central Africa

Geography

The study of Earth's physical and cultural features

The Physical World

What does it mean to say that physical geography is the study of physical features? Physical geographers want to know all about the different features found on our planet. They want to know where plains and mountain ranges are, how rivers flow across the landscape, and why different amounts of rain fall from place to place.

More importantly, however, physical geographers want to know what causes the different shapes on Earth. They want to know why mountain ranges rise up where they do and what causes rivers to flow in certain directions. They also want to know why various parts of the world have very different weather and climate patterns.

To answer these questions, physical geographers take detailed measurements. They study the heights of mountains and the temperatures of places. To track any changes that occur over time, physical geographers keep careful records of all the information they collect.

Uses of Physical Geography

Earth is made up of hundreds of types of physical features. Without a complete understanding of what these features are and the effect they have on the world's people and landscapes, we cannot fully understand our world. This is the major reason that geographers study the physical world—to learn how it works.

There are also other, more specific reasons for studying physical geography, though. Studying the changes that take place on our planet can help us prepare to live with those changes. For example, knowing what causes volcanoes to erupt can help us predict eruptions. Knowing what causes terrible storms can help us prepare for them. In this way, the work of physical geographers helps us adjust to the dangers and changes of our world.

READING CHECK **Analyzing** What are some features in your area that a physical geographer might study?

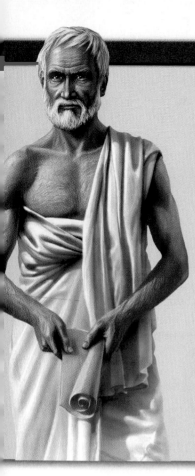

Eratosthenes
(c. 276–c. 194 BC)

Did you know that geography is over 2,000 years old? Actually, the study of the world is even older than that, but the first person ever to use the word *geography* lived then. His name was Eratosthenes (er-uh-TAHS-thuh-neez), and he was a Greek scientist and librarian. With no modern instruments of any kind, Eratosthenes figured out how large Earth is. He also drew a map that showed all of the lands that the Greeks knew about. Because of his many contributions to the field, Eratosthenes has been called the Father of Geography.

Generalizing Why is Eratosthenes called the Father of Geography?

Human Geography

The physical world is only one part of the puzzle of geography. People are also part of the world. **Human geography is the study of the world's people, communities, and landscapes.** It is the second major branch of geography.

The Human World

Put simply, human geographers study the world's people, past and present. They look at where people live and why. They ask why some parts of the world have more people than others, and why some places have almost no people at all.

Human geographers also study what people do. What jobs do people have? What crops do they grow? What makes them move from place to place? These are the types of questions that geographers ask about people around the world.

Because people's lives are so different around the world, no one can study every aspect of human geography. As a result, human geographers often specialize in a smaller area of study. Some may choose to study only the people and landscapes in a certain region. For example, a geographer may study only the lives of people who live in West Africa.

Other geographers choose not to limit their studies to one place. Instead, they may choose to examine only one aspect of people's lives. For example, a geographer could study only economics, politics, or city life. However, that geographer may compare economic patterns in various parts of the world to see how they differ.

Uses of Human Geography

Although every culture is different, people around the world have some common needs. All people need food and water. All people need shelter. All people need to deal with other people in order to survive.

Human geographers study how people in various places address their needs. They look at the foods people eat and the types of governments they form. The knowledge they gather can help us better understand people in other cultures. Sometimes this type of understanding can help people improve their landscapes and situations.

On a smaller scale, human geographers can help people design their cities and towns. By understanding where people go and what they need, geographers can help city planners place roads, shopping malls, and schools. Geographers also study the effect people have on the world. As a result, they often work with private groups and government agencies who want to protect the environment.

READING CHECK **Summarizing** What do human geographers study?

Other Fields of Geography

Physical geography and human geography are the two largest branches of the subject, but they are not the only ones. Many other fields of geography exist, each one devoted to studying one aspect of the world.

Most of these fields are smaller, more specialized areas of either physical or human geography. For example, economic geography—the study of how people make and spend money—is a branch of human geography. Another specialized branch of human geography is urban geography, the study of cities and how people live in them. Physical geography also includes many fields, such as the study of climates. Other fields of physical geography are the studies of soils and plants.

Cartography

One key field of geography is **cartography, the science of making maps.** You have already seen how important maps are to the study of geography. Without maps, geographers would not be able to study where things are in the world.

In the past, maps were always drawn by hand. Many were not very accurate. Today, though, most maps are made using computers and satellite images. Through advances in mapmaking, we can make accurate maps on almost any scale, from the whole world to a single neighborhood, and keep them up to date. These maps are not only used by geographers. For example, road maps are used by people who are planning long trips.

CONNECTING TO Technology

Computer Mapping

In the past, maps were drawn by hand. Making a map was a slow process. Even the simplest map took a long time to make. Today, however, cartographers have access to tools people in the past—even people who lived just 50 years ago—never imagined. The most important of these tools are computers.

Computers allow us to make maps quickly and easily. In addition, they let us make new types of maps that people could not make in the past.

The map shown here, for example, was drawn on a computer. It shows the number of computer users in the United States who were connected to the Internet on a particular day. Each of the lines that rises off of the map represents a city in which people were using the Internet. The color of the line indicates the number of computer users in that city. As you can see, this data resulted in a very complex map.

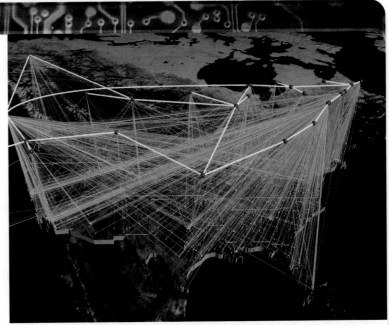

Making such a map required cartographers to sort through huge amounts of complex data. Such sorting would not have been possible without computers.

Contrasting How are today's maps different from those created in the past?

Meteorology is the study of weather. This meteorologist is using computers to follow and predict the movement of a powerful storm.

Meteorology

Have you ever seen the weather report on television? If so, you have seen the results of another branch of geography. This branch is called **meteorology**, the study of weather and what causes it.

Meteorologists study weather patterns in a particular area. Then they use the information to predict what the weather will be like in the coming days. Their work helps people plan what to wear and what to do on any given day. At the same time, their work can save lives by predicting the arrival of terrible storms. These predictions are among the most visible ways in which the work of geographers affects our lives every day.

READING CHECK Finding Main Ideas What are some major branches of geography?

Hydrology

Another important branch of geography is hydrology, the study of water on Earth. Geographers in this field study the world's river systems and rainfall patterns. They study what causes droughts and floods and how people in cities can get safe drinking water. They also work to measure and protect the world's supply of water.

FOCUS ON READING

What do you already know about drinking water?

SUMMARY AND PREVIEW In this section, you learned about two main branches of geography, physical and human. In the next chapter, you will learn more about the physical features that surround us and the processes that create them.

go.hrw.com
Online Quiz
KEYWORD: SK7 HP1

Section 3 Assessment

Reviewing Ideas, Terms, and Places

1. a. **Define** What is **physical geography**?
   b. **Explain** Why do we study physical geography?
2. a. **Identify** What are some things that people study as part of **human geography**?
   b. **Summarize** What are some ways in which the study of human geography can influence our lives?
   c. **Evaluate** Which do you think would be more interesting to study, physical geography or human geography? Why?
3. a. **Identify** What are two specialized fields of geography?
   b. **Analyze** How do cartographers contribute to the work of other geographers?

Critical Thinking

4. **Comparing and Contrasting** Draw a diagram like the one shown here. In the left circle, list three features of physical geography from your notes. In the right circle, list three features of human geography. Where the circles overlap, list one feature they share.

Physical   Human

FOCUS ON WRITING

5. **Choosing a Branch** Your job description should point out to people that there are many branches of geography. How will you note that?

Chapter Review

Geography's Impact
video series
Review the video to answer the closing question:
Why do you think it might be valuable to know the absolute location of a place?

Visual Summary

Use the visual summary below to help you review the main ideas of the chapter.

QUICK FACTS

Physical geography—the study of the world's physical features—is one main branch of geography.

Human geography—the study of the world's people and how they live—is the second main branch.

Geographers use many tools to study the world. The most valuable of these tools are maps.

Reviewing Vocabulary, Terms, and Places

Match the words in the columns with the correct definitions listed below.

1. geography

2. physical geography

3. human geography

4. element

5. meteorology

6. region

7. cartography

8. map

9. landscape

10. globe

a. a part of the world that has one or more common features that make it different from surrounding areas

b. a flat drawing of part of Earth's surface

c. a part

d. a spherical model of the planet

e. the study of the world's physical features

f. the study of weather and what causes it

g. the study of the world, its people, and the landscapes they create

h. the science of making maps

i. the physical and human features that define an area and make it different from other places

j. the study of people and communities

Comprehension and Critical Thinking

SECTION 1 *(Pages 4–9)*

11. a. Identify What are three levels at which a geographer might study the world? Which of these levels covers the largest area?

b. Compare and Contrast How are maps and globes similar? How are they different?

c. Elaborate How might satellite images and computers help geographers improve their knowledge of the world?

SECTION 2 *(Pages 10–14)*

12. a. Define What do geographers mean when they discuss an area's landscape?

b. Explain Why did geographers create the five themes and the six essential elements?

c. Predict How might the five themes and six essential elements help you in your study of geography?

SECTION 3 *(Pages 16–20)*

13. a. Identify What are the two main branches of geography? What does each include?

b. Summarize How can physical geography help people adjust to the dangers of the world?

c. Elaborate Why do geographers study both physical and human characteristics of places?

Using the Internet

go.hrw.com
KEYWORD: SK7 CH1

14. Activity: Using Maps What does your town or community look like? What can be found there? Maps can help you understand your community and learn about its features. Enter the activity keyword to learn more about maps and how they can help you better understand your community. Then search the Internet to find a map of your community. Use the map to find the locations of at least five important features. For example, you might locate your school, the library, a park, or major highways. Be creative and find other places you think your classmates should be aware of.

Social Studies Skills

Analyzing Satellite Images *Use the satellite images of Italy from the Social Studies Skills lesson in this chapter to answer the following questions.*

15. On which image do forests appear more clearly, the true-color or the infrared image?

16. What color do you think represents mountains on the infrared satellite image?

17. Why might geographers use satellite images like these while making maps of Italy?

18. Using Prior Knowledge Create a chart with three columns. In the first column list what you knew about geography before you read the chapter. In the second column list what you learned in the chapter. In the third column list questions that you now have about geography.

19. Writing Your Job Description Review your notes on the different jobs geographers do. Then write your job description. You should begin your description by explaining why the job is important. Then identify the job's tasks and responsibilities. Finally, tell what kind of person might do well as a geographer.

Map Activity

20. Sketch Map Draw a map that shows your school and the surrounding neighborhood. Your map does not have to be complicated, but you should include major features like streets and buildings. Use the map shown here as an example.

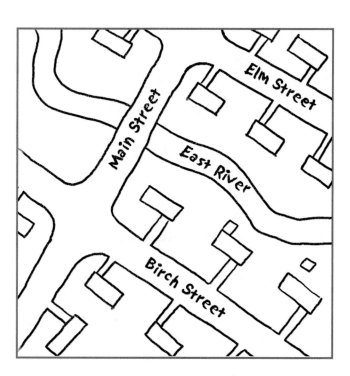

Standardized Test Practice

DIRECTIONS: Read questions 1 through 7 and write the letter of the best response. Then read question 8 and write your own well-constructed response.

1 **Which of the following subjects would a human geographer study the most?**

   A mountains

   B populations

   C rivers

   D volcanoes

2 **The study of weather is called**

   A meteorology.

   B hydrology.

   C social science.

   D cartography.

3 **A region is an area that has**

   A one or more common features.

   B no people living in it.

   C few physical features.

   D set physical boundaries.

4 **How many essential elements of geography have geographers identified?**

   A two

   B four

   C six

   D eight

5 **The physical and human characteristics that define an area are its**

   A landscape.

   B location.

   C region.

   D science.

The United States

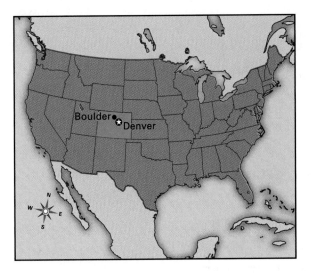

6 **Which of the five themes of geography would a geographer most likely study using this map?**

   A movement

   B location

   C human-environment interaction

   D landscape

7 **The smallest level at which a geographer might study a place is**

   A microscopic.

   B local.

   C regional.

   D global.

8 **Extended Response** Look at the map of the United States above. Do you think this map is more likely to be used by a physical geographer or by a human geographer? Give two reasons for your answer. Then write two statements about what a geographer could find on this map.

CHAPTER 2

Planet Earth

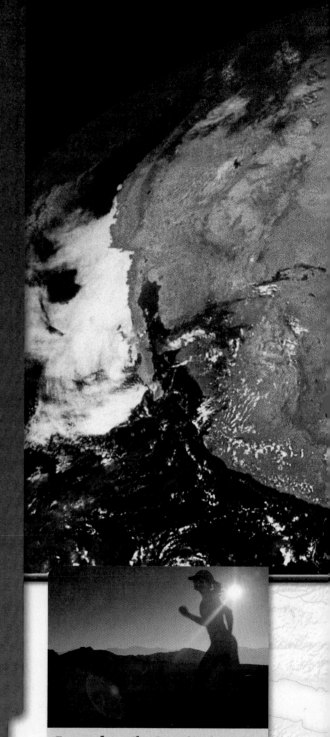

What You Will Learn...

In this chapter you will learn about important processes on planet Earth. You will discover how Earth's movements affect the energy we receive from the sun, how water affects life, and how Earth's landforms were made.

SECTION 1
Earth and the Sun's Energy **26**

SECTION 2
Water on Earth **30**

SECTION 3
The Land **35**

FOCUS ON READING AND WRITING

Using Word Parts Sometimes you can figure out the meaning of a word by looking at its parts. A root is the base of the word. A prefix attaches to the beginning, and a suffix attaches to the ending. When you come across a word you don't know, check to see whether you recognize its parts. **See the lesson, Using Word Parts, on page 669.**

Writing a Haiku Join the poets who have celebrated our planet for centuries. Write a haiku, a short poem, about planet Earth. As you read the chapter, gather information about changes in the sun's energy, Earth's water supply, and shapes on the land. Then choose the most intriguing information to include in your haiku.

Energy from the Sun The planet's movement creates differences in the amount of energy Earth receives from the sun.

Many of Earth's features are visible from space. This photo, taken from a satellite orbiting the planet, shows part of the North American continent.

Which of Earth's features are visible in this photo?

HOLT

Geography's Impact
video series
Watch the video to understand the impact of water on Earth.

Land Forces on and under Earth's surface have shaped the different landforms on our planet. Geographers study how mountains and other landforms were made.

Water on Earth Water is essential for life on Earth. Much of the planet's water supply is stored in Earth's oceans and ice caps.

Earth and the Sun's Energy

What You Will Learn...

Main Ideas

1. Earth's movement affects the amount of energy we receive from the sun.
2. Earth's seasons are caused by the planet's tilt.

The Big Idea

Earth's movement and the sun's energy interact to create day and night, temperature changes, and the seasons.

Key Terms

solar energy, *p. 26*
rotation, *p. 26*
revolution, *p. 27*
latitude, *p. 27*
tropics, *p. 29*

TAKING NOTES As you read, take notes on Earth's movement and the seasons. Use a chart like the one below to organize your notes.

Earth's Movement	The Seasons

If YOU lived there...

You live in Chicago and have just won an exciting prize—a trip to Australia during winter vacation in January. As you prepare for the trip, your mother reminds you to pack shorts and a swimsuit. You are confused. In January you usually wear winter sweaters and a heavy jacket.

Why is the weather so different in Australia?

BUILDING BACKGROUND Seasonal differences in weather are an important result of Earth's constant movement. As the planet moves, we experience changes in the amount of energy we receive from the sun. Geographers study and explain why different places on Earth receive differing amounts of energy from the sun.

Earth's Movement

Energy from the sun helps crops grow, provides light, and warms Earth. It even influences the clothes we wear, the foods we eat, and the sports we play. All life on Earth requires **solar energy, or energy from the sun**, to survive. The amount of solar energy Earth receives changes constantly. Earth's rotation, revolution, and tilt, as well as latitude, all affect the amount of solar energy the planet receives from the sun.

Rotation

Imagine that Earth has a rod running through it from the North Pole to the South Pole. This rod represents Earth's axis—an imaginary line around which a planet turns. As Earth spins on its axis, different parts of the planet face the sun. It takes Earth 24 hours, or one day, to complete this rotation. A **rotation is one complete spin of Earth on its axis**. As Earth rotates during this 24-hour period, it appears to us that the sun moves across the sky. The sun seems to rise in the east and set in the west. The

Solar Energy

Earth's tilt and rotation cause changes in the amount of energy we receive from the sun. As Earth rotates on its axis, energy from the sun creates periods of day and night. Earth's tilt causes some locations, especially those close to the equator, to receive more direct solar energy than others.

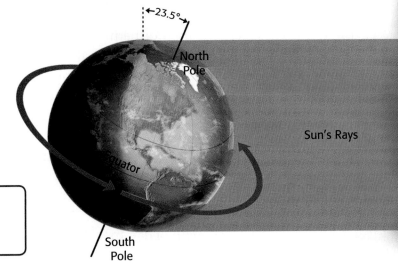

←23.5°

North Pole

Equator

Sun's Rays

South Pole

ANALYSIS SKILL **ANALYZING VISUALS**

Is the region north or south of the equator receiving more solar energy? How can you tell?

---

sun, however, does not move. It is actually Earth's rotation that creates the sense of the sun's movement.

Earth's rotation also explains why day changes to night. As you can see in the illustration, solar energy strikes only the half of Earth facing the sun. Warmth and light from the sun create daytime. At the same time, the half of the planet facing away from the sun experiences the cooler temperatures and darkness of night. Earth's rotation causes regular shifts from day to night. As a result, levels of solar energy on Earth constantly change.

Revolution

As Earth spins on its axis, it also follows a path, or orbit, around the sun. Earth's orbit around the sun is not a perfect circle. Sometimes the orbit takes Earth closer to the sun, and at other times the orbit takes it farther away. It takes 365¼ days for Earth to complete one **revolution, or trip around the sun.** We base our calendar year on the time it takes Earth to complete its orbit around the sun. To allow for the fraction of a day, we add an extra day—February 29—to our calendar every four years.

Tilt and Latitude

Another **factor** affecting the amount of solar energy we receive is the planet's tilt. As the illustration shows, Earth's axis is not straight up and down. It is actually tilted at an angle of 23½ degrees from vertical. At any given time of year, some locations on Earth are tilting away from the sun, and others are tilting toward it. Places tilting toward the sun receive more solar energy and experience warmer temperatures. Those tilting away from the sun receive less solar energy and experience cooler temperatures.

A location's **latitude, the distance north or south of Earth's equator,** also affects the amount of solar energy it receives. Low-latitude areas, those near the equator like Hawaii, receive direct rays from the sun all year. These direct rays are more intense and produce warmer temperatures. Regions with high latitudes, like Antarctica, are farther from the equator. As a result, they receive indirect rays from the sun and have colder temperatures.

READING CHECK Finding Main Ideas What factors affect the solar energy Earth receives?

ACADEMIC VOCABULARY

factor cause

The Seasons

FOCUS ON READING
The prefix *hemi-* means half. What does the word *hemisphere* mean?

Does the thought of snow in July or 100-degree temperatures in January seem odd to you? It might if you live in the Northern Hemisphere, where cold temperatures are common in January, not July. The planet's changing seasons explain why we often connect certain weather with specific times of the year, like snow in January. Seasons are periods during the year that are known for a particular type of weather. Many places on Earth experience four seasons—winter, spring, summer, and fall. These seasons are based on temperature and length of day. In some parts of the world, however, seasons are based on the amount of rainfall.

Winter and Summer

The change in seasons is created by Earth's tilt. As you can see in the illustration below, while one of Earth's poles tilts away from the sun, the other tilts toward it. During winter part of Earth is tilted away from the sun, causing less direct solar energy, cool temperatures, and less daylight. Summer occurs when part of Earth is tilted toward the sun. This creates more direct solar energy, warmer temperatures, and longer periods of daylight.

Because of Earth's tilt, the Northern and Southern hemispheres experience opposite seasons. As the North Pole tilts toward the sun in summer, the South Pole tilts away

The Seasons: Northern Hemisphere

As Earth orbits the sun, the tilt of its axis toward and away from the sun causes the seasons to change. Seasons in the Northern Hemisphere change at about the same time every year.

ANALYZING VISUALS As the Northern Hemisphere experiences winter, what season is it in the Southern Hemisphere?

March 21 SPRING

Winter and Spring The North Pole tilts away from the sun in winter, causing cooler temperatures. In the spring, temperatures gradually rise as the North Pole begins to point toward the sun.

December 21 WINTER

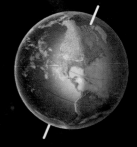

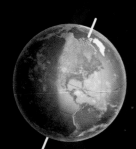

June 21 SUMMER

Summer and Fall Summer's warm temperatures are the result of the North Pole's tilt toward the sun. As we move away from the sun in the fall, temperatures slowly decline.

North Pole
Tropic of Cancer
Equator
Tropic of Capricorn
South Pole

September 22 FALL

from it. As a result, the Southern Hemisphere experiences winter. Likewise, when it is spring in the Northern Hemisphere, it is fall in the Southern Hemisphere.

Spring and Fall

As Earth orbits the sun, there are periods when the poles tilt neither toward nor away from the sun. These periods mark spring and fall. During the spring, as part of Earth begins to tilt toward the sun, solar energy increases. Temperatures slowly start to rise, and days grow longer. In the fall the opposite occurs as winter approaches. Solar energy begins to decrease, causing cooler temperatures and shorter days.

Rainfall and Seasons

Some regions on Earth have seasons marked by rainfall rather than temperature. This is true in the **tropics**, regions close to the equator. At certain times of year, winds bring either dry or moist air to the tropics, creating wet and dry seasons. In India, for example, seasonal winds called monsoons bring heavy rains from June to October and dry air from November to January.

READING CHECK Identifying Cause and Effect What causes the seasons to change?

FOCUS ON CULTURE

The Midnight Sun

Can you imagine going to sleep late at night with the sun shining in the sky? People who live near the Arctic and Antarctic circles experience this every summer, when they can receive up to 24 hours of sunlight a day. The time-lapse photo below shows a typical sunset during this period—except the sun never really sets! This phenomenon is known as the midnight sun. For locations like Tromso, Norway, this means up to two months of constant daylight each summer. People living near Earth's poles often use the long daylight hours to work on outdoor projects in preparation for winter, when they can receive 24 hours of darkness a day.

Predicting How might people's daily lives be affected by the midnight sun?

SUMMARY AND PREVIEW Solar energy is crucial for all life on the planet. Earth's position and movements affect the amount of energy we receive from the sun and determine our seasons. Next, you will learn about Earth's water supply and its importance to us.

go.hrw.com
Online Quiz
KEYWORD: SK7 HP2

Section 1 Assessment

Reviewing Ideas, Terms, and Places

1. **a. Identify** What is **solar energy**, and how does it affect Earth?
   b. Analyze How do **rotation** and tilt each affect the amount of solar energy Earth receives?
   c. Predict What might happen if Earth received less solar energy than it currently does?
2. **a. Describe** Name and describe Earth's seasons.
   b. Contrast How are seasons different in the Northern and Southern hemispheres?
   c. Elaborate How might the seasons affect human activities?

Critical Thinking

3. **Identifying Cause and Effect** Use your notes and the diagram to identify the causes of seasons.

Cause	→	
		Effect: Earth's changing seasons
Cause	→	

FOCUS ON WRITING

4. **Describing the Seasons** What are the seasons like where you live? In your notebook, jot down a few notes that describe the changing seasons.

Water on Earth

What You Will Learn...

Main Ideas

1. Salt water and freshwater make up Earth's water supply.
2. In the water cycle, water circulates from Earth's surface to the atmosphere and back again.
3. Water plays an important role in people's lives.

The Big Idea

Water is a dominant feature on Earth's surface and is essential for life.

Key Terms

freshwater, *p. 31*
glaciers, *p. 31*
surface water, *p. 31*
precipitation, *p. 31*
groundwater, *p. 32*
water vapor, *p. 32*
water cycle, *p. 33*

TAKING NOTES As you read, take notes about Earth's water, the water cycle, and how water affects our lives. Use a diagram like the one below to organize your notes.

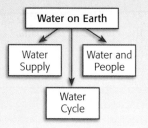

Water on Earth
Water Supply
Water and People
Water Cycle

If **YOU** lived there...

You live in the desert Southwest, where heavy water use and a lack of rainfall have led to water shortages. Your city plans to begin a water conservation program that asks people to limit how much water they use. Many of your neighbors have complained that the program is unnecessary. Others support the plan to save water.

How do you feel about the city's water plan?

BUILDING BACKGROUND Although water covers much of Earth's surface, water shortages, like those in the American Southwest, are common all over the planet. Because water is vital to the survival of all living things, geographers study Earth's water supply.

Earth's Water Supply

Think of the different uses for water. We use water to cook and clean, we drink it, and we grow crops with it. Water is used for recreation, to generate electricity, and even to travel from place to place. Water is perhaps the most important and abundant resource on Earth. In fact, water covers some two-thirds of the planet. Understanding Earth's water supply and how it affects our lives is an important part of geography.

Earth's Distribution of Water

Earth's water supply is divided into two main types—salt water and freshwater. Humans, plants, and animals rely on Earth's freshwater supply for survival.

Salt Water

Although water covers much of the planet, we cannot use most of it. About 97 percent of the Earth's water is salt water. Because salt water contains high levels of salt and other minerals, it is unsafe to drink.

In general, salt water is found in Earth's oceans. Oceans are vast bodies of water covering some 71 percent of the planet's surface. Earth's oceans are made up of smaller bodies of water such as seas, gulfs, bays, and straits. Altogether, Earth's oceans cover some 139 million square miles (360 million square km) of the planet's surface.

Some of Earth's lakes contain salt water. The Great Salt Lake in Utah, for example, is a saltwater lake. As salt and other minerals have collected in the lake, which has no outlet, the water has become salty.

Freshwater

Since the water in Earth's oceans is too salty to use, we must rely on other sources for freshwater. **Freshwater**, or water without salt, makes up only about 3 percent of our total water supply. Much of that freshwater is locked in Earth's **glaciers**, **large areas of slow-moving ice**, and in the ice of the Antarctic and Arctic regions. Most of the freshwater we use everyday is found in lakes, in rivers, and under Earth's surface.

One form of freshwater is surface water. **Surface water** is water that is found in Earth's streams, rivers, and lakes. It may seem that there is a great deal of water in our lakes and rivers, but only a tiny amount of Earth's water supply—less than 1 percent—comes from surface water.

Streams and rivers are a common source of surface water. Streams form when precipitation collects in a narrow channel and flows toward the ocean. **Precipitation is water that falls to Earth's surface as rain, snow, sleet, or hail**. In turn, streams join together to form rivers. Any smaller stream or river that flows into a larger stream or river is called a tributary. For example, the Missouri River is the largest tributary of the Mississippi River.

Lakes are another important source of surface water. Some lakes were formed as rivers filled low-lying areas with water. Other lakes, like the Great Lakes along the U.S.–Canada border, were formed when glaciers carved deep holes in Earth's surface and deposited water as they melted.

Most of Earth's available freshwater is stored underground. As precipitation falls to Earth, much of it is absorbed into the ground, filling spaces in the soil and rock.

Salt Water Earth's oceans contain some 97 percent of the planet's water supply. Unfortunately, this water is too salty to drink.

Freshwater Freshwater from lakes, rivers, and streams makes up only a fraction of Earth's water supply.

Water found below Earth's surface is called **groundwater**. In some places on Earth, groundwater naturally bubbles from the ground as a spring. More often, however, people obtain groundwater by digging wells, or deep holes dug into the ground to reach the water.

READING CHECK **Contrasting** How is salt water different from freshwater?

Interactive
Close-up

The Water Cycle

Energy from the sun drives the water cycle. Surface water evaporates into Earth's atmosphere, where it condenses, then falls back to Earth as precipitation. This cycle repeats continuously, providing us with a fairly constant water supply.

go.hrw.com KEYWORD: SK7 CH2

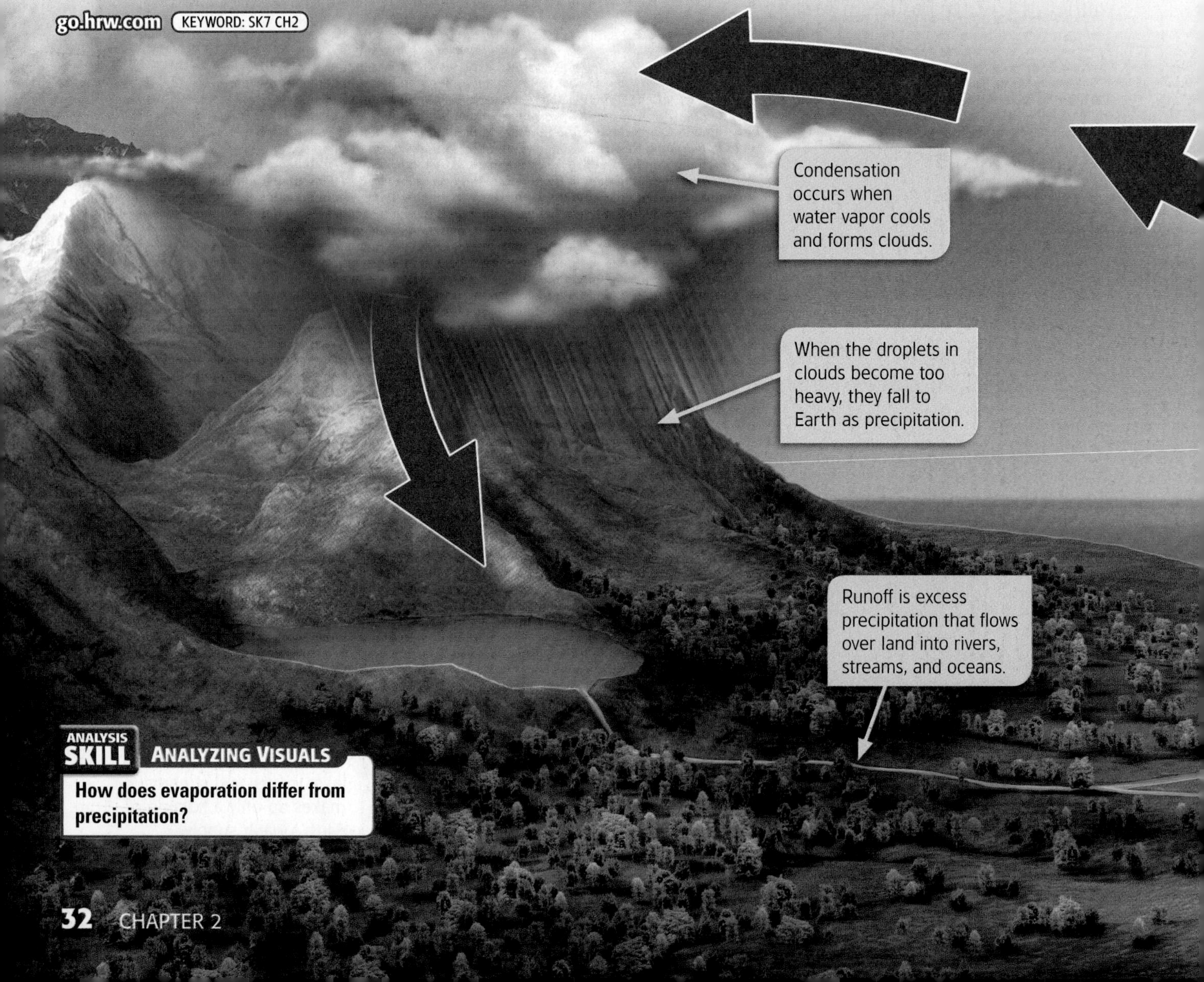

The Water Cycle

When you think of water, you probably visualize a liquid—a flowing stream, a glass of ice-cold water, or a wave hitting the beach. But did you know that water is the only substance on Earth that occurs naturally as a solid, a liquid, and a gas? We see water as a solid in snow and ice and as a liquid in oceans and rivers. Water also occurs in the air as an invisible gas called **water vapor**.

Water is always moving. As water heats up and cools down, it moves from the planet's surface to the atmosphere, or the mass of air that surrounds Earth. One of the most important processes in nature

Condensation occurs when water vapor cools and forms clouds.

When the droplets in clouds become too heavy, they fall to Earth as precipitation.

Runoff is excess precipitation that flows over land into rivers, streams, and oceans.

ANALYSIS SKILL **ANALYZING VISUALS**

How does evaporation differ from precipitation?

is the water cycle. The **water cycle** is the movement of water from Earth's surface to the atmosphere and back.

The sun's energy drives the water cycle. As the sun heats water on Earth's surface, some of that water evaporates, or turns from liquid to gas, or water vapor. Water vapor then rises into the air. As the vapor rises, it cools. The cooling causes the water vapor to condense, or change from a vapor into tiny liquid droplets. These droplets join together to form clouds. If the droplets become heavy enough, precipitation occurs—that is, the water falls back to Earth as rain, snow, sleet, or hail.

When that precipitation falls back to Earth's surface, some of the water is absorbed into the soil as groundwater. Excess water, called runoff, flows over land and collects in streams, rivers, and oceans. Because the water cycle is constantly repeating, it allows us to maintain a fairly constant supply of water on Earth.

READING CHECK Finding Main Ideas What is the water cycle?

As energy from the sun heats water on Earth's surface, the water evaporates, or turns to water vapor, and rises to the atmosphere.

Water and People

How many times a day do you think about water? Many of us rarely give it a second thought, yet water is crucial for survival. Water problems such as the lack of water, polluted water, and flooding are concerns for people all around the world. Water also provides us with <u>countless</u> benefits, such as energy and recreation.

Water Problems

One of the greatest water problems people face is a lack of available freshwater. Many places face water shortages as a result of droughts, or long periods of lower-than-normal precipitation. Another cause of water shortages is overuse. In places like the southwestern United States, where the population has grown rapidly, the heavy demand for water has led to shortages.

Even where water is plentiful, it may not be clean enough to use. If chemicals and household wastes make their way into streams and rivers, they can contaminate the water supply. Polluted water can carry diseases. These diseases may harm humans, plants, and animals.

Flooding is another water problem that affects people around the world. Heavy rains often lead to flooding, which can damage property and threaten lives. One example of dangerous flooding occurred in Bangladesh in 2004. Floods there destroyed roads and schools and left some 25 million people homeless.

Water's Benefits

Water does more than just quench our thirst. It provides us with many benefits, such as food, power, and even recreation.

Water's most important benefit is that it provides us with food to eat. Everything we eat depends on water. For example, fruits and vegetables need water to grow.

FOCUS ON READING

Look at the word *countless* in this paragraph. The suffix *-less* means unable to. What does *countless* mean?

The Benefits of Water

Many people take advantage of the recreational and agricultural benefits that water provides.

Animals also need water to live and grow. As a result, we use water to farm and raise animals so that we will have food to eat.

Water is also an important source of energy. Using dams, we harness the power of moving water to produce electricity. Electricity provides power to air-condition or heat our homes, to run our washers and dryers, and to keep our food cold.

Water also provides us with recreation. Rivers, lakes, and oceans make it possible for us to swim, to fish, to surf, or to sail a boat. Although recreation is not critical for our survival, it does make our lives richer and more enjoyable.

READING CHECK Summarizing How does water affect people's lives?

SUMMARY AND PREVIEW In this section you learned that water is essential for life on Earth. Next, you will learn about the shapes on Earth's surface.

Section 2 Assessment

Reviewing Ideas, Terms, and Places

1. **a. Describe** Name and describe the different types of water that make up Earth's water supply.
   b. Analyze Why is only a small percentage of Earth's **freshwater** available to us?
   c. Elaborate In your opinion, which is more important—**surface water** or **groundwater**? Why?
2. **a. Recall** What drives the **water cycle**?
   b. Make Inferences From what bodies of water do you think most evaporation occurs? Why?
3. **a. Define** What is a drought?
   b. Analyze How does water support life on Earth?
   c. Evaluate What water problem do you think is most critical in your community? Why?

Critical Thinking

4. **Sequencing** Draw the graphic organizer at right. Then use your notes and the graphic organizer to identify the stages in Earth's water cycle.

FOCUS ON WRITING

5. **Learning about Water** Consider what you have learned about water in this section. How might you describe water in your haiku? What words might you use to describe Earth's water supply?

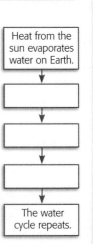

Heat from the sun evaporates water on Earth.

↓

↓

↓

The water cycle repeats.

The Land

If YOU lived there...

You live in the state of Washington. All your life, you have looked out at the beautiful, cone-shaped peaks of nearby mountains. One of them is Mount Saint Helens, an active volcano. You know that in 1980 it erupted violently, blowing a hole in the mountain and throwing ash and rock into the sky. Since then, scientists have watched the mountain carefully.

How do you feel about living near a volcano?

BUILDING BACKGROUND Over billions of years, many different forces have changed Earth's surface. Processes deep underground have built up landforms and even shifted the position of continents. Wind, water, and ice have also shaped the planet's landforms. Changes in Earth's surface continue to take place.

Landforms

Do you know the difference between a valley and a volcano? Can you tell a peninsula from a plateau? If you answered yes, then you are familiar with some of Earth's many landforms. **Landforms** are shapes on the planet's surface, such as hills or mountains. Landforms make up the landscapes that surround us, whether it's the rugged mountains of central Colorado or the flat plains of Oklahoma.

Earth's surface is covered with landforms of many different shapes and sizes. Some important landforms include:

- mountains, land that rises higher than 2,000 feet (610 m)
- valleys, areas of low land located between mountains or hills
- plains, stretches of mostly flat land
- islands, areas of land completely surrounded by water
- peninsulas, land surrounded by water on three sides

Because landforms play an important role in geography, many scientists study how landforms are made and how they affect human activity.

READING CHECK **Summarizing** What are some common landforms?

What You Will Learn...

Main Ideas

1. Earth's surface is covered by many different landforms.
2. Forces below Earth's surface build up our landforms.
3. Forces on the planet's surface shape Earth's landforms.
4. Landforms influence people's lives and culture.

The Big Idea

Processes below and on Earth's surface shape the planet's physical features.

Key Terms

landforms, *p. 35*
continents, *p. 36*
plate tectonics, *p. 36*
lava, *p. 37*
earthquakes, *p. 38*
weathering, *p. 39*
erosion, *p. 39*

TAKING NOTES As you read, use a diagram like the one below to take notes on Earth's landforms. In the circles, be sure to note how landforms are created, change, and affect people's lives.

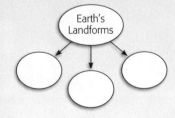

Forces below Earth's Surface

Geographers often study how landforms are made. One explanation for how landforms have been shaped involves forces below Earth's surface.

Earth's Plates

ACADEMIC
VOCABULARY
structure the
way something
is set up or
organized

To understand how these forces work, we must examine Earth's **structure**. The planet is made up of three layers. A solid inner core is surrounded by a liquid layer, or mantle. The solid outer layer of Earth is called the crust. The planet's **continents**, or large landmasses, are part of Earth's crust.

Geographers use the theory of plate tectonics to explain how forces below Earth's surface have shaped our landforms. The theory of **plate tectonics** suggests that Earth's surface is divided into a dozen or so slow-moving plates, or pieces of Earth's crust. As you can see in the image below, some plates, like the Pacific plate, are quite large. Others, like the Nazca plate, are much smaller. These plates cover Earth's entire surface. Some plates are under the ocean. These are known as ocean plates. Other plates, known as continental plates, are under Earth's continents.

Why do these plates move? Energy deep inside the planet puts pressure on Earth's crust. As this pressure builds up, it forces the plates to shift. Earth's tectonic plates all move. However, they move in different directions and at different speeds.

The Movement of Continents

Earth's tectonic plates move slowly—up to several inches per year. The continents, which are part of Earth's plates, shift as the plates move. If we could look back some 200 million years, we would see that the continents have traveled great distances. This idea is known as continental drift.

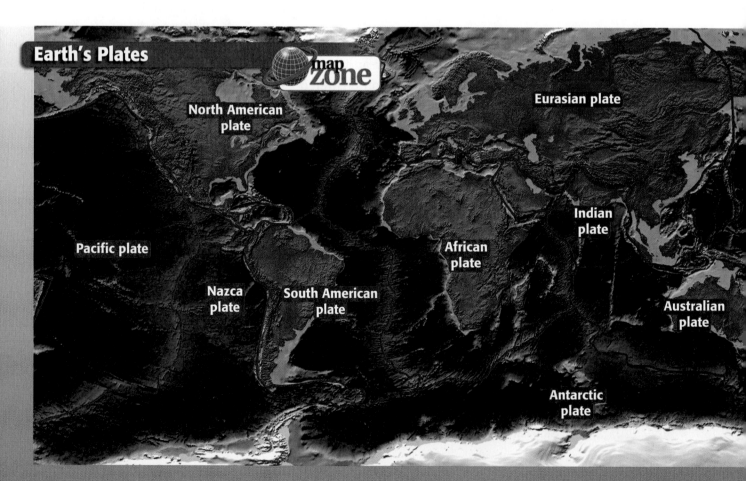

Earth's Plates

map Zone

North American plate

Eurasian plate

Pacific plate

Indian plate

African plate

Nazca plate

South American plate

Australian plate

Antarctic plate

The theory of continental drift, first developed by Alfred Wegener, states that the continents were once united in a single supercontinent. According to this theory, Earth's plates shifted over millions of years. As a result, the continents slowly separated and moved to their present positions.

Earth's continents are still moving. Some plates move toward each other and collide. Other plates separate and move apart. Still others slide past one another. Over time, colliding, separating, and sliding plates have shaped Earth's landforms.

Plates Collide

As plates collide, the energy created from their collision produces distinct landforms. The collision of different types of plates creates different shapes on Earth's surface. Ocean trenches and mountain ranges are two examples of landforms produced by the collision of tectonic plates.

BIOGRAPHY

Alfred Wegener
(1880–1930)

German scientist Alfred Wegener's fascination with the similarities between the western coast of Africa and the eastern coast of South America led to his theory of continental drift. Wegener argued that the two continents had once been joined together. Years of plate movement broke the continents apart and moved them to their current locations. It was only after Wegener's death that his ideas became a central part of the theory of plate tectonics.

The theory of plate tectonics suggests that the plates that make up Earth's crust are moving, usually only a few inches per year. As Earth's plates collide, separate, and slide past each other, they create forces great enough to shape many of Earth's landforms.

ANALYZING VISUALS Looking at the map, what evidence indicates that plates have collided or separated?

When two ocean plates collide, one plate pushes under the other. This process creates ocean trenches. Ocean trenches are deep valleys in the ocean floor. Near Japan, for example, the Pacific plate is slowly moving under other plates. This collision has created several deep ocean trenches, including the world's deepest trench, the Mariana Trench.

Ocean plates and continental plates can also collide. When this occurs, the ocean plate drops beneath the continental plate. This action forces the land above to crumple and form a mountain range. The Andes in South America, for example, were formed when the South American and Nazca plates collided.

The collision of two continental plates also results in mountain-building. When continental plates collide, the land pushes up, sometimes to great heights. The world's highest mountain range, the Himalayas, formed when the Indian plate crashed into the Eurasian plate. In fact, the Himalayas are still growing as the two plates continue to crash into each other.

Plates Separate

A second type of plate movement causes plates to separate. As plates move apart, gaps between the plates allow magma, a liquid rock from the planet's interior, to rise to Earth's crust. **Lava**, or magma that reaches Earth's surface, emerges from the gap that has formed. As the lava cools, it builds a mid-ocean ridge, or underwater mountain. For example, the separation of the North American and Eurasian plates formed the largest underwater mountain, the Mid-Atlantic Ridge. If these mid-ocean ridges grow high enough, they can rise above the surface of the ocean, forming volcanic islands. Iceland, on the boundary of the Eurasian and North American plates, is an example of such an island.

FOCUS ON READING
The suffix *–sion* means the act of. What does the word *collision* mean?

Plates Slide

Tectonic plates also slide past each other. As plates pass by one another, they sometimes grind together. This grinding produces **earthquakes**—sudden, violent movements of Earth's crust. Earthquakes often take place along faults, or breaks in Earth's crust where movement occurs. In California, for example, the Pacific plate is sliding by the edge of the North American plate. This has created the San Andreas Fault zone, an area where earthquakes are quite common.

The San Andreas Fault zone is one of many areas that lie along the boundaries of the Pacific plate. The frequent movement of this plate produces many earthquakes and volcanic eruptions along its edges. In fact, the region around the Pacific plate, called the Ring of Fire, is home to most of the world's earthquakes and volcanoes.

READING CHECK Finding Main Ideas What forces below Earth's surface shape landforms?

Plate Movement

The movement of tectonic plates has produced many of Earth's landforms. Volcanoes, islands, and mountains often result from the separation or collision of Earth's plates.

ANALYZING VISUALS What type of landform is created by the collision of two continental plates?

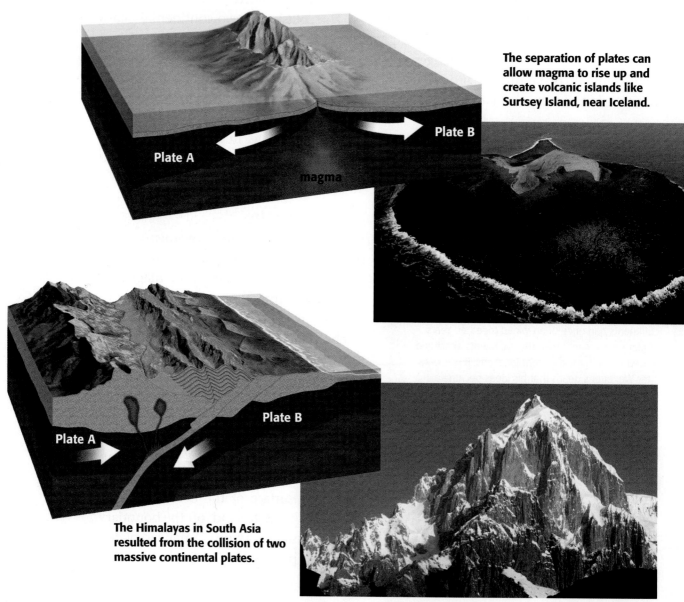

Plate A

Plate B

magma

The separation of plates can allow magma to rise up and create volcanic islands like Surtsey Island, near Iceland.

Plate A

Plate B

The Himalayas in South Asia resulted from the collision of two massive continental plates.

Forces on Earth's Surface

For millions of years, the movement of Earth's tectonic plates has been building up landforms on Earth's surface. At the same time, other forces are working to change those very same landforms.

Imagine a small pile of dirt and rock on a table. If you poured water on the pile, it would move the dirt and rock from one place to another. Likewise, if you were to blow at the pile, the rock and dirt would also move. The same process happens in nature. Weather, water, and other forces change Earth's landforms by wearing them away or reshaping them.

Weathering

One force that wears away landforms is weathering. **Weathering is the process by which rock is broken down into smaller pieces.** Several factors cause rock to break down. In desert areas, daytime heating and nighttime cooling can cause rocks to crack. Water may get into cracks in rocks and freeze. The ice then expands with a force great enough to break the rock. Even the roots of trees can pry rocks apart.

Regardless of which weathering process is at work, rocks eventually break down. These small pieces of rock are known as sediment. Once weathering has taken place, wind, ice, and water often move sediment from one place to another.

Erosion

Another force that changes landforms is the process of erosion. **Erosion is the movement of sediment from one location to another.** Erosion can wear away or build up landforms. Wind, ice, and water all cause erosion.

Powerful winds often cause erosion. Winds lift sediment into the air and carry it across great distances. On beaches and in deserts, wind can deposit large amounts of sand to form dunes. Blowing sand can also wear down rock. The sand acts like sandpaper to polish and wear away at rocks. As you can see in the photo below, wind can have a dramatic effect on landforms.

Earth's glaciers also have the power to cause massive erosion. Glaciers, or large, slow-moving sheets of ice, build up when winter snows do not melt the following summer. Glaciers can be huge. Glaciers in Greenland and Antarctica, for example, are great sheets of ice up to two miles (3 km) thick. Some glaciers flow slowly downhill like rivers of ice. As they do so, they erode the land by carving large U-shaped valleys and sharp mountain peaks. As the ice flows downhill, it crushes rock into sediment and can move huge rocks long distances.

Wind Erosion

Landforms in Utah's Canyonlands National Park have been worn away mostly by thousands of years of powerful winds.

Water is the most common cause of erosion. Waves in oceans and lakes can wear away the shore, creating jagged coastlines, like those on the coast of Oregon. Rivers also cause erosion. Over many years, the flowing water can cut through rock, forming canyons, or narrow areas with steep walls. Arizona's Horseshoe Bend and Grand Canyon are examples of canyons created in this way.

Flowing water shapes other landforms as well. When water deposits sediment in new locations, it creates new landforms. For example, rivers create floodplains when they flood their banks and deposit sediment along the banks. Sediment that is carried by a river all the way out to sea creates a delta. The sediment settles to the bottom, where the river meets the sea. The Nile and Mississippi rivers have created two of the world's largest river deltas.

READING CHECK **Comparing** How are weathering and erosion similar?

Water Erosion

For millions of years, the Colorado River has worn away the rock at Horseshoe Bend in Arizona.

Landforms Influence Life

Why do you live where you do? Perhaps your family moved to the desert to avoid harsh winter weather. Or possibly one of your ancestors settled near a river delta because its fertile soil was ideal for growing crops. Maybe your family wanted to live near the ocean to start a fishing business. As these examples show, landforms exert a strong influence on people's lives. Earth's landforms affect our settlements and our culture. At the same time, we affect the landforms around us.

Earth's landforms can influence where people settle. People sometimes settle near certain landforms and avoid others. For example, many settlements are built near fertile river valleys or deltas. The earliest urban civilization, for example, was built in the valley between the Tigris and Euphrates rivers. Other times, landforms discourage people from settling in a certain place. Tall, rugged mountains, like the Himalayas, and harsh desert climates, like the Sahara, do not usually attract large settlements.

Landforms affect our culture in ways that we may not have noticed. Landforms often influence what jobs are available in a region. For example, rich mineral deposits in the mountains of Colorado led to the development of a mining industry there. Landforms even affect language. On the island of New Guinea in Southeast Asia, rugged mountains have kept the people so isolated that more than 700 languages are spoken on the island today.

People sometimes change landforms to suit their needs. People may choose to modify landforms in order to improve their lives. For example, engineers built the Panama Canal to make travel from the Atlantic Ocean to the Pacific Ocean easier. In Southeast Asia, people who farm on steep hillsides cut terraces into the slope to

Living with Landforms

The people of Rio de Janeiro, Brazil, have learned to adapt to the mountains and bays that dominate their landscape.

ANALYZING VISUALS How have people in Rio de Janiero adapted to their landscape?

create more level space to grow their crops. People have even built huge dams along rivers to divert water for use in nearby towns or farms.

READING CHECK **Analyzing** What are some examples of humans adjusting to and changing landforms?

SUMMARY AND PREVIEW Landforms are created by actions deep within the planet's surface, and they are changed by forces on Earth's surface, like weathering and erosion. In the next chapter you will learn how other forces, like weather and climate, affect Earth's people.

Section 3 Assessment

go.hrw.com
Online Quiz
KEYWORD: SK7 HP2

Reviewing Ideas, Terms, and Places

1. **a. Describe** What are some common **landforms**?
   b. Analyze Why do geographers study landforms?
2. **a. Identify** What is the theory of **plate tectonics**?
   b. Compare and Contrast How are the effects of colliding plates and separating plates similar and different?
   c. Predict How might Earth's surface change as tectonic plates continue to move?
3. **a. Recall** What is the process of **weathering**?
   b. Elaborate How does water affect sediment?
4. **a. Recall** How do landforms affect life on Earth?
   b. Predict How might people adapt to life in an area with steep mountains?

Critical Thinking

5. **Analyzing** Use your notes and the chart below to identify the different factors that alter Earth's landforms and the changes that they produce.

Factor	Change in Landform

FOCUS ON WRITING

6. **Writing about Earth's Land** Think of some vivid words you could use to describe Earth's landforms. As you think of them, add them to your notebook.

The Ring of Fire

Essential Elements

The World in Spatial Terms
Places and Regions
Physical Systems
Human Systems
Environment and Society
The Uses of Geography

Background Does "the Ring of Fire" sound like the title of a fantasy novel? It's actually the name of a region that circles the Pacific Ocean known for its fiery volcanoes and powerful earthquakes. The Ring of Fire stretches from the tip of South America all the way up to Alaska, and from Japan down to the islands east of Australia. Along this belt, the Pacific plate moves against several other tectonic plates. As a result, thousands of earthquakes occur there every year, and dozens of volcanoes erupt.

The Eruption of Mount Saint Helens One of the best-known volcanoes in the Ring of Fire is Mount Saint Helens in Washington State. Mount Saint Helens had been dormant, or quiet, since 1857. Then in March 1980, it began spitting out puffs of steam and ash. Officials warned people to leave the area. Scientists brought in equipment to measure the growing bulge in the mountainside. Everyone feared the volcano might erupt at any moment.

On May 18, after a sudden earthquake, Mount Saint Helens let loose a massive explosion of rock and lava. Heat from the blast melted snow on the mountain, which

Ring of Fire

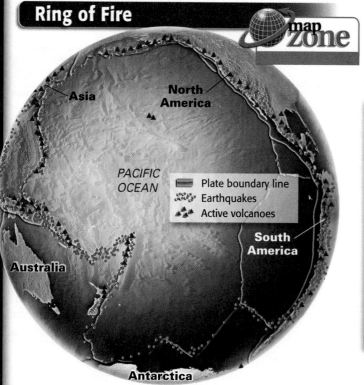

map zone

Asia

North America

PACIFIC OCEAN

- Plate boundary line
- Earthquakes
- Active volcanoes

South America

Australia

Antarctica

WORLD ALMANAC Facts about the World

Major Eruptions in the Ring of Fire

Volcano	Year
Tambora, Indonesia	1815
Krakatau, Indonesia	1883
Mount Saint Helens, United States	**1980**
Nevado del Ruiz, Colombia	1985
Mount Pinatubo, Philippines	1991

go.hrw.com KEYWORD: SK7 CH2

Mount Saint Helens, 1980

The 1980 eruption of Mount Saint Helens blew ash and hot gases miles into the air. Today, scientists study the volcano to learn more about predicting eruptions.

mixed with ash to create deadly mudflows. As the mud quickly poured downhill, it flattened forests, swept away cars, and destroyed buildings. Clouds of ash covered the land, killing crops, clogging waterways, and blanketing towns as far as 200 miles (330 km) away. When the volcano finally quieted down, 57 people had died. Damage totaled nearly $1 billion. If it were not for the early evacuation of the area, the destruction could have been much worse.

What It Means By studying Mount Saint Helens, scientists learned a great deal about stratovolcanoes. These are tall, steep, cone-shaped volcanoes that have violent eruptions. Stratovolcanoes often form in areas where tectonic plates collide.

Because stratovolcanoes often produce deadly eruptions, scientists try to predict when they might erupt. The lessons learned from Mount Saint Helens helped scientists

warn people about another stratovolcano, Mount Pinatubo in the Philippines. That eruption in 1991 was the second-largest of the 1900s. It was far from the deadliest, however. Careful observation and timely warnings saved thousands of lives.

The Ring of Fire will always remain a threat. However, the better we understand its volcanoes, the better prepared we'll be when they erupt.

Geography for Life Activity

1. How did the eruption of Mount Saint Helens affect the surrounding area?

2. Why do scientists monitor volcanic activity?

3. **Investigating the Effects of Volcanoes** Some volcanic eruptions affect environmental conditions around the world. Research the eruption of either Mount Saint Helens or the Philippines' Mount Pinatubo to find out how its eruption affected the global environment.

Using a Physical Map

Learn

Physical maps show important physical features, like oceans and mountains, in a particular area. They also indicate an area's elevation, or the height of the land in relation to sea level.

When you use a physical map, there are important pieces of information you should always examine.

- Identify physical features. Natural features, such as mountains, rivers, and lakes, are labeled on physical maps. Read the labels carefully to identify what physical features are present.

- Read the legend. On physical maps, the legend indicates scale as well as elevation. The different colors in the elevation key indicate how far above or below sea level a place is.

Practice

Use the physical map of India at right to answer the questions below.

❶ What landforms and bodies of water are indicated on the map?

❷ What is the highest elevation in India? Where is it located?

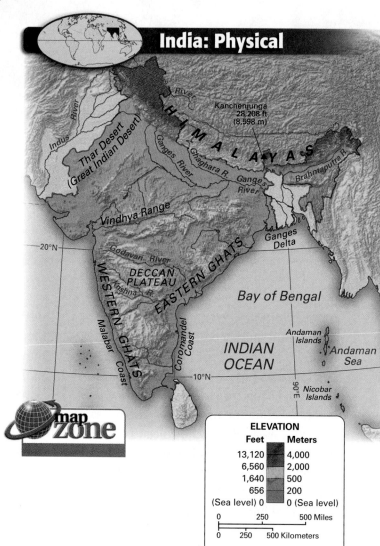

India: Physical

ELEVATION	
Feet	Meters
13,120	4,000
6,560	2,000
1,640	500
656	200
(Sea level) 0	0 (Sea level)

0   250   500 Miles
0   250   500 Kilometers

Projection: Lambert Conformal Conic

Apply

Locate the physical map of Africa in the atlas in the back of the book. Use the map to answer the questions below.

1. Which region has the highest elevation?

2. What bodies of water surround Africa?

3. What large island is located off the east coast of Africa?

Chapter Review

Geography's Impact
video series
Review the video to answer the closing question:
What are some reasons for water shortages, and what can be done to solve this problem?

Visual Summary

Use the visual summary below to help you review the main ideas of the chapter.

QUICK FACTS

The amount of solar energy Earth receives changes based on Earth's movement and position.

Water is crucial to life on Earth. Our abundant water supply is stored in oceans, in lakes, and underground.

Earth's various landforms are shaped by complex processes both under and on the planet's surface.

Reviewing Vocabulary, Terms, and Places

For each statement below, write T if it is true and F if it is false. If the statement is false, write the correct term that would make the sentence a true statement.

1. **Weathering** is the movement of sediment from one location to another.

2. Because high **latitude** areas receive indirect rays from the sun, they have cooler temperatures.

3. Most of our **groundwater** is stored in Earth's streams, rivers, and lakes.

4. It takes 365¼ days for Earth to complete one **rotation** around the sun.

5. Streams are formed when **precipitation** collects in narrow channels.

6. **Earthquakes** cause erosion as they flow downhill, carving valleys and mountain peaks.

7. The planet's tilt affects the amount of **erosion** Earth receives from the sun.

Comprehension and Critical Thinking

SECTION 1 *(Pages 26–29)*

8. **a. Identify** What factors influence the amount of energy Earth receives from the sun?

   b. Analyze Why do the Northern and Southern hemispheres experience opposite seasons?

   c. Predict What might happen to the amount of solar energy we receive if Earth's axis were straight up and down?

SECTION 2 *(Pages 30–34)*

9. **a. Describe** What different sources of water are available on Earth?

   b. Draw Conclusions How does the water cycle keep Earth's water supply relatively constant?

   c. Elaborate What water problems affect people around the world? What solutions can you think of for one of those problems?

SECTION 3 *(Pages 35–41)*

10. a. Define What is a landform? What are some common types of landforms?

b. Analyze Why are Earth's landforms still changing?

c. Elaborate What physical features dominate the landscape in your community? How do they affect life there?

Using the Internet

go.hrw.com
KEYWORD: SK7 CH2

11. Activity: Researching Earth's Seasons Earth's seasons not only affect temperatures, they also affect how much daylight is available during specific times of the year. Enter the activity keyword to research Earth's seasons and view animations to see how seasons change. Then use the interactive worksheet to answer some questions about what you learned.

FOCUS ON READING AND WRITING

Using Word Parts *Use what you learned about prefixes, suffixes, and word roots to answer the questions below.*

12. Examine the word *separation*. What is the suffix? What is the root? What does separation mean?

13. The prefix *in-* means not. What do the words *invisible* and *inactive* mean?

14. The suffix *-ment* means action or process. What does the word *movement* mean?

Writing a Haiku *Use your notes and the directions below to write a haiku.*

15. Look back through the notes you made about planet Earth. Choose one aspect of Earth to describe in a haiku. Haikus are short, three-line poems. Traditional haikus consist of only 17 syllables—five in the first line, seven in the second line, and five in the third line. You may choose to write a traditional haiku, or you may choose to write a haiku with a different number of syllables. Be sure to use descriptive words to paint a picture of planet Earth.

Social Studies Skills

Using a Physical Map *Examine the physical map of the United States in the back of this book. Use it to answer the questions below.*

16. What physical feature extends along the Gulf of Mexico?

17. What mountain range in the West lies above 6,560 feet?

18. Where does the elevation drop below sea level?

Map Activity ★ Interactive

Physical Map *Use the map below to answer the questions that follow.*

19. Which letter indicates a river?

20. Which letter on the map indicates the highest elevation?

21. The lowest elevation on the map is indicated by which letter?

22. An island is indicated by which letter?

23. Which letter indicates a large body of water?

24. Which letter indicates an area of land between 1,640 feet and 6,560 feet above sea level?

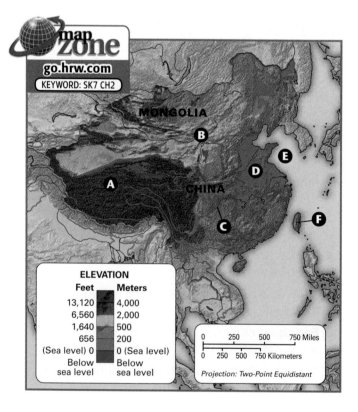

Standardized Test Practice

DIRECTIONS: Read questions 1 through 7 and write the letter of the best response. Then read question 8 and write your own well-constructed response.

1 Which regions on Earth have seasons tied to the amount of rainfall?

A polar regions

B the tropics

C the Northern Hemisphere

D high latitudes

2 Most of Earth's water supply is made up of

A groundwater.

B water vapor.

C freshwater.

D salt water.

3 The theory of continental drift explains how

A Earth's continents have moved thousands of miles.

B Earth's axis has moved to its current position.

C mountains and valleys are formed.

D sediment moves from one place to another.

4 Which of the following is a cause of erosion?

A evaporation

B ice

C plate collisions

D Earth's tilt

5 Changes in solar energy that create day and night are a result of

A the movement of tectonic plates.

B Earth's rotation.

C the revolution of Earth around the sun.

D Earth's tilt.

The Water Cycle

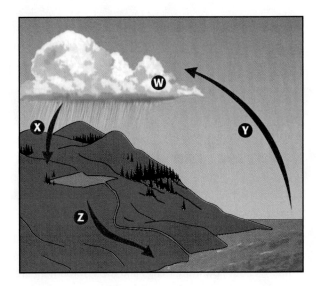

6 In the illustration above, which letter *best* reflects the process of evaporation?

A W

B X

C Y

D Z

7 Which of the following is *most likely* a cause of water pollution?

A River water is used to produce electricity.

B Heavy rainfall causes a river to overflow its banks.

C Chemicals from a factory seep into the local water supply.

D Groundwater is used faster than it can be replaced.

8 **Extended Response Question** Use the water cycle diagram above to explain how Earth's water cycle affects our water supply.

Climate, Environment, and Resources

What You Will Learn...

In this chapter you will learn about weather and climate. Climate is the weather conditions over a long period of time. You will also learn about how living things and the environment are connected and about the importance of Earth's natural resources.

SECTION 1
Weather and Climate **50**

SECTION 2
World Climates **55**

SECTION 3
Natural Environments **62**

SECTION 4
Natural Resources **68**

FOCUS ON READING AND VIEWING

Understanding Cause and Effect A cause makes something happen. An effect is the result of a cause. Words such as *because, result, since,* and *therefore* can signal causes or effects. As you read, look for causes and effects to understand how things relate. **See the lesson, Understanding Cause and Effect, on page 670.**

Presenting and Viewing a Weather Report You have likely seen a TV weather report, which tells the current weather conditions and predicts future conditions. After reading this chapter, prepare a weather report for a season and place of your choosing. Present your report to the class and then view your classmates' reports.

Climate Earth has many climates, such as the dry climate of the region shown here.

This photo shows a severe thunderstorm. These storms produce violent weather, such as heavy rainfall and strong winds, which affects people's lives.

How do you think this storm might have affected the people who lived in this area?

HOLT

Geography's Impact
video series
Watch the video to understand the impact of weather.

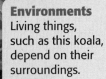

Environments Living things, such as this koala, depend on their surroundings.

Natural Resources Earth provides many valuable and useful natural resources, such as oil.

49

Weather and Climate

What You Will Learn...

Main Ideas

1. While weather is short term, climate is a region's average weather over a long period.
2. The amount of sun at a given location is affected by Earth's tilt, movement, and shape.
3. Wind and water move heat around Earth, affecting how warm or wet a place is.
4. Mountains influence temperature and precipitation.

The Big Idea

The sun, location, wind, water, and mountains affect weather and climate.

Key Terms

weather, *p. 50*
climate, *p. 50*
prevailing winds, *p. 51*
ocean currents, *p. 52*
front, *p. 53*

TAKING NOTES As you read, use a chart like the one here to take notes about the factors that affect weather and climate.

Sun and Location	Wind and Water	Mountains

If YOU lived there...

You live in Buffalo, New York, at the eastern end of Lake Erie. One evening in January, you are watching the local TV news. The weather forecaster says, "A huge storm is brewing in the Midwest and moving east. As usual, winds from this storm will drop several feet of snow on Buffalo as they blow off Lake Erie."

Why will winds off the lake drop snow on Buffalo?

BUILDING BACKGROUND All life on Earth depends on the sun's energy and on the cycle of water from the land to the air and back again. In addition, sun and water work with other forces, such as wind, to create global patterns of weather and climate.

Understanding Weather and Climate

" Climate is what you expect; weather is what you get. "
—Robert Heinlein, from *Time Enough for Love*

What is it like outside right now where you live? Is it hot, sunny, wet, cold? Is this what it is usually like outside for this time of year? The first two questions are about **weather**, the short-term changes in the air for a given place and time. The last question is about **climate**, a region's average weather conditions over a long period.

Weather is the temperature and precipitation from hour to hour or day to day. "Today is sunny, but tomorrow it might rain" is a statement about weather. Climate is the expected weather for a place based on data and experience. "Summer here is usually hot and muggy" is a statement about climate. The factors that shape weather and climate include the sun, location on Earth, wind, water, and mountains.

READING CHECK Finding Main Ideas How are weather and climate different from each other?

Sun and Location

Energy from the sun heats the planet. Different locations receive different amounts of sunlight, though. Thus, some locations are warmer than others. The differences are due to Earth's tilt, movement, and shape.

You have learned that Earth is tilted on its axis. The part of Earth tilted toward the sun receives more solar energy than the part tilted away from the sun. As the Earth revolves around the sun, the part of Earth that is tilted toward the sun changes during the year. This process creates the seasons. In general, temperatures in summer are warmer than in winter.

Earth's shape also affects the amount of sunlight different locations receive. Look at the diagram of Earth at right. You can see that Earth is a sphere, or wider in the middle. For this reason, the sun's rays directly strike the equator but only somewhat strike the poles.

As a result, areas near the equator, called the lower latitudes, are mainly hot year-round. Areas near the poles, called the higher latitudes, are cold year-round. Areas about halfway between the equator and poles have more seasonal change. In general, the farther from the equator, or the higher the latitude, the colder the climate.

READING CHECK **Summarizing** How does Earth's tilt on its axis affect climate?

Wind and Water

Heat from the sun moves across Earth's surface. The reason is that air and water warmed by the sun are constantly on the move. You might have seen a gust of wind or a stream of water carrying dust or dirt. In a similar way, wind and water carry heat from place to place. As a result, they make different areas of Earth warmer or cooler.

Global Wind Systems

Prevailing winds blow in circular belts across Earth. These belts occur at about every 30° of latitude.

ANALYZING VISUALS Which direction do the prevailing winds blow across the United States?

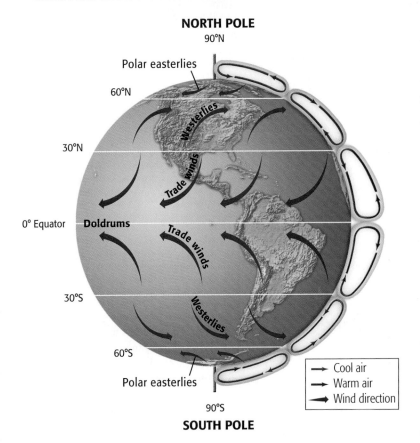

NORTH POLE
90°N
Polar easterlies
60°N
Westerlies
30°N
Trade winds
0° Equator   Doldrums
Trade winds
30°S
Westerlies
60°S
Polar easterlies
90°S
SOUTH POLE

→ Cool air
→ Warm air
→ Wind direction

Global Winds

Wind, or the sideways movement of air, blows in great streams around the planet. **Prevailing winds** are winds that blow in the same direction over large areas of Earth. The diagram above shows the patterns of Earth's prevailing winds.

To understand Earth's wind patterns, you need to think about the weight of air. Although you cannot feel it, air has weight. This weight changes with the temperature. Cold air is heavier than warm air. For this reason, when air cools, it gets heavier and sinks. When air warms, it gets lighter and rises. As warm air rises, cooler air moves in to take its place, creating wind.

Major Ocean Currents

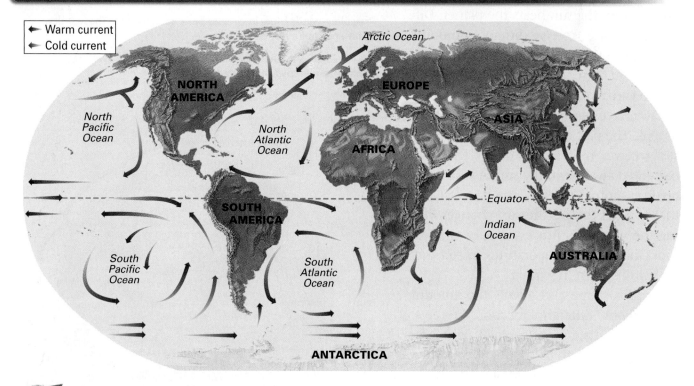

Warm current
Cold current

Arctic Ocean

NORTH AMERICA

EUROPE

ASIA

North Pacific Ocean

North Atlantic Ocean

AFRICA

SOUTH AMERICA

Equator

Indian Ocean

AUSTRALIA

South Pacific Ocean

South Atlantic Ocean

ANTARCTICA

map zone
Geography Skills

Movement Ocean currents carry warm water from the equator toward the poles and cold water from the poles toward the equator. The currents affect temperature.

1. **Use the Map** Does a warm or cold ocean current flow along the lower west coast of North America?
2. **Explain** How do ocean currents move heat between warmer and colder areas of Earth?

FOCUS ON READING
What is the effect of Earth's rotation on prevailing winds?

On a global scale, this rising, sinking, and flowing of air creates Earth's prevailing wind patterns. At the equator, hot air rises and flows toward the poles. At the poles, cold air sinks and flows toward the equator. Meanwhile, Earth is rotating. Earth's rotation causes prevailing winds to curve east or west rather than flowing directly north or south.

Depending on their source, prevailing winds make a region warmer or colder. In addition, the source of the winds can make a region drier or wetter. Winds that form from warm air or pass over lots of water often carry moisture. In contrast, winds that form from cold air or pass over lots of land often are dry.

Ocean Currents

Like wind, ocean currents—large streams of surface seawater—move heat around Earth. Winds drive these currents. The map above shows how Earth's ocean currents carry warm or cool water to different areas. The water's temperature affects air temperature near it. Warm currents raise temperatures; cold currents lower them.

The Gulf Stream is a warm current that flows north along the U.S. East Coast. It then flows east across the Atlantic to become the North Atlantic Drift. As the warm current flows along northwestern Europe, it heats the air. Westerlies blow the warmed air across Europe. This process makes Europe warmer than it otherwise would be.

Large Bodies of Water

Large bodies of water, such as an ocean or sea, also affect climate. Water heats and cools more slowly than land does. For this reason, large bodies of water make the temperature of the land nearby milder. Thus, coastal areas, such as the California coast, usually do not have as wide temperature ranges as inland areas.

As an example, the state of Michigan is largely surrounded by the Great Lakes. The lakes make temperatures in the state milder than other places as far north.

Wind, Water, and Storms

If you watch weather reports, you will hear about storms moving across the United States. Tracking storms is important to us because the United States has so many of them. As you will see, some areas of the world have more storms than others do.

Most storms occur when two air masses collide. An air mass is a large body of air. The place where two air masses of different temperatures or moisture content meet is a front. Air masses frequently collide in regions like the United States, where the westerlies meet the polar easterlies.

Fronts can produce rain or snow as well as severe weather such as thunderstorms and icy blizzards. Thunderstorms produce rain, lightning, and thunder. In the United States, they are most common in spring and summer. Blizzards produce strong winds and large amounts of snow and are most common during winter.

Thunderstorms and blizzards can also produce tornadoes, another type of severe storm. A tornado is a small, rapidly twisting funnel of air that touches the ground. Tornadoes usually affect a limited area and last only a few minutes. However, they can be highly destructive, uprooting trees and tossing large vehicles through the air. Tornadoes can be extremely deadly as well.

In 1925 a tornado that crossed Missouri, Illinois, and Indiana left 695 people dead. It is the deadliest U.S. tornado on record.

The largest and most destructive storms, however, are hurricanes. These large, rotating storms form over tropical waters in the Atlantic Ocean, usually from late summer to fall. Did you know that hurricanes and typhoons are the same? Typhoons are just hurricanes that form in the Pacific Ocean.

Extreme Weather

Severe weather is often dangerous and destructive. In the top photo, rescuers search for people during a flood in Yardley, Pennsylvania. Below, a tornado races across a wheat field.

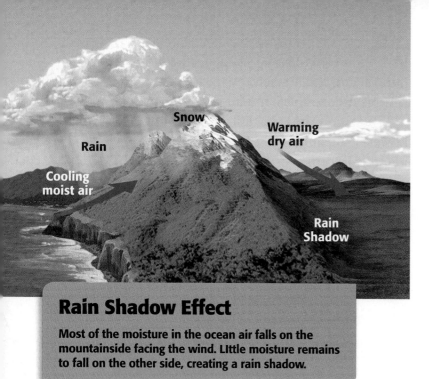

Rain Shadow Effect

Most of the moisture in the ocean air falls on the mountainside facing the wind. LIttle moisture remains to fall on the other side, creating a rain shadow.

Mountains

Mountains can influence an area's climate by affecting both temperature and precipitation. Many high mountains are located in warm areas yet have snow at the top all year. How can this be? The reason is that temperature decreases with elevation—the height on Earth's surface above sea level.

Mountains also create wet and dry areas. Look at the diagram at left. A mountain forces air blowing against it to rise. As it rises, the air cools and precipitation falls as rain or snow. Thus, the side of the mountain facing the wind is often green and lush. However, little moisture remains for the other side. This effect creates a rain shadow, a dry area on the mountainside facing away from the direction of the wind.

READING CHECK Finding Main Ideas How does temperature change with elevation?

SUMMARY AND PREVIEW As you can see, the sun, location on Earth, wind, water, and mountains affect weather and climate. In the next section you will learn what the world's different climate regions are like.

Hurricanes produce drenching rain and strong winds that can reach speeds of 155 miles per hour (250 kph) or more. This is more than twice as fast as most people drive on highways. In addition, hurricanes form tall walls of water called storm surges. When a storm surge smashes into land, it can wipe out an entire coastal area.

READING CHECK Analyzing Why do coastal areas have milder climates than inland areas?

Section 1 Assessment

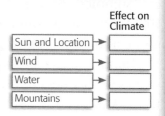

go.hrw.com
Online Quiz
KEYWORD: SK7 HP3

Reviewing Ideas, Terms, and Places

1. **a. Recall** What shapes **weather** and **climate**?
   b. Contrast How do weather and climate differ?
2. **a. Identify** What parts of Earth receive the most heat from the sun?
   b. Explain Why do the poles receive less solar energy than the equator does?
3. **a. Describe** What creates wind?
   b. Summarize How do **ocean currents** and large bodies of water affect climate?
4. **a. Define** What is a rain shadow?
   b. Explain Why might a mountaintop and a nearby valley have widely different temperatures?

Critical Thinking

5. **Identifying Cause and Effect** Draw a chart like this one. Use your notes to explain how each factor affects climate.

	Effect on Climate
Sun and Location →	
Wind →	
Water →	
Mountains →	

FOCUS ON VIEWING

6. **Writing about Weather and Climate** Jot down information to include in your weather report. For example, you might want to include a term such as *fronts* or describe certain types of storms such as hurricanes or tornadoes.

World Climates

If YOU lived there...

You live in Colorado and are on your first serious hike in the Rocky Mountains. Since it is July, it is hot in the campground in the valley. But your guide insists that you bring a heavy fleece jacket. By noon, you have climbed to 11,000 feet. You are surprised to see patches of snow in shady spots. Suddenly, you are very happy that you brought your jacket!

Why does it get colder as you climb higher?

BUILDING BACKGROUND While weather is the day-to-day changes in a certain area, climate is the average weather conditions over a long period. Earth's different climates depend partly on the amount of sunlight a region receives. Differences in climate also depend on factors such as wind, water, and elevation.

Major Climate Zones

In January, how will you dress for the weekend? In some places, you might get dressed to go skiing. In other places, you might head out in a swimsuit to go to the beach. What the seasons are like where you live depends on climate.

Earth is a patchwork of climates. Geographers identify these climates by looking at temperature, precipitation, and native plant life. Using these items, we can divide Earth into five general climate zones—tropical, temperate, polar, dry, and highland.

The first three climate zones relate to latitude. Tropical climates occur near the equator, in the low latitudes. Temperate climates occur about halfway between the equator and the poles, in the middle latitudes. Polar climates occur near the poles, in the high latitudes. The last two climate zones occur at many different latitudes. In addition, geographers divide some climate zones into more specific climate regions. The chart and map on the next two pages describe the world's climate regions.

READING CHECK Drawing Inferences Why do you think geographers consider native plant life when categorizing climates?

What You Will Learn...

Main Ideas

1. Geographers use temperature, precipitation, and plant life to identify climate zones.
2. Tropical climates are wet and warm, while dry climates receive little or no rain.
3. Temperate climates have the most seasonal change.
4. Polar climates are cold and dry, while highland climates change with elevation.

The Big Idea

Earth's five major climate zones are identified by temperature, precipitation, and plant life.

Key Terms

monsoons, p. 58
savannas, p. 58
steppes, p. 59
permafrost, p. 61

TAKING NOTES As you read, use a chart like the one here to help you note the characteristics of Earth's major climate zones.

Climate Zone	Characteristics

55

World Climate Regions

To explore the world's climate regions, start with the chart below. After reading about each climate region, locate the places on the map that have that climate. As you locate climates, look for patterns. For example, places near the equator tend to have warmer climates than places near the poles. See if you can identify some other climate patterns.

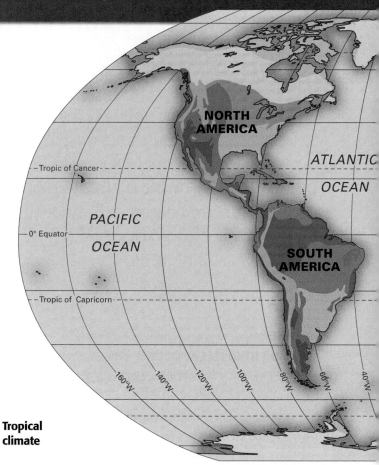

Tropical climate

Climate		Where is it?	What is it like?	Plants
Tropical	HUMID TROPICAL	On and near the equator	Warm with high amounts of rain year-round; in a few places, monsoons create extreme wet seasons	Tropical rain forest
	TROPICAL SAVANNA	Higher latitudes in the tropics	Warm all year; distinct rainy and dry seasons; at least 20 inches (50 cm) of rain during the summer	Tall grasses and scattered trees
Dry	DESERT	Mainly center on 30° latitude; also in middle of continents, on west coasts, or in rain shadows	Sunny and dry; less than 10 inches (25 cm) of rain a year; hot in the tropics; cooler with wide daytime temperature ranges in middle latitudes	A few hardy plants, such as cacti
	STEPPE	Mainly bordering deserts and interiors of large continents	About 10–20 inches (25–50 cm) of precipitation a year; hot summers and cooler winters with wide temperature ranges during the day	Shorter grasses; some trees and shrubs by water
Temperate	MEDITERRANEAN	West coasts in middle latitudes	Dry, sunny, warm summers; mild, wetter winters; rain averages 15–20 inches (30–50 cm) a year	Scrub woodland and grassland
	HUMID SUBTROPICAL	East coasts in middle latitudes	Humid with hot summers and mild winters; rain year-round; in paths of hurricanes and typhoons	Mixed forest
	MARINE WEST COAST	West coasts in the upper-middle latitudes	Cloudy, mild summers and cool, rainy winters; strong ocean influence	Evergreen forests
	HUMID CONTINENTAL	East coasts and interiors of upper-middle latitudes	Four distinct seasons; long, cold winters and short, warm summers; average precipitation varies	Mixed forest

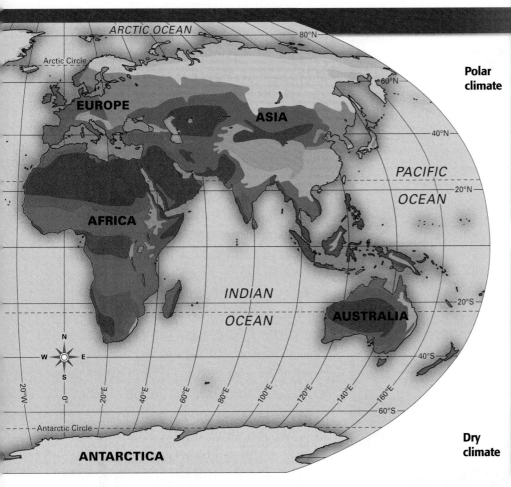

Polar climate

Dry climate

Climate		Where is it?	What is it like?	Plants
Polar	SUBARCTIC	Higher latitudes of the interior and east coasts of continents	Extremes of temperature; long, cold winters and short, warm summers; little precipitation	Northern evergreen forests
	TUNDRA	Coasts in high latitudes	Cold all year; very long, cold winters and very short, cool summers; little precipitation; permafrost	Moss, lichens, low shrubs
	ICE CAP	Polar regions	Freezing cold; snow and ice; little precipitation	No vegetation
Highland	HIGHLAND	High mountain regions	Wide range of temperatures and precipitation amounts, depending on elevation and location	Ranges from forest to tundra

map zone
Geography Skills

Regions Note how Earth's climate regions relate to different locations.

1. Locate Which climates are found mainly in the Northern Hemisphere?

2. Identify What climate does most of northern Africa have?

3. Make Generalizations Where are many of the world's driest climates found on Earth?

4. Interpreting Charts Examine the chart. Which two climates have the least amount of vegetation?

go.hrw.com   KEYWORD: SK7 CH3

The Tuareg of the Sahara.

In the Sahara, the world's largest desert, temperatures can top 130°F (54°C). Yet the Tuareg (TWAH-reg) of North and West Africa call the Sahara home—and prefer it. The Tuareg have raised camels and other animals in the Sahara for more than 1,000 years. The animals graze on sparse desert plants. When the plants are gone, the Tuareg move on.

In camp, Tuareg families live in tents made from animal skins. Some wealthier Tuareg live in adobe homes. The men traditionally wear blue veils wrapped around their face and head. The veils help protect against windblown desert dust.

Summarizing How have the Tuareg adapted to life in a desert?

Tropical and Dry Climates

Are you the type of person who likes to go to extremes? Then tropical and dry climates might be for you. These climates include the wettest, driest, and hottest places on Earth.

Tropical Climates

Our tour of Earth's climates starts at the equator, in the heart of the tropics. This region extends from the Tropic of Cancer to the Tropic of Capricorn. Look back at the map to locate this region.

Humid Tropical Climate At the equator, the hot, damp air hangs like a thick, wet blanket. Sweat quickly coats your body.

Welcome to the humid tropical climate. This climate is warm, muggy, and rainy year-round. Temperatures average about 80°F (26°C). Showers or storms occur almost daily, and rainfall ranges from 70 to more than 450 inches (180 to 1,140 cm) a year. In comparison, only a few parts of the United States average more than 70 inches (180 cm) of rain a year.

Some places with a humid tropical climate have **monsoons,** seasonal winds that bring either dry or moist air. During one part of the year, a moist ocean wind creates an extreme wet season. The winds then shift direction, and a dry land wind creates a dry season. Monsoons affect several parts of Asia. For example, the town of Mawsynram, India, receives on average more than 450 inches (1,140 cm) of rain a year—all in about six months! That is about 37 feet (11 m) of rain. As you can imagine, flooding during wet seasons is common and can be severe.

The humid tropical climate's warm temperatures and heavy rainfall support tropical rain forests. These lush forests contain more types of plants and animals than anywhere else on Earth. The world's largest rain forest is in the Amazon River basin in South America. There you can find more than 50,000 species, including giant lily pads, poisonous tree frogs, and toucans.

Tropical Savanna Climate Moving north and south away from the equator, we find the tropical savanna climate. This climate has a long, hot, dry season followed by short periods of rain. Rainfall is much lower than at the equator but still high. Temperatures are hot in the summer, often as high as 90°F (32°C). Winters are cooler but rarely get cold.

This climate does not receive enough rainfall to support dense forests. Instead, it supports **savannas**—areas of tall grasses and scattered trees and shrubs.

Dry Climates

Leaving Earth's wettest places, we head to its driest. These climates are found in a number of locations on the planet.

Desert Climate Picture the sun baking down on a barren wasteland. This is the desert, Earth's hottest and driest climate. Deserts receive less than 10 inches (25 cm) of rain a year. Dry air and clear skies produce high daytime temperatures and rapid cooling at night. In some deserts, highs can top 130°F (54°C)! Under such conditions, only very hardy plants and animals can live. Many plants grow far apart so as not to compete for water. Others, such as cacti, store water in fleshy stems and leaves.

Steppe Climate Semidry grasslands or prairies—called **steppes** (STEPS)—often border deserts. Steppes receive slightly more rain than deserts do. Short grasses are the most common plants, but shrubs and trees grow along streams and rivers.

> **READING CHECK** **Contrasting** What are some ways in which tropical and dry climates differ?

Temperate Climates

If you enjoy hot, sunny days as much as chilly, rainy ones, then temperate climates are for you. *Temperate* means "moderate" or "mild." These mild climates tend to have four seasons, with warm or hot summers and cool or cold winters.

Temperate climates occur in the middle latitudes, the regions halfway between the equator and the poles. Air masses from the tropics and the poles often meet in these regions, which creates a number of different temperate climates. You very likely live in one, because most Americans do.

Mediterranean Climate Named for the region of the Mediterranean Sea, this sunny, pleasant climate is found in many popular vacation areas. In a Mediterranean climate, summers are hot, dry, and sunny. Winters are mild and somewhat wet. Plant life includes shrubs and short trees with scattered larger trees. The Mediterranean climate occurs mainly in coastal areas. In the United States, much of California has this climate.

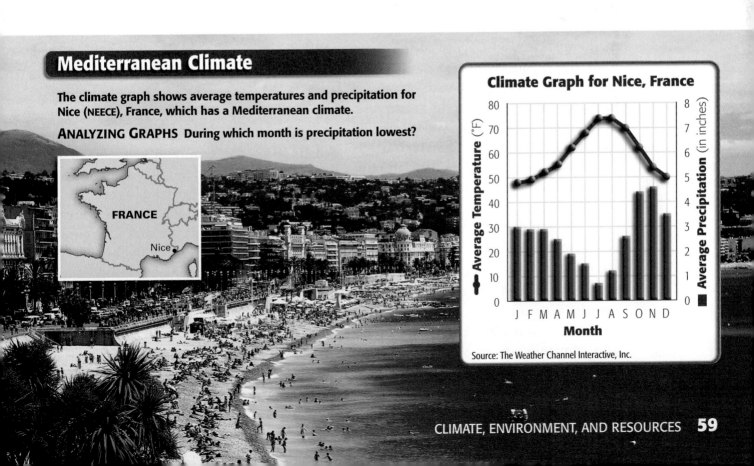

Mediterranean Climate

The climate graph shows average temperatures and precipitation for Nice (NEECE), France, which has a Mediterranean climate.

ANALYZING GRAPHS During which month is precipitation lowest?

FRANCE

Nice

Climate Graph for Nice, France

Source: The Weather Channel Interactive, Inc.

Highland Climates

Mount Kilimanjaro is the tallest mountain in Africa. Although Kilimanjaro is only about 200 miles (320 km) south of the equator, snow blankets its highest peak year-round.

Kilimanjaro rises to 19,341 ft (5,895 m). The snow-covered summit has an ice cap climate.

Climate and plant life ranges from rain forest, to steppe, to desert, to tundra.

A tropical savanna climate is found around the base of Mount Kilimanjaro.

ANALYSIS SKILL **ANALYZING VISUALS**

Which type of tropical climate is found on Mount Kilimanjaro?

Humid Subtropical Climate The southeastern United States is an example of the humid subtropical climate. This climate occurs along east coasts near the tropics. In these areas, warm, moist air blows in from the ocean. Summers are hot and muggy. Winters are mild, with occasional frost and snow. Storms occur year-round. In addition, hurricanes can strike, bringing violent winds, heavy rain, and high seas.

A humid subtropical climate supports mixed forests. These forests include both deciduous trees, which lose their leaves each fall, and coniferous trees, which are green year-round. Coniferous trees are also known as evergreens.

Marine West Coast Climate Parts of North America's Pacific coast and of western Europe have a marine west coast climate. This climate occurs on west coasts where winds carry moisture in from the seas.

The moist air keeps temperatures mild year-round. Winters are foggy, cloudy, and rainy, while summers can be warm and sunny. Dense evergreen forests thrive in this climate.

Humid Continental Climate Closer to the poles, in the upper–middle latitudes, many inland and east coast areas have a humid continental climate. This climate has four **distinct** seasons. Summers are short and hot. Spring and fall are mild, and winters are long, cold, and in general, snowy.

This climate's rainfall supports vast grasslands and forests. Grasses can grow very tall, such as in parts of the American Great Plains. Forests contain both deciduous and coniferous trees, with coniferous forests occurring in the colder areas.

ACADEMIC VOCABULARY

distinct
clearly different and separate

READING CHECK **Categorizing** Which of the temperate climates is too dry to support forests?

Polar and Highland Climates

Get ready to feel the chill as we end our tour in the polar and highland climates. The three polar climates are found in the high latitudes near the poles. The varied highland climate is found on mountains.

Subarctic Climate The subarctic climate and the tundra climate described below occur mainly in the Northern Hemisphere south of the Arctic Ocean. In the subarctic climate, winters are long and bitterly cold. Summers are short and cool. Temperatures stay below freezing for about half the year. The climate's moderate rainfall supports vast evergreen forests called taiga (TY-guh).

Tundra Climate The tundra climate occurs in coastal areas along the Arctic Ocean. As in the subarctic climate, winters are long and bitterly cold. Temperatures rise above freezing only during the short summer. Rainfall is light, and only plants such as mosses, lichens, and small shrubs grow.

In parts of the tundra, soil layers stay frozen all year. Permanently frozen layers of soil are called **permafrost**. Frozen earth absorbs water poorly, which creates ponds and marshes in summer. This moisture causes plants to burst forth in bloom.

Ice Cap Climate The harshest places on Earth may be the North and South poles. These regions have an ice cap climate. Temperatures are bone-numbingly cold, and lows of more than –120°F (–84°C) have been recorded. Snow and ice remain year-round, but precipitation is light. Not surprisingly, no vegetation grows. However, mammals such as penguins and polar bears thrive.

Highland Climates The highland climate includes polar climates plus others. In fact, this mountain climate is actually several climates in one. As you go up a mountain, the climate changes. Temperatures drop, and plant life grows sparser. Going up a mountain can be like going from the tropics to the poles. On very tall mountains, ice coats the summit year-round.

FOCUS ON READING

What is the effect of elevation on climate?

READING CHECK **Comparing** How are polar and highland climates similar?

SUMMARY AND PREVIEW As you can see, Earth has many climates, which we identify based on temperature, precipitation, and native plant life. In the next section you will read about how nature and all living things are connected.

Section 2 Assessment

go.hrw.com
Online Quiz
KEYWORD: SK7 HP3

Reviewing Ideas, Terms, and Places

1. **a. Recall** Which three major climate zones occur at certain latitudes?
   b. Summarize How do geographers categorize Earth's different climates?
2. **a. Define** What are **monsoons**?
   b. Make Inferences In which type of dry climate do you think the fewest people live, and why?
3. **a. Identify** What are the four temperate climates?
   b. Draw Conclusions Why are places with a Mediterranean climate popular vacation spots?
4. **a. Describe** What are some effects of **permafrost**?
   b. Explain How are highland climates unique?

Critical Thinking

5. **Categorizing** Create a chart like the one below for each climate region. Then use your notes to describe each climate region's average temperatures, precipitation, and native plant life.

Climate Region	→	Temperature	Precipitation	Plant Life

FOCUS ON VIEWING

6. **Discussing World Climates** Add information about the climate of the place you have selected, such as average temperature and precipitation.

Natural Environments

What You Will Learn...

Main Ideas

1. The environment and life are interconnected and exist in a fragile balance.
2. Soils play an important role in the environment.

The Big Idea

Plants, animals, and the environment, including soil, interact and affect one another.

Key Terms

environment, *p. 62*
ecosystem, *p. 63*
habitat, *p. 64*
extinct, *p. 64*
humus, *p. 65*
desertification, *p. 65*

TAKING NOTES As you read, use a chart like the one below to help you take notes on the main topics in this section.

Limits and Connections in Nature	
Changes to Environments	
Soil and the Environment	

If YOU lived there...

When your family moved to the city, you were sure you would miss the woods and pond near your old house. Then one of your new friends at school told you there's a large park only a few blocks away. You wondered how interesting a city park could be. But you were surprised at the many plants and animals that live there.

What environments might you see in the park?

BUILDING BACKGROUND No matter where you live, you are part of a natural environment. From a desert to a rain forest to a city park, every environment is home to a unique community of plant and animal life. These plants and animals live in balance with nature.

The Environment and Life

If you saw a wild polar bear outside your school, you would likely be shocked. In most parts of the United States, polar bears live only in zoos. This is because plants and animals must live where they are suited to the **environment**, or surroundings. Polar bears are suited to very cold places with lots of ice, water, and fish. As you will see, living things and their environments are connected and affect each other in many ways.

Limits on Life

The environment limits life. As our tour of the world's climates showed, factors such as temperature, rainfall, and soil conditions limit where plants and animals can live. Palm trees cannot survive at the frigid North Pole. Ferns will quickly wilt and die in deserts, but they thrive in tropical rain forests.

At the same time, all plants and animals are adapted to specific environments. For example, kangaroo rats are adapted to dry desert environments. These small rodents can get all the water they need from food, so they seldom have to drink water.

Connections in Nature

The interconnections between living things and the environment form ecosystems. An **ecosystem** is a group of plants and animals that depend on each other for survival and the environment in which they live. Ecosystems can be any size and can occur wherever air, water, and soil support life. A garden pond, a city park, a prairie, and a rain forest are all examples of ecosystems.

The diagram below shows a forest ecosystem. Each part of this ecosystem fills a certain role. The sun provides energy to the plants, which use the energy to make their own food. The plants then serve as food, either directly or indirectly, for all other life in the forest. When the plants and animals die, their remains break down and provide nutrients for the soil and new plant growth. Thus, the cycle continues.

Close-up

A Forest Ecosystem

A forest is one type of ecosystem. The plants and animals in the forest depend on one another and the forest environment for survival.

1 Sunlight is the source of energy for most living things.

2 Plants use the energy in sunlight to make food. They serve as the basis for other life in the ecosystem.

3 Animals such as rabbits eat plants and gain some of their energy.

4 Predators, such as wolves and hawks, eat rabbits and other prey for energy.

5 Larger predators, such as mountain lions, compete for the prey that is available.

ANALYSIS SKILL ANALYZING VISUALS

What might happen in the forest ecosystem above if the number of rabbits fell significantly?

Changes to Environments

The interconnected parts of an ecosystem exist in a fragile balance. For this reason, a small change to one part can affect the whole system. A lack of rain in the forest ecosystem could kill off many of the plants that feed the rabbits. If the rabbits die, there will be less food for the wolves and mountain lions. Then they too may die.

Many actions can affect ecosystems. For example, people need places to live and food to eat, so they clear land for homes and farms. Clearing land has **consequences**, however. It can cause the soil to erode. In addition, the plants and animals that live in the area might be left without food and shelter.

Actions such as clearing land and polluting can destroy habitats. A **habitat** is the place where a plant or animal lives. The most diverse habitats on Earth are tropical rain forests. People are clearing Earth's rain forests for farmland, lumber, and other reasons, though. As a result, these diverse habitats are being lost.

ACADEMIC VOCABULARY

consequences the effects of a particular event or events

FOCUS ON READING

What are some causes of habitat destruction?

Extreme changes in ecosystems can cause species to die out, or become **extinct**. As an example, flightless birds called dodos once lived on Mauritius (maw-RI-shuhs), an island in the Indian Ocean. When people began settling on the island, their actions harmed the dodos' habitat. First seen in 1507, dodos were extinct by 1681.

Recognizing these problems, many countries are working to balance people's needs with the needs of the environment. The United States, for example, has passed many laws to limit pollution, manage forests, and protect valuable ecosystems.

READING CHECK Drawing Inferences How might one change affect an entire ecosystem?

Soil and the Environment

As you know, plants are the basis for all food that animals eat. Soils help determine what plants will grow and how well. Because soils support plant life, they play an important role in the environment.

CONNECTING TO Science

Soil Factory

The next time you see a fallen tree in the forest, do not think of it as a dead log. Think of it as a soil factory. A fallen tree is buzzing with the activity of countless insects, bacteria, and other organisms. These organisms invade the fallen log and start to break the wood down.

As the tree decays and crumbles, it turns into humus. Humus is a rich blend of organic material. The humus mixes with the soil and broken rock material. These added nutrients then enrich the soil, making it possible for new trees and plants to grow. Fallen trees provide as much as one-third of the organic material in forest soil.

Summarizing What causes a fallen tree to change into soil?

Fertile soils are rich in minerals and **humus** (HYOO-muhs), decayed plant or animal matter. These soils can support abundant plant life. Like air and water, fertile soil is essential for life. Without it, we could not grow much of the food we eat.

Soils can lose fertility in several ways. Erosion from wind or water can sweep topsoil away. Planting the same crops over and over can also rob soil of its fertility. When soil becomes worn out, it cannot support as many plants. In fragile dry environments this can lead to the spread of desertlike conditions, or **desertification**. The spread of desertlike conditions is a serious problem in many parts of the world.

READING CHECK Analyzing What do fertile soils contain, and why are these soils important?

SUMMARY AND PREVIEW Living things and the environment are connected, but changes can easily upset the balance in an ecosystem. Because they support plant life, soils are important parts of ecosystems. In the next section you will learn about Earth's many resources.

Soil Layers

The three layers of soil are the topsoil, subsoil, and broken rock. The thickness of each layer depends on the conditions in a specific location.

ANALYZING VISUALS In which layer of soil are most plant roots and insects found?

Topsoil

Subsoil

Broken Rock

Solid Rock

go.hrw.com
Online Quiz
KEYWORD: SK7 HP3

Section 3 Assessment

Reviewing Ideas, Terms, and Places

1. **a. Define** What is an **ecosystem**, and what are two examples of ecosystems?
   b. Summarize How do nature and people change ecosystems?
   c. Elaborate Why can plants and animals not live everywhere?
2. **a. Recall** What is **humus**, and why is it important to soil?
   b. Identify Cause and Effect What actions can cause **desertification**, and what might be some possible effects?
   c. Elaborate Why it is important for geographers and scientists to study soils?

Critical Thinking

3. **Identifying Cause and Effect** Review your notes. Then use a chart like this one to identify some of the causes and effects of changes to ecosystems.

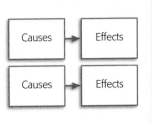

FOCUS ON VIEWING

4. **Writing about Natural Environments** Jot down ideas about how different types of weather might affect the environment of the place you chose. For example, how might lack of rain affect the area?

Earth's Changing Environments

PANGAEA

Pangaea About 250 million years ago, all of Earth's continents were connected, forming one giant landmass called Pangaea.

What was North America like 74 million years ago, when dinosaurs roamed Earth? You might be surprised to learn that it was a very different place. Earth's environments are always changing. The map at right shows North America in the age of dinosaurs. Back then, the climate was warm and humid, and large inland seas covered much of the land. The region's plants and animals were completely different. Slowly, however, things changed. Some major event, possibly an asteroid impact, wiped out the dinosaurs. Over time, North America's environments changed into the ones that exist today.

What Survived Dinosaurs, such as the plant-eating ceratopsian at left, are long gone. But insects, such as cockroaches and dragonflies, are still around.

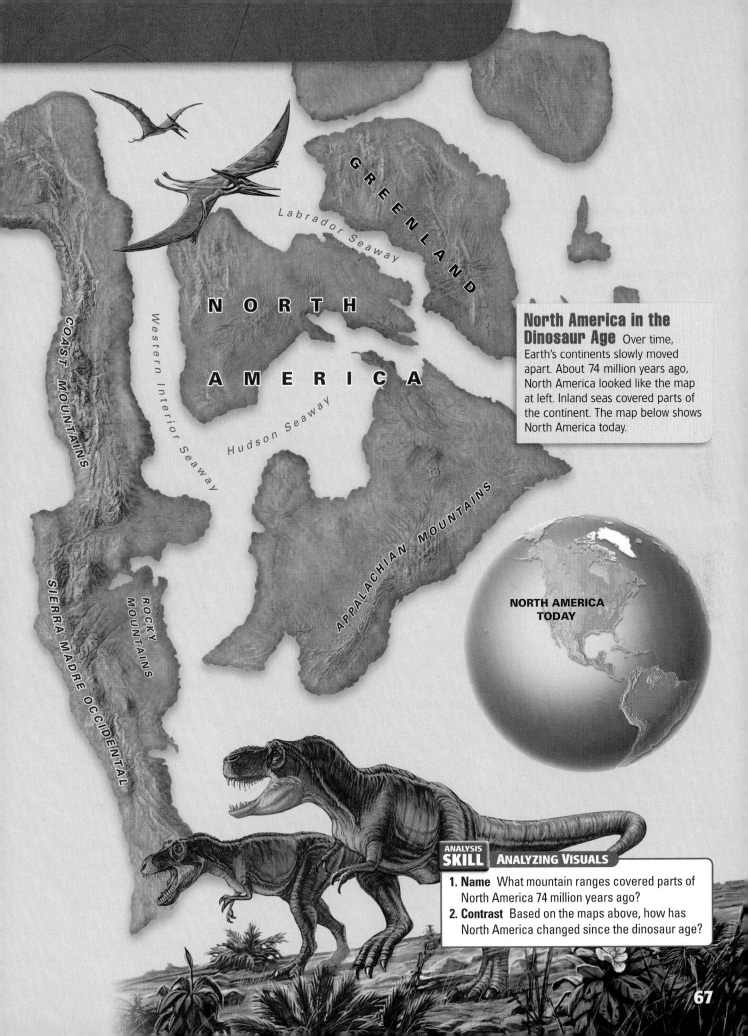

GREENLAND

Labrador Seaway

NORTH

AMERICA

COAST MOUNTAINS

Western Interior Seaway

Hudson Seaway

APPALACHIAN MOUNTAINS

SIERRA MADRE OCCIDENTAL

ROCKY MOUNTAINS

North America in the Dinosaur Age
Over time, Earth's continents slowly moved apart. About 74 million years ago, North America looked like the map at left. Inland seas covered parts of the continent. The map below shows North America today.

NORTH AMERICA TODAY

ANALYSIS SKILL **ANALYZING VISUALS**

1. **Name** What mountain ranges covered parts of North America 74 million years ago?
2. **Contrast** Based on the maps above, how has North America changed since the dinosaur age?

Natural Resources

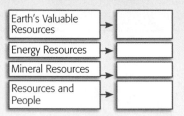

Earth's Valuable Resources	
Energy Resources	
Mineral Resources	
Resources and People	

If YOU lived there...

You live in Southern California, where the climate is warm and dry. Every week, you water the grass around your house to keep it green. Now the city has declared a "drought emergency" because of a lack of rain. City officials have put limits on watering lawns and on other uses of water.

How can you help conserve scarce water?

BUILDING BACKGROUND In addition to plant and animal life, other resources in the environment greatly influence people. In fact, certain vital resources, such as water, soils, and minerals, may determine whether people choose to live in a place or how wealthy people are.

Earth's Valuable Resources

Think about the materials in nature that you use. You have learned about the many ways we use sun, water, and land. They are just a start, though. Look at the human-made products around you. They all required the use of natural materials in some way. We use trees to make paper for books. We use petroleum, or oil, to make plastics for cell phones. We use metals to make machines, which we then use to make many items. Without these materials, our lives would change drastically.

Using Natural Resources

Trees, oil, and metals are all examples of natural resources. A **natural resource** is any material in nature that people use and value. Earth's most important natural resources include air, water, soils, forests, and minerals.

Understanding how and why people use natural resources is an important part of geography. We use some natural resources just as they are, such as wind. Usually, though, we change natural resources to make something new. For example, we change metals to make products such as bicycles and watches. Thus, most natural resources are raw materials for other products.

Reforestation

Members of the Green Belt Movement plant trees in Kenya. Although trees are a renewable resource, some forests are being cut down faster than new trees can replace them. Reforestation helps protect Earth's valuable forestlands.

ANALYZING VISUALS How does reforestation help the environment?

BIOGRAPHY

Wangari Maathai
(1940–)

Can planting a tree improve people's lives? Wangari Maathai thinks so. Born in Kenya in East Africa, Maathai wanted to help people in her country, many of whom were poor. She asked herself what Kenyans could do to improve their lives. "Planting a tree was the best idea that I had," she says. In 1977 Maathai founded the Green Belt Movement to plant trees and protect forestland. The group has now planted more than 30 million trees across Kenya! These trees provide wood and prevent soil erosion. In 2004 Maathai was awarded the Nobel Peace Prize. She is the first African woman to receive this famous award.

Types of Natural Resources

We group natural resources into two types, those we can replace and those we cannot. **Renewable resources** are resources Earth replaces naturally. For example, when we cut down a tree, another tree can grow in its place. Renewable resources include water, soil, trees, plants, and animals. These resources can last forever if used wisely.

Other natural resources will run out one day. These **nonrenewable resources** are resources that cannot be replaced. For example, coal formed over millions of years. Once we use the coal up, it is gone.

Managing Natural Resources

People need to manage natural resources to protect them for the future. Consider how your life might change if we ran out of forests, for example. Although forests are renewable, we can cut down trees far faster than they can grow. The result is the clearing of trees, or **deforestation**.

By managing resources, however, we can repair and prevent resource loss. For example, some groups are engaged in **reforestation**, planting trees to replace lost forestland.

READING CHECK Contrasting How do renewable and nonrenewable resources differ?

Energy Resources

Every day you use plants and animals from the dinosaur age—in the form of energy resources. These resources power vehicles, produce heat, and generate electricity. They are some of our most important and valuable natural resources.

Nonrenewable Energy Resources

Most of the energy we use comes from **fossil fuels**, nonrenewable resources that formed from the remains of ancient plants and animals. The most important fossil fuels are coal, petroleum, and natural gas.

Coal has long been a reliable energy source for heat. However, burning coal causes some problems. It pollutes the air and can harm the land. For these reasons, people have used coal less as other fuel options became available.

FOCUS ON READING

In the second sentence on this page, what cause does the word *because* signal? What is the effect of this cause?

Today we use coal mainly to create electricity at power plants, not to heat single buildings. Because coal is plentiful, people are looking for cleaner ways to burn it.

Petroleum, or oil, is a dark liquid used to make fuels and other products. When first removed from the ground, petroleum is called crude oil. This oil is shipped or piped to refineries, factories that process the crude oil to make products. Fuels made from oil include gasoline, diesel fuel, and jet fuel. Oil is also used to make petrochemicals, which are processed to make products such as plastics and cosmetics.

As with coal, burning oil-based fuels can pollute the air and land. In addition, oil spills can harm wildlife. Because we are so dependent on oil for energy, however, it is an extremely valuable resource.

The cleanest-burning fossil fuel is natural gas. We use it mainly for heating and cooking. For example, your kitchen stove may use natural gas. Some vehicles run on natural gas as well. These vehicles cause less pollution than those that run on gasoline.

Renewable Energy Resources

Unlike fossil fuels, renewable energy resources will not run out. They also are generally better for the environment. On the other hand, they are not available everywhere and can be costly.

The main alternative to fossil fuels is **hydroelectric power**—the production of electricity from waterpower. We obtain energy from moving water by damming rivers. The dams harness the power of moving water to generate electricity.

Hydroelectric power has both pros and cons. On the positive side, it produces power without polluting and lessens our use of fossil fuels. On the negative side, dams create lakes that replace existing resources, such as farmland, and disrupt wildlife habitats.

Another renewable energy source is wind. People have long used wind to power windmills. Today we use wind to power wind turbines, a type of modern windmill. At wind farms, hundreds of turbines create electricity in windy places.

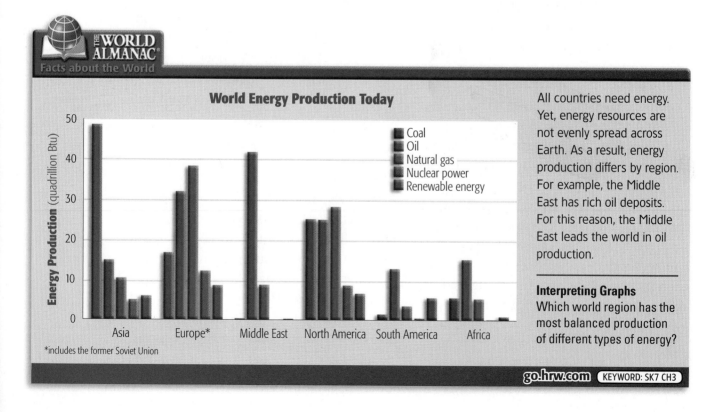

THE WORLD ALMANAC®
Facts about the World

World Energy Production Today

Energy Production (quadrillion Btu)

- Coal
- Oil
- Natural gas
- Nuclear power
- Renewable energy

Asia   Europe*   Middle East   North America   South America   Africa

*includes the former Soviet Union

All countries need energy. Yet, energy resources are not evenly spread across Earth. As a result, energy production differs by region. For example, the Middle East has rich oil deposits. For this reason, the Middle East leads the world in oil production.

Interpreting Graphs
Which world region has the most balanced production of different types of energy?

go.hrw.com   KEYWORD: SK7 CH3

A third source of renewable energy is heat from the sun and Earth. We can use solar power, or power from the sun, to heat water or homes. Using special solar panels, we turn solar energy into electricity. We can also use geothermal energy, or heat from within Earth. Geothermal power plants use steam and hot water located within Earth to create electricity.

Nuclear Energy

A final energy source is nuclear energy. We obtain this energy by splitting atoms, small particles of matter. This process uses the metal uranium, so some people consider nuclear energy a nonrenewable resource. Nuclear power does not pollute the air, but it does produce dangerous wastes. These wastes must be stored for thousands of years before they are safe. In addition, an accident at a nuclear power plant can have terrible effects.

READING CHECK Drawing Inferences Why might people look for alternatives to fossil fuels?

Mineral Resources

Like energy resources, mineral resources can be quite valuable. These resources include metals, salt, rocks, and gemstones.

Minerals fulfill countless needs. Look around you to see a few. Your school building likely includes steel, made from iron. The outer walls might be granite or limestone. The window glass is made from quartz, a mineral in sand. From staples to jewelry to coins, metals are everywhere.

Minerals are nonrenewable, so we need to conserve them. Recycling items such as aluminum cans will make the supply of these valuable resources last longer.

READING CHECK Categorizing What are the major types of mineral resources?

From the Ground to the Air

The photo shows a bauxite mine. Bauxite is used to make aluminum. This metal is used in many products, such as jet planes. See how many other products made from aluminum you can name.

Resources and People

Natural resources vary from place to place. The resources available in a region can shape life and wealth for the people there.

Resources and Daily Life

The natural resources available to people affect their lifestyles and needs. In the United States we have many different kinds of natural resources. We can choose among many different ways to dress, eat, live, travel, and entertain ourselves. People in places with fewer natural resources will likely have fewer choices and different needs than Americans.

For example, people who live in remote rain forests depend on forest resources for most of their needs. These people may craft containers by weaving plant fibers together. They may make canoes by hollowing out tree trunks. Instead of being concerned about money, they might be more concerned about food.

Products from Petroleum

This Ohio family shows some common products made from petroleum, or oil.

ANALYZING VISUALS What petroleum-based products can you identify in this photo?

Resources and Wealth

The availability of natural resources affects countries' economies as well. For example, the many natural resources available in the United States have helped it become one of the world's wealthiest countries. In contrast, countries with few natural resources often have weak economies.

Some countries have one or two valuable resources but few others. For example, Saudi Arabia is rich in oil but lacks water for growing food. As a result, Saudi Arabia must use its oil profits to import food.

READING CHECK **Identifying Cause and Effect** How can having few natural resources affect life and wealth in a region or country?

SUMMARY AND PREVIEW You can see that Earth's natural resources have many uses. Important natural resources include air, water, soils, forests, fuels, and minerals. In the next chapter you will read about the world's people and cultures.

go.hrw.com
Online Quiz
KEYWORD: SK7 HP3

Section 4 Assessment

Reviewing Ideas, Terms, and Places

1. **a. Define** What are **renewable resources** and **nonrenewable resources**?
   b. Explain Why is it important for people to manage Earth's natural resources?
   c. Develop What are some things you can do to help manage and conserve natural resources?
2. **a. Define** What are **fossil fuels**, and why are they significant?
   b. Summarize What are three examples of renewable energy resources?
   c. Predict How do you think life might change as we begin to run out of petroleum?
3. **a. Recall** What are the main types of mineral resources?
   b. Analyze What are some products that we get from mineral resources?

4. **a. Describe** How do resources affect people?
   b. Make Inferences How might a country with only one valuable resource develop its economy?

Critical Thinking

5. **Categorizing** Draw a chart like this one. Use your notes to identify and evaluate each energy resource.

Fossil Fuels	Renewable Energy	Nuclear Energy
Pros	Pros	Pros
Cons	Cons	Cons

FOCUS ON VIEWING

6. **Noting Details about Natural Resources** What natural resources does the place you chose have? Note ways to refer to some of these resources (or the lack of them) in your weather report.

from
The River

by Gary Paulsen

About the Reading *In the novel* The River, *a teenager named Brian has already proven his ability to survive in the wilderness. On this trip into the wilderness, he is accompanied by a man who wants to learn survival skills from him. With only a pocket knife and a transistor radio as tools, the two men meet challenges that at first appear too difficult to overcome. In the following passage from the novel, the men have just arrived in the wilderness.*

AS YOU READ Notice how Brian uses his senses to predict how some natural resources can help him survive.

He didn't just hear birds singing, not just a background sound of birds, but each bird. He listened to each bird. Located it, knew where it was by the sound, listened for the sound of alarm. He didn't just see clouds, but light clouds, scout clouds that came before the heavier clouds that could mean rain and maybe wind. ❶ The clouds were coming out of the northwest, and that meant that weather would come with them. Not could, but would. There would be rain. Tonight, late, there would be rain.

His eyes swept the clearing. . . There was a stump there that probably held grubs; hardwood there for a bow, and willows there for arrows; a game trail, . . . porcupines, raccoons, bear, wolves, moose, skunk would be moving on the trail and into the clearing. ❷ He flared his nostrils, smelled the air, pulled the air along the sides of his tongue in a hissing sound and tasted it, but there was nothing. Just summer smells. The tang of pines, soft air, some mustiness from rotting vegetation. No animals. ❸

GUIDED READING

WORD HELP

grubs soft, thick wormlike forms of insects
flared widened
tang sharp, biting smell
mustiness damp, stale smell

❶ Scout clouds are clouds that appear to be searching for other clouds to come.

❷ Brian notes that the stump likely holds grubs. He can eat the grubs for food.

❸ Brian does not smell any animals nearby.

Why might Brian want to know if animals are around?

Connecting Literature to Geography

1. **Predicting** Brian observes the clouds and can tell from their appearance and movement that rain is coming. What might be some ways that he can use his environment to prepare for the rain?

2. **Finding Main Ideas** The environment provides many resources that we can use, from wood to plants to animals. What resources does Brian identify around him that he can use to survive?

73

Social Studies Skills

Chart and Graph | **Critical Thinking** | **Geography** | **Study**

Analyzing a Bar Graph

Learn

Bar graphs are drawings that use bars to show data in a clear, visual format. Use these guidelines to analyze bar graphs.

- Read the title to identify the graph's subject and purpose.

- Read the graph's other labels. Note what the graph is measuring and the units of measurement being used. For example, this bar graph is measuring precipitation by climate. The unit of measurement is inches. If the graph uses colors, note their purpose.

- Analyze and compare the data. As you do, note any increases or decreases and look for trends or changes over time.

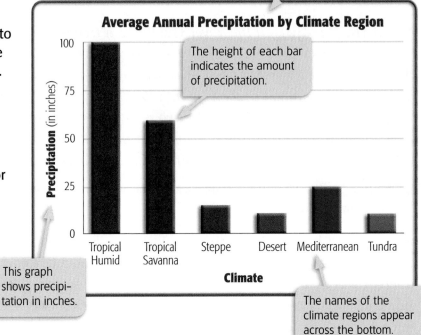

This bar graph compares the average annual precipitation of six climate regions.

Average Annual Precipitation by Climate Region

The height of each bar indicates the amount of precipitation.

This graph shows precipitation in inches.

The names of the climate regions appear across the bottom.

Practice

❶ On the bar graph above, which climate region has the highest average annual precipitation?

❷ Which two climate regions have about the same amount?

❸ Which climate region receives an average of between 50 and 75 inches of precipitation each year?

Apply

Examine the World Energy Production Today bar graph in Section 4. Then use the graph to answer the following questions.

1. Which region produces the most oil?

2. Which three regions produce little or no nuclear power?

3. Based on the graph, what type of energy resource do most Asian countries likely use?

Geography's Impact
video series
Review the video to answer the closing question:
How are climate and weather different, and how does the influence they have differ?

Visual Summary

Use the visual summary below to help you review the main ideas of the chapter.

QUICK FACTS

Earth has a wide range of climates, which we identify by precipitation, temperature, and native plant life.

Plants, animals, and the environment are interconnected and affect one another in many ways.

Earth's valuable natural resources, such as air, water, forests, and minerals, have many uses and affect people's lives.

Reviewing Vocabulary, Terms, and Places

Unscramble each group of letters below to spell a term that matches the given definition.

1. **usumh**—decayed plant or animal matter
2. **tahrewe**—changes or conditions in the air at a certain time and place
3. **netorietfaosr**—planting trees where forests were
4. **neticxt**—completely died out
5. **estpep**—semidry grassland or prairie
6. **sifeticatorined**—spread of desertlike conditions
7. **laitemc**—an area's weather patterns over a long period of time
8. **arsmofrtpe**—permanently frozen layers of soil
9. **snonomo**—winds that change direction with the seasons and create wet and dry periods
10. **vansanas**—areas of tall grasses and scattered shrubs and trees

Comprehension and Critical Thinking

SECTION 1 *(Pages 50–54)*

11. **a. Identify** What five factors affect climate?

    b. Analyze Is average annual precipitation an example of weather or climate?

    c. Evaluate Of the five factors that affect climate, which one do you think is the most important? Why?

SECTION 2 *(Pages 55–61)*

12. **a. Recall** What are the five major climate zones?

    b. Explain How does latitude relate to climate?

    c. Elaborate Why do you think the study of climate is important in geography?

SECTION 3 *(Pages 62–65)*

13. **a. Define** What is an ecosystem, and why does it exist in a fragile balance?

SECTION 3 (continued)

b. Explain Why are plants an important part of the environment?

c. Predict What might be some results of desertification?

SECTION 4 (Pages 68–72)

14. a. Define What are minerals?

b. Contrast How do nonrenewable resources and renewable resources differ?

c. Elaborate How might a scarcity of natural resources affect life in a region?

Using the Internet

go.hrw.com
KEYWORD: SK7 CH3

15. Activity: Experiencing Extremes Could you live in a place where for part of the year it is always dark and temperatures plummet to –104°F? What if you had to live in a place where it is always wet and stormy? Enter the activity keyword to learn more about some of the world's extreme climates. Then create a poster that describes some of those climates and the people, animals, and plants that live in them.

FOCUS ON READING AND VIEWING

Understanding Cause and Effect *Answer the following questions about causes and effects.*

16. What causes desertification?

17. What are the effects of abundant natural resources on a country's economy?

Presenting and Viewing a Weather Report *Use your weather report notes to complete the activity below.*

18. Select a place and a season. Then write a script for a weather report for that place during that season. Describe the current weather and predict the upcoming weather. During your presentation, use a professional, friendly tone of voice and make frequent eye contact with your audience. Then view your classmates' weather reports. Be prepared to give feedback on the content and their presentation techniques.

Social Studies Skills

Analyzing a Bar Graph *Examine the bar graph titled Average Annual Precipitation by Climate Region in the Social Studies Skills for this chapter. Then use the bar graph to answer the following questions.*

19. Which climate region receives an average of 100 inches of precipitation a year?

20. Which climate region receives an average of 25 inches of precipitation a year?

21. What is the difference in average annual precipitation between tropical humid climates and Mediterranean climates?

Map Activity ★Interactive

22. Prevailing Winds On a separate sheet of paper, match the letters on the map with their correct labels.

| equator | South Pole | westerly |
| North Pole | trade wind | |

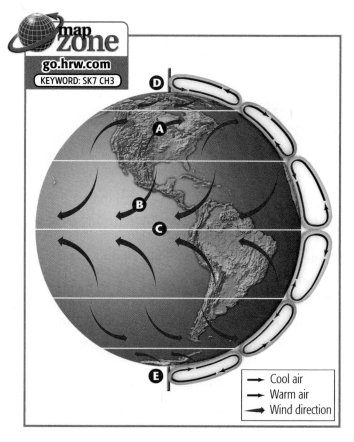

go.hrw.com
KEYWORD: SK7 CH3

→ Cool air
→ Warm air
→ Wind direction

Standardized Test Practice

DIRECTIONS: Read questions 1 through 7 and write the letter of the best response. Then read question 8 and write your own well-constructed response.

1 The cold winds that flow away from the North and South poles are the

   A doldrums.

   B polar easterlies.

   C trade winds.

   D westerlies.

2 Which climate zone occurs only in the upper latitudes?

   A highland

   B temperate

   C tropical

   D polar

3 Where are the most diverse habitats on Earth found?

   A steppe

   B tropical rain forest

   C tropical savanna

   D tundra

4 What is the cleanest burning fossil fuel?

   A coal

   B natural gas

   C oil

   D petroleum

5 Which renewable energy source uses the heat of Earth's interior to generate power?

   A geothermal energy

   B hydroelectric energy

   C nuclear energy

   D solar energy

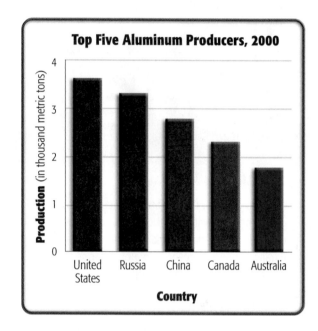

6 Based on the graph above, which country produced about 2,750 metric tons of aluminum in 2000?

   A Australia

   B China

   C Russia

   D United States

7 Which of the following form over tropical waters and are Earth's largest and most destructive storms?

   A blizzards

   B hurricanes

   C thunderstorms

   D tornadoes

8 **Extended Response** Forces such as the sun, latitude, wind, and water shape climate. Examine the World Climate Regions map in Section 2. Describe two climate patterns that you see on the map and explain how various forces combine to create the two patterns.

CHAPTER 4

The World's People

What You Will Learn...

In this chapter you will learn what culture is and how it changes over time. You will also study population and the different types of governments and economic systems used around the world. Finally, you will discover how global connections are bringing the world's people closer together.

SECTION 1
Culture 80

SECTION 2
Population 86

SECTION 3
Government and Economy 91

SECTION 4
Global Connections 97

FOCUS ON READING AND WRITING

Understanding Main Ideas A main idea is the central idea around which a paragraph or passage is organized. As you read, ask yourself what each paragraph is about. Look for a sentence or two that summarizes the main point of the entire paragraph. **See the lesson, Understanding Main Ideas, on page 671.**

Creating a Poster Think of some great posters you have seen—at the movies, in bus stations, or in the halls of your school. They likely all had a colorful image that captured your attention and a few words that explained the main idea. Read this chapter about the world's people. Then create a poster that includes words and images that summarize the chapter's main ideas.

Culture Thousands of different cultures make up our world. Clothing, language, and music are just some parts of culture.

ANALYSIS
SKILL **ANALYZING VISUALS**

Many of the world's people come together every four years to compete in the Olympics.

What indicates that some of the people in this photo are from different parts of the world?

HOLT

Geography's Impact

video series
Watch the video to learn the impact of culture.

Global Connections
Technology allows people in remote places around the world to communicate.

Population Geographers study human populations like this one in India to learn where and why people live in certain places.

Culture

What You Will Learn...

Main Ideas

1. Culture is the set of beliefs, goals, and practices that a group of people share.
2. The world includes many different culture groups.
3. New ideas and events lead to changes in culture.

The Big Idea

Culture, a group's shared practices and beliefs, differs from group to group and changes over time.

Key Terms

culture, *p. 80*
culture trait, *p. 81*
culture region, *p. 82*
ethnic group, *p. 83*
cultural diversity, *p. 83*
cultural diffusion, *p. 85*

TAKING NOTES As you read, take notes on culture. Use a web diagram like the one below to organize your notes.

If **YOU** lived there...

You live in New York City, and your young cousin from out of state has come to visit. As you take her on a tour of the city, you point out the different cultural neighborhoods, like Chinatown, Little Italy, Spanish Harlem, and Koreatown. Your cousin isn't quite sure what culture means or why these neighborhoods are so different.

How can you explain what culture is?

BUILDING BACKGROUND For hundreds of years, immigrants from around the world have moved to the United States to make a new home here. They have brought with them all the things that make up culture—language, religion, beliefs, traditions, and more. As a result, the United States has one of the most diverse cultures in the world.

What Is Culture?

If you traveled around the world, you would experience many different sights and sounds. You would probably hear unique music, eat a variety of foods, listen to different languages, see distinctive landscapes, and learn new customs. You would see and take part in the variety of cultures that exist in our world.

A Way of Life

What exactly is culture? **Culture is the set of beliefs, values, and practices that a group of people has in common.** Culture includes many aspects of life, such as language and religion, that we may share with people around us. Everything in your day-to-day life is part of your culture, from the clothes you wear to the music you hear to the foods you eat.

On your world travels, you might notice that all societies share certain cultural features. All people have some kind of government, educate their children in some way, and create some type of art or music. However, not all societies practice their culture in the same way. For example, in Japan the school year begins in the spring, and students wear school uniforms. In the United States, however, the school year begins in the late

Culture Traits

These students in Japan and Kenya have some culture traits in common, like eating lunch at school. Other culture traits are different.

ANALYZING VISUALS What culture traits do these students share? Which are different?

summer, and most schools do not require uniforms. Differences like these are what make each culture unique.

Culture Traits

Cultural features like starting the school year in the spring or wearing uniforms are types of culture traits. A **culture trait** is an activity or behavior in which people often take part. The language you speak and the sports you play are some of your culture traits. Sometimes a culture trait is shared by people around the world. For example, all around the globe people participate in the game of soccer. In places as different as Germany, Nigeria, and Saudi Arabia, many people enjoy playing and watching soccer.

While some culture traits are shared around the world, others change from place to place. One example of this is how people around the world eat. In China most people use chopsticks to eat their food. In Europe, however, people use forks and spoons. In Ethiopia, many people use bread or their fingers to scoop their food.

Development of Culture

How do cultures develop? Culture traits are often learned or passed down from one generation to the next. Most culture traits develop within families as traditions, foods, or holiday customs are handed down over the years. Laws and moral codes are also passed down within societies. Many laws in the United States, for example, can be traced back to England in the 1600s and were brought by colonists to America.

Cultures also develop as people learn new culture traits. Immigrants who move to a new country, for example, might learn to speak the language or eat the foods of their adopted country.

Other factors, such as history and the environment, also affect how cultures develop. For example, historical events changed the language and religion of much of Central and South America. In the 1500s, when the Spanish conquered the region, they introduced their language and Roman Catholic faith. The environment in which we live can also shape culture.

FOCUS ON READING
What is the main idea of this paragraph?

For example, the desert environment of Africa's Sahara influences the way people who live there earn a living. Rather than grow crops, they herd animals that have adapted to the harsh environment. As you can see, history and the environment affect how cultures develop.

READING CHECK **Finding Main Ideas** What practices and customs make up culture?

Culture Groups

Earth is home to thousands of different cultures. People who share similar culture traits are members of the same culture group. Culture groups can be based on a variety of factors, such as age, language, or religion. American teenagers, for example, can be said to form a culture group based on location and age. They share similar tastes in music, clothing, and sports.

Culture Regions

When we refer to culture groups, we are speaking of people who share a common culture. At other times, however, we need to refer to the area, or region, where the culture group is found. A **culture region** is an area in which people have many shared culture traits.

In a specific culture region, people share certain culture traits, such as religious beliefs, language, or lifestyle. One well-known culture region is the Arab world. As you can see at right, an Arab culture region spreads across Southwest Asia and North Africa. In this region, most people write and speak Arabic and are Muslim. They also share other traits, such as foods, music, styles of clothing, and architecture.

Occasionally, a single culture region dominates an entire country. In Japan, for example, one primary culture dominates the country. Nearly everyone in Japan speaks the same language and follows the same practices. Many Japanese bow to their elders as a sign of respect and remove their shoes when they enter a home.

A single country may also include more than one culture region within its borders. Mexico is one of many countries that is made up of different culture regions. People in northern Mexico and southern Mexico, for example, have different culture traits. The culture of northern Mexico tends to be more modern, while traditional culture remains strong in southern Mexico.

A culture region may also stretch across country borders. As you have already learned, an Arab culture region dominates much of Southwest Asia and North Africa. Another example is the Kurdish culture region, home to the Kurds, a people that live throughout Turkey, Iran, and Iraq.

Arab Culture Region

Culture regions are based on shared culture traits. Southwest Asia and North Africa make up an Arab culture region based on ethnic heritage, a common language, and religion. Most people in this region are Arab, speak and write Arabic, and practice Islam.

Cultural Diversity

As you just learned, countries may contain several culture regions within their borders. Often, these culture regions are based on ethnic groups. An **ethnic group** is a group of people who share a common culture and ancestry. Members of ethnic groups often share certain culture traits such as religion, language, and even special foods.

Some countries are home to a variety of ethnic groups. For example, more than 100 different ethnic groups live in the East African country of Tanzania. Countries with many ethnic groups are culturally diverse. **Cultural diversity** is the state of having a variety of cultures in the same area. While cultural diversity creates an interesting mix of ideas, behaviors, and practices, it can also lead to conflict.

In some countries, ethnic groups have been in conflict. In Canada, for example, some French Canadians want to separate from the rest of Canada to preserve their language and culture. In the 1990s ethnic conflict in the African country of Rwanda led to extreme violence and bloodshed.

Although ethnic groups have clashed in some culturally diverse countries, they have cooperated in others. In the United States, for example, many different ethnic groups live side by side. Cities and towns often celebrate their ethnic heritage with festivals and parades, like the Saint Patrick's Day Parade in Boston or Philadelphia's Puerto Rican Festival.

READING CHECK **Making Inferences** Why might cultural diversity cause conflict?

Many people share Arab culture traits. An Algerian boy, above, and Palestinian girls, at left, share the same language and religion.

ANALYZING VISUALS What culture traits do you see in the photos?

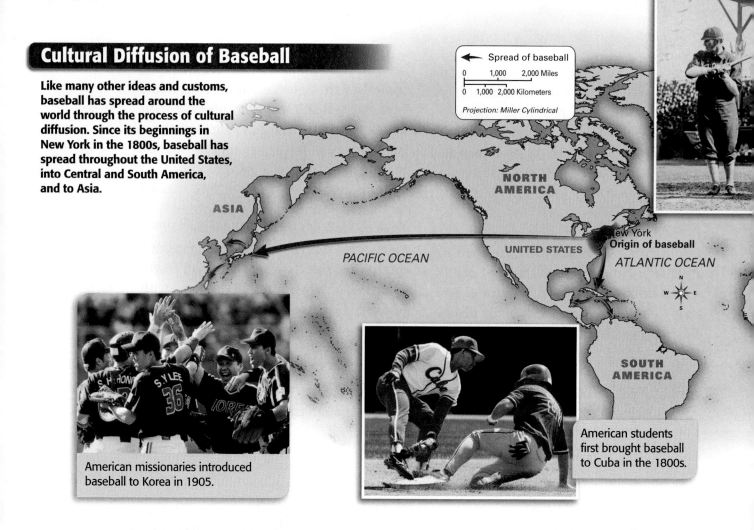

Cultural Diffusion of Baseball

Like many other ideas and customs, baseball has spread around the world through the process of cultural diffusion. Since its beginnings in New York in the 1800s, baseball has spread throughout the United States, into Central and South America, and to Asia.

Spread of baseball

0   1,000   2,000 Miles

0 1,000 2,000 Kilometers

Projection: Miller Cylindrical

ASIA

NORTH AMERICA

PACIFIC OCEAN

UNITED STATES

New York
Origin of baseball

ATLANTIC OCEAN

SOUTH AMERICA

American missionaries introduced baseball to Korea in 1905.

American students first brought baseball to Cuba in the 1800s.

Changes in Culture

You've read books or seen movies set in the time of the Civil War or in the Wild West of the late 1800s. Think about how our culture has changed since then. Clothing, food, music—all have changed drastically. When we study cultural change, we try to find out what caused the changes and how those changes spread from place to place.

How Cultures Change

Cultures change constantly. Some changes happen rapidly, while others take many years. What causes cultures to change? **Innovation** and contact with other people are two key causes of cultural change.

New ideas often bring about cultural changes. For example, when Alexander Graham Bell invented the telephone, it changed how people communicate with each other. Other innovations, such as motion pictures, changed how people spend their free time. More recently, the creation of the Internet dramatically altered the way people find information, communicate, and shop.

Cultures also change as societies come into contact with each other. For example, when the Spanish arrived in the Americas, they introduced firearms and horses to the region, changing the lifestyle of some Native American groups. At the same time, the Spaniards learned about new foods like potatoes and chocolate. These foods then became an important part of Europeans' diet. The Chinese had a similar influence on Korea and Japan, where they introduced Buddhism and written language.

ACADEMIC VOCABULARY

innovation
a new idea or way of doing something

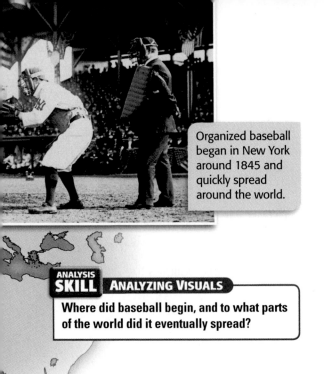

Organized baseball began in New York around 1845 and quickly spread around the world.

ANALYSIS SKILL **ANALYZING VISUALS**

Where did baseball begin, and to what parts of the world did it eventually spread?

How Ideas Spread

You have probably noticed that a new slang word might spread from teenager to teenager and state to state. In the same way, clothing styles from New York or Paris might become popular all over the world. More serious cultural traits spread as well. Religious beliefs or ideas about government may spread from place to place. The spread of culture traits from one region to another is called **cultural diffusion**.

Cultural diffusion often occurs when people move from one place to another. For example, when Europeans settled in the Americas, they brought their culture along with them. As a result, English, French, Spanish, and Portuguese are all spoken in the Americas. American culture also spread as pioneers moved west, taking with them their form of government, religious beliefs, and customs.

Cultural diffusion also takes place as new ideas spread from place to place. As you can see on the map above, the game of baseball first began in New York, then spread throughout the United States. As

more and more people learned the game, it spread even faster and farther. Baseball eventually spread around the world. Wearing blue jeans became part of our culture in a similar way. Blue jeans originated in the American West in the mid-1800s. They gradually became popular all over the country and the world.

READING CHECK **Finding Main Ideas** How do cultures change over time?

SUMMARY AND PREVIEW In this section you learned about the role that culture plays in our lives and how our cultures change. Next, you will learn about human populations and how we keep track of Earth's changing population.

go.hrw.com
Online Quiz
KEYWORD: SK7 HP4

Section 1 Assessment

Reviewing Ideas, Terms, and Places

1. **a. Define** What is **culture**?
   b. Analyze What influences the development of culture?
   c. Elaborate How might the world be different if we all shared the same culture?

2. **a. Identify** What are the different types of **culture regions**?
   b. Analyze How does **cultural diversity** affect societies?

3. **a. Describe** How does **cultural diffusion** take place?
   b. Make Inferences How can the spread of new ideas lead to cultural change?
   c. Evaluate Do you think that cultural diffusion has a positive or a negative effect? Explain your answer.

Critical Thinking

4. **Finding Main Ideas** Using your notes and a chart like the one here, explain the main idea of each aspect of culture in your own words.

Culture Traits	Culture Groups	Cultural Change

FOCUS ON WRITING

5. **Writing about Culture** What key words about culture can you include on your poster? What images might you include? Jot down your ideas in your notebook.

Population

What You Will Learn...

Main Ideas

1. The study of population patterns helps geographers learn about the world.
2. Population statistics and trends are important measures of population change.

The Big Idea

Population studies are an important part of geography.

Key Terms

population, *p. 86*
population density, *p. 86*
birthrate, *p. 88*
migration, *p. 89*

TAKING NOTES As you read, take notes on population. Use a graphic organizer like the one below to organize your notes on population patterns and population change.

If **YOU** lived there...

You live in Mexico City, one of the largest and most crowded cities in the world. You realize just how crowded it is whenever you ride the subway at rush hour! You love the excitement of living in a big city. There is always something interesting to do. At the same time, the city has a lot of crime. Heavy traffic pollutes the air.

What do you like and dislike about living in a large city?

BUILDING BACKGROUND An important part of geographers' work is the study of human populations. Many geographers are interested in where people live, how many people live there, and what effects those people have on resources and the environment.

Population Patterns

How many people live in your community? Do you live in a small town, a huge city, or somewhere in between? Your community's **population**, or the total number of people in a given area, determines a great deal about the place in which you live. Population influences the variety of businesses, the types of transportation, and the number of schools in your community.

Because population has a huge impact on our lives, it is an important part of geography. Geographers who study human populations are particularly interested in patterns that emerge over time. They study such information as how many people live in an area, why people live where they do, and how populations change. Population patterns like these can tell us much about our world.

Population Density

Some places on Earth are crowded with people. Others are almost empty. One statistic geographers use to examine populations is **population density**, a measure of the number of people living in an area. Population density is expressed as persons per square mile or square kilometer.

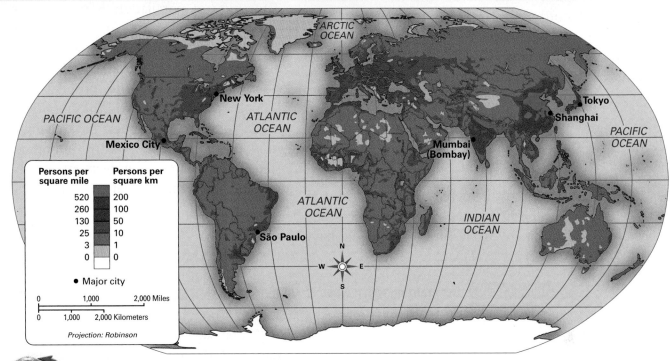

Geography Skills

Location While low population densities are common throughout much of the world, South and East Asia are two of the world's most densely populated regions.

1. **Identify** Which continent is the most densely populated? Which is the least densely populated?
2. **Making Inferences** Why might the population density of far North America be so low?

go.hrw.com KEYWORD: SK7 CH4

Population density provides us with important information about a place. The more people per square mile in a region, the more crowded, or dense, it is. Japan, for example, has a population density of 873 people per square mile (340 per square km). That is a high population density. In many parts of Japan, people are crowded together in large cities, and space is very limited. In contrast, Australia has a very low population density. Only 7 people per square mile (3 per square km) live there. Australia has many wide-open spaces with very few people.

How do you think population density affects life in a particular place? In places with high population densities, the land is often expensive, roads are crowded, and buildings tend to be taller. On the other hand, places with low population densities tend to have more open spaces, less traffic, and more available land.

Where People Live

Can you tell where most of the world's people live by examining the population density map above? The reds and purples on the map indicate areas of very high population density, while the light yellow areas indicate sparse populations. When an area is thinly populated, it is often because the land does not provide a very good life. These areas may have rugged mountains or harsh deserts where people cannot grow crops. Some areas may be frozen all year long, making survival there very difficult.

For these reasons, very few people live in parts of far North America, Greenland, northern Asia, and Australia.

Notice on the map that some areas have large clusters of population. Such clusters can be found in East and South Asia, Europe, and eastern North America. Fertile soil, reliable sources of water, and a good agricultural climate make these good regions for settlement. For example, the North China Plain in East Asia is one of the most densely populated regions in the world. The area's plentiful agricultural land, many rivers, and mild climate have made it an ideal place to settle.

READING CHECK **Generalizing** What types of information can population density provide?

CONNECTING TO Math

Calculating Population Density

Population density measures the number of people living in an area. To calculate population density, divide a place's total population by its area in square miles (or square kilometers). For example, if your city has a population of 100,000 people and an area of 100 square miles, you would divide 100,000 by 100. This would give you a population density of 1,000 people per square mile (100,000 ÷ 100 = 1,000).

Analyzing If a city had a population of 615,000 and a total land area of 250 square miles, what would its population density be?

City	Population	Total Area (square miles)	Population Density (people per square mile)
Adelaide, Australia	1,032,585	336	3,073
Lima, Peru	8,043,521	1,029	7,816
Nairobi, Kenya	2,143,254	266	8,057

$6 + x^3 - 54 \div 102 = 9y \div \pi^\circ 8 + x^2 - 33 \times 157 + x^3 - ab - 102 - 8$

Population Change

The study of population is much more important than you might realize. The number of people living in an area affects all elements of life—the availability of housing and jobs, whether hospitals and schools open or close, even the amount of available food. Geographers track changes in populations by examining important statistics, studying the movement of people, and analyzing population trends.

Tracking Population Changes

Geographers examine three key statistics to learn about population changes. These statistics are important for studying a country's population over time.

Three key statistics—birthrate, death rate, and the rate of natural increase—track changes in population. Births add to a population. Deaths subtract from it. The annual number of births per 1,000 people is called the **birthrate**. Similarly, the death rate is the annual number of deaths per 1,000 people. The birthrate minus the death rate equals the percentage of natural increase, or the rate at which a population is changing. For example, Japan has a rate of natural increase of .07%. This means it has slightly more births than deaths and a very slight population increase.

Population growth rates differ from one place to another. In some countries, populations are growing very slowly or even shrinking. Many countries in Europe and North America have very low rates of natural increase. In Russia, for example, the birthrate is about 9.6 and the death rate is 15.2. The result is a negative rate of natural increase and a shrinking population.

In most countries around the world, however, populations are growing. Mali, for example, has a rate of natural increase of almost 3 percent. While that may sound

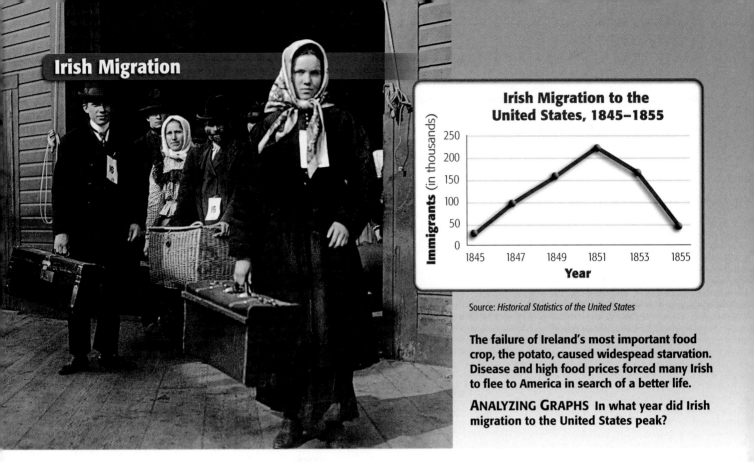

Irish Migration

Irish Migration to the United States, 1845–1855

Source: *Historical Statistics of the United States*

The failure of Ireland's most important food crop, the potato, caused widespread starvation. Disease and high food prices forced many Irish to flee to America in search of a better life.

ANALYZING GRAPHS In what year did Irish migration to the United States peak?

small, it means that Mali's population is expected to double in only 23 years! High population growth rates can pose some challenges, as governments try to provide enough jobs, education, and medical care for their rapidly growing populations.

Migration

A common cause of population change is migration. **Migration** is the process of moving from one place to live in another. As one country loses citizens as a result of migration, its population can decline. At the same time, another country may gain population as people settle there.

People migrate for many reasons. Some factors push people to leave their country, while other factors pull, or attract, people to new countries. Warfare, a lack of jobs, or a lack of good farmland are common push factors. For example, during the Irish potato famine of the mid-1800s, poverty and disease forced some 1.5 million people

to leave Ireland. Opportunities for a better life often pull people to new countries. For example, in the 1800s and early 1900s thousands of British citizens migrated to Australia in search of cheap land.

World Population Trends

In the last 200 years Earth's population has exploded. For thousands of years world population growth was low and relatively steady. About 2,000 years ago, the world had some 300 million people. By 1800 there were almost 1 billion people. Since 1800, better health care and improved food production have supported tremendous population growth. In 1999 the world's population passed 6 billion people.

Population trends are an important part of the study of the world's people. Two important population trends are clear today. The first trend indicates that the population growth in some of the more industrialized nations has begun to slow.

FOCUS ON READING
What is the main idea of this paragraph? What facts are used to support that idea?

World Population Growth

Advances in food production and health care have dramatically lowered death rates. As a result, the global population has seen incredible growth over the last 200 years.

ANALYZING GRAPHS By how much did the world's population increase between 1800 and 2000?

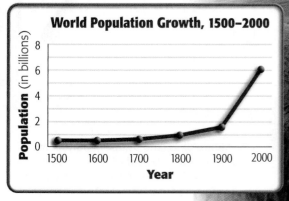

World Population Growth, 1500–2000

Source: *Atlas of World Population History*

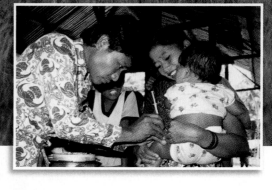

For example, Germany and France have low rates of natural increase. A second trend indicates that less industrialized countries, like Nigeria and Bangladesh, often have high growth rates. These trends affect a country's workforce and government aid.

READING CHECK **Summarizing** What population statistics do geographers study? Why?

SUMMARY AND PREVIEW In this section you have learned where people live, how crowded places are, and how population affects our world. Geographers study past and present population patterns in order to plan for the future. In the next section, you will learn how governments and economies affect people on Earth.

Section 2 Assessment

go.hrw.com
Online Quiz
KEYWORD: SK7 HP4

Reviewing Ideas, Terms, and Places

1. **a. Identify** What regions of the world have the highest levels of **population density**?
   b. Draw Conclusions What information can be learned by studying population density?
   c. Evaluate Would you prefer to live in a region with a dense or a sparse population? Why?
2. **a. Describe** What is natural increase? What can it tell us about a country?
   b. Analyze What effect does **migration** have on human populations?
   c. Predict What patterns do you think world population might have in the future?

Critical Thinking

3. **Summarizing** Draw a graphic organizer like the one here. Use your notes to write a sentence that summarizes each aspect of the study of population.

   Population Patterns

   Population Change

FOCUS ON WRITING

4. **Discussing Population** What effect does population have on our world? Write down some words and phrases that you might use on your poster to explain the importance of population.

Government and Economy

If YOU lived there...

You live in Raleigh, North Carolina. Your class at school is planning a presentation about life in the United States for a group of visitors from Japan. Your teacher wants you to discuss government and economics in the United States. As you prepare for your speech, you wonder what you should say.

How do government and economics affect your life?

BUILDING BACKGROUND Although you probably don't think about them every day, your country's government and economy have a big influence on your life. That is true in every country in every part of the world. Governments and economic systems affect everything from a person's rights to the type of job he or she has.

Governments of the World

Can you imagine what life would be like if there were no rules? Without ways to establish order and ensure justice, life would be chaotic. That explains why societies have governments. Our governments make and enforce laws, regulate business and trade, and provide aid to people. Governments help shape the culture and economy of a country as well as the daily lives of the people who live there.

Democratic Governments

Many countries—including the United States, Canada, and Mexico—have democratic governments. A **democracy** is a form of government in which the people elect leaders and rule by majority. In most democratic countries, citizens are free to choose representatives to make and enforce the laws. Voters in the United States, for example, elect members of Congress, who make the laws, and the president, who enforces those laws.

What You Will Learn...

Main Ideas

1. The governments of the world include democracy, monarchy, dictatorship, and communism.
2. Different economic activities and systems exist throughout the world.
3. Geographers group the countries of the world based on their level of economic development.

The Big Idea

The world's countries have different governments and levels of economic development.

Key Terms

democracy, *p. 91*
communism, *p. 92*
market economy, *p. 94*
command economy, *p. 94*
gross domestic product (GDP), *p. 95*
developed countries, *p. 95*
developing countries, *p. 95*

TAKING NOTES As you read, use a chart like this one to take notes on the different types of governments and economies.

Government	Economy

Governments of the World

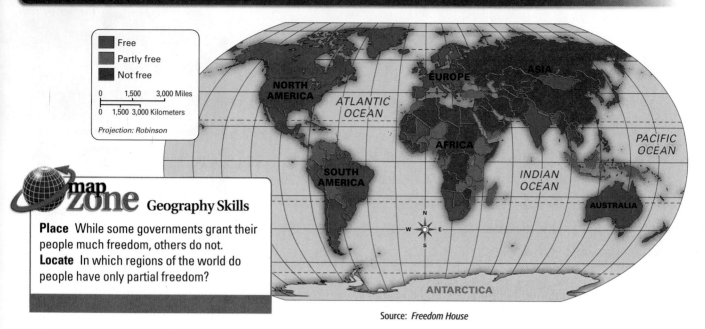

▮	Free
▮	Partly free
▮	Not free

0   1,500   3,000 Miles

0   1,500 3,000 Kilometers

Projection: Robinson

map zone Geography Skills

Place While some governments grant their people much freedom, others do not.
Locate In which regions of the world do people have only partial freedom?

Source: *Freedom House*

FOCUS ON READING

Main ideas are not always stated in the first sentence. Which sentence in this paragraph states the main idea?

Most democratic governments in the world work to protect the freedoms and rights of their people, such as the freedom of speech and the freedom of religion. Other democracies, however, restrict the rights and freedoms of their people. Not all democratic governments in the world are completely free.

Other Types of Government

Not all of the world's countries, however, are democracies. Several other types of government are found in the world today, including monarchies, dictatorships, and Communist states.

Monarchy is one of the oldest types of government in the world. A monarchy is ruled directly by a king or queen, the head of a royal family. Saudi Arabia is an example of a monarchy. The Saudi king has executive, legislative, and judicial powers. In some monarchies, power is in the hands of just one person. As a result, the people have little say in their government. Other monarchies, however, like Norway and Spain, use many democratic practices.

Dictatorship is a type of government in which a single, powerful ruler has total control. This leader, called a dictator, often rules by force. Iraq under Saddam Hussein was an example of a dictatorship. People who live under a dictatorship are not free. They have few rights and no say in their own government.

Yet another form of government is communism. **Communism** is a political system in which the government owns all property and dominates all aspects of life in a country. Leaders of most Communist governments are not elected by citizens. Rather, they are chosen by the Communist Party or by Communist leaders. In most Communist states, like Cuba and North Korea, the government strictly controls the country's economy and the daily life of its people. As a result, people in Communist states often have restricted rights and very little freedom.

READING CHECK **Supporting a Point of View** Why might people prefer to live in a democracy as opposed to a dictatorship?

Economies of the World

One important function of government is to monitor a country's economy. The economy is a system that includes all of the activities that people and businesses do to earn a living. Countries today use a mix of different economic activities and systems.

Economic Activity

Every country has some level of economic activity. Economic activities are ways in which people make a living. Some people farm, others manufacture goods, while still others provide services, such as driving a taxi or designing skyscrapers. Geographers divide these economic activities into four different levels.

The first level of economic activity, the primary industry, uses natural resources or raw materials. People in these industries earn a living by providing raw materials to others. Farming, fishing, and mining are all examples of primary industries. These activities provide raw materials such as grain, seafood, and coal for others to use.

Secondary industries perform the next step. They use natural resources or raw materials to manufacture other products. Manufacturing is the process in which raw materials are changed into finished goods. For example, people who make furniture might take wood and make products such as tables, chairs, or desks. Automobile manufacturers use steel, plastic, glass, and rubber to put together trucks and cars.

In the third level of activity, or tertiary industry, goods and services are exchanged. People in tertiary industries sell the furniture, automobiles, or other products made in secondary industries. Other people, like health care workers or mechanics, provide services rather than goods. Teachers, store clerks, doctors, and TV personalities are all engaged in this level of economic activity.

Economic Activity

Primary Industry
Primary industries use natural resources to make money. Here a farmer sells milk from dairy cows to earn a living.

Secondary Industry
Secondary economic activities use raw materials to produce or manufacture something new. In this case, the milk from dairy cows is used to make cheese.

Tertiary Industry
Tertiary economic activities provide services to people and businesses. This grocer selling cheese in a market is involved in a tertiary activity.

Quaternary Industry
Quaternary industries process and distribute information. Skilled workers research and gather information. Here, inspectors examine and test the quality of cheese.

The highest level of economic activity, quaternary industry, involves the research and distribution of information. People making a living at this level work with information rather than goods, and often have specialized knowledge and skills. Architects, lawyers, and scientists all work in quaternary industries.

Economic Systems

Just as economic activities are organized into different types, so are our economic systems. Economic systems can be divided into three types: traditional, market, and command. Most countries today use a mix of these economic systems.

One economic system is a **traditional** economy, a system in which people grow their own food and make their own goods. Trade may take place through barter, or the exchange of goods without the use of money. Rural and remote communities often have a mostly traditional economy.

ACADEMIC VOCABULARY
traditional
customary, time-honored

The most common economic system used around the world today is a market economy. A **market economy** is a system based on private ownership, free trade, and competition. Individuals and businesses are free to buy and sell what they wish. Prices are determined by the supply and demand for goods. This is sometimes called capitalism. The United States is one of many countries that use this system.

A third system is a **command economy**, a system in which the central government makes all economic decisions. The government decides what goods to produce, how much to produce, and what prices will be. While no country has a purely command economy, the economies of North Korea and Cuba are close to it. The Communist governments of these nations own and control most businesses.

READING CHECK **Summarizing** What economic systems are used in the world today?

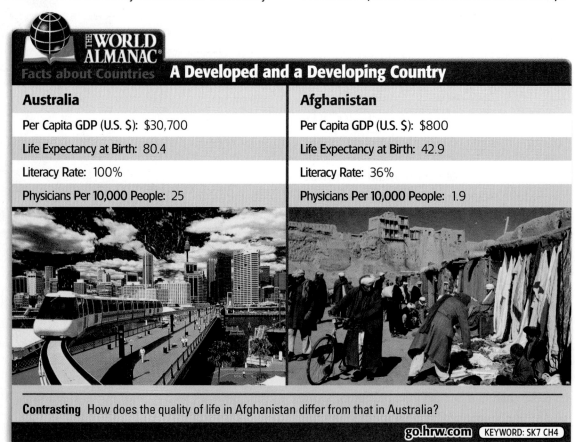

THE WORLD ALMANAC
Facts about Countries
A Developed and a Developing Country

Australia	Afghanistan
Per Capita GDP (U.S. $): $30,700	Per Capita GDP (U.S. $): $800
Life Expectancy at Birth: 80.4	Life Expectancy at Birth: 42.9
Literacy Rate: 100%	Literacy Rate: 36%
Physicians Per 10,000 People: 25	Physicians Per 10,000 People: 1.9

Contrasting How does the quality of life in Afghanistan differ from that in Australia?

go.hrw.com KEYWORD: SK7 CH4

Economic Development

Economic systems and activities affect a country's economic development, or the level of economic growth and quality of life. Geographers often group countries into two basic categories—developed and developing countries—based on their level of economic development.

Economic Indicators

Geographers use economic indicators, or measures of a country's wealth, to decide if a country is developed or developing. One such measure is gross domestic product. **Gross domestic product (GDP)** is the value of all goods and services produced within a country in a single year. Another indicator is a country's per capita GDP, or the total GDP divided by the number of people in a country. As you can see in the chart, per capita GDP allows us to compare incomes among countries. Other indicators include the level of industrialization and overall quality of life. In other words, we look at the types of industries and technology a country has, in addition to its level of health care and education.

Developed and Developing Countries

Many of the world's wealthiest and most powerful nations are **developed countries**, countries with strong economies and a high quality of life. Developed countries like Germany and the United States have a high per capita GDP and high levels of industrialization. Their health care and education systems are among the best in the world. Many people in developed countries have access to technology.

The world's poorer nations are known as **developing countries**, countries with less productive economies and a lower quality of life. Almost two-thirds of the people in the world live in developing countries. These developing countries have a lower per capita GDP than developed countries. Most of their citizens work in farming or other primary industries. Although these countries typically have large cities, much of their population still lives in rural areas. People in developing countries usually have less access to health care or technology. Guatemala, Nigeria, and Afghanistan are all developing countries.

READING CHECK **Analyzing** What factors separate developed and developing countries?

SUMMARY AND PREVIEW The world's countries have different governments, economies, and levels of development. Next, you will learn how people are linked in a global community.

Section 3 Assessment

Reviewing Ideas, Terms, and Places

1. **a. Identify** What are some different types of government?
   b. Elaborate Under which type of government would you most want to live? Why?
2. **a. Describe** What are the levels of economic activity?
   b. Evaluate Which economic system do you think is best? Explain your answer.
3. **a. Define** What is **gross domestic product**?
   b. Contrast In what ways do **developed countries** differ from **developing countries**?

Critical Thinking

4. **Categorizing** Draw a chart like the one here. Use the chart and your notes to identify the different governments, economies, and levels of economic development in the world today.

Types of Government	Economic Systems	Economic Development

FOCUS ON WRITING

5. **Thinking about Government and Economy** What kind of images and words might you use to present the main ideas behind the world's governments and economies?

Organizing Information

Learn

Remembering new information is easier if you organize it clearly. As you read and study, try to organize what you are learning. One way to do this is to create a graphic organizer. Follow these steps to create a graphic organizer as you read.

- Identify the main idea of the passage. Write the main idea in a circle at the top of your page.

- As you read, look for subtopics under the main idea. On your paper, draw a row of circles below the main idea, one for each subtopic. Write the subtopics in the circles.

- Below each subtopic, draw a big box. Look for facts and supporting details for each subtopic. List them in the box below the subtopic.

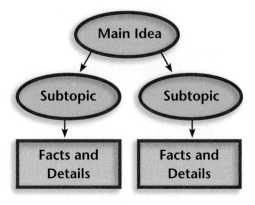

Practice

Read the passage below carefully. Then use the graphic organizer above to organize the information from the passage.

Cultures change slowly over time. New ideas and new people can often lead to cultural change.

Cultures often change as new ideas are introduced to a society. New ways of doing things, new inventions, and even new beliefs can all change a culture. One example of this is the spread of computer technology. As people adopted computers, they learned a new language and new work habits.

Cultures also change when new people introduce their culture traits to a society. For example, as immigrants settle in the United States, they add new culture traits, like food, music, and clothing, to American culture.

Apply

Turn to Section 1 and read the passage titled Culture Regions. Draw a graphic organizer like the one above. Then follow the steps to organize the information you have read. The passage will have two or more subtopics. Add additional circles and rectangles for each additional subtopic you find.

Global Connections

If YOU lived there...

You live in Louisville, Kentucky, and you have never traveled out of the United States. However, when you got ready for school this morning, you put on a T-shirt made in Guatemala and jeans made in Malaysia. Your shoes came from China. You rode to school on a bus with parts manufactured in Mexico. At school, your class even took part in a discussion with students in Canada.

What makes your global connections possible?

> **BUILDING BACKGROUND** Trade and technology have turned the world into a "global village." People around the world wear clothes, eat foods, and use goods made in other countries. Global connections are bringing people around the world closer than ever before.

Globalization

In just seconds an e-mail message sent by a teenager in India beams all the way to a friend in London. A band in Seattle releases a new CD that becomes popular in China. People from New York to Singapore respond to a crisis in Brazil. These are all examples of **globalization, the process in which countries are increasingly linked to each other through culture and trade.**

What caused globalization? Improvements in transportation and communication over the past 100 years have brought the world closer together. Airplanes, telecommunications, and the Internet allow us to communicate and travel the world with ease. As a result, global culture and trade have increased.

Popular Culture

What might you have in common with a teenager in Japan? You probably have more in common than you think. You may use similar technology, wear similar clothes, and watch many of the same movies. You share the same global popular culture.

What You Will Learn...

Main Ideas

1. Globalization links the world's countries together through culture and trade.
2. The world community works together to solve global conflicts and crises.

The Big Idea

Fast, easy global connections have made cultural exchange, trade, and a cooperative world community possible.

Key Terms

globalization, *p. 97*
popular culture, *p. 98*
interdependence, *p. 99*
United Nations (UN), *p. 99*
humanitarian aid, *p. 100*

TAKING NOTES As you read, take notes on globalization and the world community. Use a graphic organizer like the one below to take notes.

Globalization	World Community

More and more, people around the world are linked through popular culture. **Popular culture** refers to culture traits that are well known and widely accepted. Food, sports, music, and movies are all examples of our popular culture.

The United States has great influence on global popular culture. For example, American soft drinks are sold in almost every country in the world. Many popular American television shows are broadcast internationally. English has become the major global language. One-quarter of the world's people speak English. It has become the main language for international music, business, science, and education.

At the same time, the United States is influenced by global culture. Martial arts movies from Asia attract large audiences in the United States. Radio stations in the United States play music by African, Latin American, and European musicians. We even adopt many foreign words, like *sushi* and *plaza*, into English.

Close-up

A Global Economy

The growth of the global economy has affected many businesses, especially the automobile industry. Automakers can now buy parts from countries all around the world, depending on where they can get the best price.

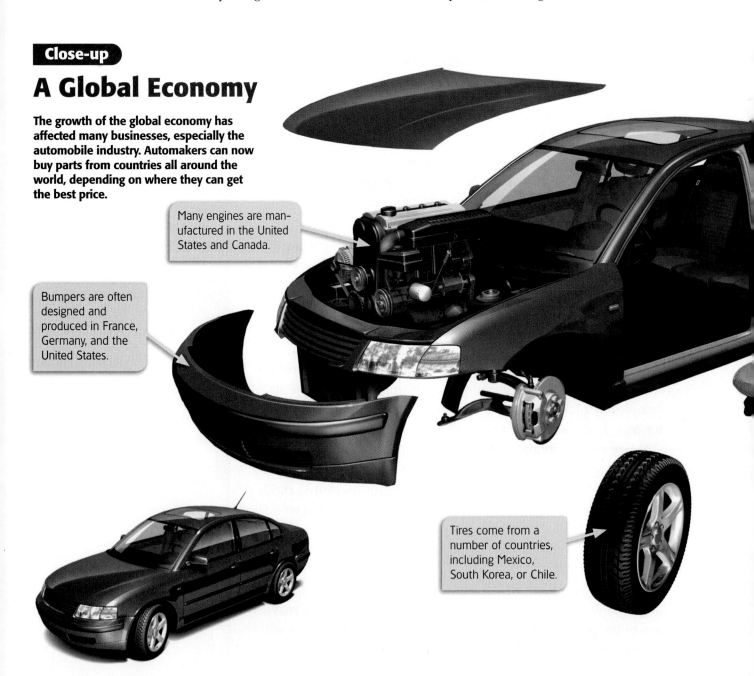

Many engines are man-ufactured in the United States and Canada.

Bumpers are often designed and produced in France, Germany, and the United States.

Tires come from a number of countries, including Mexico, South Korea, or Chile.

Global Trade

Globalization not only links the world's people, but it also connects businesses and affects trade. For centuries, societies have traded with each other. Improvements in transportation and communication have made global trade quicker and easier. For example, a shoe retailer in Chicago can order the sneakers she needs on a Web site from a company in China. The order can be flown to Chicago the next day and sold to customers that afternoon.

The expansion of global trade has increased interdependence among the world's countries. **Interdependence** is a relationship between countries in which they rely on one another for resources, goods, or services. Many companies in one country often rely on goods and services produced in another country. For example, automakers in Europe might purchase auto parts made in the United States or Japan. Consumers also rely on goods produced elsewhere. For example, American shoppers buy bananas from Ecuador and tomatoes from Mexico. Global trade gives us access to goods from around the world.

READING CHECK Finding Main Ideas How has globalization affected the world?

Many cars feature windows manufactured in Venezuela or the United States.

Seats are sometimes assembled in Japan from covers sewn in Mexico.

ANALYSIS SKILL ANALYZING VISUALS

From what different countries do automotive parts often originate?

A World Community

Some people call our world a global village. What do you think this means? Because of globalization, the world seems smaller. Places are more connected. What happens in one part of the world can affect the entire planet. Because of this, the world community works together to promote cooperation among countries in times of conflict and crisis.

The world community encourages cooperation by working to resolve global conflicts. From time to time, conflicts erupt among the countries of the world. Wars, trade disputes, and political disagreements can threaten the peace. Countries often join together to settle such conflicts. In 1945, for example, 51 nations created the United Nations. The **United Nations (UN)** is an organization of the world's countries that promotes peace and security around the globe.

The world community also promotes cooperation in times of crisis. A disaster may leave thousands of people in need.

FOCUS ON READING
What is the main idea of this paragraph? What facts are used to support that idea?

HISTORIC DOCUMENT
The Charter of the United Nations

Created in 1945, the United Nations is an organization of the world's countries that works to solve global problems. The Charter of the United Nations outlines the goals of the UN, some of which are included here.

We the Peoples of the United Nations Determined ...

to save succeeding generations from the scourge [terror] of war ...

to practice tolerance and live together in peace with one another as good neighbors, and

to unite our strength to maintain international peace and security, and

to ensure ... that armed forces shall not be used, save [except] in the common interest, and

to employ international machinery [systems] for the promotion of the economic and social advancement of all peoples,

Have Resolved to Combine our Efforts to Accomplish these Aims.

—from the Charter of the United Nations

ANALYSIS SKILL **ANALYZING PRIMARY SOURCES**

What are some of the goals of the United Nations?

Earthquakes, floods, and drought can cause crises around the world. Groups from many nations often come together to provide **humanitarian aid**, or assistance to people in distress.

Organizations representing countries around the globe work to help in times of crisis. For example, in 2004 a tsunami, or huge tidal wave, devastated parts of Southeast Asia. Many organizations, like the United Nations Children's Fund (UNICEF) and the International Red Cross, stepped in to provide humanitarian aid to the victims of the tsunami. Some groups lend aid to refugees, or people who have been forced to flee their homes. Groups like Doctors Without Borders give medical aid to those in need around the world.

READING CHECK **Analyzing** How has globalization promoted cooperation?

SUMMARY In this section you learned how globalization links the countries of the world through shared culture and trade. Globalization allows organizations around the world to work together. They often solve conflicts and provide humanitarian aid.

Section 4 Assessment

go.hrw.com
Online Quiz
KEYWORD: SK7 HP4

Reviewing Ideas, Terms, and Places
1. **a. Describe** What is **globalization**?
   b. Make Inferences How has **popular culture** influenced countries around the world?
   c. Evaluate In your opinion, has globalization hurt or helped the people of the world?
2. **a. Define** What is **humanitarian aid**?
   b. Draw Conclusions How has globalization promoted cooperation among countries?
   c. Predict What types of problems might lead to international cooperation?

Critical Thinking
3. **Identifying Cause and Effect** Use your notes and the graphic organizer at right to identify the effects that globalization has on our world.

Globalization → Effects, Effects, Effects

FOCUS ON WRITING
4. **Writing about Global Connections** What aspects of globalization might you include in your poster? Jot down your ideas in your notebook.

Geography's Impact
video series
Review the video to answer the closing question:
Why do you think some peoples must work to preserve their cultures in the modern world?

Visual Summary

Use the visual summary below to help you review the main ideas of the chapter.

QUICK FACTS

The world has many different cultures, or shared beliefs and practices.

The world's people practice different economic activities and systems.

Globalization brings people around the world closer than ever before.

Reviewing Vocabulary, Terms, and Places

Choose one word from each word pair to correctly complete each sentence below.

1. Members of a/an _____ often share the same religion, traditions, and language. **(ethnic group/population)**

2. People in a _____ are free to buy and sell goods as they please. **(command economy/ market economy)**

3. Organizations like the International Red Cross provide _____ to people in need around the world. **(humanitarian aid/cultural diffusion)**

4. _____, the process of moving from one place to live in another, is a cause of population change. **(Population density/Migration)**

5. A country with a strong economy and a high standard of living is considered a _____. **(developed country/developing country)**

Comprehension and Critical Thinking

SECTION 1 *(Pages 80–85)*

6. **a. Describe** What is cultural diversity?

   b. Analyze What causes cultures to change over time?

   c. Elaborate Describe some of the culture traits practiced by people in your community.

SECTION 2 *(Pages 86–90)*

7. **a. Describe** What does population density tell us about a place?

   b. Draw Conclusions Why do certain areas attract large populations?

   c. Elaborate Why do you think it is important for geographers to study population trends?

SECTION 3 *(Pages 91–95)*

8. **a. Recall** What is a command economy?

SECTION 3 (continued)

b. Make Inferences Why might developing countries have only primary and secondary economic activities?

c. Evaluate Do you think government is important in our everyday lives? Why or why not?

SECTION 4 (Pages 97–100)

9. a. Describe How have connections among the world's countries improved?

b. Analyze What impact has globalization had on world trade and culture?

c. Evaluate What do you think has been the most important result of globalization? Why?

Social Studies Skills

10. Organizing Information Practice organizing information by creating a graphic organizer for Section 3. Use the main ideas on the first page of the section for your large circles. Then write the subtopics under each main idea. Finally, identify supporting details for each subtopic.

FOCUS ON READING AND WRITING

Understanding Main Ideas *Read the paragraph below carefully, then write out the main idea of the paragraph.*

11. The ancient Greeks were the first to practice democracy. Since then many countries have adopted democratic government. The United Kingdom, South Korea, and Ghana all practice democracy. Democracy is the most widely used government in the world today.

Creating a Poster *Use your notes and the instructions below to help you create a poster.*

12. Review your notes about the world's cultures, populations, governments, and economies. Then select a subject for your poster. On a large sheet of paper, write a title that identifies your topic. Decorate your poster with illustrations that relate to your main idea. Write a short caption explaining each image. Be sure to use words and images that will grab your audience's attention and clearly express your main idea.

Using the Internet

go.hrw.com
KEYWORD: SK7 CH4

13. Activity: Writing a Report Population changes have a huge impact on the world around us. Countries around the world must deal with shrinking populations, growing populations, and other population issues. Enter the activity keyword and explore the issues surrounding global population. Then imagine you have been asked to report on global population trends to the United Nations. Write a report in which you identify world population trends and their impact on the world today.

Map Activity ✱Interactive

Population Density *Use the map below to answer the questions that follow.*

14. What letter on the map indicates the least crowded area?

15. What letter on the map indicates the most densely crowded area?

16. Which letter indicates a region with 260–520 people per square mile (100–200 people per square km)?

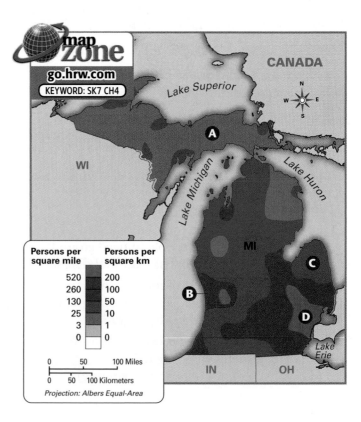

Standardized Test Practice

DIRECTIONS: Read questions 1 through 7 and write the letter of the best response. Then read question 8 and write your own well-constructed response.

1 Which of the following is *most likely* a culture trait?

A religion

B population density

C interdependence

D cultural diffusion

2 What developments led to the rapid increase in world population in the last 200 years?

A a decline in migration

B improvements in technology and communication

C a decrease in standard of living

D improvements in health care and agriculture

3 Which economic system is used in the United States?

A market economy

B command economy

C traditional economy

D domestic economy

4 A government in which a single, powerful ruler exerts complete control is a

A Communist state.

B democracy.

C dictatorship.

D republic.

5 Global connections have improved as a result of

A population growth.

B cultural diversity.

C the spread of democratic government.

D improvements in technology.

Developed and Developing Countries

Country	Per Capita GDP (U.S. $)	Life Expectancy at Birth	TVs per 1,000 People
Cameroon	$1,900	50.9	34
Singapore	$27,800	81.6	341
Ukraine	$6,300	69.7	433
Uruguay	$14,500	76.1	531

6 Which of the countries in the chart above is *most likely* a developed country?

A Cameroon

B Singapore

C Ukraine

D Uruguay

7 Which of the following is an example of economic interdependence?

A Cattle ranchers in Oklahoma sell beef to grocery stores in Maryland.

B Students in Germany use the Internet to communicate with scientists in Brazil.

C Construction companies in Canada build skyscrapers with steel imported from the United States.

D Immigrants from Russia settle in London.

8 **Extended Response** Using the data in the chart above, write a paragraph in which you compare and contrast the standard of living in Ukraine and Singapore.

Explaining a Process

How does soil renewal work? How do cultures change? Often the first question we ask about something is how it works or what process it follows. One way we can answer these questions is by writing an explanation.

Assignment

Write a paper explaining one of these topics:

- how water recycles on Earth
- how agriculture developed

1. Prewrite

Choose a Process

- Choose one of the topics above to write about.
- Turn your topic into a big idea, or thesis. For example, your big idea might be "Water continually circulates from Earth's surface to the atmosphere and back."

> **TIP** **Organizing Information** Explanations should be in a logical order. You should arrange the steps in the process in chronological order, the order in which the steps take place.

Gather and Organize Information

- Look for information about your topic in your textbook, in the library, or on the Internet.
- Start a plan to organize support for your big idea. For example, look for the individual steps of the water cycle.

2. Write

Use a Writer's Framework

> **A Writer's Framework**
>
> **Introduction**
> - Start with an interesting fact or question.
> - Identify your big idea.
>
> **Body**
> - Create at least one paragraph for each point supporting the big idea. Add facts and details to explain each point.
> - Use chronological order or order of importance.
>
> **Conclusion**
> - Summarize your main points in your final paragraph.

3. Evaluate and Revise

Review and Improve Your Paper

- Re-read your paper and make sure you have followed the framework.
- Make the changes needed to improve your paper.

Evaluation Questions for an Explanation of a Process

1. Do you begin with an interesting fact or question?
2. Does your introduction identify your big idea? Does it provide any background information your readers might need?
3. Do you have at least one paragraph for each point you are using to support the big idea?
4. Do you include facts and details to explain and illustrate each point?
5. Do you use chronological order or order of importance to organize your main points?

4. Proofread and Publish

Give Your Explanation the Finishing Touch

- Make sure you have capitalized the first word in every sentence.
- Check for punctuation at the end of every sentence.
- Think of a way to share your explanation.

5. Practice and Apply

Use the steps and strategies outlined in this workshop to write your explanation. Share your paper with others and find out whether the explanation makes sense to them.

Southwest and Central Asia

The Caspian Sea

The vast Caspian Sea, which is the world's largest inland body of water, contains valuable resources like oil.

Great Mountains

In Central Asia, high mountain ranges such as the Tian Shan separate the region from other parts of Asia.

Huge Deserts

Southwest Asia is home to huge deserts such as the Rub' al-Khali, or "Empty Quarter," which is virtually uninhabited.

Southwest and Central Asia

Explore the Satellite Image
Vast deserts, high mountains, and large rivers stand out clearly on this satellite image of Southwest and Central Asia. How do you think these features influence life in the region?

The Satellite's Path

>44'56.08

>>>>>>>>>665.00'87<

567·476·348

+355

+794
+803

+966

456.094.

Southwest and Central Asia: Physical

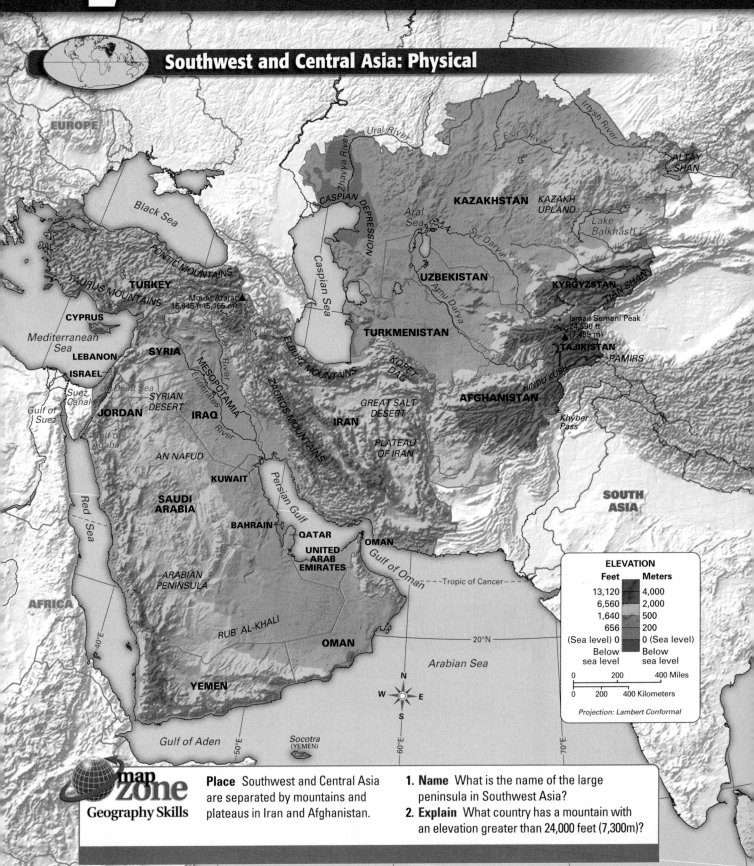

EUROPE

Black Sea

Ural River

Zhayva River

CASPIAN DEPRESSION

Esil River

Irtysh River

ALTAY SHAN

KAZAKHSTAN

KAZAKH UPLAND

Aral Sea

Lake Balkhash

Ili River

PONTIC MOUNTAINS

TAURUS MOUNTAINS

TURKEY

Mount Ararat
16,945 ft (5,165 m)

Caspian Sea

ELBURZ MOUNTAINS

Syr Darya

UZBEKISTAN

Amu Darya

KYRGYZSTAN

TIAN SHAN

CYPRUS

Mediterranean Sea

LEBANON

ISRAEL

Suez Canal

Dead Sea

Gulf of Suez

JORDAN

Gulf of Aqaba

SYRIA

Tigris River

MESOPOTAMIA

Euphrates River

SYRIAN DESERT

IRAQ

AN NAFUD

TURKMENISTAN

KOPET DAG

ZAGROS MOUNTAINS

GREAT SALT DESERT

IRAN

PLATEAU OF IRAN

Ismail Semani Peak
24,590 ft (7,495 m)

TAJIKISTAN

PAMIRS

HINDU KUSH

AFGHANISTAN

Khyber Pass

SOUTH ASIA

KUWAIT

Persian Gulf

SAUDI ARABIA

BAHRAIN

QATAR

UNITED ARAB EMIRATES

OMAN

Gulf of Oman

AFRICA

Red Sea

ARABIAN PENINSULA

RUB' AL-KHALI

OMAN

40°E

20°N

Tropic of Cancer

Arabian Sea

YEMEN

N
W   E
S

50°E

Gulf of Aden

Socotra
(YEMEN)

60°E

70°E

ELEVATION

Feet	Meters
13,120	4,000
6,560	2,000
1,640	500
656	200
(Sea level) 0	0 (Sea level)
Below sea level	Below sea level

0   200   400 Miles

0   200   400 Kilometers

Projection: Lambert Conformal

map zone
Geography Skills

Place Southwest and Central Asia are separated by mountains and plateaus in Iran and Afghanistan.

1. **Name** What is the name of the large peninsula in Southwest Asia?

2. **Explain** What country has a mountain with an elevation greater than 24,000 feet (7,300m)?

Southwest and Central Asia

THE WORLD ALMANAC®
Facts about the World

Geographical Extremes: Southwest and Central Asia

Longest River	Euphrates River, Turkey/Syria/Iraq: 1,700 miles (2,735 km)
Highest Point	Qullai Ismoili Somoni, Tajikistan: 24,590 feet (7,495 m)
Lowest Point	Dead Sea, Israel/Jordan: 1,348 feet (411 m) below sea level
Highest Recorded Temperature	Tirat Tsvi, Israel: 129°F (53.9°C)
Driest Place	Aden, Yemen: 1.8 inches (4.6 cm) average precipitation per year
Largest Country	Kazakhstan: 1,049,155 square miles (2,717,311 square km)
Smallest Country	Bahrain: 257 square miles (666 square km)
Saltiest Lake	Dead Sea, Israel/Jordan: 33 percent salt content
Most Powerful Earthquake	Erzincan, Turkey, 1939: 8.0 magnitude

A high salt content keeps people afloat in the Dead Sea.

go.hrw.com   KEYWORD: SK7 UN2

Size Comparison: The United States and Southwest and Central Asia

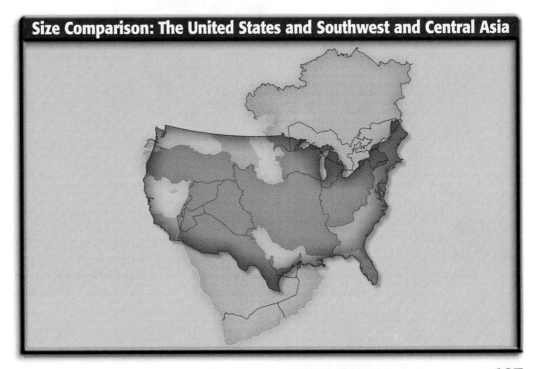

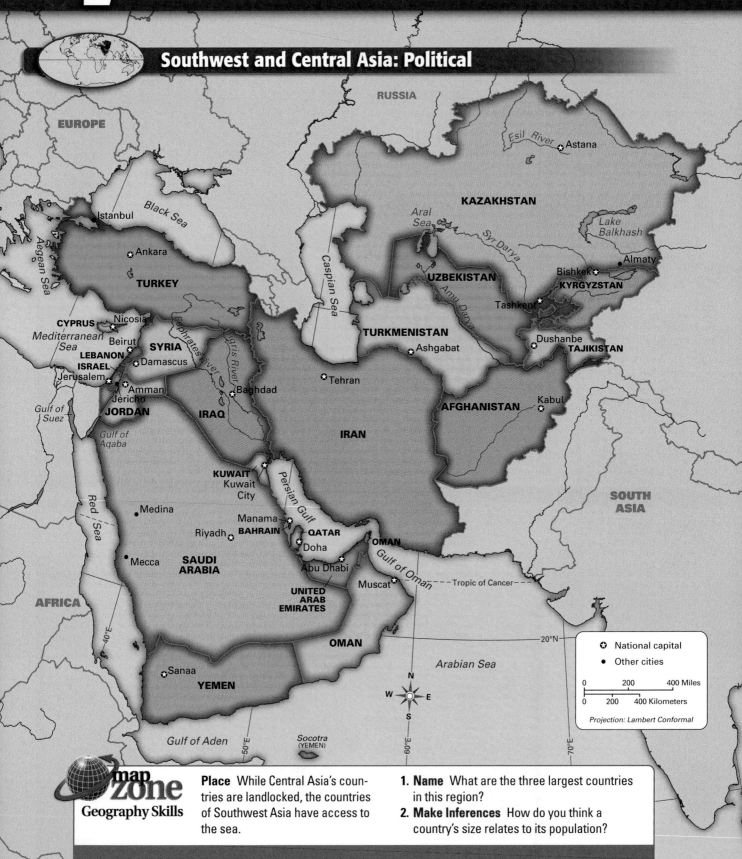

Southwest and Central Asia: Political

EUROPE

RUSSIA

Istanbul

Black Sea

Esil River · Astana

Aegean Sea

Ankara

KAZAKHSTAN

Aral Sea

Lake Balkhash

TURKEY

Caspian Sea

Syr Darya

Bishkek · Almaty

UZBEKISTAN

KYRGYZSTAN

CYPRUS Nicosia

Amu Darya

Tashkent

Mediterranean Sea

Euphrates River

SYRIA

TURKMENISTAN

Dushanbe

TAJIKISTAN

LEBANON Beirut

Tigris River

Ashgabat

ISRAEL Damascus

Jerusalem

Baghdad

Tehran

Kabul

Amman

Jericho

Gulf of Suez

JORDAN

IRAQ

AFGHANISTAN

Gulf of Aqaba

IRAN

SOUTH ASIA

Red Sea

KUWAIT

Kuwait City

Persian Gulf

Medina

Manama

Riyadh BAHRAIN QATAR

OMAN

Doha

Gulf of Oman

Mecca

SAUDI ARABIA

Abu Dhabi

Muscat

Tropic of Cancer

AFRICA

UNITED ARAB EMIRATES

20°N

National capital

Other cities

OMAN

Arabian Sea

0   200   400 Miles

Sanaa

0   200   400 Kilometers

YEMEN

N W E S

Projection: Lambert Conformal

Gulf of Aden

Socotra (YEMEN)

40°E   50°E   60°E   70°E

map zone

Geography Skills

Place While Central Asia's countries are landlocked, the countries of Southwest Asia have access to the sea.

1. **Name** What are the three largest countries in this region?
2. **Make Inferences** How do you think a country's size relates to its population?

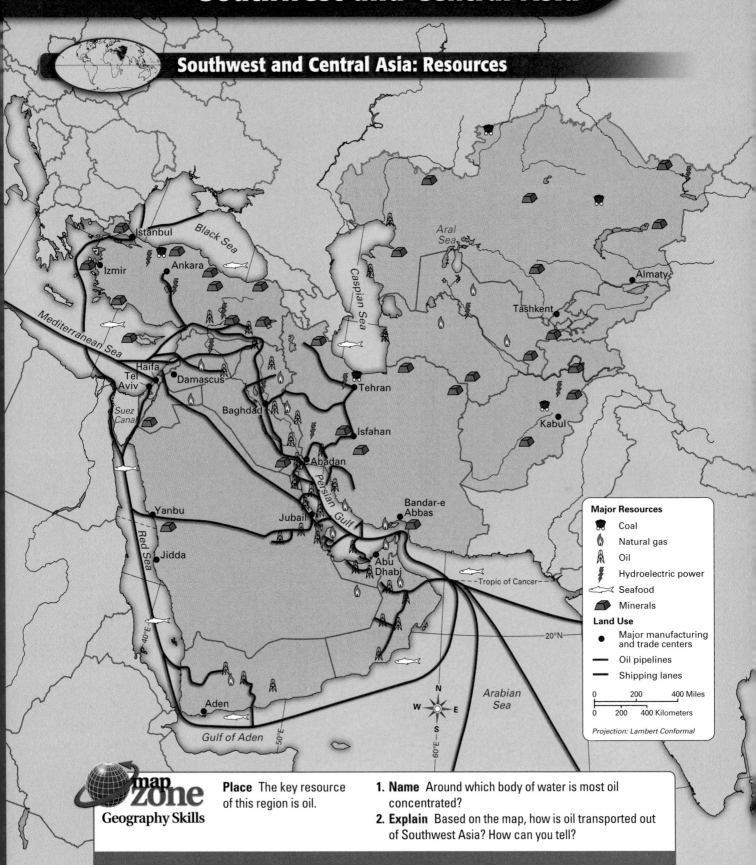

Southwest and Central Asia: Resources

Major Resources

- 🪨 Coal
- 🔥 Natural gas
- ⚗ Oil
- ⚡ Hydroelectric power
- 🐟 Seafood
- 🔷 Minerals

Land Use

- ● Major manufacturing and trade centers
- ─── Oil pipelines
- ━━━ Shipping lanes

0 200 400 Miles
0 200 400 Kilometers

Projection: Lambert Conformal

map Zone
Geography Skills

Place The key resource of this region is oil.

1. **Name** Around which body of water is most oil concentrated?
2. **Explain** Based on the map, how is oil transported out of Southwest Asia? How can you tell?

Southwest and Central Asia: Population

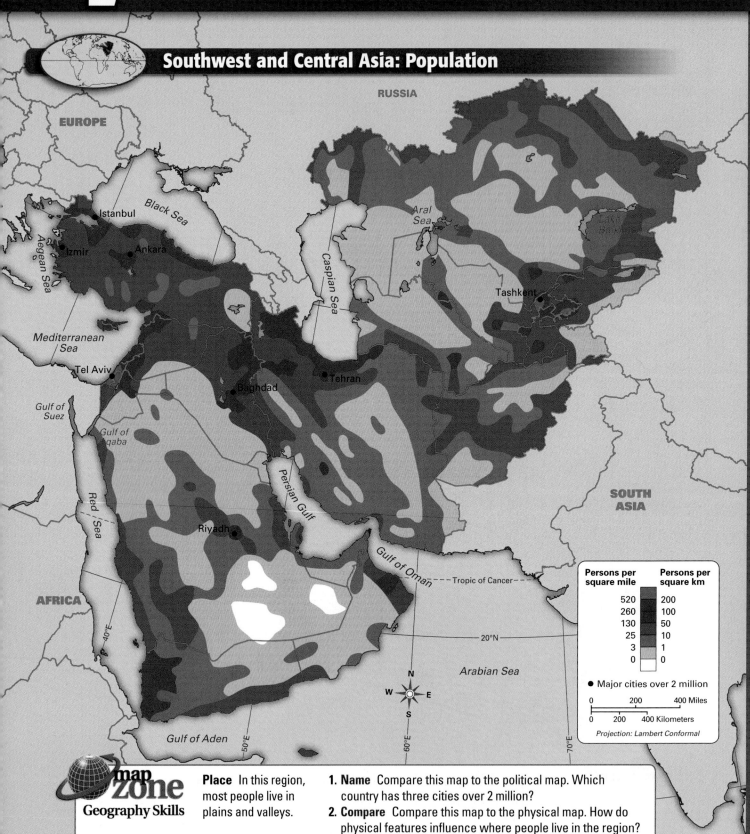

RUSSIA

EUROPE

Black Sea

Istanbul

Izmir   Ankara

Aegean Sea

Aral Sea

Lake Balkash

Caspian Sea

Tashkent

Mediterranean Sea

Tel Aviv

Baghdad

Tehran

Gulf of Suez

Gulf of Aqaba

Red Sea

Persian Gulf

Riyadh

SOUTH ASIA

AFRICA

40°E

Gulf of Oman

Tropic of Cancer

20°N

Arabian Sea

50°E

60°E

70°E

Gulf of Aden

Persons per square mile	**Persons per square km**
520	200
260	100
130	50
25	10
3	1
0	0

● Major cities over 2 million

| 0 | 200 | 400 Miles |
| 0 | 200 | 400 Kilometers |

Projection: Lambert Conformal

map zone
Geography Skills

Place In this region, most people live in plains and valleys.

1. **Name** Compare this map to the political map. Which country has three cities over 2 million?

2. **Compare** Compare this map to the physical map. How do physical features influence where people live in the region?

Southwest and Central Asia

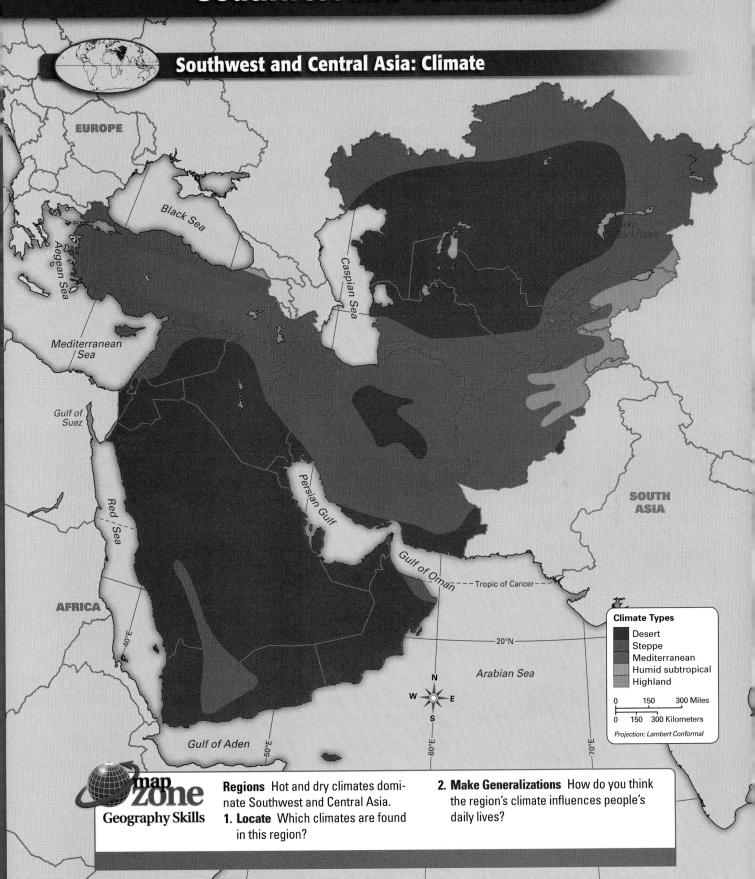

EUROPE

Black Sea

Aegean Sea

Caspian Sea

Lake Balkhash

Mediterranean Sea

Gulf of Suez

Gulf of Aqaba

Red Sea

Persian Gulf

Gulf of Oman

SOUTH ASIA

AFRICA

40°E

Tropic of Cancer

20°N

Arabian Sea

Gulf of Aden

50°E

60°E

70°E

N
W   E
S

Climate Types
- Desert
- Steppe
- Mediterranean
- Humid subtropical
- Highland

0   150   300 Miles

0   150   300 Kilometers

Projection: Lambert Conformal

map zone

Geography Skills

Regions Hot and dry climates dominate Southwest and Central Asia.

1. Locate Which climates are found in this region?

2. Make Generalizations How do you think the region's climate influences people's daily lives?

Southwest and Central Asia

COUNTRY Capital	FLAG	POPULATION	AREA (sq mi)	PER CAPITA GDP (U.S. $)	LIFE EXPECTANCY AT BIRTH	TVS PER 1,000 PEOPLE
Afghanistan Kabul		29.9 million	250,001	$800	42.9	14
Bahrain Manama		688,300	257	$19,200	74.2	446
Cyprus Nicosia		780,100	3,571	$20,300	77.7	154
Iran Tehran		68 million	636,296	$7,700	70.0	154
Iraq Baghdad		26.1 million	168,754	$3,500	68.7	82
Israel Jerusalem		6.3 million	8,019	$20,800	79.3	328
Jordan Amman		5.8 million	35,637	$4,500	78.2	83
Kazakhstan Astana		15.2 million	1,049,155	$7,800	66.6	240
Kuwait Kuwait City		2.3 million	6,880	$21,300	77.0	480
Kyrgyzstan Bishkek		5.1 million	76,641	$1,700	68.2	49
Lebanon Beirut		3.8 million	4,015	$5,000	72.6	355
Oman Muscat		3 million	82,031	$13,100	73.1	575
Qatar Doha		863,100	4,416	$23,200	73.7	866
Saudi Arabia Riyadh		26.4 million	756,985	$12,000	75.5	263
Syria Damascus		18.4 million	71,498	$3,400	70.0	68
United States Washington, D.C.		295.7 million	3,718,710	$40,100	77.7	844

COUNTRY Capital	FLAG	POPULATION	AREA (sq mi)	PER CAPITA GDP (U.S. $)	LIFE EXPECTANCY AT BIRTH	TVS PER 1,000 PEOPLE
Tajikistan Dushanbe		7.2 million	55,251	$1,100	64.6	328
Turkey Ankara		69.7 million	301,383	$7,400	72.4	328
Turkmenistan Ashgabat		5 million	188,456	$5,700	61.4	198
United Arab Emirates Abu Dhabi		2.6 million	32,000	$25,200	75.2	309
Uzbekistan Tashkent		26.9 million	172,742	$1,800	64.2	280
Yemen Sanaa		20.7 million	203,850	$800	61.8	286
United States Washington, D.C.		295.7 million	3,718,710	$40,100	77.7	844

ANALYSIS SKILL **ANALYZING TABLES**

1. How does the per capita GDP of countries in this region compare to the per capita GDP of the United States?
2. Based on the table, which countries seem to have the highest standard of living?

Oil Giants

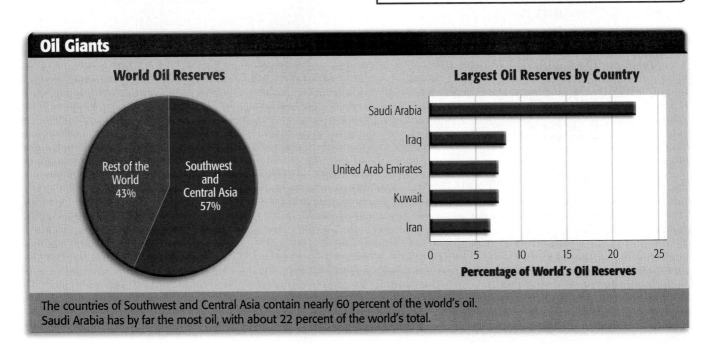

The countries of Southwest and Central Asia contain nearly 60 percent of the world's oil. Saudi Arabia has by far the most oil, with about 22 percent of the world's total.

History of the Fertile Crescent

7000–500 BC

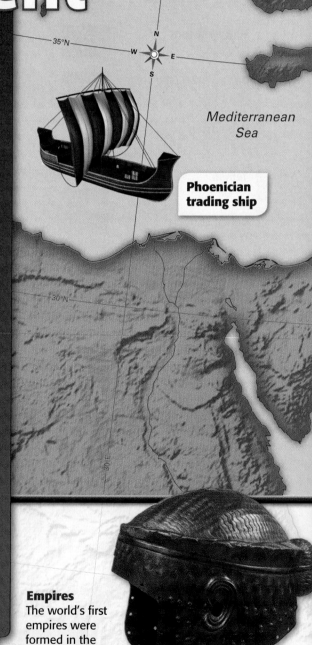

Phoenician trading ship

Mediterranean Sea

What You Will Learn...

In this chapter you will learn about the world's oldest civilizations. These civilizations developed in the region of Mesopotamia, part of a larger area known as the Fertile Crescent.

SECTION 1
Geography of the Fertile Crescent . 116

SECTION 2
The Rise of Sumer..................... 122

SECTION 3
Sumerian Achievements............ 127

SECTION 4
Later Peoples of the
Fertile Crescent 132

FOCUS ON READING AND WRITING

Paraphrasing One way to be sure you understand a passage of text is to paraphrase it, or restate it in your own words. Practice paraphrasing sentences and whole paragraphs as you read this chapter. **See the lesson, Paraphrasing, on page 672.**

Creating a Poster Most elementary students have not read or heard much about ancient Mesopotamia. As you read this chapter, you can gather information about that land. Then you can create a colorful poster to share some of what you have learned with a young child.

Empires
The world's first empires were formed in the Fertile Crescent. Soldiers from these empires wore bronze helmets like this one.

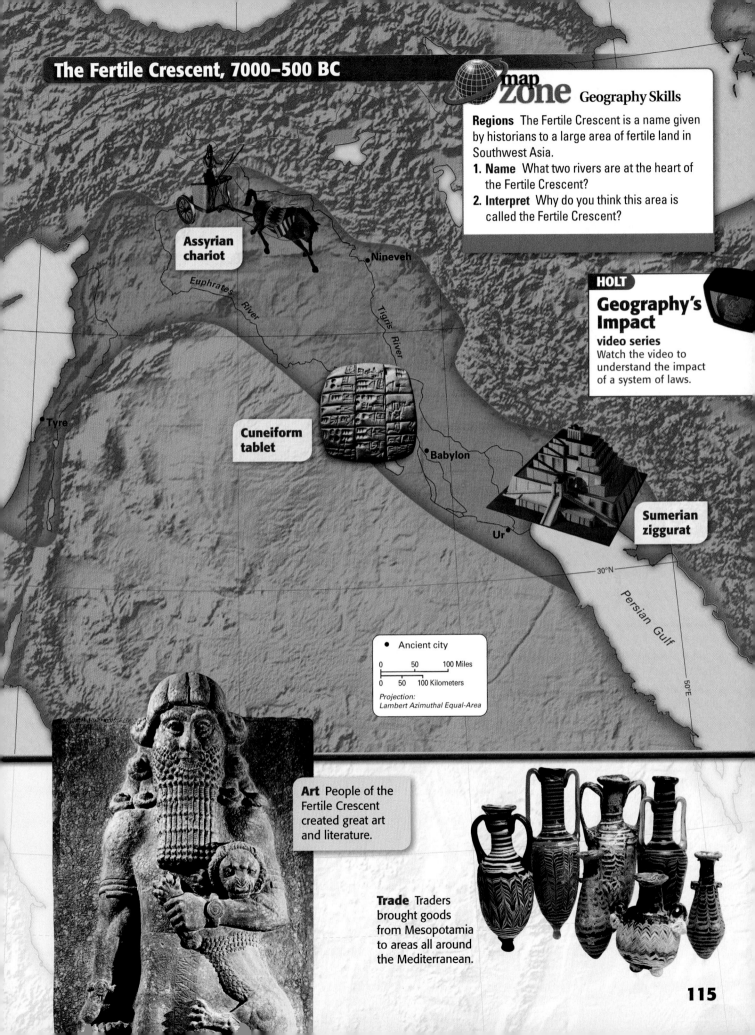

The Fertile Crescent, 7000–500 BC

map zone Geography Skills

Regions The Fertile Crescent is a name given by historians to a large area of fertile land in Southwest Asia.

1. **Name** What two rivers are at the heart of the Fertile Crescent?
2. **Interpret** Why do you think this area is called the Fertile Crescent?

Assyrian chariot

Nineveh

Euphrates River

Tigris River

HOLT

Geography's Impact
video series
Watch the video to understand the impact of a system of laws.

Cuneiform tablet

Babylon

Sumerian ziggurat

Tyre

Ur

30°N

Persian Gulf

50°E

• Ancient city

0   50   100 Miles
0   50   100 Kilometers

*Projection:
Lambert Azimuthal Equal-Area*

Art People of the Fertile Crescent created great art and literature.

Trade Traders brought goods from Mesopotamia to areas all around the Mediterranean.

115

Geography of the Fertile Crescent

What You Will Learn...

Main Ideas

1. The rivers of Southwest Asia supported the growth of civilization.
2. New farming techniques led to the growth of cities.

The Big Idea

The valleys of the Tigris and Euphrates rivers were the site of the world's first civilizations.

Key Terms

Fertile Crescent, *p. 117*
silt, *p. 117*
irrigation, *p. 118*
canals, *p. 118*
surplus, *p. 118*
division of labor, *p. 118*

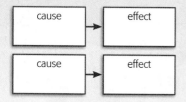
TAKING NOTES As you read, take notes on the cause-and-effect relationship between river valleys and the civilizations that developed around it. Use a graphic organizer like this one to list causes and effects.

cause	→	effect

cause	→	effect

If **YOU** lived there...

You are a farmer in Southwest Asia about 6,000 years ago. You live near a slow-moving river that has many shallow lakes and marshes. The river makes the land in the valley rich and fertile, so you can grow wheat and dates. But in the spring, raging floods spill over the riverbanks, destroying your fields. In the hot summers, you are often short of water.

How can you control the waters of the river?

BUILDING BACKGROUND In several parts of the world, bands of hunter-gatherers began to settle down in farming settlements. They domesticated plants and animals. Gradually, their cultures became more complex. Most early civilizations grew up along rivers, where people learned to work together to irrigate fields and control floods.

Rivers Support the Growth of Civilization

Early peoples settled where crops would grow. Crops usually grew well near rivers, where water was available and regular floods made the soil rich. One region in Southwest Asia was especially well suited for farming. It lay between two rivers.

The Land between the Rivers

The Tigris and Euphrates rivers are the most important physical features of the region sometimes known as Mesopotamia (mes-uh-puh-TAY-mee-uh). Mesopotamia means "between the rivers" in Greek.

As you can see on the map, the region called Mesopotamia lies between Asia Minor and the Persian Gulf. The region is part of the **Fertile Crescent**, a large arc of rich, or fertile, farmland. As you can see on the map, the Fertile Crescent extends from the Persian Gulf to the Mediterranean Sea.

In ancient times, Mesopotamia was made of two parts. Northern Mesopotamia was a plateau bordered on the north and the east by mountains. The southern part of Mesopotamia was a flat plain. The Tigris and Euphrates rivers flowed down from the hills into this low-lying plain.

The Rise of Civilization

Hunter-gatherer groups first settled in Mesopotamia more than 12,000 years ago. Over time, these people learned how to plant crops to grow their own food. Every year, floods on the Tigris and Euphrates rivers brought **silt**, a mixture of rich soil and tiny rocks, to the land. The fertile silt made the land ideal for farming.

The first farm settlements were formed in Mesopotamia as early as 7000 BC. There, farmers grew wheat, barley, and other types of grain. Livestock, birds, and fish were also good sources of food. Plentiful food led to population growth, and villages formed. Eventually, these early villages developed into the world's first civilization.

READING CHECK **Summarizing** What made civilization possible in Mesopotamia?

FOCUS ON READING
Make sure you understand this paragraph by restating it in your own words.

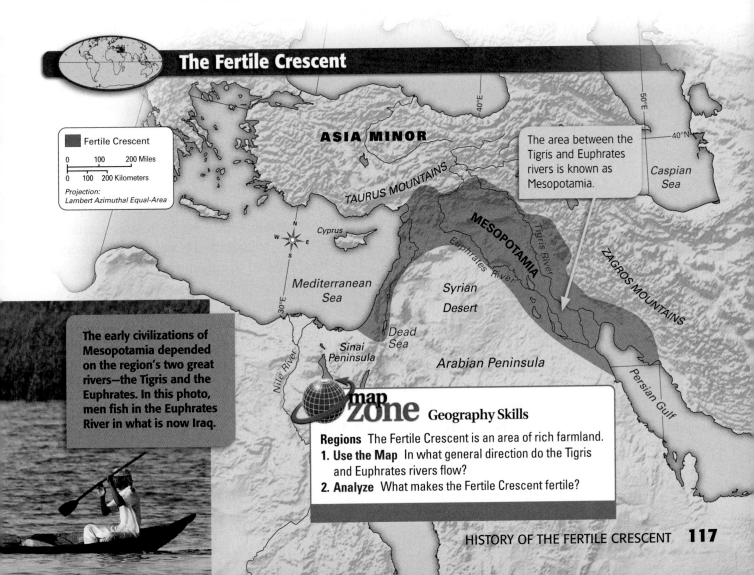

The Fertile Crescent

Fertile Crescent

0   100   200 Miles
0   100   200 Kilometers
Projection: Lambert Azimuthal Equal-Area

ASIA MINOR

The area between the Tigris and Euphrates rivers is known as Mesopotamia.

Caspian Sea

TAURUS MOUNTAINS

Cyprus

MESOPOTAMIA

Euphrates River

Tigris River

ZAGROS MOUNTAINS

Mediterranean Sea

Syrian Desert

Dead Sea

Sinai Peninsula

Nile River

Arabian Peninsula

Persian Gulf

The early civilizations of Mesopotamia depended on the region's two great rivers—the Tigris and the Euphrates. In this photo, men fish in the Euphrates River in what is now Iraq.

map zone Geography Skills

Regions The Fertile Crescent is an area of rich farmland.
1. **Use the Map** In what general direction do the Tigris and Euphrates rivers flow?
2. **Analyze** What makes the Fertile Crescent fertile?

Irrigation and Civilization

Early farmers faced the challenge of learning how to control the flow of river water to their fields in both rainy and dry seasons.

1 Early settlements in Mesopotamia were located near rivers. Water was not controlled, and flooding was a continual problem.

2 Later, people built canals to protect houses from flooding and to move water to their fields.

Farming and Cities

Although Mesopotamia had fertile soil, farming wasn't easy there. The region received little rain. This meant that water levels in the Tigris and Euphrates rivers depended on rainfall in eastern Asia Minor where the two rivers began. When a great amount of rain fell, water levels got very high. This flooding destroyed crops, killed livestock, and washed away homes. When water levels were too low, crops dried up. Farmers knew that they needed to develop a way to control the rivers' flow.

Controlling Water

To solve their problems, Mesopotamians used **irrigation**, a way of supplying water to an area of land. To irrigate their land, they dug out large storage basins to catch rainwater that fell to the north. Then they dug **canals**, human-made waterways, that connected these basins to a network of ditches. These ditches brought water to the fields. To protect their fields from flooding, farmers built up the rivers' banks. These built-up banks held back floodwaters even when river levels were high.

THE IMPACT TODAY

People still build dikes, or earthen walls along rivers or shorelines, to hold back water.

Food Surpluses

Irrigation increased the amount of food farmers were able to grow. In fact, farmers could produce a food **surplus**, or more than they needed. Farmers also used irrigation to water grazing areas for cattle and sheep. As a result, Mesopotamians ate a variety of foods. Fish, meat, wheat, barley, and dates were plentiful.

Because irrigation made farmers more productive, fewer people needed to farm. Some people became free to do other jobs. As a result, new occupations developed. For the first time, people became crafters, religious leaders, and government workers. The type of arrangement in which each worker specializes in a particular task or job is called a **division of labor**.

Having people available to work on different jobs meant that society could accomplish more. Large projects, such as raising buildings and digging irrigation systems, required specialized workers, managers, and organization. To complete these types of projects, the Mesopotamians needed structure and rules. These could be provided by laws and government.

③ With irrigation, the people of Mesopotamia were able to grow more food.

④ Food surpluses allowed some people to stop farming and concentrate on other jobs, like making clay pots or tools.

Appearance of Cities

Over time, Mesopotamian settlements grew both in size and complexity. They gradually developed into cities between 4000 and 3000 BC.

Despite the growth of cities, society in Mesopotamia was still based on agriculture. Most people still worked in farming jobs. However, cities were becoming important places. People traded goods there, and cities provided leaders with power bases.

Cities were the political, religious, cultural, and economic centers of civilization.

READING CHECK **Analyzing** Why did the Mesopotamians create irrigation systems?

SUMMARY AND PREVIEW Mesopotamia's rich, fertile lands supported productive farming, which led to the development of cities. In Section 2 you will learn about some of the first city builders.

go.hrw.com
Online Quiz
KEYWORD: SK7 HP5

Section 1 Assessment

Reviewing Ideas, Terms, and Places

1. **a. Identify** Where was Mesopotamia?
   b. Explain How did the **Fertile Crescent** get its name?
   c. Evaluate What was the most important factor in making Mesopotamia's farmland fertile?
2. **a. Describe** Why did farmers need to develop a system to control their water supply?
   b. Explain In what ways did a **division of labor** contribute to the growth of the Mesopotamian civilization?
   c. Elaborate How might managing large projects prepare people for running a government?

Critical Thinking

3. **Identifying Cause and Effect** Farmers who used the rivers for irrigation were part of a cause-effect chain. Use a chart like this one to show that chain.

| River levels were uneven. | | | People enjoy many foods. |

FOCUS ON WRITING

4. **Understanding Geography** Think of the images you might use on your poster. Would you want to show an image of the canals or rivers? Can you find pictures to show important features?

HISTORY OF THE FERTILE CRESCENT   **11**

River Valley Civilizations

All of the world's earliest civilizations had at least one thing in common—they arose in river valleys that were good locations for farming. Three key factors made river valleys good for farming. First, the fields that bordered the rivers were flat, which made it easier for farmers to plant crops. Second, the soils were nourished by flood deposits and silt, which made them very fertile. Finally, the river provided the water farmers needed for irrigation.

Natural Highways River travel allowed early civilizations to trade goods and ideas. These people are traveling on the Euphrates River, one of the two main rivers of ancient Mesopotamia.

Caspian Se

M e d i t e r r a n e a n   S e a

MESOPOTAMIA

Tigris River

Euphrates River

A F R I C A

Memphis

Ur

EGYPT

Red Sea

Nile River

A R A B I A N   P E N I N S U L A

From Village to City With the development of agriculture, people settled into farming villages. Over time, some of these villages grew into large cities. These ancient ruins are near Memphis, Egypt.

Gifts of the River River water was key to farming in early civilizations. This farmer is using water from the Huang He (Yellow River) in China to water her crops.

A S I A

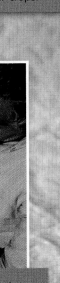

New Activities Food surpluses allowed people to pursue other activities, like crafts, art, and writing. This tile designer lives in the Indus Valley.

Huang He (Yellow River)

CHINA

• Harappa

H I M A L A Y A S

Indus River

Ganges River

Chang Jiang (Yangzi River)

Mohenjo Daro •

INDUS VALLEY

INDIA

Arabian Sea

Bay of Bengal

map zone Geography Skills

Human-Environment Interaction Four of the world's earliest civilizations arose on the banks of large rivers.

1. **Locate** Where were the four earliest river valley civilizations located?

2. **Explain** Why did the world's first civilizations all develop in river valleys?

N
W ✦ E
S

River valley

0 ———— 500 ———— 1,000 Miles
0 —— 500 —— 1,000 Kilometers

INDIAN OCEAN

SECTION 2

The Rise of Sumer

What You Will Learn...

Main Ideas

1. The Sumerians created the world's first advanced society.
2. Religion played a major role in Sumerian society.

The Big Idea

The Sumerians developed the first civilization in Mesopotamia.

Key Terms and Places

Sumer, *p. 122*
city-state, *p. 122*
empire, *p. 123*
polytheism, *p. 124*
priests, *p. 125*
social hierarchy, *p. 125*

TAKING NOTES As you read, use a chart like the one below to take notes on the Sumerian civilization.

Characteristics	Notes
Cities	
Government	
Religion	
Society	

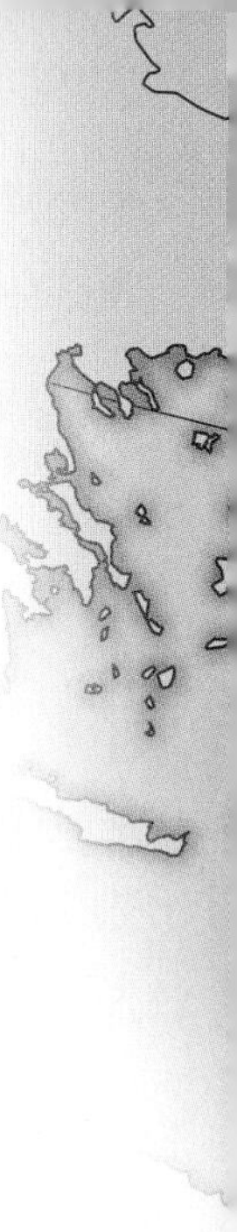

If YOU lived there...

You are a crafter living in one of the cities of Sumer. Thick walls surround and protect your city, so you feel safe from the armies of other city-states. But you and your neighbors are fearful of other beings—the many gods and spirits that you have been taught are everywhere. They can bring illness or sandstorms or bad luck.

How might you protect yourself from gods and spirits?

BUILDING BACKGROUND As civilizations developed along rivers, their societies and governments became more advanced. Religion became a main characteristic of these ancient cultures. Kings claimed to rule with the approval of the gods, and ordinary people wore charms and performed rituals to avoid bad luck.

An Advanced Society

In southern Mesopotamia, a people known as the Sumerians (SOO-MER-ee-unz) developed the world's first civilization. No one knows where they came from or when they moved into the region. All we know is that by 3000 BC, several hundred thousand Sumerians had settled in Mesopotamia, in a land they called **Sumer** (SOO-muhr). There they built an advanced society.

City-States of Sumer

Most people in Sumer were farmers. They lived mainly in rural, or countryside, areas. The centers of Sumerian society, however, were the urban, or city, areas. The first cities in Sumer had about 10,000 residents. Over time, the cities grew. Historians think that by 2000 BC, some of Sumer's largest cities had more than 100,000 residents.

As a result, the basic political unit of Sumer combined the two parts. This unit was the city-state. A **city-state** consisted of a central city and all the countryside around it. The amount of farmland controlled by a city-state depended on its military strength. Stronger city-states controlled larger areas.

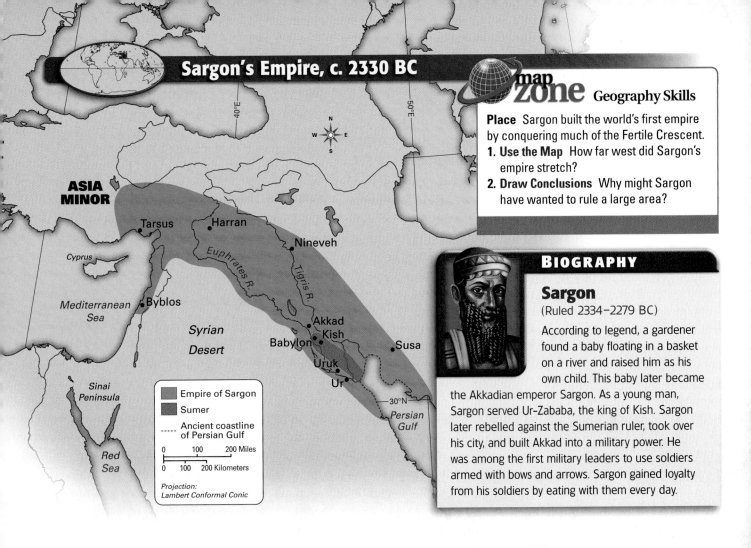

ASIA MINOR

Tarsus

Harran

Nineveh

Cyprus

Euphrates R.

Tigris R.

Mediterranean Sea

Byblos

Syrian Desert

Akkad

Kish

Babylon

Susa

Uruk

Ur

Sinai Peninsula

30°N

Persian Gulf

Red Sea

Empire of Sargon

Sumer

Ancient coastline of Persian Gulf

0   100   200 Miles

0   100   200 Kilometers

Projection: Lambert Conformal Conic

map zone Geography Skills

Place Sargon built the world's first empire by conquering much of the Fertile Crescent.
1. **Use the Map** How far west did Sargon's empire stretch?
2. **Draw Conclusions** Why might Sargon have wanted to rule a large area?

BIOGRAPHY

Sargon
(Ruled 2334–2279 BC)

According to legend, a gardener found a baby floating in a basket on a river and raised him as his own child. This baby later became the Akkadian emperor Sargon. As a young man, Sargon served Ur-Zababa, the king of Kish. Sargon later rebelled against the Sumerian ruler, took over his city, and built Akkad into a military power. He was among the first military leaders to use soldiers armed with bows and arrows. Sargon gained loyalty from his soldiers by eating with them every day.

City-states in Sumer fought each other to gain more farmland. As a result of these conflicts, the city-states built up strong armies. Sumerians also built strong, thick walls around their cities for protection.

Individual city-states gained and lost power over time. By 3500 BC, a city-state known as Kish had become quite powerful. Over the next 1,000 years, the city-states of Uruk and Ur fought for dominance. One of Uruk's kings, known as Gilgamesh, became a legendary figure in Sumerian literature.

Rise of the Akkadian Empire

In time, another society developed along the Tigris and Euphrates. This society was built by the Akkadians (uh-KAY-dee-uhns). They lived just north of Sumer, but they were not Sumerians. They even spoke a different language than the Sumerians.

In spite of their differences, however, the Akkadians and the Sumerians lived in peace for many years.

That peace was broken in the 2300s BC when Sargon sought to extend Akkadian territory. He built a new capital, Akkad (A-kad), on the Euphrates River, near what is now the city of Baghdad. Sargon was the first ruler to have a permanent army. He used that army to launch a series of wars against neighboring kingdoms.

Sargon's soldiers defeated all the city-states of Sumer. They also conquered northern Mesopotamia, finally bringing the entire region under his rule. With these conquests, Sargon established the world's first **empire**, or land with different territories and peoples under a single rule. Sargon's huge empire stretched from the Persian Gulf to the Mediterranean Sea.

Sargon was emperor, or ruler of his empire, for more than 50 years. However, the empire lasted only a century after his death. Later rulers could not keep the empire safe from invaders. Hostile tribes from the east raided and captured Akkad. A century of chaos followed.

Eventually, however, the Sumerian city-state of Ur rebuilt its strength and conquered the rest of Mesopotamia. Political stability was restored. The Sumerians once again became the most powerful civilization in the region.

READING CHECK **Summarizing** How did Sargon build an empire?

Religion Shapes Society

Religion was very important in Sumerian society. In fact, it played a **role** in nearly every aspect of life. In many ways, religion was the basis for all of Sumerian society.

Sumerian Religion

The Sumerians practiced **polytheism, the worship of many gods**. Among the gods they worshipped were Enlil, lord of the air; Enki, god of wisdom; and Inanna, goddess of love and war. The sun and moon were represented by the gods Utu and Nanna. Each city-state considered one god to be its special protector.

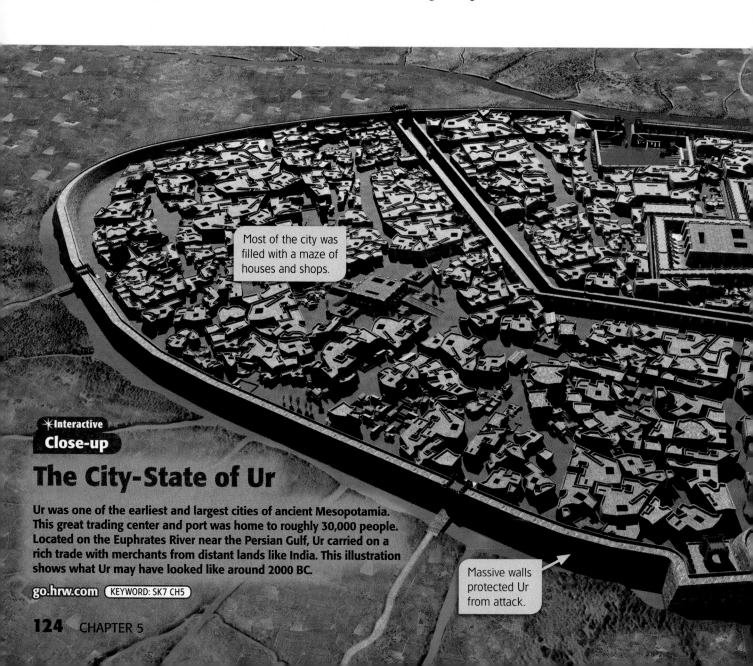

Most of the city was filled with a maze of houses and shops.

★ Interactive
Close-up

The City-State of Ur

Ur was one of the earliest and largest cities of ancient Mesopotamia. This great trading center and port was home to roughly 30,000 people. Located on the Euphrates River near the Persian Gulf, Ur carried on a rich trade with merchants from distant lands like India. This illustration shows what Ur may have looked like around 2000 BC.

go.hrw.com KEYWORD: SK7 CH5

Massive walls protected Ur from attack.

The Sumerians believed that their gods had enormous powers. Gods could bring good harvests or disastrous floods. They could bring illness, or they could bring good health and wealth. The Sumerians believed that success in life depended on pleasing the gods. Every Sumerian had to serve and worship the gods.

Priests, people who performed or led religious ceremonies, had great status in Sumer. People relied on them to help gain the gods' favor. Priests interpreted the wishes of the gods and made offerings to them. These offerings were made in temples, special buildings where priests performed their religious ceremonies.

Sumerian Social Order

Because of their status, priests occupied a high level in Sumer's **social hierarchy, the division of society by rank or class**. In fact, priests were just below kings. The kings of Sumer claimed that they had been chosen by the gods to rule.

Below the priests were Sumer's skilled craftspeople, merchants, and traders. Trade had a great **impact** on Sumerian society. Traders traveled to faraway places and exchanged grain for gold, silver, copper, lumber, and precious stones.

Below traders, farmers and laborers made up the large working class. Slaves were at the bottom of the social order.

ACADEMIC VOCABULARY

impact effect, result

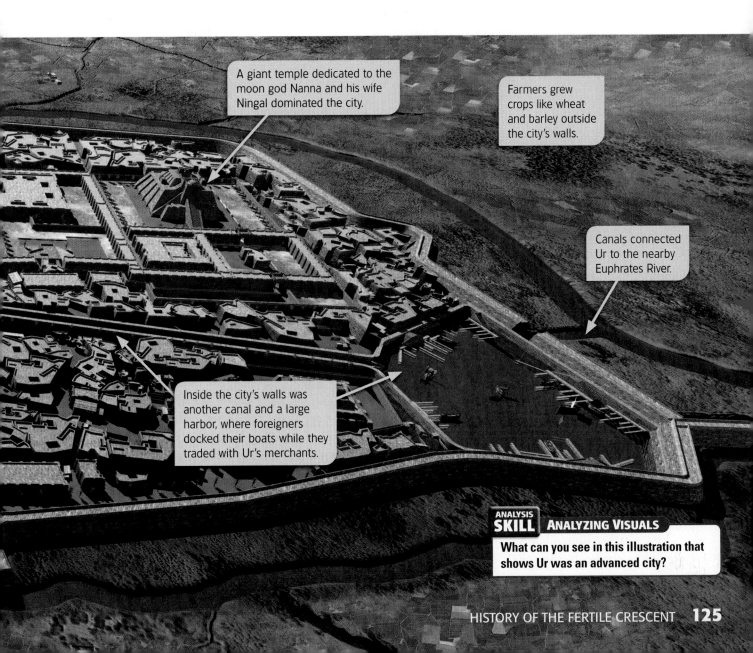

A giant temple dedicated to the moon god Nanna and his wife Ningal dominated the city.

Farmers grew crops like wheat and barley outside the city's walls.

Canals connected Ur to the nearby Euphrates River.

Inside the city's walls was another canal and a large harbor, where foreigners docked their boats while they traded with Ur's merchants.

ANALYSIS SKILL | **ANALYZING VISUALS**

What can you see in this illustration that shows Ur was an advanced city?

Sumerian Society

Sumerian society was divided into different groups. This ancient artifact shows Sumerian leaders celebrating a military victory while a musician plays an instrument.

Men and Women in Sumer

Sumerian men and women had different roles. In general, men held political power and made laws while women took care of the home and children. Education was usually reserved for men, but some upper-class women were educated as well.

Some educated women were priestesses in Sumer's temples. They helped shape Sumerian culture. One, Enheduanna, the daughter of Sargon, wrote hymns to the goddess Inanna. The first known female writer in history, she wrote these verses:

" My Queen,
[all] the Anunna, the great gods,
Fled before you like fluttering bats,
Could not stand before your awesome face, "
–Enheduanna, from *Adoration of Inanna of Ur*

READING CHECK **Analyzing** How did trade affect Sumerian society?

SUMMARY AND PREVIEW In this section you learned about Sumerian city-states, religion, and society. In Section 3, you will read about Sumerian achievements.

Section 2 Assessment

go.hrw.com
Online Quiz
KEYWORD: SK7 HP5

Reviewing Ideas, Terms, and Places

1. **a. Recall** What was the basic political unit of Sumer?
   b. Explain What steps did Sumerian **city-states** take to protect themselves from their rivals?
   c. Elaborate How do you think that Sargon's creation of an **empire** changed the later history of Mesopotamia? Defend your answer.
2. **a. Identify** What is **polytheism**?
   b. Draw Conclusions Why do you think **priests** were so influential in ancient Sumerian society?
   c. Elaborate Why would rulers benefit if they claimed to be chosen by the gods?

Critical Thinking

3. **Summarizing** In the right column of your note-taking chart, write a summary sentence for each of the four characteristics. Then add a box at the bottom of the chart and write a sentence summarizing the Sumerian civilization.

Characteristics	Notes
Cities	
Government	
Religion	
Society	

Summary Sentence:

FOCUS ON WRITING

4. **Gathering Information about Sumer** You will need some pictures of Sumerian society on your poster. Note two or three things to add.

Sumerian Achievements

If YOU lived there...

You are a student at a school for scribes in Sumer. Learning all the symbols for writing is very hard. Your teacher assigns you lessons to write on your clay tablet, but you can't help making mistakes. Then you have to smooth out the surface and try again. Still, being a scribe can lead to important jobs for the king. You could make your family proud.

Why would you want to be a scribe?

BUILDING BACKGROUND Sumerian society was advanced in terms of religion and government organization. The Sumerians were responsible for many other achievements, which were passed down to later civilizations.

Invention of Writing

The Sumerians made one of the greatest cultural advances in history. They developed **cuneiform** (kyoo-NEE-uh-fohrm), the world's first system of writing. The Sumerians did not have pens, pencils, or paper, though. Instead, they used sharp tools called styluses to make wedge-shaped symbols on clay tablets.

What You Will Learn...

Main Ideas

1. The Sumerians invented the world's first writing system.
2. Advances and inventions changed Sumerian lives.
3. Many types of art developed in Sumer.

The Big Idea

The Sumerians made many advances that helped their society develop.

Key Terms

cuneiform, *p. 127*
pictographs, *p. 128*
scribe, *p. 128*
epics, *p. 128*
architecture, *p. 130*
ziggurat, *p. 130*

TAKING NOTES Create a chart like the one below. As you read, list the achievements and advances made by the Sumerians.

Sumerian Advances and Achievements

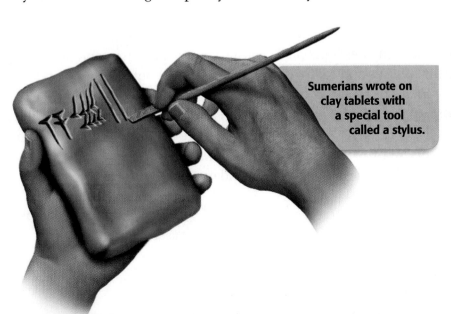

Sumerians wrote on clay tablets with a special tool called a stylus.

Development of Writing

	3300 BC	2800 BC	2400 BC	1800 BC
Heaven				
Grain				
Fish				
Bird				
Water				

Sumerian writing developed from early symbols called pictographs. Writers used clay tablets to record business deals, like this tablet that describes a number of sheep and goats.

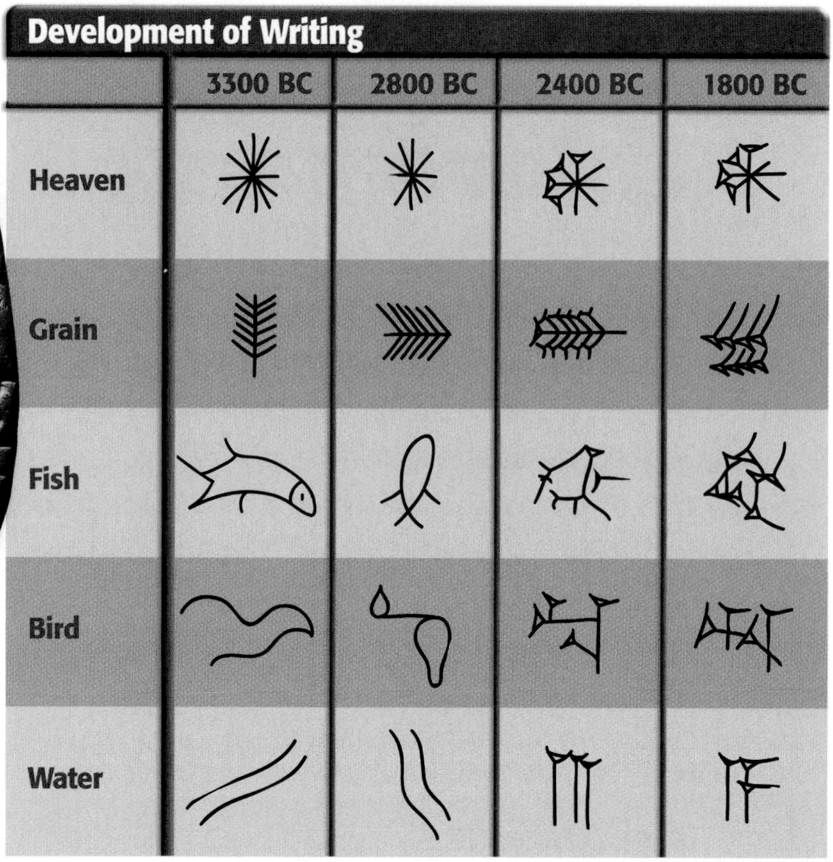

Earlier written communication had used **pictographs**, or picture symbols. Each pictograph represented an object, such as a tree or an animal. In cuneiform, symbols could also represent syllables, or basic parts of words. As a result, Sumerian writers could combine multiple symbols to express more **complex** ideas such as "joy" or "powerful."

Sumerians first used cuneiform to keep business records. A **scribe**, or writer, would be hired to keep track of the items people traded. Government officials and temples also hired scribes to keep their records. Becoming a scribe was a way to move up in social class.

Sumerian students went to school to learn to read and write. Like today, though, some students did not want to study. A Sumerian story tells of a father who urged his son to do his schoolwork:

ACADEMIC VOCABULARY

complex
difficult, not simple

" Go to school, stand before your 'school-father,' recite your assignment, open your schoolbag, write your tablet … After you have finished your assignment and reported to your monitor [teacher], come to me, and do not wander about in the street. "

–Sumerian essay quoted in *History Begins at Sumer*, by Samuel Noah Kramer

In time, Sumerians put their writing skills to new uses. They wrote works on history, law, grammar, and math. They also created works of literature. Sumerians wrote stories, proverbs, and songs. They wrote poems about the gods and about military victories. Some of these were **epics**, long poems that tell the stories of heroes. Later, people used some of these poems to create *The Epic of Gilgamesh*, the story of a legendary Sumerian king.

READING CHECK **Generalizing** How was cuneiform first used in Sumer?

Advances and Inventions

Writing was not the only great Sumerian invention. These early people made many other advances and discoveries.

Technical Advances

One of the Sumerians' most important developments was the wheel. They were the world's first people to build wheeled vehicles, such as carts. Using the wheel, Sumerians invented a device that spins clay as a craftsperson shapes it into bowls. This device is called a potter's wheel.

The plow was another important Sumerian invention. Pulled by oxen, plows broke through the hard clay soil of Sumer to prepare it for planting. This technique greatly increased farm production. The Sumerians also invented a clock that used falling water to measure time.

Sumerian advances improved daily life. Sumerians built sewers under city streets. They used bronze to make strong tools and weapons. They even produced makeup and glass jewelry.

Math and Science

Another area in which Sumerians excelled was math. In fact, they developed a math system based on the number 60. Based on this system, they divided a circle into 360 degrees. Dividing a year into 12 months—a factor of 60—was another Sumerian idea. Sumerians also calculated the areas of rectangles and triangles.

Sumerian scholars studied science, too. They wrote long lists to record their study of the natural world. These tablets included the names of thousands of animals, plants, and minerals.

The Sumerians also made advances in medicine. Using ingredients from animals, plants, and minerals, they produced many healing drugs. Among the items used in these medicines were milk, turtle shells, figs, and salt. The Sumerians catalogued their medical knowledge, listing treatments according to symptoms and body parts.

READING CHECK **Categorizing** What areas of life were improved by Sumerian inventions?

THE IMPACT TODAY
We still use a base-60 system when we talk about 60 seconds in a minute and 60 minutes in an hour.

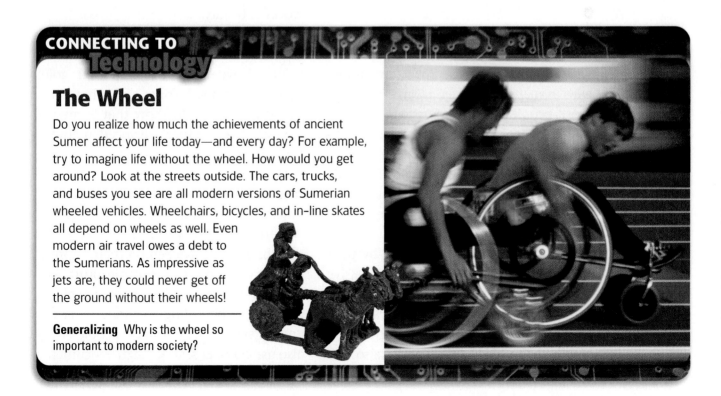

CONNECTING TO Technology

The Wheel

Do you realize how much the achievements of ancient Sumer affect your life today—and every day? For example, try to imagine life without the wheel. How would you get around? Look at the streets outside. The cars, trucks, and buses you see are all modern versions of Sumerian wheeled vehicles. Wheelchairs, bicycles, and in-line skates all depend on wheels as well. Even modern air travel owes a debt to the Sumerians. As impressive as jets are, they could never get off the ground without their wheels!

Generalizing Why is the wheel so important to modern society?

Sumerian Achievements

The Sumerians' artistic achievements included beautiful works of gold, wood, and stone.

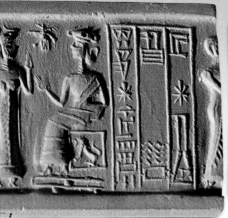

Cylinder seals like this one were carved into round stones and then rolled over clay to leave their mark.

This stringed musical instrument is called a lyre. It features a cow's head and is made of silver decorated with shell and stone.

The Arts of Sumer

The Sumerians' skills in the fields of art, metalwork, and **architecture**—the science of building—are well known to us. The ruins of great buildings and fine works of art have provided us with many examples of the Sumerians' creativity.

Architecture

Most Sumerian rulers lived in large palaces. Other rich Sumerians had two-story homes with as many as a dozen rooms. However, most people lived in smaller, one-story houses. These homes had six or seven rooms arranged around a small courtyard. Large and small houses stood side by side along the narrow, unpaved streets of the city. Bricks made of mud were the houses' main building blocks.

City centers were dominated by their temples, the largest and most impressive buildings in Sumer. A **ziggurat**, a pyramid-shaped temple, rose high above each city. Outdoor staircases led to a platform and a shrine at the top. Some temples also had columns to make them more attractive.

FOCUS ON READING

What was a cylinder seal? Describe one in your own words.

The Arts

Sumerian sculptors produced many fine works. Among them are the statues of gods created for temples. Sumerian artists also sculpted small objects out of ivory and rare woods. Sumerian pottery is better known for its quantity than its quality. Potters turned out many items, but few were works of beauty.

Jewelry was a popular item in Sumer. The jewelers of the region made many beautiful works out of imported gold, silver, and gems. Earrings and other items found in the region show that Sumerian jewelers knew rather advanced methods for putting gold pieces together.

Cylinder seals are perhaps Sumer's most famous works of art. These small objects were stone cylinders engraved with designs. When rolled over clay, the designs would leave behind their imprint. Each seal left its own distinct imprint. As a result, a person could show ownership of a container by rolling a cylinder over the container's wet clay surface. People could also use cylinder seals to "sign" documents or to decorate other clay objects.

The Sumerians were the first people in Mesopotamia to build large temples called ziggurats.

This gold dagger was found in a royal tomb. The bull's head is made of gold and silver.

ANALYSIS SKILL ANALYZING VISUALS
What animal is shown in two of these works?

Some cylinder seals showed battle scenes. Others displayed worship rituals. Some were highly decorative, covered with hundreds of carefully cut gems.

The Sumerians also enjoyed music. Kings and temples hired musicians to play on special occasions. Sumerian musicians played reed pipes, drums, tambourines, and harplike stringed instruments called lyres. Children learned songs in school. People sang hymns to gods and kings. Music and dance provided entertainment in marketplaces and homes.

READING CHECK Drawing Inferences What might historians learn from cylinder seals?

SUMMARY AND PREVIEW The Sumerians greatly enriched their society. Next, you will learn about the later peoples who lived in Mesopotamia.

Section 3 Assessment

go.hrw.com
Online Quiz
KEYWORD: SK7 HP5

Reviewing Ideas, Terms, and Places

1. **a. Identify** What is **cuneiform**?
   b. Analyze Why do you think writing is one of history's most important cultural advances?
   c. Elaborate What current leader would you choose to write an **epic** about, and why?
2. **a. Recall** What were two early uses of the wheel?
   b. Explain Why do you think the invention of the plow was so important to the Sumerians?
3. **a. Describe** What was the basic Sumerian building material?
   b. Make Inferences Why do you think cylinder seals developed into works of art?

Critical Thinking

4. **Identifying Effects** In a chart like this one, identify the effect of each Sumerian advance you listed in your notes.

Advance/Achievement	Effect

FOCUS ON WRITING

5. **Evaluating Information** What will you include on your poster to show Sumerian achievements? A ziggurat? A piece of jewelry? A musical instrument? Make a list of the pictures you think would be most interesting to elementary students.

Later Peoples of the Fertile Crescent

What You Will Learn...

Main Ideas

1. The Babylonians conquered Mesopotamia and created a code of law.
2. Invasions of Mesopotamia changed the region's culture.
3. The Phoenicians built a trading society in the eastern Mediterranean region.

The Big Idea

After the Sumerians, many cultures ruled parts of the Fertile Crescent.

Key Terms and Places

Babylon, *p. 132*
Hammurabi's Code, *p. 133*
chariot, *p. 134*
alphabet, *p. 137*

TAKING NOTES As you read, use a diagram like the one below to keep track of the later empires of the Fertile Crescent.

If YOU lived there...

You are a noble in ancient Babylon, an adviser to the great king Hammurabi. One of your duties is to collect all the laws of the kingdom. They will be carved on a tall block of black stone and placed in the temple. The king asks your opinion about the punishments for certain crimes. For example, should common people be punished more harshly than nobles?

How will you advise the king?

BUILDING BACKGROUND Many peoples invaded Mesopotamia. A series of kings conquered the lands between the rivers. Each new culture inherited the earlier achievements of the Sumerians. Some of the later invasions of the region also introduced new skills and ideas that still influence civilization today, such as a written law code.

The Babylonians Conquer Mesopotamia

Although Ur rose to glory after the death of Sargon, repeated foreign attacks drained its strength. By 2000 BC, Ur lay in ruins. With Ur's power gone, several waves of invaders battled to gain control of Mesopotamia.

Rise of Babylon

Babylon was home to one such group. That city was located on the Euphrates near what is now Baghdad, Iraq. Babylon had once been a Sumerian town. By 1800 BC, however, it was home to a powerful government of its own. In 1792 BC, Hammurabi (ham-uh-RAHB-ee) became Babylon's king. He would become the city's greatest ruler.

Hammurabi's Code

Hammurabi was a brilliant war leader. His armies fought many battles to expand his power. Eventually, Hammurabi brought all of Mesopotamia into his empire, called the Babylonian Empire after his capital city.

Hammurabi was not only skilled on the battlefield, though. He was also an able ruler who could govern a huge empire. He used tax money to pay for building and irrigation projects. He also brought wealth through increased trade. Hammurabi is best known, however, for his code of laws.

Hammurabi's Code was a set of 282 laws that dealt with almost every part of daily life. There were laws on everything from trade, loans, and theft to marriage, injury, and murder. It contained some ideas that are still found in laws today.

Under Hammurabi's Code, each crime brought a specific penalty. However, social class did matter. For example, injuring a rich man brought a greater penalty than injuring a poor man.

Hammurabi's Code was important not only for how thorough it was but also because it was written down for all to see. People all over the empire could read exactly what was against the law.

Hammurabi ruled for 42 years. During his reign, Babylon became the major city in Mesopotamia. However, after his death, Babylonian power declined. The kings that followed faced invasions from the people Hammurabi had conquered. Before long, the Babylonian Empire came to an end.

READING CHECK **Analyzing** What was Hammurabi's most important accomplishment?

Primary Source

HISTORIC DOCUMENT
Hammurabi's Code

The Babylonian ruler Hammurabi is credited with putting together the earliest known written collection of laws. The code set down rules for both criminal and civil law and informed citizens about what was expected of them.

196. If a man put out the eye of another man, his eye shall be put out.

197. If he break another man's bone, his bone shall be broken.

198. If he put out the eye of a freed man, or break the bone of a freed man, he shall pay one gold mina.

199. If he put out the eye of a man's slave, or break the bone of a man's slave, he shall pay one-half of its value.

221. If a physican heal the broken bone or diseased soft part of a man, the patient shall pay the physician five shekels in money.

222. If he were a freed man he shall pay three shekels.

223. If he were a slave his owner shall pay the physician two shekels.

–Hammurabi, from *The Code of Hammurabi*, translated by L. W. King

ANALYSIS SKILL **ANALYZING PRIMARY SOURCES**

How do you think Hammurabi's code of laws affected citizens of that time?

Invasions of Mesopotamia

Several other civilizations developed in and around the Fertile Crescent. As their armies battled for land, control of the region passed from one empire to another.

Hittites and Kassites

FOCUS ON READING
Make sure you understand this paragraph by restating it in your own words.

A people known as the Hittites built a strong kingdom in Asia Minor, in what is today Turkey. Their success came, in part, from two key military advantages they had over rivals. First, the Hittites were among the first people to master ironworking. This meant they could make stronger weapons than their foes. Second, the Hittite army skillfully used the **chariot**, a wheeled, horse-drawn cart used in battle. Chariots allowed Hittite soldiers to move quickly around a battlefield. Archers riding in the chariots fired arrows at the enemy.

Using these advantages, Hittite forces captured Babylon around 1595 BC. Hittite rule did not last long, however. Soon after taking Babylon, the Hittite king was killed by an assassin. The kingdom plunged into chaos. The Kassites, a people who lived north of Babylon, captured the city and ruled for almost 400 years.

Assyrians

Later, in the 1200s BC, a group called the Assyrians (uh-SIR-ee-unz) from northern Mesopotamia briefly gained control of Babylon. However, their empire was soon overrun by invaders. After this defeat, the Assyrians took about 300 years to recover their strength. Then, starting about 900 BC, they began to conquer all of the Fertile Crescent. They even took over parts of Asia Minor and Egypt.

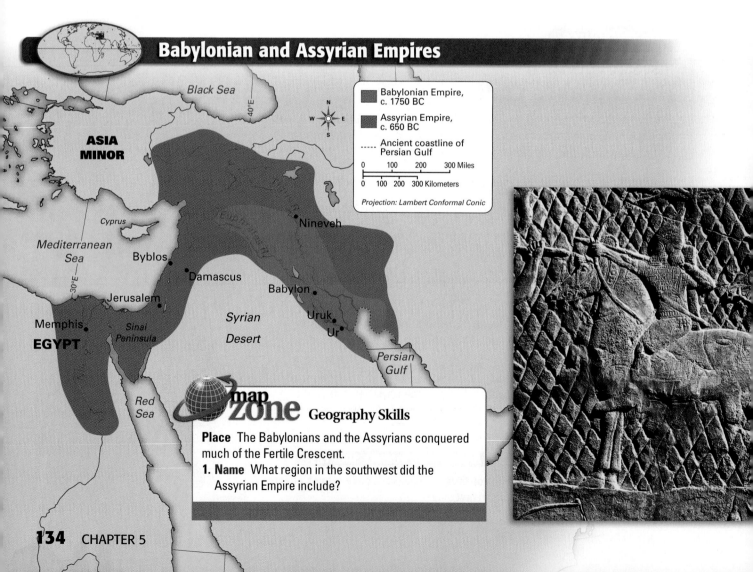

Babylonian and Assyrian Empires

■ Babylonian Empire, c. 1750 BC
■ Assyrian Empire, c. 650 BC
---- Ancient coastline of Persian Gulf

0   100   200   300 Miles
0   100   200   300 Kilometers

Projection: Lambert Conformal Conic

Black Sea

ASIA MINOR

Cyprus

Mediterranean Sea

Byblos

Damascus

Jerusalem

Memphis

EGYPT

Sinai Peninsula

Red Sea

Nineveh

Babylon

Syrian Desert

Uruk

Ur

Persian Gulf

map zone Geography Skills

Place The Babylonians and the Assyrians conquered much of the Fertile Crescent.

1. Name What region in the southwest did the Assyrian Empire include?

The key to the Assyrians' success was their strong army. Like the Hittites, the Assyrians used iron weapons and chariots. The army was very well organized, and every soldier knew his role.

The Assyrians were fierce in battle. Before attacking, they spread terror by looting villages and burning crops. Anyone who still dared to resist them was killed.

After conquering the Fertile Crescent, the Assyrians ruled from their capital city, Nineveh (NI-nuh-vuh). They demanded heavy taxes from across the empire. Areas that resisted the Assyrians' demands were harshly punished.

Assyrian kings ruled their large empire through local leaders. Each governed a small area, collected taxes, enforced laws, and raised troops for the army. Roads were built to link distant parts of the empire. Messengers on horseback were sent to deliver orders to faraway officials.

Chaldeans

In 652 BC a series of wars broke out in the Assyrian Empire over who should rule. These wars greatly weakened the empire.

Sensing this weakness, the Chaldeans (kal-DEE-unz), a group from the Syrian Desert, led other peoples in an attack on the Assyrians. In 612 BC, they destroyed Nineveh and the Assyrian Empire.

In its place, the Chaldeans set up a new empire of their own. Nebuchadnezzar (neb-uh-kuhd-NEZ-uhr), the most famous Chaldean king, rebuilt Babylon into a beautiful city. According to legend, his grand palace featured the famous Hanging Gardens. Trees and flowers grew on its terraces and roofs. From the ground the gardens seemed to hang in the air.

The Chaldeans greatly admired the ideas and culture of the Sumerians. They studied the Sumerian language and built temples to Sumerian gods.

At the same time, Babylon became a center for astronomy. Chaldeans charted the positions of the stars and kept track of economic, political, and weather events. They also created a calendar and solved complex problems of geometry.

READING CHECK **Sequencing** List in order the peoples who ruled Mesopotamia.

The Assyrian Army

The Assyrian army was the most powerful fighting force the world had ever seen. Large and well organized, it featured iron weapons, war chariots, and giant war machines used to knock down city walls.

ANALYZING VISUALS What kinds of weapons can you see in this carving?

SPAIN

ATLANTIC OCEAN

Strait of Gibraltar

ATLAS MOUNTAINS

The Phoenicians sailed throughout the Mediterranean, seeking trade goods and founding new cities.

The Phoenicians

At the western end of the Fertile Crescent, along the Mediterranean Sea, was a land known as Phoenicia (fi-NI-shuh). It was not home to a great military power and was often ruled by foreign governments. Nevertheless, the Phoenicians created a wealthy trading society.

Geography of Phoenicia

Today the nation of Lebanon occupies most of what was Phoenicia. Mountains border the region to the north and east. To the west lies the Mediterranean.

The Phoenicians were largely an urban people. Among their chief cities were Tyre, Sidon, and Byblos. These three cities, like many Phoenician cities, still exist today.

Phoenicia had few resources. One thing it did have, however, was cedar. Cedar trees were prized for their timber, a valuable trade item. But Phoenicia's overland trade routes were blocked by mountains and hostile neighbors. Phoenicians had to look to the sea for a way to trade.

THE IMPACT TODAY

Because so many cedar trees have been cut down in Lebanon's forests over the years, very few trees remain.

Expansion of Trade

Motivated by a desire for trade, the people of Phoenicia became expert sailors. They built one of the world's finest harbors at the city of Tyre. Fleets of fast Phoenician trading ships sailed to ports all around the Mediterranean Sea. Traders traveled to Egypt, Greece, Italy, Sicily, and Spain. They even passed through the Strait of Gibraltar to reach the Atlantic Ocean.

The Phoenicians founded several new colonies along their trade routes. Carthage (KAHR-thij), located on the northern coast of Africa, was the most famous of these. It later became one of the most powerful cities on the Mediterranean.

Phoenicia grew wealthy from its trade. Besides lumber, the Phoenicians traded silverwork, ivory carvings, and slaves. They also made and sold beautiful glass items. In addition, the Phoenicians made purple dye from a type of shellfish. They then traded cloth that had been dyed with this purple color. Phoenician purple fabric was very popular with rich people all around the Mediterranean.

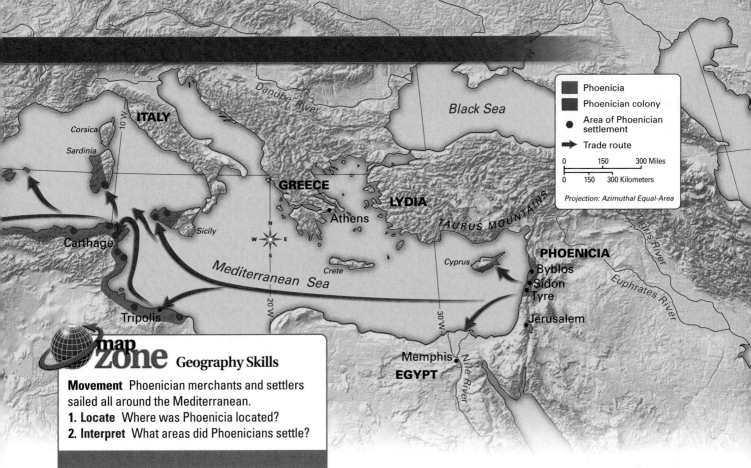

Phoenicia
Phoenician colony
• Area of Phoenician settlement
→ Trade route

0   150   300 Miles
0   150   300 Kilometers

Projection: Azimuthal Equal-Area

map zone **Geography Skills**

Movement Phoenician merchants and settlers sailed all around the Mediterranean.
1. **Locate** Where was Phoenicia located?
2. **Interpret** What areas did Phoenicians settle?

The Phoenicians' most important achievement, however, wasn't a trade good. To record their activities, Phoenician traders developed one of the world's first alphabets. An **alphabet** is a set of letters that can be combined to form words. This development made writing much easier. It had a major impact on the ancient world and on our own. In fact, the alphabet we use today is based on the Phoenicians'.

READING CHECK **Finding Main Ideas** What were the Phoenicians' main achievements?

SUMMARY AND PREVIEW Many peoples ruled in the Fertile Crescent after the Sumerians. Some made contributions that are still valued today. Next, you will learn about two religions that developed in the Fertile Crescent and are still alive today—Judaism and Christianity.

go.hrw.com
Online Quiz
KEYWORD: SK7 HP5

Section 4 Assessment

Reviewing Ideas, Terms, and Places
1. **a. Identify** Where was **Babylon** located?
   b. Analyze What does **Hammurabi's Code** reveal about Babylonian society?
2. **a. Describe** What two advantages did Hittite soldiers have over their opponents?
   b. Rank Which empire discussed in this section do you feel contributed the most to modern-day society? Why?
3. **a. Identify** For what trade goods were the Phoenicians known? For what else were they known?
   b. Analyze How did Phoenicia grow wealthy?

Critical Thinking
4. **Categorizing** Use your note-taking diagram with the names of the empires. List at least one advance or achievement made by each empire.

Fertile Crescent Empires

FOCUS ON WRITING

5. **Gathering Information about Later Peoples** Several different peoples contributed to civilization in the Fertile Crescent after the Sumerians. Which ones, if any, will you include on your poster? What will you show?

Sequencing and Using Time Lines

Learn

When you are reading about events in the past, it is important to learn their sequence, or the order in which the events occurred. If you do not know the sequence in which events happen, history will not make any sense.

One way to examine the sequence of events is to construct a time line. A time line is a visual display showing events in the order in which they happened. Events on the left side of the time line occurred first. Events farther to the right occurred later.

Practice

Use the time line below to answer the following questions.

❶ Around what year did Hammurabi issue his code of laws?

❷ Which happened earlier, the formation of Sargon's empire or the beginning of Phoenician trade?

❸ About how many years after Hammurabi issued his law code did the Assyrians conquer Babylon?

Major Events in the Fertile Crescent

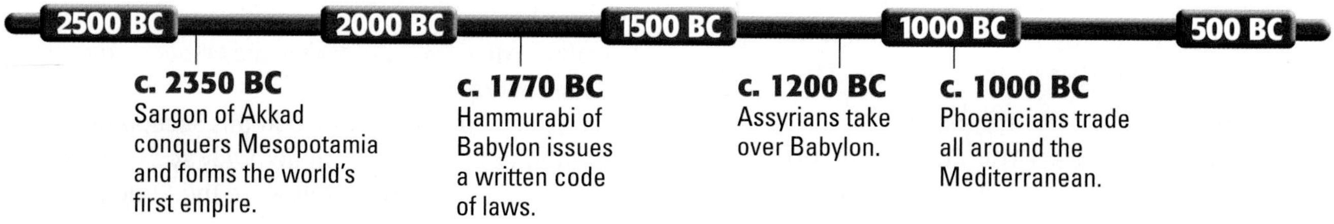

2500 BC   2000 BC   1500 BC   1000 BC   500 BC

c. 2350 BC
Sargon of Akkad conquers Mesopotamia and forms the world's first empire.

c. 1770 BC
Hammurabi of Babylon issues a written code of laws.

c. 1200 BC
Assyrians take over Babylon.

c. 1000 BC
Phoenicians trade all around the Mediterranean.

Apply

Think about a typical school day. What time do you wake up? What classes do you have? When do you get home? Make a list of events that occur on a typical day. Once you have made your list, rearrange it so that the events are listed in sequence. Then use your list to draw a time line of your day.

Chapter Review

Geography's Impact
video series
Review the video to answer the closing question:
What would life in America be like today without a written code of laws?

Visual Summary

Use the visual summary below to help you review the main ideas of the chapter.

QUICK FACTS

The early Mesopotamians developed irrigation to grow food. As a result, they were able to form cities.

Sumerian advances included ziggurats, the wheel, and the world's first writing system, cuneiform.

Later peoples created the first written laws and the first empires. They also formed great trading networks.

Reviewing Vocabulary, Terms, and Places

Using your own paper, complete the sentences below by providing the correct term for each blank.

1. Mesopotamian farmers built _____ to irrigate their fields.

2. The art and science of building is known as _____.

3. The people of Sumer practiced _____, the worship of many gods.

4. Instead of using pictographs, Sumerians developed a type of writing called _____.

5. Horse-drawn _____ gave the Hittites an advantage during battle.

6. _____ was Hammurabi's capital and one of Mesopotamia's greatest cities.

7. _____ ideas are not simple.

8. Sumerian society was organized in _____, which consisted of a city and the surrounding lands.

Comprehension and Critical Thinking

SECTION 1 *(Pages 116–119)*

9. **a. Describe** Where was Mesopotamia, and what does the name mean?

b. Analyze How did Mesopotamian irrigation systems allow civilization to develop?

c. Elaborate Do you think a division of labor is necessary for civilization to develop? Why or why not?

SECTION 2 *(Pages 122–126)*

10. **a. Identify** Who built the world's first empire, and what land did that empire include?

b. Analyze Politically, how was early Sumerian society organized? How did that organization affect society?

c. Elaborate Why did the Sumerians consider it everyone's responsibility to keep the gods happy?

SECTION 3 *(Pages 127–131)*

11. a. Identify What was the Sumerian writing system called, and why is it so significant?

b. Compare and Contrast What were two ways in which Sumerian society was similar to our society today? What were two ways in which it was different?

c. Evaluate Other than writing and the wheel, which Sumerian invention do you think is most important? Why?

SECTION 4 *(Pages 132–137)*

12. a. Describe What were two developments of the Phoenicians?

b. Draw Conclusions Why do you think several peoples banded together to fight the Assyrians?

c. Evaluate Do you think Hammurabi was more effective as a ruler or as a military leader? Why?

FOCUS ON READING AND WRITING

Paraphrasing *Read the paragraph below carefully. Then rewrite the paragraph in your own words, taking care to include all the main ideas.*

13. Mesopotamia was the home of many ancient civilizations. The first of these civilizations was the Sumerians. They lived in Mesopotamia by 3000 BC. There they built cities, created a system of writing, and invented the wheel.

Creating a Poster *Use your notes and the instructions below to help you create a poster.*

14. Using a large poster board, create a poster on the Fertile Crescent. From your list, select 5 or 6 pictures to show. Remember that your audience is young children and think about what would interest them.

Begin by collecting pictures or drawings from magazines or the Internet. Then make a plan for your poster. Decide where you will place each picture and what you will say about each. After you have arranged the pictures, create a title for the poster and center it at the top. Write a one- or two-sentence introduction for your poster. You will also have to create a label or short caption for each picture.

Social Studies Skills

15. Sequencing and Using Time Lines Create a time line that shows the various people who ruled the Fertile Crescent. Remember that the people should appear on your time line in order.

Using the Internet

go.hrw.com
KEYWORD: SK7 CH5

16. Activity: Looking at Writing The Sumerians made one of the greatest cultural advances in history by developing the world's first system of writing. Enter the activity keyword and research the evolution of language and its written forms. Look at one of the newest methods of writing: text messaging. Then write a paragraph explaining why writing is important using abbreviations and symbols used in text messaging.

Map Activity ★Interactive

17. The Fertile Crescent On a separate sheet of paper, match the letters on the map with their correct labels.

Babylon   Euphrates River

Phoenicia   Tigris River

Sumer

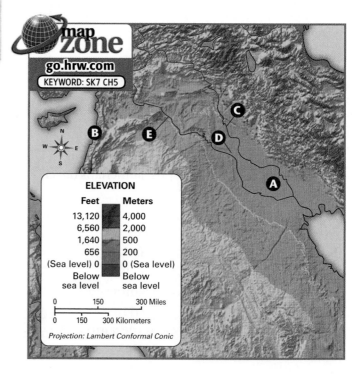

map zone
go.hrw.com
KEYWORD: SK7 CH5

ELEVATION

Feet	Meters
13,120	4,000
6,560	2,000
1,640	500
656	200
(Sea level) 0	0 (Sea level)
Below sea level	Below sea level

0   150   300 Miles
0   150   300 Kilometers
Projection: Lambert Conformal Conic

DIRECTIONS: *Read questions 1 through 7 and write the letter of the best response. Then read question 8 and write your own well-constructed response.*

1 The first people to develop a civilization in Mesopotamia were the

  A Akkadians.

  B Babylonians.

  C Egyptians.

  D Sumerians.

2 Which of the following statements about the first writing system is false?

  A It was developed by the Babylonians.

  B It began with the use of pictures to represent syllables and objects.

  C It was recorded on tablets made of clay.

  D It was first used to keep business records.

3 In Sumerian society, people's social class or rank depended on their wealth and their

  A appearance.

  B religion.

  C location.

  D occupation.

4 Which of the following was the subject of a great Sumerian epic?

  A Cuneiform

  B Ziggurat

  C Gilgamesh

  D Babylon

5 What was the most important contribution of the Phoenicians to our civilization?

  A purple dye

  B their alphabet

  C founding of Carthage

  D sailing ships

Mesopotamia

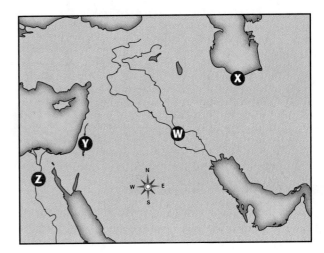

6 The region known as Mesopotamia is indicated on the map above by the letter

  A W.

  B X.

  C Y.

  D Z.

7 Hammurabi's Code is important in world history because it was an early

  A form of writing that could be used to record important events.

  B written list of laws that controlled people's daily life and behavior.

  C record-keeping system that enabled the Phoenicians to become great traders.

  D set of symbols that allowed the Sumerians to communicate with other peoples.

8 **Extended Response** The early civilizations of Mesopotamia developed in the valleys of two rivers. Look back at the map in the Geography and History feature. All around the world, river valleys were the home of early civilizations. Why was this? Write a short paragraph in which you give at least two reasons why early civilizations were often formed in river valleys.

CHAPTER 6

Judaism and Christianity
2000 BC–AD 1453

What You Will Learn...

In this chapter you will learn about the origins and spread of two major world religions—Judaism and Christianity. You will also learn how Christianity changed as it spread from the western Roman Empire into the Byzantine Empire.

SECTION 1
Origins of Judaism **144**

SECTION 2
Origins of Christianity **152**

SECTION 3
The Byzantine Empire................**160**

FOCUS ON READING AND WRITING

Understanding Implied Main Ideas Sometimes the main idea of a paragraph or passage is not directly stated; instead it is implied, or suggested. In those cases, you have to examine the details the author provides. Then you decide what point, or main idea, the author is expressing with those details. **See the lesson, Understanding Implied Main Ideas, on page 673.**

Writing a Letter As you read this chapter, think about what it would have been like to witness the events that occurred as Judaism and Christianity developed and spread. Then you will write a letter from the point of view of someone who actually witnessed an event.

Early Christian church

Judaism Jews pray at the Western Wall in Jerusalem. The wall is part of the Second Temple, which was built by ancient Hebrews.

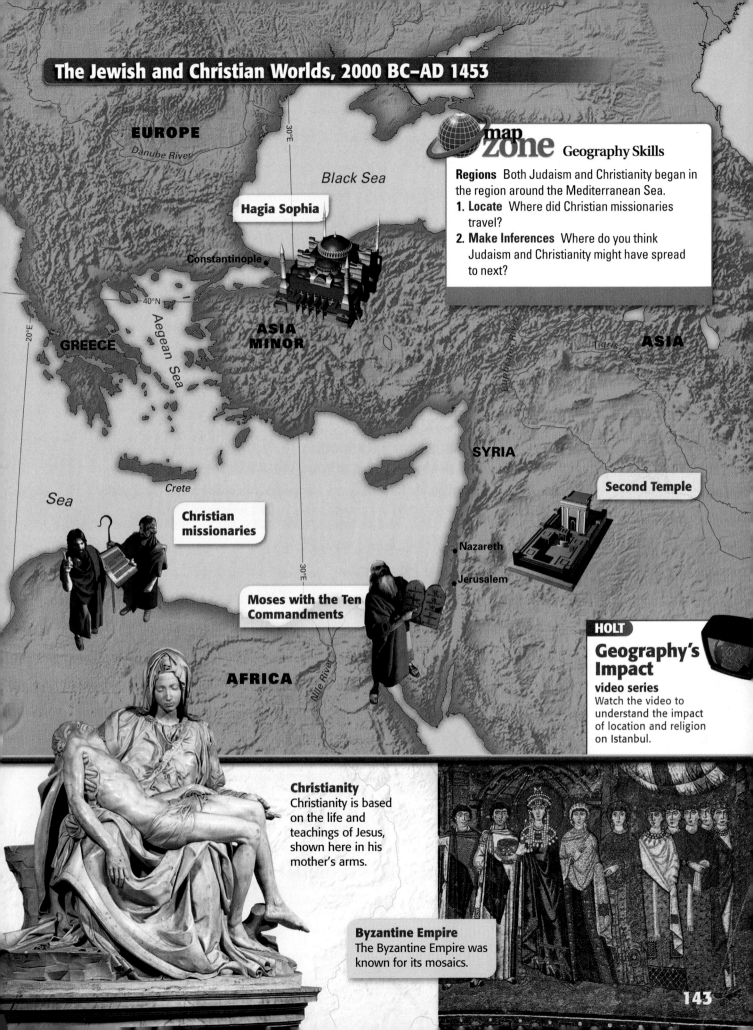

The Jewish and Christian Worlds, 2000 BC–AD 1453

EUROPE

Danube River

Black Sea

Hagia Sophia

Constantinople

30°E

40°N

GREECE

Aegean Sea

ASIA MINOR

ASIA

Tigris River

Euphrates River

SYRIA

Crete

Second Temple

Sea

Christian missionaries

Nazareth

Jerusalem

30°E

Moses with the Ten Commandments

20°E

AFRICA

Nile River

map zone — Geography Skills

Regions Both Judaism and Christianity began in the region around the Mediterranean Sea.

1. **Locate** Where did Christian missionaries travel?
2. **Make Inferences** Where do you think Judaism and Christianity might have spread to next?

HOLT

Geography's Impact

video series
Watch the video to understand the impact of location and religion on Istanbul.

Christianity
Christianity is based on the life and teachings of Jesus, shown here in his mother's arms.

Byzantine Empire
The Byzantine Empire was known for its mosaics.

143

Origins of Judaism

If YOU lived there...

You and your family are herders, looking after large flocks of sheep. Your grandfather, the leader of your tribe, is very rich, so your life is easy. One day, your grandfather says that your whole family will be moving to a new country. The trip will be very long, and people there may not welcome you.

How do you feel about going to a new land?

BUILDING BACKGROUND Like the family described above, the early Hebrews moved to new lands several times. From the beginning, the Hebrews were wanderers. According to Hebrew tradition, their history began with a search for a new home.

Early History

Sometime between 2000 and 1500 BC a new people appeared in Southwest Asia. They were the Hebrews (HEE-brooz). Most of what is known about early Hebrew history comes from the work of archaeologists and from accounts written by Hebrew scribes. These accounts describe the Hebrews' early history and the laws of **Judaism** (JOO-dee-i-zuhm), the Hebrews' religion. In time these accounts became the Hebrew Bible.

What You Will Learn...

Main Ideas

1. The Hebrews' early history began in Canaan and ended when the Romans forced them out of Israel.
2. Jewish beliefs in God, justice, and law anchor their society.
3. Jewish sacred texts describe the laws and principles of Judaism.
4. Traditions and holy days celebrate the history and religion of the Jewish people.

The Big Idea

The Hebrews formed a great kingdom in Israel and started a religion called Judaism.

Key Terms and Places

Judaism, *p. 144*
Canaan, *p. 145*
Exodus, *p. 145*
monotheism, *p. 147*
Torah, *p. 148*
rabbis, *p. 149*

TAKING NOTES As you read, use a graphic organizer like this one to organize your notes on the origins of Judaism.

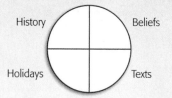

History · Beliefs
Holidays · Texts

Time Line

Early Hebrew History

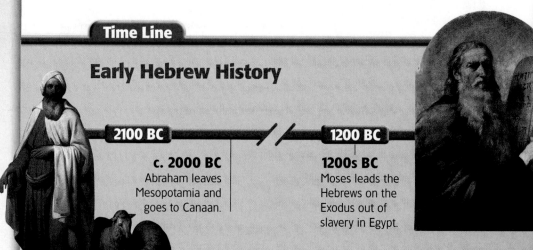

2100 BC

c. 2000 BC
Abraham leaves Mesopotamia and goes to Canaan.

1200 BC

1200s BC
Moses leads the Hebrews on the Exodus out of slavery in Egypt.

Beginnings in Canaan and Egypt

The Bible traces the Hebrews back to a man named Abraham. One day, the Bible says, God told Abraham to leave his home in Mesopotamia. He was to take his family on a long journey to the west. God promised to lead Abraham to a new land and make his descendants into a mighty nation.

Abraham left Mesopotamia and settled in **Canaan** (kay-nuhn) on the Mediterranean Sea. His descendants—the Hebrews—lived in Canaan for many years. Later, however, some Hebrews moved to Egypt, perhaps because of famine in Canaan.

The Hebrews lived well in Egypt, and their population grew. This growth worried Egypt's ruler, the pharaoh. He feared that the Hebrews might soon take over Egypt. To stop this from happening, the pharaoh made the Hebrews slaves.

The Exodus

According to the Bible, a leader named Moses appeared among the Hebrews in Egypt. In the 1200s BC, God told Moses to lead the Hebrews out of Egypt. Moses went to the pharaoh and demanded that he free the Hebrews. The pharaoh refused. Soon afterward a series of terrible plagues, or disasters, struck Egypt.

The plagues frightened the pharaoh so much that he agreed to free the Hebrews. Overjoyed with the news of their release, Moses led his people out of Egypt in a journey called the **Exodus**. To the Hebrews, the release from slavery proved that God was protecting and watching over them.

For years after their release, the Hebrews wandered through the desert, trying to return to Canaan. On their journey, they reached a mountain called Sinai. The Bible says that while Moses was on the mountain, God gave him two stone tablets. On the tablets was written a code of moral laws known as the Ten Commandments. These laws have since shaped Hebrew society.

Once the Hebrews reached Canaan, they had to fight to gain control of the land. After they conquered Canaan and settled down on the land, the Hebrews became known as the Israelites.

A Series of Invasions

The Israelites soon faced more threats to their land. Invaders swept through the region in the mid-1000s BC. For a while, strong kings kept Israel together. Israel even grew rich through trade and expanded its territory. With their riches, the Israelites built a great temple to God in Jerusalem.

Some years later when one king died, the Israelites could not agree on who would be the next king. This conflict caused Israel to split into two kingdoms, one called Israel and one called Judah (joo-duh). The people of Judah became known as Jews.

FOCUS ON READING
Study the details in the first paragraph under A Series of Invasions. What is the main idea?

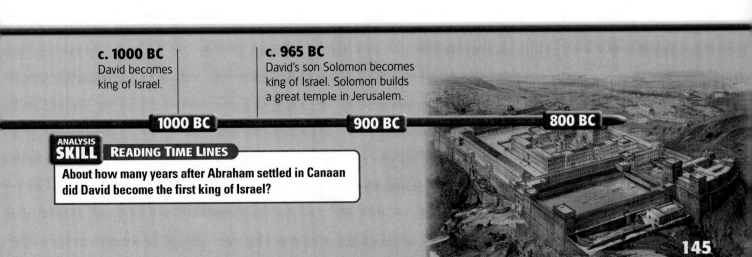

c. 1000 BC
David becomes king of Israel.

c. 965 BC
David's son Solomon becomes king of Israel. Solomon builds a great temple in Jerusalem.

1000 BC   900 BC   800 BC

ANALYSIS SKILL **READING TIME LINES**

About how many years after Abraham settled in Canaan did David become the first king of Israel?

145

The two new kingdoms lasted for a few centuries. Israel eventually fell to invaders about 722 BC. Judah lasted until 586 BC, when invaders captured Jerusalem and destroyed Solomon's temple. They sent the Jews out of Jerusalem as slaves. When these invaders were themselves conquered, some Jews returned home. Others moved to other places in Southwest Asia. Scholars call the scattering of Jews outside of Israel and Judah the Diaspora (dy-AS-pruh).

The Jews who returned to Jerusalem ruled themselves for about 100 years. They even rebuilt Solomon's temple. Eventually, however, they were conquered by the Romans. The Jews revolted against the Romans, but most gave up after the Romans destroyed their temple. As punishment for the rebellion, the Romans killed or enslaved much of Jerusalem's population. Thousands of Jews fled Jerusalem. Over the next centuries, Jews moved all around the world. Often they were forced to move by other religious groups who discriminated against them.

READING CHECK **Identifying Cause and Effect** How did invasions affect the Hebrews?

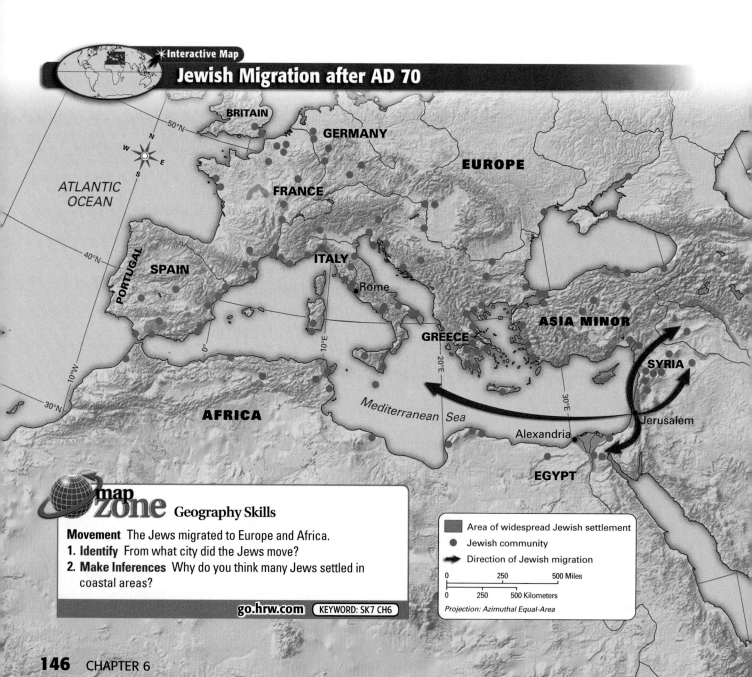

Jewish Migration after AD 70

★Interactive Map

map zone Geography Skills

Movement The Jews migrated to Europe and Africa.
1. **Identify** From what city did the Jews move?
2. **Make Inferences** Why do you think many Jews settled in coastal areas?

go.hrw.com KEYWORD: SK7 CH6

Area of widespread Jewish settlement
● Jewish community
➔ Direction of Jewish migration

0   250   500 Miles
0   250   500 Kilometers
Projection: Azimuthal Equal-Area

Jewish Beliefs

Wherever Jews live around the world, their religion is the foundation upon which they base their whole society. In fact, much of Jewish culture is based directly on Jewish beliefs. The central beliefs of Judaism are belief in one God, in justice and righteousness, and in law.

Belief in One God

Most importantly, Jews believe in one God. The belief in one and only one God is called **monotheism**. Many people believe that Judaism was the world's first monotheistic religion.

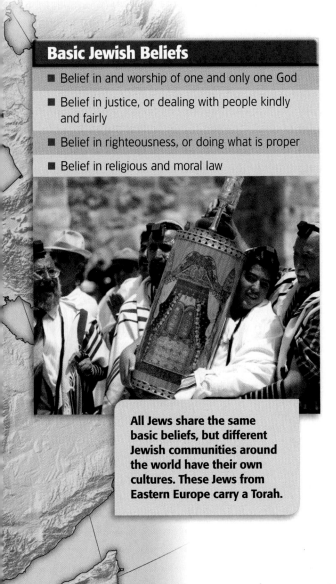

Basic Jewish Beliefs

- Belief in and worship of one and only one God
- Belief in justice, or dealing with people kindly and fairly
- Belief in righteousness, or doing what is proper
- Belief in religious and moral law

All Jews share the same basic beliefs, but different Jewish communities around the world have their own cultures. These Jews from Eastern Europe carry a Torah.

In the ancient world where most people worshipped many gods, the Jews' worship of only one God set them apart. This worship shaped Jewish society. The Jews believed they were God's chosen people. They believed that God had guided their history through relationships with Abraham, Moses, and other leaders.

Belief in Justice and Righteousness

Also central to the Jews' religion are the ideas of justice and righteousness. To Jews, justice means kindness and fairness in dealing with other people. Everyone deserves justice, even strangers and criminals. Jews are expected to give aid to those who need it, including the poor, the sick, and orphans. Jews are also expected to be fair in business dealings.

Righteousness refers to doing what is proper. Jews are supposed to behave properly, even if others around them do not. For the Jews, righteous behavior is more important than rituals, or ceremonies.

Belief in Law

Closely related to the ideas of justice and righteousness is obedience to the law. Jews believe that God gave them religious and moral laws to follow. The most important Jewish laws are the Ten Commandments. The commandments require that Jews worship only one God. They also do not allow Jews to do bad things like murder, steal, or lie.

The commandments are only one part of Jewish law. Jews believe that Moses recorded a system of laws, now called Mosaic law, that God had set down for them. Mosaic laws guide many areas of Jews' daily lives, such as how people pray and observe holy days.

READING CHECK **Generalizing** What are the most important beliefs of Judaism?

The Torah

Using a special pointer called a *yad*, this girl is reading aloud from the Torah. The Torah is the most sacred of Hebrew texts. It plays a central role in many Jewish ceremonies.

ANALYSIS SKILL **ANALYZING VISUALS**

How does the Torah look different from the Hebrew Bible and the commentaries?

ACADEMIC VOCABULARY

principle basic belief, rule, or law

Jewish Texts

The laws and **principles** of Judaism are described in several sacred texts. Among the main texts are the Torah, the Hebrew Bible, and the commentaries.

The Torah

The ancient Jews recorded most of their laws in five books. Together, these books are called the Torah. The Torah is the most sacred text of Judaism. In addition to laws, it includes a history of the Jewish people until the death of Moses. Jews believe the contents of the Torah were revealed to Moses by God.

Readings from the Torah are central to Jewish religious services today. Nearly every synagogue (si-nuh-gawg), or Jewish house of worship, has at least one Torah. Out of respect for the Torah, readers do not touch it. They use special pointers to mark their places in the text.

The Hebrew Bible

The Torah is the first of three parts of a group of writings called the Hebrew Bible, or Tanach (tah-NAHK). The second part is made up of eight books that describe the messages of Hebrew prophets. Prophets are people who are said to receive messages from God to be taught to others.

The final part of the Hebrew Bible is 11 books of poetry, songs, stories, lessons, and history. Many of these stories are told by Jews to show the power of faith.

Also in the final part of the Hebrew Bible are the Proverbs, short expressions of Hebrew wisdom. For example, one Proverb says, "A good name is to be chosen rather than great riches." In other words, it is better to be seen as a good person than to be rich and not respected.

The third part of the Hebrew Bible also includes the Book of Psalms. The Book of Psalms is a collection of short and long songs of praise to God.

The Hebrew Bible
These beautifully decorated pages are from a Hebrew Bible. The Hebrew Bible, sometimes called the Tanach, includes the Torah and other ancient writings.

The Commentaries
The Talmud is a collection of commentaries and discussions about the Torah and the Hebrew Bible. The Talmud is a rich source of information for discussion and debate. Religious scholars like these young men study the Talmud to learn about Jewish history and laws.

The Commentaries

For centuries rabbis, or religious teachers, and scholars have studied the Torah and Jewish laws. Because some laws are hard to understand, scholars write commentaries to explain them. Many explanations can be found in the Talmud (TAHL-moohd), a set of laws, commentaries, stories, and folklore. The writings of the Talmud were produced between AD 200 and 600. Many Jews consider them second only to the Hebrew Bible in significance to Judaism.

READING CHECK Analyzing What texts do Jews consider sacred?

Traditions and Holy Days

Jews feel that understanding their history will help them better follow the Jewish teachings. Their traditions and holy days help Jews connect with their past and celebrate their history.

Hanukkah

One Jewish tradition is celebrated by Hanukkah, which falls in December. It honors a historical event. The ancient Jews wanted to celebrate a victory that had convinced their rulers to let them keep their religion. According to legend, though, the Jews did not have enough lamp oil to celebrate at the temple. Miraculously, the oil they had—enough for only one day—burned for eight full days.

Today Jews celebrate this event by lighting candles in a special candleholder called a menorah (muh-nohr-uh). Its eight branches represent the eight days through which the oil burned. Many Jews also exchange gifts on each of the eight nights.

Passover

More important to Jews than Hanukkah, Passover is celebrated in March or April. During Passover Jews honor the Exodus, the journey of the Hebrews out of slavery.

A Passover Meal

During a special Passover meal called a seder, participants reflect on the events of the Exodus.

ANALYZING VISUALS How can you tell this is a special meal?

High Holy Days

The two most sacred of all Jewish holidays are the High Holy Days. They take place in September or October. The first two days of celebration, Rosh Hashanah (rahsh uh-SHAH-nuh), celebrate the start of a new year in the Jewish calendar.

On Yom Kippur (yohm ki-poohr), which falls soon afterward, Jews ask God to forgive their sins. Jews consider Yom Kippur to be the holiest day of the entire year. Because it is so holy, Jews do not eat or drink anything all day. They also perform ancient ceremonies that help many Jews feel more connected with their past.

READING CHECK Finding Main Ideas What are the two most important Jewish holidays?

According to Jewish tradition, the Hebrews left Egypt so quickly that bakers did not have time to let their bread rise. Therefore, during Passover Jews eat only matzo, a flat, unrisen bread. They also celebrate the holiday with ceremonies.

SUMMARY AND PREVIEW Judaism was the world's first monotheistic religion. Jewish culture and traditions are rooted in the history of the Hebrew people. Next, you will read about a religion that is related to Judaism—Christianity.

Section 1 Assessment

Reviewing Ideas, Terms, and Places

1. **a. Identify** Who first led the Jews to **Canaan**?
   b. Evaluate Why was the **Exodus** a significant event in Hebrew history?
2. **a. Define** What is **monotheism**?
   b. Explain What is the Jewish view of justice and righteousness?
3. **a. Identify** What are the main sacred texts of Judaism?
   b. Elaborate Why do you think the Commentaries are so significant to many Jews?
4. **a. Identify** What event in Hebrew history does Passover celebrate?
   b. Elaborate How do you think celebrating traditions and holy days helps Jews connect to their past?

Critical Thinking

5. **Sequencing** Review your notes on Hebrew history. Then draw a diagram like this one and fill in important events from Hebrew history in the order they occurred. You may add as many boxes as you need for the information.

Abraham settles in Canaan.

FOCUS ON WRITING

6. **Noting Main Events in the Origins of Judaism** Look back over this section and imagine what it would have been like to witness some of these events. Identify one or two events that you might describe in your letter.

Interpreting a Route Map

Learn

A route map shows movement from one place to another. Usually, different routes are shown with different colored arrows. Look at the legend to see what the different arrows represent.

Practice

Use the map of Possible Routes of the Exodus to answer the following questions.

1. How many possible Exodus routes does the map show?

2. Where did the Exodus begin?

3. Which possible route would have been the longest?

4. Which route would have passed closest to the Mediterranean Sea?

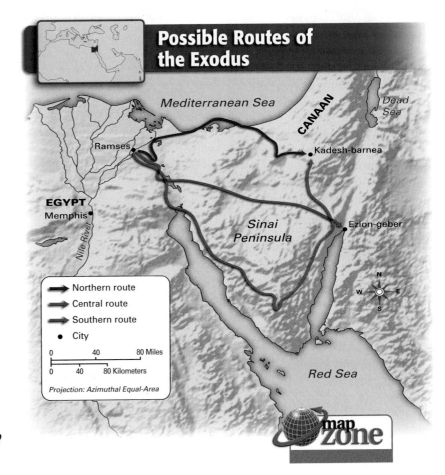

Possible Routes of the Exodus

Mediterranean Sea

CANAAN

Dead Sea

Ramses

Kadesh-barnea

EGYPT

Memphis

Sinai Peninsula

Ezion-geber

Nile River

Red Sea

→ Northern route
→ Central route
→ Southern route
• City

0   40   80 Miles
0   40   80 Kilometers

Projection: Azimuthal Equal-Area

map zone

Apply

Find a map of your city either in an atlas or on the Internet. You will need to draw on the map, so either print it, copy it, or draw a map on your own paper using the information. On the city map, draw the route you take from your home to school. Then draw another route you could take to get to school. Be sure to create a legend to show what your route lines mean.

Origins of Christianity

SECTION 2

What You Will Learn...

Main Ideas

1. The life and death of Jesus of Nazareth inspired a new religion called Christianity.
2. Christians believe that Jesus's acts and teachings focused on love and salvation.
3. Jesus's followers taught others about Jesus's life and teachings.
4. Christianity spread throughout the Roman Empire by 400.

The Big Idea

Christianity, a religion based on the life and teachings of Jesus of Nazareth, spread throughout the Roman Empire.

Key Terms and Places

Messiah, *p. 152*
Christianity, *p. 152*
Bible, *p. 152*
Bethlehem, *p. 152*
Resurrection, *p. 153*
disciples, *p. 153*
saint, *p. 156*

TAKING NOTES As you read, use a graphic organizer like this one to take notes on Jesus and the spread of Christianity.

Life and Death	
Acts and Teachings	
Jesus's Followers	
Spread of Christianity	

If YOU lived there...

You are a fisher in Judea, bringing in the day's catch. As you reach the shore, you see a large crowd. They are listening to a man tell stories. A man in the crowd whispers to you that the speaker is a teacher with some new ideas about religion. You are eager to get your fish to the market, but you are also curious.

What might convince you to stay and listen?

BUILDING BACKGROUND In the first century AD, Roman soldiers occupied Judea, but the Jews living there held firmly to their own beliefs and customs. During that time, one religious teacher began to attract large followings among the people of Judea. That teacher was Jesus of Nazareth.

Jesus of Nazareth

Jesus of Nazareth was the man many people believed was the **Messiah**—a great leader the ancient Jews predicted would come to restore the greatness of Israel. Jesus was a great leader and one of the most influential figures in world history. Jesus's life and teachings form the basis of a religion called **Christianity**. However, we know relatively little about his life. Everything we do know is contained in the **Bible**, the holy book of Christianity.

The Christian Bible is made up of two parts. The first part, the Old Testament, is largely the same as the Hebrew Bible. The second part, the New Testament, is an account of the life and teachings of Jesus and of the early history of Christianity.

The Birth of Jesus

According to the Bible, Jesus was born in a small town called **Bethlehem** (BETH-li-hem) at the end of the first century BC. Jesus's mother, Mary, was married to a carpenter named Joseph. But Christians believe God, not Joseph, was Jesus's father.

As a young man Jesus lived in the town of Nazareth and probably studied with Joseph to become a carpenter. Like many young Jewish men of the time, Jesus also studied the laws and teachings of Judaism. By the time he was about 30, Jesus had begun to travel and teach. Stories of his teachings and actions from this time make up the beginning of the New Testament.

The Crucifixion

As a teacher, Jesus drew many followers with his ideas. But at the same time, his teachings challenged the authority of political and religious leaders. According to the Bible, they arrested Jesus while he was in Jerusalem in or around AD 30.

Shortly after his arrest, Jesus was tried and executed. He was killed by crucifixion (kroo-suh-FIK-shuhn), a type of execution in which a person was nailed to a cross. In fact, the word *crucifixion* comes from the Latin word for "cross." After he died, Jesus's followers buried him.

The Resurrection

According to Christian beliefs, Jesus rose from the dead and vanished from his tomb three days after he was crucified. Now Christians refer to Jesus's rise from the dead as the **Resurrection** (re-suh-REK-shuhn).

Christians further believe that after the Resurrection, Jesus appeared to some groups of his **disciples** (di-SY-puhls), or followers. Jesus stayed with these disciples for the next 40 days, teaching them and giving them instructions about how to pass on his teachings. Then Jesus rose up into heaven.

Early Christians believed that the Resurrection was a sign that Jesus was the Messiah and the son of God. Some people began to call him Jesus Christ, from the Greek word for Messiah, *Christos*. It is from this word that the words *Christian* and *Christianity* eventually developed.

READING CHECK **Summarizing** What do Christians believe happened after Jesus died?

Jesus of Nazareth

The Bible says that Jesus was born in Bethlehem but grew up in Nazareth. The famous artist Giotto (1266–1336) painted this scene from Jesus's childhood.

ANALYZING VISUALS How does the artist imply that Jesus was important?

Mediterranean Sea

Nazareth

Sea of Galilee

Jordan River

Jerusalem

Bethlehem

Dead Sea

JUDEA

Christian Holidays

For centuries, Christians have honored key events in Jesus's life. Some of these events inspired holidays that Christians celebrate today.

The most sacred holiday for Christians is Easter, which is celebrated each spring. Easter is a celebration of the Resurrection. On Easter Christians usually attend church services. Many people also celebrate by dyeing eggs because eggs are seen as a symbol of new life.

Another major Christian holiday is Christmas. It honors Jesus's birth and is celebrated every December 25. Although no one knows on what date Jesus was actually born, Christians have placed Christmas in December since the 200s. Today people celebrate with church services and the exchange of gifts. Some people reenact scenes of Jesus's birth.

Drawing Conclusions Why do you think people celebrate events in Jesus's life?

Jesus's Acts and Teachings

During his lifetime, Jesus traveled from village to village spreading his message among the Jewish people. As he traveled, he attracted many followers. These early followers later became the first Christians.

Miracles

According to the New Testament, many people became Jesus's followers after they saw him perform miracles. A miracle is an event that cannot normally be performed by a human. For example, the books of the New Testament tell of times when Jesus healed people who were sick or injured. One passage also describes how Jesus once fed an entire crowd with just a few loaves of bread and a few fish. Although there should not have been enough food for everyone, people ate their fill and even had food to spare.

Parables

The Bible says that miracles drew followers to Jesus and convinced them that he was the son of God. Once Jesus had attracted followers, he began to teach them. One way he taught was through parables, or stories that teach lessons about how people should live. Parables are similar to fables, but they usually teach religious lessons. The New Testament includes many of Jesus's parables.

Through his parables, Jesus linked his beliefs and teachings to people's everyday lives. The parables explained complicated ideas in ways that most people could understand. For example, in one parable, Jesus compared people who lived sinfully to a son who had left his home and his family. Just as the son's father would joyfully welcome him home, Jesus said, God would forgive sinners when they turned away from sin.

In another parable, Jesus compared society to a wheat field. In this story, a farmer plants wheat seed, but an enemy comes and plants weeds among the wheat. The farmer lets the weeds and wheat grow in the field together. At harvest time, he gathers the wheat in his barn, but he burns the weeds. Jesus explained this parable by comparing the wheat and weeds to good people and evil people who must live together. However, in the end, Jesus said, the good people would be rewarded and the evil people would be punished.

Jesus's Message

Much of Jesus's message was rooted in older Jewish traditions. For example, he emphasized two rules that were also in the Torah: love God and love other people.

Jesus expected his followers to love all people, not just friends and family. He encouraged his followers to be generous to the poor and the sick. He told people that they should even love their enemies. The way people treated others, Jesus said, showed how much they loved God.

Another important theme in Jesus's teachings was salvation, or the rescue of people from sin. Jesus taught that people who were saved from sin would enter the Kingdom of God when they died. Many of his teachings dealt with how people could reach God's kingdom.

Over the many centuries since Jesus lived, people have interpreted his teachings in different ways. As a result, many different denominations of Christians have developed. A denomination is a group of people who hold mostly the same beliefs. Despite their differences, however, Christians around the world share many basic beliefs about Jesus.

READING CHECK **Summarizing** What were the main ideas in Jesus's message?

The Sermon on the Mount

The Bible says that Jesus attracted many followers. One day he led his followers onto a mountainside to give a religious speech. In this speech, called the Sermon on the Mount, Jesus said that people who love God will be blessed. An excerpt of this sermon appears below.

When Jesus saw the crowds, he went up the mountain; and after he sat down, his disciples came to him. Then he began to speak, and taught them, saying:

"Blessed are the poor in spirit, for theirs is the kingdom of heaven.

"Blessed are those who mourn, for they will be comforted.

"Blessed are the meek, for they will inherit the earth.

"Blessed are those who hunger and thirst for righteousness, for they will be filled.

"Blessed are the merciful, for they will receive mercy.

"Blessed are the pure in heart, for they will see God.

"Blessed are the peacemakers, for they will be called children of God.

"Blessed are those who are persecuted for righteousness' sake, for theirs is the kingdom of heaven.

"Blessed are you when people revile you and persecute you and utter all kinds of evil against you falsely on my account. Rejoice and be glad, for your reward is great in heaven, for in the same way they persecuted the prophets who were before you."

—Matthew 5:1–12, New Revised Standard Version

Scholars believe this location was the site of Jesus's Sermon on the Mount.

ANALYSIS SKILL **ANALYZING INFORMATION**

What kinds of qualities does Jesus say will be rewarded?

The Last Supper

This famous painting by Italian artist Leonardo da Vinci shows Jesus and his Apostles sharing their last meal before Jesus was arrested.

ANALYZING VISUALS What kind of mood do people appear to be in?

Jesus's Followers

Shortly after the Resurrection, the Bible says, Jesus's followers traveled throughout the Roman world telling about Jesus and his teachings. Among the people to pass on Jesus's teachings were 12 chosen disciples called Apostles (uh-PAHS-uhlz) and a man called Paul.

The Apostles

The Apostles were 12 men whom Jesus chose to receive special teaching. During Jesus's lifetime they were among his closest followers and knew him very well. Jesus frequently sent the Apostles to spread his teachings. After the Resurrection, the Apostles continued this task.

One of the Apostles, Peter, became the leader of the group after Jesus died. Peter traveled to a few Roman cities and taught about Jesus in the Jewish communities there. Eventually, he went to live in Rome, where he had much authority among Jesus's followers. In later years after the Christian Church was more organized, many people looked back to Peter as its first leader.

ACADEMIC VOCABULARY

ideals ideas or goals that people try to live up to

The Gospels

Some of Jesus's disciples wrote accounts of his life and teachings. These accounts are called the Gospels. Four Gospels are found in the New Testament of the Bible.

The Gospels were written by men known as Matthew, Mark, Luke, and John. All the men's accounts differ slightly from one another, but together they make up the best source we have on Jesus's life. Historians and religious scholars depend on these stories for information about Jesus's life and teachings. The Gospels tell of miracles Jesus performed. They also contain the parables he told.

Paul

Probably the most important person in the spread of Christianity after Jesus's death was Paul of Tarsus. Although he had never met Jesus, Paul did more to spread Christian beliefs and **ideals** than anyone else. He had so much influence that many people think of him as another Apostle. After Paul died, he was named a **saint**, a person known and admired for his or her holiness.

Like most of Jesus's early followers, Paul was born Jewish. At first he did not like Jesus's ideas, which he considered a threat to Judaism. For a time, Paul even worked to prevent followers of Jesus from spreading their message.

According to the Bible, though, something happened to Paul one day as he traveled on the road to Damascus. He saw a blinding light and heard the voice of Jesus calling out to him. Soon after that event, Paul became a Christian.

After his conversion, Paul traveled widely, spreading Christian teachings. As you can see on the map, he visited many of the major cities along the eastern coast of the Mediterranean. In addition, he wrote long letters to communities throughout the Roman world. These letters helped explain and elaborate on Jesus's teachings.

In his letters Paul wrote at length about the Christian belief in the Resurrection and about salvation. He also mentioned the idea of the Trinity. The Trinity is a central Christian belief that God is made up of three persons. They are God the Father, Jesus the Son, and the Holy Spirit. This belief holds that, even though there are three persons, there is still only one God.

Paul's teachings attracted both Jews and non-Jews to Christianity in many areas around the Mediterranean. In time, this growing number of Christians helped the Christian Church break away from its Jewish roots. People began to recognize Christianity as a separate religion.

READING CHECK **Finding Main Ideas**
What did Jesus's followers do to help spread Christianity?

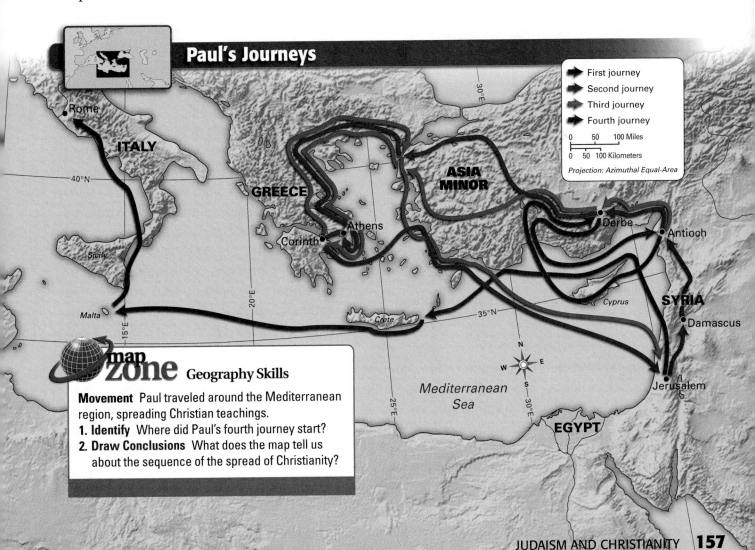

Paul's Journeys

First journey
Second journey
Third journey
Fourth journey

0   50   100 Miles
0   50   100 Kilometers
Projection: Azimuthal Equal-Area

Rome
ITALY
40°N
Sicily
Malta
GREECE
Athens
Corinth
Crete
ASIA MINOR
Derbe
Antioch
Cyprus
SYRIA
Damascus
Jerusalem
Mediterranean Sea
EGYPT
30°E
20°E
15°E
25°E
35°N
30°E

map zone **Geography Skills**

Movement Paul traveled around the Mediterranean region, spreading Christian teachings.
1. **Identify** Where did Paul's fourth journey start?
2. **Draw Conclusions** What does the map tell us about the sequence of the spread of Christianity?

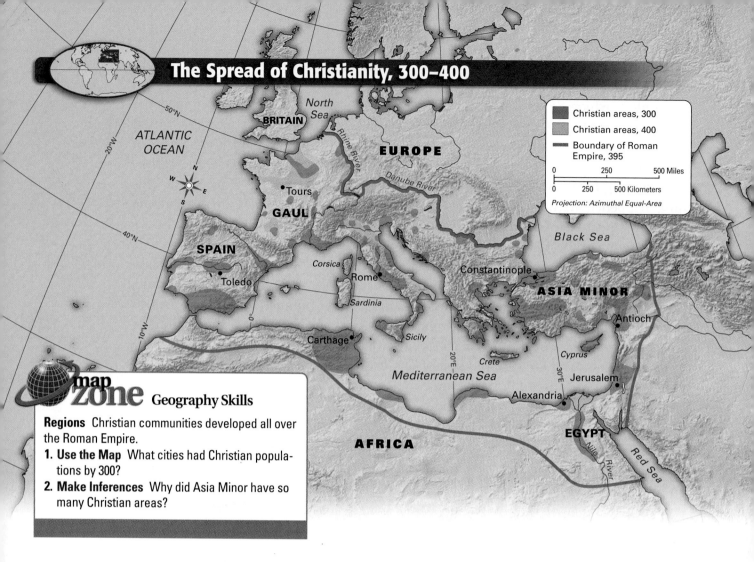

The Spread of Christianity, 300–400

BRITAIN
North Sea
EUROPE
ATLANTIC OCEAN
Rhine River
Danube River
Tours
GAUL
SPAIN
Corsica
Toledo
Rome
Sardinia
Carthage
Sicily
Mediterranean Sea
Crete
AFRICA
Black Sea
Constantinople
ASIA MINOR
Antioch
Cyprus
Jerusalem
Alexandria
EGYPT
Nile River
Red Sea

Christian areas, 300
Christian areas, 400
Boundary of Roman Empire, 395

0   250   500 Miles
0   250   500 Kilometers
Projection: Azimuthal Equal-Area

map zone Geography Skills

Regions Christian communities developed all over the Roman Empire.
1. **Use the Map** What cities had Christian populations by 300?
2. **Make Inferences** Why did Asia Minor have so many Christian areas?

The Spread of Christianity

Early Christians like Paul wanted to share their message about Jesus with the world. To do that, Christians began to write down parts of Jesus's message, including the Gospels. They distributed copies of the Gospels and other writings to strengthen people's faith. Because of their efforts, Christianity spread quickly in Roman communities.

Persecution

As Christianity became more popular, some Roman leaders became concerned. They looked for ways to put an end to this new religion. Sometimes local officals challenged the Christians trying to spread their beliefs. Some of these officials even arrested and killed Christians who refused

to worship the gods of Rome. Many of the leaders of the early Christians, including Peter and Paul, were killed for their efforts in spreading Christian teachings.

Most of Rome's emperors let Christians worship as they pleased. However, a few emperors in the 200s and 300s feared that the Christians could cause unrest in the empire. To prevent such unrest, these emperors banned Christianity. Christians were often forced to meet in secret.

Growth of the Church

Because the early church usually had to meet in secret, it did not have any single leader to govern it. Instead, bishops, or local Christian leaders, led each Christian community. Most of these early bishops lived in cities.

By the late 100s Christians were looking to the bishops of large cities for guidance. These bishops had great influence, even over other bishops. The most honored of all the empire's bishops was the bishop of Rome, or the pope. Gradually, the pope's influence grew and many people in the West came to see him as the head of the whole Christian Church. As the church grew, so did the influence of the pope.

Acceptance of Christianity

As the pope's influence grew, Christianity continued to spread throughout Rome even though it was banned. Then an event changed things for Christians in Rome. The emperor himself became a Christian.

The emperor who became a Christian was Constantine (KAHN-stuhn-teen). According to legend, Constantine was preparing for battle against a rival when he saw a cross in the sky. He thought that this vision meant he would win the battle if he converted to Christianity. Constantine did convert, and he won the battle. As a result of his victory, he became the new emperor of Rome.

As emperor, Constantine removed bans against the practice of Christianity. He also called together a council of Christian leaders from around the empire to try to clarify Christian teachings. Almost 60 years after Constantine died, another emperor banned all non-Christian religious practices in the empire. Christianity eventually spread from Rome all around the world.

FOCUS ON READING
What is the implied main idea of this paragraph?

READING CHECK Analyzing What difficulties did early Christians face in practicing and spreading their religion?

SUMMARY AND PREVIEW The life and teachings of Jesus of Nazareth inspired a new religion among the Jews. This religion was Christianity. Next, you will learn about how Christianity and other factors influenced culture in the eastern part of the Roman Empire.

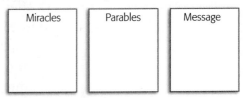

go.hrw.com
Online Quiz
KEYWORD: SK7 HP6

Section 2 Assessment

Reviewing Ideas, Terms, and Places

1. **a. Define** In Christian teachings, what was the **Resurrection**?
   b. Elaborate Why do you think Christians use the cross as a symbol of their religion?
2. **a. Identify** What did Jesus mean by salvation?
   b. Explain How have differing interpretations of Jesus's teachings affected Christianity?
3. **a. Define** What is a **saint**?
   b. Summarize How did Paul influence early Christianity?
4. **a. Recall** What was the role of bishops in the early Christian Church?
   b. Explain Why were some Roman leaders worried about the growing popularity of Christianity?
   c. Predict What do you think might have happened to Christianity if Constantine had not become a Christian?

Critical Thinking

5. **Making Generalizations** Review your notes on Jesus's acts and teachings. Then make generalizations about the topics shown in the graphic organizer.

Acts and Teachings of Jesus of Nazareth

Miracles	Parables	Message

FOCUS ON WRITING

6. **Identifying Events Related to Christianity** What events from this time can you imagine witnessing? Identify at least one or two events that you might describe in your letter.

The Byzantine Empire

What You Will Learn...

Main Ideas

1. Eastern emperors ruled from Constantinople and tried but failed to reunite the whole Roman Empire.
2. The people of the eastern empire created a new society that was very different from society in the west.
3. Byzantine Christianity was different from religion in the west.

The Big Idea

The Roman Empire split into two parts, and the eastern Roman Empire prospered for hundreds of years after the western empire fell.

Key Terms and Places

Constantinople, *p. 160*
Byzantine Empire, *p. 162*
mosaics, *p. 163*

TAKING NOTES As you read, use a chart like the one below to record your notes on the western Roman Empire and the Byzantine, or eastern, Empire.

Western Roman Empire	Byzantine Empire

If YOU lived there...

You are a trader visiting Constantinople. You have traveled to many cities but have never seen anything so magnificent. The city has huge palaces and stadiums for horse races. In the city center you enter a church and stop, speechless with amazement. Above you is a vast, gold dome lit by hundreds of candles.

How does the city make you feel about its rulers?

BUILDING BACKGROUND The Roman emperor Constantine moved the capital of the empire from Rome east to Constantinople. Power shifted to the eastern part of the empire. Soon, political problems and invasions caused the end of the western Roman Empire.

Emperors Rule from Constantinople

Constantinople was built on the site of an ancient Greek trading city called Byzantium (buh-ZAN-shuhm). Its location between two seas protected the city from attack and let the city control trade between Europe and Asia. Constantinople was in an ideal place to grow in wealth and power.

Justinian

After Rome fell in 476, the emperors of the eastern Roman Empire dreamed of taking it back and reuniting the old Roman Empire. For Justinian (juh-STIN-ee-uhn), an emperor who ruled from 527 to 565, reuniting the empire was a passion. He sent his army to retake Italy. In the end this army conquered not only Italy but also much land around the Mediterranean.

Justinian's other passions were the law and the church. He ordered officials to remove any out-of-date or unchristian laws. He then organized the laws into a legal system called Justinian's Code. By simplifying Roman law, the code helped guarantee fairer treatment for all.

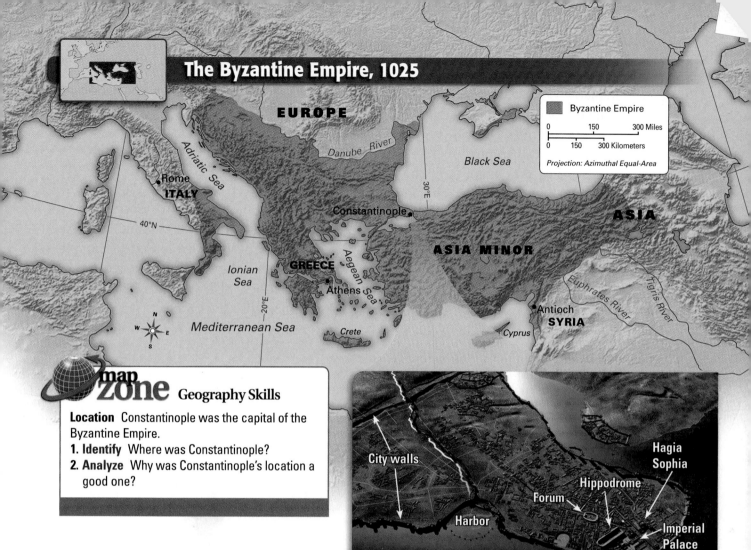

The Byzantine Empire, 1025

EUROPE

Danube River

Adriatic Sea

Rome
ITALY

40°N

Black Sea

Constantinople

ASIA

ASIA MINOR

Ionian
Sea

GREECE

Aegean Sea

Athens

30°E

Euphrates River

Tigris River

Antioch
SYRIA

20°E

Mediterranean Sea

Crete

Cyprus

N
W   E
S

Byzantine Empire

0   150   300 Miles
0   150   300 Kilometers

Projection: Azimuthal Equal-Area

Geography Skills

Location Constantinople was the capital of the Byzantine Empire.
1. **Identify** Where was Constantinople?
2. **Analyze** Why was Constantinople's location a good one?

City walls

Harbor

Forum

Hippodrome

Hagia
Sophia

Imperial
Palace

Constantinople was strategically located where Europe and Asia meet. As a result, the city was in a perfect location to control trade routes between the two continents.

Despite his achievements, Justinian made many enemies. Two groups of these enemies joined together and tried to overthrow him in 532. These groups led riots in the streets. Scared for his life, Justinian prepared to leave Constantinople.

Justinian was stopped from leaving by his wife, Theodora (thee-uh-DOHR-uh). She convinced Justinian to stay in the city. Smart and powerful, Theodora helped her husband rule effectively. With her advice, he found a way to end the riots. Justinian's soldiers killed all the rioters—some 30,000 people—and saved the emperor's throne.

The Empire after Justinian

After the death of Justinian in 565, the eastern empire began to decline. Faced with invasions by barbarians, Persians, and Muslims, later emperors lost all the land

Justinian had gained. The eastern empire remained a major power in the world for hundreds of years, but it never regained its former strength.

The eastern Roman Empire finally ended nearly 900 years after the death of Justinian. In 1453 a group called the Ottoman Turks swept in and captured Constantinople. With this defeat the 1,000-year history of the eastern Roman Empire came to an end.

FOCUS ON READING
What is the implied main idea of the last paragraph?

READING CHECK Drawing Conclusions
Why did Justinian reorganize Roman law?

A New Society

In many ways Justinian was the last Roman emperor of the eastern empire. After he died, non-Roman influences took hold throughout the empire. People began to speak Greek, the language of the eastern empire, rather than Latin. Scholars studied Greek, not Roman, philosophy. Gradually, the empire lost its ties to the old Roman Empire, and a new society developed.

The people who lived in this society never stopped thinking of themselves as Romans. But modern historians have given their society a new name. They call the society that developed in the eastern Roman Empire after the west fell the **Byzantine** (BI-zuhn-teen) **Empire**, named after the Greek city of Byzantium.

Outside Influence

One reason eastern and western Roman societies were different was the Byzantines' interaction with other groups. This interaction was largely a result of trade. Because Constantinople's location was ideal for trading between Europe and Asia, it became the greatest trading city in Europe.

Merchants from all around Europe, Asia, and Africa traveled to Constantinople to trade. Over time Byzantine society began to reflect these outside influences as well as its Roman and Greek roots.

Government

The forms of government in the two empires were also different. Byzantine emperors had more power than western

Close-up

The Glory of Constantinople

Constantinople was a crossroads for traders, a center of Christianity, and the capital of an empire. It was a magnificent city filled with great buildings, palaces, and churches. The city's rulers led processions, or ceremonial walks, to show their wealth and power.

This procession went from the church to the royal palace. The procession showed the power and importance of the emperor as head of the church.

emperors did. Eastern emperors also liked to show off their great power. For example, people could not stand while they were in the presence of the eastern emperor. They had to crawl on their hands and knees to talk to him.

The power of an eastern emperor was greater, in part, because the emperor was considered the head of the church as well as the political ruler. The Byzantines thought the emperor had been chosen by God to lead both the empire and the church. In contrast, the emperor in the west was limited to political power. Popes and other bishops were the leaders of the church.

READING CHECK **Contrasting** What were two ways in which eastern and western Roman societies were different?

Byzantine Christianity

Christianity was central to the Byzantines' lives, just as it was to the lives of people in the west. Nearly everyone who lived in the Byzantine Empire was Christian.

To show their devotion to God and the Christian Church, Byzantine artists created beautiful works of religious art. Among the grandest works were **mosaics**, pictures made with pieces of colored stone or glass. Some mosaics sparkled with gold, silver, and jewels.

The procession began at Hagia Sophia, the Byzantines' famous church.

Citizens and visitors crowded the square to see the royal rulers pass by.

ANALYSIS SKILL **ANALYZING VISUALS**

Where did the procession begin and end? What was the significance of this beginning and ending?

163

The Western Roman and Byzantine Empires

In the Western Roman Empire . . .

- Popes and bishops led the church, and the emperor led the government.
- Latin was the main language.

In the Byzantine Empire . . .

- Emperors led the church and the government.
- Greek was the main language.

THE IMPACT TODAY

The Orthodox Church is still the main religion in Russia, Greece, and other parts of Eastern Europe.

Even more magnificent than their mosaics were Byzantine churches, especially Hagia Sophia (HAH-juh soh-FEE-uh). Built by Justinian in the 530s, its huge domes rose high above Constantinople. According to legend, when Justinian saw the church he exclaimed in delight:

❝ Glory to God who has judged me worthy of accomplishing such a work as this! O Solomon, I have outdone you! ❞

–Justinian, quoted in *The Story of the Building of the Church of Santa Sophia*

As time passed, people in the east and west began to interpret and practice Christianity differently. For example, eastern priests could get married, while priests in the west could not. Religious services were performed in Greek in the east. In the west they were held in Latin.

For hundreds of years, church leaders from the east and west worked together peacefully despite their differences. However, the differences between their ideas continued to grow. In time the differences led to divisions within the Christian Church. In the 1000s the split between east and west became official. Eastern Christians formed what became known as the Orthodox Church. As a result, eastern and western Europe were divided by religion.

READING CHECK **Contrasting** What led to a split in the Christian Church?

SUMMARY AND PREVIEW The Roman Empire and the Christian Church both divided into two parts. The Orthodox Church became a major force in the Byzantine Empire. Before long, though, Orthodox Christians encountered members of a religious group they had never met before, the Muslims.

Section 3 Assessment

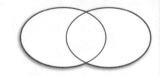
go.hrw.com
Online Quiz
KEYWORD: SK7 HP6

Reviewing Ideas, Terms, and Places

1. **a. Describe** Where was **Constantinople** located?
   b. Summarize What were two of Justinian's major accomplishments?
   c. Elaborate What do you think Theodora's role in the government says about women in the eastern empire?
2. **a. Identify** What was one major difference between the powers of emperors in the east and the west?
   b. Explain How did contact with other cultures help change the **Byzantine Empire**?
3. **a. Define** What is a **mosaic**?
   b. Make Generalizations What led to the creation of two different Christian societies in Europe?

Critical Thinking

4. **Comparing and Contrasting** Draw a diagram like this one. Using your notes and the diagram, compare and contrast Christianity in the western Roman Empire with Christianity in the Byzantine Empire.

FOCUS ON WRITING

5. **Choosing Events of the Byzantine Empire** What are the most important events that occurred in the Byzantine Empire? Identify one or two you could write about.

Chapter Review

Geography's Impact
video series
Review the video to answer the closing question:
Why has Istanbul remained a major cultural center throughout its history?

Visual Summary

Use the visual summary below to help you review the main ideas of the chapter.

QUICK FACTS

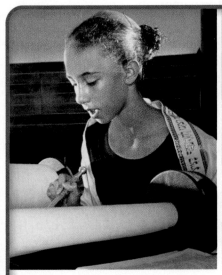

Jews read the Torah to learn about Jewish history and traditions.

Christianity is based on the life and teachings of Jesus of Nazareth.

Byzantine society was greatly influenced by Christianity.

Reviewing Vocabulary, Terms, and People

Match each "I" statement with the person, place, or thing that might have made the statement.

a. Messiah
b. Bible
c. saint
d. rabbi
e. Bethlehem
f. principle
g. disciple
h. Constantinople
i. monotheism
j. Torah

1. "I am the town where Jesus of Nazareth was born."
2. "I was the capital of the Byzantine Empire."
3. "I am the holy book of Christianity."
4. "I was a promised leader who was to appear among the Jews."
5. "I am a person known and admired for my holiness."
6. "I am a basic belief, rule, or law."
7. "I am the most sacred text of Judaism."
8. "I am a follower."
9. "I am a Jewish religious teacher."
10. "I am the belief in only one God."

Comprehension and Critical Thinking

SECTION 1 *(Pages 144–150)*

11. **a. Identify** What are the basic beliefs of Judaism?

   b. Analyze What do the various sacred Jewish texts contribute to Judaism?

   c. Elaborate How are Jewish ideas reflected in modern western society today?

SECTION 2 *(Pages 152–159)*

12. **a. Describe** According to the Bible, what were the crucifixion and Resurrection?

   b. Analyze Why do you think Jesus's teachings appealed to many people in the Roman Empire?

SECTION 2 (continued)

c. Evaluate Why do you think Paul is considered one of the most important people in the history of Christianity?

SECTION 3 (Pages 160–164)

13. a. Identify Who were Justinian and Theodora, and what did they accomplish?

b. Contrast In what ways was the Byzantine Empire different from the western Roman Empire?

c. Elaborate Would Constantinople have been an exciting place to visit in the 500s? Why or why not?

Using the Internet

go.hrw.com
KEYWORD: SK7 CH6

14. Activity: Creating Maps Within 400 years of Jesus's death, Christianity had grown from a small group of Jesus's disciples into the only religion practiced in the Roman Empire. Although 400 years sounds like a long time, to a historian it is practically the blink of an eye. What explains the rapid growth of Christianity? Enter the activity keyword. Then research the key figures, events, and factors in the spread of Christianity. Use what you learn to create an illustrated and annotated map of the spread of Christianity.

Social Studies Skills

Interpreting Route Maps *Use the map of Paul's Journeys in Section 2 to answer the following questions.*

15. How many journeys did Paul take?

16. From where did he start his third journey?

17. On which journeys did Paul visit the cities of Corinth and Athens?

18. What was the last city Paul traveled to?

---

19. Understanding Implied Main Ideas Look back at the beginning of Section 3 of this chapter. For each paragraph under the heading "Justinian," write a statement that you think is the implied main idea of the paragraph.

20. Writing a Letter Your letter is from an eyewitness to the event to a good friend. Look back over your notes and choose one important event for your letter. Your letter should answer the Who? What? Where? How? questions about the event. Write the letter as though you are writing to a good friend and are excited about having seen this event.

Map Activity

21. The Jewish and Christian Worlds On a separate sheet of paper, match the letters on the map with their correct labels.

Rome   Constantinople

Jerusalem

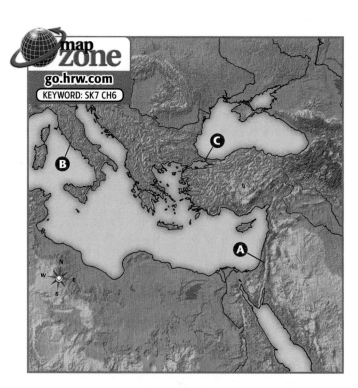

go.hrw.com
KEYWORD: SK7 CH6

Standardized Test Practice

DIRECTIONS: Read questions 1 through 6 and write the letter of the best response. Then read question 7 and write your own well-constructed response.

1 **The Jewish holiday that celebrates the start of a new year in the Jewish calendar is**

A Hanukkah.

B Passover.

C Rosh Hashanah.

D Yom Kippur.

2 **Many people believe that the Hebrews were the first people to practice**

A monotheism.

B rituals.

C religion.

D law.

3 **Which of the following statements about Jesus is false?**

A Some people believed Jesus was the Messiah that Jewish prophets had predicted.

B Some Jewish and Roman leaders viewed Jesus as a threat to their power.

C Jesus fought in a rebellion of the Jews against the Romans.

D Jesus taught people to love God and to be kind to each other.

4 **Which of the following statements correctly describes how the Byzantine Empire differed from the western Roman Empire?**

A People practiced Christianity in the Byzantine Empire but not in the western Roman Empire.

B Byzantine emperors had more power than western emperors did.

C Popes and bishops were leaders of the church in the Byzantine Empire, while rabbis led the church in the western Roman Empire.

D The western Roman Empire was known more for its mosaics than the Byzantine Empire was.

*"*Honor your father and mother. You will then live long on the land that God your Lord is giving you.

Do not commit murder.

Do not commit adultery.

Do not steal.

Do not testify as a false witness against your neighbor.

Do not be envious of your neighbor's house. *"*

–Exodus 20:12–14, from *The Living Torah*

5 **The passage above is a selection from the Ten Commandments. Based on this passage, which aspect of Judaism has a basis in the Ten Commandments?**

A celebration of Hanukkah

B belief in doing what is proper

C explanation of the Commentaries

D respect for the Talmud

6 **Which of Jesus's followers knew Jesus well and received special teaching?**

A Paul

B Constantine

C the Gospels

D the Apostles

7 **Extended Response** Consider what you have learned in this chapter as well as the selection above from the Ten Commandments. Use this information to write a brief essay explaining one or two ideas and beliefs that are shared by Jews and Christians. Your essay should also explain why you think these two religions share some ideas and beliefs.

History of the Islamic World

AD 550–1650

Scholars at Córdoba

Córdoba

A Muslim trader

What You Will Learn...

In this chapter you will learn about a religion called Islam. First taught by a man named Muhammad, Islam is now one of the largest religions in the world. Throughout history, Muslims, or people who practice Islam, have ruled empires and made great advances in many fields.

SECTION 1
Origins of Islam 170

SECTION 2
Islamic Beliefs and Practices 174

SECTION 3
Muslim Empires 180

SECTION 4
Cultural Achievements................186

FOCUS ON READING AND WRITING

Sequencing When you read, it is important to keep track of the sequence, or order, in which events happen. Look for dates and other clues to help you figure out the proper sequence. **See the lesson, Sequencing, on page 674.**

Designing A Web Site You have been asked to design a Web site to teach children about Islam and the history of the Muslim people. As you read this chapter, you will collect information about Islam and the Muslim empires. Then you will use that information to design your Web site.

Islam One of the world's largest religions, Islam is practiced by people all around the world. These Muslims, or people who practice Islam, are in India.

The Islamic World, AD 550–1650

map zone Geography Skills

Regions For hundreds of years, Islam was the major religion in a huge region that stretched from Spain to India.
1. **Locate** In what city was the Blue Mosque?
2. **Draw Conclusions** How do you think Islam helped tie people in such distant locations together?

The Blue Mosque in Constantinople

Caspian Sea

Black Sea

Constantinople

20°E

Mediterranean Sea

The Taj Mahal

Baghdad • Esfahan •

Persian Gulf

Tropic of Cancer

20°N

70°E

Medina •

Mecca •

The Great Mosque, Mecca

0   300   600 Miles
0   300   600 Kilometers
Projection: Lambert Azimuthal Equal-Area

40°E

60°E

50°E

10°N

INDIAN OCEAN

HOLT

Geography's Impact
video series
Watch the video to understand the impact of Mecca on Islam.

Empires The Muslims formed powerful empires in many parts of the world.

Achievements All through history, Muslims have made great achievements in many fields. This device, an astrolabe, was invented by Muslim scholars.

169

Origins of Islam

What You Will Learn...

Main Ideas

1. Arabia is mostly a desert land, where two ways of life, nomadic and sedentary, developed.
2. A new religion called Islam, founded by the prophet Muhammad, spread throughout Arabia in the 600s.

The Big Idea

In the harsh desert climate of Arabia, Muhammad, a merchant from Mecca, introduced a major world religion called Islam.

Key Terms and Places

Mecca, *p. 172*
Islam, *p. 172*
Muslim, *p. 172*
Qur'an, *p. 172*
Medina, *p. 173*
mosque, *p. 173*

 **TAKING NOTES** As you read, take notes on key places, people, and events in the origins of Islam. Organize your notes in a series of boxes like the ones below.

Places	People	Events

If YOU lived there...

You live in a town in Arabia, in a large merchant family. Your family has grown rich from selling goods brought by traders crossing the desert. Your house is larger than most others in town, and you have servants to wait on you. Although many townspeople are poor, you have always taken such differences for granted. Now you hear that some people are saying the rich should give money to the poor.

How might your family react to this idea?

BUILDING BACKGROUND For thousands of years, traders have crossed the deserts of Arabia to bring goods to market. Scorching temperatures and lack of water have made the journey difficult. However, Arabia not only developed into a thriving trade center, it also became the birthplace of a new religion.

Life in a Desert Land

The Arabian Peninsula, or Arabia, is located in the southwest corner of Asia. It lies near the intersection of Africa, Europe, and Asia. For thousands of years Arabia's location, physical features, and climate have shaped life in the region.

Physical Features and Climate

Arabia lies in a region with hot and dry air. With a blazing sun and clear skies, summer temperatures in the interior parts of the peninsula reach 100°F (38°C) daily. This climate has created a band of deserts across Arabia and northern Africa. Sand dunes, or hills of sand shaped by the wind, can rise to 800 feet (240 m) high and stretch across hundreds of miles!

Arabia's deserts have a very limited amount of water. What water there is exists mainly in scattered oases. An oasis is a wet, fertile area in a desert. Oases have long been key stops along Arabia's overland trade routes.

Two Ways of Life

To live in Arabia's harsh deserts, people developed two main ways of life. Nomads lived in tents and raised herds of sheep, goats, and camels. The animals provided milk, meat, wool, and leather. The camels also carried heavy loads. Nomads traveled with their herds across the desert in search of food and water for their animals.

Among the nomads, water and land belonged to tribes. Membership in a tribe, a group of related people, offered safety from desert dangers.

While nomads moved around, other Arabs lived a more settled life. They made their homes in oases where they could farm. These settlements, particularly the ones along trade routes, became towns.

Towns became centers of trade. There, nomads traded animal products and herbs for goods like cooking supplies and clothes. Merchants sold spices, gold, leather, and other goods brought by caravans.

READING CHECK **Categorizing** What two ways of life were common in Arabia?

Close-up
Life in Arabia

The city of Mecca in Arabia is shown here as it might have looked in the late 500s. Nomads from the desert and merchants from distant lands came to trade in Mecca. As a result of this trade, many Meccan merchants became very wealthy.

> Nomads traveled across Arabia, moving their animals as the seasons changed.

> Towns were centers of trade for both nomads and townspeople. They traded goods like food and cloth.

ANALYSIS SKILL **ANALYZING VISUALS**

Which figures in this image do you think are nomads? Which are townspeople? How can you tell?

171

A New Religion

In early times, Arabs worshipped many gods. That changed, however, when a man named Muhammad brought a new religion to Arabia. Historians know little about Muhammad. What they do know comes from religious writings.

Muhammad Becomes a Prophet

FOCUS ON READING

What clues in this paragraph can help you track the sequence of events?

Muhammad was born into an important family in the city of **Mecca** around 570. As a small child, he traveled with his uncle's caravans. Once he was grown, he managed a caravan business owned by a wealthy woman named Khadijah (ka-DEE-jah). At age 25, Muhammad married Khadijah.

The caravan trade made Mecca a rich city, but most of the wealth belonged to just a few people. Traditionally, wealthy people in Mecca had helped the poor. As Muhammad was growing up, though, many rich merchants ignored the needy.

Concerned about these changes, Muhammad often went to the hills to pray and meditate. One day, when he was about 40 years old, he went to meditate in a cave. According to religious writings, an angel spoke to Muhammad, telling him to "Recite! Recite!" Muhammad asked what he should recite. The angel answered:

"Recite in the name of your Lord who created, created man from clots of blood! Recite! Your Lord is the Most Bountiful One, Who by the pen taught man what he did not know."
— From *The Koran*, translated by N. J. Dawood

Muslims believe that God had spoken to Muhammad through the angel and had made him a prophet, a person who tells of messages from God. The messages that Muhammad received form the basis of the religion called **Islam**. In Arabic, the word *Islam* means "to submit to God."

Muslims, or people who follow Islam, believe that God chose Muhammad to be his messenger to the world. They also believe that Muhammad continued to receive messages from God for the rest of his life. Eventually, these messages were collected in the **Qur'an** (kuh-RAN), the holy book of Islam.

Muhammad's Teachings

In 613 Muhammad began to talk about his messages. He taught that there was only one God, Allah, which means "the God" in Arabic. Like Judaism and Christianity, Islam is monotheistic, or based on the belief in one God. Although people of all three religions believe in one God, their beliefs about God are not the same.

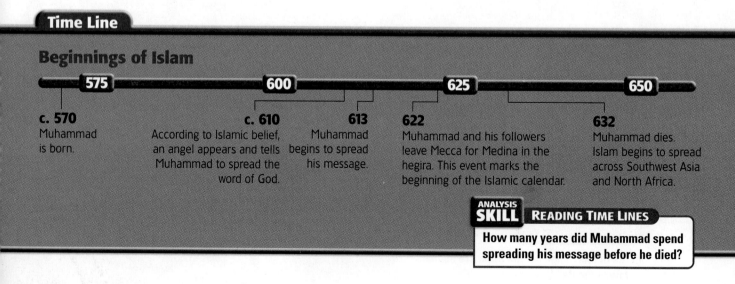

Time Line

Beginnings of Islam

575 — 600 — 625 — 650

c. 570 Muhammad is born.

c. 610 According to Islamic belief, an angel appears and tells Muhammad to spread the word of God.

613 Muhammad begins to spread his message.

622 Muhammad and his followers leave Mecca for Medina in the hegira. This event marks the beginning of the Islamic calendar.

632 Muhammad dies. Islam begins to spread across Southwest Asia and North Africa.

ANALYSIS SKILL **READING TIME LINES**

How many years did Muhammad spend spreading his message before he died?

Muhammad's teachings also dealt with how people should live. He taught that all people who believed in Allah were bound together like members of a family. As a result, he said, people should help those who are less fortunate. For example, he thought that people who had money should use that money to help the poor.

Islam Spreads in Arabia

At first Muhammad had few followers. Slowly, more people began to listen to his ideas. As Islam spread, Mecca's rulers grew worried. They threatened Muhammad and even planned to kill him.

A group of people living north of Mecca invited Muhammad to move to their city. So in 622 Muhammad and many of his followers went to **Medina** (muh-DEE-nuh). The name *Medina* means "the Prophet's city" in Arabic. Muhammad's departure from Mecca is called the hegira (hi-JY-ruh), or journey. It is so important a date in the history of Islam that Muslims made 622 the first year of the Islamic calendar.

Muhammad became a spiritual and political leader in Medina. His house became the first **mosque** (MAHSK), or building for Muslim prayer.

As the Muslim community in Medina grew stronger, other Arab tribes began to accept Islam. Conflict with the Meccans, however, increased. In 630, after several years of fighting, the people of Mecca gave in. They accepted Islam as their religion.

Before long, most people in Arabia had accepted Muhammad as their spiritual and political leader and become Muslims. Muhammad died in 632, but the religion he taught would soon spread far beyond the Arabian Peninsula.

READING CHECK **Summarizing** How did Islam spread in Arabia?

Over many centuries, Islam spread from Arabia into other parts of the world. These Muslim children in the African country of Nigeria are studying the Qur'an.

SUMMARY AND PREVIEW In the early 600s Islam was introduced to Arabia by Muhammad. In the next section, you will learn more about the main Islamic teachings and beliefs.

Section 1 Assessment

Reviewing Ideas, Terms, and Places

1. a. Define What is an oasis?
   b. Make Generalizations Where did towns develop? Why?
   c. Predict Do you think life would have been better for nomads or townspeople in early Arabia? Explain.

2. a. Identify What is the **Qur'an**?
   b. Explain According to Islamic belief, what was the source of Islamic teachings?
   c. Elaborate Why did Muhammad move from **Mecca** to **Medina**? What did he accomplish there?

Critical Thinking

3. Sequencing Draw a time line like the one below. Using your notes on Muhammad, identify the key dates in

FOCUS ON WRITING

4. Thinking about Muhammad and Is read about Muhammad and the might you organize these two Write down some notes.

Islamic Beliefs and Practices

What You Will Learn...

Main Ideas

1. The Qur'an guides Muslims' lives.
2. The Sunnah tells Muslims of important duties expected of them.
3. Islamic law is based on the Qur'an and the Sunnah.

The Big Idea

Sacred texts called the Qur'an and the Sunnah guide Muslims in their religion, daily life, and laws.

Key Terms

jihad, *p. 175*
Sunnah, *p. 175*
Five Pillars of Islam, *p. 176*

TAKING NOTES As you read, take notes on the most important beliefs and practices of Islam. You can organize your notes in a table like this one.

Religious beliefs	Daily life	Laws

If YOU lived there...

Your family owns an inn in Mecca. Usually business is pretty calm, but this week your inn is packed. Travelers have come from all over the world to visit your city. One morning you leave the inn and are swept up in a huge crowd of these visitors. They speak many different languages, but everyone is wearing the same white robes. They are headed to the mosque.

What might draw so many people to your city?

BUILDING BACKGROUND One basic Islamic belief is that everyone who can must make a trip to Mecca sometime during his or her lifetime. More Islamic teachings can be found in Islam's holy books—the Qur'an and the Sunnah.

The Qur'an

During Muhammad's life, his followers memorized his messages from God along with his words and deeds. After Muhammad's death, they collected his teachings and wrote them down to form the book known as the Qur'an. Muslims consider the Qur'an to be the exact word of God as it was told to Muhammad.

Beliefs

The central teaching in the Qur'an is that there is only one God—Allah—and that Muhammad is his prophet. The Qur'an says people must obey Allah's commands. Muslims learned of these commands from Muhammad.

Islam teaches that the world had a definite beginning and will end one day. Muhammad said that on the final day God will judge all people. Those who have obeyed his orders will be granted life in paradise. According to the Qur'an, paradise is a beautiful garden full of fine food and drink. People who have not obeyed God, however, will suffer.

Studying the Qur'an

The Qur'an, pictured below, plays a central role in the lives of many Muslims. Both children and adults study and memorize verses from the Qur'an at home, at Islamic schools, and in mosques.

ANALYZING VISUALS Where do you think these children are studying the Qur'an?

Guidelines for Behavior

Like holy books of other religions, the Qur'an describes Muslim acts of worship, guidelines for moral behavior, and rules for social life.

Some of these guidelines for life are stated **explicitly**. For example, the Qur'an clearly describes how a person should prepare for worship. Muslims must wash themselves before praying so they will be pure before Allah. The Qur'an also tells Muslims what they should not eat or drink. Muslims are not allowed to eat pork or drink alcohol.

Other guidelines for behavior are not stated directly but are **implicit** in the Qur'an. Even though they are not written directly, many of these ideas altered early Arabian society. For example, the Qur'an does not expressly forbid the practice of slavery, which was common in early Arabia. It does, however, imply that slavery should be abolished. Based on this implication, many Muslim slaveholders chose to free their slaves.

Another important subject in the Qur'an has to do with **jihad** (ji-HAHD), which means "to make an effort, or to struggle." Jihad refers to the inner struggle people go through in their effort to obey God and behave according to Islamic ways. Jihad can also mean the struggle to defend the Muslim community, or, historically, to convert people to Islam. The word has also been translated as "holy war."

READING CHECK **Analyzing** Why is the Qur'an important to Muslims?

The Sunnah

The Qur'an is not the only source for the teachings of Islam. Muslims also study the hadith (huh-DEETH), the written record of Muhammad's words and actions. It is also the basis for the Sunnah. The **Sunnah** (SOOH-nuh) refers to the way Muhammad lived, which provides a model for the duties and the way of life expected of Muslims. The Sunnah guides Muslims' behavior.

ACADEMIC VOCABULARY

explicit fully revealed without vagueness

ACADEMIC VOCABULARY

implicit understood though not clearly put into words

The Five Pillars of Islam

Saying "There is no god but God, and Muhammad is his prophet"

Praying five times a day

Giving to the poor and needy

Fasting during the holy month of Ramadan

Traveling to Mecca at least once on a hajj

Which of the five pillars shows how Muslims are supposed to treat other people?

The Five Pillars of Islam

The first duties of a Muslim are known as the **Five Pillars of Islam**, which are five acts of worship required of all Muslims. The first pillar is a statement of faith. At least once in their lives, Muslims must state their faith by saying, "There is no god but God, and Muhammad is his prophet." Muslims say this when they accept Islam. They also say it in their daily prayers.

The second pillar of Islam is daily prayer. Muslims must pray five times a day: before sunrise, at midday, in late afternoon, right after sunset, and before going to bed. At each of these times, a call goes out from a mosque, inviting Muslims to come pray. Muslims try to pray together at a mosque. They believe prayer is proof that someone has accepted Allah.

The third pillar of Islam is a yearly donation to charity. Muslims must pay part of their wealth to a religious official. This money is used to help the poor, build mosques, or pay debts. Helping and caring for others is important in Islam.

The fourth pillar of Islam is fasting— going without food and drink. Muslims fast during the holy month of Ramadan (RAH-muh-dahn). The Qur'an says Allah began his revelations to Muhammad in this month. Throughout Ramadan, most Muslims will not eat or drink anything between dawn and sunset. Muslims believe fasting is a way to show that God is more important than one's own body. Fasting also reminds Muslims of people in the world who struggle to get enough food.

The fifth pillar of Islam is the hajj (HAJ), a pilgrimage to Mecca. All Muslims must travel to Mecca at least once in their lives if they can. The Kaaba, in Mecca, is Islam's most sacred place.

The Sunnah and Daily Life

Besides the five pillars, the Sunnah has other examples of Muhammad's actions and teachings. These form the basis for rules about how to treat others. According to Muhammad's example, people should treat guests with generosity.

The Sunnah also provides guidelines for how people should conduct their relations in business and government. For example, one Sunnah rule says that it is bad to owe someone money. Another rule says that people should obey their leaders.

READING CHECK **Generalizing** What do Muslims learn from the Sunnah?

Islamic Law

Together, the Qur'an and the Sunnah are important guides for how Muslims should live. They also form the basis of Islamic law, or Shariah (shuh-REE-uh). Shariah uses both Islamic sources and human reason to judge the rightness of actions a person or community might take. All actions fall on a scale ranging from required to accepted to disapproved to forbidden. Islamic law makes no distinction between religious beliefs and daily life, so Islam affects all aspects of Muslims' lives.

Shariah sets rewards for good behavior and punishments for crimes. It also describes limits of authority. It was the basis for law in Muslim countries until modern times.

Sources of Islamic Beliefs		
Qur'an	**Sunnah**	**Shariah**
Holy book that includes all the messages Muhammad received from God	Muhammad's example for the duties and way of life expected of Muslims	Islamic law, based on interpretations of the Qur'an Sunnah

Today, though, most Muslim countries blend Islamic law with legal systems like those in the United States or western Europe.

Islamic law is not found in one book. Instead, it is a set of opinions and writings that have changed over the centuries. As a result, different ideas about Islamic law are found in different Muslim regions.

READING CHECK **Finding Main Ideas** What is the purpose of Islamic law?

SUMMARY AND PREVIEW The Qur'an, the Sunnah, and Shariah teach Muslims how to live. In the next chapter, you will learn more about Muslim culture and the spread of Islam from Arabia to other lands in Europe, Africa, and Asia.

Section 2 Assessment

Reviewing Ideas, Terms, and Places

1. **a. Recall** What is the central teaching of the Qur'an?
   b. Explain How does the Qur'an help Muslims obey God?
2. **a. Recall** What are the **Five Pillars of Islam**?
   b. Make Generalizations Why do Muslims fast during Ramadan?
3. **a. Identify** What is Islamic law called?
   b. Make Inferences How is Islamic law different from law in the United States?
   c. Elaborate What is one possible reason that opinions and writings about Islamic law have changed over the centuries?

Critical Thinking

4. **Categorizing** Draw a chart like the one to the right. Use your notes to list three key teachings from the Qur'an and three teachings from the Sunnah.

Qur'an	Sunnah

FOCUS ON WRITING

5. **Describing Islam** What information would you include on your Web site about the beliefs and practices of Islam? Note how you might organize one page of your Web site about this topic.

The Hajj

Every year, as many as 2 million Muslims make a religious journey, or pilgrimage, to Mecca, Saudi Arabia. This journey, called the hajj, is one of the Five Pillars of Islam—all Muslims are expected to make the journey at least once in their lifetime if they can.

Mecca is the place where Muhammad lived and taught more than 1,300 years ago. As a result, it is the holiest city in Islam. The pilgrims who travel to Mecca each year serve as a living reminder of the connection between history and geography.

On the Road to Mecca

- Before entering Mecca, pilgrims undergo a ritual cleansing and put on special white garments.

- At Mecca, guides help pilgrims through religious rituals.

- One important ritual is the "Standing" on Mount Arafat, near Mecca. Pilgrims stand for hours, praying, at a place where Muhammad is said to have held his last sermon.

- Pilgrims then participate in a three-day ritual of "Stoning," in which they throw pebbles at three pillars.

- Finally, pilgrims complete their journey by returning to the Grand Mosque in Mecca, where a great feast is held.

Europe and the Americas Many countries in Europe and the Americas have a Muslim population. These pilgrims are from Germany.

Africa Pilgrims also come from Africa. These pilgrims are from Nigeria, just one of the African countries that is home to a large Muslim population.

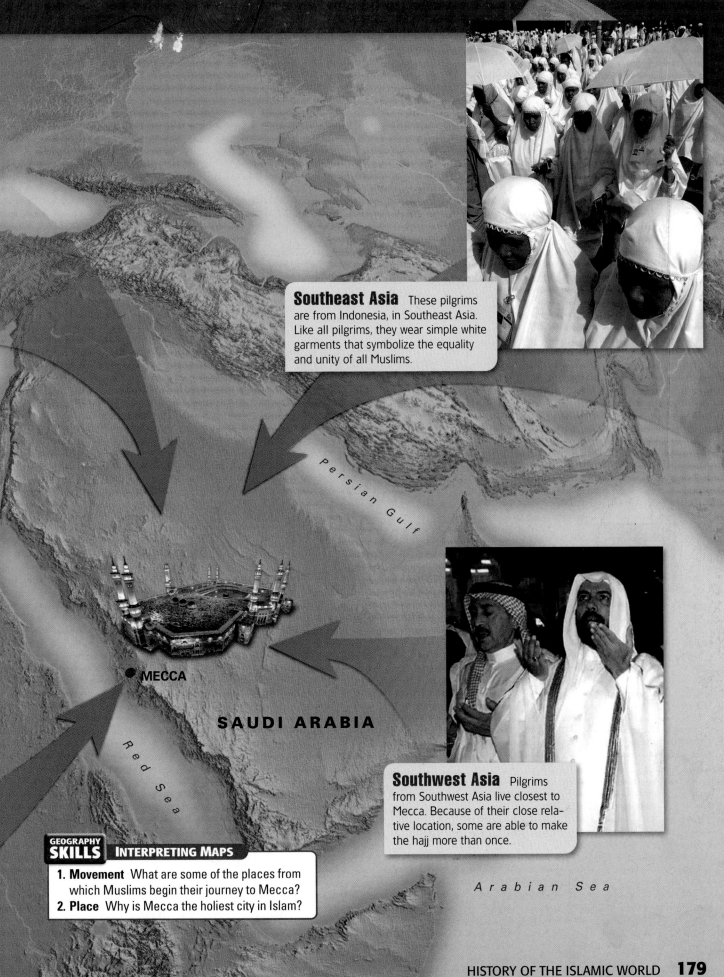

Southeast Asia These pilgrims are from Indonesia, in Southeast Asia. Like all pilgrims, they wear simple white garments that symbolize the equality and unity of all Muslims.

Persian Gulf

MECCA

SAUDI ARABIA

Red Sea

Southwest Asia Pilgrims from Southwest Asia live closest to Mecca. Because of their close relative location, some are able to make the hajj more than once.

Arabian Sea

GEOGRAPHY
SKILLS **INTERPRETING MAPS**

1. **Movement** What are some of the places from which Muslims begin their journey to Mecca?
2. **Place** Why is Mecca the holiest city in Islam?

Muslim Empires

If **YOU** lived there...

You are a farmer living in a village on the coast of India. For centuries, your people have raised cotton and spun its fibers into a soft fabric. One day, a ship arrives in the harbor carrying Muslim traders from far away. They bring interesting goods you have never seen before. They also bring new ideas.

What ideas might you learn from the traders?

BUILDING BACKGROUND For years traders traveled from Arabia to markets far away. As they traveled, they picked up new goods and ideas, and they introduced these to the people they met. Some of the new ideas the traders spread were Islamic ideas.

Muslim Armies Conquer Many Lands

After Muhammad's death his followers quickly chose Abu Bakr (uh-boo BAK-uhr), one of Muhammad's first converts, to be the next leader of Islam. He was the first **caliph** (KAY-luhf), a title that Muslims use for the highest leader of Islam. In Arabic, the word *caliph* means "successor." As Muhammad's successors, the caliphs had to follow the prophet's example. This meant that they had to rule according to the Qur'an. Unlike Muhammad, however, the early caliphs were not religious leaders.

Beginnings of an Empire

Abu Bakr directed a series of battles to unite Arabia. By his death in 634, he had made Arabia into a unified Muslim state. With Arabia united, Muslim leaders turned their attention elsewhere. Their armies, strong after their battles in Arabia, won many stunning victories. They defeated the Persian and Byzantine empires, which were weak from many years of fighting.

When the Muslims conquered lands, they set certain rules for non-Muslims living there. For example, some non-Muslims could not build churches in Muslim cities or dress like Muslims. However, Christians and Jews could continue to practice their own religion. They were not forced to convert to Islam.

What You Will Learn...

Main Ideas

1. Muslim armies conquered many lands into which Islam slowly spread.
2. Trade helped Islam spread into new areas.
3. Three Muslim empires controlled much of Europe, Asia, and Africa from the 1400s to the 1800s.

The Big Idea

After the early spread of Islam, three large Muslim empires formed—the Ottoman, Safavid, and Mughal empires.

Key Terms and Places

caliph, *p. 180*
tolerance, *p. 182*
Baghdad, *p. 182*
Córdoba, *p. 182*
janissaries, *p. 182*
Istanbul, *p. 182*
Esfahan, *p. 184*

TAKING NOTES As you read, take notes on the spread of Islam and three large Muslim empires. Organize your notes in a chart like the one below.

Spread of Islam	Muslim Empires

Growth of the Empire

Many early caliphs came from one family, the Umayyad (oom-EYE-yuhd) family. The Umayyads moved the capital to Damascus, in Muslim-conquered Syria, and continued to expand the empire. They took over lands in Central Asia and in northern India. The Umayyads also gained control of trade in the eastern Mediterranean and conquered parts of North Africa.

The Berbers, the native people of North Africa, resisted Muslim rule at first. After years of fighting, however, many Berbers converted to Islam.

In 711 a combined Arab and Berber army invaded Spain and quickly conquered it. Next, the army moved into what is now France, but it was stopped by a Christian army near the city of Tours (TOOR). Despite this defeat, Muslims called Moors ruled parts of Spain for the next 700 years.

A new Islamic dynasty, the Abbasids (uh-BAS-idz), came to power in 749. They reorganized the government to make it easier to rule such a large region.

READING CHECK **Analyzing** What role did armies play in spreading Islam?

Trade Helps Islam Spread

Islam gradually spread through areas the Muslims conquered. Trade also helped spread Islam. Along with their goods, Arab merchants took Islamic beliefs to India, Africa, and Southeast Asia. Though Indian kingdoms remained Hindu, coastal trading cities soon had large Muslim communities. In Africa, many leaders converted to Islam. As a result, societies often had both African and Muslim customs. Between 1200 and 1600, Muslim traders carried Islam even farther east. Muslim communities grew up in what are now Malaysia and Indonesia.

Trade also brought new products to Muslim lands. For example, Arabs learned from the Chinese how to make paper and use gunpowder. New crops such as cotton, rice, and oranges arrived from India, China, and Southeast Asia.

Many Muslim merchants traveled to African market towns, too. They wanted African products such as ivory, cloves, and slaves. In return they offered fine white pottery called porcelain from China, cloth goods from India, and iron from Europe and Southwest Asia. Arab traders grew wealthy from trade between regions.

FOCUS ON READING

As you read this page, look for words that give clues to the sequence of events.

THE IMPACT TODAY

Indonesia now has the world's largest Muslim population.

The City of Córdoba

By the early 900s, Córdoba, Spain, was one of the wealthiest cities in Europe and a center of Islamic learning. Rich examples of Islamic architecture can still be seen in the city.

181

A Mix of Cultures

As Islam spread, Arabs came into contact with people who had different beliefs and lifestyles than they did. Muslims generally practiced **tolerance**, or **acceptance**, with regard to the people they conquered. For example, Muslims did not ban all other religions in their lands. Because they shared some beliefs with Muslims, Christians and Jews in particular kept many of their rights. They did, however, have to pay a special tax. Christians and Jews were also forbidden from converting anyone to their religions.

Many people conquered by the Arabs converted to Islam. These people often adopted other parts of Arabic culture, including the Arabic language. The Arabs, in turn, adopted some customs from the people they conquered. This cultural blending changed Islam from a mostly Arab religion into a religion that included many other cultures. However, the Arabic language and shared religion helped unify the different groups of the Islamic world.

Growth of Cities

The growing cities of the Muslim world reflected the blending of cultures. Trade had brought people together and created wealth, which supported great cultural development in Muslim cities.

Baghdad, in what is now Iraq, became the capital of the Islamic Empire in 762. Trade and farming made Baghdad one of the world's richest cities. The caliphs there supported science and the arts. The city was a center of culture and learning.

Córdoba (KAWR-doh-bah), a great city in Spain, became another showplace of Muslim civilization. By the early 900s Córdoba was the largest and most advanced city in western Europe.

READING CHECK **Finding the Main Idea** How did trade affect the spread of Islam?

Three Muslim Empires

The great era of Arab Muslim expansion lasted until the 1100s. Afterward, three non-Arab Muslim groups built large, powerful empires that took control of much of Europe, Asia, and Africa.

The Ottoman Empire

In the mid-1200s Muslim Turkish warriors known as Ottomans began to take territory from the Christian Byzantine Empire. They eventually ruled land from eastern Europe to North Africa and Arabia.

The key to the empire's expansion was the Ottoman army. The Ottomans trained Christian boys from conquered towns to be soldiers. These slave soldiers, called janissaries, converted to Islam and became fiercely loyal warriors. The Ottomans also benefitted from their use of new weapons, especially gunpowder.

In 1453 Ottomans led by Mehmed II used huge cannons to conquer the city of Constantinople. With the city's capture, Mehmed defeated the Byzantine Empire. He became known as the Conqueror. Mehmed made Constantinople, which the Ottomans called **Istanbul**, his capital. He also turned the Byzantines' great church, Hagia Sophia, into a mosque.

After Mehmed's death, another ruler, or sultan, continued his conquests. This sultan expanded the empire to the east through the rest of Anatolia, another name for Asia Minor. His armies also conquered Syria and Egypt. The holy cities of Mecca and Medina then accepted Ottoman rule.

The Ottoman Empire reached its height under Suleyman I (soo-lay-MAHN), "the Magnificent." During his rule from 1520 to 1566, the Ottomans took control of the eastern Mediterranean and pushed farther into Europe, areas they would control until the early 1800s.

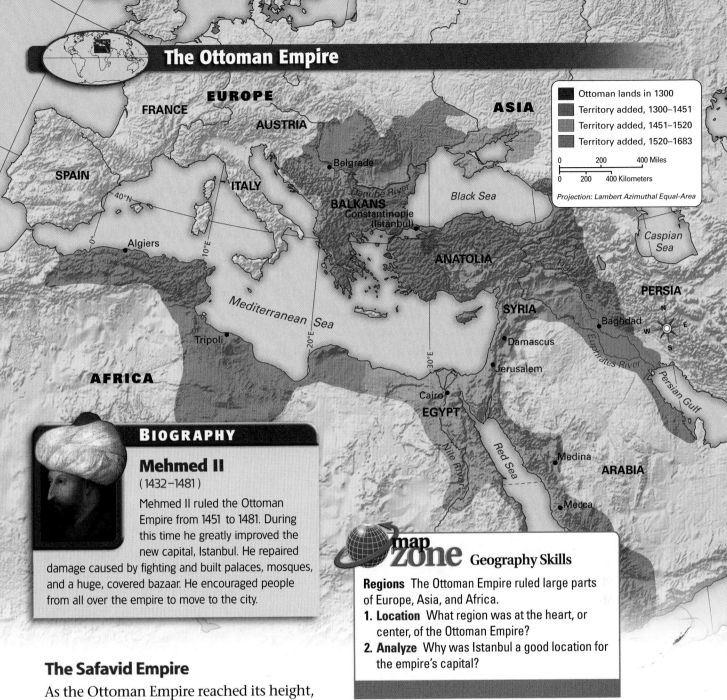

The Ottoman Empire

EUROPE

FRANCE

AUSTRIA

SPAIN

ITALY

Belgrade

Danube River

BALKANS

Constantinople
(Istanbul)

40°N

Algiers

Mediterranean Sea

Tripoli

AFRICA

Cairo

EGYPT

Nile River

Red Sea

ASIA

Black Sea

Caspian
Sea

ANATOLIA

PERSIA

SYRIA

Baghdad

Damascus

Jerusalem

Euphrates River

Persian Gulf

Medina

ARABIA

Mecca

	Ottoman lands in 1300
	Territory added, 1300–1451
	Territory added, 1451–1520
	Territory added, 1520–1683

0   200   400 Miles
0   200   400 Kilometers

Projection: Lambert Azimuthal Equal-Area

BIOGRAPHY

Mehmed II
(1432–1481)

Mehmed II ruled the Ottoman Empire from 1451 to 1481. During this time he greatly improved the new capital, Istanbul. He repaired damage caused by fighting and built palaces, mosques, and a huge, covered bazaar. He encouraged people from all over the empire to move to the city.

map zone Geography Skills

Regions The Ottoman Empire ruled large parts of Europe, Asia, and Africa.

1. **Location** What region was at the heart, or center, of the Ottoman Empire?
2. **Analyze** Why was Istanbul a good location for the empire's capital?

The Safavid Empire

As the Ottoman Empire reached its height, a group of Persian Muslims, the Safavids (sah-FAH-vuhds), was gaining power to the east, in the area of present-day Iran. Before long, the Safavids came into conflict with the Ottomans and other Muslims.

The conflict arose from an old dispute among Muslims about who should be caliph. In the mid-600s, Islam split into two groups. The two groups were the Shia (SHEE-ah) and the Sunni (SOO-nee). Shia Muslims thought only Muhammad's descendants could become caliphs. The Sunni did not think caliphs had to be related to Muhammad. The Ottomans were Sunni, and the Safavid leaders were Shia.

The Safavid Empire began in 1501 when a strong Safavid leader named Esma'il (is-mah-EEL) conquered Persia. He took the ancient Persian title of shah, or king.

Esma'il made Shiism—the beliefs of the Shia—the official religion of the empire. But he wanted to spread Shiism farther.

THE IMPACT TODAY

Most Muslims today belong to the Sunni branch of Islam.

He tried to gain more Muslim lands and convert more Muslims to Shiism. He fought the Uzbek people, but he suffered a major defeat by the Ottomans in 1514.

In 1588 the greatest Safavid leader, 'Abbas, became shah. He strengthened the military and gave his soldiers gunpowder weapons. Copying the Ottomans, 'Abbas trained foreign slave boys to be soldiers. Under 'Abbas's rule the Safavids defeated the Uzbeks and took back land that had been lost to the Ottomans.

The Safavids blended many Persian and Muslim traditions. They grew wealthy from trade and built glorious mosques in their capital, **Esfahan** (es-fah-HAHN). The Safavid Empire lasted until the mid-1700s.

The Mughal Empire

East of the Safavid Empire, in northern India, lay the Mughal (MOO-guhl) Empire. The Mughals were Turkish Muslims from Central Asia. Their empire was established by a leader named Babur (BAH-boohr), or "tiger." He tried for years to build an empire in Central Asia. When he did not succeed there, he decided to create an empire in northern India instead. The result was the Mughal Empire, created in 1526.

In the mid-1500s an emperor named Akbar conquered many new lands and worked to strengthen the government of the empire. He also instituted a tolerant religious policy. Akbar believed members of all religions could live and work together.

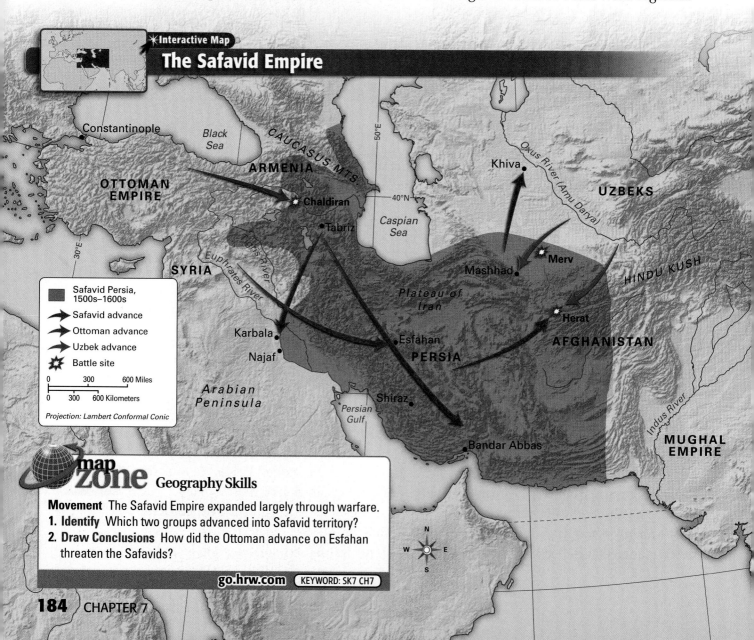

★ Interactive Map

The Safavid Empire

Legend
- Safavid Persia, 1500s–1600s
- Safavid advance
- Ottoman advance
- Uzbek advance
- ✸ Battle site

0 300 600 Miles
0 300 600 Kilometers
Projection: Lambert Conformal Conic

map zone Geography Skills

Movement The Safavid Empire expanded largely through warfare.
1. **Identify** Which two groups advanced into Safavid territory?
2. **Draw Conclusions** How did the Ottoman advance on Esfahan threaten the Safavids?

go.hrw.com KEYWORD: SK7 CH7

Akbar's tolerance allowed Muslims and Hindus in the empire to live in peace. In time, cooperation between the two groups helped create a unique Mughal culture. It blended Persian, Islamic, and Hindu elements. The Mughals became known for their monumental works of architecture. One famous example of this architecture is the Taj Mahal, a tomb built in the 1600s by emperor Shah Jahan for his wife. Its graceful domes and towers are a symbol of India today.

In the late 1600s, an emperor reversed Akbar's tolerant policies. He destroyed many Hindu temples, and violent revolts broke out. The Mughal Empire fell apart.

READING CHECK **Analyzing** How did the Ottomans gain land for their empire?

SUMMARY AND PREVIEW Islam spread beyond Arabia through warfare and trade. The Ottomans, Safavids, and Mughals built empires and continued the spread of Islam. In Section 4, you will learn about the cultural achievements of the Islamic world.

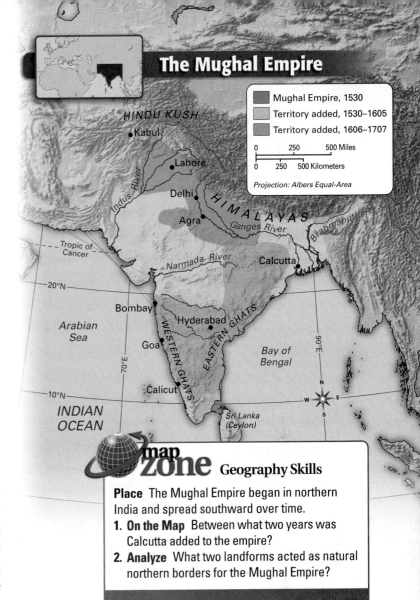

The Mughal Empire

▨	Mughal Empire, 1530
▢	Territory added, 1530–1605
▨	Territory added, 1606–1707

0   250   500 Miles
0   250   500 Kilometers
Projection: Albers Equal-Area

map **zone** Geography Skills

Place The Mughal Empire began in northern India and spread southward over time.
1. **On the Map** Between what two years was Calcutta added to the empire?
2. **Analyze** What two landforms acted as natural northern borders for the Mughal Empire?

go.hrw.com
Online Quiz
KEYWORD: SK7 HP7

Section 3 Assessment

Reviewing Ideas, Terms, and Places

1. **a. Define** What is a **caliph**?
   b. Evaluate Do you think the rules that Muslims made for conquered non-Muslims were fair? Why or why not?
2. **a. Identify** Name three places Islam spread to through trade.
   b. Explain How did trade help spread Islam?
   c. Elaborate What was life in **Córdoba** like?
3. **a. Recall** Who were the **janissaries**?
   b. Contrast How did Sunni and Shia beliefs about caliphs differ?
   c. Evaluate Which of the Muslim empires do you think made the greatest achievements? Why?

Critical Thinking

4. **Comparing and Contrasting** Draw a chart like the one below. Use your notes to compare and contrast the Ottoman, Safavid, and Mughal empires.

	Ottomans	Safavids	Mughals
Leaders			
Location			
Religious policy			

FOCUS ON WRITING

5. **Collecting Information about Empires** You will need one Web page on Muslim empires. Note one or two points you will make about each empire.

Cultural Achievements

What You Will Learn...

Main Ideas

1. Muslim scholars made lasting contributions to the fields of science and philosophy.
2. In literature and the arts, Muslim achievements included beautiful poetry, memorable short stories, and splendid architecture.

The Big Idea

Muslim scholars and artists made important contributions to science, art, and literature.

Key Terms

Sufism, *p. 187*
minarets, *p. 189*
calligraphy, *p. 189*

TAKING NOTES As you read, take notes on the achievements and advances the Muslims made in various fields. In each outer circle of this word web, describe one achievement or advance. You may need to add more circles.

Achievements and Advances

If YOU lived there...

You are a servant in the court of a powerful Muslim ruler. Your life at court is comfortable, though not one of luxury. Now the ruler is sending your master to explore unknown lands and distant kingdoms. The dangerous journey will take him across seas and deserts. He can take only a few servants with him. He has not ordered you to come but has given you a choice.

Would you join the expedition? Why or why not?

BUILDING BACKGROUND Muslim explorers traveled far and wide to learn about new places. They used what they learned to make maps. Their contributions to geography were just one way Muslim scholars made advancements in science and learning.

Science and Philosophy

The empires of the Islamic world made great advances in many fields—astronomy, geography, philosophy, math, and science. Scholars at Baghdad and Córdoba translated ancient writings on these subjects into Arabic. Scholars all over the Arabic world then used these ancient writings as the bases for their own works.

Islamic Achievements

Astronomy

The use of observatories allowed Muslim scientists to make other significant advances in astronomy. This observatory was built in the 1700s in Delhi, the capital of Mughal India.

Astronomy

Many Muslim cities had observatories. In these observatories, Muslim scientists worked to increase their knowledge of astronomy. Their study of the sky had practical benefits as well. For example, scientists used astronomy to improve their understanding of time, which let them build better clocks. They also improved the astrolabe, a device that allowed people to calculate their location on Earth.

Geography

Studying astronomy also helped Muslims explore the world. As people learned to use the stars to calculate time and location, merchants and explorers began to travel widely. The explorer Ibn Battutah traveled to Africa, India, China, and Spain in the 1320s. As a result of such travels, Muslim geographers made more accurate maps than were available before.

Philosophy

Many great thinkers lived in the Muslim world. Some studied **classical** writings and, like the ancient Greeks, believe in the importance of reason. Other philosophies taught that religion was more important than science. One of these philosophies was **Sufism** (SOO-fi-zuhm), which taught people they could find God's love by having a personal relationship with God.

Math

Muslim scholars also made advances in mathematics. For example, in the 800s they combined the ancient Indian system of numbers—including the use of zero—with Greek mathematical ideas. The results of these Muslim advances still affect how we think of math today. The number system we use is based on ancient Muslim writings. In addition, the field of algebra, an advanced type of mathematics, was first developed by Muslim scholars.

Medicine

The greatest of all Muslim achievements may have come in medicine. They based their medical skills on ancient Greek and Indian knowledge and added many new discoveries of their own.

Muslim doctors started the world's first pharmacy school to teach people how to make medicine. They built hospitals and learned to cure many serious diseases, such as smallpox. A Muslim doctor known in the West as Avicenna (av-uh-SEN-uh) recorded medical knowledge in an encyclopedia. It was used throughout Europe until the 1600s and is one of the most famous books in the history of medicine.

READING CHECK Drawing Conclusions
How did Muslims influence the fields of science and medicine?

THE IMPACT TODAY
We still call the numerals 0, 1, 2, 3, 4, 5, 6, 7, 8, and 9 Arabic or Hindu-Arabic numerals.

ACADEMIC VOCABULARY
classical referring to the cultures of ancient Greece or Rome

Geography
Muslim travelers collected much information about the world, some of which was used to make this map. New and better maps led to even more travel and a greater understanding of the world's geography.

Medicine
Muslim doctors made medicines from plants like this mandrake plant, which was used to treat pain and illnesses. They developed better ways to prevent, diagnose, and treat many diseases.

The Blue Mosque

The Blue Mosque in Istanbul was built in the early 1600s for an Ottoman sultan. It upset many people at the time it was built because they thought its six minarets—instead of the usual four—were an attempt to make it as great as the mosque in Mecca.

Domes are a common feature of Islamic architecture. Huge columns support the center of this dome, and more than 250 windows let light into the mosque.

The mosque gets its name from its beautiful blue Iznik tiles.

Tall towers called minarets are found outside many mosques.

The most sacred part of a mosque is the mihrab, the niche that points the way to Mecca. These men are praying facing the mihrab.

ANALYSIS SKILL **ANALYZING VISUALS**

Why do you think the decoration of the Blue Mosque is so elaborate?

Literature and the Arts

In addition to scientific achievements, the Muslims made great advances in the arts. Some of these artistic advances can be seen in literature and the visual arts.

Literature

Literature, especially poetry, was popular in the Muslim world. Much of this poetry was influenced by Sufism. Sufi poets often wrote about their loyalty to God. One of the most famous Sufi poets was Omar Khayyám (oh-mahr ky-AHM). In a book of poems known as the *Rubáiyát*, he wrote about faith, hope, and other emotions.

Muslims also enjoyed reading short stories. Many stories are collected in *The Thousand and One Nights*. This collection includes tales about legendary characters such as Sinbad, Aladdin, and Ali Baba.

Visual Arts

Of the visual arts, architecture was the most important in the Muslim world. Rich Muslim rulers used their wealth to have beautiful mosques built to honor God and inspire religious followers. Many mosques feature large domes and graceful **minarets**, **tall towers from where Muslims are called to prayer.**

Muslim architects also built palaces, marketplaces, and libraries. Many of these buildings have complicated domes and arches, colored bricks, and decorated tiles.

Although most Muslim buildings were highly decorated, most Muslim art does not show any people or animals. Muslims think only God can create humans and animals or their images. Instead, Muslim artists created complex geometric patterns. Muslim artists also turned to **calligraphy, or decorative writing.** They made sayings from the Qur'an into works of art to decorate mosques and other buildings.

Muslim art and literature combined Islamic influences with regional traditions of the places Muslims conquered. This mix of Islam with cultures from Asia, Africa, and Europe gave literature and the arts a unique style and character.

READING CHECK Generalizing What were two Muslim artistic achievements?

SUMMARY AND PREVIEW The Muslims made great advances in science and art. In the next chapter, you'll learn about an area where many of these advances were made—the Eastern Mediterranean.

go.hrw.com
Online Quiz
KEYWORD: SK3 HP7

Section 4 Assessment

Reviewing Ideas, Terms, and Places

1. **a. Identify** Who traveled to India, Africa, China, and Spain and contributed his knowledge to the study of geography?
   b. Explain How did Muslim scholars help preserve learning from the ancient world?
   c. Rank In your opinion, what was the most important Muslim scientific achievement? Why?
2. **a. Describe** What function do **minarets** serve in mosques?
   b. Explain How did Muslim artists create art without showing humans or animals?

Critical Thinking

3. **Analyzing** Using your notes, complete a chart like the one below. For each category in the first column, list one important achievement or advance the Muslims made.

Category	Achievement or Advance
Astronomy	
Geography	
Math	
Medicine	
Philosophy	

FOCUS ON WRITING

4. **Describing Muslim Achievements and Advances** Review your notes on key Muslim achievements and advances in science, philosophy, literature, and the arts. Now decide what information about each of these topics you will include on your Web site.

Social Studies Skills

Chart and Graph | Critical Thinking | Geography | Study

Outlining

Learn

The chapters in your textbooks are full of facts and ideas. Sometimes keeping track of all the information that you read can be overwhelming. At these times, it may help you to construct an outline of what you are reading.

An outline lists the main ideas of a chapter and the details that support those main ideas. The most important ideas are labeled with Roman numerals (I, II, III, and so on). Supporting ideas are listed below the main ideas, indented and labeled with capital letters (A, B, C, and so on). Less important details are indented farther and labeled with numbers (1, 2, 3) and lowercase letters (a, b, c). By arranging the ideas in an outline, you can see which are most important and how various ideas are related.

Practice

To the right is a partial outline of the discussion titled Literature and the Arts in Section 4 of this chapter. Study the outline and then answer the following questions.

1 What are the major ideas on this outline? How are they marked?

2 What details were listed to support the first main idea?

3 How are the heads on the outline related to the heads in the text of the discussion?

Literature and the Arts
I. Literature
  A. Poetry
  B. Short Stories
    1. *The Thousand and One Nights*
    2. Stories about Sinbad, Aladdin, Ali Baba
II. Visual Arts
  A. Architecture
    1. Mosques
      a. Large domes
      b. Minarets
    2. Palaces, marketplaces, libraries
  B. Art
    1. No people or animals
    2. Complex geometric patterns
    3. Calligraphy

Apply

Read back over the discussion titled Science and Philosophy in Section 4 of this chapter. Create an outline of this discussion. Before you write your outline, decide what you will use as your main heads. Then fill in the details below each of the heads.

Chapter Review

Geography's Impact
video series
Review the video to answer the closing question:
Why might the pilgrimage to Mecca mean so much to the Muslims who go there?

Visual Summary

Use the visual summary below to help you review the main ideas of the chapter.

QUICK FACTS

Islam was first taught by Muhammad. Its teachings are found in the Qur'an and the Sunnah.

From Arabia, Islam spread into many parts of the world. Muslims ruled great empires in Asia, Europe, and Africa.

Muslim scholars and artists made great achievements in science, medicine, math, philosophy, and the arts.

Reviewing Vocabulary, Terms, and Places

For each statement below, write T if it is true and F if it is false. If the statement is false, write the correct term that would make the sentence a true statement.

1. Muslims gather to pray at a **jihad**.

2. Traders often traveled in **caravans** to take their goods to markets.

3. An **Islam** is a person who submits to God and follows the teachings of Muhammad.

4. According to Islamic belief, God's messages to Muhammad during his lifetime make up the **Sunnah**.

5. A **caliph** is a journey to a sacred place.

6. A **minaret** is a tower from where Muslims are called to prayer.

7. **Janissaries** converted to Islam and became fierce warriors in the Ottoman army.

8. The hajj is a pilgrimage to Islam's most sacred city, **Baghdad**.

Comprehension and Critical Thinking

SECTION 1 *(pages 170–173)*

9. **a. Recall** According to Muslim belief, how was Islam revealed to Muhammad?

b. Analyze How did Muhammad encourage people to treat each other?

c. Evaluate What are some possible benefits to a nomadic lifestyle, and what are some possible benefits to a town lifestyle?

SECTION 2 *(pages 174–177)*

10. **a. Define** What is the hajj?

b. Contrast Both the Qur'an and the Sunnah have guided Muslims' behavior for centuries. Apart from discussing different topics, how do these two differ?

c. Predict Which of the Five Pillars of Islam do you think would be the most difficult to perform? Why?

SECTION 3 *(pages 180–185)*

11. a. Identify Who was Abu Bakr, and why is he important in the history of Islam?

b. Analyze Why did the Safavids come into conflict with the Ottomans?

c. Evaluate In your opinion, was conquest or trade more effective in spreading Islam? Why?

SECTION 4 *(pages 186–189)*

12. a. Describe What are two elements often found in Muslim architecture?

b. Draw Conclusions How did having a common language help scholars in the Islamic world?

c. Elaborate Why might a ruler want to use his or her wealth to build a mosque?

Social Studies Skills

13. Outlining Find the discussion titled The Five Pillars of Islam in Section 2 of this chapter. If you were going to outline this discussion, what would you use as your main ideas?

FOCUS ON READING AND WRITING

14. Sequencing Arrange the following list of events in the order in which they happened. Then write a brief paragraph describing the events, using clue words such as after, *then,* and *later* to show the proper sequence.

- Muhammad moves to Medina.
- Muhammad begins to teach about Islam.
- Muhammad works as a merchant.
- Muhammad becomes a political leader.

15. Creating Your Web Site You have now collected information on Muhammad, the religion of Islam, major Muslim empires, and Muslim achievements. Create a home page and one Web page on each of these topics. You can write about your topics in paragraph form or in a list of bullet points. You may design the pages either online or on sheets of paper. Remember that your audience is children, so you should keep your sentences simple.

Using the Internet

go.hrw.com
KEYWORD: SK7 CH7

16. Activity: Researching Muslim Achievements Muslim advances in science, math, and art were spread around the world both by explorers and by traders. Enter the activity keyword and learn about some of these advances. Choose an object created by Muslim scholars in the 600s or 700s and write a paragraph that explains its roots and how it spread to other cultures. End your paragraph with a discussion of how the object is used in modern times.

Map Activity ⚐Interactive

17. The Islamic World On a separate sheet of paper, match the letters on the map with their correct labels.

Mecca   Medina   Red Sea

Persian Gulf   Arabian Sea

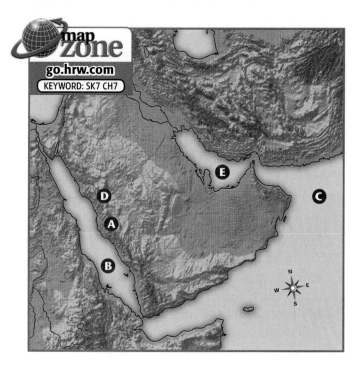

DIRECTIONS: Read questions 1 through 7 and write the letter of the best response. Then read question 8 and write your own well-constructed response.

1 During the month of Ramadan, Muslims
   A fast.
   B do not pray.
   C travel to Medina.
   D hold feasts.

2 The teachings of Muhammad are found mainly in the Qur'an and the
   A Commentaries.
   B Sunnah.
   C Analects.
   D Torah.

3 Which Muslim empire was located in India?
   A Ottoman Empire
   B Mughal Empire
   C Safavid Empire
   D Córdoba Empire

4 Muslim scholars are credited with developing
   A geometry.
   B algebra.
   C calculus.
   D physics.

5 The most sacred city in Islam is
   A Baghdad.
   B Mecca.
   C Medina.
   D Esfahan.

Travels in Asia and Africa

"From Tabuk the caravan travels with great speed night and day, for fear of this desert. Halfway through is the valley of al-Ukhaydir . . . One year the pilgrims suffered terribly here from the samoom-wind; the water-supplies dried up and the price of a single drink rose to a thousand dinars, but both seller and buyer perished. Their story is written on a rock in the valley."

—Ibn Battutah, from *The Travels*

6 From the passage above, you can conclude that the climate near Tabuk is
   A mild and sunny.
   B cold and wet.
   C hot and dry.
   D cool and pleasant.

7 Which of the following people was known as a great traveler and geographer?
   A Abu Bakr
   B 'Abbas
   C Omar Khayyám
   D Ibn Battutah

8 **Extended Response** Look back at Section 3 of this chapter and read the discussion of the Ottoman and Safavid empires again. Write a paragraph that notes one way in which the Ottoman Empire and the Safavid Empire were similar. Then describe two ways in which they were different.

CHAPTER 8

The Eastern Mediterranean

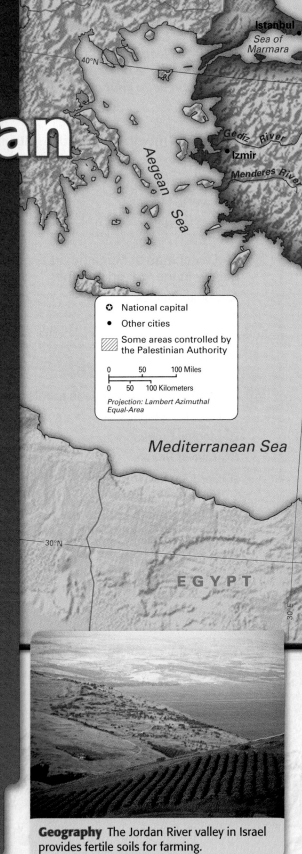

EUROPE

Istanbul
Sea of Marmara
40°N
Aegean Sea
Izmir
Gediz River
Menderes River

National capital
Other cities
Some areas controlled by the Palestinian Authority

0   50   100 Miles
0   50   100 Kilometers

Projection: Lambert Azimuthal Equal-Area

Mediterranean Sea

30°N

EGYPT

30°E

What You Will Learn...

In this chapter you will learn about the countries of the Eastern Mediterranean region—Turkey, Israel, Syria, Lebanon, and Jordan. You will study their physical geography, history, government, economy, and culture.

SECTION 1
Physical Geography 196

SECTION 2
Turkey 200

SECTION 3
Israel 204

SECTION 4
Syria, Lebanon, and Jordan 210

FOCUS ON READING AND WRITING

Setting a Purpose Good readers often set a purpose before they read. Ask yourself, "Why am I reading this chapter?" For example, you might want to learn about the geography of a country. Keeping your purpose in mind will help you focus on what is important. **See the lesson, Setting a Purpose, on page 675.**

Writing a Description As your read this chapter, you will collect information about the lands and people in this region. Later you will write a description of these lands and people. You will be writing for readers who have not read the chapter or visited the region.

Geography The Jordan River valley in Israel provides fertile soils for farming.

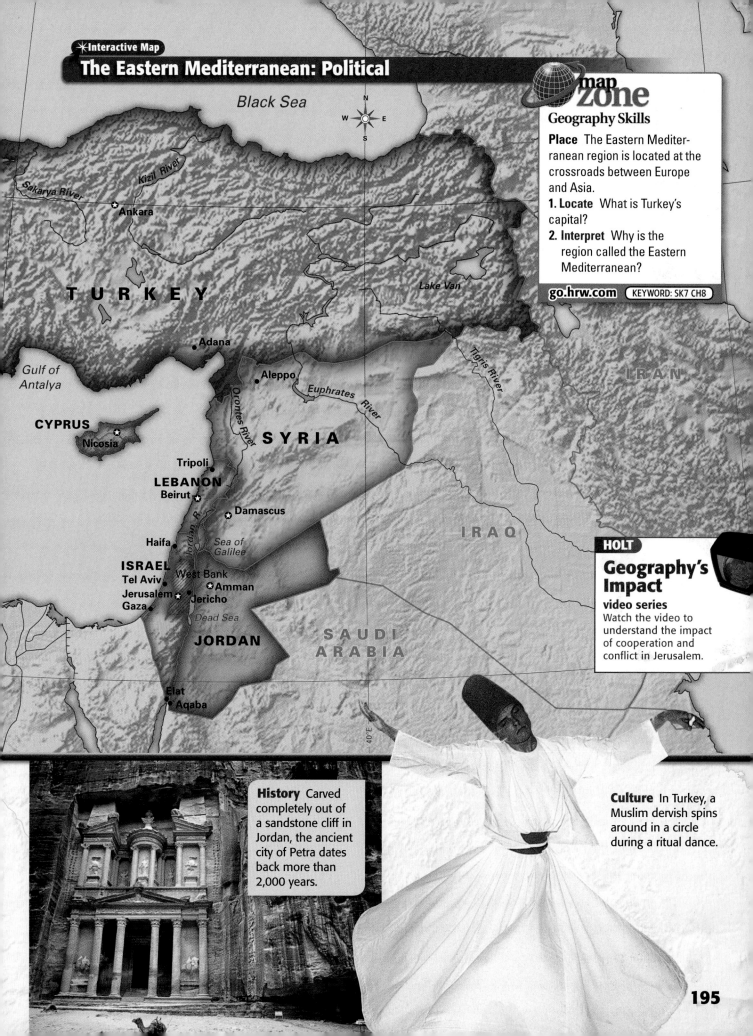

The Eastern Mediterranean: Political

Black Sea

map zone

Geography Skills

Place The Eastern Mediterranean region is located at the crossroads between Europe and Asia.

1. Locate What is Turkey's capital?

2. Interpret Why is the region called the Eastern Mediterranean?

go.hrw.com | KEYWORD: SK7 CH8

Sakarya River

Kizii River

Ankara

T U R K E Y

Lake Van

Adana

Aleppo

Euphrates River

Tigris River

I R A N

Gulf of Antalya

Orontes River

S Y R I A

CYPRUS

Nicosia

Tripoli

LEBANON

Beirut

Damascus

Jordan R.

Haifa

Sea of Galilee

I R A Q

ISRAEL

Tel Aviv

West Bank

Jerusalem

Amman

Gaza

Jericho

Dead Sea

JORDAN

S A U D I A R A B I A

40°E

Elat

Aqaba

HOLT

Geography's Impact

video series

Watch the video to understand the impact of cooperation and conflict in Jerusalem.

History Carved completely out of a sandstone cliff in Jordan, the ancient city of Petra dates back more than 2,000 years.

Culture In Turkey, a Muslim dervish spins around in a circle during a ritual dance.

Physical Geography

What You Will Learn...

Main Ideas

1. The Eastern Mediterranean's physical features include the Bosporus, the Dead Sea, rivers, mountains, deserts, and plains.
2. The region's climate is mostly dry with little vegetation.
3. Important natural resources in the Eastern Mediterranean include valuable minerals and the availability of water.

The Big Idea

The Eastern Mediterranean, a region with a dry climate and valuable resources, sits in the middle of three continents.

Key Terms and Places

Dardanelles, *p. 196*
Bosporus, *p. 196*
Jordan River, *p. 197*
Dead Sea, *p. 197*
Syrian Desert, *p. 198*

 TAKING NOTES As you read, take notes on the physical features, climate and vegetation, and natural resources of the region.

Physical Features	Climate and Vegetation	Natural Resources

If **YOU** lived there...

You live in Izmir, Turkey, on the Aegean Sea, but are traveling into the far eastern part of the country called eastern Anatolia. At home you are used to a warm, dry Mediterranean climate. You are surprised by the colder and wetter climate you're experiencing. Two mountain ranges come together here, and you notice that the peaks are covered with snow.

How does geography affect climate in these two places?

BUILDING BACKGROUND The Eastern Mediterranean region lies at the crossroads of Europe, Africa, and Asia. In ancient times, Greek colonists settled here, and it was later part of the Roman Empire. Geographically, however, it is almost entirely in Southwest Asia.

The countries of the Eastern Mediterranean make up part of a larger region called Southwest Asia. This region is sometimes referred to as the Middle East. Europeans first called the region the Middle East to distinguish it from the Far East, which included China and Japan.

Physical Features

As you can see on the physical map on the next page, a narrow waterway separates Europe from Asia. This waterway is made up of the **Dardanelles** (dahrd-uhn-ELZ), the **Bosporus** (BAHS-puh-ruhs), and the **Sea of Marmara** (MAHR-muh-ruh). Large ships travel through the waterway, which connects the Black Sea to the Mediterranean Sea. The Bosporus also splits the country of Turkey into two parts, a small part lies in Europe and the rest in Asia. The Asian part of Turkey includes the large peninsula called Anatolia (a-nuh-TOH-lee-uh).

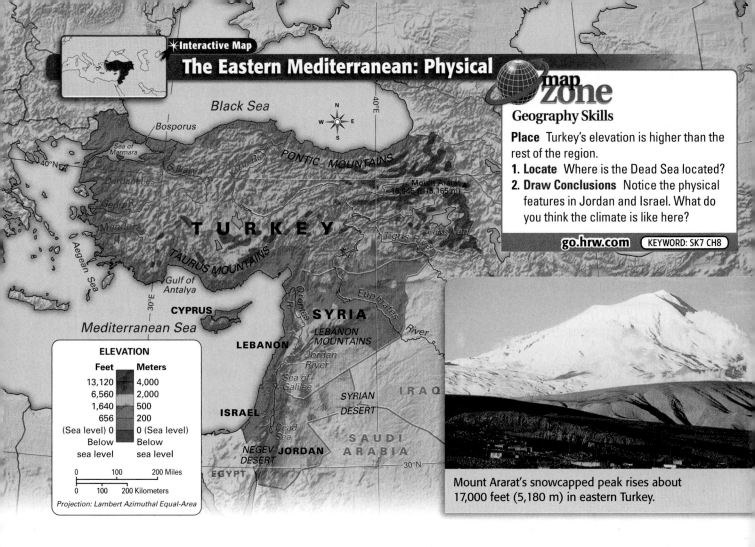

The Eastern Mediterranean: Physical

map **zone**

Geography Skills

Place Turkey's elevation is higher than the rest of the region.
1. **Locate** Where is the Dead Sea located?
2. **Draw Conclusions** Notice the physical features in Jordan and Israel. What do you think the climate is like here?

go.hrw.com | KEYWORD: SK7 CH8

ELEVATION

Feet	Meters
13,120	4,000
6,560	2,000
1,640	500
656	200
(Sea level) 0	0 (Sea level)
Below sea level	Below sea level

0   100   200 Miles
0   100   200 Kilometers
Projection: Lambert Azimuthal Equal-Area

Mount Ararat's snowcapped peak rises about 17,000 feet (5,180 m) in eastern Turkey.

Rivers and Lakes

The **Jordan River** begins in Syria and flows south through Israel and Jordan. The river finally empties into a large lake called the **Dead Sea**. As its name suggests, the Dead Sea contains little life. Only bacteria lives in the lake's extremely salty water. The world's saltiest lake, its surface is 1,312 feet (400 m) below sea level—the lowest point on any continent.

Mountains and Plains

As you can see on the map, two mountain systems stretch across Turkey. The Pontic Mountains run east–west along the northern edge. The Taurus Mountains run east–west along the southern edge.

Heading south from Turkey and into Syria lies a narrow plain. The Euphrates River flows southeast from Turkey through the plains to Syria and beyond.

Dead Sea

Because of its high salt content, swimmers do not sink in the Dead Sea.

ANALYZING VISUALS
What appears on the shore of the Dead Sea?

FOCUS ON READING

Set a purpose for reading the paragraphs under Climate and Vegetation.

Farther inland lies plateaus, hills, and valleys. A rift valley that begins in Africa extends northward into Syria. Hills rise on both sides of the rift. Two main mountain ridges run north–south. One runs from southwestern Syria through western Jordan. The other, closer to the coast, runs through Lebanon and Israel.

READING CHECK **Summarizing** What are the region's main physical features?

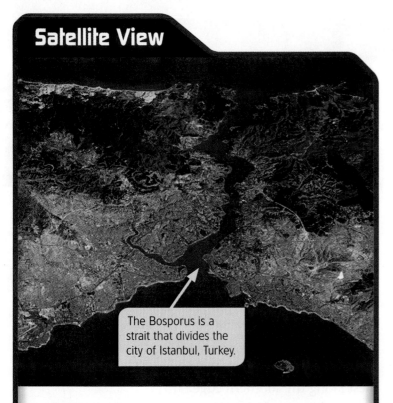

The Bosporus is a strait that divides the city of Istanbul, Turkey.

Istanbul and the Bosporus

Throughout history, geography has almost always determined the location of a city. Istanbul, Turkey, which sits between Europe and Asia, is no exception. In this satellite image the city of Istanbul appears light brown and white. The body of water that cuts through the city is a strait called the Bosporus. It separates the Sea of Marmara in the south with the Black Sea in the north. Historically, the Bosporus has served as a prized area for empires that have controlled the city. Today, the strait is a major shipping route.

Drawing Conclusions Why do you think the Bosporus was seen as a strategic location?

Climate and Vegetation

The Eastern Mediterranean is a mostly dry region. However, there are important variations. As you can see on the map on the next page, Turkey's Black Sea coast and the Mediterranean coast all the way to northern Israel have a Mediterranean climate. Much of interior Turkey experiences a steppe climate. Central Syria and lands farther south have a desert climate. A small area of northeastern Turkey has a humid subtropical climate.

The region's driest areas are its deserts. Much of Syria and Jordan is covered by the **Syrian Desert**. This desert of rock and gravel usually receives less than five inches (12.7 cm) of rainfall a year. Another desert, the Negev (NE-gev), lies in southern Israel. Here the temperatures can reach as high as 114°F (46°C), and annual rainfall totals barely two inches.

In such dry conditions, only shrubs grow scattered throughout the region's deserts. However, in other areas vegetation is plentiful. In Israel, more than 2,800 species of plants thrive throughout the country's various environments.

READING CHECK **Generalizing** What are climates like in the Eastern Mediterranean?

Natural Resources

Because the Eastern Mediterranean is so dry, water is a valuable resource. The people of this region are mostly farmers. The region lacks oil resources, but does have valuable minerals.

Land and Water

In this dry region the limited availability of water limits how land is used. Commercial farms can only grow crops where rain or irrigation provides enough water.

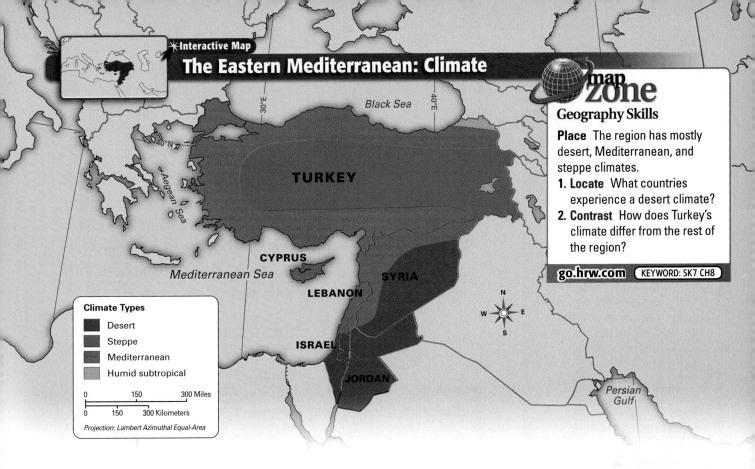

The Eastern Mediterranean: Climate

Black Sea

TURKEY

Aegean Sea

CYPRUS

Mediterranean Sea

SYRIA

LEBANON

ISRAEL

JORDAN

Persian Gulf

Climate Types
- Desert
- Steppe
- Mediterranean
- Humid subtropical

0   150   300 Miles
0   150   300 Kilometers
Projection: Lambert Azimuthal Equal-Area

map zone

Geography Skills

Place The region has mostly desert, Mediterranean, and steppe climates.

1. **Locate** What countries experience a desert climate?
2. **Contrast** How does Turkey's climate differ from the rest of the region?

go.hrw.com KEYWORD: SK7 CH8

In drier areas, subsistence farming and livestock herding are common. In the desert areas, available water supports a few nomadic herders, but no farming.

Mineral Resources

The region's resources include many minerals, including sulfur, mercury, and copper. Syria, Jordan, and Israel all produce phosphates—mineral salts that contain the element phosphorus. Phosphates are used to make fertilizers. This region also produces asphalt—the dark tarlike material used to pave streets.

READING CHECK **Drawing Conclusions** How do people use the region's mineral resources?

SUMMARY AND PREVIEW In this section you learned about the physical geography of the Eastern Mediterranean. Next, you will learn about Turkey.

Section 1 Assessment

go.hrw.com
Online Quiz
KEYWORD: SK7 HP8

Reviewing Ideas, Terms, and Places

1. **a. Describe** What makes the **Dead Sea** unusual?
   b. Explain What physical features separate Europe and Asia?
2. **a. Recall** What desert covers much of Syria and Jordan?
   b. Make Generalizations What is the climate of the Eastern Mediterranean like?
3. **a. Identify** What mineral resource is produced by Syria, Jordan, and Israel?
   b. Draw Conclusions Why must farmers in the region rely on irrigation?

Critical Thinking

4. **Summarizing** Using your notes, summarize the physical geography of Israel and Turkey. Use this chart to organize your notes.

Physical Features	
Turkey	Israel

FOCUS ON WRITING

5. **Describing the Physical Geography** What physical features would you include in your description? How would you describe the climate? Note your ideas.

THE EASTERN MEDITERRANEAN **199**

Turkey

What You Will Learn...

Main Ideas

1. Turkey's history includes invasion by the Romans, rule by the Ottomans, and a twentieth-century democracy.
2. Turkey's people are mostly ethnic Turks, and its culture is a mixture of modern and traditional.
3. Today, Turkey is a democratic nation seeking economic opportunities as a future member of the European Union.

The Big Idea

Although Turkey has historically been more Asian than European, its leaders are seeking to develop closer economic ties to Europe.

Key Terms and Places

Ankara, *p. 202*
Istanbul, *p. 203*
secular, *p. 203*

TAKING NOTES As you read, use a diagram like the one below to take notes on Turkey.

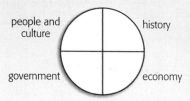

people and culture | history

government | economy

If YOU lived there...

Your cousins from central Turkey are coming to visit your hometown, Istanbul. You think your city is both beautiful and interesting. You like to stroll in the Grand Bazaar and smell the spices for sale. You admire the architecture of the Blue Mosque, whose walls are lined with thousands of tiny tiles. You also like to visit the elegant Topkapi Palace, where sultans once lived.

What sights will you show your cousins?

BUILDING BACKGROUND Many sites in Turkey reflect the country's long and diverse cultural history. Throughout the country you will find the ruins of ancient Greek temples and Roman palaces. You can also see magnificent early Christian buildings and art, as well as the palaces and mosques of Ottoman rulers.

Close-up

Early Farming Village

The village of Çatal Hüyük in modern Turkey is one of the earliest farming villages discovered. Around 8,000 years ago, the village was home to about 5,000–6,000 people living in more than 1,000 houses. Villagers farmed, hunted and fished, traded with distant lands, and worshipped gods in special shrines.

Villagers used simple channels to move water to their fields.

Wheat, barley, and peas were some of the main crops grown outside the village.

History

Around 8,000 years ago the area that is now Turkey was home to one of the world's earliest farming villages. For centuries invasions from powerful empires shaped the region. By the 1920s Turkey was a democratic nation.

Invasions

When the Romans invaded the area, they captured the city of Byzantium and later renamed it Constantinople. Its location at the crossroads between Europe and Asia made Constantinople an important trading port. After the fall of Rome, Constantinople became the capital of the Byzantine Empire.

In the AD 1000s a nomadic people from central Asia called the Seljuk Turks invaded the area. In 1453 another Turkish people, the Ottoman Turks, captured the city of Constantinople and made it the capital of their Islamic empire.

BIOGRAPHY

Kemal Atatürk
(1881–1938)

Known as the Father of the Turks, Kemal Atatürk was Turkey's first president. As president, he modernized Turkey, which dramatically changed Turkish way of life. Atatürk separated all aspects of Islam from Turkey's government. He even closed Islamic schools. Turkey's people were also encouraged to wear Western dress and adopt surnames.

Generalizing How did Atatürk change Turkey's government?

The Ottoman Empire

During the 1500s and 1600s the Ottoman Empire was very powerful. The empire controlled territory in northern Africa, southwestern Asia, and southeastern Europe.

In World War I the Ottomans fought on the losing side. When the war ended, they lost most of their territory.

Houses were made of wood covered with mud. Since they didn't have doors, people entered on ladders through rooftop openings.

Inside their houses, villagers made the earliest-known wooden bowls and cups, pottery, and mirrors.

Some houses were built as shrines and had small statues of goddesses and large sculpted bulls' heads.

ANALYSIS SKILL **ANALYZING VISUALS**

How did farmers get water to their fields?

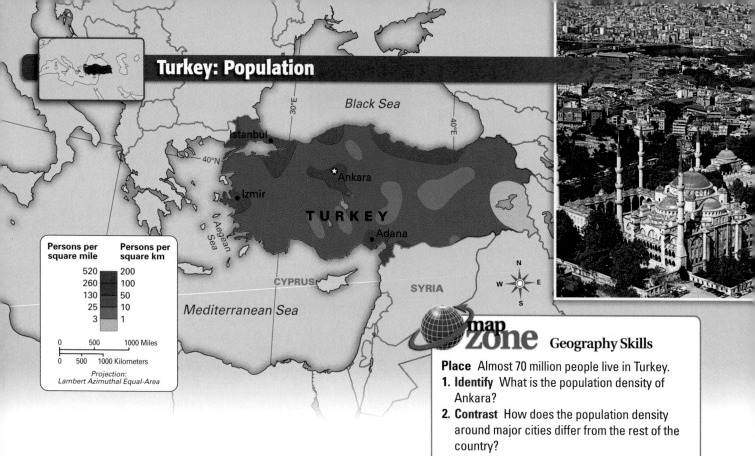

Turkey: Population

Black Sea

Istanbul

Ankara

Izmir

TURKEY

Adana

Persons per square mile / **Persons per square km**

Persons per square mile	Persons per square km
520	200
260	100
130	50
25	10
3	1

0 500 1000 Miles
0 500 1000 Kilometers

Projection:
Lambert Azimuthal Equal-Area

CYPRUS

Mediterranean Sea

SYRIA

map zone Geography Skills

Place Almost 70 million people live in Turkey.
1. **Identify** What is the population density of Ankara?
2. **Contrast** How does the population density around major cities differ from the rest of the country?

FOCUS ON READING

Set a purpose for reading by asking yourself what you want to learn about Turkey's people and culture.

Military officers then took over the government, led by a war hero, Mustafa Kemal. He later adopted the name Kemal Atatürk, which means Father of Turks. Atatürk created the democratic nation of Turkey and moved the capital to **Ankara** from Constantinople, which was renamed Istanbul.

Modern Turkey

ACADEMIC VOCABULARY

method a way of doing something

Atatürk believed Turkey needed to modernize and adopt Western **methods** in order to be a strong nation. For example, he banned the fez, the traditional hat of Turkish men, and required that they wear European-style hats. Reforms urged women to stop wearing traditional veils. Women were also encouraged to vote, work, and hold office. Other ways Atatürk modernized Turkey included replacing the Arabic alphabet with the Latin alphabet, and adopting the metric system.

READING CHECK **Finding Main Ideas** How did Atatürk modernize Turkey?

People and Culture

Most of Turkey's people are mostly ethnic Turks. Kurds are the largest minority and make up 20 percent of the population.

Turkey's culture today is a reflection of some of Kemal Atatürk's changes. He created a cultural split between Turkey's urban middle class and rural villagers. The lifestyle and attitudes of middle-class Turks have much in common with those of the European middle class. In contrast, most rural Turks are more traditional. Islam strongly influences their attitudes on matters such as the role of women.

Turkish cooking features olives, vegetables, cheese, yogurt, and bread. Kebabs—grilled meats on a skewer—are a favorite Turkish dish.

READING CHECK **Contrasting** How are urban Turks different from rural Turks?

With about 9 million people, Istanbul, shown here, is Turkey's largest city.

Turkey Today

Turkey's government meets in the capital of Ankara, but **Istanbul** is Turkey's largest city. Istanbul's location will serve as an economic bridge to Europe as Turkey plans to join the European Union.

Government

Turkey's legislature is called the National Assembly. A president and a prime minister share executive power.

Although most of its people are Muslim, Turkey is a secular state. **Secular** means that religion is kept separate from government. For example, the religion of Islam allows a man to have up to four wives. However, by Turkish law a man is permitted to have just one wife. In recent years Islamic political parties have attempted to increase Islam's role in Turkish society.

Economy and Resources

As a member of the European Union, Turkey's economy and people would benefit by increased trade with Europe. Turkey's economy includes modern factories as well as village farming and craft making.

Among the most important industries are textiles and clothing, cement, and electronics. About 40 percent of Turkey's labor force works in agriculture. Grains, cotton, sugar beets, and hazelnuts are major crops.

Turkey is rich in natural resources, which include oil, coal, and iron ore. Water is also a valuable resource in the region. Turkey has spent billions of dollars building dams to increase its water supply. On one hand, these dams provide hydroelectricity. On the other hand, some of these dams have restricted the flow of river water into neighboring countries.

READING CHECK **Finding Main Ideas** What kind of government does Turkey have?

SUMMARY AND PREVIEW In this section you learned about Turkey's history, people, government, and economy. Next, you will learn about Israel.

Section 2 Assessment

go.hrw.com
Online Quiz
KEYWORD: SK7 HP8

Reviewing Ideas, Terms, and Places

1. **a. Recall** What city did both the Romans and Ottoman Turks capture?
   b. Explain In what ways did Atatürk try to modernize Turkey?
2. **a. Recall** What ethnic group makes up 20 percent of Turkey's population?
   b. Draw Conclusions What makes Turkey **secular**?
   c. Elaborate Why do you think Turkey wants to be a member of the European Union?

Critical Thinking

3. **Summarizing** Using the information in your notes, summarize Turkey's history and Turkey today.

Turkey's History	Turkey Today

FOCUS ON WRITING

4. **Describing Turkey** A description of Turkey might include details about its people, culture, government, and economy. Take notes on the details you think are important and interesting.

Israel

What You Will Learn...

Main Ideas

1. Israel's history includes the ancient Hebrews and the creation of the nation of Israel.
2. In Israel today, Jewish culture is a major part of daily life.
3. The Palestinian Territories are areas within Israel controlled partly by Palestinian Arabs.

The Big Idea

Israel and the Palestinian Territories are home to Jews and Arabs who continue to struggle over the region's land.

Key Terms and Places

Diaspora, *p. 204*
Jerusalem, *p. 204*
Zionism, *p. 205*
kosher, *p. 206*
kibbutz, *p. 206*
Gaza, *p. 207*
West Bank, *p. 207*

TAKING NOTES As you read, take notes on Israel and the Palestinian Territories. Use the chart below to organize your notes.

Israel	Palestinian Territories

If **YOU** lived there...

When you were only six years old, your family moved to Israel from Russia. You are learning Hebrew in school, but your parents and grandparents still speak Russian at home. When you first moved here, your parents worked in an office building, but you now live on a farm where you grow oranges and tomatoes.

What do you like about living in Israel?

BUILDING BACKGROUND Modern Israel was formed in 1948. Since then immigrants from many parts of the world have made the population of Israel very diverse. Many Jews emigrated to Israel from Russia and Eastern European countries.

History

Do you know that Israel is often referred to as the Holy Land? Some people call Israel the Holy Land because it is home to sacred sites for three major religions—Judaism, Christianity, and Islam. According to the Bible, many events in Jewish history and in the life of the Jesus happened in Israel.

The Holy Land

The Hebrews, the ancestors of the Jews, first established the kingdom of Israel about 3,000 years ago. It covered roughly the same area as the modern state of Israel. In the 60s BC the Roman Empire conquered the region, which they called Palestine. After several Jewish revolts, the Romans forced many Jews to leave the region. This scattering of the Jewish population is known as the **Diaspora**.

Arabs conquered Palestine in the mid-600s. However, from the late 1000s to the late 1200s, Christians from Europe launched a series of invasions of Palestine called the Crusades. The Crusaders captured the city of **Jerusalem** in 1099. In time the Crusaders were pushed out of the area. Palestine then became part of the Ottoman Empire. After World War I, it came under British control.

Creation of Israel

Zionism, a movement calling for Jews to establish a country or community in Palestine, began in Europe in the late 1800s. Tens of thousands of Jews from around the world began moving to the region.

In 1948 Jewish leaders declared Palestine the nation of Israel. Arab Palestinians living in Palestine and the Arab countries that surrounded Israel were opposed to the new nation. The Arab countries invaded Israel with their armies. In a very short war, the Israelis defeated the Arab armies.

After the Jewish victory, many Palestinians fled to neighboring Arab countries. Israel and Arab countries have fought each other in several wars since then. Disputes between the two sides continue today.

READING CHECK **Summarizing** What two groups played a large role in Israel's history?

Israel Today

Jews from all over the world have settled in Israel hoping to find peace and stability. Yet, they have faced continual conflicts with neighboring countries. In spite of these problems, Israelis have built a modern, democratic country.

HISTORIC DOCUMENT
The Dead Sea Scrolls

Written by Jews about 2,000 years ago, the Dead Sea Scrolls include prayers, commentaries, letters, and passages from the Hebrew Bible. Hidden in caves near the Dead Sea, these scrolls were not found until 1947. Here are two passages from a prayer written on one of the scrolls.

❝With knowledge shall I sing out my music, only for the glory of God, my harp, my lyre for His holiness established; the flute of my lips will I lift, His law its tuning fork.**❞**

❝When first I begin campaign or journey, His name shall I bless; when first I set out or turn to come back; when I sit down or rise up, when I spread my bed, then shall I rejoice in Him.**❞**

ANALYSIS SKILL **ANALYZING PRIMARY SOURCES**

What does this prayer from the Dead Sea Scrolls reveal about the people who wrote it?

Government and Economy

Israel has a prime minister and a parliament—the Knesset. There are two major political parties and many smaller parties.

Israel's government has built a strong military. At age 18 most Israeli men and women must serve at least one year.

FOCUS ON READING

What do you want to find out about Israel today?

Jerusalem

The city of Jerusalem is sacred to three world religions—Judaism, Islam, and Christianity.

Israel's economy is modern and diverse. Items like high-technology equipment and cut diamonds are important exports. Israel has increased food production by irrigating farmland. Israel's economy also benefits from the millions of visitors who come to Israel to see the country's historic sites.

Cities, Diversity, and Languages

Most of Israel's population lives in cities. With about 2 million people, Tel Aviv is Israel's largest city.

About 80 percent of Israel's population is Jewish. The rest of the country's people are mostly Arab. About three-fourths of Israeli Arabs are Muslim, but some are Christian. Israel's Jewish population includes Jews from all parts of the world. Many arrive not knowing Hebrew, one of Israel's official languages. To assist these new citizens, the government provides language classes. Israeli Arabs speak Arabic, Israel's other official language.

Culture and Rural Settlements

Israeli Jewish culture is rich in holidays and special foods. For Jews, the Sabbath, or Saturday, is a holy day. Yom Kippur, a very important Jewish holiday, is celebrated in the fall. Passover, in the spring, celebrates the Hebrews' escape from captivity in ancient Egypt.

Because food is such a large part of Jewish culture, there are religious laws stating what Jews should not eat. These laws are ancient and appear in the Hebrew Bible. **Kosher**, which means "proper" in Hebrew, is the term used to refer to Jewish dietary laws. Jews eating a kosher diet do not eat pork or shellfish. They also do not mix meat and milk products.

About 100,000 Israeli Jews live in rural settlements. Each settlement, or **kibbutz** (ki-BOOHTS), is a large farm where people share everything in common. Israeli Jews live in more than 250 kibbutzim.

READING CHECK **Generalizing** What is Jewish culture in Israel like?

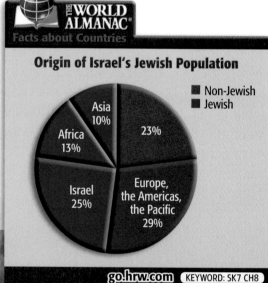

THE **WORLD ALMANAC**
Facts about Countries

Origin of Israel's Jewish Population

- Non-Jewish
- Jewish

Asia 10%
Africa 13%
Israel 25%
Europe, the Americas, the Pacific 29%
23%

go.hrw.com KEYWORD: SK7 CH8

Jews from all over the world have settled in Israel. The graph above shows the percentages of Jews who migrated from different places. Non-Jews in Israel include Arabs, Christians, and Druze. This photo shows a Jewish teenager celebrating his bar mitzvah—a ceremony that acknowledges 13-year-old Jewish boys as adults in the Jewish community.

ANALYZING VISUALS What part of the world did the majority of Israel's Jewish population emigrate from?

Israel's Population

The Palestinian Territories

In 1967 during the Six-Day War, Israel captured areas inhabited by Palestinian Arabs—Gaza, the West Bank, and East Jerusalem. Since then, Jews and Arabs have fought over the right to live in these areas.

Gaza

Gaza is a small, crowded piece of coastal land where more than a million Palestinians live. The area has almost no resources. However, citrus fruit is grown in irrigated fields. Unemployment is a problem for the Palestinians living in Gaza. Many travel to Israel each day to work.

West Bank

The **West Bank** is much larger than Gaza and has a population of about 2.4 million. It is mostly rural, but the territory has three large cities—Nablus, Hebron, and Ramallah. The West Bank's economy is mostly based on agriculture. Farmers rely on irrigation to grow their crops.

Since Israel took control of the West Bank, tens of thousands of Jews have moved into settlements there. However, the Palestinians consider the Jewish settlements an invasion of their land. This conflict over land causes the greatest tension and violence between Arabs and Israelis.

East Jerusalem

Other disputed land includes Israel's capital, Jerusalem. Control of Jerusalem is a difficult and often emotional question for Jews, Muslims, and Christians. The city contains sites that are holy to all three religions. Areas of the old city are divided into Jewish, Muslim, and Christian neighborhoods.

Palestinians claimed East Jerusalem as their rightful capital. However, Israel annexed East Jerusalem in 1980. Even before this, the Israeli government had moved its capital to Jerusalem from Tel Aviv. Most foreign countries have chosen not to recognize this transfer.

The Future of the Territories

In the 1990s Israel agreed to turn over parts of the territories to the Palestinians. In return, the Palestinian leadership—the Palestinian Authority—agreed to work for peace. In 2005 Israelis transferred Gaza to the Palestinian Authority.

Israel and the Palestinian Territories

map zone Geography Skills

Place Some areas of Israel are controlled by the Palestinian Authority.
1. **Name** What Palestinian territory is located on the Mediterranean Sea?
2. **Interpret** Who controls parts of the West Bank?

Some areas controlled by the Palestinian Authority

0   25   50 Miles

0   25   50 Kilometers

Projection: Cassini-Soldner Transverse Cylindrical

Israeli Teens for Peace

Peace between Israeli Jews and Palestinian Arabs has not been easy in the past. Moreover, some believe peace in the region might be impossible ever to accomplish. But don't tell that to a group of 200 Jewish and Arab teenagers who are making a difference in Israel. These teens belong to an organization called Seeds of Peace. To learn more about each other's cultures and thus understand each other better, these teens meet regularly. For example, Jews teach Arabs Hebrew and Arabs teach Jews Arabic. They also participate in community service projects.

By bridging the gap between their two cultures, these teens hope they can one day live peacefully together. A Palestinian boy in the group expressed his hope for the future. He explained, "I realize that peace is not a dream when you truly get to know who you are making peace with."

Drawing Conclusions How are Jewish and Arab teenagers in Israel working toward peace?

The future of the peace process is uncertain. Some Palestinian groups have continued to commit acts of terrorism. Jewish Israelis fear they would be open to attack if they withdrew from the territories.

READING CHECK **Analyzing** Why have the Palestinian Territories been a source of conflict?

SUMMARY AND PREVIEW In this section you learned about Israel's history, people, government and economy, and the future of the Palestinian Territories. In the next section you will learn about the history and culture of Israel's neighbors—Syria, Lebanon, and Jordan.

Section 3 Assessment

go.hrw.com
Online Quiz
KEYWORD: SK7 HP8

Reviewing Ideas, Terms, and Places

1. **a. Define** What is the **Diaspora**?
   b. Explain How did **Zionism** help create the nation of Israel?
2. **a. Recall** What are Jewish dietary laws called?
   b. Draw Conclusions Why have Israeli leaders built up a strong military?
   c. Elaborate Why do you think Jews from around the world migrate to Israel?
3. **a. Identify** What are the two territories partly controlled by Palestinians?
   b. Make Inferences How might giving land to the Palestinians help in achieving peace in Israel?

Critical Thinking

4. **Categorizing** Use the chart below to separate your notes on Israel into categories.

Israel Today

Government	
Economy	
Diversity and Languages	
Jewish Culture	

FOCUS ON WRITING

5. **Describing Israel** What features make Israel unique? Take notes on how you might describe these features for your readers.

Social Studies Skills

Chart and Graph   Critical Thinking   Geography   Study

Analyzing a Cartogram

Learn

For statistical information like population figures, geographers sometimes create a special map called a cartogram. A cartogram displays information about countries by the size shown for each country. In contrast, a political map like the one on the right reflects countries' actual physical size. Here are some guidelines for reading and analyzing a cartogram.

- Read the title of the map to determine the subject area covered.

- Compare the political map to the cartogram. Notice how some countries are much different in size on the cartogram compared to the map.

- Read the cartogram's legend and think about what the information means.

Practice

❶ Which country has the largest population?

❷ How is the size of Saudi Arabia's land area different from the size of its population?

❸ Using the cartogram legend, what is the approximate population of Lebanon?

Apply

Draw your own cartogram using the gross domestic product, or GDP, of each country in Southwest and Central Asia. Use a reference source or the Internet to find these statistics. Then determine the scale for sizing each country by GDP. For example, you might use one square unit of area per $10 billion or $100 billion. Countries with a high GDP should appear larger than countries with a low GDP.

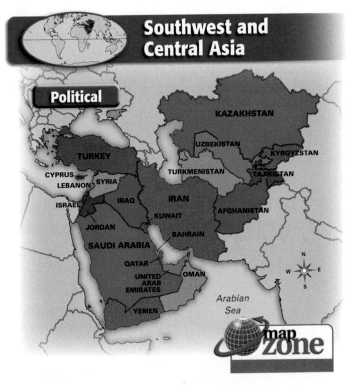

Southwest and Central Asia

Political

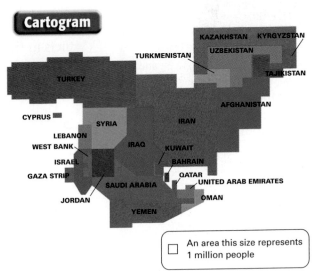

Cartogram

An area this size represents 1 million people

Syria, Lebanon, and Jordan

What You Will Learn...

Main Ideas

1. Syria, once part of the Ottoman Empire, is an Arab country ruled by a powerful family.
2. Lebanon is recovering from civil war and its people are divided by religion.
3. Jordan has few resources and is home to Bedouins and Palestinian refugees.

The Big Idea

Syria, Lebanon, and Jordan are Arab nations coping with religious diversity.

Key Terms and Places

Damascus, *p. 210*
Beirut, *p. 212*
Bedouins, *p. 212*
Amman, *p. 213*

TAKING NOTES As you read, take notes on Syria, Lebanon, and Jordan. Use a chart like the one below to organize your notes.

Syria	
Lebanon	
Jordan	

If **YOU** lived there...

You live in Beirut, Lebanon. Your grandparents often tell you about the years before civil wars destroyed the heart of Beirut. The city then had wide boulevards, parks, and elegant shops. It was popular with tourists. Even though much of Beirut has been rebuilt, you find it hard to imagine what the city used to look like.

What hopes do you have for your country?

BUILDING BACKGROUND The histories of Lebanon, Syria, and Jordan have been tangled together since the countries gained independence in the 1940s. Syria is a large nation with a strong military. Syria has often dominated Lebanon's political life. Other conflicts in the region have also spilled over into Lebanon.

Look again at the map at the beginning of this chapter. Notice that Syria, Lebanon, and Jordan all border Israel. Because of their location near Israel, these countries have been involved in conflicts in the region. In addition, Syria, Lebanon, and Jordan also share a similar history, religion, and culture.

Syria

The capital of Syria, **Damascus**, is believed to be the oldest continuously inhabited city in the world. For centuries it was a leading regional trade center. Syria became part of the Ottoman Empire in the 1500s. After World War I, France controlled Syria. Syria finally became independent in the 1940s.

History and Government

From 1971 to 2000, the Syrian government was led by a dictator, Hafiz al-Assad. As president, Assad increased the size of Syria's military. He wanted to match Israel's military strength and protect his rule from his political enemies within Syria. After Assad's death in 2000, his son, Bashar, was elected president. Bashar al-Assad's goals during his seven-year term include improving Syria's economy.

Syria has a socialist government, which owns the country's oil refineries, larger electrical plants, railroads, and some factories. Syria's key manufactured goods are textiles, food products, and chemicals. Agriculture remains important. Syria has only small deposits of oil and natural gas. It is rich in iron ore, basalt, and phosphates.

Syria's People

Syria's population of more than 18 million is about 90 percent Arab. The other 10 percent includes Kurds and Armenians. About 74 percent of Syrians are Sunni Muslim. Another 16 percent are Druze and Alawites, members of small branches of Islam. About 10 percent of Syrians are Christian. There are also small Jewish communities in some cities.

READING CHECK **Analyzing** How is Syria's economy organized?

Lebanon

Lebanon is a small, mountainous country on the Mediterranean coast. It is home to several different groups of people. At times these different groups have fought.

Lebanon's History and People

During the Ottoman period, many religious and ethnic minority groups settled in Lebanon. After World War I, France controlled Lebanon and Syria. Lebanon finally gained independence in the 1940s. Even so, some aspects of French culture influenced Lebanese culture. For example, in addition to Arabic, many Lebanese also speak French.

Lebanon's people are overwhelmingly Arab, but they are divided by religion. Most Lebanese are either Muslim or Christian. Each of those groups is divided into several smaller groups. Muslims are divided into Sunni, Shia, and Druze.

The Maronites are the largest of the Christian groups in the country. Over time, however, Muslims have become Lebanon's majority religious group.

FOCUS ON READING

Look at the headings under Lebanon to set your purpose for reading these paragraphs.

Ancient Syria

In Syria today, ruins of an ancient Roman trading center still stand. The Romans called the city Palmyra, meaning "city of palm trees."

People of Syria, Lebanon, and Jordan

The people of Syria, Lebanon, and Jordan share many cultural traits. For example, most people living in this region are Arab and practice Islam.

ANALYZING VISUALS What can you see in these photos that tells you about daily life in the region?

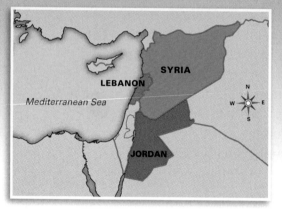

Syria In Syria drinking tea is an important part of Arab culture. Many Syrians, like this carpet seller, drink tea every day with family and friends.

Lebanon's Civil War and Today

For some decades after independence, Christian and Muslim politicians managed to share power. Certain government positions were held by different religious groups. For example, the president was always a Maronite. However, over time this cooperation broke down. Tensions between Christians and Muslims mounted.

Adding to these political divisions was the presence of hundreds of thousands of Palestinian refugees living in Lebanon. Ethnic and religious groups armed themselves, and in the 1970s fighting broke out. Warfare between Lebanese groups lasted until 1990. Tens of thousands of people died. The capital, **Beirut**, was badly damaged. Today Beirut is finally rebuilding.

During the 1990s the Lebanese economy slowly recovered from the civil war. Today, Lebanon's industries include food processing, textiles, cement, chemicals, and jewelry making.

READING CHECK Drawing Conclusions
What has caused divisions in Lebanese society?

Jordan

Jordan's short history has been full of conflict. The country has few resources and several powerful neighbors.

Jordan's History and Government

The country of Jordan was created after World War I. The British controlled the area and named an Arab prince as the monarch of the new country. In the 1940s the country became fully independent.

At the time of its independence, Jordan's population was small. Most Jordanians lived a nomadic or semi-nomadic life. Hundreds of thousands of Palestinian Arab refugees fled Israel and came to live in Jordan. From 1952 to 1999 Jordan was ruled by King Hussein. The king enacted some democratic reforms in the 1990s.

Jordan's People and Resources

Many of Jordan's people are **Bedouins**, or Arabic-speaking nomads who mostly live in the deserts of Southwest Asia. Jordan produces phosphates, cement, and potash. Tourism and banking are becoming impor-

Lebanon After more than two decades of civil war, Lebanon's people are rebuilding their capital, Beirut. The city's people now enjoy a new public square.

Jordan Jordan's people value education and equal rights for women. Jordanian teenagers like these girls are required to attend school until age 15.

tant industries. Jordan depends on economic aid from the oil-rich Arab nations and the United States. **Amman**, the capital, is Jordan's largest city. Jordanian farmers grow fruits and vegetables and raise sheep and goats. A shortage of water is a crucial resource issue for Jordan.

READING CHECK **Summarizing** How did King Hussein affect Jordan's history?

SUMMARY AND PREVIEW In this section you learned about the history, government, and people of Syria, Lebanon, and Jordan. In the next chapter you will learn about Iraq and Iran and the countries of the Arabian Peninsula—Saudi Arabia, Kuwait, Bahrain, Qatar, the United Arab Emirates, Yemen, and Oman.

Section 4 Assessment

Reviewing Ideas, Terms, and Places

1. a. **Recall** What is the capital of Syria?
   b. **Explain** What does Syria's government own?
   c. **Elaborate** Why did Hafiz al-Assad want to increase the size of Syria's military?
2. a. **Identify** What European country ruled Lebanon after World War I?
   b. **Analyze** How was **Beirut** damaged?
   c. **Elaborate** What is the history of political divisions between religious groups in Lebanon's government?
3. a. **Define** Who are the **Bedouins**?
   b. **Summarize** Who provides economic aid to Jordan?

Critical Thinking

4. **Comparing and Contrasting** Use your notes to identify similarities and differences among the people in the three countries.

	Similarities	Differences
Syria		
Lebanon		
Jordan		

FOCUS ON WRITING

5. **Describing Syria, Lebanon, and Jordan** If you could only include two details about these countries, what would they be?

THE EASTERN MEDITERRANEAN **213**

from
Red Brocade

by Naomi Shihab Nye

Drinking tea with guests is a traditional Arab custom.

GUIDED READING

WORD HELP

pine nuts a small sweet edible seed of some pine trees

brocade a heavy fabric of silk, cotton, or wool woven with a raised design, often using metallic threads

mint a plant with aromatic leaves that grows in northern temperate regions and is often used for flavoring

❶ Arabs are a cultural group that speak Arabic. They live mostly in Southwest Asia and North Africa.

❷ When entertaining, Arabs often sit on pillows on the floor.

About the Poem *In "Red Brocade," Arab-American writer Naomi Shihab Nye tells about an Arab custom. As a part of this custom, strangers are given a special welcome by those who meet them at the door. Since the poet is Arab-American, she is suggesting that we go "back to that" way of accepting new people.*

AS YOU READ Identify the special way that Arab people in Southwest Asia greet strangers at their door.

The Arabs ❶ used to say,
When a stranger appears at your
    door,
feed him for three days
before asking who he is,
where he's come from,
where he's headed.
That way, he'll have strength
enough to answer.
Or, by then you'll be
such good friends
you don't care.

Let's go back to that.
Rice? Pine nuts?

Here, take the red brocade
    pillow. ❷
My child will serve water
to your horse.

No, I was not busy when you
    came!
I was not preparing to be busy.
That's the armor everyone put
    on to pretend they had a purpose
in the world.

I refuse to be claimed.
Your plate is waiting.
We will snip fresh mint
into your tea.

Connecting Literature to Geography

1. **Describing** What details in the second verse show us that the Arab speaker is extending a warm welcome to the stranger?

2. **Comparing and Contrasting** Do you think this poem about greeting a stranger at the door would be different if it had taken place in another region of the world? Explain your answer.

Chapter Review

Geography's Impact
video series
Review the video to answer the closing question:
Why do you think the conflict in Jerusalem today is difficult to solve?

Visual Summary

Use the visual summary below to help you review the main ideas of the chapter.

QUICK FACTS

The eastern Mediterranean is a dry region, and water is a key resource.

The region's history includes conflict between three major religions.

Most people living in the eastern Mediterranean are Arab Muslims.

Reviewing Vocabulary, Terms, and Places

Fill in the blanks with the correct term or place from this chapter.

1. The_____ is the lowest point on any continent and the world's saltiest body of water.

2. A desert located in southern Israel is called the_____.

3. A_____is a way of doing something.

4. Turkey's largest city is_____.

5. _____means that religion is kept separate from government.

6. The scattering of the Jewish population is known as_____.

7. A_____is a large farm where people share everything in common.

8. _____is Lebanon's capital that was badly damaged during the country's civil war.

Comprehension and Critical Thinking

SECTION 1 *(Pages 196–199)*

9. **a. Describe** How is the Eastern Mediterranean considered a part of the Middle East?

   b. Draw Conclusions How would the region's dry climates affect where people lived?

   c. Predict What would happen if the region's people did not have access to water?

SECTION 2 *(Pages 200–203)*

10. **a. Recall** How was control of Constantinople important?

    b. Make Inferences How did modernization change Turkey?

    c. Elaborate Why do you think Turkey wants to be a member of the European Union?

SECTION 3 *(Pages 204–208)*

11. **a. Define** What is Zionism?

SECTION 3 (continued)

b. Make Inferences Why does Israel need a strong military?

c. Elaborate How has Israel's history affected the country today?

SECTION 4 (Pages 210–213)

12. a. Identify What is the capital of Syria? Why is it historically significant?

b. Analyze Why did Lebanon have a civil war?

c. Evaluate How do you think Jordan survives with so few resources?

Using the Internet

go.hrw.com
KEYWORD: SK7 CH8

13. Creating an Exhibit Jerusalem is a city rich in tradition, history, and culture dating back thousands of years. Enter the activity keyword and travel back in time to historic Jerusalem. Explore its history, archaeology, buildings, daily life, food, and more. Then create a museum exhibit to highlight the artifacts, information, and stories you encounter in your journey through Jerusalem's past. Some things you may want to include are artifacts, models, time lines, maps, small placards providing information, and an exhibit guide for viewers.

Social Studies Skill

Analyzing a Cartogram *Use the cartogram and political map of Southwest and Central Asia on this chapter's Social Studies Skills page to answer the following questions.*

14. Why do you think Turkey's size on the political map is similar to its size on the cartogram?

15. How does the cartogram show the high population density of Israel and the Palestinian Territories?

16. From looking at the cartogram, is the population density of Kazakhstan, high or low? Explain your answer.

FOCUS ON READING AND WRITING

Setting a Purpose *Use the information in this chapter to answer the following questions.*

17. How does setting a purpose before you read help you become a better reader?

18. How is your purpose in reading this chapter different from your purpose when you read a newspaper comic strip?

19. How can looking at headings and main idea statements help you set a purpose for reading?

20. Writing a Description Look over your notes and choose one Eastern Mediterranean country to describe. Organize your notes by topic—physical features, people, culture and government. Then, write a one-to two-paragraph description of the country. Include information you think would be interesting to someone who knows nothing about the country. Add details that will help your readers picture the country.

Map Activity ★Interactive

21. The Eastern Mediterranean On a separate sheet of paper, match the letters on the map with their correct labels.

Bosporus · Negev

Jordan River · Euphrates River

Dead Sea

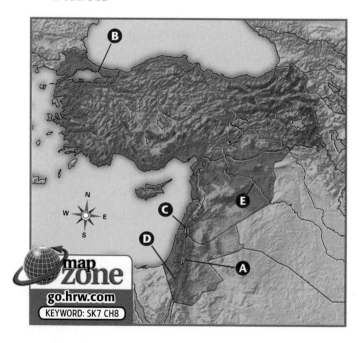

map zone

go.hrw.com
KEYWORD: SK7 CH8

DIRECTIONS: Read questions 1 through 7 and write the letter of the best response. Then read question 8 and write your own well-constructed response.

1 **The climate of most of Israel, Jordan, and Syria is**

A desert.

B steppe.

C humid subtropical.

D Mediterranean.

2 **Turkey's government wants to be more like countries on what continent?**

A Asia

B South America

C Australia

D Europe

3 **Jews and Palestinian Arabs make up most of what country's population?**

A Jordan

B Israel

C Turkey

D Lebanon

4 **What city is sacred to Jews, Muslims, and Christians?**

A Istanbul

B Tel Aviv

C Jerusalem

D Damascus

5 **Most people living in Syria, Lebanon, and Jordan are**

A Arabs.

B Jews.

C European.

D Christians.

Turkey: Physical Geography

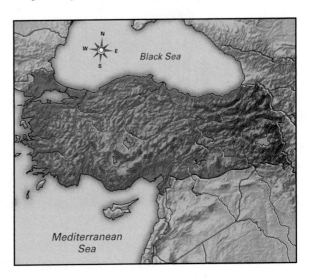

6 **Based on the map above, what physical features surround most of Turkey?**

A mountains

B seas

C plateaus

D lakes

7 **Many of Jordan's people are**

A Bedouins.

B Lebanese.

C Jewish.

D Turkish.

8 **Extended Response** Based on the map above and your knowledge of the region, write a brief essay explaining how Turkey's location has influenced its history and the country today.

The Arabian Peninsula, Iraq, and Iran

What You Will Learn...

In this chapter you will learn about the Arabian Peninsula, which includes Saudi Arabia, Kuwait, Bahrain, Qatar, United Arab Emirates, Oman, and Yemen. You will also learn about the history and people of Iraq and Iran.

SECTION 1
Physical Geography 220

SECTION 2
The Arabian Peninsula 224

SECTION 3
Iraq 230

SECTION 4
Iran 234

FOCUS ON READING AND WRITING

Re-Reading Sometimes a single reading is not enough to fully understand a passage of text. If you feel like you do not fully understand something you have read, it may help to re-read the passage more slowly. **See the lesson, Re-Reading, on page 676.**

Creating a Geographer's Log You are a geographer taking a journey of discovery through the Arabian Peninsula, Iraq, and Iran. As you travel from place to place, create a geographer's log, a written record of what you see on your journey.

Mediterranean Sea

20°E

AFRICA

- ⊙ National capital
- ● Other cities

0   150   300 Miles
0 150 300 Kilometers

Projection: Lambert Conformal Conic

map zone
Geography Skills

Place The countries of the Arabian Peninsula, Iraq, and Iran are centered around the Persian Gulf.

1. **Locate** What is the capital of Saudi Arabia?
2. **Analyze** Approximately how many miles would you have to travel from Baghdad to Kuwait City?

go.hrw.com   KEYWORD: SK7 CH9

Culture Islam is a major part of the culture in every country in the region. These women pray at a mosque in Mecca, Saudi Arabia.

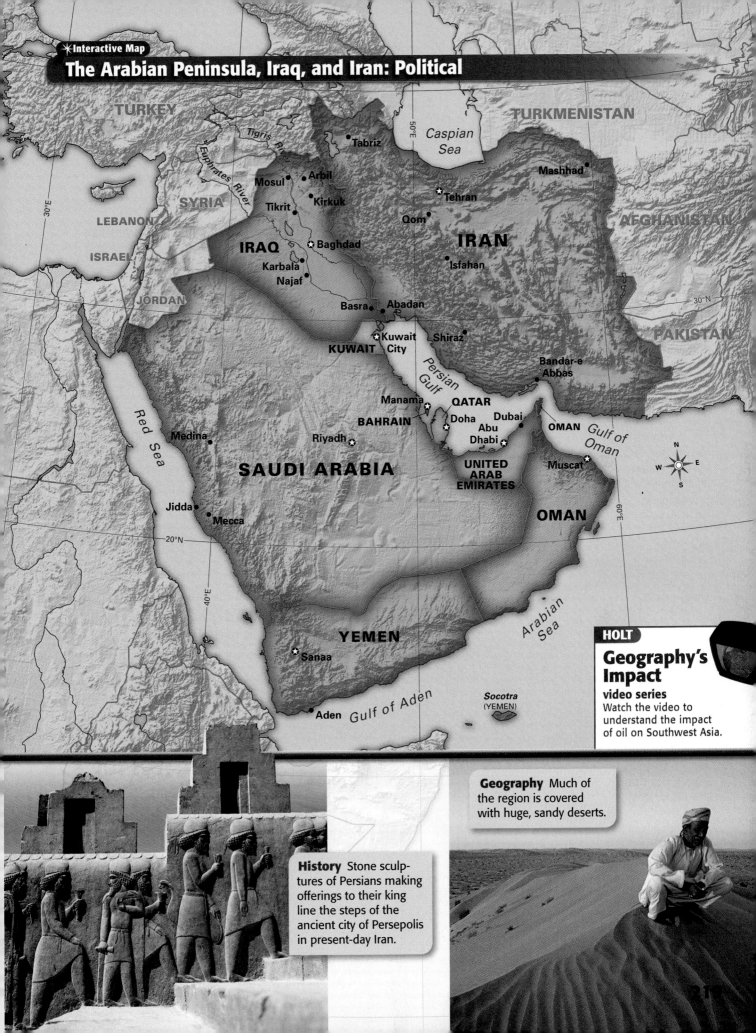

The Arabian Peninsula, Iraq, and Iran: Political

TURKEY

TURKMENISTAN

Tigris River

Tabriz

Caspian Sea

Mashhad

50°E

Euphrates River

Mosul • Arbil

Tehran

SYRIA

Tikrit • Kirkuk

Qom

IRAN

AFGHANISTAN

LEBANON

Isfahan

ISRAEL

IRAQ • Baghdad

30°N

JORDAN

Karbala
Najaf

Basra • Abadan

PAKISTAN

Kuwait City

Shiraz

KUWAIT

Bandar-e Abbas

Persian Gulf

Manama

QATAR

BAHRAIN

Doha   Dubai

OMAN

Gulf of Oman

Medina

Riyadh

Abu Dhabi

N

Red Sea

SAUDI ARABIA

UNITED ARAB EMIRATES

Muscat

W   E

S

Jidda

Mecca

OMAN

60°E

20°N

40°E

Arabian Sea

YEMEN

Sanaa

HOLT

Geography's Impact
video series
Watch the video to understand the impact of oil on Southwest Asia.

Socotra (YEMEN)

Aden   Gulf of Aden

Geography Much of the region is covered with huge, sandy deserts.

History Stone sculptures of Persians making offerings to their king line the steps of the ancient city of Persepolis in present-day Iran.

Physical Geography

If **YOU** lived there...

You are in a plane flying over the vast desert areas of the Arabian Peninsula. As you look down, you see some tents of desert nomads around trees of an oasis. Sometimes you can see a truck or a line of camels crossing the dry, rocky terrain. A shiny oil pipeline stretches for miles in the distance.

What is life like for people in the desert?

BUILDING BACKGROUND Iran, Iraq, and the countries of the Arabian Peninsula are part of a region sometimes called the "Middle East." This region lies at the intersection of Africa, Asia, and Europe. Much of the region is dry and rugged.

Physical Features

Did you know that not all deserts are made of sand? The **Arabian Peninsula** has the largest sand desert in the world. But it also has huge expanses of desert covered with bare rock or gravel. These wide desert plains are a common landscape in the region that includes the Arabian Peninsula, Iraq, and Iran.

The countries of this region appear on the map in sort of a semicircle, with the **Persian Gulf** in the center. The Arabian Peninsula is also bounded by the Gulf of Oman, the Arabian Sea, and the Red Sea. The Caspian Sea borders Iran to the north.

The region contains four main landforms: rivers, plains, plateaus, and mountains. The **Tigris** (TY-gruhs) and **Euphrates** (yooh-FRAY-teez) rivers flow across a low, flat plain in Iraq. They join together before they reach the Persian Gulf. The Tigris and Euphrates are what are known as exotic rivers, or rivers that begin in humid regions and then flow through dry areas. The rivers create a narrow fertile area, which in ancient times was called Mesopotamia, or the "land between the rivers." The Arabian Peninsula has no permanent rivers.

The vast, dry expanse of the Arabian Peninsula is covered by plains in the east. The peninsula's desert plains are covered with sand in the south and volcanic rock in the north. As you can see on the map, the surface of the peninsula rises gradually from the Persian Gulf to the Red Sea. Near the Red Sea the landscape becomes one of plateaus and mountains, with almost no coastal plain. The highest point on the peninsula is in the mountains of Yemen.

Plateaus and mountains also cover most of Iran. In fact, Iran is one of the world's most mountainous countries. In the west, the land climbs sharply to form the Zagros Mountains. The Elburz Mountains and the Kopet-Dag lie in the north. Historically, this mountainous landscape has kept towns there isolated from each other.

FOCUS ON READING
After you read this paragraph, re-read it to make sure you understand Iran's landscape.

READING CHECK Summarizing What are the major physical features of this area?

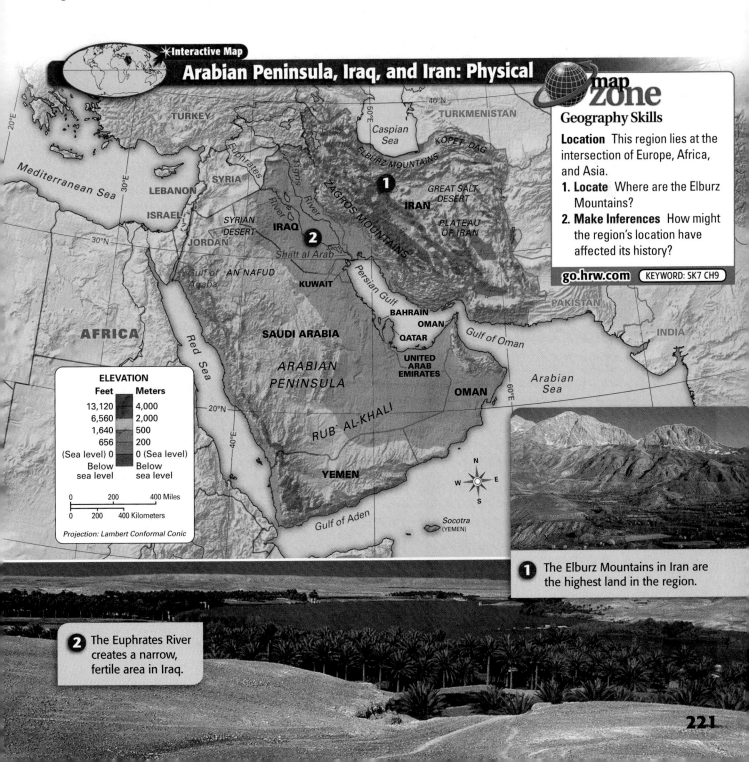

Arabian Peninsula, Iraq, and Iran: Physical

Interactive Map

map zone

Geography Skills

Location This region lies at the intersection of Europe, Africa, and Asia.
1. **Locate** Where are the Elburz Mountains?
2. **Make Inferences** How might the region's location have affected its history?

go.hrw.com   KEYWORD: SK7 CH9

TURKEY
TURKMENISTAN
Caspian Sea
KOPET-DAG
ELBURZ MOUNTAINS
Euphrates
Tigris
Mediterranean Sea
SYRIA
LEBANON
ZAGROS MOUNTAINS
GREAT SALT DESERT
IRAN
ISRAEL
SYRIAN DESERT
IRAQ
Euphrates River
Tigris River
PLATEAU OF IRAN
JORDAN
Shatt al Arab
Gulf of Aqaba
AN NAFUD
Persian Gulf
KUWAIT
PAKISTAN
AFRICA
Red Sea
BAHRAIN
OMAN
QATAR
Gulf of Oman
INDIA
SAUDI ARABIA
UNITED ARAB EMIRATES
Arabian Sea
ARABIAN PENINSULA
OMAN
RUB` AL-KHALI
YEMEN
Socotra (YEMEN)
Gulf of Aden

ELEVATION

Feet		Meters
13,120		4,000
6,560		2,000
1,640		500
656		200
(Sea level) 0		0 (Sea level)
Below sea level		Below sea level

0   200   400 Miles
0   200   400 Kilometers

Projection: Lambert Conformal Conic

N   W   E   S

① The Elburz Mountains in Iran are the highest land in the region.

② The Euphrates River creates a narrow, fertile area in Iraq.

221

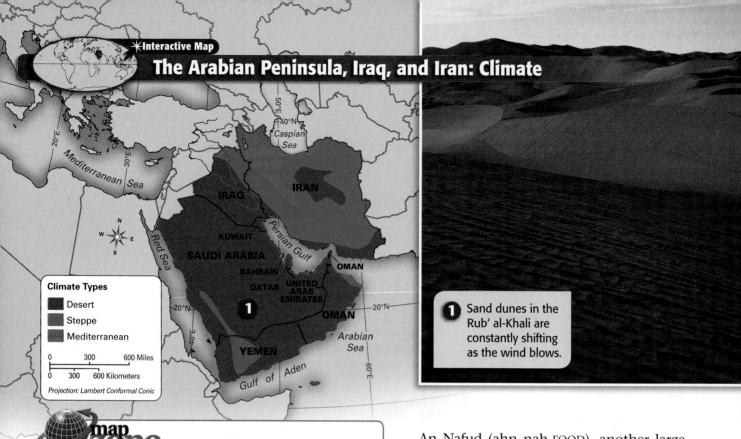

Interactive Map
The Arabian Peninsula, Iraq, and Iran: Climate

Climate Types
- Desert
- Steppe
- Mediterranean

0   300   600 Miles
0   300   600 Kilometers
Projection: Lambert Conformal Conic

1 Sand dunes in the Rub' al-Khali are constantly shifting as the wind blows.

map zone Geography Skills

Regions Most of this region has a desert climate.
1. **Identify** Which countries have only desert climates?
2. **Interpret** Look back at the physical map. How do landforms in the region influence climate?

go.hrw.com KEYWORD: SK7 CH9

Climate and Vegetation

As you have already read, most of this region has a desert climate. The desert can be both very hot and very cold. In the summer, afternoon temperatures regularly climb to over 100°F (38°C). During the night, however, the temperature may drop quickly. Nighttime temperatures in the winter sometimes dip below freezing.

The world's largest sand desert, the Rub' al-Khali (ROOB ahl-KAH-lee), covers much of southern Saudi Arabia. *Rub' al-Khali* means "Empty Quarter," a name given to the area because there is so little life there. Sand dunes in the desert can rise to 800 feet (245 m) high and stretch for nearly 200 miles! In northern Saudi Arabia is the

An Nafud (ahn nah-FOOD), another large desert. These deserts are among the driest places in the world. The Rub' al-Khali receives an average of less than 4 inches (10 cm) of rainfall each year.

Some plateau and mountain areas do get winter rains or snow. These higher areas generally have semiarid steppe climates. Some mountain peaks receive more than 50 inches (130 cm) of rain per year.

Rainfall supports vegetation in some parts of the region. Trees are common in mountain regions and in scattered desert oases. An **oasis** is a wet, fertile area in a desert that forms where underground water bubbles to the surface. Most desert plants have adapted to survive without much rain. For example, the shrubs and grasses that grow on the region's dry plains have roots that either grow deep or spread out far to capture as much water as possible. Still, some places in the region are too dry or too salty to support any vegetation.

READING CHECK Finding the Main Idea
What climate dominates this region?

Resources

Water is one of the region's two most valuable resources. However, this resource is very scarce. In some places in the desert, springs provide water. At other places, water can come from wells dug into dry streambeds called **wadis**. Modern wells can reach water deep underground, but the groundwater in these wells is often fossil water. **Fossil water** is water that is not being replaced by rainfall. Wells that pump fossil water will eventually run dry.

While water is scarce, the region's other important resource, oil, is plentiful. Oil exports bring great wealth to the countries that have oil fields. Most of the oil fields are located near the shores of the Persian Gulf. However, although oil is plentiful now, it cannot be replaced once it is taken from Earth. Too much drilling for oil now may cause problems in the future because most countries of the region are not rich in other resources. Iran is an exception with its many mineral deposits.

READING CHECK **Summarizing** What are the region's important resources?

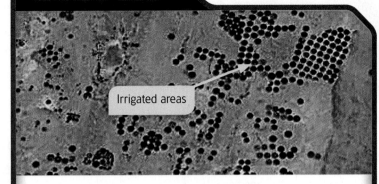

Satellite View

Irrigated areas

Pivot-Irrigated Fields

This satellite image shows how fossil water has converted desert land into farmers' fields. Each circular plot of land has a water source at its center. An irrigation device extends out and pivots around the center.

Drawing Inferences Why are the fields circular?

SUMMARY AND PREVIEW The Arabian Peninsula, Iraq, and Iran form a desert region with significant oil resources. Next, you will learn more about the countries of the Arabian Peninsula.

go.hrw.com
Online Quiz
KEYWORD: SK7 HP9

Section 1 Assessment

Reviewing Ideas, Terms, and Places

1. a. **Describe** Where was Mesopotamia?
   b. **Explain** Where are the region's mountains?
   c. **Elaborate** Why do you think the **Tigris** and **Euphrates** rivers were so important in history?
2. a. **Recall** What parts of the region receive the most rainfall?
   b. **Explain** How have desert plants adapted to their environment?
3. a. **Define** What is **fossil water**?
   b. **Make Inferences** How do you think resources in the region influence where people live?
   c. **Predict** What might happen to the oil-rich countries if their oil was used up or if people found a new energy source to replace oil?

Critical Thinking

4. **Comparing and Contrasting** Using your notes and a graphic organizer like the one here, note physical characteristics unique to each area. Then list characteristics shared by all three areas.

Arabian Peninsula	Iraq	Iran
All		

FOCUS ON WRITING

5. **Describing Physical Geography** Take notes on the physical features, climate and vegetation, and resources that you could record in your log. What would you see and feel if you were in this region?

The Arabian Peninsula

What You Will Learn...

Main Ideas

1. Islamic culture and an economy greatly based on oil influence life in Saudi Arabia.
2. Most other Arabian Peninsula countries are monarchies influenced by Islamic culture and oil resources.

The Big Idea

Most countries of the Arabian Peninsula share three main characteristics: Islamic religion and culture, monarchy as a form of government, and valuable oil resources.

Key Terms

Shia, *p. 224*
Sunni, *p. 224*
OPEC, *p. 225*

TAKING NOTES As you read, use a chart like the one here to take notes on the countries on the Arabian Peninsula.

Saudi Arabia	
Kuwait	
Bahrain	
Qatar	
United Arab Emirates	
Oman	
Yemen	

If **YOU** lived there...

You are a financial adviser to the ruler of Oman. Your country has been making quite a bit of money from oil exports. However, you worry that your economy is too dependent on oil. You think Oman's leaders should consider expanding the economy. Oman is a small country, but it has beautiful beaches, historic palaces and mosques, and colorful markets.

How would you suggest expanding the economy?

BUILDING BACKGROUND Oman and all the countries of the Arabian Peninsula have valuable oil resources. In addition to oil, these countries share two basic characteristics: Islamic religion and monarchy as a form of government. The largest country, and the one with the most influence in the region, is Saudi Arabia.

Saudi Arabia

Saudi Arabia is by far the largest of the countries of the Arabian Peninsula. It is also a major religious and cultural center and has one of the region's strongest economies.

People and Customs

Nearly all Saudis are Arabs and speak Arabic. Their culture is strongly influenced by Islam, a religion founded in Saudi Arabia by Muhammad. Islam is based on submitting to God and on messages Muslims believe God gave to Muhammad. These messages are written in the Qur'an, the holy book of Islam.

Nearly all Saudis follow one of two main branches of Islam. **Shia** Muslims believe that true interpretation of Islamic teaching can only come from certain religious and political leaders called imams. **Sunni** Muslims believe in the ability of the majority of the community to interpret Islamic teachings. About 85 percent of Saudi Muslims are Sunni.

CONNECTING TO Math

Muslim Contributions to Math

During the early centuries of the Middle Ages, European art, literature, and science declined. However, during this same period, Muslim scholars made important advances in literature, art, medicine, and mathematics.

Our familiar system of numerals, which we call Arabic, was first created in India. However, it was Muslim thinkers who introduced that system to Europe. They also developed algebra and made advances in geometry. Muslims used math to advance the study of astronomy and physics. Muslim geographers calculated distances between cities, longitudes and latitudes, and the direction from one city to another. Muslim scientists even defined ratios and used mathematics to explain the appearance of rainbows.

Drawing Inferences Why do we need math to study geography?

Islam influences Saudi Arabia's culture in many ways. For example, in part because Islam requires modesty, Saudi clothing keeps arms and legs covered. Men usually wear a long, loose shirt. They often wear a cotton headdress held in place with a cord. Saudi women traditionally wear a black cloak and veil in public, although some now wear Western-style clothing.

Saudi laws and customs limit women's activities. For example, a woman rarely appears in public without her husband or a male relative. Also, women are not allowed to drive cars. However, women can own and run businesses in Saudi Arabia.

Government and Economy

Saudi Arabia is a monarchy. Members of the Saud family have ruled Saudi Arabia since 1932. Most government officials are relatives of the king. The king may ask members of his family, Islamic scholars, and tribal leaders for advice on decisions.

The country has no elected legislature. Local officials are elected, but only men are allowed to vote.

Saudi Arabia's economy is based on oil. In fact, Saudi Arabia has the world's largest reserves, or supplies, of oil and is the world's leading exporter of oil. Because it controls so much oil, Saudi Arabia is an influential member of the Organization of Petroleum Exporting Countries, or OPEC. **OPEC is an international organization whose members work to influence the price of oil on world markets by controlling the supply.**

Oil has brought wealth to Saudi Arabia. The country has a sizable middle class, and the government provides free health care and education to its citizens. Even so, Saudi Arabia faces economic challenges. For example, it must import much of its food because freshwater needed for farming is scarce. The country uses desalination plants to remove salt from seawater, but this requires an extremely expensive **procedure**.

ACADEMIC VOCABULARY
procedure a series of steps taken to accomplish a task

FOCUS ON READING
After you read this paragraph, re-read it to make sure you understand Saudi Arabia's economic challenges.

Another economic challenge for Saudi Arabia is its high unemployment rate. One reason for the lack of jobs is the high population growth rate. More than 40 percent of Saudis are younger than 15. Another reason for unemployment is that many young Saudis choose to study religion instead of the technical subjects their economy requires.

READING CHECK **Finding Main Ideas** What religion influences Saudi Arabia's culture?

Other Countries of the Arabian Peninsula

Saudi Arabia shares the Arabian Peninsula with six smaller countries. Like Saudi Arabia, these countries are all influenced by Islam. Also like Saudi Arabia, most have monarchies and economies based on oil.

Kuwait

Oil was discovered in Kuwait in the 1930s. Since then it has made Kuwait very rich. In 1990 Iraq invaded Kuwait to try to control its oil, starting the Persian Gulf War. The United States and other countries defeated Iraq, but the war caused major destruction to Kuwait's oil fields.

Although Kuwait's government is dominated by a royal family, the country did elect a legislature in 1992. Only men from certain families—less than 15 percent of Kuwait's population—had the right to vote in these elections. However, Kuwait recently gave women the right to vote.

Bahrain and Qatar

Bahrain is a group of islands in the Persian Gulf. It is a monarchy with a legislature. Bahrain is a rich country. Most people there live well in big, modern cities. Oil made Bahrain wealthy, but in the 1990s the country began to run out of oil. Now banking and tourism are major industries.

Qatar occupies a small peninsula in the Persian Gulf. Like Bahrain, Qatar is ruled by a powerful monarch. In 2003 men and women in Qatar voted to approve a new constitution that would give more power to elected officials. Qatar is a wealthy country. Its economy relies on its oil and natural gas.

Oil Wealth

Big, modern cities such as Dubai, UAE, were built with money from oil exports. Many people in the region's cities can afford to buy luxury items.

ANALYZING VISUALS What kind of luxury items is this man selling?

The United Arab Emirates

The United Arab Emirates, or UAE, consists of seven tiny kingdoms. Profits from oil and natural gas have created a modern, comfortable lifestyle for the people of the UAE. Partly because it is so small, the UAE depends on foreign workers. In fact, it has more foreign workers than citizens.

Oman and Yemen

Oman covers most of the southeastern part of the Arabian Peninsula. Oman's economy is also based on oil. However, Oman does not have the great oil wealth of Kuwait or the UAE. Therefore, the government is attempting to develop new industries.

Yemen is located on the southwestern part of the Arabian Peninsula. The country has an elected government, but it has suffered from corruption. Oil was not discovered in Yemen until the 1980s. Oil and coffee generate much of the national income, but Yemen is still the poorest country on the Arabian Peninsula.

READING CHECK **Summarizing** How has oil affected the countries of the Arabian Peninsula?

Yemen
Yemen's architecture is an important part of its culture. These buildings are more than 2,000 years old.

SUMMARY AND PREVIEW Islam is a major influence on the people and culture of Saudi Arabia and the other countries of the Arabian Peninsula. The other major influence in the region is oil. Oil has brought wealth to most countries on the peninsula. In the next section you will learn about Iraq, a neighboring country with similar influences.

Section 2 Assessment

go.hrw.com
Online Quiz
KEYWORD: SK7 HP9

Reviewing Ideas, Terms, and Places

1. **a. Define** What is **OPEC**?
   b. Compare and Contrast How are **Sunni** and **Shia** Muslims similar, and how are they different from each other?
   c. Elaborate What do you think Saudi Arabia would be like if it did not have such huge oil reserves?
2. **a. Identify** What resource is the most important to the economies of countries on the Arabian Peninsula?
   b. Analyze How does its small size affect the United Arab Emirates?
   c. Predict How might Yemen change now that oil is a major part of its economy?

Critical Thinking

3. **Summarizing** Look at your notes on the countries of the Arabian Peninsula. Then copy the graphic organizer here and for each topic, write a one-sentence summary about the region.

	Summary
Culture	
Government	
Economy	

FOCUS ON WRITING

4. **Writing about the Arabian Peninsula** If you were traveling through these lands, what would you see or experience? Write some notes in your journal.

Oil in Saudi Arabia

Background Try to imagine your life without oil. You would probably walk or ride a horse to school. You would heat your home with coal or wood. You would never fly in a plane, walk in rubber-soled shoes, or even drink out of a plastic cup.

Our society depends on oil. However, oil is a nonrenewable resource. This means that supplies are limited, and we may one day run out of oil. In fact, the United States no longer produces enough oil to satisfy its own needs. We now depend on foreign countries, such as Saudi Arabia, for oil.

Oil Reserves in Saudi Arabia

Saudi Arabia has the world's largest supply of oil. This important resource, found naturally in the environment, has had a huge impact on Saudi Arabia's society.

Before the discovery of oil there in the 1930s, Saudi Arabia was a poor country. But income from oil exports has given the government money to invest in improvements such as new apartments, communications systems, airports, oil pipelines, and roads.

Saudi Arabia's Oil Fields

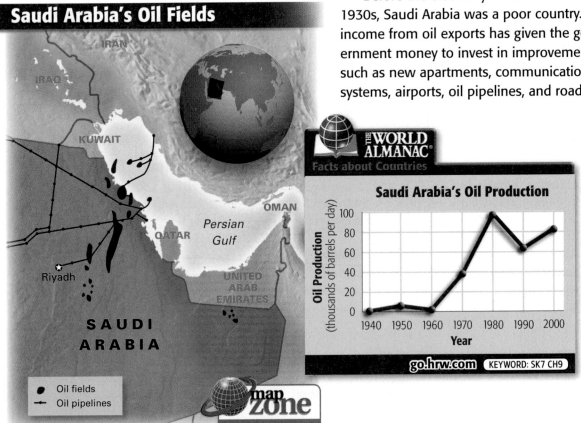

IRAN

IRAQ

KUWAIT

OMAN

Persian Gulf

QATAR

★ Riyadh

UNITED ARAB EMIRATES

SAUDI ARABIA

- ● Oil fields
- ┅ Oil pipelines

map Zone

THE WORLD ALMANAC®
Facts about Countries

Saudi Arabia's Oil Production

Oil Production (thousands of barrels per day)

100
80
60
40
20
0

1940 1950 1960 1970 1980 1990 2000

Year

go.hrw.com   KEYWORD: SK7 CH9

Saudi Arabia's oil is pumped through pipelines to tankers that ship the oil around the world. The oil industry has made Saudi Arabia a rich country.

THE WORLD ALMANAC
Facts about Countries

Saudi Arabia's Exports

Oil Products 91%

Other 9%

go.hrw.com KEYWORD: SK7 CH9

For example, in 1960 Saudi Arabia had only about 1,000 miles (1,600 km) of roads. By 2005 it had over 94,000 miles (152,000 km) of roads. These improvements have helped modernize Saudi Arabia's economy.

Oil exports have also affected Saudi society. Rising incomes have given many people there more money to spend on consumer goods. New stores and restaurants have opened, and new schools have been built throughout the country. Education is now available to all citizens. Increased education means the literacy rate has increased also—from about 3 percent when oil was discovered to about 79 percent today. Health care there has also improved.

The oil industry has also increased Saudi Arabia's importance in the world. Since it is a member of the Organization of Petroleum Exporting Countries (OPEC), Saudi Arabia influences the price of oil on the world market. Countries around the world want to have good relations with Saudi Arabia because of its vast oil reserves.

What It Means Today Saudi Arabia's government has a lot of money. This wealth has come almost entirely from the sale of oil. However, since the world's oil supplies are limited, Saudi Arabia's economy may be at risk in the future. Many countries are beginning to research other types of energy that can one day be used in place of oil. Until then, the many countries buying oil from Saudi Arabia will continue to pump wealth into Saudi society.

Geography for Life Activity

1. How has oil changed Saudi Arabia's society?
2. What are some advantages and disadvantages for a society that relies on oil?
3. **Other Types of Energy** Research other types of energy we can get from the environment. Based on your findings, do you think other types of energy will replace oil in the near future? Why or why not?

Iraq

What You Will Learn...

Main Ideas

1. Iraq's history includes rule by many conquerors and cultures, as well as recent wars.
2. Most of Iraq's people are Arabs, and Iraqi culture includes the religion of Islam.
3. Iraq today must rebuild its government and economy, which have suffered from years of conflict.

The Big Idea

Iraq, a country with a rich culture and natural resources, faces the challenge of rebuilding after years of conflict.

Key Terms and Places

embargo, *p. 231*
Baghdad, *p. 233*

TAKING NOTES Draw two boxes like the ones below. As you read, fill in the box on the left with your notes on Iraq's history. In the box on the right, take notes on Iraq today.

Iraq's History and Culture	Iraq Today

If YOU lived there...

You are a student in a school in Iraq's capital, Baghdad. During the war, your school and its library were badly damaged. Since then, you and your friends have had few books to read. Now your teachers and others are organizing a project to rebuild your library. They want to include books from all countries of the world as well as computers so students can use the Internet.

What would you like to have in the new library?

BUILDING BACKGROUND In spite of its generally harsh climate, the area that is now Iraq was one of the ancient cradles of civilization. Mesopotamia—the "land between the rivers"—was part of the "Fertile Crescent." Thousands of years ago, people there developed farming, domesticated animals, and organized governments.

History

Did you know that the world's first civilization was located in Iraq? Thousands of years ago people known as Sumerians settled in Mesopotamia—a region that is part of Iraq today. The country's recent history includes wars and a corrupt leader.

Early Civilization

Throughout Mesopotamia's history, different cultures and empires conquered the region. As you can see on the map on the next page, the Sumerians settled in southern Mesopotamia. By about 3000 BC, the Sumerians built the world's first known cities there. The Persians then conquered Mesopotamia in the 500s BC. By 331 BC Alexander the Great made it part of his empire. In the AD 600s Arabs conquered Mesopotamia, and the people gradually converted to Islam.

In the 1500s Mesopotamia became part of the Ottoman Empire. During World War I Great Britain took over the region. The British set up the kingdom of Iraq in 1932 and placed a pro-British ruler in power. In the 1950s a group of Iraqi army officers overthrew this government.

Saddam Takes Power

In 1968, after several more changes in Iraq's government, the Baath (BAHTH) Party took power. In 1979, a Baath leader named Saddam Hussein became Iraq's president. Saddam Hussein was a harsh ruler. He controlled Iraq's media, restricted personal freedoms, and killed an unknown number of political enemies.

Invasions of Iran and Kuwait

Under Saddam's leadership, Iraq invaded Iran in 1980. The Iranians fought back, and the Iran-Iraq War dragged on until 1988. Both countries' economies were seriously damaged, and many people died.

In 1990 Iraq invaded Kuwait, Iraq's oil-rich neighbor to the south. This event shocked and worried many world leaders. They were concerned that Iraq might gain control of the region's oil. In addition, they worried about Iraq's supply of weapons of mass destruction, including chemical and biological weapons.

War and Its Effects

In 1991, an alliance of countries led by the United States forced the Iraqis out of Kuwait. This six week event was called the Persian Gulf War. Saddam, who remained in power after the war, would not accept all the United Nations' (UN) terms for peace. In response, the UN placed an **embargo, or limit on trade**, on Iraq. As a result, Iraq's economy suffered.

Soon after the fighting ended, Saddam faced two rebellions from Shia Muslims and Kurds. He brutally put down these uprisings. In response, the UN forced Iraq to end all military activity. The UN also required that Iraq allow inspectors into the country. They wanted to make sure that Saddam had destroyed the weapons of mass destruction. Iraq later refused to cooperate completely with the UN.

Ten years after the Persian Gulf War, the terrorist attacks of September 11, 2001, led to new tensions between the United States and Iraq. U.S. government officials believed that Iraq aided terrorists. In March 2003, President George W. Bush, ordered U.S. forces to attack Iraqi targets. Within a few weeks the Iraqi army was defeated and Saddam's government was crushed. Saddam went into hiding. Eight months later, U.S. soldiers found Saddam hiding in an underground hole in rural Iraq. Saddam was arrested for his crimes.

READING CHECK Summarizing What are some key events in Iraq's history?

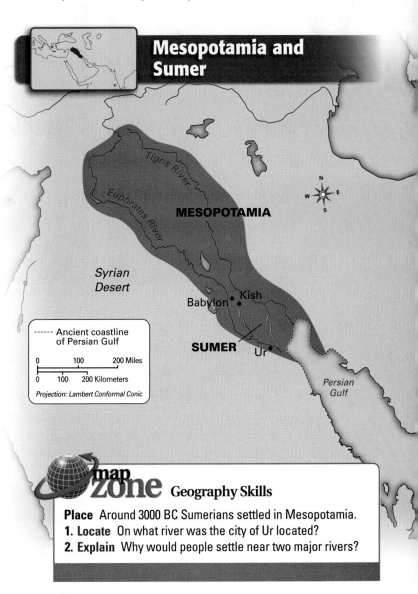

Mesopotamia and Sumer

Tigris River

Euphrates River

MESOPOTAMIA

Syrian Desert

Babylon • Kish

SUMER Ur

Persian Gulf

------ Ancient coastline of Persian Gulf

0   100   200 Miles

0   100   200 Kilometers

Projection: Lambert Conformal Conic

map zone Geography Skills

Place Around 3000 BC Sumerians settled in Mesopotamia.
1. **Locate** On what river was the city of Ur located?
2. **Explain** Why would people settle near two major rivers?

With the help of the United States, Iraqis are hopeful that they can rebuild their country.

ANALYZING VISUALS How is the United States helping Iraq's people today?

Iraqi boys enjoy their summer at soccer camp in Baghdad.

During Iraq's first democratic elections, an Iraqi woman shows her ink-stained finger that she used to vote.

A U.S. soldier passes out school supplies to Iraqi schoolchildren.

People and Culture

Iraq is about the size of California, with a population of about 26 million. Most Iraqis live in cities. Ethnic identity, religion, and food are all important elements of Iraqi culture.

Ethnic Groups

Most of Iraq's people belong to two major ethnic groups—Arabs and Kurds. Arabs are the largest group and make up more than 75 percent of Iraq's population. Iraqi Arabs speak the country's official language, Arabic. The smaller group, the Kurds, make up some 15 to 20 percent of the population. The Kurds are mostly farmers and live in a large region of northern Iraq. Most Iraqi Kurds speak Kurdish in addition to Arabic.

Religion

Like ethnic identity, religion plays a large role in the lives of most Iraqis. Nearly all Iraqis, both Arab and Kurdish, are Muslim. Within Iraq, the two different branches of Islam—Shia and Sunni—are practiced. About 60 percent of Iraqis are Shia and live in the south. Some 35 percent of Iraqis are Sunnis and live in the north.

READING CHECK **Summarizing** What ethnic groups do most Iraqis belong to?

Iraq Today

After years of war, Iraq is slowly rebuilding. Signs of progress include a new government and a recovering economy.

Rebuilding Baghdad

Iraq's capital, **Baghdad**, was severely damaged in the recent war. Baghdad's 6 million people suffered from a lack of electricity and running water, both of which were lost during the war. U.S. military and private contractors helped the Iraqis restore electricity and water throughout the city. They have also helped the Iraqis rebuild homes, businesses, and schools.

Government and Economy

In January 2005 Iraqis participated in democracy for the first time ever. Millions of Iraqis went to the polls to elect members to the National Assembly. One of the assembly members' main tasks was to draft Iraq's new constitution.

As Iraqis create a new government, they are also trying to recover their once strong economy. In the 1970s Iraq was the world's second-largest oil exporter. Time will tell if Iraq can again be a major oil producer.

Oil isn't Iraq's only resource. From earliest times, Iraq's wide plains and fertile soils have produced many food crops. Irrigation from the Tigris and Euphrates rivers allows farmers to grow barley, cotton, and rice.

After decades of a harsh government and wars, Iraq's future remains uncertain. Rebuilding schools, hospitals, roads, and making other improvements may take years. Even with help from the United States, Iraq faces an even bigger challenge of creating a free and prosperous society.

READING CHECK **Drawing Conclusions** What happened to Iraq's oil industry?

FOCUS ON READING
Do you understand everything you just read? If not, try re-reading the paragraphs that you do not understand.

SUMMARY AND PREVIEW In this section, you have learned about Iraq's ancient history, rich culture, and the progress made toward a new government and economy. Next, you will learn about Iran, which also has an ancient history but has a much different culture, government, and economy.

Section 3 Assessment

go.hrw.com
Online Quiz
KEYWORD: SK7 HP9

Reviewing Ideas, Terms, and Places

1. **a. Recall** Where was the world's first civilization located?
   b. Sequence What events led to the **embargo** on Iraq by the United Nations?
2. **a. Identify** What are two major ethnic groups in Iraq?
   b. Contrast What is one difference between Shia Muslims and Sunni Muslims?
3. **a. Describe** How was **Baghdad** damaged by war?
   b. Draw Conclusions What natural resource may help Iraq's economy recover?
   c. Predict What kind of country do you think Iraq will be in five years?

Critical Thinking

4. **Summarizing** Use your notes on Iraq today to fill in this table by summarizing what you have learned about Baghdad and Iraq's government and economy.

Baghdad	Government	Economy

FOCUS ON WRITING

5. **Writing about Iraq** Add details about Iraq's people, culture, and the country today to your notes. What sights have you seen that you might record in your log?

Iran

What You Will Learn...

Main Ideas

1. Iran's history includes great empires and an Islamic republic.
2. In Iran today, Islamic religious leaders restrict the rights of most Iranians.

The Big Idea

Islam is a huge influence on government and daily life in Iran.

Key Terms and Places

shah, *p. 235*
revolution, *p. 235*
Tehran, *p. 235*
theocracy, *p. 236*

TAKING NOTES As you read, take notes on Iran's history and life in the country today. Use the chart below to organize your notes.

Iran	
History	Today

If YOU lived there...

You are a student in Tehran, the capital of Iran. In school, you are taught that the way of life in the West—countries of Europe and the Americas—is bad. News reports and newspapers are filled with negative propaganda about Western countries. Yet you know that some of your friends secretly listen to Western popular music and watch American television programs that they catch using illegal satellite dishes at home. This makes you very curious about Western countries.

What would you like to know about life in other countries?

BUILDING BACKGROUND Like Iraqis, Iranians have a proud and ancient history. While most people living in the Arabian Peninsula and Iraq are Arabs, the majority of Iranians are Persian. They have a distinct culture and language.

History

The early history of the country we now call Iran includes the Persian Empire and a series of Muslim empires. Iran's recent history includes an Islamic revolution. Today Iran is an Islamic republic, which limits the rights of many Iranians.

Persian Empire

Beginning in the 500s BC, the Persian Empire ruled the region around present-day Iran. For centuries Persia was a great center of art and learning. The Persian Empire was known for its spectacular paintings, carpets, metalwork, and architecture. In the empire's capital, Persepolis, walls and statues throughout the city glittered with gold, silver, and precious jewels.

The Persian Empire was later conquered by several Muslim empires. Muslims converted the Persians to Islam, but most people retained their Persian culture. They built beautiful mosques with colorful tiles and large domes.

The Shah and Islamic Revolution

In 1921 an Iranian military officer took power and encouraged change in Iran's government. He claimed the old Persian title of **shah**, or king. In 1941 the shah's son took control. This shah became an ally of the United States and Great Britain and tried to modernize Iran. His programs were unpopular with many Iranians.

In 1978 Iranians began a revolution. A **revolution** is a drastic change in a country's government and way of life. By 1979, Iranians overthrew the shah and set up an Islamic republic. This type of government follows strict Islamic law.

Soon after Iran's Islamic Revolution began, relations with the United States broke down. A mob of students attacked the U.S. Embassy in Iran's capital, **Tehran**. With the approval of Iran's government, the students took Americans working at the embassy hostage. More than 50 Americans were held by force for over a year.

READING CHECK **Drawing Conclusions** How did Iran's history lead to the Islamic Revolution?

Iran Today

Iranian culture differs from many other cultures of Southwest Asia. Unlike most of the Arab peoples living in the region, more than half of all Iranians are Persian. They speak Farsi, the Persian language.

People and Culture

With about 68 million people, Iran has one of the largest populations in Southwest Asia. Iran's population is very young. Over 35 million Iranians are younger than 25 years old. It is also ethnically diverse. Iranian ethnic groups other than the Persian majority include Azerbaijanis, Kurds, Arabs, and Turks.

Most Iranians belong to the Shia branch of Islam. Only about 10 percent are Sunni Muslim. The rest of Iran's people practice Christianity, Judaism, or other religions.

In addition to the Islamic holy days, Iranians celebrate Nowruz—the Persian New Year. Iranians tend to spend this holiday outdoors. As a part of this celebration, they display goldfish in their homes to symbolize life.

FOCUS ON READING

Re-read the paragraphs under The Shah and Islamic Revolution to better understand important parts of Iran's recent history.

Yazd, Iran

In the ancient city of Yazd, spectacular tilework covers the dome of an Islamic mausoleum built in the 1300s.

Iranian culture also includes close-knit families and respect for elders. Most family gatherings in Iran are centered around Persian food, which includes rice, bread, vegetables, fruits, lamb, and tea.

Economy and Government

Huge oil reserves, which are among the largest in the world, make Iran a wealthy country. In addition to oil, the production of beautiful woven carpets contributes to Iran's economy. The country's strong agricultural sector employs nearly one-third of the Iranian workforce.

The current government of Iran is a **theocracy**—a government ruled by religious leaders. These religious leaders, or *ayatollahs,* control Iran's government. The head of the *ayatollahs,* or supreme leader, has unlimited power. Even though religious leaders control Iran, its government has an elected president and parliament.

Life in Iran and the United States

Iran	United States
Daily Life	**Daily Life**
■ An Iranian woman has to cover her head and most of her body with clothing in public.	■ Americans are free to wear any type of clothing.
■ Iranians are forbidden to view most Western Web sites, and Internet use is monitored by the government.	■ Americans are free to surf the Internet and view most Web sites.
■ Boys and girls have separate schools, and they can not be alone with each other without adult supervision.	■ Boys and girls can attend the same school.
Government	**Government**
■ Iran is a theocracy.	■ The United States is a democracy.
■ A supreme religious leader rules Iran.	■ A president is the leader of our country.
■ Only candidates approved by the government can run for political office.	■ Any U.S. citizen can run for political office.
Basic Rights	**Basic Rights**
■ Freedom of speech, religion, and the press is limited.	■ Freedom of speech, religion, and the press is allowed.

Iranian teenagers can shop for computers, but a girl must wear clothing that covers most of her body.

Unlike Iranians, Americans are free to speak in public. Here a teenager speaks on the steps of the Texas State Capitol in Austin.

Contrasting In what ways does Iran's government differ from the U.S. government?

Iran's government has supported many hard-line policies. For example, it has called for the destruction of Israel. It has also supported terrorist groups in other countries. With a newly elected president in 1997, some signs indicated that Iran's government might adopt democratic reforms. This government attempted to improve Iran's economy and rights for women.

However, in 2005 Iranians moved away from democratic reforms by electing Mahmoud Ahmadinejad (mah-MOOD ah-mah-di-nee-ZHAHD) president. He wants Iranians to follow strict Islamic law. After the election, a reporter asked the new president if he had any plans for reforms. He responded, "We did not have a revolution in order to have a democracy."

Iran's government and future as a peaceful nation remains uncertain. The United States and many European countries are concerned that Iran might try to build nuclear weapons. The U.S. and other nations see Iran's nuclear program as a threat to world security.

READING CHECK **Analyzing** What are Iran's government and people like?

SUMMARY AND PREVIEW In this section you learned about Iran's history, people, culture, economy, and government. In the next chapter, you will learn about the countries of Central Asia that lie to the north and east of Iran.

Section 4 Assessment

Reviewing Ideas, Terms, and Places

1. **a. Define** What is a **revolution**?
   b. Explain What was the Persian Empire known for?
   c. Elaborate What changes were made in Iran after the Islamic Revolution?
2. **a. Recall** What kind of leaders have authority over their people in a **theocracy**?
   b. Compare In what ways does Iran's culture differ from cultures in other countries of Southwest Asia?
   c. Predict How do you think the United States and other nations will deal with Iran's nuclear weapons program?

Critical Thinking

3. **Finding Main Ideas** Use your notes on Iran today to fill in this diagram with the main ideas of Iran's people, culture, economy, and government.

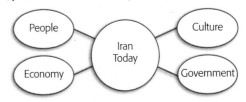

People — Iran Today — Culture
Economy — Iran Today — Government

FOCUS ON WRITING

4. **Writing about Iran** Record details about Iran in your log. What types of things would you see if you were traveling around Iran?

Social Studies Skills

Chart and Graph	Critical Thinking	Geography	Study

Analyzing Tables and Statistics

Learn

Tables provide an organized way of presenting statistics, or data. The data are usually listed side by side for easy reference and comparison. Use the following guidelines to analyze a table:

- Read the table's title to determine its subject.
- Note the headings and labels of the table's columns and rows. This will tell you how the data are organized.
- Locate statistics where rows and columns intersect by reading across rows and down columns.
- Use critical thinking skills to compare and contrast data, identify relationships, and note trends.

Literacy Rates in Southwest Asia			
	Literacy Rate (%)		
Country	Male	Female	Total
Iran	85.6	73.0	79.4
Iraq	55.9	24.4	40.4
Oman	83.1	76.2	75.8
Qatar	81.4	85.0	82.5
Saudi Arabia	84.7	70.8	78.8

Source: Central Intelligence Agency, *World Factbook 2005*

Practice

Use the table here to answer the following questions.

1. Which country has the highest total literacy rate? Which country has the lowest?

2. Which country has the largest difference between the literacy rate among men and the literacy rate among women?

3. What inference, or educated guess, can you make about education in these countries?

Apply

Using the Internet, an encyclopedia, or an almanac, locate information on the population density, birthrate, and death rate for each country listed in the table above. Then create your own table to show this information.

Chapter Review

Geography's Impact
video series
Review the video to answer the closing question:
Why is it important for countries to prepare for possible oil shortages?

Visual Summary

Use the visual summary below to help you review the main ideas of the chapter.

QUICK FACTS

Deserts cover much of the Arabian Peninsula. The region also has a lot of valuable oil reserves.

Many people in Iraq, such as this woman, enjoy more freedoms now that the country has a new government.

Iran is a theocracy. Islam is an important part of the country's government and culture.

Reviewing Vocabulary, Terms, and Places

Match the words in the columns with the correct definitions listed below.

1. wadis
2. revolution
3. embargo
4. procedure

5. fossil water
6. OPEC
7. shah
8. theocracy

a. the Persian title for a king

b. dry streambeds

c. a limit on trade

d. a series of steps taken to accomplish a task

e. water that is not being replaced by rainfall

f. an organization whose members try to influence the price of oil on world markets

g. a drastic change in a country's government

h. a government ruled by religious leaders

Comprehension and Critical Thinking

SECTION 1 *(Pages 220–223)*

9. a. Identify Through what country do the Tigris and Euphrates rivers flow?

b. Analyze Based on the landforms and climate, where do you think would be the best place in the region to live? Explain your answer.

c. Evaluate Do you think oil or water is a more important resource in the region? Explain your answer.

SECTION 2 *(Pages 224–227)*

10. a. Describe What kind of government does Saudi Arabia have?

b. Analyze In what ways does religion affect Saudi Arabia's culture and economy?

c. Elaborate What challenges might countries on the Arabian Peninsula face in attempting to create new industries in addition to oil?

SECTION 3 (Pages 230–233)

11. a. Recall What is the region of Mesopotamia known for?

   b. Draw Conclusions Why did Iraq invade Kuwait in 1990?

   c. Elaborate How did war damage Baghdad?

SECTION 4 (Pages 234–237)

12. a. Describe What occurred at the U.S. Embassy in Tehran after the Islamic Revolution?

   b. Compare and Contrast How is Iran similar to or different from the United States?

   c. Predict Do you think Iran's government will ever become more democratic? Why or why not?

FOCUS ON READING AND WRITING

13. Re-Reading Read the passage titled Resources in Section 1. After you read, write down the main ideas of the passage. Then go back and re-read the passage carefully. Identify at least one thing you learned from the passage when you re-read it and add the new information to your list of main ideas.

14. Creating a Geographer's Log Imagine that you began your journey in Saudi Arabia. Write the name of the country at the top of your log. Under the country's name, record details about what you saw as you traveled through the country. Then choose another country and do the same until you have created an entry in your log for each country. Look at your notes to help you remember what you saw. Be sure to include descriptions of the land and people.

Social Studies Skills

Analyzing Tables and Statistics *Use the Facts about Countries table at the beginning of the unit to answer the following questions.*

15. What is the population of Iraq?

16. What country is the smallest?

17. How many TVs per thousand people are there in Qatar?

Using the Internet

go.hrw.com KEYWORD: SK7 CH9

18. Activity: Charting Democracy Some countries face challenges as they work to promote a more democratic form of government. Iraq, for example, has struggled as it has begun to develop its own form of democracy. How does democracy in Iraq, or elections in Saudi Arabia and Lebanon, affect the people of those countries? How do the media, literature, and arts in those areas reflect life in a democratic society? In what ways does democracy in Southwest Asia differ from democracy in the United States? Enter the activity keyword. Then create a chart or diagram that compares democratic life in the United States with democratic life in Iraq and other countries of Southwest Asia.

Map Activity ★Interactive

19. The Arabian Peninsula, Iraq, and Iran On a separate sheet of paper, match the letters on the map with their correct labels.

Rub' al-Khali   Tehran, Iran

Persian Gulf   Riyadh, Saudi Arabia

Tigris River   Euphrates River

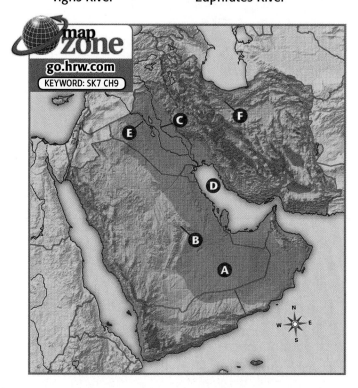

Standardized Test Practice

DIRECTIONS: Read questions 1 through 7 and write the letter of the best response. Then read question 8 and write your own well-constructed response.

1 Dry streambeds in the desert are known as

A wadis.

B salty rivers.

C exotic rivers.

D disappearing rivers.

2 What kind of government does Saudi Arabia have?

A monarchy

B legislature

C democracy

D republic

3 Iraq's official language is

A Persian.

B Arabic.

C French.

D Kurdish.

4 Saddam Hussein is the former president of

A Saudi Arabia.

B Iraq.

C Oman.

D Iran.

5 Iran's government is a theocracy ruled by

A Islamic religious leaders.

B priests.

C Christian ministers.

D democratic leaders.

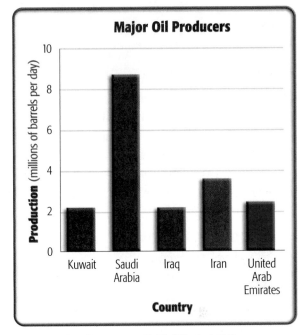

Source: Central Intelligence Agency, *The World Factbook 2005*

6 Based on the graph above, about how many barrels of oil per day does Saudi Arabia produce?

A 9

B 9 thousand

C 9 million

D 9 billion

7 Based on the graph above, which country produces the second largest amount of oil in the region?

A United Arab Emirates

B Iran

C Iraq

D Kuwait

8 **Extended Response** Based on the graph above and your knowledge of the region, write a paragraph explaining the influence oil has on the region. Identify at least two ways in which oil affects the region.

CHAPTER 10
Central Asia

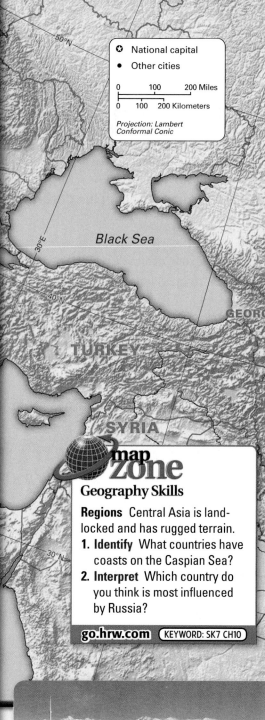

What You Will Learn...

In this chapter you will learn about the rugged physical geography of Central Asia. This physical geography has affected the region's history. You will also learn about the many influences on Central Asia throughout history. Finally, you will see how these influences have affected the region's culture, governments, and economies today.

SECTION 1
Physical Geography 244

SECTION 2
History and Culture..................248

SECTION 3
Central Asia Today 253

FOCUS ON READING AND VIEWING

Using Context Clues As you read, you may come across words in your textbook that you do not know. When this happens, look for context clues that restate the unknown word in other words that you know. **See the lesson, Using Context Clues, on page 677.**

Giving a Travel Presentation You work for a travel agency, and you are going to give a presentation encouraging people to visit Central Asia. Gather information from the chapter to help you prepare your presentation. Later you will view your classmates' presentations and provide feedback to them.

National capital
Other cities

0   100   200 Miles
0 100   200 Kilometers

Projection: Lambert Conformal Conic

Black Sea

GEORGIA

TURKEY

SYRIA

map zone
Geography Skills

Regions Central Asia is land-locked and has rugged terrain.
1. **Identify** What countries have coasts on the Caspian Sea?
2. **Interpret** Which country do you think is most influenced by Russia?

go.hrw.com KEYWORD: SK7 CH10

Geography Much of Central Asia's land is rugged. Here, mountains rise behind the city of Almaty, Kazakhstan.

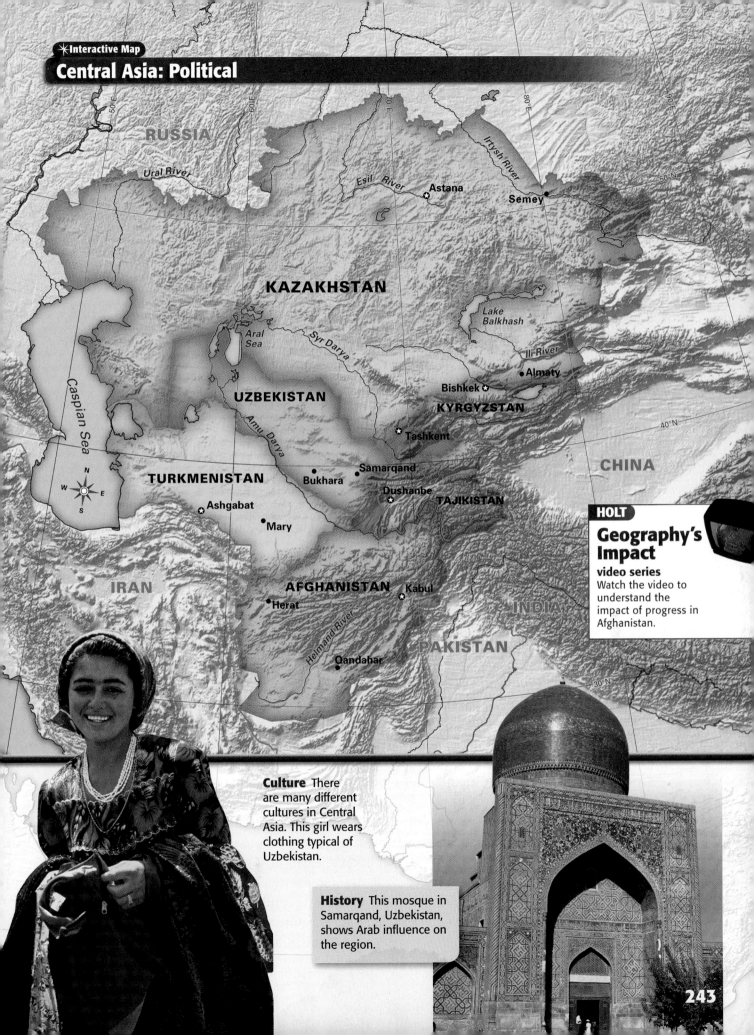

RUSSIA

Ural River

Esil River

Irtysh River

Astana

Semey

50°N

KAZAKHSTAN

Aral Sea

Syr Darya

Lake Balkhash

Ili River

UZBEKISTAN

Almaty

Bishkek

KYRGYZSTAN

Caspian Sea

Tashkent

40°N

CHINA

TURKMENISTAN

Bukhara

Samarqand

Amu Darya

Dushanbe

TAJIKISTAN

N
W E
S

Ashgabat

Mary

HOLT

Geography's Impact

video series
Watch the video to understand the impact of progress in Afghanistan.

IRAN

AFGHANISTAN

Kabul

Herat

Helmand River

INDIA

PAKISTAN

30°N

Qandahar

Culture There are many different cultures in Central Asia. This girl wears clothing typical of Uzbekistan.

History This mosque in Samarqand, Uzbekistan, shows Arab influence on the region.

Physical Geography

What You Will Learn...

Main Ideas

1. Key physical features of land-locked Central Asia include rugged mountains.
2. Central Asia has a harsh, dry climate that makes it difficult for vegetation to grow.
3. Key natural resources in Central Asia include water, oil and gas, and minerals.

The Big Idea

Central Asia, a dry, rugged, landlocked region, has oil and other valuable mineral resources.

Key Terms and Places

landlocked, *p. 244*
Pamirs, *p. 244*
Fergana Valley, *p. 245*
Kara-Kum, *p. 246*
Kyzyl Kum, *p. 246*
Aral Sea, *p. 247*

 TAKING NOTES As you read, use a chart like the one below to help you organize your notes on the physical geography of Central Asia.

Physical Features	
Climate and Vegetation	
Natural Resources	

If YOU lived there...

You are flying in a plane low over the mountains of Central Asia. You look down and notice that the area below you looks as if a giant hand has crumpled the land into steep mountains and narrow valleys. Icy glaciers fill some of the valleys. A few silvery rivers flow out of the mountains and across a green plain. This plain is the only green spot you can see in this rugged landscape.

How would this landscape affect people?

BUILDING BACKGROUND The physical geography of Central Asia affects the lives of the people who live there. This region has been shaped throughout its history by its isolated location, high mountains, dry plains, and limited resources.

Physical Features

As the name suggests, Central Asia lies in the middle of Asia. All of the countries in this region are landlocked. **Landlocked means completely surrounded by land with no direct access to the ocean.** This isolated location is just one challenge presented by the physical features of the region.

Mountains

Much of Central Asia has a rugged landscape. In the south, many high mountain ranges, such as the Hindu Kush, stretch through Afghanistan. Tajikistan and Kyrgyzstan are also very mountainous. Large glaciers are common in high mountains such as the **Pamirs**.

Like its landlocked location, Central Asia's rugged terrain presents a challenge for the region. Throughout history, the mountains have made travel and communication difficult and have contributed to the region's isolation. In addition, tectonic activity causes frequent earthquakes there.

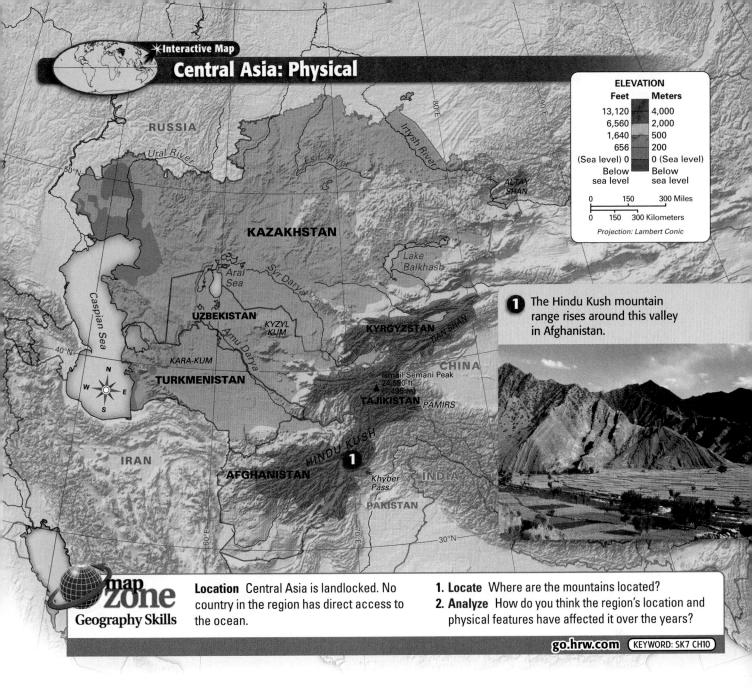

ELEVATION

Feet		Meters
13,120		4,000
6,560		2,000
1,640		500
656		200
(Sea level) 0		0 (Sea level)
Below sea level		Below sea level

0   150   300 Miles

0   150   300 Kilometers

Projection: Lambert Conic

RUSSIA

Ural River

Esil River

Irysh River

ALTAY SHAN

KAZAKHSTAN

Aral Sea

Syr Darya

Lake Balkhash

UZBEKISTAN

KYZYL KUM

KYRGYZSTAN

TIAN SHAN

CHINA

Caspian Sea

Amu Darya

KARA-KUM

Ismail Semani Peak 24,590 ft (7,495 m) ▲

TURKMENISTAN

TAJIKISTAN — PAMIRS

IRAN

HINDU KUSH

1

AFGHANISTAN

Khyber Pass

INDIA

PAKISTAN

1 The Hindu Kush mountain range rises around this valley in Afghanistan.

map zone
Geography Skills

Location Central Asia is landlocked. No country in the region has direct access to the ocean.

1. Locate Where are the mountains located?
2. Analyze How do you think the region's location and physical features have affected it over the years?

go.hrw.com KEYWORD: SK7 CH10

Plains and Plateaus

From the mountains in the east, the land gradually slopes toward the west. There, near the Caspian Sea, the land is as low as 95 feet (29 m) below sea level. The central part of the region, between the mountains and the Caspian Sea, is covered with plains and low plateaus.

The plains region is the site of the fertile **Fergana Valley**. This large valley has been a major center of farming in the region for thousands of years.

Rivers and Lakes

The Fergana Valley is fertile because of two rivers that flow through it—the Syr Darya (sir duhr-YAH) and the Amu Darya (uh-MOO duhr-YAH). These rivers flow from eastern mountains into the Aral Sea, which is really a large lake. Another important lake, Lake Balkhash, has freshwater at one end and salty water at the other end.

READING CHECK **Generalizing** What challenges do the mountains present to this region?

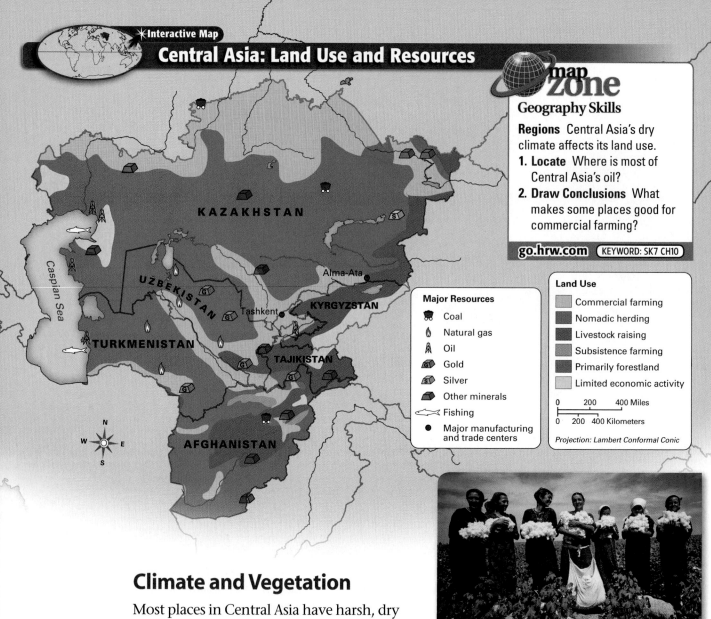

Interactive Map
Central Asia: Land Use and Resources

map zone
Geography Skills

Regions Central Asia's dry climate affects its land use.
1. **Locate** Where is most of Central Asia's oil?
2. **Draw Conclusions** What makes some places good for commercial farming?

go.hrw.com KEYWORD: SK7 CH10

Major Resources
- Coal
- Natural gas
- Oil
- Gold
- Silver
- Other minerals
- Fishing
- Major manufacturing and trade centers

Land Use
- Commercial farming
- Nomadic herding
- Livestock raising
- Subsistence farming
- Primarily forestland
- Limited economic activity

0   200   400 Miles
0 200 400 Kilometers

Projection: Lambert Conformal Conic

Cotton is the main crop in Central Asia. Students often have to take time off from school to help harvest the cotton.

Climate and Vegetation

Most places in Central Asia have harsh, dry climates. Extreme temperature ranges and limited rainfall make it difficult for plants to grow there.

One area with harsh climates in the region is the mountain area in the east. The high peaks in this area are too cold, dry, and windy for vegetation.

West of the mountains and east of the Caspian Sea is another harsh region. Two deserts—the **Kara-Kum** (kahr-uh-KOOM) in Turkmenistan and the **Kyzyl Kum** (ki-ZIL KOOM) in Uzbekistan and Kazakhstan— have extremely high temperatures in the summer. Rainfall is limited, though both deserts contain several settlements. Rivers crossing this dry region make settlements possible, because they provide water for irrigation. Irrigation is a way of supplying water to an area of land.

The only part of Central Asia with a milder climate is the far north. There, temperature ranges are not so extreme and rainfall is heavy enough for grasses and trees to grow.

READING CHECK **Generalizing** Why is it hard for plants to grow in much of Central Asia?

246 CHAPTER 10

Natural Resources

In this dry region, water is one of the most valuable resources. Although water is scarce, or limited, the countries of Central Asia do have oil and other resources.

Water

The main water sources in southern Central Asia are the Syr Darya and Amu Darya rivers. Since water is so scarce there, different ideas over how to use the water from these rivers have led to conflict between Uzbekistan and Turkmenistan.

Today farmers use river water mostly to irrigate cotton fields. Cotton grows well in Central Asia's sunny climate, but it requires a lot of water. Irrigation has taken so much water from the rivers that almost no water actually reaches the **Aral Sea** today. The effect of this irrigation has been devastating to the Aral Sea. It has lost more than 75 percent of its water since 1960. Large areas of seafloor are now exposed.

In addition to water for irrigation, Central Asia's rivers supply power. Some countries have built large dams on the rivers to generate hydroelectricity.

Oil and Other Resources

The resources that present the best economic opportunities for Central Asia are oil and gas. Uzbekistan, Kazakhstan, and Turkmenistan all have huge reserves of oil and natural gas.

However, these oil and gas reserves cannot benefit the countries of Central Asia unless they can be exported. Since no country in the region has an ocean port, the only way to transport the oil and gas efficiently is through pipelines. But the rugged mountains, along with economic and political turmoil in some surrounding countries, make building and maintaining pipelines difficult.

In addition to oil and gas, some parts of Central Asia are rich in other minerals. They have deposits of gold, silver, copper, zinc, uranium, and lead. Kazakhstan, in particular, has many mines with these minerals. It also has large amounts of coal.

FOCUS ON READING
What context clues give you a restatement of the term *scarce*?

READING CHECK **Categorizing** What are three types of natural resources in Central Asia?

SUMMARY AND PREVIEW In this section you learned about Central Asia's rugged terrain, dry climate, and limited resources. In the next section you will learn about the history and culture of Central Asia.

Section 1 Assessment

go.hrw.com
Online Quiz
KEYWORD: SK7 HP10

Reviewing Ideas, Terms, and Places

1. **a. Identify** What fertile area has been a center of farming in Central Asia for many years?
   b. Make Inferences How does Central Asia's terrain affect life there?
2. **a. Describe** Where do people find water in the deserts?
   b. Make Generalizations What is the climate like in most of Central Asia?
3. **a. Recall** What mineral resources does Central Asia have?
   b. Explain How have human activities affected the **Aral Sea**?
   c. Elaborate What kinds of situations would make it easier for countries of Central Asia to export oil and gas?

Critical Thinking

4. **Finding Main Ideas** Look at your notes on this section. Then, using a chart like the one here, write a main idea statement about each topic.

	Main Idea
Physical Features	
Climate and Vegetation	
Natural Resources	

FOCUS ON VIEWING

5. **Describing Physical Geography** Note information about physical features, climates, and resources of this region. Highlight information to include in your presentation.

History and Culture

What You Will Learn...

Main Ideas

1. Throughout history, many different groups have conquered Central Asia.
2. Many different ethnic groups and their traditions influence culture in Central Asia.

The Big Idea

The countries of Central Asia share similar histories and traditions, but particular ethnic groups give each country a unique culture.

Key Terms and Places

Samarqand, *p. 248*
nomads, *p. 250*
yurt, *p. 250*

TAKING NOTES As you read, use a chart like the one here to organize your notes on the history and culture of Central Asia. Be sure to pay attention to the different peoples that influenced the region.

History	Culture

If YOU lived there...

Your family has always farmed a small plot of land. Most days you go to school and work in the fields. One day you get news that invaders have taken over your country. They don't look like you and they speak a different language, but now they are in charge.

How do you think your life will change under the new rulers?

BUILDING BACKGROUND You may have noticed that the names of the countries in this region all end with *stan*. In the language of the region, *stan* means "land of." So, for example, Kazakhstan means "land of the Kazakhs." However, throughout history many different groups have ruled these lands.

History

Central Asia has been somewhat of a crossroads for traders and invaders for hundreds of years. As these different peoples have passed through Central Asia, they have each left their own unique and lasting influences on the region.

Trade

At one time, the best trade route between Europe and India ran through Afghanistan. The best route between Europe and China ran through the rest of Central Asia. Beginning in about 100 BC, merchants traveled along the China route to trade European gold and wool for Chinese spices and silk. As a result, this route came to be called the Silk Road. Cities along the road, such as **Samarqand** and Bukhara, grew rich from the trade.

By 1500 the situation in Central Asia had changed, however. When Europeans discovered they could sail to East Asia through the Indian Ocean, trade through Central Asia declined. The region became more isolated and poor.

Central Asia Today

If YOU lived there...

Your country, Kyrgyzstan, has just had an election. You listen to the radio with your brother, anxiously awaiting the results of the election. When the radio announcer says that the same president has won again, your brother is very angry. He says the election was unfair, and he is going to protest outside the president's palace. He expects there to be a big crowd.

Will you join your brother? Why or why not?

BUILDING BACKGROUND Political protests have been fairly common in some Central Asian countries in recent years. Political instability is just one of the challenges facing Central Asia today as the region learns to deal with independence.

Central Asia Today

A history of invasions and foreign rule has made an impact on Central Asia. Because of years of fighting and changes in the region, today many countries of Central Asia face similar issues in building stable governments and strong economies.

Afghanistan

The situation in Afghanistan today is in many ways a result of a long war with the Soviet Union in the 1980s. The Soviets left in 1989. However, turmoil continued under an alliance of Afghan groups. In the mid-1990s a radical Muslim group known as the **Taliban** arose. The group's leaders took over most of the country, including the capital, **Kabul**.

The Taliban used a strict interpretation of Islamic teachings to rule Afghanistan. For example, the Taliban severely limited the role of women in society. They forced women to wear veils and to stop working outside the home. They also banned all music and dancing. Although most Muslims sharply disagreed with the Taliban's policies, the group remained in power for several years.

What You Will Learn...

Main Ideas

1. The countries of Central Asia are working to develop their economies and to improve political stability in the region.
2. The countries of Central Asia face issues and challenges related to the environment, the economy, and politics.

The Big Idea

Central Asian countries are mostly poor, but they are working to create stable governments and sound economies.

Key Terms and Places

Taliban, *p. 253*
Kabul, *p. 253*
dryland farming, *p. 255*
arable, *p. 255*

TAKING NOTES As you read, use a chart like the one below to help you take notes on governments, economies, and challenges in Central Asia today.

Afghanistan	
Kazakhstan	
Kyrgyzstan	
Tajikistan	
Turkmenistan	
Uzbekistan	

Reforms in Afghanistan

Some reforms have taken place in Afghanistan since the end of Taliban rule. However, the country still faces many challenges.

ANALYZING VISUALS What opportunities might education create for this girl?

Since the End of Taliban Rule . . .

- Afghanistan has a new constitution and an elected president.
- Many people are registered to vote.
- Afghanistan's rules are written and accessible to citizens for the first time.
- New clinics and trained doctors provide more people with access to health care.
- Women can work outside the home.
- Girls can attend school.

Eventually, the Taliban came into conflict with the United States. Investigation of the September 11, 2001, terrorist attacks on New York City and Washington, D.C., led to terrorist leader Osama bin Laden and his al Qaeda network, based in Afghanistan. U.S. and British forces attacked Taliban and al Qaeda targets and toppled Afghanistan's Taliban government.

Since the fall of the Taliban, Afghanistan's government has changed in many ways. The country has a new constitution. Also, all men and women age 18 and older can vote for the president and for the members of a national assembly. Some members of the assembly are appointed by the president, and the constitution requires that half of these appointees be women.

FOCUS ON READING

What is a restatement of *factions*?

Many Afghans hope their government will be stable. However, political factions, or opposing groups, disagree with some of the recent changes. These groups threaten violence, which may make Afghanistan's new government less stable.

Kazakhstan

Kazakhstan was the first part of Central Asia to be conquered by Russia. As a result, Russian influence remains strong in that country today. About one-third of Kazakhstan's people are ethnic Russians. Kazakh and Russian are both official languages. Many ethnic Kazakhs grow up speaking Russian at home and have to learn Kazakh in school.

Kazakhstan's economy was once tied to the former Soviet Union's. It was based on manufacturing. When the Soviet Union collapsed, the economy suffered. However, due to its valuable oil reserves and quick adaptation to the free market, Kazakhstan's economy is now growing steadily. The country is the richest in Central Asia.

Kazakhstan also has one of the more stable governments in Central Asia. The country is a democratic republic with an elected president and parliament. In 1998 Kazakhstan moved its capital from Almaty to Astana, which is closer to Russia.

Kyrgyzstan

The word *kyrgyz* means "forty clans." Throughout history, clan membership has been an important part of Kyrgyzstan's social, political, and economic life. Many people still follow nomadic traditions.

Many other people in Kyrgyzstan are farmers. Fertile soils there allow a mix of irrigated crops and **dryland farming, or farming that relies on rainfall instead of irrigation.** Farming is the most important industry in Kyrgyzstan. However, it does not provide much income for the country.

Although the standard of living in Kyrgyzstan is low, the economy shows signs of strengthening. Tourism might also help Kyrgyzstan's economy. The country has a Muslim pilgrimage site as well as the beautiful Lake Issyk-Kul.

Kyrgyzstan's government is changing. The country has been fairly stable for some years. However, protests in 2005 over what some people thought were unfair elections could signal that times are changing.

Tajikistan

Like other countries in Central Asia, Tajikistan is struggling to overcome its problems. In the mid-1990s the country's Communist government fought against a group of reformers. Some reformers demanded democracy. Others called for a government that ruled by Islamic law. The groups came together and signed a peace agreement in 1997. As a result, Tajikistan is now a republic with an elected president.

Years of civil war damaged Tajikistan's economy. Both industrial and agricultural production declined. Even with the decline in agricultural production, Tajikistan still relies on cotton farming for much of its income. However, only 5 to 6 percent of the country's land is **arable, or suitable for growing crops.** Lack of arable land makes progress there difficult.

Turkmenistan

Turkmenistan's president holds all power in the country. He was voted president for life by the country's parliament. He has used his power to name a month of the year after himself, and his face appears on almost everything in Turkmenistan.

The Turkmen government supports Islam and has ordered schools to teach Islamic principles. However, it also views Islam with caution. It does not want Islam to become a political movement.

Tajikistan's economy is based on oil, gas, and cotton. Although the country is a desert, about half of it is planted with cotton fields. Farming is possible because Turkmenistan has the longest irrigation channel in the world.

FOCUS ON CULTURE

Turkmen Carpets

Decorative carpets are an essential part of a nomad's home. They are also perhaps the most famous artistic craft of Turkmenistan. Carpet factories operate in cities all through Turkmenistan, but some women still weave carpets by hand. These weavers memorize hundreds of intricate designs so they can make rugs that look the same. Each of several different Turkmen tribes has its own rug design.

Analyzing Why are carpets good for a nomadic way of life?

Uzbekistan

Uzbekistan has the largest population of the Central Asian countries. It also has the largest cities in the region. Two cities—Bukhara and Samarqand—are famous for their mosques and monuments.

As in Turkmenistan, Uzbekistan's elected president holds all the political power. The United States has criticized the government for not allowing political freedom or respecting human rights.

The government also closely controls the economy. Uzbekistan's economy, based on oil, gold, and cotton, is fairly stable even though it is growing only very slowly.

READING CHECK **Drawing Inferences** How does physical geography affect the economies of Kyrgyzstan and Tajikistan?

Issues and Challenges

As you have read, the countries of Central Asia face similar issues and challenges. Their greatest challenges are in the areas of environment, economy, and politics.

Environment

One of the most serious environmental problems is the shrinking of the Aral Sea. Winds sweep the dry seafloor and blow dust, salt, and pesticides hundreds of miles. Also, towns that once relied on fishing are now dozens of miles from the shore.

Another problem is the damage caused by Soviet military practices. The Soviets tested nuclear bombs in Central Asia. Now people there suffer poor health because of radiation left over from the tests.

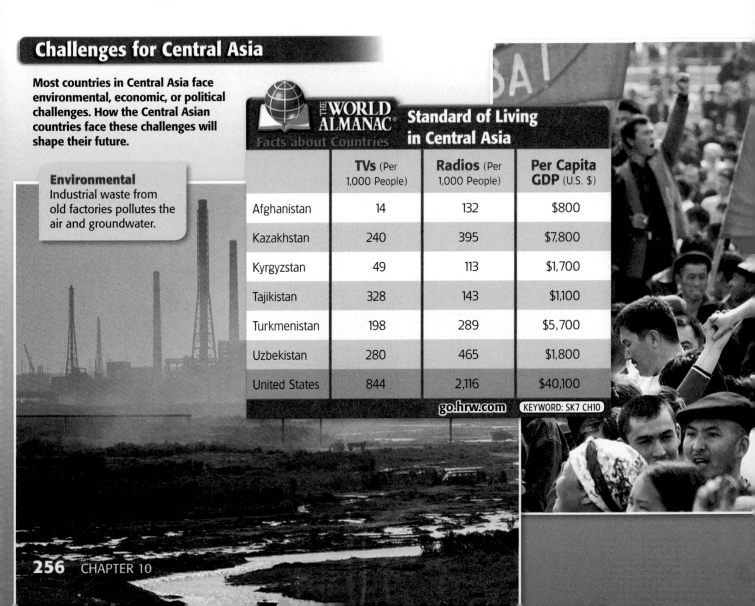

Challenges for Central Asia

Most countries in Central Asia face environmental, economic, or political challenges. How the Central Asian countries face these challenges will shape their future.

Environmental
Industrial waste from old factories pollutes the air and groundwater.

THE WORLD ALMANAC Facts about Countries
Standard of Living in Central Asia

	TVs (Per 1,000 People)	Radios (Per 1,000 People)	Per Capita GDP (U.S. $)
Afghanistan	14	132	$800
Kazakhstan	240	395	$7,800
Kyrgyzstan	49	113	$1,700
Tajikistan	328	143	$1,100
Turkmenistan	198	289	$5,700
Uzbekistan	280	465	$1,800
United States	844	2,116	$40,100

go.hrw.com   KEYWORD: SK7 CH10

Another environmental problem has been caused by the overuse of chemicals to increase crop production. These chemicals have ended up ruining some farmlands. Instead of increasing crop production, the chemicals have hurt the economy.

Economy

Many of Central Asia's economic problems are due to reliance on one crop—cotton. Suitable farmland is limited, so employment in the cotton industry is limited. Also, the focus on cotton has not encouraged countries to develop manufacturing.

Some countries have oil and gas reserves that may someday make them rich. For now, though, outdated equipment, lack of funds, and poor transportation systems slow development in Central Asia.

Politics

The other main challenge in Central Asia today is lack of political stability. In some countries, such as Kyrgyzstan, people do not agree on the best kind of government. People who are dissatisfied with their government sometimes turn to violence. These countries today are often faced with terrorist threats from different political groups within their own countries.

READING CHECK **Summarizing** What environmental challenges does Central Asia face?

SUMMARY Central Asia is recovering from a history of foreign rule. The region is struggling to develop sound economies and stable governments.

Political Protesters show their opposition to the government in Kyrgyzstan.

ANALYSIS SKILL ANALYZING VISUALS

What do you think could be done to improve the environment in Central Asia?

go.hrw.com
Online Quiz
KEYWORD: SK7 HP10

Section 3 Assessment

Reviewing Ideas, Terms, and Places

1. **a. Describe** How did the **Taliban** affect Afghanistan?
   b. Contrast What are some major differences between Afghanistan and Kazakhstan?
   c. Elaborate What is one way a country might create more **arable** land?
2. **a. Identify** What three types of challenges does Central Asia face today?
   b. Make Generalizations Why does much of Central Asia face political instability?

Critical Thinking

3. **Categorizing** Using your notes and a chart like the one here, categorize your information on each Central Asian country. You will have to add more lines as needed.

	Government	Economy
Afghanistan		
Kazakhstan		

FOCUS ON VIEWING

4. **Describing Central Asia Today** Write notes about each country in Central Asia. Which countries will you suggest listeners visit? What details will encourage them?

Geography and History

The Aral Sea

In 1960 the Aral Sea was the world's fourth-largest lake. However, human activities over the years have caused the Aral Sea to shrink drastically. The lake's former seafloor is now a desert of sand and salt. Also, towns that once benefited from their lakeside location are now left without access to the water. Area governments have built dams to control the flow of water into the lake, but so far their efforts have been unsuccessful.

Cause
Farmers have taken water from the Amu Darya and the Syr Darya to irrigate cotton fields. Now, less water flows into the sea than evaporates from it.

Effect
Stranded boats are a reminder of the fishing industry once based near the Aral Sea. The fishing industry is dying with the sea.

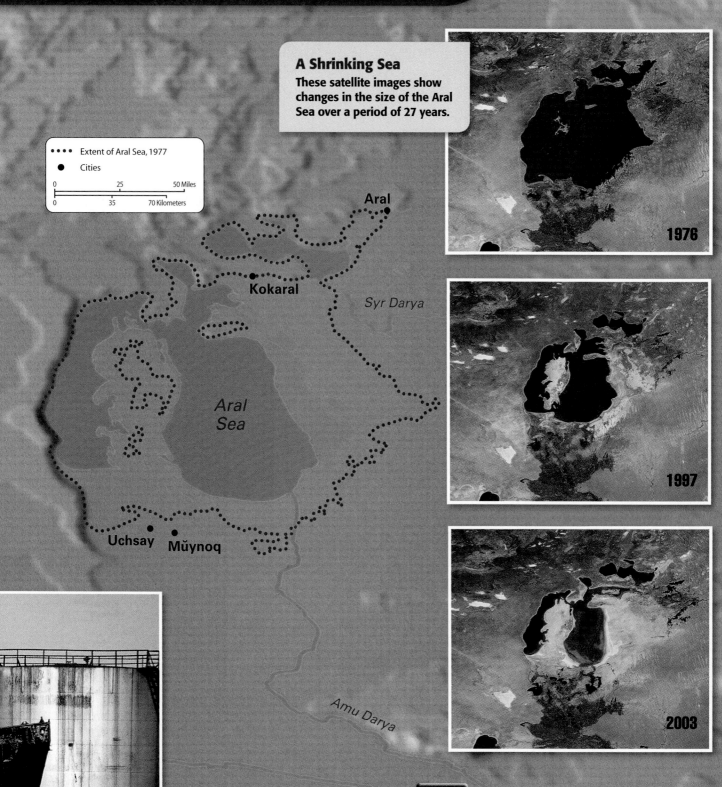

A Shrinking Sea
These satellite images show changes in the size of the Aral Sea over a period of 27 years.

•••• Extent of Aral Sea, 1977
● Cities

0   25   50 Miles
0   35   70 Kilometers

Aral

Kokaral

Syr Darya

Aral Sea

Uchsay   Mŭynoq

Amu Darya

1976

1997

2003

ANALYSIS SKILL ANALYZING VISUALS

1. **Place** How much wider was the Aral Sea in 1977 than it is now?
2. **Human-Environment Interaction** How might the shrinking of the sea affect towns that were once on its shore?

Using Scale

Learn

Mapmakers use scales to represent distances between points on a map. On each map legend in this book, you will notice some lines marked to measure miles and kilometers. These lines are the map's scale.

To find the distance between two points on a map, place a piece of paper so that the edge connects the two points. Mark the location of each point on the paper with a line or dot. Then compare the distance between the two dots with the map's scale.

Practice

Use the maps here to practice using scale and to answer the following questions.

❶ Which map shows a larger region?

❷ About how many miles does one inch represent on the map of Kyrgyzstan? on the map of Bishkek?

❸ How far is it from Dubovy Park to Victory Square in Bishkek?

Apply

Use the map of Southwest Asia in the unit opener to answer the following questions.

1. How many miles does one inch represent? How many kilometers does one inch represent?

2. How long is the Caspian Sea from north to south?

3. How far is Turkmenistan from the Persian Gulf?

Bishkek

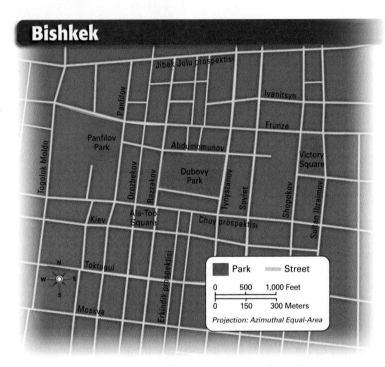

Kyrgyzstan

Chapter Review

Geography's Impact
video series
Review the video to answer the closing question:
What challenges do the people of Afghanistan face today?

Visual Summary

Use the visual summary below to help you review the main ideas of the chapter.

QUICK FACTS

Central Asia is a dry, rugged region. However, people use irrigation to grow cotton, the region's main crop.

Some people follow traditional ways of life. For example, nomads move their yurts from place to place.

Central Asia faces environmental, economic, and political challenges. One challenge is the Aral Sea's shrinking.

Reviewing Vocabulary, Terms, and Places

Unscramble each group of letters below to spell a term that matches the given definition.

1. **mnodsa**—people who move often from place to place

2. **yrddnal mrignaf**—farming that relies on rainfall, not irrigation

3. **tryu**—a moveable round house of wool felt mats hung over a wood frame

4. **ssblhieat**—to set up or create

5. **fgrenaa vlyela**—fertile region that has been a center of farming for thousands of years

6. **tlbania**—a radical Muslim group

7. **dknadclleo**—completely surrounded by land with no direct access to the ocean

8. **kluba**—the capital of Afghanistan

9. **aabler**—suitable for growing crops

10. **aalr sae**—body of water that is shrinking because of use of water for irrigation

Comprehension and Critical Thinking

SECTION 1 *(Pages 244–247)*

11. **a. Describe** How are farmers able to grow crops in Central Asia's dry landscapes?

    b. Analyze What factors make it difficult for the countries of Central Asia to export their oil and gas resources?

    c. Evaluate Do you think Central Asia's location or its mountains do more to keep the region isolated? Explain your answer.

SECTION 2 *(Pages 248–252)*

12. **a. Describe** How did life in Central Asia change under Russian and Soviet rule?

    b. Analyze In what ways do the people of Central Asia show their pride in their past and their culture?

    c. Evaluate Why do you think many former nomads now live in cities? Why do you think other people still choose to live as nomads?

SECTION 3 *(Pages 253–257)*

13. a. Identify What are some reforms that have taken place in Afghanistan since the fall of the Taliban?

b. Analyze How does having a limited amount of arable land affect Tajikistan's economy?

c. Elaborate How do you think political and environmental challenges in Central Asia affect the region's economy?

Using the Internet

go.hrw.com
KEYWORD: SK7 CH10

14. Activity: Writing Home For thousands of years, nomads have traveled the lands of Central Asia. They move their herds to several different pasture areas as the seasons change. Enter the activity keyword and join a caravan of nomads. Find out what it is like to pack up your house, clothes, and all you own as you move from place to place. Then create a postcard to share your adventures with friends and family back home in the United States.

FOCUS ON READING AND VIEWING

15. Using Context Clues Look through your book for examples of restatement. Note one or two examples of restatement for each section of the chapter.

16. Giving a Travel Presentation Review your notes and select one country in Central Asia your audience might want to visit. Look for pictures of at least five locations in that country: buildings, monuments, or other interesting places. As you plan your presentation, create a brief introduction, a brief description of each location and its picture, and a conclusion. Hold up each picture and point out important features as you make your presentation.

As you watch your classmates' presentations, view and listen carefully. Make note of their eye contact with the audience, use of gestures to add interest, use of interesting pictures, and persuasiveness.

Social Studies Skills

Using Scale *Use the physical map of Central Asia in Section 1 to answer the following questions.*

17. How many miles does one inch represent?

18. How far is it from the shore of the Caspian Sea to Ismail Samani Peak?

19. How many kilometers long is the Amu Darya?

Map Activity

20. Central Asia On a separate sheet of paper, match the letters on the map with their correct labels.

Aral Sea	Pamirs
Caspian Sea	Astana, Kazakhstan
Afghanistan	Tashkent, Uzbekistan

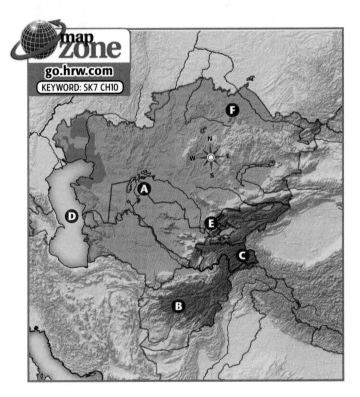

go.hrw.com
KEYWORD: SK7 CH10

Standardized Test Practice

DIRECTIONS: Read questions 1 through 7 and write the letter of the best response. Then read question 8 and write your own well-constructed response.

1 **What is the main crop grown in Central Asia?**

A wheat

B olives

C cotton

D corn

2 **Which of the following descriptions *best* describes the landscape of Central Asia?**

A dry and rugged

B dry and flat

C humid and landlocked

D humid and cold

3 **How did the Arabs influence Central Asia in the 700s and 800s?**

A separated ethnic groups

B destroyed cities and irrigation systems

C built railroads and expanded oil production

D introduced Islam

4 **Which of the following statements about the nomadic lifestyle is false?**

A Nomads move their herds depending on the season.

B Nomads decorate their yurts with carpets.

C It is a symbol of the region's heritage.

D Nomads often move from one dwelling to another.

5 **What country did the Taliban rule?**

A Kazakhstan

B Afghanistan

C Kyrgyzstan

D Uzbekistan

Farmland in Central Asia

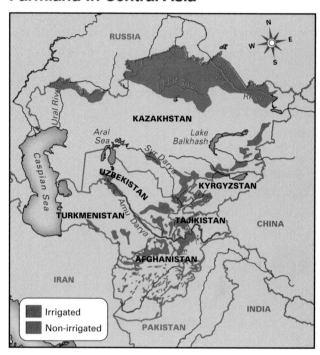

6 **Based on the map above, what country has the most non-irrigated farmland?**

A Kazakhstan

B Tajikistan

C Turkmenistan

D Uzbekistan

7 **Based on the map above and your knowledge of the physical geography of Central Asia, what is the main reason there is little farmland in eastern Kyrgyzstan and Tajikistan?**

A There are too many rivers.

B Most people live as nomads there.

C The area is too mountainous.

D The area is a desert.

8 **Extended Response** Using the map and your knowledge of Central Asia, write a brief essay explaining how irrigation has affected the region.

Compare and Contrast

How are two countries alike? How are they different? Comparing the similarities and contrasting the differences between countries can teach us more than we can learn by studying them separately.

Assignment

Write a paper comparing and contrasting two countries from this unit. Consider physical geography, government, and/or culture.

1. Prewrite

Choose a Topic

- Choose two countries to write about.
- Create a big idea, or thesis, about the two counries. For example, your big idea might be "Iran and Iraq both have oil-based economies, but they also have many differences."

> **TIP** **Organizing Information** A Venn diagram (two overlapping circles) can help you plan your paper. Write similarities in the overlapping area and differences in the areas that do not overlap.

Gather and Organize Information

- Identify at least three similarities or differences between the countries.
- Decide whether to write about each country one at a time or to discuss each point of similarity or difference one at a time.

2. Write

Use a Writer's Framework

A Writer's Framework

Introduction
- Start with a fact or question relating to both countries.
- Identify your big idea.

Body
- Write at least one paragraph for each country or each point of similarity or difference. Include facts and details to help explain each point.
- Use block style or point-by-point style.

Conclusion
- Summarize the process in your final paragraph.

3. Evaluate And Revise

Review and Improve Your Paper

- Re-read your draft, then ask yourself the questions below to see if you have followed the framework.
- Make any changes needed to improve your comparison and conrast paper.

Evaluation Questions for a Compare and Contrast Paper

1. Do you begin with an interesting fact or question that relates to both countries?
2. Does your first paragraph clearly state your big idea and provide background information?
3. Do you discuss at least three similarities and differences between the countries?
4. Do you include facts and details to explain each similarity or difference?
5. Is your paper clearly organized by country or by similarities and differences?

4. Proofread And Publish

Give Your Explanation the Finishing Touch

- Make sure you have capitalized the names of countries and cities.
- Check for punctuation around transitional words and phrases like and, but, or similarly.
- Share your compare-and-contrast paper by reading it aloud in class or in small groups.

5. Practice And Apply

Use the steps outlined in this workshop to write a compare-and-contrast paper. Compare and contrast your paper to those of your classmates.

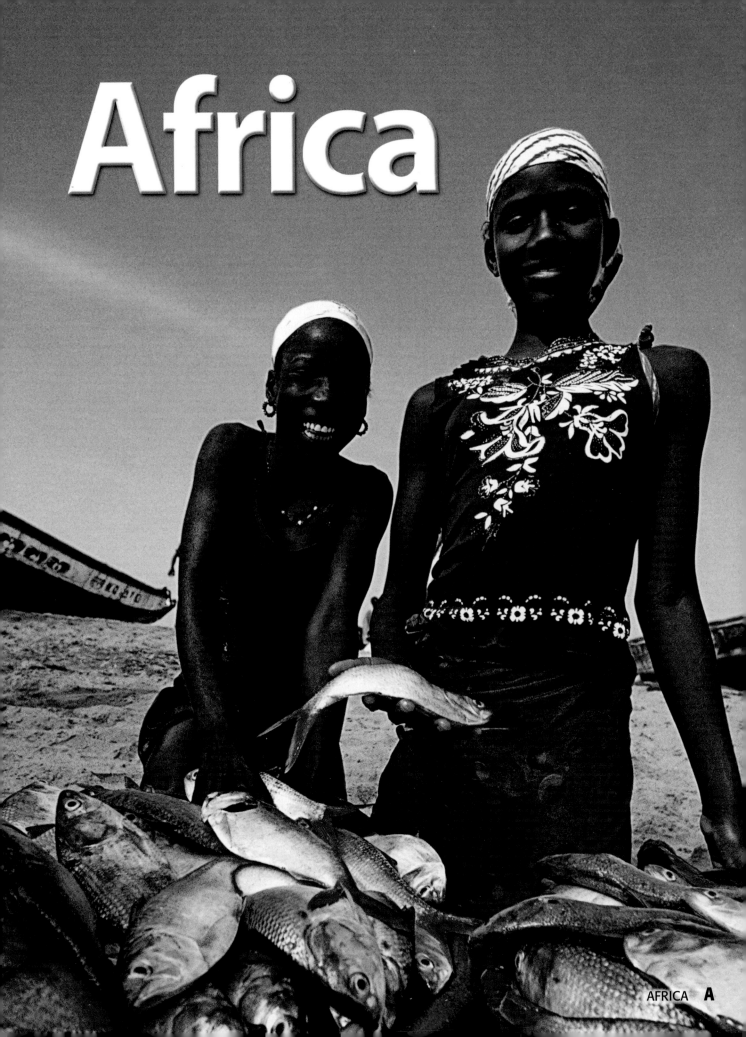

Africa

The Sahara

The world's largest desert, the Sahara, dominates land and life in North Africa.

Savannas

Grassy plains called savannas stretch across large parts of the continent and are home to much African wildlife.

Africa

Rift Valleys

In East Africa, Earth's crust is slowly being pulled apart. This causes hills, long lakes, and wide "rift valleys" to form.

Explore the Satellite Image
A huge continent, Africa is home to many different kinds of physical features. Based on this satellite image, how would you describe Africa's physical geography?

The Satellite's Path

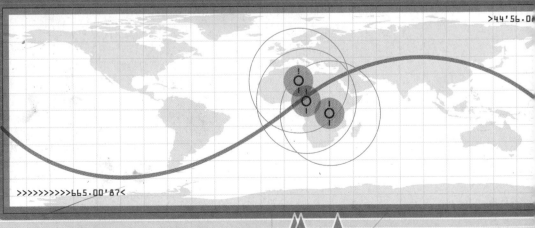

>44'56.0

>>>>>>>>665.00'87<

567.476.348

456.094.

+803
+799

+966

+355

Africa: Physical

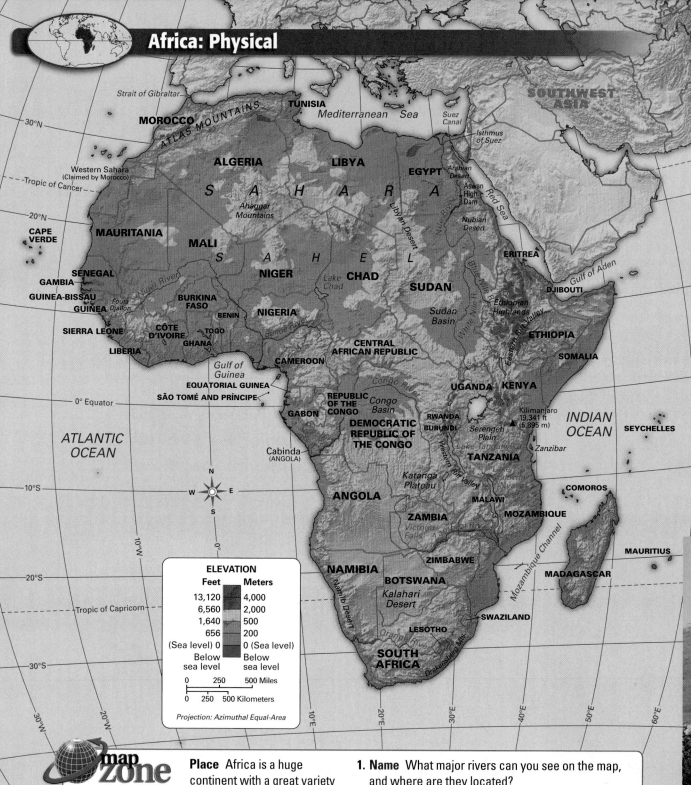

SOUTHWEST ASIA

Strait of Gibraltar
MOROCCO
TUNISIA
Mediterranean Sea
Suez Canal
Isthmus of Suez
30°N
ATLAS MOUNTAINS
ALGERIA
LIBYA
EGYPT
Arabian Desert
Aswan High Dam
Western Sahara (Claimed by Morocco)
Tropic of Cancer
S A H A R A
Ahaggar Mountains
Libyan Desert
Nubian Desert
Red Sea
20°N
CAPE VERDE
MAURITANIA
MALI
S A H E L
NIGER
Lake Chad
CHAD
SUDAN
ERITREA
Gulf of Aden
SENEGAL
GAMBIA
GUINEA-BISSAU
GUINEA
Fouta Djallon
Niger River
BURKINA FASO
BENIN
NIGERIA
Sudan Basin
Blue Nile R.
Ethiopian Highlands
DJIBOUTI
SIERRA LEONE
CÔTE D'IVOIRE
TOGO
GHANA
Benue River
CENTRAL AFRICAN REPUBLIC
White Nile R.
Eastern Rift Valley
ETHIOPIA
SOMALIA
LIBERIA
Gulf of Guinea
CAMEROON
EQUATORIAL GUINEA
SÃO TOMÉ AND PRÍNCIPE
0° Equator
REPUBLIC OF THE CONGO
GABON
Congo River
Congo Basin
UGANDA
KENYA
Kilimanjaro 19,341 ft (5,895 m)
INDIAN OCEAN
SEYCHELLES
DEMOCRATIC REPUBLIC OF THE CONGO
RWANDA
BURUNDI
Serengeti Plain
Lake Tanganyika
Zanzibar
TANZANIA
ATLANTIC OCEAN
Cabinda (ANGOLA)
Western Rift Valley
Lake Nyasa
10°S
Katanga Plateau
MALAWI
COMOROS
ANGOLA
Victoria Falls
ZAMBIA
MOZAMBIQUE
MAURITIUS
Mozambique Channel
ZIMBABWE
MADAGASCAR

ELEVATION

Feet		Meters
13,120		4,000
6,560		2,000
1,640		500
656		200
(Sea level) 0		0 (Sea level)
Below sea level		Below sea level

NAMIBIA
BOTSWANA
Kalahari Desert
20°S
Tropic of Capricorn
Namib Desert
SWAZILAND
LESOTHO
Orange River
Drakensberg Mts.
30°S
SOUTH AFRICA

0 250 500 Miles
0 250 500 Kilometers

Projection: Azimuthal Equal-Area

N W E S

map zone
Geography Skills

Place Africa is a huge continent with a great variety of physical features.

1. **Name** What major rivers can you see on the map, and where are they located?
2. **Compare** How does the average elevation of southern Africa compare to that of northern Africa?

Africa

THE WORLD ALMANAC — Facts about the World

Geographical Extremes: Africa

Longest River	Nile River, Egypt: 4,160 miles (6,693 km)	**Driest Place**	Wadi Halfa, Sudan: .1 inches (.3 cm) average precipitation per year
Highest Point	Mount Kilimanjaro, Tanzania: 19,340 feet (5,895 m)	**Largest Country**	Sudan: 967,498 square miles (2,505,820 square km)
Lowest Point	Lake Assal, Djibouti: 512 feet (156 m) below sea level	**Smallest Country**	Seychelles: 176 square miles (456 square km)
Highest Recorded Temperature	El Azizia, Libya: 136°F (57.8°C)	**Largest Desert**	Sahara: 3,500,000 square miles (9,065,000 square km)
Lowest Recorded Temperature	Ifrane, Morocco: -11°F (-23.9°C)	**Largest Island**	Madagascar: 226,658 square miles (587,044 square km)
Wettest Place	Debundscha, Cameroon: 405 inches (1,028.7 cm) average precipitation per year	**Highest Waterfall**	Tugela, South Africa: 2,014 feet (614 m)

go.hrw.com KEYWORD: SK7 UN3

Size Comparison: The United States and Africa

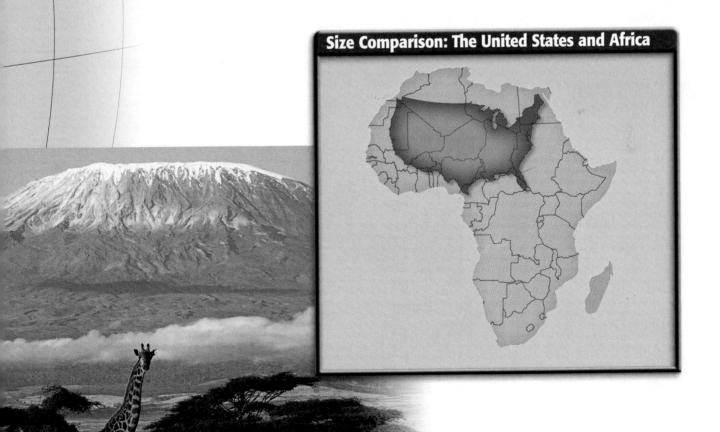

Mount Kilimanjaro, Tanzania

Africa: Political

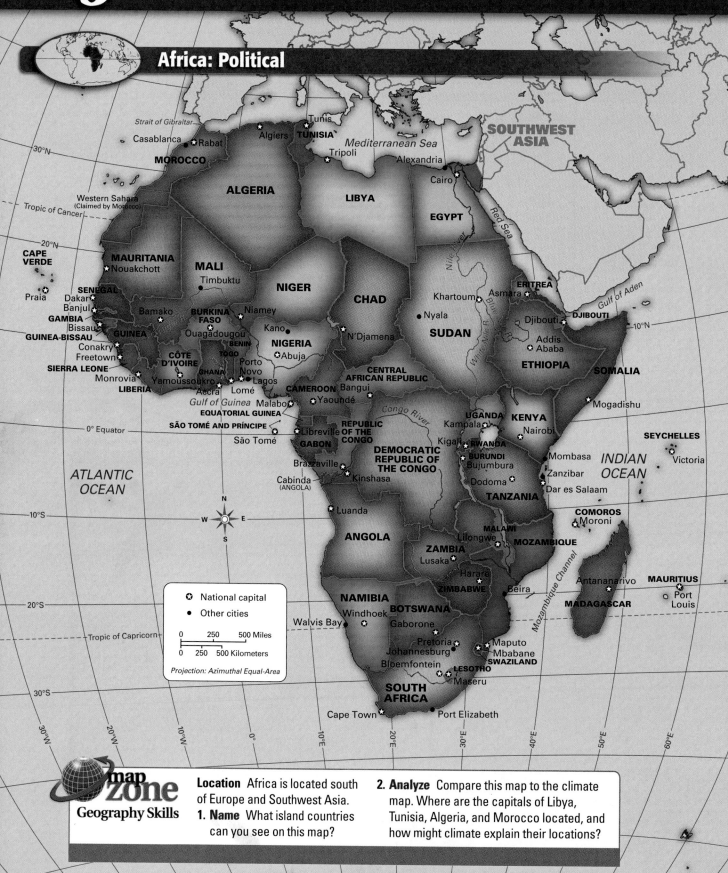

Strait of Gibraltar
Casablanca • Rabat
MOROCCO
Tunis
Algiers • TUNISIA
Tripoli
Mediterranean Sea
Alexandria
SOUTHWEST ASIA

30°N

ALGERIA
LIBYA
Cairo
EGYPT

Western Sahara
(Claimed by Morocco)
Tropic of Cancer

Red Sea

20°N
CAPE VERDE
MAURITANIA
Nouakchott
MALI
Timbuktu
NIGER
CHAD
Khartoum
Nyala
ERITREA
Asmara
Gulf of Aden
DJIBOUTI
Djibouti
10°N

Praia
SENEGAL
Dakar
Banjul
GAMBIA
Bissau
GUINEA-BISSAU
Conakry
Freetown
SIERRA LEONE
Monrovia
LIBERIA
Bamako
BURKINA FASO
Ouagadougou
GUINEA
CÔTE D'IVOIRE
GHANA
Yamoussoukro
Accra
Niamey
Kano
BENIN
TOGO
NIGERIA
Abuja
Porto Novo
Lomé
Lagos
N'Djamena
CENTRAL AFRICAN REPUBLIC
Bangui
SUDAN
Nile River
Blue Nile R.
White Nile R.
Addis Ababa
ETHIOPIA
SOMALIA
Mogadishu

CAMEROON
Yaoundé
Malabo
EQUATORIAL GUINEA
Gulf of Guinea
SÃO TOMÉ AND PRÍNCIPE
São Tomé
Libreville
GABON
REPUBLIC OF THE CONGO
Brazzaville
Cabinda (ANGOLA)
Kinshasa
Congo River
DEMOCRATIC REPUBLIC OF THE CONGO
Kampala
UGANDA
Kigali
RWANDA
BURUNDI
Bujumbura
Dodoma
KENYA
Nairobi
Mombasa
Zanzibar
Dar es Salaam
TANZANIA
SEYCHELLES
Victoria
INDIAN OCEAN

0° Equator

ATLANTIC OCEAN

Luanda
10°S
ANGOLA
ZAMBIA
Lusaka
MALAWI
Lilongwe
MOZAMBIQUE
COMOROS
Moroni
Mozambique Channel
Antananarivo
MAURITIUS
Port Louis

N
W E
S

Harare
ZIMBABWE
Beira
MADAGASCAR

20°S
NAMIBIA
Windhoek
Walvis Bay
BOTSWANA
Gaborone
Pretoria
Johannesburg
Maputo
Mbabane
SWAZILAND
Bloemfontein
LESOTHO
Maseru
Tropic of Capricorn

○ National capital
• Other cities

0   250   500 Miles
0   250   500 Kilometers
Projection: Azimuthal Equal-Area

SOUTH AFRICA
Cape Town
Port Elizabeth

30°S

30°W 20°W 10°W 0° 10°E 20°E 30°E 40°E 50°E 60°E

map zone
Geography Skills

Location Africa is located south of Europe and Southwest Asia.
1. Name What island countries can you see on this map?

2. Analyze Compare this map to the climate map. Where are the capitals of Libya, Tunisia, Algeria, and Morocco located, and how might climate explain their locations?

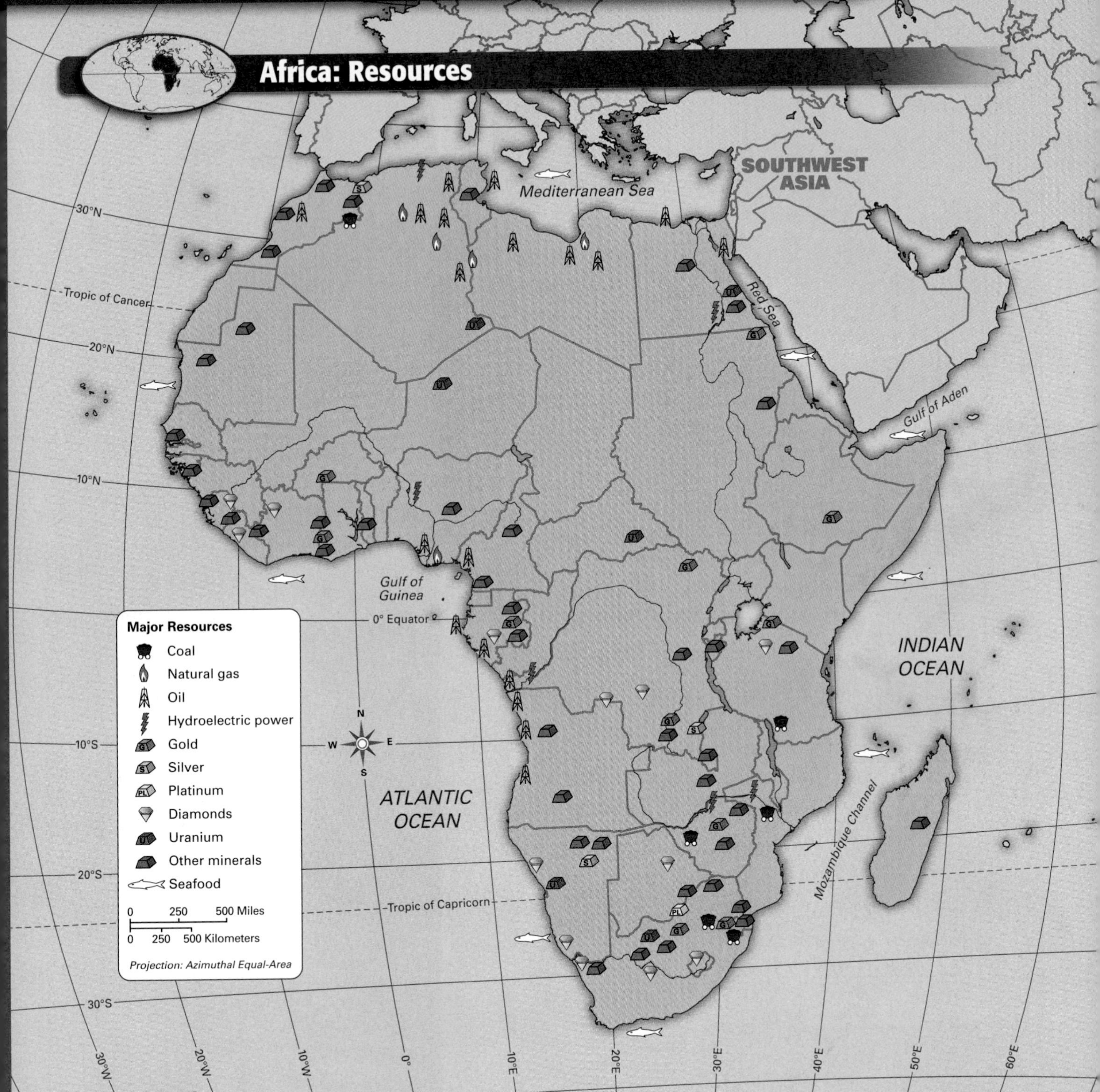

Africa: Resources

SOUTHWEST ASIA

Mediterranean Sea

Red Sea

Gulf of Aden

Tropic of Cancer

30°N

20°N

10°N

Gulf of
Guinea

0° Equator

INDIAN
OCEAN

Major Resources

- Coal
- Natural gas
- Oil
- Hydroelectric power
- Gold
- Silver
- Platinum
- Diamonds
- Uranium
- Other minerals
- Seafood

| 0 | 250 | 500 Miles |
| 0 | 250 | 500 Kilometers |

Projection: Azimuthal Equal-Area

10°S

ATLANTIC
OCEAN

Mozambique Channel

20°S

Tropic of Capricorn

30°S

30°W 20°W 10°W 0° 10°E 20°E 30°E 40°E 50°E 60°E

map zone
Geography Skills

Place The African continent is rich in resources.

1. Identify What are some of the key resources in southern Africa?

2. Make Generalizations Where in Africa are oil resources found? How do you think oil affects the economies of these regions?

Africa: Population

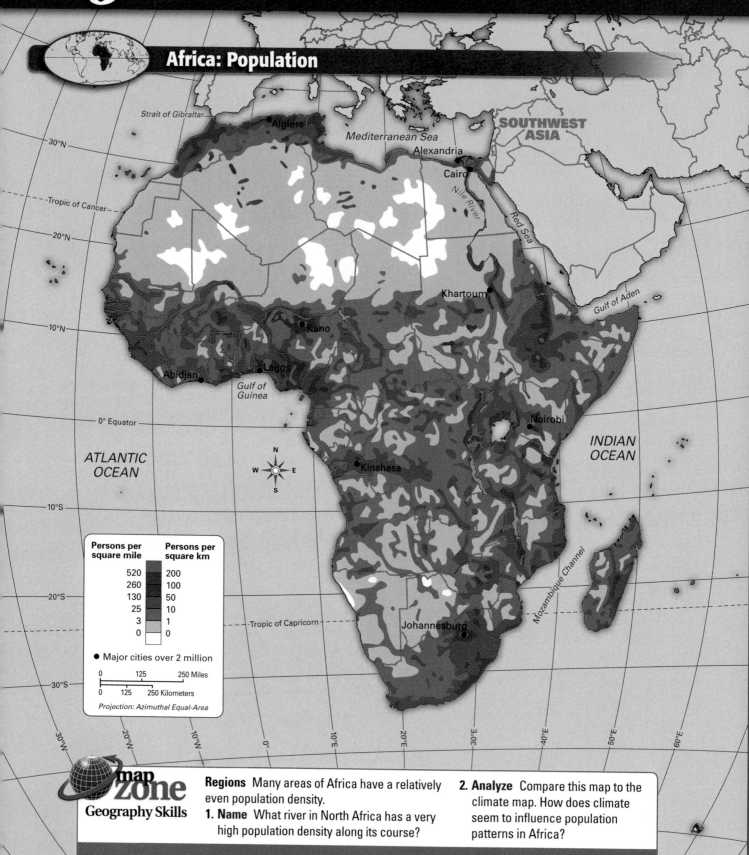

Strait of Gibraltar

Algiers

Mediterranean Sea

SOUTHWEST ASIA

Alexandria

Cairo

30°N

Tropic of Cancer

20°N

Nile River

Red Sea

Khartoum

Gulf of Aden

10°N

Kano

Abidjan

Lagos

Gulf of Guinea

Nairobi

0° Equator

ATLANTIC OCEAN

INDIAN OCEAN

Kinshasa

N
W   E
S

10°S

Persons per square mile | **Persons per square km**

520	200
260	100
130	50
25	10
3	1
0	0

20°S

Tropic of Capricorn

Mozambique Channel

Johannesburg

● Major cities over 2 million

0   125   250 Miles
0   125   250 Kilometers

Projection: Azimuthal Equal-Area

30°S

30°W 20°W 10°W 0° 10°E 20°E 30°E 40°E 50°E 60°E

map zone
Geography Skills

Regions Many areas of Africa have a relatively even population density.

1. Name What river in North Africa has a very high population density along its course?

2. Analyze Compare this map to the climate map. How does climate seem to influence population patterns in Africa?

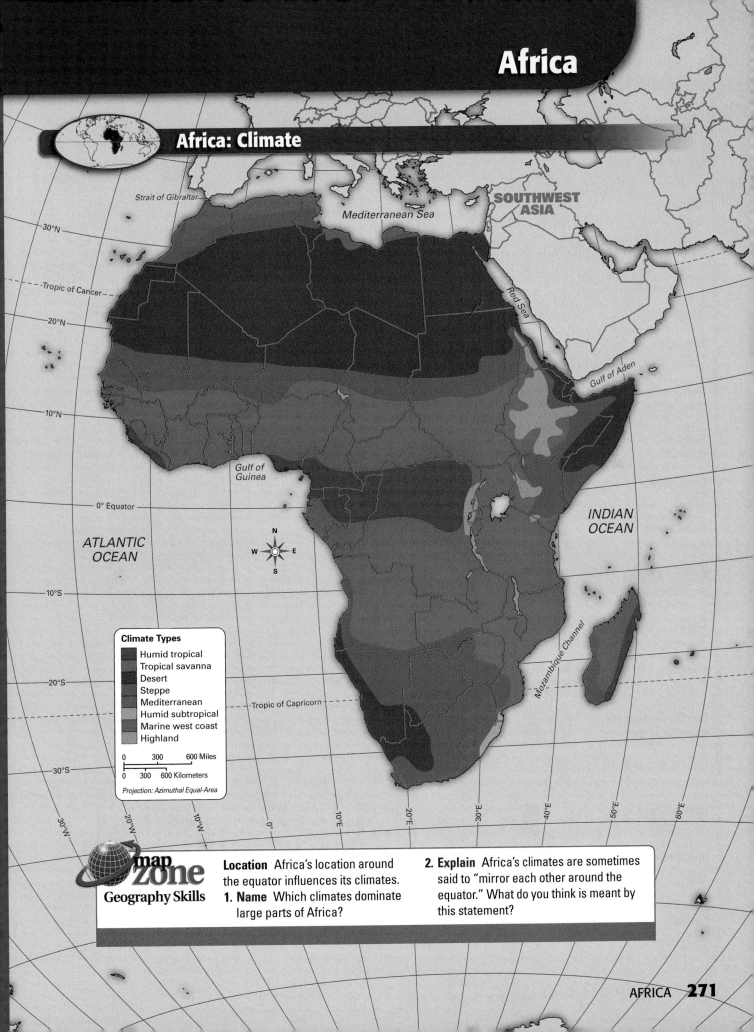

Africa: Climate

SOUTHWEST ASIA

Strait of Gibraltar

Mediterranean Sea

30°N

Tropic of Cancer

20°N

Red Sea

Gulf of Aden

10°N

Gulf of Guinea

0° Equator

ATLANTIC OCEAN

N
W   E
S

INDIAN OCEAN

10°S

Mozambique Channel

Climate Types

- Humid tropical
- Tropical savanna
- Desert
- Steppe
- Mediterranean
- Humid subtropical
- Marine west coast
- Highland

20°S

Tropic of Capricorn

0   300   600 Miles
0   300   600 Kilometers

Projection: Azimuthal Equal-Area

30°S

30°W   20°W   10°W   0°   10°E   20°E   30°E   40°E   50°E   60°E

map zone
Geography Skills

Location Africa's location around the equator influences its climates.

1. Name Which climates dominate large parts of Africa?

2. Explain Africa's climates are sometimes said to "mirror each other around the equator." What do you think is meant by this statement?

Africa

COUNTRY Capital	FLAG	POPULATION	AREA (sq mi)	PER CAPITA GDP (U.S. $)	LIFE EXPECTANCY AT BIRTH	TVS PER 1,000 PEOPLE
Algeria Algiers		32.5 million	919,595	$6,600	73.0	107
Angola Luanda		11.2 million	481,353	$2,100	38.4	15
Benin Porto-Novo		7.5 million	43,483	$1,200	52.7	44
Botswana Gaborone		1.6 million	231,804	$9,200	33.9	21
Burkina Faso Ouagadougou		13.9 million	105,869	$1,200	48.5	11
Burundi Bujumbura		6.4 million	10,745	$600	50.3	15
Cameroon Yaoundé		16.4 million	183,568	$1,900	50.9	34
Cape Verde Praia		418,200	1,557	$1,400	70.5	5
Central African Republic; Bangui		3.8 million	240,535	$1,100	43.4	6
Chad N'Djamena		9.8 million	495,755	$1,600	47.2	1
Comoros Moroni		671,200	838	$700	62.0	4
Congo, Democratic Republic of the; Kinshasa		60.1 million	905,567	$700	51.1	2
Congo, Republic of the; Brazzaville		3 million	132,047	$800	52.3	13
Côte d'Ivoire Yamoussoukro		17.3 million	124,502	$1,500	48.6	65
Djibouti Djibouti		476,700	8,880	$1,300	43.1	48
United States Washington, D.C.		295.7 million	3,718,710	$40,100	77.7	844

COUNTRY Capital	FLAG	POPULATION	AREA (sq mi)	PER CAPITA GDP (U.S. $)	LIFE EXPECTANCY AT BIRTH	TVS PER 1,000 PEOPLE
Egypt Cairo		77.5 million	386,662	$4,200	71.0	170
Equatorial Guinea Malabo		535,900	10,831	$2,700	49.7	116
Eritrea Asmara		4.6 million	46,842	$900	58.5	16
Ethiopia Addis Ababa		73.1 million	435,186	$800	48.8	5
Gabon Libreville		1.4 million	103,347	$5,900	55.0	251
Gambia Banjul		1.6 million	4,363	$1,800	53.8	3
Ghana Accra		21 million	92,456	$2,300	58.5	115
Guinea Conakry		9.5 million	94,926	$2,100	49.4	47
Guinea-Bissau Bissau		1.4 million	13,946	$700	46.6	43
Kenya Nairobi		33.8 million	224,962	$1,100	48.0	22
Lesotho Maseru		1.9 million	11,720	$3,200	34.5	16
Liberia Monrovia		3.5 million	43,000	$900	38.9	26
Libya Tripoli		5.8 million	679,362	$6,700	76.5	139
Madagascar Antananarivo		18 million	226,657	$800	57.0	23
Malawi Lilongwe		12.2 million	45,745	$600	41.4	3
United States Washington, D.C.		295.7 million	3,718,710	$40,100	77.7	844

COUNTRY / Capital	FLAG	POPULATION	AREA (sq mi)	PER CAPITA GDP (U.S. $)	LIFE EXPECTANCY AT BIRTH	TVS PER 1,000 PEOPLE
Mali / Bamako		12.3 million	478,766	$900	48.6	13
Mauritania / Nouakchott		3.1 million	397,955	$1,800	52.7	95
Mauritius / Port Louis		1.2 million	788	$12,800	72.4	248
Morocco / Rabat		32.7 million	172,414	$4,200	70.7	165
Mozambique / Maputo		19.4 million	309,496	$1,200	40.3	5
Namibia / Windhoek		2 million	318,696	$7,300	43.9	38
Niger / Niamey		11.7 million	489,191	$900	43.5	15
Nigeria / Abuja		128.8 million	356,669	$1,000	46.7	69
Rwanda / Kigali		8.4 million	10,169	$1,300	47.0	0.09
São Tomé and Príncipe; São Tomé		187,400	386	$1,200	67.0	229
Senegal / Dakar		11.1 million	75,749	$1,700	58.9	41
Seychelles / Victoria		81,200	176	$7,800	71.8	214
Sierra Leone / Freetown		6 million	27,699	$600	39.9	13
Somalia / Mogadishu		8.6 million	246,201	$600	48.1	14
United States / Washington, D.C.		295.7 million	3,718,710	$40,100	77.7	844

COUNTRY Capital	FLAG	POPULATION	AREA (sq mi)	PER CAPITA GDP (U.S. $)	LIFE EXPECTANCY AT BIRTH	TVS PER 1,000 PEOPLE
South Africa; Pretoria, Cape Town, Bloemfontein		44.3 million	471,010	$11,100	43.3	138
Sudan Khartoum		40.2 million	967,498	$1,900	58.5	173
Swaziland Mbabane		1.2 million	6,704	$5,100	33.2	112
Tanzania Dar es Salaam, Dodoma		36.8 million	364,900	$700	45.2	21
Togo Lomé		5.7 million	21,925	$1,600	57.0	22
Tunisia Tunis		10.1 million	63,170	$7,100	74.9	190
Uganda Kampala		27.3 million	91,136	$1,500	51.6	28
Zambia Lusaka		11.3 million	290,586	$900	39.7	145
Zimbabwe Harare		12.7 million	150,804	$1,900	39.1	35
United States Washington, D.C.		295.7 million	3,718,710	$40,100	77.7	844

Africa's Growing Population

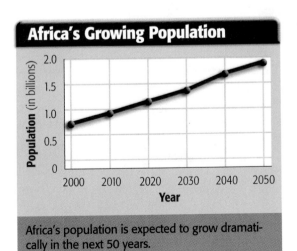

Africa's population is expected to grow dramatically in the next 50 years.

Africa and the World

	Average Age	Life Expec- tancy at Birth	Per Capita GDP (in U.S. $)
Africa	19.1 years	52.6	$2,700
Rest of the World	28.9 years	70.6	$9,819

Compared to the rest of the world, Africa's population is younger, has a shorter life expectancy, and has less money.

ANALYSIS SKILL ANALYZING INFORMATION

1. Based on the information above, what do you think are some key challenges in Africa today?

CHAPTER 11

History of Ancient Egypt

4500–500 BC

What You Will Learn...

In this chapter you will learn about the fascinating civilization of ancient Egypt and how it developed along the Nile River.

SECTION 1
Geography and Early Egypt 278

SECTION 2
The Old Kingdom 283

SECTION 3
The Middle and New Kingdoms.....291

SECTION 4
Egyptian Achievements..............298

FOCUS ON READING AND WRITING

Categorizing A good way to make sense of what you read is to separate facts and details into groups called categories. For example, you could sort facts about ancient Egypt into categories like geography, history, and culture. As you read this chapter, look for ways to categorize the information you are learning. **See the lesson, Categorizing, on page 678.**

Writing a Riddle In this chapter you will read about the civilization of the ancient Egyptians. In ancient times a sphinx, an imaginary creature like the sculpture in Egypt shown on the next page, was supposed to have demanded the answer to a riddle. People died if they did not answer the riddle correctly. After you read this chapter, you will write a riddle. The answer to your riddle should be "Egypt."

Geography Skills

Location The civilization of ancient Egypt developed along the fertile Nile River.
1. **Name** What other bodies of water are near Egypt?
2. **Make Inferences** Based on the land around ancient Egypt, why do you think the Nile was so important to life?

0	75	150 Miles
0	75	150 Kilometers

Projection: Lambert Equal-Area

The Gift of the Nile The fertile land along the Nile River drew early people to the region. Cities are still found along the Nile today.

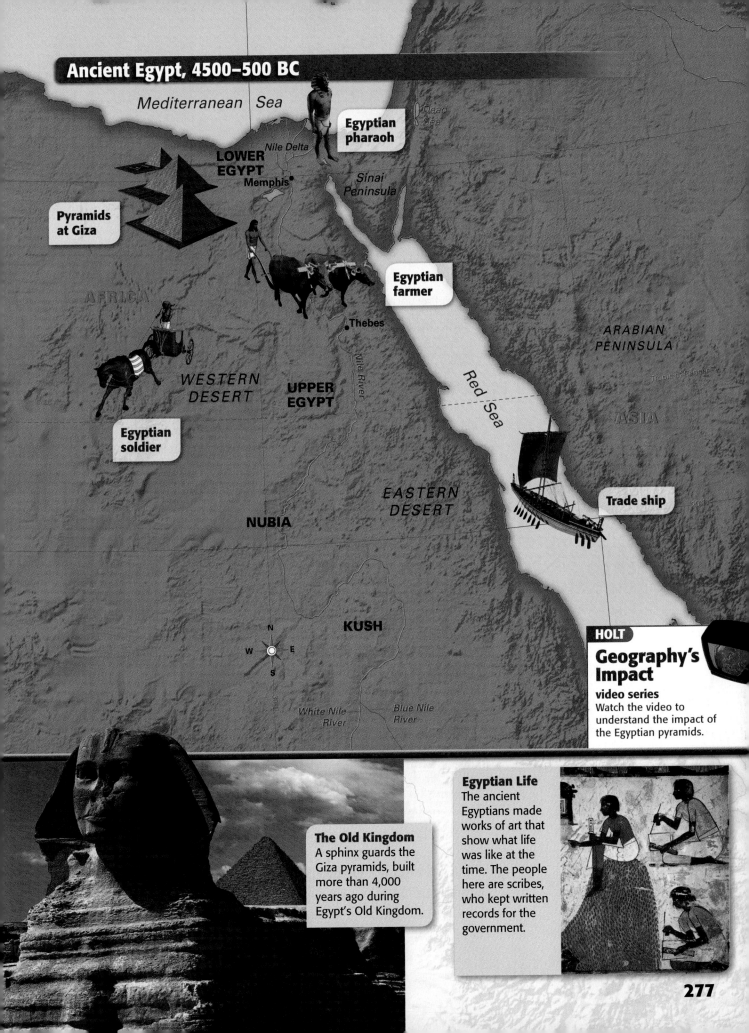

Ancient Egypt, 4500–500 BC

Mediterranean Sea

Egyptian pharaoh

Dead Sea

LOWER EGYPT

Nile Delta

Memphis

Sinai Peninsula

Pyramids at Giza

AFRICA

Egyptian farmer

Thebes

ARABIAN PENINSULA

Tropic of Cancer

WESTERN DESERT

UPPER EGYPT

Nile River

Red Sea

ASIA

Egyptian soldier

EASTERN DESERT

Trade ship

NUBIA

KUSH

N
W · E
S

White Nile River

Blue Nile River

HOLT

Geography's Impact

video series
Watch the video to understand the impact of the Egyptian pyramids.

The Old Kingdom
A sphinx guards the Giza pyramids, built more than 4,000 years ago during Egypt's Old Kingdom.

Egyptian Life
The ancient Egyptians made works of art that show what life was like at the time. The people here are scribes, who kept written records for the government.

Geography and Early Egypt

What You Will Learn...

Main Ideas

1. Egypt was called the gift of the Nile because the Nile River was so important.
2. Civilization developed after people began farming along the Nile River.
3. Strong kings unified all of ancient Egypt.

The Big Idea

The water and fertile soils of the Nile Valley enabled a great civilization to develop in Egypt.

Key Terms and Places

Nile River, *p. 278*
Upper Egypt, *p. 278*
Lower Egypt, *p. 278*
cataracts, *p. 279*
delta, *p. 279*
pharaoh, *p. 281*
dynasty, *p. 281*

TAKING NOTES As you read, take notes on the characteristics of the Nile River and on the way in which it affected Egypt. Use a chart like this one to organize your notes.

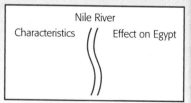

Nile River

Characteristics | Effect on Egypt

If YOU lived there...

Your family farms in the Nile Valley. Each year when the river's floodwaters spread rich soil on the land, you help your father plant barley. When you are not in the fields, you spin fine linen thread from flax you have grown. Sometimes your family goes on an outing to the river, where your father hunts birds in the tall grasses.

Why do you like living in the Nile Valley?

BUILDING BACKGROUND In ancient times, the fertile land in the Nile River Valley drew people to live in the area. Over time, a farming civilization developed that became ancient Egypt. This civilization would be stable and long-lasting.

The Gift of the Nile

Geography played a key role in the development of Egyptian civilization. The **Nile River** brought life to Egypt and enabled it to thrive. The river was so important to people in this region that the Greek historian Herodotus (hi-RAHD-uh-tuhs) called Egypt the gift of the Nile.

Location and Physical Features

The Nile is the longest river in the world. It begins in central Africa and runs north through Egypt to the Mediterranean Sea, a distance of over 4,000 miles. The civilization of ancient Egypt developed along a 750-mile stretch of the Nile.

Ancient Egypt included two regions, a southern region and a northern region. The southern region was called **Upper Egypt**. It was so named because it was located upriver in relation to the Nile's flow. **Lower Egypt**, the northern region, was located downriver. The Nile sliced through the desert of Upper Egypt. There, it created a fertile river valley about 13 miles wide. On either side of the Nile lay hundreds of miles of bleak desert sands.

As you can see on the map, the Nile flowed through rocky, hilly land to the south of Egypt. At several points, this rough terrain caused **cataracts**, or rapids, to form. The first cataract was located 720 miles south of the Mediterranean Sea. This cataract, shown by a red bar on the map, marked the southern border of Upper Egypt. Five more cataracts lay farther south. These cataracts made sailing on that portion of the Nile very difficult.

In Lower Egypt, the Nile divided into several branches that fanned out and flowed into the Mediterranean Sea. These branches formed a **delta, a triangle-shaped area of land made from soil deposited by a river.** At the time of ancient Egypt, swamps and marshes covered much of the Nile Delta. Some two-thirds of Egypt's fertile farmland was located in the Nile Delta.

The Floods of the Nile

Because little rain fell in the region, most of Egypt was desert. Each year, however, rain fell far to the south of Egypt in the highlands of East Africa. This rainfall caused the Nile River to flood. Almost every year, the Nile flooded Upper Egypt in mid-summer and Lower Egypt in the fall.

The Nile's flooding coated the land around it with a rich silt. This silt made the soil ideal for farming. The silt also made the land a dark color. That is why Egyptians called their country the black land. They called the dry, lifeless desert beyond the river valley the red land.

Each year, Egyptians eagerly awaited the flooding of the Nile River. For them, the river's floods were a life-giving miracle. Without the Nile's regular flooding, people never could have farmed in Egypt. The Nile truly was a gift to Egypt.

READING CHECK **Finding Main Ideas** Why was Egypt called the gift of the Nile?

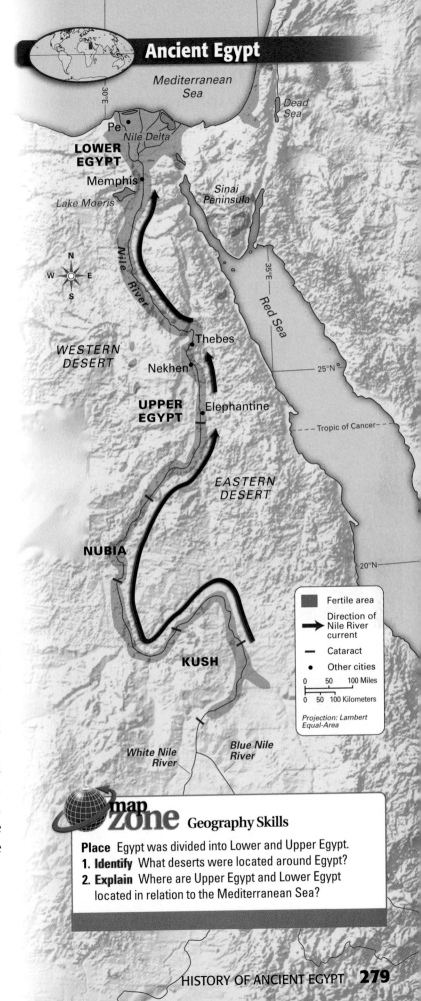

Ancient Egypt

map**zone** **Geography Skills**

Place Egypt was divided into Lower and Upper Egypt.
1. **Identify** What deserts were located around Egypt?
2. **Explain** Where are Upper Egypt and Lower Egypt located in relation to the Mediterranean Sea?

Civilization Develops in Egypt

The Nile provided both water and fertile soil for farming. Over time, scattered farms grew into villages and cities. Eventually, an Egyptian civilization developed.

Increased Food Production

Hunter-gatherers first moved into the Nile Valley more than 12,000 years ago. They found plants, wild animals, and fish there to eat. In time, these people learned how to farm, and they settled along the Nile. By 4500 BC, farmers living in small villages grew wheat and barley.

Over time, farmers in Egypt developed an irrigation system. This system consisted of a series of canals that directed the Nile's flow and carried water to the fields.

The Nile provided Egyptian farmers with an abundance of food. Farmers in Egypt grew wheat, barley, fruits, and vegetables. They also raised cattle and sheep. The river provided many types of fish, and hunters trapped wild geese and ducks along its banks. With these many sources of food, the Egyptians enjoyed a varied diet.

Two Kingdoms

In addition to a stable food supply, Egypt's location offered another advantage. It had natural barriers, which made it hard to invade Egypt. To the west, the desert was too big and harsh to cross. To the north, the Mediterranean Sea kept many enemies away. To the east, more desert and the Red Sea provided protection. Finally, to the south, cataracts in the Nile made it difficult for invaders to sail into Egypt that way.

Farming in Egypt

Protected from invaders, the villages of Egypt grew. Wealthy farmers emerged as village leaders. In time, strong leaders gained control over several villages. By 3200 BC, villages had grown and banded together to create two kingdoms—Lower Egypt and Upper Egypt.

Each kingdom had its own capital city where its ruler was based. The capital city of Lower Egypt was Pe, located in the Nile Delta. There, wearing a red crown, the king of Lower Egypt ruled. The capital city of Upper Egypt was Nekhen, located on the Nile's west bank. In this southern kingdom, the king wore a cone-shaped white crown. For centuries, Egyptians referred to their country as the two lands.

READING CHECK **Summarizing** What attracted early settlers to the Nile Valley?

Farmers in ancient Egypt learned how to grow wheat and barley. The tomb painting at left shows a couple harvesting their crop. As the photo above shows, people in Egypt still farm along the Nile.

ANALYZING VISUALS Based on the above photo, what methods do Egyptian farmers use today?

Kings Unify Egypt

According to tradition, around 3100 BC Menes (MEE-neez) rose to power in Upper Egypt. Some historians think Menes is a myth and that his accomplishments were really those of other ancient kings named Aha, Scorpion, or Narmer.

Menes wanted to unify the kingdoms of Upper and Lower Egypt. He had his armies invade Lower Egypt and take control of it. Menes then married a princess from Lower Egypt to strengthen his control over the newly unified country.

Menes wore both the white crown of Upper Egypt and the red crown of Lower Egypt to symbolize his leadership over the two kingdoms. Later, he combined the two crowns into a double crown, as you can see on the next page.

Many historians consider Menes to be Egypt's first **pharaoh** (FEHR-oh), the title used by the rulers of ancient Egypt. The title *pharaoh* means "great house." Menes also founded Egypt's first **dynasty**, or series of rulers from the same family.

Menes built a new capital city at the southern tip of the Nile Delta. The city was later named Memphis. It was near where Lower Egypt met Upper Egypt, close to what is now Cairo, Egypt. For centuries, Memphis was the political and cultural center of Egypt. Many government offices were located there, and the city bustled with artistic activity.

Egypt's First Dynasty was a theocracy that lasted for about 200 years. A theocracy is a government ruled by religious leaders such as priests or a monarch thought to be divine.

Over time, Egypt's rulers extended Egyptian territory southward along the Nile River and into Southwest Asia. They also improved irrigation and trade, making Egypt wealthier.

FOCUS ON READING

Identify two or three categories that you could use to organize the information under Kings Unify Egypt.

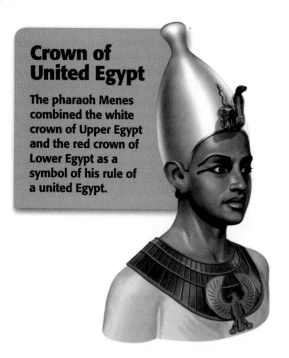

Crown of United Egypt

The pharaoh Menes combined the white crown of Upper Egypt and the red crown of Lower Egypt as a symbol of his rule of a united Egypt.

Eventually, however, rivals arose to challenge Egypt's First Dynasty for power. These challengers took over Egypt and established the Second Dynasty. In time, some 30 dynasties would rule ancient Egypt over a span of more than 2,500 years.

READING CHECK **Drawing Inferences** Why do you think Menes wanted to rule over both kingdoms of Egypt?

SUMMARY AND PREVIEW As you have read, ancient Egypt began in the fertile Nile River Valley. Two kingdoms developed in this region. The two kingdoms were later united under one ruler, and Egyptian territory grew. In the next section you will learn how Egypt continued to grow and change under later rulers in a period known as the Old Kingdom.

Section 1 Assessment

go.hrw.com
Online Quiz
KEYWORD: SK7 HP11

Reviewing Ideas, Terms, and Places

1. **a. Identify** Where was the Egyptian kingdom of **Lower Egypt** located?
   b. Analyze Why was the **delta** of the **Nile River** well suited for settlement?
   c. Predict How might the Nile's **cataracts** have both helped and hurt Egypt?
2. **a. Describe** What foods did the Egyptians eat?
   b. Analyze What role did the Nile play in supplying Egyptians with the foods they ate?
   c. Elaborate How did the desert on both sides of the Nile help ancient Egypt?
3. **a. Identify** Who was the first **pharaoh** of Egypt?
   b. Draw Conclusions Why did the pharaohs of the First Dynasty wear a double crown?

Critical Thinking

4. **Categorizing** Create a chart like the one shown here. Use your notes to provide information for each category in the chart.

Development along Nile	Two Kingdoms	United Kingdoms

FOCUS ON WRITING

5. **Thinking about Geography and Early History** In your riddle, what clues could you include related to Egypt's geography and early history? For example, you might include the Nile River or pharaohs as clues. Add some ideas to your notes.

The Old Kingdom

If **YOU** lived there...

You are a farmer in ancient Egypt. To you, the pharaoh is the god Horus as well as your ruler. You depend on his strength and wisdom. For part of the year, you are busy planting crops in your fields. But at other times of the year, you work for the pharaoh. You are helping to build a great tomb so that your pharaoh will be comfortable in the afterlife.

How do you feel about working for the pharaoh?

BUILDING BACKGROUND As in other ancient cultures, Egyptian society was based on a strict order of social classes. A small group of royalty and nobles ruled Egypt. They depended on the rest of the population to supply food, crafts, and labor. Few people questioned this arrangement of society.

Life in the Old Kingdom

The First and Second Dynasties ruled ancient Egypt for about four centuries. Around 2700 BC, though, a new dynasty rose to power in Egypt. Called the Third Dynasty, its rule began a period in Egyptian history known as the Old Kingdom.

Early Pharaohs

The **Old Kingdom** was a period in Egyptian history that lasted for about 500 years, from about 2700 to 2200 BC. During this time, the Egyptians continued to develop their political system. The system they developed was based on the belief that Egypt's pharaoh, or ruler, was both a king and a god.

The ancient Egyptians believed that Egypt belonged to the gods. The Egyptians believed the pharaoh had come to Earth in order to manage Egypt for the rest of the gods. As a result, he had absolute power over all the land and people in Egypt.

But the pharaoh's status as both king and god came with many responsibilities. People blamed him if crops did not grow well or if disease struck. They also demanded that the pharaoh make trade profitable and prevent wars.

What You Will Learn...

Main Ideas

1. Life in the Old Kingdom was influenced by pharaohs, roles in society, and trade.
2. Religion shaped Egyptian life.
3. The pyramids were built as tombs for Egypt's pharaohs.

The Big Idea

Egyptian government and religion were closely connected during the Old Kingdom.

Key Terms

Old Kingdom, *p. 283*
nobles, *p. 284*
afterlife, *p. 286*
mummies, *p. 286*
elite, *p. 287*
pyramids, *p. 288*
engineering, *p. 288*

TAKING NOTES As you read, take notes on government and religion during Egypt's Old Kingdom. Use a chart like the one below to record your notes.

Government	Religion

Egyptian Society

Pharaoh

The pharaoh ruled Egypt as a god.

Nobles

Officials and priests helped run the government and temples.

Scribes and Craftspeople

Scribes and craftspeople wrote and produced goods.

Farmers, Servants, and Slaves

Most Egyptians were farmers, servants, or slaves.

ANALYSIS SKILL **ANALYZING VISUALS**

Which group helped run the government and temples?

The most famous pharaoh of the Old Kingdom was Khufu (KOO-foo), who ruled in the 2500s BC. Even though he is famous, we know relatively little about Khufu's life. Egyptian legend says that he was cruel, but historical records tell us that the people who worked for him were well fed. Khufu is best known for the monuments that were built to him.

Society and Trade

ACADEMIC VOCABULARY

acquire (uh-KWYR) to get

By the end of the Old Kingdom, Egypt had about 2 million people. As the population grew, social classes appeared. The Egyptians believed that a well-ordered society would keep their kingdom strong.

At the top of Egyptian society was the pharaoh. Just below him were the upper classes, which included priests and key government officials. Many of these priests and officials were **nobles**, or people from rich and powerful families.

Next in society was the middle class. This class included lesser government officials, scribes, and a few rich craftspeople.

The people in Egypt's lower class, more than 80 percent of the population, were mostly farmers. During flood season, when they could not work in the fields, farmers worked on the pharaoh's building projects. Servants and slaves also worked hard.

As society developed during the Old Kingdom, Egypt traded with some of its neighbors. Traders traveled south along the Nile to Nubia to **acquire** gold, copper, ivory, slaves, and stone for building. Trade with Syria provided Egypt with wood for building and for fire.

Egyptian society grew more complex during this time. It continued to be organized, disciplined, and highly religious.

READING CHECK **Generalizing** How was society structured in the Old Kingdom?

Religion and Egyptian Life

Worshipping the gods was a part of daily life in Egypt. But the Egyptian focus on religion extended beyond people's lives. Many customs focused on what happened after people died.

The Gods of Egypt

The Egyptians practiced polytheism. Before the First Dynasty, each village worshipped its own gods. During the Old Kingdom, however, Egyptian officials expected everyone to worship the same gods, though how people worshipped the gods might differ from place to place.

The Egyptians built temples to the gods all over the kingdom. Temples collected payments from both worshippers and the government. These payments enabled the temples to grow more influential.

Over time, certain cities became centers for the worship of certain gods. In the city of Memphis, for example, people prayed to Ptah, the creator of the world.

The Egyptians worshipped many gods besides Ptah. They had gods for nearly everything, including the sun, the sky, and Earth. Many gods blended human and animal forms. For example, Anubis, the god of the dead, had a human body but a jackal's head. Other major gods included

- Re, or Amon-Re, the sun god
- Osiris, the god of the underworld
- Isis, the goddess of magic
- Horus, a sky god; god of the pharaohs
- Thoth, the god of wisdom
- Geb, the Earth god

Egyptian families also worshipped household gods at shrines in their homes.

FOCUS ON READING

How is the text under the heading Religion and Egyptian Life categorized?

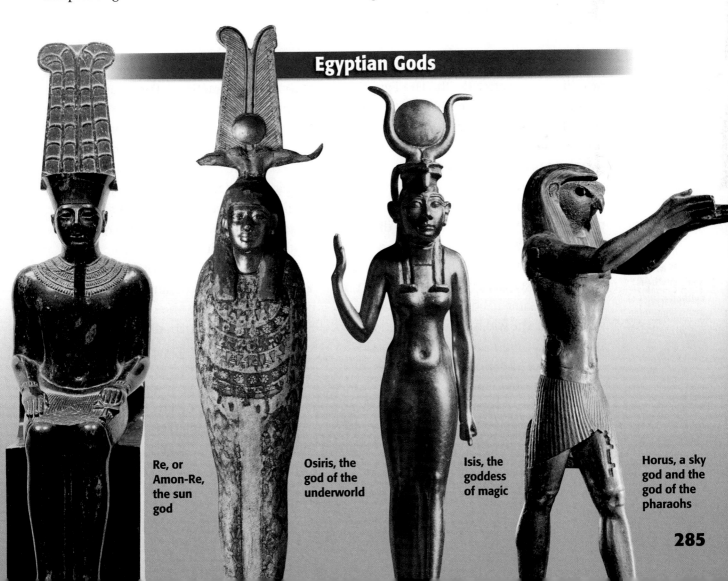

Egyptian Gods

Re, or Amon-Re, the sun god

Osiris, the god of the underworld

Isis, the goddess of magic

Horus, a sky god and the god of the pharaohs

Mummies and the Afterlife

Osiris, god of the underworld, waited to judge the dead person's soul.

The god Anubis weighed the dead person's heart against the feather of truth. If they weighed the same amount, the person was allowed into the underworld.

Emphasis on the Afterlife

Much of Egyptian religion focused on the **afterlife**, or life after death. The Egyptians believed that the afterlife was a happy place. Paintings from Egyptian tombs show the afterlife as an ideal world where all the people are young and healthy.

The Egyptian belief in the afterlife stemmed from their idea of *ka* (KAH), or a person's life force. When a person died, his or her *ka* left the body and became a spirit. The *ka* remained linked to the body and could not leave its burial site. However, it had all the same needs that the person had when he or she was living. It needed to eat, sleep, and be entertained.

To fulfill the *ka*'s needs, people filled tombs with objects for the afterlife. These objects included furniture, clothing, tools, jewelry, and weapons. Relatives of the dead were expected to bring food and beverages to their loved ones' tombs so the *ka* would not be hungry or thirsty.

ACADEMIC VOCABULARY

method a way of doing something

Burial Practices

Egyptian ideas about the afterlife shaped their burial practices. For example, the Egyptians believed that a body had to be prepared for the afterlife before it could be placed in a tomb. This meant the body had to be preserved. If the body decayed, its spirit could not recognize it. That would break the link between the body and spirit. The *ka* would then be unable to receive the food and drink it needed.

To help the *ka*, Egyptians developed a **method** called embalming to preserve bodies and to keep them from decaying. Egyptians preserved bodies as **mummies**, specially treated bodies wrapped in cloth. Embalming preserves a body for many, many years. A body that was not embalmed decayed far more quickly.

Embalming was a complex process that took several weeks to complete. In the first step, embalmers cut open the body and removed all organs except for the heart.

1 Only the god Anubis was allowed to perform the first steps in preparing a mummy.

2 The body's organs were preserved in special jars and kept next to the mummy.

Embalmers stored the removed organs in special jars. Next, the embalmers used a special substance to dry out the body. They later applied some special oils. The embalmers then wrapped the dried-out body with linen cloths and bandages, often placing special charms inside the cloth wrappings.

Wrapping the body was the last step in the mummy-making process. Once it was completely wrapped, a mummy was placed in a coffin called a sarcophagus, such as the one shown at right.

Only royalty and other members of Egypt's **elite** (AY-leet), or people of wealth and power, could afford to have mummies made. Peasant families did not need the process. They buried their dead in shallow graves at the edge of the desert. The hot, dry sand preserved the bodies naturally.

3 The body was preserved as a mummy and kept in a case called a sarcophagus.

READING CHECK **Analyzing** How did religious beliefs affect Egyptian burial practices?

ANALYSIS
SKILL **ANALYZING VISUALS**

How did gods participate in the afterlife?

The Pyramids

The Egyptians believed that burial sites, especially royal tombs, were very important. For this reason, they built spectacular monuments in which to bury their rulers. The most spectacular were the **pyramids**— huge, stone tombs with four triangle-shaped sides that met in a point on top.

The Egyptians built the first pyramids during the Old Kingdom. Some of the largest pyramids were built during that time.

Many of these huge Egyptian pyramids are still standing. The largest is the Great Pyramid of Khufu near the town of Giza. It covers more than 13 acres at its base and stands 481 feet (146 m) high. This one pyramid took thousands of workers and more than 2 million limestone blocks to build. Like all the pyramids, it is an amazing example of Egyptian **engineering**, the application of scientific knowledge for practical purposes.

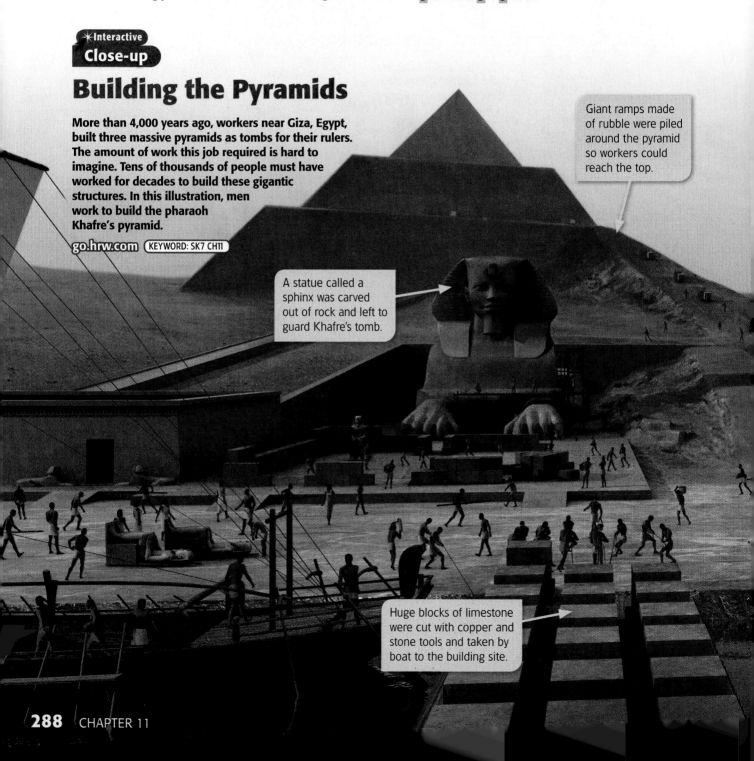

★Interactive Close-up

Building the Pyramids

More than 4,000 years ago, workers near Giza, Egypt, built three massive pyramids as tombs for their rulers. The amount of work this job required is hard to imagine. Tens of thousands of people must have worked for decades to build these gigantic structures. In this illustration, men work to build the pharaoh Khafre's pyramid.

go.hrw.com (KEYWORD: SK7 CH11)

Giant ramps made of rubble were piled around the pyramid so workers could reach the top.

A statue called a sphinx was carved out of rock and left to guard Khafre's tomb.

Huge blocks of limestone were cut with copper and stone tools and taken by boat to the building site.

Building the Pyramids

The earliest pyramids did not have the smooth sides we usually imagine when we think of pyramids. The Egyptians began building the smooth-sided pyramids we usually see around 2700 BC. The steps of these pyramids were filled and covered with limestone. The burial chamber was located deep inside the pyramid. After the pharaoh's burial, workers sealed the passages to this room with large blocks.

Historians do not know for certain how the ancient Egyptians built the pyramids. What is certain is that such massive projects required a huge labor force. As many as 100,000 workers may have been needed to build just one pyramid. The government paid the people working on the pyramids.

Inside the Great Pyramid, tunnels led to the pharaoh's burial chamber, which was sealed off with rocks.

Teams of workers dragged the stones on wooden sleds to the pyramid.

ANALYSIS SKILL **ANALYZING VISUALS**

How did workers get their stone blocks to the pyramids?

Wages for working on construction projects were paid in goods such as grain instead of money, however.

For years, scholars have debated how the Egyptians moved the massive stones used to build the pyramids. Some scholars think that during the Nile's flooding, builders floated the stones downstream directly to the construction site. Most historians believe that workers used brick ramps and strong wooden sleds to drag the stones up the pyramid once at the building site.

Significance of the Pyramids

Burial in a pyramid showed a pharaoh's importance. Both the size and shape of the pyramid were symbolic. Pointing to the sky above, the pyramid symbolized the pharaoh's journey to the afterlife. The Egyptians wanted the pyramids to be spectacular because they believed the pharaoh, as their link to the gods, controlled everyone's afterlife. Making the pharaoh's spirit happy was a way of ensuring happiness in one's own afterlife.

To ensure that the pharaohs remained safe after death, the Egyptians sometimes wrote magical spells and hymns on tombs.

Together, these spells and hymns are called Pyramid Texts. The first such text, addressed to Re, the sun god, was carved into the pyramid of King Unas (OO-nuhs). He was a pharaoh of the Old Kingdom.

> "Re, this Unas comes to you,
> A spirit indestructible . . .
> Your son comes to you, this Unas . . .
> May you cross the sky united in the dark,
> May you rise in lightland, [where] you shine!"
> –from Pyramid Text, Utterance 217

The builders of Unas's pyramid wanted the god Re to look after their leader's spirit. Even after death, the Egyptians' pharaoh was important to them.

READING CHECK **Identifying Points of View** Why were pyramids important to the ancient Egyptians?

SUMMARY AND PREVIEW As you have read, during the Old Kingdom, new political and social orders were created in Egypt. Religion was important, and many pyramids were built for pharaohs. In the next section you will learn about Egypt's Middle and New Kingdoms.

Section 2 Assessment

go.hrw.com
Online Quiz
KEYWORD: SK7 HP11

Reviewing Ideas, Terms, and Places

1. **a. Define** To what Egyptian period does the phrase **Old Kingdom** refer?
   b. Analyze Why did Egyptians never question the pharaoh's authority?
   c. Elaborate Why do you think pharaohs might have wanted the support of **nobles**?
2. **a. Define** What did Egyptians mean by the **afterlife**?
   b. Analyze Why was embalming important to Egyptians?
3. **a. Describe** What is **engineering**?
   b. Elaborate What does the building of the **pyramids** tell us about Egyptian society?

Critical Thinking

4. **Generalizing** Using your notes, complete this graphic organizer by listing three facts about the relationship between government and religion in the Old Kingdom.

Government and Religion
1.
2.
3.

FOCUS ON WRITING

5. **Noting Characteristics of the Old Kingdom** The Old Kingdom has special characteristics of government, society, and religion. Write down details about any of those characteristics that you might want to include as one of the clues in your Egypt riddle.

The Middle and New Kingdoms

If YOU lived there...

You are a servant to Hatshepsut, the ruler of Egypt. You admire her, but some people think a woman should not rule. She calls herself king and dresses like a pharaoh—even wearing a fake beard. That was your idea! But you want to help more.

What could Hatshepsut do to show her authority?

BUILDING BACKGROUND The power of the pharaohs expanded during the Old Kingdom. Society was orderly, based on great differences between social classes. But rulers and dynasties changed, and Egypt changed with them. In time, these changes led to new eras in Egyptian history, eras called the Middle and New Kingdoms.

The Middle Kingdom

At the end of the Old Kingdom, the wealth and power of the pharaohs declined. Building and maintaining pyramids cost a lot of money. Pharaohs could not collect enough taxes to keep up with their expenses. At the same time, ambitious nobles used their government positions to take power from pharaohs.

In time, nobles gained enough power to challenge Egypt's pharaohs. By about 2200 BC the Old Kingdom had fallen. For the next 160 years, local nobles ruled much of Egypt. During this period, the kingdom had no central ruler.

What You Will Learn...

Main Ideas

1. The Middle Kingdom was a period of stable government between periods of disorder.
2. The New Kingdom was the peak of Egyptian trade and military power, but its greatness did not last.
3. Work and daily life differed among Egypt's social classes.

The Big Idea

During the Middle and New Kingdoms, order and greatness were restored in Egypt.

Key Terms and Places

Middle Kingdom, *p. 292*
New Kingdom, *p. 292*
Kush, *p. 292*
trade routes, *p. 293*

TAKING NOTES As you read, use a chart like the one here to take notes on the Middle and New Kingdoms and on work and life in ancient Egypt.

Middle Kingdom	New Kingdom	Work and Life

Time Line

Periods of Egyptian History

3000 BC	2000 BC	1000 BC
c. 2700–2200 BC Old Kingdom	c. 2050–1750 BC Middle Kingdom	c. 1550–1050 BC New Kingdom

Finally, around 2050 BC, a powerful pharaoh defeated his rivals. Once again all of Egypt was united. His rule began the **Middle Kingdom**, a period of order and stability that lasted to about 1750 BC. Toward the end of the Middle Kingdom, however, Egypt began to fall into disorder once again.

Around 1750 BC, a group from Southwest Asia called the Hyksos (HIK-sohs) invaded. The Hyksos used horses, chariots, and advanced weapons to conquer Lower Egypt. The Hyksos then ruled the region as pharaohs for 200 years.

The Egyptians eventually fought back. In the mid-1500s BC, Ahmose (AHM-ohs) of Thebes declared himself king and drove the Hyksos out of Egypt. Ahmose then ruled all of Egypt.

READING CHECK **Summarizing** What caused the end of the Middle Kingdom?

BIOGRAPHY

Queen Hatshepsut
(Ruled c. 1503–1482 BC)

Hatshepsut was married to the pharaoh Thutmose II, her half brother. He died young, leaving the throne to Thutmose III, his son by another woman. Because Thutmose III was still very young, Hatshepsut took over power. Many people did not think women should rule, but Hatshepsut dressed as a man and called herself king. After she died, her stepson took back power and vandalized all the monuments she had built.

Identifying Cause and Effect
What do you think caused Hatshepsut to dress like a man?

The New Kingdom

Ahmose's rise to power marked the start of Egypt's eighteenth dynasty. More importantly, it was the start of the **New Kingdom**, the period during which Egypt reached the height of its power and glory. During the New Kingdom, which lasted from about 1550 to 1050 BC, conquest and trade brought wealth to the pharaohs.

Building an Empire

After battling the Hyksos, Egypt's leaders feared future invasions. To prevent such invasions from occurring, they decided to take control of all possible invasion routes into the kingdom. In the process, these leaders turned Egypt into an empire.

Egypt's first target was the homeland of the Hyksos. After taking over that area, the army continued north and conquered Syria. As you can see from the map, Egypt took over the entire eastern shore of the Mediterranean and the kingdom of **Kush**, south of Egypt. By the 1400s BC, Egypt was the leading military power in the region. Its empire extended from the Euphrates River to southern Nubia.

Military conquests made Egypt rich as well as powerful. The kingdoms that Egypt conquered regularly sent gifts and treasure to their Egyptian conquerors. For example, the kingdom of Kush in Nubia sent yearly payments of gold, precious stones, and leopard skins to the pharaohs. In addition, Assyrian, Babylonian, and Hittite kings sent expensive gifts to Egypt in an effort to maintain good relations.

Growth and Effects of Trade

As Egypt's empire expanded, so did its trade. Conquest brought Egyptian traders into contact with more distant lands. Many of these lands had valuable resources for trade. The Sinai Peninsula is one example.

It had valuable supplies of turquoise and copper. Profitable **trade routes**, or paths followed by traders, developed from Egypt to these lands, as the map shows.

One of Egypt's rulers who worked to increase trade was Queen Hatshepsut. She sent Egyptian traders south to trade with the kingdom of Punt on the Red Sea and north to trade with people in Asia Minor and Greece.

Hatshepsut and later pharaohs used the money they gained from trade to support the arts and architecture. Hatshepsut in particular is remembered for the many impressive monuments and temples built during her reign. The best known of these structures was a magnificent temple built for her near the city of Thebes.

Invasions of Egypt

Despite its military might, Egypt still faced threats to its power. In the 1200s BC the pharaoh Ramses (RAM-seez) II, or Ramses the Great, fought the Hittites, who came from Asia Minor. The two powers fought fiercely for years, but neither one could defeat the other.

Egypt faced threats in other parts of its empire as well. To the west, a people known as the Tehenu invaded the Nile Delta. Ramses fought them off and built a series of forts to strengthen the western frontier. This proved to be a wise decision because the Tehenu invaded again a century later. Faced with Egypt's strengthened defenses, the Tehenu were defeated once again.

Soon after Ramses the Great died, invaders called the Sea Peoples sailed into Southwest Asia. Little is known about these people. Historians are not even sure who they were. All we know is that they were strong warriors who had crushed the Hittites and destroyed cities in Southwest Asia. Only after 50 years of fighting were the Egyptians able to turn them back.

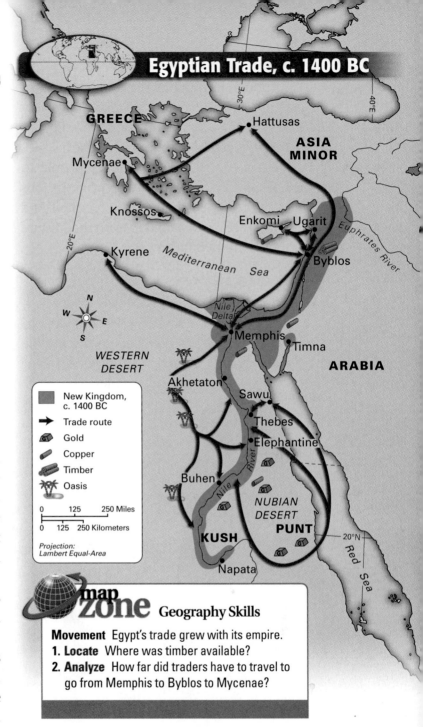

Egyptian Trade, c. 1400 BC

map zone Geography Skills

Movement Egypt's trade grew with its empire.
1. **Locate** Where was timber available?
2. **Analyze** How far did traders have to travel to go from Memphis to Byblos to Mycenae?

Egypt survived, but its empire in Asia was gone. Shortly after the invasions of the Hittites and the Sea Peoples, the New Kingdom came to an end. Ancient Egypt fell into a period of violence and disorder. Egypt would never regain its power.

READING CHECK **Identifying Cause and Effect**
What caused Egypt's growth of trade during the New Kingdom?

Work and Daily Life

FOCUS ON READING
What categories of jobs made up the society of ancient Egypt?

Although Egyptian dynasties rose and fell, daily life for Egyptians did not change very much. But as the population grew, Egypt's society became even more complex.

A complex society requires people to take on different jobs. In Egypt, these jobs were often passed on within families. At a young age, boys started to learn their future jobs from their fathers.

Scribes

After the priests and government officials, scribes were the most respected people in ancient Egypt. As members of the middle class, scribes worked for the government and the temples. This work involved keeping records and accounts. Scribes also wrote and copied religious and literary texts.

Because of their respected position, scribes did not have to pay taxes. For this reason, many scribes became wealthy.

Artisans, Artists, and Architects

Another group in society was made up of artisans whose jobs required advanced skills. Among the artisans who worked in Egypt were sculptors, builders, carpenters, jewelers, metalworkers, and leatherworkers. Artisans made items such as statues, furniture, jewelry, pottery, and shoes. Most artisans worked for the government or for temples. Egypt's artisans were admired and often paid fairly well.

Architects and artists were admired in Egypt as well. Architects designed the temples and royal tombs for which Egypt is famous. Talented architects could rise to become high government officials. Artists often worked for the state or for temples.

Daily Life in Egypt

Most Egyptians spent their days in the fields, plowing and harvesting their crops.

Queen Nefertiti, shown here, and other Egyptian queens wore makeup, jewelry, and perfume.

Egyptian artists produced many different types of works. Many artists worked in the deep burial chambers of the pharaohs' tombs painting detailed pictures.

Merchants and Traders

Although trade was important to Egypt, only a small group of Egyptians became merchants and traders. Some traveled long distances to buy and sell goods. On their journeys, merchants were usually accompanied by soldiers, scribes, and laborers.

Soldiers

After the wars of the Middle Kingdom, Egypt established a professional army. The military offered people a chance to rise in social status. Soldiers received land as payment and could also keep any treasure they captured in war. Soldiers who excelled could be promoted to officer positions.

Farmers and Other Peasants

As in the society of the Old Kingdom, Egyptian farmers and other peasants were toward the bottom of Egypt's social scale. These hardworking people made up the vast majority of Egypt's population.

Egyptian farmers grew crops to support their families. These farmers depended on the Nile's regular floods to grow their crops. Farmers used wooden hoes or plows pulled by cows to prepare the land before the flood. After the floodwaters had drained away, farmers planted seeds for crops such as wheat and barley. At the end of the growing season, Egypt's farmers worked together to gather the harvest.

Farmers had to give some of their crops to the pharaoh as taxes. These taxes were intended to pay the pharaoh for use of the land. Under Egyptian law, the pharaoh controlled all land in the kingdom.

This jar probably held perfume, a valuable trade item.

Servants worked for Egypt's rulers and nobles and did many jobs, like preparing food.

ANALYSIS SKILL ANALYZING VISUALS

What were some luxury goods used by Egypt's queens and rulers?

ACADEMIC
VOCABULARY
contracts binding
legal agreements

All peasants, including farmers, were also subject to special duty. Under Egyptian law, the pharaoh could demand at any time that people work on projects, such as building pyramids, mining gold, or fighting in the army. The government paid the workers in grain.

Slaves

The few slaves in Egyptian society were considered lower than farmers. Many slaves were convicted criminals or prisoners captured in war. These slaves worked on farms, on building projects, in workshops, and in private households. Unlike most slaves in history, however, slaves in Egypt had some legal rights. Also, in some cases, they could earn their freedom.

Family Life in Egypt

Family life was very important in Egyptian society. Most Egyptian families lived in their own homes. Sometimes unmarried female relatives lived with them, but men were expected to marry young so that they could start having children.

Most Egyptian women were devoted to their homes and families. Some women, however, did have jobs outside the home.

A few women served as priestesses, and some worked as royal officials, administrators, or artisans. Unlike most women in ancient times, Egyptian women had a number of legal rights. They could own property, make **contracts**, and divorce their husbands. They could even keep their property after a divorce.

Children's lives were not as structured as adults' lives were. Children played with toys such as dolls, tops, and clay animal figurines. Children also played ballgames and hunted. Most children, boys and girls, received some education. At school they learned morals, writing, math, and sports. At age 14 most boys left school to enter their father's profession. At that time, they took their place in Egypt's social structure.

READING CHECK Categorizing What types of jobs existed in ancient Egypt?

SUMMARY AND PREVIEW Pharaohs faced many challenges to their rule. After the defeat of the Hyksos, Egypt grew in land and wealth. People in Egypt worked at many jobs. In the next section you will learn about Egyptian achievements.

go.hrw.com
Online Quiz
KEYWORD: SK7 HP11

Section 3 Assessment

Reviewing Ideas, Terms, and Places

1. **a. Define** What was the **Middle Kingdom**?
   b. Analyze How did Ahmose manage to become king of all Egypt?
2. **a. Recall** What two things brought wealth to the pharaohs during the **New Kingdom**?
   b. Explain What did Hatshepsut do as pharaoh of Egypt?
3. **a. Identify** What job employed the majority of the people in Egypt?
   b. Analyze What rights did Egyptian women have?
   c. Elaborate Why do you think scribes were so honored in Egyptian society?

Critical Thinking

4. **Categorizing** Draw pyramids like the ones shown. Using your notes, fill in the pyramids with the political and military factors that led to the rise and fall of the Middle and New Kingdoms.

Rise — Fall — Rise — Fall
Middle Kingdom   New Kingdom

FOCUS ON WRITING

5. **Developing Ideas from the Middle and New Kingdoms** Your riddle should contain information about these periods. Decide which key ideas you should include and add them to your list.

Ramses the Great

How could a ruler achieve fame that would last 3,000 years?

When did he live? late 1300s and early 1200s BC

Where did he live? As pharaoh, Ramses lived in a city he built on the Nile Delta. The city's name, Pi-Ramesse, means the "house of Ramses."

What did he do? From a young age, Ramses was trained as a ruler and a fighter. Made an army captain at age 10, he began military campaigns even before he became pharaoh. During his reign, Ramses greatly increased the size of his kingdom.

Why is he important? Many people consider Ramses the last great Egyptian pharaoh. He accomplished great things, but the pharaohs who followed could not maintain them. Both a great warrior and a great builder, he is known largely for the massive monuments he built. The temples at Karnak, Luxor, and Abu Simbel stand as 3,000-year-old symbols of the great pharaoh's power.

Drawing Conclusions Why do you think Ramses built monuments all over Egypt?

KEY IDEAS

Ramses had a poem praising him carved into the walls of five temples, including Karnak. One verse of the poem praises Ramses as a great warrior and the defender of Egypt.

" Gracious lord and bravest
    king, savior–guard
Of Egypt in the battle, be our
    ward;
Behold we stand alone, in the
    hostile Hittite ring,
Save for us the breath of life,
Give deliverance from the
    strife,
Oh! protect us Ramses Miamun!
Oh! save us, mighty king! "

–Pen-ta-ur, quoted in *The World's Story*, edited by Eva March Tappan

This copy of an ancient painting shows Ramses the Great on his chariot in battle against the Hittites.

Egyptian Achievements

What You Will Learn...

Main Ideas

1. Egyptian writing used symbols called hieroglyphics.
2. Egypt's great temples were lavishly decorated.
3. Egyptian art filled tombs.

The Big Idea

The Egyptians made lasting achievements in writing, art, and architecture.

Key Terms

hieroglyphics, *p. 298*
papyrus, *p. 298*
Rosetta Stone, *p. 299*
sphinxes, *p. 300*
obelisk, *p. 300*

TAKING NOTES As you read, use a chart like this one to take notes on the achievements of the ancient Egyptians. In each column, identify Egyptian achievements in the appropriate field.

Writing	Architecture	Art

If YOU lived there...

You are an artist in ancient Egypt. A powerful noble has hired you to decorate the walls of his family tomb. You are standing inside the new tomb, studying the bare, stone walls that you will decorate. No light reaches this chamber, but your servant holds a lantern high. You've met the noble only briefly but think that he is someone who loves his family, the gods, and Egypt.

What will you include in your painting?

BUILDING BACKGROUND The ancient Egyptians had a rich and varied history. Today, though, most people remember them for their cultural achievements. Egyptian art, such as the tomb paintings mentioned above, and Egypt's unique writing system are admired by millions of tourists in museums around the world.

Egyptian Writing

If you were reading a book and saw pictures of folded cloth, a leg, a star, a bird, and a man holding a stick, would you know what it meant? You would if you were an ancient Egyptian. In the Egyptian writing system, or **hieroglyphics** (hy-ruh-GLIH-fiks), those five symbols together meant "to teach." Egyptian hieroglyphics were one of the world's first writing systems.

Writing in Ancient Egypt

The earliest known examples of Egyptian writing are from around 3300 BC. These early Egyptian writings were carved in stone or on other hard materials. Later, Egyptians learned how to make **papyrus** (puh-PY-ruhs), a long-lasting, paperlike material made from reeds. The Egyptians made papyrus by pressing layers of reeds together and pounding them into sheets. These sheets were tough and durable, yet could be rolled into scrolls. Scribes wrote on papyrus using brushes and ink.

Egyptian Writing

Egyptian hieroglyphics used picture symbols to represent sounds.

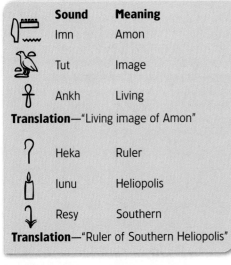

	Sound	Meaning
	Imn	Amon
	Tut	Image
	Ankh	Living

Translation—"Living image of Amon"

	Heka	Ruler
	Iunu	Heliopolis
	Resy	Southern

Translation—"Ruler of Southern Heliopolis"

ANALYSIS SKILL **ANALYZING VISUALS**

What does the symbol for ruler look like?

The hieroglyphic writing system used more than 600 symbols, mostly pictures of objects. Each symbol represented one or more sounds in the Egyptian language. For example, a picture of an owl represented the same sound as our letter M.

Hieroglyphics could be written either horizontally or vertically. They could be written from right to left or from left to right. These options made hieroglyphics flexible to write but difficult to read. The only way to tell which way a text is written is to look at individual symbols.

The Rosetta Stone

Historians and archaeologists have known about hieroglyphics for centuries. For a long time, though, historians did not know how to read them. In fact, it was not until 1799 that a lucky discovery by a French soldier gave historians the key they needed to read ancient Egyptian writing.

That key was the **Rosetta Stone**, a huge, stone slab inscribed with hieroglyphics. In addition to the hieroglyphics, the Rosetta Stone had text in Greek and a later form of Egyptian. Because the message in all three languages was the same, scholars who knew Greek were able to figure out what the hieroglyphics said.

Egyptian Texts

Because papyrus did not decay in Egypt's dry climate, many ancient Egyptian texts still survive. These texts include government records, historical records, science texts, and medical manuals. In addition, many literary works have survived. Some of them, such as *The Book of the Dead,* tell about the afterlife. Others tell stories about gods and kings.

READING CHECK **Comparing** How is our writing system similar to hieroglyphics?

THE IMPACT TODAY

An object that helps solve a difficult mystery is sometimes now called a Rosetta Stone.

Egypt's Great Temples

In addition to their writing system, the ancient Egyptians are famous for their magnificent architecture. You have already read about the Egyptians' most famous structures, the pyramids. But the Egyptians also built massive temples. Those that survive are among the most spectacular sites in Egypt today.

The Egyptians believed that temples were the homes of the gods. People visited the temples to worship, offer the gods gifts, and ask for favors.

Many Egyptian temples shared some similar features. Rows of stone **sphinxes**—imaginary creatures with the bodies of lions and the heads of other animals or humans—lined the path leading to the entrance. That entrance itself was a huge, thick gate. On either side of the gate might stand an **obelisk** (AH-buh-lisk), a tall, four-sided pillar that is pointed on top.

Inside, Egyptian temples were lavishly decorated, as you can see in the drawing of the Temple of Karnak. Huge columns supported the temple's roof. These columns were often covered with paintings and hieroglyphics, as were the temple walls. Statues of gods and pharaohs often stood along the walls as well. The sanctuary, the most sacred part of the building, was at the far end of the temple.

The Temple of Karnak is only one of Egypt's great temples. Other temples were built by Ramses the Great at Abu Simbel and Luxor. The temple at Abu Simbel is especially known for the huge statues that stand next to its entrance. The 66-foot-tall statues are carved out of sandstone cliffs and show Ramses the Great as pharaoh. Nearby are smaller statues of his family.

THE IMPACT TODAY

The Washington Monument, in Washington, D.C., is an obelisk.

READING CHECK **Generalizing** What were some features of ancient Egyptian temples?

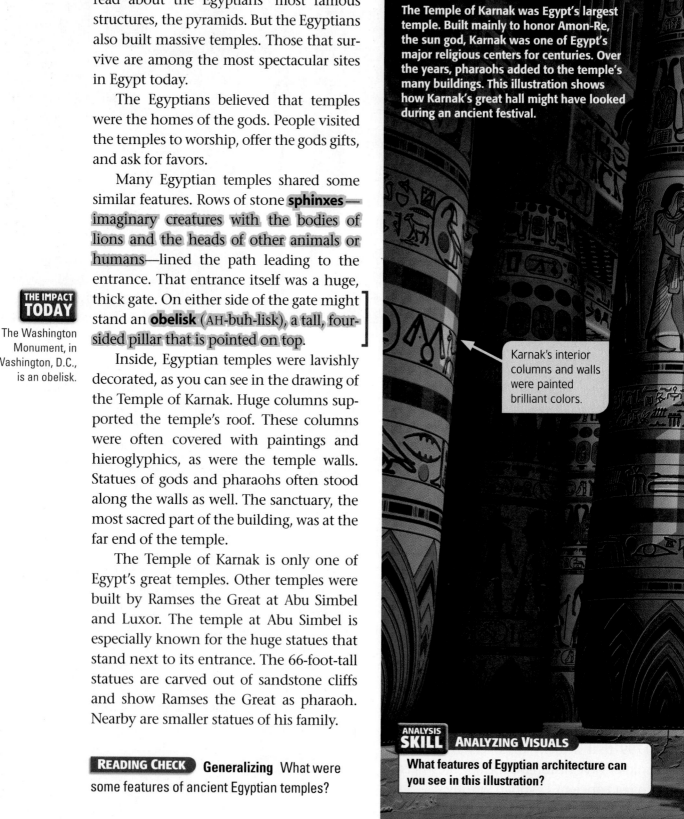

Close-up

The Temple of Karnak

The Temple of Karnak was Egypt's largest temple. Built mainly to honor Amon-Re, the sun god, Karnak was one of Egypt's major religious centers for centuries. Over the years, pharaohs added to the temple's many buildings. This illustration shows how Karnak's great hall might have looked during an ancient festival.

Karnak's interior columns and walls were painted brilliant colors.

ANALYSIS SKILL **ANALYZING VISUALS**

What features of Egyptian architecture can you see in this illustration?

Treasures of King Tut's Tomb

In 1922 the archaeologist Howard Carter discovered the tomb of King Tut. Unlike most Egyptian tombs, it had never been robbed and was still filled with treasures, some of which are shown here.

Howard Carter examining King Tut's coffin in 1922

Egyptian Art

FOCUS ON READING

What categories could you use to organize the information under Egyptian Art?

One reason Egypt's temples are so popular with tourists is the art they contain. The ancient Egyptians were masterful artists. Many of their greatest works were created to fill the tombs of pharaohs and other nobles. The Egyptians took great care in making these items because they believed the dead could enjoy them in the afterlife.

Paintings

Egyptian art was filled with lively, colorful scenes. Detailed works covered the walls of temples and tombs. Artists also painted on canvas, papyrus, pottery, plaster, and wood. Most Egyptians never saw these paintings, however. Only kings, priests, and important people could enter temples and tombs, and even they rarely entered the tombs.

The subjects of Egyptian paintings vary widely. Some of the paintings show important historical events, such as the crowning of a new king or the founding of a temple.

Others show major religious rituals. Still other paintings show scenes from everyday life, such as farming or hunting.

Egyptian painting has a distinctive style. People, for example, are drawn in a certain way. In Egyptian paintings, people's heads and legs are always seen from the side, but their upper bodies and shoulders are shown straight on. In addition, people do not all appear the same size. Important figures such as pharaohs appear huge in comparison to others, especially servants or conquered people. In contrast, Egyptian animals were usually drawn realistically.

Carvings and Jewelry

Painting was not the only art form Egyptians practiced. The Egyptians were also skilled stoneworkers. Many tombs included huge statues and detailed carvings.

In addition, the Egyptians made lovely objects out of gold and precious stones. They made jewelry for both men and women.

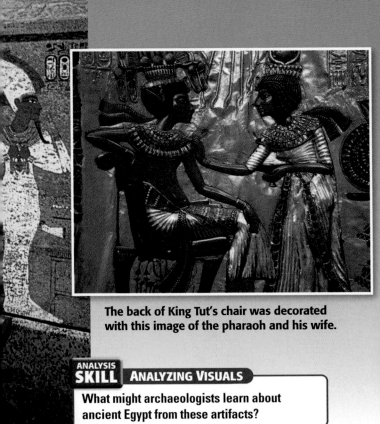

The back of King Tut's chair was decorated with this image of the pharaoh and his wife.

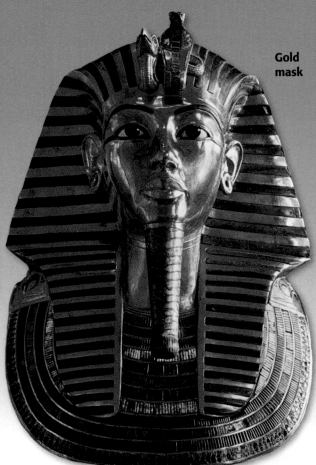

Gold mask

ANALYSIS
SKILL **ANALYZING VISUALS**

What might archaeologists learn about ancient Egypt from these artifacts?

This jewelry included necklaces, bracelets, and collars. The Egyptians also used gold to make burial items for their pharaohs.

Over the years, treasure hunters emptied many pharaohs' tombs. At least one tomb, however, was not disturbed. In 1922 some archaeologists found the tomb of King Tutankhamen (too-tang-KAHM-uhn), or King Tut. The tomb was filled with many treasures, including boxes of jewelry, robes, a burial mask, and ivory statues. King Tut's treasures have taught us much about Egyptian burial practices and beliefs.

READING CHECK **Summarizing** What types of artwork were contained in Egyptian tombs?

SUMMARY AND PREVIEW The Egyptians developed one of the best-known cultures of the ancient world. Next, you will learn about a culture that developed in the shadow of Egypt—Kush.

go.hrw.com
Online Quiz
KEYWORD: SK7 HP11

Section 4 Assessment

Reviewing Ideas, Terms, and Places

1. **a. Define** What are **hieroglyphics**?
   b. Contrast How was hieroglyphic writing different from our writing today?
   c. Evaluate Why was the **Rosetta Stone** important?
2. **a. Describe** What were two ways the Egyptians decorated their temples?
   b. Evaluate Why do you think pharaohs like Ramses the Great built huge temples?
3. **Recall** Why did Egyptians fill tombs with art, jewelry, and other treasures?

Critical Thinking

4. **Summarizing** Draw a chart like the one below. In each column, write a statement that summarizes Egyptian achievements in the listed category.

Writing	Architecture	Art

FOCUS ON WRITING

5. **Considering Egyptian Achievements** For your riddle, note some Egyptian achievements in writing, architecture, and art that make Egypt different from other places.

Analyzing Primary and Secondary Sources

Learn

Primary sources are materials created by people who lived during the times they describe. Examples include letters, diaries, and photographs. *Secondary sources* are accounts written later by someone who was not present. They often teach about or discuss a historical topic. This chapter is an example of a secondary source.

By studying both types, you can get a better picture of a historical period or event. However, not all sources are accurate or reliable. Use these checklists to judge which sources are reliable.

Checklist for Primary Sources

• Who is the author? Is he or she trustworthy?

• Was the author present at the event described in the source? Might the author have based his or her writing on rumor, gossip, or hearsay?

• How soon after the event occurred was the source written? The more time that passed, the greater the chance for error.

• What is the purpose? Authors can have reasons to exaggerate—or even lie—to suit their own purposes. Look for evidence of emotion, opinion, or bias in the source. They can affect the accuracy.

• Can the information in the source be verified in other primary or secondary sources?

Checklist for Secondary Sources

• Who is the author? What are his or her qualifications? Is he or she an authority on the subject?

• Where did the author get his or her information? Good historians always tell you where they got their information.

• Has the author drawn valid conclusions?

Practice

" The Egyptians quickly extended their military and commercial influence over an extensive [wide] region that included the rich provinces of Syria … and the numbers of Egyptian slaves grew swiftly. "

–C. Warren Hollister, from *Roots of the Western Tradition*

" Let me tell you how the soldier fares … how he goes to Syria, and how he marches over the mountains. His bread and water are borne [carried] upon his shoulders like the load of [a donkey]; … and the joints of his back are bowed [bent] … When he reaches the enemy, … he has no strength in his limbs. "

–from *Wings of the Falcon: Life and Thought of Ancient Egypt*, translated by Joseph Kaster

1 Which of the above passages is a primary source, and which is a secondary source?

2 Is there evidence of opinion, emotion, or bias in the second passage? Why, or why not?

3 Which passage would be better for learning about what life was like for Egyptian soldiers, and why?

Apply

Refer to the Ramses the Great biography in this chapter to answer the following questions.

1. Identify the primary source in the biography.

2. What biases or other issues might affect the reliability or accuracy of this primary source?

Geography's Impact
video series
Review the video to answer the closing question:
What do the pyramids of ancient Egypt tell you about the people of that civilization?

Visual Summary

Use the visual summary below to help you review the main ideas of the chapter.

QUICK FACTS

Egyptian civilization developed along the Nile River, which provided water and fertile soil for farming.

Egypt's kings were considered gods, and Egyptians made golden burial masks and pyramids in their honor.

Egyptian cultural achievements included beautiful art and the development of a hieroglyphic writing system.

Reviewing Vocabulary, Terms, and Places

Imagine these terms are answers to items in a crossword puzzle. Write the clues for the answers. Then make the puzzle with answers down and across.

1. cataract
2. Nile River
3. pharaoh
4. nobles
5. mummy
6. acquire
7. contract
8. pyramids
9. hieroglyphics
10. sphinxes

Comprehension and Critical Thinking

SECTION 1 *(Pages 278–282)*

11. **a. Identify** Where was most of Egypt's fertile land located?

b. Make Inferences Why did Memphis become a political and social center of Egypt?

c. Predict How might history have been different if the Nile had not flooded every year?

SECTION 2 *(Pages 283–290)*

12. **a. Describe** Who were the pharaohs, and what responsibilities did they have?

b. Analyze How were beliefs about the afterlife linked to items placed in tombs?

c. Elaborate What challenges, in addition to moving stone blocks, do you think the pyramid builders faced?

SECTION 3 *(Pages 291–296)*

13. **a. Describe** What did a scribe do, and what benefits did a scribe receive?

b. Analyze When was the period of the New Kingdom, and what two factors contributed to Egypt's wealth during that period?

c. Evaluate Ramses the Great was a powerful pharaoh. Do you think his military successes or his building projects are more important to evaluating his greatness? Why?

14. **a. Describe** For what was papyrus used?

   b. Contrast How are the symbols in Egyptian hieroglyphics different from the symbols used in our writing system?

   c. Elaborate How does the Egyptian style of painting people reflect their society?

Social Studies Skills

Analyzing Primary and Secondary Sources *Each of the questions below lists two sources that a historian might consult to answer a question about ancient Egypt. For each question, decide which source is likely to be more accurate or reliable and why. Then indicate whether that source is a primary or secondary source.*

15. What were Egyptian beliefs about the afterlife?

   a. Egyptian tomb inscriptions

   b. writings by a priest who visited Egypt in 1934

16. Why did the Nile flood every year?

   a. songs of praise to the Nile River written by Egyptian priests

   b. a book about the rivers of Africa written by a modern geographer

17. What kinds of goods did the Egyptians trade?

   a. ancient Egyptian trade records

   b. an ancient Egyptian story about a trader

18. What kind of warrior was Ramses the Great?

   a. a poem in praise of Ramses

   b. a description of a battle in which Ramses fought, written by an impartial observer

Using the Internet

go.hrw.com
KEYWORD: SK7 CH11

19. **Activity: Creating Egyptian Art** The Egyptians excelled in the arts. Egyptian artwork included beautiful paintings, carvings, and jewelry. Egyptian architecture included huge pyramids and temples. Enter the activity keyword and research Egyptian art and architecture. Then imagine you are an Egyptian. Create a work of art for the pharaoh's tomb. Provide hieroglyphics telling the pharaoh about your art.

FOCUS ON READING AND WRITING

20. **Categorizing** Create a chart with three columns. Title the chart "Egyptian Pharaohs." Label the three chart columns "Position and Power," "Responsibilities," and "Famous Pharaohs." Then list facts and details from the chapter under each category in the chart.

21. **Writing a Riddle** Choose five details about Egypt. Then write a sentence about each detail. Each sentence of your riddle should be a statement ending with "me." For example, if you were writing about the United States, you might say, "People come from all over the world to join me." After you have written your five sentences, end your riddle with "Who am I?" The answer to your riddle must be "Egypt."

Map Activity ★Interactive

22. **Ancient Egypt** On a separate sheet of paper, match the letters on the map with their correct labels.

   Lower Egypt   Red Sea

   Mediterranean Sea   Sinai Peninsula

   Nile River   Upper Egypt

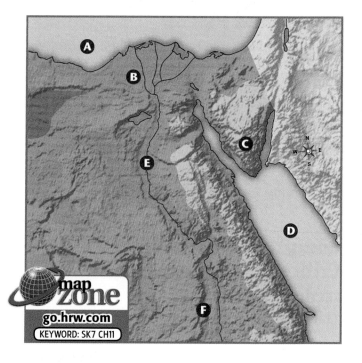

map zone
go.hrw.com
KEYWORD: SK7 CH11

DIRECTIONS: Read questions 1 through 8 and write the letter of the best response. Then read question 9 and write your own well-constructed response.

1 **Which statement about how the Nile helped civilization develop in Egypt is false?**

A It provided a source of food and water.

B It enabled farming in the area.

C Its flooding enriched the soil along its banks.

D It protected against invasion from the west.

2 **The most fertile soil in Egypt was located in the**

A Nile Delta.

B deserts.

C cataracts.

D far south.

3 **The high position that priests held in Egyptian society shows that**

A the pharaoh was a descendant of a god.

B government was large and powerful.

C religion was important in Egyptian life.

D the early Egyptians worshipped many gods.

4 **The Egyptians are probably *best* known for building**

A pyramids.

B irrigation canals.

C cataracts.

D deltas.

5 **During which period did ancient Egypt reach the height of its power and glory?**

A First Dynasty

B Old Kingdom

C Middle Kingdom

D New Kingdom

Oh great god and ruler, the gift of Amon-Re, god of the Sun.
Oh great protector of Egypt and its people.
Great one who saved us from the Tehenu.
You, who have fortified our western border to protect us from our enemies.
You, who honored the gods with mighty temples at Abu Simbel and Luxor.
We bless you, oh great one.
We worship and honor you, oh great and mighty pharaoh.

6 **A tribute such as the one above might have been written in honor of which Egyptian ruler?**

A Menes

B Ramses the Great

C King Tutankhamen

D Queen Hatshepsut

7 **What discovery gave historians the key they needed to read Egyptian hieroglyphics?**

A obelisk

B papyrus

C Rosetta Stone

D sphinx

8 **Which of the following groups made up the majority of people in ancient Egypt?**

A artisans and artists

B farmers and other peasants

C merchants and traders

D scribes

9 **Extended Response** Examine the diagram titled Egyptian Society in Section 2. Based on the information in the diagram and in the text, explain the role that religion and government played in the organization of Egyptian society.

CHAPTER 12

History of Ancient Kush

2300 BC–AD 350

What You Will Learn...

In this chapter you will learn about the geography and history of ancient Kush. You will also discover the connections between Egypt and Kush. Finally, you will study the culture of ancient Kush and the reasons for its decline.

SECTION 1
Kush and Egypt 310

SECTION 2
Later Kush 315

FOCUS ON READING AND WRITING

Asking Questions As you read new information, asking questions can help you learn and remember. After you read a paragraph or section, ask yourself who or what the passage is about, when and where the events took place, and why the information is important. **See the lesson, Asking Questions, on p. 679.**

Writing a Fictional Narrative In this chapter you will read about events surrounding the rise and fall of Kush. Then you will write a short story about fictional characters who lived through these events. The main character in your story will be from Kush. Other main characters could be from Egypt, Assyria, or Aksum.

Trade caravan in the Sahara

0   125   250 Miles
0   125   250 Kilometers

Projection: Lambert Azimuthal Equal-Area

Kushite Pyramids Egyptian culture greatly influenced the people of Kush. Kush's leaders built pyramids like these as tombs.

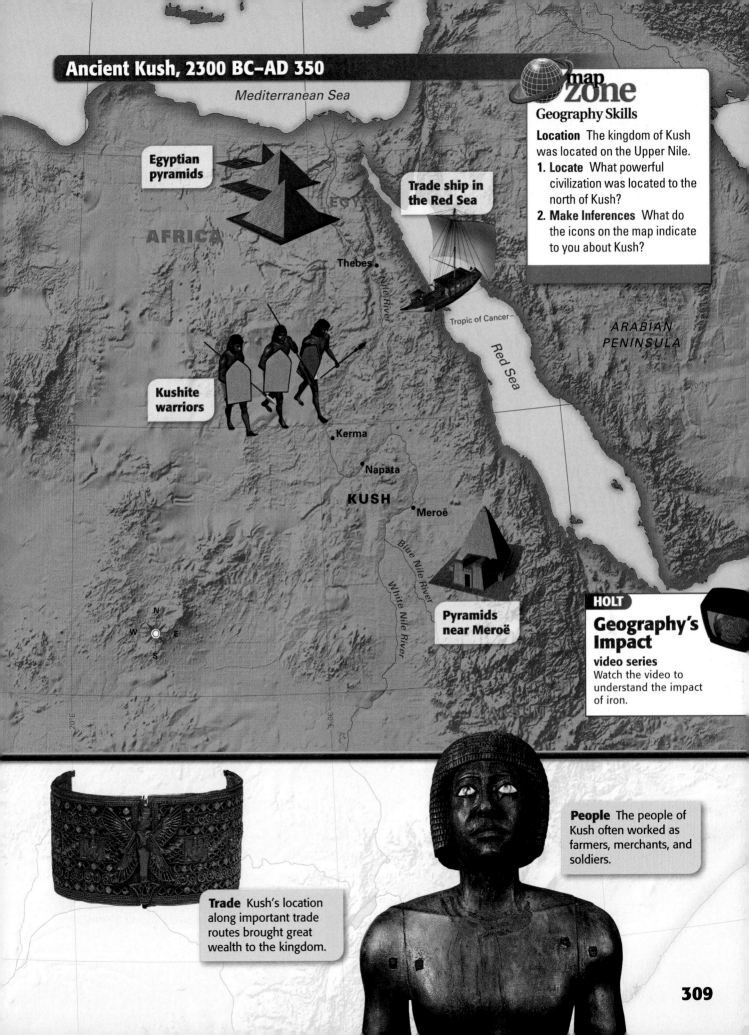

Ancient Kush, 2300 BC–AD 350

Mediterranean Sea

Egyptian pyramids

AFRICA

EGYPT

Thebes

Trade ship in the Red Sea

Tropic of Cancer

Red Sea

ARABIAN PENINSULA

Kushite warriors

Nile River

Kerma

Napata

KUSH

Meroë

Blue Nile River

White Nile River

Pyramids near Meroë

AKSUM

N W E S

map zone

Geography Skills

Location The kingdom of Kush was located on the Upper Nile.

1. **Locate** What powerful civilization was located to the north of Kush?

2. **Make Inferences** What do the icons on the map indicate to you about Kush?

HOLT

Geography's Impact

video series
Watch the video to understand the impact of iron.

People The people of Kush often worked as farmers, merchants, and soldiers.

Trade Kush's location along important trade routes brought great wealth to the kingdom.

309

Kush and Egypt

What You Will Learn...

Main Ideas

1. Geography helped early Kush civilization develop in Nubia.
2. Egypt controlled Kush for about 450 years.
3. After winning its independence, Kush ruled Egypt and set up a new dynasty there.

The Big Idea

The kingdom of Kush, in the region of Nubia, was first conquered by Egypt but later conquered and ruled Egypt.

Key Terms and Places

Nubia, *p. 310*
ebony, *p. 312*
ivory, *p. 312*

TAKING NOTES As you read, take notes on the important events in the early history of the kingdom of Kush. Use a chart like the one below to identify significant events, their dates, and their importance.

Event	Date	Importance

If YOU lived there...

You live along the Nile River, where it moves quickly through rapids. A few years ago, armies from the powerful kingdom of Egypt took over your country. Some Egyptians have moved to your town. They bring new customs, which many people are beginning to imitate. Now your sister has a new baby and wants to give it an Egyptian name! This upsets many people in your family.

How do you feel about following Egyptian customs?

BUILDING BACKGROUND Egypt dominated the lands along the Nile, but it was not the only ancient culture to develop along the river. Another kingdom, called Kush, arose to the south of Egypt. Through trade, conquest, and political dealings, the histories of Egypt and Kush became closely tied together.

Geography and Early Kush

South of Egypt along the Nile, a group of people settled in the region we now call Nubia. These Africans established the first large kingdom in the interior of Africa. We know this kingdom by the name the ancient Egyptians gave it—Kush. Development of Kushite civilization was greatly influenced by the geography and resources of the region.

The Land of Nubia

Nubia is a region in northeast Africa. It lies on the Nile River south of Egypt. Today desert covers much of Nubia, located in the present-day country of Sudan. In ancient times, however, the region was much more fertile. Heavy rainfall flooded the Nile every year. These floods provided a rich layer of fertile soil to nearby lands. The kingdom of Kush developed in this area.

In addition to having fertile soil, ancient Nubia was rich in valuable minerals such as gold, copper, and stone. These natural resources contributed to the region's wealth and played a major role in its history.

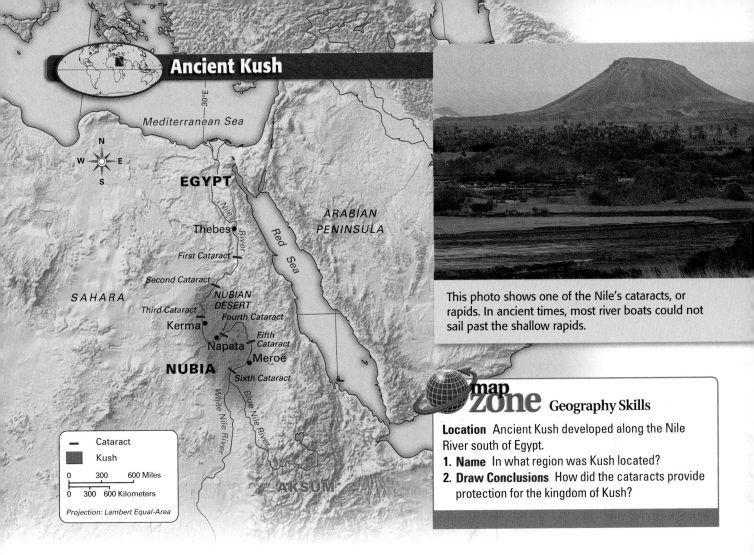

Ancient Kush

This photo shows one of the Nile's cataracts, or rapids. In ancient times, most river boats could not sail past the shallow rapids.

map zone Geography Skills

Location Ancient Kush developed along the Nile River south of Egypt.
1. **Name** In what region was Kush located?
2. **Draw Conclusions** How did the cataracts provide protection for the kingdom of Kush?

Early Civilization in Nubia

Like all early civilizations, the people of Nubia depended on agriculture for their food. Fortunately for them, the Nile's floods allowed the Nubians to plant both summer and winter crops. Among the crops they grew were wheat, barley, and other grains. In addition to farmland, the banks of the river provided grazing land for cattle and other livestock. As a result, farming villages thrived all along the Nile by about 3500 BC.

Over time some farmers became richer and more successful than others. These farmers became leaders of their villages. Sometime around 2000 BC, one of these leaders took control of other villages and made himself king of the region. His new kingdom was called Kush.

The early kings of Kush ruled from their capital at Kerma (KAR-muh). This city was located on the Nile just south of a cataract, or stretch of rapids. Cataracts made travel through some parts of the Nile extremely difficult. As a result, the cataracts were natural barriers against invaders. For many years the cataracts kept Kush safe from the powerful Egyptian kingdom to the north.

As time passed, Kushite society grew more complex. In addition to farmers and herders, some people of Kush became priests or artisans. Early on, civilizations to the south greatly influenced the kingdom of Kush. Later, however, Egypt played a greater role in the kingdom's history.

FOCUS ON READING
What is this paragraph about? Why is it important?

READING CHECK Finding Main Ideas How did geography help civilization grow in Nubia?

Egypt Controls Kush

Kush and Egypt were neighbors. At times the neighbors lived in peace with each other and helped each other prosper. For example, Kush became a supplier of slaves and raw materials to Egypt. The Kushites sent materials such as gold, copper, and stone to Egypt. The Kushites also sent the Egyptians **ebony**, a type of dark, heavy wood, and **ivory**, a white material taken from elephant tusks.

Egypt's Conquest of Kush

Relations between Kush and Egypt were not always peaceful. As Kush grew wealthy from trade, its army grew stronger as well. Egypt's rulers soon feared that Kush would grow even stronger. They were afraid that a powerful Kush might attack Egypt.

To prevent such an attack, the pharaoh Thutmose I sent an army to take control of Kush around 1500 BC. The pharaoh's army conquered all of Nubia north of the Fifth Cataract. As a result, the kingdom of Kush became part of Egypt.

After his army's victory, the pharaoh destroyed the Kushite palace at Kerma. Later pharaohs—including Ramses the Great—built huge temples in what had been Kushite territory.

Effects of the Conquest

Kush remained an Egyptian territory for about 450 years. During that time, Egypt's influence over Kush grew tremendously. Many Egyptians settled in Kush. Egyptian became the language of the region. Many Kushites used Egyptian names and wore Egyptian-style clothing. They also adopted Egyptian religious practices.

A Change in Power

In the mid-1000s BC the New Kingdom in Egypt was ending. As the power of Egypt's pharaohs declined, Kushite leaders regained control of Kush. Kush once again became independent.

READING CHECK **Identifying Cause and Effect**
How did Egyptian rule change Kush?

Kush and Egypt

Early in its history, Egypt dominated Kush, forcing Kushites to give tribute to Egypt.

Kush Rules Egypt

We know almost nothing about the history of the Kushites for about 200 years after they regained independence from Egypt. Kush is not mentioned in any historical records until the 700s BC, when armies from Kush swept into Egypt and conquered it.

The Conquest of Egypt

By around 850 BC, Kush had regained its strength. It was once again as strong as it had been before it was conquered by Egypt. Because the Egyptians had captured the old capital at Kerma, the kings of Kush ruled from the city of Napata. Napata was located on the Nile, about 100 miles southeast of Kerma.

As Kush was growing stronger, Egypt was losing power. A series of weak pharaohs left Egypt open to attack. In the 700s BC a Kushite king, Kashta, took advantage of Egypt's weakness. Kashta attacked Egypt. By about 751 BC he had conquered Upper Egypt. He then established relations with Lower Egypt.

BIOGRAPHY

Piankhi
(c. 751–716 BC)

Also known as Piye, Piankhi was among Kush's most successful military leaders. A fierce warrior on the battlefield, the king was also deeply religious. Piankhi's belief that he had the support of the gods fueled his passion for war against Egypt. His courage inspired his troops on the battlefield. Piankhi loved his horses and was buried with eight of them.

Drawing Conclusions How did Piankhi's belief that he was supported by the gods help him in the war against Egypt?

After Kashta died, his son Piankhi (PYANG-kee) continued to attack Egypt. The armies of Kush captured many cities, including Egypt's ancient capital. Piankhi fought the Egyptians because he believed that the gods wanted him to rule all of Egypt. By the time he died in about 716 BC, Piankhi had accomplished this task. His kingdom extended north from Napata all the way to the Nile Delta.

Later, as Kush's power increased, its warriors invaded and conquered Egypt. This photo shows Kushite and Egyptian warriors.

After conquering Egypt, Kush established a new dynasty. This sculpture shows one of Kush's pharaohs kneeling before an Egyptian god.

ANALYSIS SKILL **ANALYZING VISUALS**

What did Kushites give to Egypt as tribute?

When the Assyrians invaded Egypt with their iron weapons, they forced Kush's rulers out of Egypt and south into Nubia.

The Kushite rulers of Egypt built new temples to Egyptian gods and restored old ones. They also worked to preserve many Egyptian writings. As a result, Egyptian culture thrived during the Kushite dynasty.

The End of Kushite Rule in Egypt

The Kushite dynasty remained strong in Egypt for about 40 years. In the 670s BC, however, the powerful army of the Assyrians from Mesopotamia invaded Egypt. The Assyrians' iron weapons were better than the Kushites' bronze weapons, and the Kushites were slowly pushed out of Egypt. In just 10 years the Assyrians had driven the Kushite forces completely out of Egypt.

READING CHECK **Sequencing** How did the leaders of Kush gain control over Egypt?

SUMMARY AND PREVIEW Kush was conquered by Egypt, but later the Kushites controlled Egypt. In the next section, you will learn how the civilization of Kush developed after the Kushites were forced out of Egypt by the Assyrians.

The Kushite Dynasty

After Piankhi died, his brother Shabaka (SHAB-uh-kuh) took control of the kingdom and declared himself pharaoh. His declaration marked the beginning of Egypt's Twenty-fifth, or Kushite, Dynasty.

Shabaka and later rulers of his dynasty tried to restore many old Egyptian cultural practices. Some of these practices had died out during Egypt's period of weakness. For example, Shabaka was buried in a pyramid. The Egyptians had stopped building pyramids for their rulers centuries earlier.

Section 1 Assessment

Reviewing Ideas, Terms, and Places

1. **a. Identify** On which river did Kush develop?
   b. Analyze How did **Nubia**'s natural resources influence the early history of Kush?

2. **a. Describe** What is **ivory**?
   b. Explain How did Egypt's conquest of Kush affect the people of Kush?
   c. Evaluate Why do you think Thutmose I destroyed the Kushite palace at Kerma?

3. **a. Describe** What territory did Piankhi conquer?
   b. Make Inferences Why is the Twenty-fifth Dynasty significant in the history of Egypt?
   c. Predict What might have happened in Kush and Egypt if Kush had developed iron weapons earlier?

Critical Thinking

4. **Sequencing** Use a time line like the one below to show the sequence and dates of important events in the early history of the kingdom of Kush.

2000 BC   680 BC

FOCUS ON WRITING

5. **Planning Characters and Plot** Make a chart with two columns. Label one "Characters," and take notes on the main characters and their interactions. Label the other "Plot," and note major events and sources of conflict among the characters.

Later Kush

If **YOU** lived there...

You live in Meroë, the capital of Kush, in 250 BC. Your father is a skilled ironworker. From him you've learned to shape iron tools and weapons. Everyone expects that you will carry on his work. If you do become an ironworker, you will likely make a good living. But you are restless. You'd like to travel down the Nile to see Egypt and the great sea beyond it. Now a neighbor who is a trader has asked you to join his next trading voyage.

Will you leave Meroë to travel? Why or why not?

BUILDING BACKGROUND The Assyrians drove the Kushites out of Egypt in the 600s BC, partly through their use of iron weapons. Although the Kushites lost control of Egypt, their kingdom did not disappear. In fact, they built up another empire in the African interior, based on trade and their own iron industry.

Kush's Economy Grows

After they lost control of Egypt, the people of Kush devoted themselves to improving agriculture and trade. They hoped to make their country rich again. Within a few centuries, Kush had indeed become a rich and powerful kingdom once more.

What You Will Learn...

Main Ideas

1. Kush's economy grew because of its iron industry and trade network.
2. Some elements of Kushite society and culture were borrowed from other cultures while others were unique to Kush.
3. The decline and defeat of Kush was caused by both internal and external factors.

The Big Idea

Although Kush developed an advanced civilization, it eventually declined.

Key Terms and Places
Meroë, *p. 316*
trade network, *p. 316*
merchants, *p. 316*
exports, *p. 316*
imports, *p. 316*

TAKING NOTES As you read, take notes about the civilization of Kush and how it finally declined. Organize your notes in a diagram like the one below.

Kush	
Economy	Society

Decline

Kushite Metalwork

Kush's craftspeople made iron spear-heads and gold jewelry like you see here.

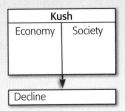

PHOTOGRAPH © 2004
MUSEUM OF FINE ARTS, BOSTON

315

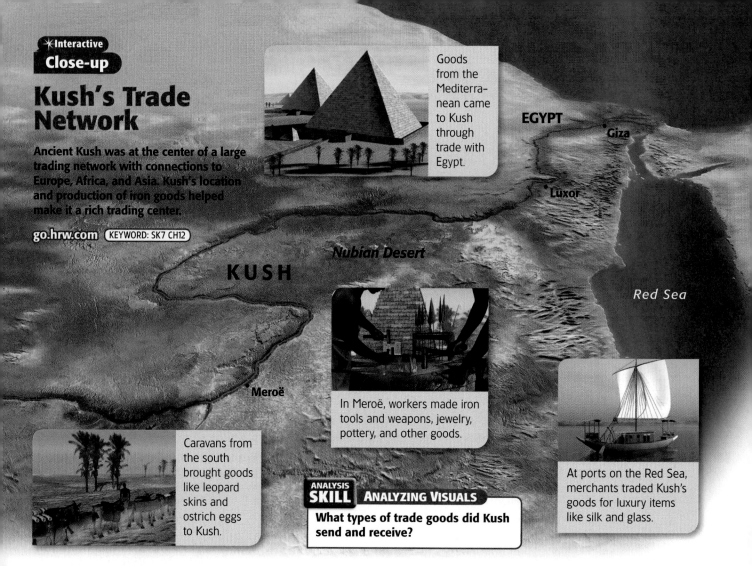

Kush's Trade Network

Ancient Kush was at the center of a large trading network with connections to Europe, Africa, and Asia. Kush's location and production of iron goods helped make it a rich trading center.

go.hrw.com KEYWORD: SK7 CH12

Goods from the Mediterranean came to Kush through trade with Egypt.

EGYPT

Giza

Luxor

Nubian Desert

KUSH

Meroë

Red Sea

In Meroë, workers made iron tools and weapons, jewelry, pottery, and other goods.

Caravans from the south brought goods like leopard skins and ostrich eggs to Kush.

At ports on the Red Sea, merchants traded Kush's goods for luxury items like silk and glass.

ANALYSIS SKILL ANALYZING VISUALS

What types of trade goods did Kush send and receive?

Kush's Iron Industry

FOCUS ON READING

As you read, ask yourself these questions: Where was Meroë? Why was it important?

During this period, the economic center of Kush was **Meroë** (MER-oh-wee), the new Kushite capital. Meroë's location on the east bank of the Nile helped Kush's economy. Gold could be found nearby, as could forests of ebony and other wood. More importantly, the area around Meroë was rich in deposits of iron ore.

In this location the Kushites developed an iron industry. Because resources such as iron ore and wood for furnaces were easily available, the industry grew quickly.

Expansion of Trade

In time, Meroë became the center of a large **trade network**, a system of people in different lands who trade goods back and forth.

The Kushites sent goods down the Nile to Egypt. From there, Egyptian and Greek **merchants**, or traders, carried goods to ports on the Mediterranean and Red seas and to southern Africa. These goods may have eventually reached India and China.

Kush's **exports**—items sent to other regions for trade—included gold, pottery, iron tools, slaves, and ivory. Merchants from Kush also exported leopard skins, ostrich feathers, and elephants. In return, Kushites received **imports**—goods brought in from other regions—such as jewelry and other luxury items from Egypt, Asia, and lands around the Mediterranean Sea.

READING CHECK Drawing Inferences What helped Kush's iron industry grow?

Society and Culture

As Kushite trade grew, merchants came into contact with people from many other cultures. As a result, the people of Kush combined customs from other cultures with their own unique culture.

Kushite Culture

The most obvious influence on the culture of Kush was Egypt. Many buildings in Meroë, especially temples, resembled those in Egypt. Many people in Kush worshipped Egyptian gods and wore Egyptian clothing. Like Egyptian rulers, Kush's rulers used the title *pharaoh* and were buried in pyramids.

Many elements of Kushite culture were unique and not borrowed from anywhere else. For example, Kushite daily life and houses were different from those in other places. One Greek geographer noted some of these differences.

"The houses in the cities are formed by interweaving split pieces of palm wood or of bricks . . . They hunt elephants, lions, and panthers. There are also serpents, which encounter elephants, and there are many other kinds of wild animals."

–Strabo, from *Geography*

In addition to Egyptian gods, Kushites worshipped their own gods. For example, their most important god was the lion-headed god Apedemek. The people of Kush also developed their own written language, known today as Meroitic. Unfortunately, historians have not yet been able to interpret the Meroitic language.

Women in Kushite Society

Unlike women in other early societies, Kushite women were expected to be active in their society. Like Kushite men, women worked long hours in the fields. They also raised children, cooked, and performed other household tasks. During times of war, many women fought alongside men.

Some Kushite women rose to positions of **authority**, especially religious authority. For example, King Piankhi made his sister a powerful priestess. Later rulers followed his example and made other princesses priestesses as well. Other women from royal families led the ceremonies in which new kings were crowned.

Some Kushite women had even more power. These women served as co-rulers with their husbands or sons. A few Kushite women, such as Queen Shanakhdakheto (shah-nahk-dah-KEE-toh), even ruled the empire alone. Several other queens ruled Kush later, helping increase the strength and wealth of the kingdom. Throughout most of its history, however, Kush was ruled by kings.

READING CHECK Analyzing In what ways were the society and culture of Kush unique?

ACADEMIC VOCABULARY

authority power or influence

THE IMPACT TODAY

More than 50 ancient Kushite pyramids still stand near the ruins of Meroë in present-day Sudan.

BIOGRAPHY

Queen Shanakhdakheto
(Ruled 170–150 BC)

Historians believe Queen Shanakhdakheto was the first woman to rule Kush. But because we can't understand Meroitic writing, we know very little about Queen Shanakhdakheto. Most of what we know about her comes from carvings found in her tomb, one of the largest pyramids at Meroë. Based on these carvings, many historians think she probably gained power after her father or husband died.

Drawing Inferences What information do you think the carvings in the queen's tomb contained?

Rulers of Kush

Like the Egyptians, the people of Kush considered their rulers to be gods. Kush's culture was similar to Egypt's, but there were also important differences.

Like the Egyptians, Kush's rulers built pyramids. Kushite pyramids, however, were much smaller and the style was different.

Kush was at times ruled by powerful queens. Queens seem to have been more important in Kush than in Egypt.

Stone carvings were made to commemorate important buildings and events, just like in Egypt. Kush's writing system was similar to Egyptian hieroglyphics, but scholars have been unable to understand most of it.

ANALYSIS SKILL **ANALYZING VISUALS**

What can you see in the illustration that is similar to Egyptian culture?

Decline and Defeat

The Kushite kingdom centered at Meroë reached its height in the first century BC. Four centuries later, Kush had collapsed. Developments both inside and outside the empire led to its downfall.

Loss of Resources

A series of problems within Kush weakened its economic power. One possible problem was that farmers allowed their cattle to overgraze the land. When the cows ate all the grass, there was nothing to hold the soil down. As a result, wind blew the soil away. Without this soil, farmers could not produce enough food for Kush's people.

In addition, ironmakers probably used up the forests near Meroë. As wood became scarce, furnaces shut down. Kush could no longer produce enough weapons or trade goods. As a result, Kush's military and economic power declined.

Trade Rivals

Kush was also weakened by a loss of trade. Foreign merchants set up new trade routes that went around Kush. For example, a new trade route bypassed Kush in favor of a nearby kingdom, Aksum (AHK-soom).

Rise of Aksum

Aksum was located southeast of Kush on the Red Sea, in present-day Ethiopia and Eritrea. In the first two centuries AD, Aksum grew wealthy from trade. But Aksum's wealth and power came at the expense of Kush. As Kush's power declined, Aksum became the most powerful state in the region.

By the AD 300s, Kush had lost much of its wealth and military might. Seeing that the Kushites were weak, the king of Aksum sent an army to conquer his former trade rival. In about AD 350, the army of Aksum's King Ezana (AY-zah-nah) destroyed Meroë and took over the kingdom of Kush.

In the late 300s, the rulers of Aksum became Christian. Their new religion reshaped culture throughout Nubia, and the last influences of Kush disappeared.

READING CHECK **Summarizing** What internal problems caused Kush's power to decline?

THE IMPACT TODAY

Much of the population of Ethiopia, which includes what used to be Aksum, is still Christian.

SUMMARY AND PREVIEW In this section you learned about the rise and fall of a powerful Kushite kingdom centered in Meroë. Next, you will learn about the rise of strong empires in West Africa.

---

go.hrw.com
Online Quiz
KEYWORD: SK7 HP12

Section 2 Assessment

Reviewing Ideas, Terms, and Places

1. **a. Recall** What were some of Kush's **exports**?
   b. Analyze Why was **Meroë** in a good location?
2. **a. Identify** Who was Queen Shanakhdakheto?
   b. Compare How were Kushite and Egyptian cultures similar?
   c. Elaborate How does our inability to understand Meroitic affect our knowledge of Kush's culture?
3. **a. Identify** What kingdom conquered Kush in about AD 350?
   b. Summarize What was the impact of new trade routes on Kush?

Critical Thinking

4. **Identifying Causes** Review your notes to identify causes of the rise and the fall of the Kushite kingdom centered at Meroë. Use a chart like this one to record the causes.

Causes of rise	Causes of fall

FOCUS ON WRITING

5. **Adding Details** Add details to your chart. What were your characters' lives like? What events caused Kush to change over time? Note events that your characters might take part in during your story.

Social Studies Skills

Identifying Bias

Learn

Occasionally when you read you may come across bias. Bias is a one-sided idea about someone or something. It is based solely on opinion, not facts. An opinion is a personal belief or judgment.

Bias can be either favorable or unfavorable. For example, when voting for a cheerleader, you might have a favorable bias toward your best friend. Identifying bias is an important part of understanding the issues, events, and people in our world. Read the tips below for information on how to identify bias.

- Look for information about the author. What is the author's background? How might that affect his or her attitude?

- Look at the language the author uses. Is the information fact or opinion? Does the author present only one side or point of view?

- Look at the author's sources. Do they favor one side over another?

Practice

As you read the text in this passage, look for possible bias. Then answer the following questions.

❶ What facts are presented in the passage? What opinions are presented?

❷ The author was a famous Greek historian. Why might he be biased regarding the Kushites?

❸ What words or phrases indicate that the author is biased? Do those words indicate a favorable or unfavorable bias?

> "The mode of life of the [Kushites] is wretched; they are for the most part naked, and wander from place to place with their flocks. Their flocks and herds are small in size . . . They [the Kushites] live on millet and barley . . . Some feed even upon grass, the tender twigs of trees, the lotus, or the roots of reeds."
>
> –Strabo, from *Geography*

Apply

Practice identifying bias. Locate a news article that you feel has favorable or unfavorable bias. Then answer the questions below.

1. What issue, event, or person is the article about?

2. What is the author's bias? What words or phrases indicate that?

3. What is the author's background? What opinions does the author give? What sources does he or she cite?

Geography's Impact
video series
Review the video to answer the closing question:
How might the manufacture of iron have led to Meroë's downfall?

Visual Summary

Use the visual summary below to help you review the main ideas of the chapter.

QUICK FACTS

Egypt dominated early Kush and forced the Kushites to pay tribute.

After Kush conquered Egypt, invaders forced the Kushites to move south to their ancient homeland.

Kush's advanced civilization blended unique Kushite traits with culture traits from Egypt and other parts of Africa.

Reviewing Terms and Places

Match the words in the columns with the correct definitions listed below.

1. authority
2. ebony
3. export
4. import
5. merchant
6. Meroë
7. Nubia
8. trade network

a. an item sent to other regions for trade
b. a region along the Nile River in present-day Sudan
c. a trader
d. dark, heavy wood
e. groups of people in different lands who trade goods back and forth
f. the capital and economic center of the kingdom of Kush
g. item brought in for purchase from other regions
h. power or influence

Comprehension and Critical Thinking

SECTION 1 *(Pages 310–314)*

9. **a. Describe** How did the physical geography of Nubia affect civilization in the region?

b. Analyze Why did the relationship between Kush and Egypt change more than once over the centuries?

c. Predict If an archaeologist found an artifact near the Fourth Cataract, why might he or she have difficulty deciding how to display it in a museum?

SECTION 2 *(Pages 315–319)*

10. **a. Identify** Who was Queen Shanakhdakheto? Why don't we know more about her?

b. Compare and Contrast What are some features that Kushite and Egyptian cultures had in common? How were they different?

c. Evaluate What do you think was the most important cause of Kush's decline? Why?

Social Studies Skills

Identifying Bias *Read the passage below, then answer the questions that follow.*

> According to inscriptions found at Napata, Piankhi led his army to Egypt to stop a rebellion. As Piankhi approached, his enemies were gripped by fear. His finest troops conquered the cities of the Nile Valley one by one. In victory, the great Pianhki was merciful.

11. What issue, event, or person is this passage about?

12. What sources does the author cite?

13. What words from the passage indicate that the author is biased? Is the author's bias favorable or unfavorable?

FOCUS ON READING AND WRITING

14. Asking Questions Read the passage below. As you read, ask yourself questions to help you learn and remember the information. On a separate sheet of paper, write down the questions you created and the answers to those questions.

> After Egypt attacked Napata in the 500s BC, Kush's rulers moved their capital to Meroë. Located about 150 miles (240 km) southeast of Napata, Meroë served as the center for trade and government in Kush.

15. Writing a Fictional Narrative Use the notes you have taken to write your short story about a character from Kush. First, identify the setting, or where the story takes place. Then tell about the problem, or conflict, your main character is facing. Relate the series of events that occur from the beginning of the conflict until it ends. Introduce the other characters as they become involved in the plot of the story.

You can help bring the story to life by adding some dialogue as your character deals with the events in the story. At the end of the story, resolve the central problem or conflict. Make sure the ending is believable and satisfies your readers' curiosity about what happened.

Using the Internet

go.hrw.com
KEYWORD: SK7 CH12

16. Activity: Researching Life in Ancient Nubia
Would you like to travel back in time to ancient Nubia and explore the wonders of that era? Enter the activity keyword. Then find out about the people, their customs, and their homes. Finally, imagine that you are a person living in ancient Kush. Take notes about the home, activities, and religion you might have experienced. Write a journal entry to show what you have learned. In your journal entry, specify which parts of your life have Egyptian influences.

Map Activity

17. Ancient Kush On a separate sheet of paper, match the letters on the map with their correct labels.

Aksum	Meroë
Blue Nile	Napata
Egypt	Red Sea
Mediterranean Sea	Sahara

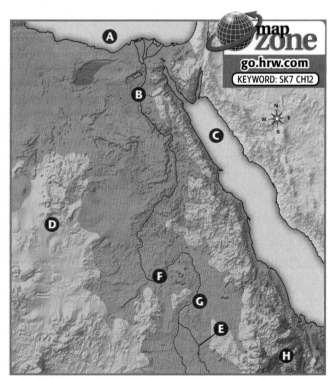

go.hrw.com
KEYWORD: SK7 CH12

DIRECTIONS: Read questions 1 through 7 and write the letter of the best response. Then read question 8 and write your own well-constructed response.

1 **Which of the following statements regarding women in ancient Kush is *true*?**

A Some Kushite women served as religious and political leaders.

B Kushite women had more rights and opportunities than Kushite men.

C Kushite women were forbidden to leave their homes.

D Many Kushite women were wealthy merchants.

2 **Around 1500 BC, Egyptian rulers conquered Kush because**

A they wanted to expand their territory.

B they wanted revenge for a Kushite attack on Thebes.

C they feared Kush's growing military power.

D they hoped to gain valuable resources.

3 **Which of the following was an advantage of the Kushite capital of Meroë?**

A It had rich deposits of iron ore and gold nearby.

B It was close to Mediterranean trade routes.

C It was the religious center of Kush.

D It was protected from invasion by the Sahara.

4 **How did cataracts on the Nile River benefit Kush?**

A They allowed Kushite farmers to plant both summer and winter crops.

B Because they were highly prized by Egyptians, Kush gained wealth from trade.

C Because they were difficult to pass through, they provided protection against invaders.

D They allowed the Kushites to build a powerful army.

*"*It is surrounded on the side of Libya by great hills of sand, and on that of Arabia by continuous precipices. In the higher parts on the south, it is bounded by . . . Astapa [the White Nile], and Astasobas [the Blue Nile]. On the north is the continuous course of the Nile to Egypt . . . *"*

–Strabo, from *Geography*

5 **What region is the author describing in the quote above?**

A Assyria

B Egypt

C Mesopotamia

D Nubia

6 **How did Egyptian culture influence Kush?**

A Egyptians taught Kushites how to raise cattle.

B Kushites adopted Christianity.

C Egyptians taught Kushites how to make iron.

D Kushites modeled their pyramids after Egyptian pyramids.

7 **Which of the following is a cause of the decline of Kush?**

A Egypt conquered Kush hoping to steal the technology for making iron.

B Piankhi's armies were crushed when they attempted to invade Aksum.

C Invaders from West Africa defeated the Kushites in about 350 BC.

D The loss of trade severely damaged Kush's economy.

8 **Extended Response** Using the information in the Close-up illustration, Rulers of Kush, in Section 2, write a short paragraph describing the culture of Kush.

History of West Africa
500 BC–AD 1650

What You Will Learn...

In this chapter you will learn about the great empires of West Africa, which grew rich from trade. You will also learn about the traditions of West Africa, which include storytelling, art, music, and dance.

SECTION 1
Empire of Ghana **326**

SECTION 2
Mali and Songhai **334**

SECTION 3
Historical and Artistic Traditions **340**

FOCUS ON READING AND WRITING

Understanding Cause and Effect When you read about history, it is important to recognize causes and effects. A cause is an action or event that makes something else happen. An effect is the result of a cause. For example, when you read this chapter you could identify the causes that led to great empires in West Africa. Then you can identify the empires' effects. **See the lesson, Understanding Cause and Effect, on page 680.**

A Journal Entry Many people feel that recording their lives in journals helps them to understand their own experiences. Writing a journal entry from someone else's point of view can help you to understand what that person's life is, or would have been, like. In this chapter, you will read about the land, people, and culture of early Africa. Then you will imagine a character and write a journal entry from his or her point of view.

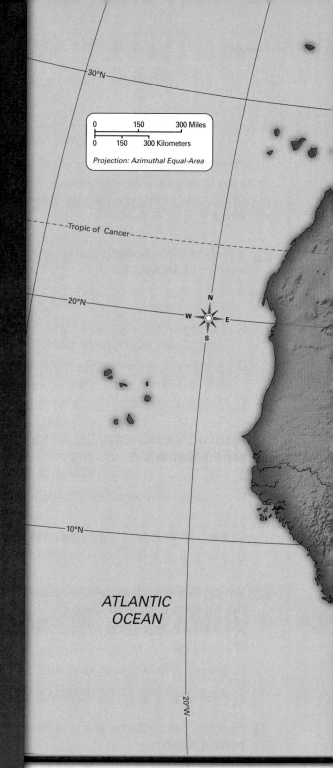

Trade West Africa's salt mines were a great source of wealth. Camels carried salt from the mines of the Sahara to the south to trade for gold.

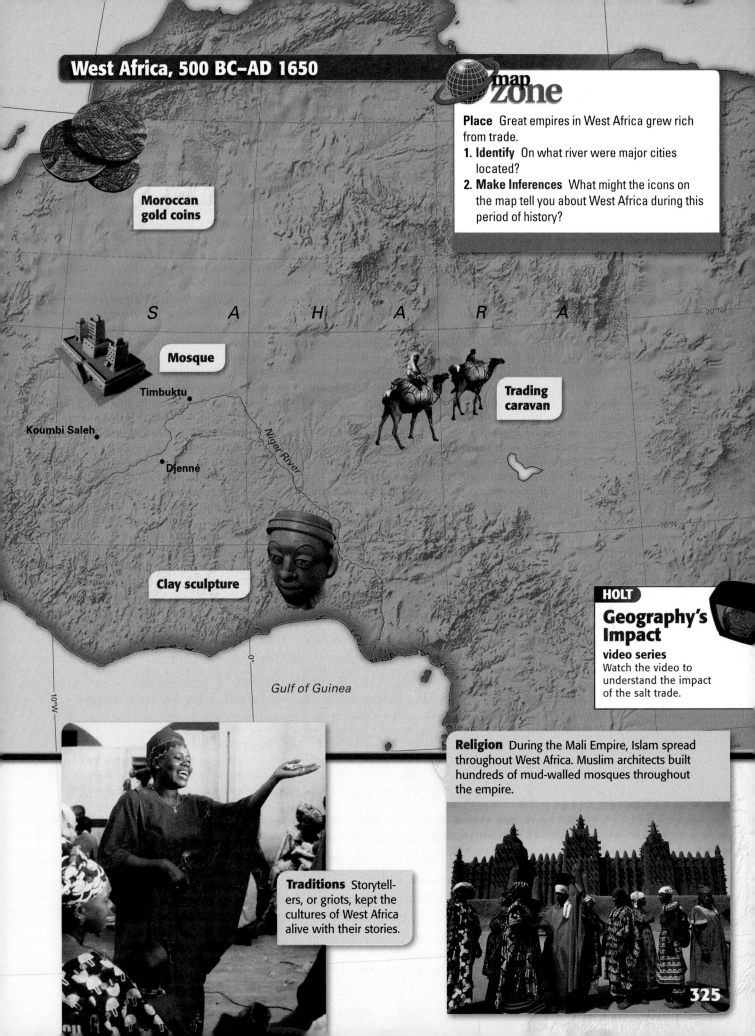

West Africa, 500 BC–AD 1650

map zone

Place Great empires in West Africa grew rich from trade.

1. Identify On what river were major cities located?

2. Make Inferences What might the icons on the map tell you about West Africa during this period of history?

Moroccan gold coins

S A H A R A

20°N

Mosque

Timbuktu

Trading caravan

Koumbi Saleh

Niger River

Djenné

10°N

Clay sculpture

0°

10°W

Gulf of Guinea

HOLT

Geography's Impact

video series
Watch the video to understand the impact of the salt trade.

Traditions Storytellers, or griots, kept the cultures of West Africa alive with their stories.

Religion During the Mali Empire, Islam spread throughout West Africa. Muslim architects built hundreds of mud-walled mosques throughout the empire.

325

Empire of Ghana

What You Will Learn...

Main Ideas

1. Ghana controlled trade and became wealthy.
2. Through its control of trade, Ghana built an empire.
3. Attacking invaders, overgrazing, and the loss of trade caused Ghana's decline.

The Big Idea

The rulers of Ghana built an empire by controlling the salt and gold trade.

Key Terms

silent barter, *p. 328*

TAKING NOTES As you read, make a list of important events from the beginning to the end of the empire of Ghana. Keep track of these events using a diagram like this one.

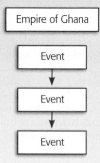

Empire of Ghana

Event

Event

Event

If **YOU** lived there...

You are a trader, traveling in a caravan from the north into West Africa in about 1000. The caravan carries many goods, but the most precious is salt. Salt is so valuable that people trade gold for it! You have never met the mysterious men who trade you the gold. You wish you could talk to them to find out where they get it.

Why do you think the traders are so secretive?

BUILDING BACKGROUND The various regions of Africa provide people with different resources. West Africa, for example, was rich in both fertile soils and minerals, especially gold and iron. Other regions had plentiful supplies of other resources, such as salt. Over time, trade developed between regions with different resources. This trade led to the growth of the first great empire in West Africa.

Ghana Controls Trade

For hundreds of years, trade routes crisscrossed West Africa. For most of that time, West Africans did not profit much from the Saharan trade because the routes were run by Berbers from northern Africa. Eventually, that situation changed. Ghana (GAH-nuh), an empire in West Africa, gained control of the valuable routes. As a result, Ghana became a powerful state.

As you can see on the map on the following page, the empire of Ghana lay between the Niger and Senegal rivers. This location was north and west of the location of the modern nation that bears the name Ghana.

Ghana's Beginnings

Archaeology provides some clues to Ghana's early history, but we do not know much about its earliest days. Historians think the first people in Ghana were farmers. Sometime after 300 these farmers, the Soninke (soh-NING-kee), were threatened by nomadic herders. The herders wanted to take the farmers' water and pastures. For protection, groups of Soninke families began to band together. This banding together was the beginning of Ghana.

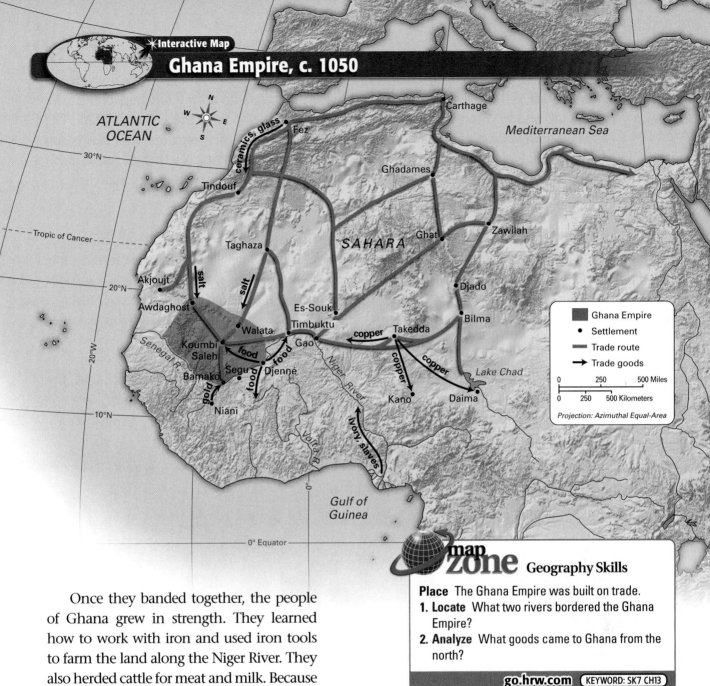

ATLANTIC OCEAN

Mediterranean Sea

Carthage

Fez

ceramics, glass

30°N

Ghadames

Tindouf

SAHARA

Ghat

Zawilah

Tropic of Cancer

Taghaza

salt

salt

Djado

Akjoujt

20°N

Bilma

Awdaghost

Es-Souk

Walata   Timbuktu   copper   Takedda

Senegal R.   Koumbi   Gao   copper

Koumbi Saleh   food   copper   Lake Chad

Bamako   Segu   Djenné   copper

food   Niger River

Niani   food

Kano   Daima

10°N

Volta R.

Ivory, slaves

Gulf of Guinea

0° Equator

Map Legend

■ Ghana Empire
• Settlement
▬ Trade route
→ Trade goods

0   250   500 Miles
0   250   500 Kilometers

Projection: Azimuthal Equal-Area

map zone Geography Skills

Place The Ghana Empire was built on trade.
1. **Locate** What two rivers bordered the Ghana Empire?
2. **Analyze** What goods came to Ghana from the north?

go.hrw.com   KEYWORD: SK7 CH13

Once they banded together, the people of Ghana grew in strength. They learned how to work with iron and used iron tools to farm the land along the Niger River. They also herded cattle for meat and milk. Because these farmers and herders could produce plenty of food, the population of Ghana increased. Towns and villages grew.

Besides farm tools, iron was also useful for making weapons. Other armies in the area had weapons made of bone, wood, and stone. These were no match for the iron spear points and blades used by Ghana's army.

Trade in Valuable Goods

Ghana lay between the vast Sahara Desert and deep forests. In this location, they were in a good position to trade in the region's most valuable resources—gold and salt. Gold came from the south, from mines near the Gulf of Guinea and along the Niger. Salt came from the Sahara in the north.

People wanted gold for its beauty. But they needed salt in their diets to survive. Salt, which could be used to preserve food, also made bland food tasty. These qualities made salt very valuable. In fact, Africans sometimes cut up slabs of salt and used the pieces as money.

ACADEMIC
VOCABULARY

process a
series of steps
by which a task
is accomplished

The exchange of gold and salt sometimes followed a **process** called silent barter. **Silent barter** is a process in which people exchange goods without ever contacting each other directly. The method made sure that the traders did business peacefully. It also kept the exact location of the gold mines secret from the salt traders.

In the silent barter process, salt traders went to a riverbank near gold fields. There they left slabs of salt in rows and beat a drum to tell the gold miners that trading had begun. Then the salt traders moved back several miles from the riverbank.

Soon afterward, the gold miners arrived by boat. They left what they considered a fair amount of gold in exchange for the salt. Then the gold miners also moved back several miles so the salt traders could return. If they were happy with the amount of gold left there, the salt traders beat the drum again, took the gold, and left. The gold miners then returned and picked up their salt. Trading continued until both sides were happy with the exchange.

Growth of Trade

As the trade in gold and salt increased, Ghana's rulers gained power. Over time, their military strength grew as well. With their armies they began to take control of this trade from the merchants who had once controlled it. Merchants from the north and south met to exchange goods in Ghana. As a result of their control of trade routes, the rulers of Ghana became wealthy.

Salt and Gold

Additional sources of wealth and trade were developed to add to Ghana's wealth. Wheat came from the north. Sheep, cattle, and honey came from the south. Local products, including leather and cloth, were also traded for wealth. Among the prized special local products were tassels made from golden thread.

As trade increased, Ghana's capital grew as well. The largest city in West Africa, Koumbi Saleh (KOOM-bee SAHL-uh) was an oasis for travelers. These travelers could find all the region's goods for sale in its markets. As a result, Koumbi Saleh gained a reputation as a great trading center.

READING CHECK **Generalizing** How did trade help Ghana develop?

Ghana's rulers became rich by controlling the trade in salt and gold. Salt came from the north in large slabs like the ones shown at left. Gold, like the woman above is wearing, came from the south.

Ghana Builds an Empire

By 800 Ghana was firmly in control of West Africa's trade routes. Nearly all trade between northern and southern Africa passed through Ghana. Traders were protected by Ghana's army, which kept trade routes free from bandits. As a result, trade became safer. Knowing they would be protected, traders were not scared to travel to Ghana. Trade increased, and Ghana's influence grew as well.

Taxes and Gold

With so many traders passing through their lands, Ghana's rulers looked for ways to make money from them. One way they raised money was by forcing traders to pay taxes. Every trader who entered Ghana had to pay a special tax on the goods he carried. Then he had to pay another tax on any goods he took with him when he left.

Traders were not the only people who had to pay taxes. The people of Ghana also had to pay taxes. In addition, Ghana conquered many small neighboring tribes, then forced them to pay tribute. Rulers used the money from taxes and tribute to support Ghana's growing army.

Not all of Ghana's wealth came from taxes and tribute. Ghana's rich mines produced huge amounts of gold. Some of this gold was carried by traders to lands as far away as England, but not all of Ghana's gold was traded. Ghana's kings kept huge stores of gold for themselves. In fact, all the gold produced in Ghana was officially the property of the king.

Knowing that rare materials are worth far more than common ones, the rulers banned anyone else in Ghana from owning gold nuggets. Common people could own only gold dust, which they used as money. This ensured that the king was richer than his subjects.

Expansion of the Empire

Ghana's kings used their great wealth to build a powerful army. With this army the kings of Ghana conquered many of their neighbors. Many of these conquered areas were centers of trade. Taking over these areas made Ghana's kings even richer.

Ghana's kings didn't think that they could rule all the territory they conquered by themselves. Their empire was quite large, and travel and communication in West Africa could be difficult. To keep order in their empire, they allowed conquered kings to retain much of their power. These kings acted as governors of their territories, answering only to the king.

The empire of Ghana reached its peak under Tunka Manin (TOOHN-kah MAH-nin). This king had a splendid court where he displayed the vast wealth of the empire. A Spanish writer noted the court's splendor.

FOCUS ON READING

How is this quotation an example of the effects of the king's wealth?

"The king adorns himself . . . round his neck and his forearms, and he puts on a high cap decorated with gold and wrapped in a turban of fine cotton. Behind the king stand ten pages holding shields and swords decorated with gold."
–al-Bakri, from *The Book of Routes and Kingdoms*

READING CHECK **Summarizing** How did the rulers of Ghana control trade?

BIOGRAPHY

Tunka Manin
(Ruled around 1068)

All we know about Tunka Manin comes from the writings of a Muslim geographer who wrote about Ghana. From his writings, we know that Tunka Manin was the nephew of the previous king, a man named Basi. Kingship and property in Ghana did not pass from father to son, but from uncle to nephew. Only the king's sister's son could inherit the throne. Once he did become king, Tunka Manin surrounded himself with finery and many luxuries.

Contrasting How was inheritance in Ghana different from inheritance in other societies you have studied?

Ghana's Decline

In the mid-1000s Ghana was rich and powerful, but by the end of the 1200s, the empire had collapsed. Three major factors contributed to its end.

Invasion

The first factor that helped bring about Ghana's end was invasion. A Muslim group called the Almoravids (al-moh-RAH-vidz) attacked Ghana in the 1060s in an effort to force its leaders to convert to Islam.

The people of Ghana fought hard against the Almoravid army. For 14 years they kept the invaders at bay. In the end, however, the Almoravids won. They destroyed the city of Koumbi Saleh.

The Almoravids didn't control Ghana for long, but they certainly weakened the empire. They cut off many trade routes through Ghana and formed new trading partnerships with Muslim leaders instead. Without this trade Ghana could no longer support its empire.

Overgrazing

A second factor in Ghana's decline was a result of the Almoravid conquest. When the Almoravids moved into Ghana, they brought herds of animals with them. These animals ate all the grass in many pastures, leaving the soil exposed to hot desert winds. These winds blew away the soil, leaving the land worthless for farming or herding. Unable to grow crops, many farmers had to leave in search of new homes.

Internal Rebellion

A third factor also helped bring about the decline of Ghana's empire. In about 1200 the people of a country that Ghana had conquered rose up in rebellion. Within a few years the rebels had taken over the entire empire of Ghana.

Overgrazing

Too many animals grazing in one area can lead to problems, such as the loss of farmland that occurred in West Africa.

❶ Animals are allowed to graze in areas with lots of grass.

❷ With too many animals grazing, however, the grass disappears, leaving the soil below exposed to the wind.

❸ The wind blows the soil away, turning what was once grassland into desert.

Once in control, however, the rebels found that they could not keep order in Ghana. Weakened, Ghana was attacked and defeated by one of its neighbors. The empire fell apart.

READING CHECK **Identifying Cause and Effect** Why did Ghana decline in the 1000s?

SUMMARY AND PREVIEW The empire of Ghana in West Africa grew rich and powerful through its control of trade routes. The empire lasted for centuries, but eventually Ghana fell. In the next section you will learn that it was replaced by a new empire, Mali.

Section 1 Assessment

Reviewing Ideas, Terms, and Places

1. **a. Identify** What were the two most valuable resources traded in Ghana?
   b. Explain How did the **silent barter** system work?
2. **a. Identify** Who was Tunka Manin?
   b. Generalize What did Ghana's kings do with the money they raised from taxes?
   c. Elaborate Why did the rulers of Ghana not want everyone to have gold?
3. **a. Identify** What group invaded Ghana in the late 1000s?
   b. Summarize How did overgrazing help cause the fall of Ghana?

Critical Thinking

4. **Identifying Causes** Draw a diagram like the one shown here. Use it to identify factors that caused Ghana's trade growth and those that caused its decline.

Growth → Ghana's Trade → Decline

FOCUS ON WRITING

5. **Gathering Information** Think about what it would have been like to live in Ghana. Whose journal would you create? Would you choose the powerful Tunka Manin? a trader? Jot down some ideas.

Crossing the Sahara

Crossing the Sahara has never been easy. Bigger than the entire continent of Australia, the Sahara is one of the hottest, driest, and most barren places on earth. Yet for centuries, people have crossed the Sahara's gravel-covered plains and vast seas of sand. Long ago, West Africans crossed the desert regularly to carry on a rich trade.

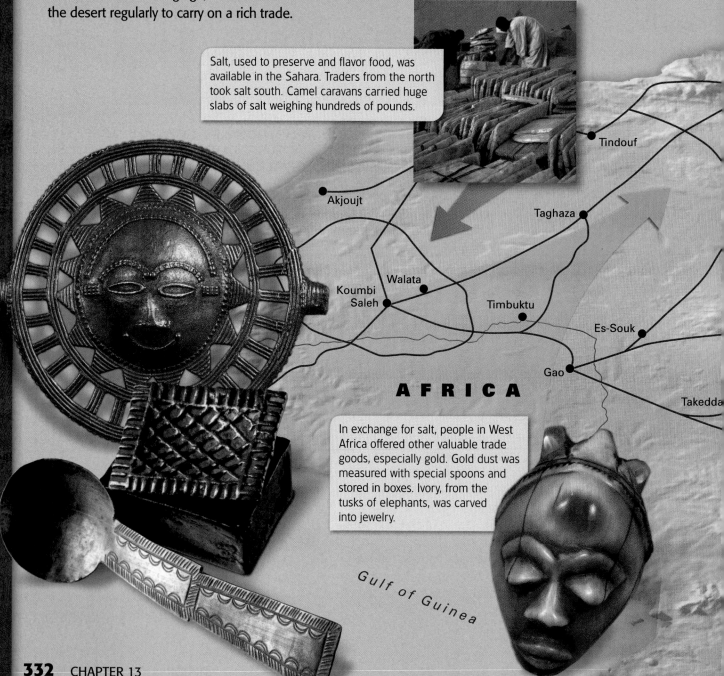

Salt, used to preserve and flavor food, was available in the Sahara. Traders from the north took salt south. Camel caravans carried huge slabs of salt weighing hundreds of pounds.

Tindouf

Akjoujt

Taghaza

Walata

Koumbi Saleh

Timbuktu

Es-Souk

Gao

Takedda

AFRICA

In exchange for salt, people in West Africa offered other valuable trade goods, especially gold. Gold dust was measured with special spoons and stored in boxes. Ivory, from the tusks of elephants, was carved into jewelry.

Gulf of Guinea

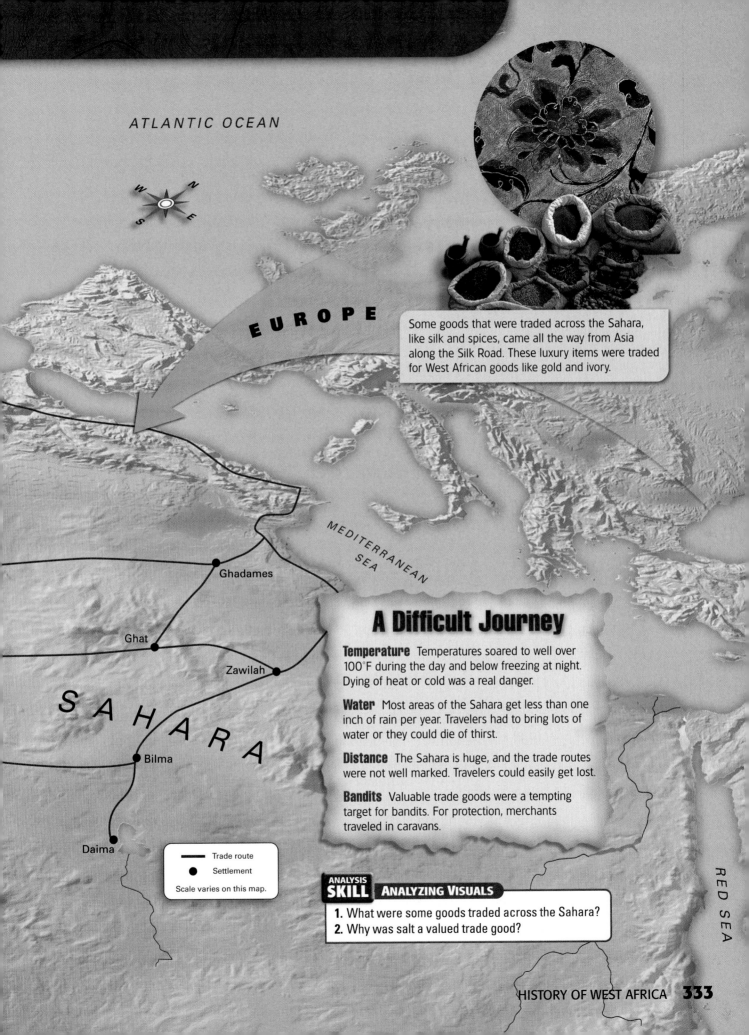

ATLANTIC OCEAN

EUROPE

Some goods that were traded across the Sahara, like silk and spices, came all the way from Asia along the Silk Road. These luxury items were traded for West African goods like gold and ivory.

MEDITERRANEAN SEA

Ghadames

Ghat

Zawilah

S A H A R A

Bilma

Daima

— Trade route
● Settlement
Scale varies on this map.

A Difficult Journey

Temperature Temperatures soared to well over 100°F during the day and below freezing at night. Dying of heat or cold was a real danger.

Water Most areas of the Sahara get less than one inch of rain per year. Travelers had to bring lots of water or they could die of thirst.

Distance The Sahara is huge, and the trade routes were not well marked. Travelers could easily get lost.

Bandits Valuable trade goods were a tempting target for bandits. For protection, merchants traveled in caravans.

RED SEA

ANALYSIS
SKILL ANALYZING VISUALS

1. What were some goods traded across the Sahara?
2. Why was salt a valued trade good?

Mali and Songhai

What You Will Learn...

Main Ideas

1. The empire of Mali reached its height under the ruler Mansa Musa, but the empire fell to invaders in the 1400s.
2. The Songhai built a new Islamic empire in West Africa, conquering many of the lands that were once part of Mali.

The Big Idea

Between 1000 and 1500 the empires of Mali and Songhai developed in West Africa.

Key Terms and Places

Niger River, *p. 334*
Timbuktu, *p. 335*
mosque, *p. 337*
Gao, *p. 337*
Djenné, *p. 338*

TAKING NOTES As you read, take notes about life in the cultures that developed in West Africa—Mali and Songhai.

West Africa

Mali   Songhai

If YOU lived there...

You are a servant of the great Mansa Musa, ruler of Mali. You've been chosen as one of the servants who will travel with him on a pilgrimage to Mecca. The king has given you all fine new clothes of silk for the trip. He will carry much gold with him. You've never left your home before. But now you will see the great city of Cairo, Egypt, and many other new places.

How do you feel about going on this journey?

BUILDING BACKGROUND Mansa Musa was one of Africa's greatest rulers, and his empire, Mali, was one of the largest in African history. Rising from the ruins of Ghana, Mali took over the trade routes of West Africa and grew into a powerful state.

Mali

Like Ghana, Mali (MAH-lee) lay along the upper **Niger River**. This area's fertile soil helped Mali grow. Mali's location on the Niger also allowed its people to control trade on the river. As a result, the empire grew rich and powerful. According to legend, Mali's rise to power began under a ruler named Sundiata (soohn-JAHT-ah).

Sundiata Makes Mali an Empire

When Sundiata was a boy, a harsh ruler conquered Mali. But as an adult, Sundiata built up an army and won back his country's independence. He then conquered nearby kingdoms, including Ghana, in the 1230s.

After Sundiata conquered Ghana, he took over the salt and gold trades. He also worked to improve agriculture in Mali. Sundiata had new farmlands cleared for beans, onions, rice, and other crops. Sundiata even introduced a new crop—cotton. People made clothing from the cotton fibers that was comfortable in the warm climate. They also sold cotton to other people.

To keep order in his prosperous kingdom, Sundiata took power away from local leaders. Each of these local leaders had the title mansa (MAHN-sah), a title Sundiata now took

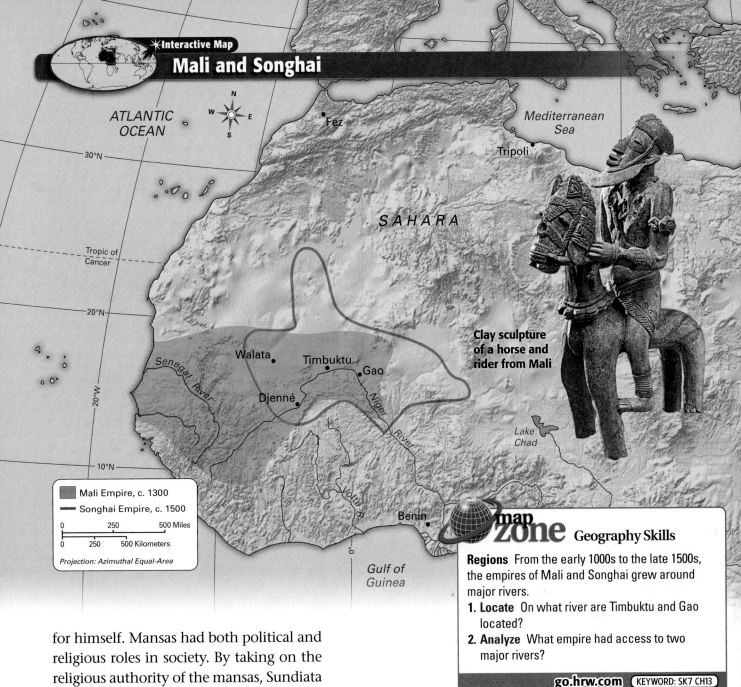

ATLANTIC
OCEAN

30°N

Fez

Mediterranean
Sea

Tripoli

SAHARA

Tropic of
Cancer

20°N

Walata

Timbuktu

Gao

**Clay sculpture
of a horse and
rider from Mali**

Senegal River

Djenné

Niger River

Lake
Chad

10°N

Volta R.

Benin

Gulf of
Guinea

Mali Empire, c. 1300

Songhai Empire, c. 1500

0   250   500 Miles

0   250   500 Kilometers

Projection: Azimuthal Equal-Area

map Zone **Geography Skills**

Regions From the early 1000s to the late 1500s, the empires of Mali and Songhai grew around major rivers.
1. **Locate** On what river are Timbuktu and Gao located?
2. **Analyze** What empire had access to two major rivers?

go.hrw.com **KEYWORD: SK7 CH13**

for himself. Mansas had both political and religious roles in society. By taking on the religious authority of the mansas, Sundiata gained even more power in Mali.

Sundiata died in 1255. Later rulers of Mali took the title of mansa. Unlike Sundiata, most of these rulers were Muslims.

Mansa Musa

Mali's most famous ruler was a Muslim named Mansa Musa (MAHN-sah moo-SAH). Under his skillful leadership, Mali reached the height of its wealth, power, and fame in the 1300s. Because of Mansa Musa's influence, Islam spread through a large part of West Africa, gaining many new believers.

Mansa Musa ruled Mali for about 25 years, from 1312 to 1337. During that time, Mali added many important trade cities to its empire, including **Timbuktu** (tim-buhk-TOO).

Religion was very important to Mansa Musa. In 1324 he left Mali on a pilgrimage to Mecca. Through his journey, Mansa Musa introduced his empire to the Islamic world. He spread Mali's fame far and wide.

Mansa Musa also supported education. He sent many scholars to study in Morocco.

Timbuktu

Timbuktu became a major trading city at the height of Mali's power under Mansa Musa. Traders came to Timbuktu from the north and south to trade for salt, gold, metals, shells, and many other goods.

Mansa Musa and later rulers built several large mosques in the city, which became a center of Islamic learning.

Winter floods allowed boats to reach Timbuktu from the Niger River.

Timbuktu's walls and build-ings were mostly built with bricks made of dried mud. Heavy rains can soften the bricks and destroy buildings.

At crowded market stalls, people traded for goods like sugar, kola nuts, and glass beads.

Camel caravans from the north brought goods like salt, cloth, books, and slaves to trade at Timbuktu.

ANALYSIS SKILL ANALYZING VISUALS

How did traders from the north bring their goods to Timbuktu?

These scholars later set up schools in Mali. Mansa Musa stressed the importance of learning to read the Arabic language so that Muslims in his empire could read the Qur'an. To spread Islam in West Africa, Mansa Musa hired Muslim architects to build mosques. A **mosque** (mahsk) is a building for Muslim prayer.

The Fall of Mali

When Mansa Musa died, his son Maghan (MAH-gan) took the throne. Maghan was a weak ruler. When raiders from the southeast poured into Mali, he couldn't stop them. The raiders set fire to Timbuktu's great schools and mosques. Mali never fully recovered from this terrible blow. The empire continued to weaken and decline.

In 1431 the Tuareg (TWAH-reg), nomads from the Sahara, seized Timbuktu. By 1500 nearly all of the lands the empire had once ruled were lost. Only a small area of Mali remained.

READING CHECK **Sequencing** What steps did Sundiata take to turn Mali into an empire?

Songhai

Even as the empire of Mali was reaching its height, a rival power was growing in the area. That rival was the Songhai (SAHNG-hy) kingdom. From their capital at **Gao**, the Songhai participated in the same trade that had made Ghana and Mali so rich.

The Building of an Empire

In the 1300s Mansa Musa conquered the Songhai, adding their lands to his empire. But as the Mali Empire weakened in the 1400s, the people of Songhai rebelled and regained their freedom.

The Songhai leaders were Muslims. So too were many of the North African Berbers who traded in West Africa. Because of this shared religion, the Berbers were willing to trade with the Songhai, who grew richer.

As the Songhai gained in wealth, they expanded their territory and built an empire. Songhai's expansion was led by Sunni Ali (SOOH-nee ah-LEE), who became ruler of the Songhai in 1464. Before he took over, the Songhai state had been disorganized and poorly run. As ruler, Sunni Ali worked to unify, strengthen, and enlarge his empire. Much of the land that he added to Songhai had been part of Mali.

As king, Sunni Ali encouraged everyone in his empire to work together. To build religious harmony, he participated in both Muslim and local religions. As a result, he brought stability to Songhai.

Askia the Great

Sunni Ali died in 1492. He was followed as king by his son Sunni Baru, who was not a Muslim. The Songhai people feared that if Sunni Baru didn't support Islam, they

THE IMPACT
TODAY

Some of the mosques built by Mansa Musa can still be seen in West Africa today.

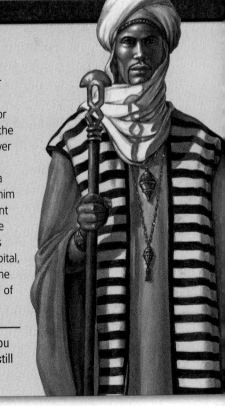

BIOGRAPHY

Askia the Great
(c. 1443–1538)

Askia the Great became the ruler of Songhai when he was nearly 50 years old. He ruled Songhai for about 35 years. During his reign the cities of Songhai gained power over the countryside.

When he was in his 80s, Askia went blind. His son Musa forced him to leave the throne. Askia was sent to live on an island. He lived there for nine years until another of his sons brought him back to the capital, where he died. His tomb is still one of the most honored places in all of West Africa.

Drawing Inferences Why do you think Askia the Great's tomb is still considered an honored place?

FOCUS ON
READING
As you read
Songhai Falls to
Morocco, identify
two causes of
Songhai's fall.

would lose their trade with Muslim lands. They rebelled against the king.

The leader of that rebellion was a general named Muhammad Ture (moo-HAH-muhd too-RAY). After overthrowing Sunni Baru, Muhammad Ture chose the title *askia*, a title of high military rank. Eventually, he became known as Askia the Great.

Askia supported education and learning. Under his rule, Timbuktu flourished, drawing thousands to its universities, schools, libraries, and mosques. The city was especially known for the University of Sankore (san-KOH-rah). People arrived there from North Africa and other places to study math, science, medicine, grammar, and law. **Djenné** was another city that became a center of learning.

Most of Songhai's traders were Muslim, and as they gained influence in the empire so did Islam. Askia, himself a devout Muslim, encouraged the growth of Islamic influence. He made many laws similar to those in other Muslim nations.

To help maintain order, Askia set up five provinces within Songhai. He appointed governors who were loyal to him. Askia also created a professional army and specialized departments to oversee tasks.

Songhai Falls to Morocco

A northern rival of Songhai, Morocco, wanted to gain control of Songhai's salt mines. So the Moroccan army set out for the heart of Songhai in 1591. Moroccan soldiers carried advanced weapons, including the terrible arquebus (AHR-kwih-buhs). The arquebus was an early form of a gun.

The swords, spears, and bows used by Songhai's warriors were no match for the Moroccans' guns and cannons. The invaders destroyed Timbuktu and Gao.

Changes in trade patterns completed Songhai's fall. Overland trade declined as port cities on the Atlantic coast became more important. Africans south of Songhai and European merchants both preferred trading at Atlantic ports to dealing with Muslim traders. Slowly, the period of great West African empires came to an end.

READING CHECK **Evaluating** What do you think was Askia's greatest accomplishment?

SUMMARY AND PREVIEW Mali was a large empire famous for its wealth and centers of learning. Songhai similarly thrived. Next, you will learn about historical and artistic traditions of West Africa.

Section 2 Assessment

go.hrw.com
Online Quiz
KEYWORD: SK7 HP13

Reviewing Ideas, Terms, and Places

1. **a. Identify** Who was Sundiata?
   b. Explain What major river was important to the people of Mali? Why?
   c. Elaborate What effects did the rule of Mansa Musa have on Mali and West Africa?
2. **a. Identify** Who led the expansion of Songhai in the 1400s?
   b. Explain How did Askia the Great's support of education affect **Timbuktu**?
   c. Elaborate What were two reasons why Songhai fell to the Moroccans?

Critical Thinking

3. **Finding Main Ideas** Use your notes to help you list three major accomplishments of Sundiata and Askia.

Sundiata	Askia

FOCUS ON WRITING

4. **Comparing and Contrasting** Whose journal could you write from the empires of Mali and Songhai? Would you create a journal for an important person, like Mansa Musa or Askia the Great? Or would you create a journal for someone who has a different role in one of the empires? List your ideas.

Mansa Musa

How could one man's travels become a major historic event?

When did he live? the late 1200s and early 1300s

Where did he live? Mali

What did he do? Mansa Musa, the ruler of Mali, was one of the Muslim kings of West Africa. He became a major figure in African and world history largely because of a pilgrimage he made to the city of Mecca.

Why is he important? Mansa Musa's spectacular journey attracted the attention of the Muslim world and of Europe. For the first time, other people's eyes turned to West Africa. During his travels, Mansa Musa gave out huge amounts of gold. His spending made people eager to find the source of such wealth. Within 200 years, European explorers would arrive on the shores of western Africa.

Identifying Points of View How do you think Mansa Musa changed people's views of West Africa?

KEY FACTS

According to chroniclers of the time, Mansa Musa was accompanied on his journey to Mecca by some 60,000 people. Of those people

- **12,000** were servants to attend to the king.

- **500** were servants to attend to his wife.

- **14,000** more were slaves wearing rich fabrics such as silk.

- **500** carried staffs heavily decorated with gold. Historians have estimated that the gold Mansa Musa gave away on his trip would be worth more than $100 million today.

This Spanish map from the 1300s shows Mansa Musa sitting on his throne.

Historical and Artistic Traditions

What You Will Learn...

Main Ideas

1. West Africans have preserved their history through storytelling and the written accounts of visitors.
2. Through art, music, and dance, West Africans have expressed their creativity and kept alive their cultural traditions.

The Big Idea

West African culture has been passed down through oral history, writings by other people, and the arts.

Key Terms

oral history, *p. 340*
griots, *p. 340*
proverbs, *p. 341*
kente, *p. 343*

TAKING NOTES As you read, take notes on West African historical and artistic traditions. Write your notes in a diagram like the one below.

West African Traditions

Historical   Artistic

If **YOU** lived there...

You are the youngest and smallest in your family. People often tease you about not being very strong. In the evenings, when work is done, your village gathers to listen to storytellers. One of your favorite stories is about the hero Sundiata. As a boy he was small and weak, but he grew to be a great warrior and hero.

How does the story of Sundiata make you feel?

BUILDING BACKGROUND Although trading empires rose and fell in West Africa, many traditions continued through the centuries. In every town and village, storytellers passed on the people's histories, legends, and wise sayings. These were at the heart of West Africa's arts and cultural traditions.

Preserving History

Writing was never very common in West Africa. In fact, none of the major early civilizations of West Africa developed a written language. Arabic was the only written language they used. The lack of a native written language does not mean that the people of West Africa didn't know their history, though. They passed along information through oral histories. An **oral history** is a spoken record of past events. The task of remembering and telling West Africa's history was entrusted to storytellers.

The Griots

The storytellers of early West Africa were called **griots** (GREE-ohz). They were highly respected in their communities because the people of West Africa were very interested in the deeds of their ancestors. Griots helped keep this history alive for each new generation.

The griots' stories were both entertaining and informative. They told of important past events and of the accomplishments of distant ancestors. For example, some stories explained the rise and fall of the West African empires. Other stories described the actions of powerful kings and warriors. Some griots made their stories more lively by acting out the events like scenes in a play.

In addition to stories, the griots recited **proverbs**, or short sayings of wisdom or truth. They used proverbs to teach lessons to the people. For example, one West African proverb warns, "Talking doesn't fill the basket in the farm." This proverb reminds people that they must work to accomplish things. It is not enough for people just to talk about what they want to do.

In order to tell their stories and proverbs, the griots memorized hundreds of names and events. Through this process the griots passed on West African history from generation to generation. However, some griots confused names and events in their heads. When this happened, the facts of some historical events became distorted. Still, the griots' stories tell us a great deal about life in the West African empires.

West African Epics

Some of the griot poems are epics—long poems about kingdoms and heroes. Many of these epic poems are collected in the *Dausi* (DAW-zee) and the *Sundiata*.

The *Dausi* tells the history of Ghana. Intertwined with historical events, though, are myths and legends. One story is about a seven-headed snake god named Bida. This god promised that Ghana would prosper if the people sacrificed a young woman to him every year. One year a mighty warrior killed Bida. As the god died, he cursed Ghana. The griots say that this curse caused the empire of Ghana to fall.

The *Sundiata* is about Mali's great ruler. According to the epic, when Sundiata was still a boy, a conqueror captured Mali and killed Sundiata's father and 11 brothers.

Oral Traditions

West African storytellers called griots had the job of remembering and passing on their people's history. Here, people gather to perform traditional dances and to listen to the stories of a griot.

Music from Mali to Memphis

Did you know that the music you listen to today may have begun with the griots? From the 1600s to the 1800s, many people from West Africa were brought to America as slaves. In America, these slaves continued to sing the way they had in Africa. They also continued to play traditional instruments such as the *kora* played by Senegalese musician Soriba Kouyaté (right), the son of a griot. Over time, this music developed into a style called the blues, made popular by such artists as B. B. King (left). In turn, the blues shaped other styles of music, including jazz and rock. So, the next time you hear a Memphis blues track or a cool jazz tune, listen for its ancient African roots!

He didn't kill Sundiata, however, because the boy was sick and didn't seem like a threat. But Sundiata grew up to be an expert warrior. Eventually he overthrew the conqueror and became king.

Visitors' Written Accounts

FOCUS ON
READING

What is one effect of visitors' written accounts of West Africa?

In addition to the oral histories told about West Africa, visitors wrote about the region. In fact, much of what we know about early West Africa comes from the writings of travelers and scholars from Muslim lands such as Spain and Arabia.

Ibn Battutah was the most famous Muslim visitor to write about West Africa. From 1353 to 1354 he traveled through the region. Ibn Battutah's account of this journey describes the political and cultural lives of West Africans in great detail.

READING CHECK Drawing Conclusions Why were oral traditions important in West Africa?

Art, Music, and Dance

Like most peoples, West Africans valued the arts. They expressed themselves creatively through sculpture, mask-making, cloth-making, music, and dance.

Sculpture

Of all the visual art forms, the sculpture of West Africa is probably the best known. West Africans made ornate statues and carvings out of wood, brass, clay, ivory, stone, and other materials.

Most statues from West Africa are of people—often the sculptor's ancestors. Usually these statues were made for religious rituals, to ask for the ancestors' blessings. Sculptors made other statues as gifts for the gods. These sculptures were kept in holy places. They were never meant to be seen by people.

Because their statues were used in religious rituals, many African artists were deeply respected. People thought artists had been blessed by the gods.

Long after the decline of Ghana, Mali, and Songhai, West African art is still admired. Museums around the world display African art. In addition, African sculpture inspired some European artists of the 1900s, including Henri Matisse and Pablo Picasso.

Masks and Clothing

In addition to statues, the artists of West Africa carved elaborate masks. Made of wood, these masks bore the faces of animals such as hyenas, lions, monkeys, and antelopes. Artists often painted the masks after carving them. People wore the masks during rituals as they danced around fires. The way firelight reflected off the masks made them look fierce and lifelike.

Many African societies were famous for the cloth they wove. The most famous of these cloths is called kente (ken-TAY). **Kente is a hand-woven, brightly colored fabric.** The cloth was woven in narrow strips that were then sewn together. Kings and queens in West Africa wore garments made of kente for special occasions.

Music and Dance

In many West African societies, music and dance were as important as the visual arts. Singing, drumming, and dancing were great entertainment, but they also helped people honor their history and mark special occasions. For example, music was played when a ruler entered a room.

Dance has long been a central part of African society. Many West African cultures used dance to celebrate specific events or ceremonies. For example, they may have performed one dance for weddings and another for funerals. In some parts of West Africa, people still perform dances similar to those performed hundreds of years ago.

READING CHECK **Summarizing** Summarize how traditions were preserved in West Africa.

SUMMARY AND PREVIEW The societies of West Africa did not have written languages, but they preserved their histories and cultures through storytelling and the arts. Next you will read about the modern region of North Africa.

---

go.hrw.com
Online Quiz
KEYWORD: SK7 HP13

Section 3 Assessment

Reviewing Ideas, Terms, and Places

1. **a. Define** What is **oral history**?
   b. Make Generalizations Why were **griots** and their stories important in West African society?
   c. Evaluate Why may an oral history provide different information than a written account of the same event?

2. **a. Identify** What were two forms of visual art popular in West Africa?
   b. Make Inferences Why do you think that the sculptures made as gifts for the gods were not meant to be seen by people?
   c. Elaborate What role did music and dance play in West African society?

Critical Thinking

3. **Summarizing** Use a chart like this one and your notes to summarize the importance of each tradition in West Africa.

Tradition	Importance
Storytelling	
Epics	
Sculpture	

FOCUS ON WRITING

4. **Identifying West African Traditions** Think about the arts and how they affected people who lived in the West African empires. Would you create a journal of one of these artists? Or would you create a journal of someone who is affected by the arts or artists?

Making Decisions

Learn

You make decisions every day. Some decisions are very easy to make and take little time. Others are much harder. Regardless of how easy or hard a decision is, it will have consequences, or results. These consequences can be either positive or negative.

Before you make a decision, consider all your possible options. Think about the possible consequences of each option and decide which will be best for you. Thinking about the consequences of your decision beforehand will allow you to make a better, more thoughtful decision.

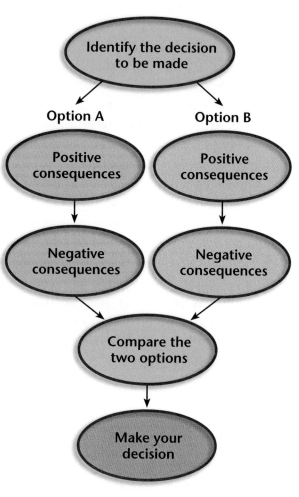

Practice

Imagine your parents have given you the option of getting a new pet. Use a graphic organizer like the one on this page to help you decide whether to get one.

1 What are the consequences of getting a pet? Which of these consequences are positive? Which are negative?

2 What are the consequences of not getting a pet? Which of them are positive? Which are negative?

3 Compare your two options. Look at the positive and negative consequences of each option. Based on these consequences, do you think you should get a pet?

Apply

Imagine that your school has just received money to build either a new art studio or a new track. School officials have asked students to vote on which of these new facilities they would prefer, and you have to decide which option you think would be better for the school. Use a graphic organizer like the one above to consider the consequences of each option. Compare your lists, and then make your decision. Write a short paragraph to explain your decision.

Chapter Review

Geography's Impact
video series
Review the video to answer the closing question:
Why was the salt trade important to African civilizations before the 1600s?

Visual Summary

Use the visual summary below to help you review the main ideas of the chapter.

QUICK FACTS

The Ghana Empire developed in West Africa and controlled the trade of salt and gold.

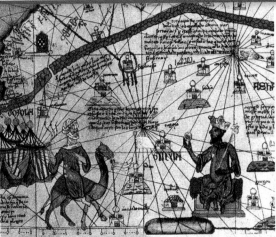

The empires of Mali and Songhai both grew powerful through trade in salt, gold, and other goods.

The history of West Africa has been preserved through story-telling, visitors' accounts, art, music, and dance.

Reviewing Vocabulary, Terms, and Places

Choose the letter of the answer that best completes each statement below.

1. The belief that natural objects have spirits is called
  a. animism.　　**c.** animalism.
  b. vegetism.　　**d.** naturalism.

2. Mali's rise to power began under a ruler named
  a. Tunka Manin.　　**c.** Ibn Battutah.
  b. Sunni Ali.　　**d.** Sundiata.

3. A spoken record of the past is
  a. a Soninke.　　**c.** a Gao.
  b. an oral history.　　**d.** an age-set proverb.

4. A West African storyteller is
  a. an Almoravid.　　**c.** an arquebus.
  b. a griot.　　**d.** a rift.

5. The Muslim leader of Mali who spread Islam and made a famous pilgrimage to Mecca was
  a. Sunni Baru.　　**c.** Mansa Musa.
  b. Askia the Great.　　**d.** Muhammad Ture.

6. A brightly colored African fabric is a
  a. kente.　　**c.** Timbuktu.
  b. mansa.　　**d.** Tuareg.

Comprehension and Critical Thinking

SECTION 1 *(pages 326–331)*

7. a. Identify What were the two major trade goods that made Ghana rich?

  b. Make Inferences Why did merchants in Ghana not want other traders to know where their gold came from?

  c. Evaluate Who do you think was more responsible for the collapse of Ghana, the people of Ghana or outsiders? Why?

SECTION 2 *(pages 334–338)*

8. **a. Describe** How did Islam influence society in Mali?

   b. Compare and Contrast How were Sundiata and Mansa Musa similar? How were they different?

   c. Evaluate Which group do you think played a larger role in Songhai, warriors or traders?

SECTION 3 *(pages 340–343)*

9. **a. Recall** What different types of information did griots pass on to their listeners?

   b. Analyze Why are the writings of visitors to West Africa so important to our understanding of the region?

   c. Evaluate Which of the various arts of West Africa do you think is most important? Why?

Social Studies Skills

10. **Making Decisions** Imagine that you are a young trader in the Ghana Empire. You have to decide which good you would prefer to trade, salt or gold. Make a list of the consequences that might result from trading each of them. Compare your lists of consequences and make a decision as to which good you will trade. You may wish to conduct research to help make your decision.

FOCUS ON READING AND WRITING

Understanding Cause and Effect *Answer the following questions about causes and effects.*

11. What caused the empire of Ghana to grow?

12. What were some effects of Mansa Musa's rule?

Writing Your Journal Entry *Use your notes and the instructions below to help you create your news report.*

13. Review your notes on possible characters for your journal entry. Choose one, and think about an experience that person, or character, might write about in his or her journal. What other people might the journal entry mention? What event or scene might you describe in your journal? What details or information would your character want to include? Write a journal entry of a paragraph or two from the point of view of your character.

Using the Internet

go.hrw.com
KEYWORD: SK6 CH13

14. **Activity: Writing a Proverb** Does the early bird get the worm? If you go outside at sunrise to check, you missed the fact that this is a proverb that means "The one that gets there first can earn something good." Griots created many proverbs that expressed wisdom or truth. Enter the activity keyword. Then use the Internet resources to write three proverbs that might have been said by griots during the time of the great West African empires. Make sure your proverbs are written from the point of view of a West African person living during those centuries.

Map Activity ✴Interactive

15. **West Africa** On a separate sheet of paper, match the letters on the map with their correct labels.

   Senegal River   Timbuktu

   Lake Chad   Niger River

   Gulf of Guinea

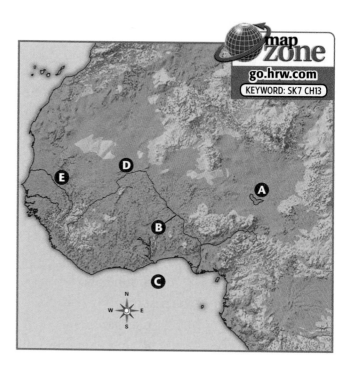

DIRECTIONS: Read questions 1 through 6 and write the letter of the best response. Then read question 7 and write your own well-constructed response.

1 **The wealth of Ghana, Mali, and Songhai was based on**

A raiding other tribes.

B the gold and salt trade.

C trade in ostriches and elephant tusks.

D making iron tools and weapons.

2 **The two rulers who were most responsible for spreading Islam in West Africa were**

A Sunni Ali and Mansa Musa.

B Sundiata and Sunni Ali.

C Ibn Battutah and Tunka Manin.

D Mansa Musa and Askia the Great.

3 **Which of the following rivers helped the development of the West African empires?**

A Niger

B Congo

C Nile

D Zambezi

4 **Griots contributed to West African societies by**

A fighting battles.

B collecting taxes.

C preserving oral history.

D trading with the Berbers.

"Well placed for the caravan trade, it was badly situated to defend itself from the Tuareg raiders of the Sahara. These restless nomads were repeatedly hammering at the gates of Timbuktu, and often enough they burst them open with disastrous results for the inhabitants. Life here was never quite safe enough to recommend it as the centre [center] of a big state."

—Basil Davidson, from *A History of West Africa*

5 **In this quote, the author is discussing why Timbuktu was**

A a good place for universities.

B not a good place for a capital city.

C a good location for trade.

D not a good location for the center of the Tuareg state.

6 **In the second sentence of the passage above, what does the phrase "hammering at the gates of Timbuktu" mean?**

A driving nails into Timbuktu's gates

B knocking on the door to get in the city

C trying to get into and conquer the city

D making noise to anger the inhabitants

7 **Extended Response** Look at the map of Ghana in Section 1. Using information from the map, explain how trade helped build a strong empire.

North Africa

What You Will Learn...

In this chapter you will learn about five countries located in the region of North Africa—Egypt, Libya, Tunisia, Algeria, and Morocco. You will learn about the importance of water in this dry region. You will also study the histories of these countries, which include ancient Egyptian civilization. In addition, you will learn about North Africa's cultures, economies, and governments.

SECTION 1
Physical Geography **350**

SECTION 2
History and Culture.....................**354**

SECTION 3
North Africa Today **361**

FOCUS ON READING AND WRITING

Summarizing To better understand what you read, it is sometimes helpful to stop and summarize the information you have read. A summary is a short restatement of important events or main ideas. As you read this chapter, stop now and then to summarize what you have read. **See the lesson, Summarizing, on page 681.**

Writing a Myth Ancient people created stories called myths to explain things about the world. For example, they created myths to explain the seasons and to explain the powers of their gods. As you read this chapter, look for information that might have seemed mysterious to ancient peoples. Later you will write your own myth to explain something about North Africa.

● National capital
● Other cities

0   150   300 Miles
0   150   300 Kilometers
Projection: Azimuthal Equal-Area

ATLANTIC
OCEAN

Strait of Gibraltar

Rabat ✪
Casablanca ●

M O R O C C O

Canary Islands
(SPAIN)

Western Sahara
(Claimed by
MOROCCO)

Tropic of Cancer

M A U R I T A N I A

Culture Most North Africans are Muslims and speak Arabic.

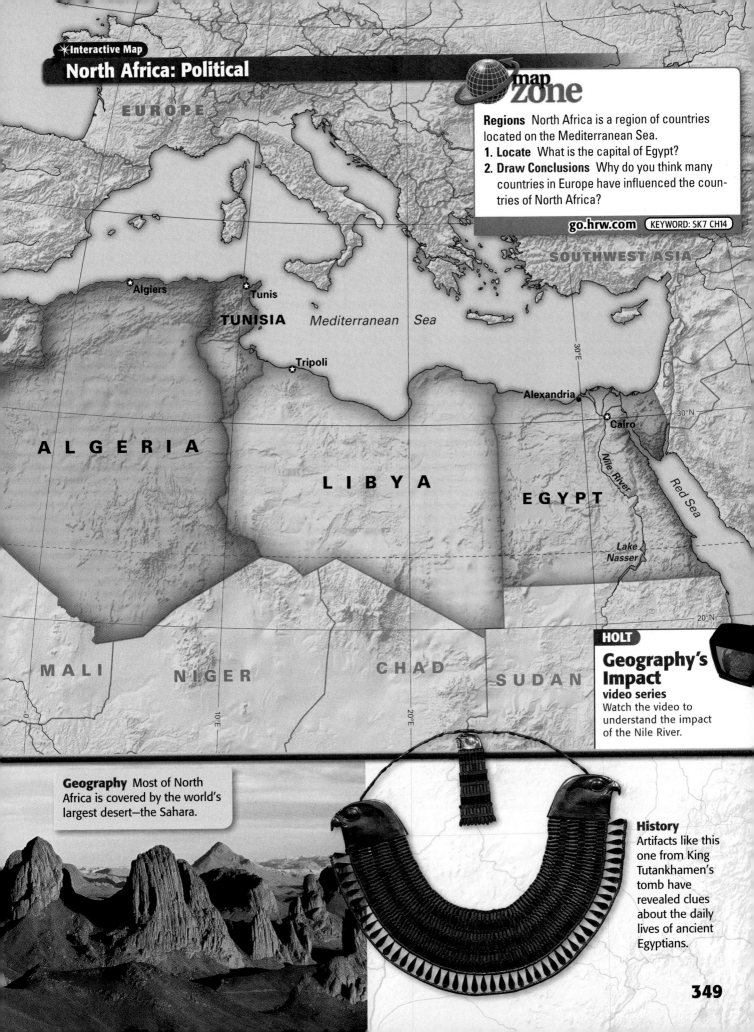

map zone

Regions North Africa is a region of countries located on the Mediterranean Sea.
1. **Locate** What is the capital of Egypt?
2. **Draw Conclusions** Why do you think many countries in Europe have influenced the countries of North Africa?

go.hrw.com KEYWORD: SK7 CH14

EUROPE

SOUTHWEST ASIA

Algiers

Tunis

TUNISIA Mediterranean Sea

Tripoli

Alexandria

Cairo

A L G E R I A

L I B Y A

E G Y P T

Nile River

Red Sea

Lake Nasser

30°E

30°N

20°N

M A L I   N I G E R   C H A D   S U D A N

0°   10°E   20°E

HOLT
Geography's Impact
video series
Watch the video to understand the impact of the Nile River.

Geography Most of North Africa is covered by the world's largest desert—the Sahara.

History Artifacts like this one from King Tutankhamen's tomb have revealed clues about the daily lives of ancient Egyptians.

Physical Geography

What You Will Learn...

Main Ideas

1. Major physical features of North Africa include the Nile River, the Sahara, and the Atlas Mountains.
2. The climate of North Africa is hot and dry, and water is the region's most important resource.

The Big Idea

North Africa is a dry region with limited water resources.

Key Terms and Places

Sahara, *p. 350*
Nile River, *p. 350*
silt, *p. 350*
Suez Canal, *p. 351*
oasis, *p. 352*
Atlas Mountains, *p. 352*

TAKING NOTES As you read, take notes on the physical geography of North Africa. Use the chart below to organize your notes.

Physical Features	
Climate	
Resources	

If YOU lived there...

As your airplane flies over Egypt, you look down and see a narrow ribbon of green—the Nile River Valley—with deserts on either side. As you fly along North Africa's Mediterranean coast, you see many towns scattered across rugged mountains and green valleys.

What are the challenges of living in a mainly desert region?

BUILDING BACKGROUND Even though much of North Africa is covered by rugged mountains and huge areas of deserts, the region is not a bare wasteland. Areas of water include wet, fertile land with date palms and almond trees.

Physical Features

The region of North Africa includes Morocco, Algeria, Tunisia, Libya, and Egypt. From east to west the region stretches from the Atlantic Ocean to the Red Sea. Off the northern coast is the Mediterranean Sea. In the south lies the **Sahara** (suh-HAR-uh), a vast desert. Both the desert sands and bodies of water have helped shape the cultures of North Africa.

The Nile

The **Nile River** is the world's longest river. It is formed by the union of two rivers, the Blue Nile and the White Nile. Flowing northward through the eastern Sahara for about 4,000 miles, the Nile finally empties into the Mediterranean Sea.

For centuries, rain far to the south caused floods along the northern Nile, leaving rich silt in surrounding fields. **Silt** is finely ground fertile soil that is good for growing crops.

The Nile River Valley is like a long oasis in the desert. Farmers use water from the Nile to irrigate their fields. The Nile fans out near the Mediterranean Sea, forming a large delta. A delta

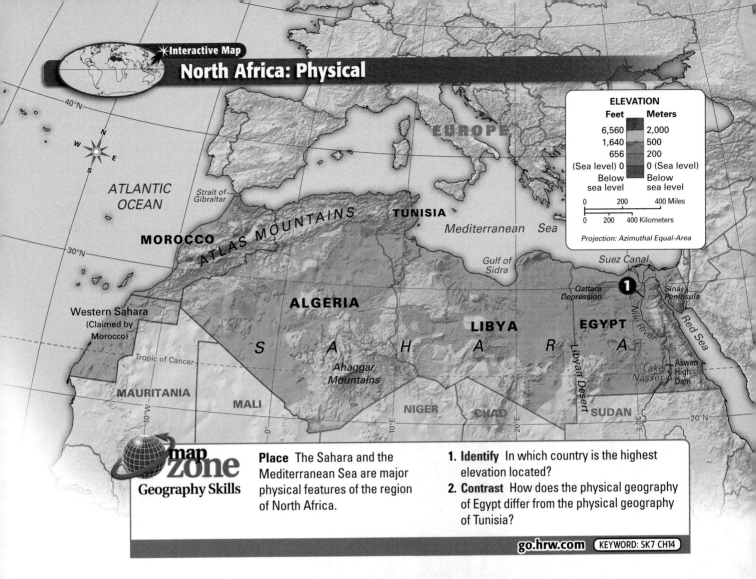

ELEVATION

Feet		Meters
6,560		2,000
1,640		500
656		200
(Sea level) 0		0 (Sea level)
Below sea level		Below sea level

0 200 400 Miles
0 200 400 Kilometers

Projection: Azimuthal Equal-Area

EUROPE

ATLANTIC OCEAN

Strait of Gibraltar

TUNISIA

Mediterranean Sea

MOROCCO

ATLAS MOUNTAINS

Gulf of Sidra

Suez Canal

Qattara Depression

Sinai Peninsula

ALGERIA

LIBYA

EGYPT

Nile River

Red Sea

Western Sahara (Claimed by Morocco)

S A H A R A

Libyan Desert

Tropic of Cancer

Ahaggar Mountains

Lake Nasser

Aswan High Dam

MAURITANIA

MALI

NIGER

CHAD

SUDAN

map zone
Geography Skills

Place The Sahara and the Mediterranean Sea are major physical features of the region of North Africa.

1. **Identify** In which country is the highest elevation located?
2. **Contrast** How does the physical geography of Egypt differ from the physical geography of Tunisia?

go.hrw.com KEYWORD: SK7 CH14

is a landform at the mouth of a river that is created by the deposit of sediment. The sediment in the Nile delta makes the area extremely fertile.

The Aswan High Dam controls flooding along the Nile. However, the dam also traps silt, preventing it from being carried downriver. Today some of Egypt's farmers must use fertilizers to enrich the soil.

The Sinai and the Suez Canal

East of the Nile is the triangular Sinai Peninsula. Barren, rocky mountains and desert cover the Sinai. Between the Sinai and the rest of Egypt is the **Suez Canal**. The French built the canal in the 1860s. It is a narrow waterway that connects the Mediterranean Sea with the Red Sea. Large cargo ships carry oil and goods through the canal.

1 Flowing for 4,132 miles, the Nile is the longest river in the world.

351

The Sahara

The Sahara, the largest desert in the world, covers most of North Africa. The name Sahara comes from the Arabic word for "desert." It has an enormous **impact** on the landscapes of North Africa.

ACADEMIC
VOCABULARY

impact effect, result

One impact of the very dry Sahara is that few people live there. Small settlements are located near a water source such as an oasis. An **oasis** is a wet, fertile area in a desert where a natural spring or well provides water.

In addition to broad, windswept gravel plains, sand dunes cover much of the Sahara. Dry streambeds are also common.

Mountains

Do you think of deserts as flat regions? You may be surprised to learn that the Sahara is far from flat. Some sand dunes and ridges rise as high as 1,000 feet (305 m). The Sahara also has spectacular mountain ranges. For example, a mountain range in southern Algeria rises to a height of 9,800 feet (3,000 m). Another range, the **Atlas Mountains** on the northwestern side of the Sahara near the Mediterranean coast, rises even higher, to 13,600 feet (4,160 m).

READING CHECK **Summarizing** What are the major physical features of North Africa?

Close-up

A Sahara Oasis

The largest desert in the world, the Sahara, spans almost 4 million square miles across North Africa. From ancient times to today, traders crossing the Sahara have relied on the desert's oases. These oases provide water and shade.

Date palms thrive on the banks of this natural spring, which provides water to travelers and irrigated fields.

By carrying supplies, camels help the nomadic Tuareg people travel from oasis to oasis.

Climate and Resources

North Africa is very dry. However, rare storms can cause flooding. In some areas these floods as well as high winds have carved bare rock surfaces out of the land.

North Africa has three main climates. A desert climate covers most of the region. Temperatures range from mild to very hot. How hot can it get? Temperatures as high as 136°F (58°C) have been recorded in Libya. However, the humidity is very low. As a result, temperatures can drop quickly after sunset. In winter temperatures can fall below freezing at night.

The second climate type in the region is a Mediterranean climate. Much of the northern coast west of Egypt has this type of climate. Winters there are mild and moist. Summers are hot and dry. Areas between the coast and the Sahara have a steppe climate.

Oil and gas are important resources, particularly for Libya, Algeria, and Egypt. Morocco mines iron ore and minerals used to make fertilizers. The Sahara has natural resources such as coal, oil, and natural gas.

READING CHECK **Generalizing** What are North Africa's major resources?

SUMMARY AND PREVIEW In this section, you learned about the physical geography of North Africa. Next, you will learn about the history and cultures of the countries of North Africa.

FOCUS ON READING
Summarize the details of what you just read about North Africa's climate.

Shelters like this one provide a place for travelers to rest.

ANALYSIS SKILL **ANALYZING VISUALS**

Why do you think an oasis would be important to people traveling through the Sahara?

Section 1 Assessment

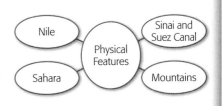
go.hrw.com
Online Quiz
KEYWORD: SK7 HP14

Reviewing Ideas, Terms, and Places

1. **a. Define** What is an **oasis**?
   b. Explain Why is the **Suez Canal** an important waterway?
   c. Elaborate Would it be possible to farm in Egypt if the **Nile River** did not exist? Explain your answer.
2. **a. Recall** What is the climate of most of North Africa?
   b. Draw Conclusions What resources of North Africa are the most valuable?

Critical Thinking

3. **Categorizing** Draw a diagram like the one shown here. Use your notes to list two facts about each physical feature of North Africa.

Nile — Physical Features — Sinai and Suez Canal — Sahara — Mountains

FOCUS ON WRITING

4. **Writing about Physical Geography** What physical feature will you choose as the subject of your myth? How will you describe this feature? Note your ideas.

History and Culture

What You Will Learn...

Main Ideas

1. North Africa's history includes ancient Egyptian civilization.
2. Islam influences the cultures of North Africa and most people speak Arabic.

The Big Idea

North Africa is rich in history and Islamic culture.

Key Terms and Places

Alexandria, *p. 355*
Berbers, *p. 357*

TAKING NOTES Draw two boxes like the ones below. As you read, take notes on the history and culture of North Africa.

North Africa	
History	Culture

If **YOU** lived there...

You live in a village in ancient Egypt in about 800 BC. Your family grows wheat and date palms along the banks of the Nile River, which brings water for your crops. You and your friends like to explore the marshy areas along the banks of the river, where many kinds of birds live in the tall reeds.

How is the Nile River important in your life?

BUILDING BACKGROUND Some of the world's earliest civilizations began in river valleys in Asia and Africa. One of these civilizations was the Nile Valley in Egypt. Egypt was called the "gift of the Nile," because the river's floods brought rich soil to the valley. The soil built up a fertile delta where the Nile emptied into the sea.

Egypt's Nile River Valley was home to some of the world's oldest civilizations. These ancient Egyptians built large monuments, participated in trade, and developed a writing system.

The Great Sphinx at Giza is one of the many monuments built by the ancient Egyptians.

North Africa's History

Sometime after 3200 BC people along the northern Nile united into one Egyptian kingdom. The ancient Egyptians built large stone monuments and developed a written system. Later Greeks and Arabs, who wanted to expand their empires, invaded North Africa.

The Ancient Egyptians

What is the first thing that comes to mind when we think of the ancient Egyptians? Most of us think of the great stone pyramids. The Egyptians built these huge monuments as tombs, or burial places, for pharaohs, or kings.

How did the Egyptians build these huge monuments? Scholars believe thousands of workers cut large blocks of stone far away and rolled them on logs to the Nile. From there the blocks were moved on barges. At the building site, the Egyptians finished carving the blocks. They built dirt and brick ramps alongside the pyramids. Then they hauled the blocks up the ramps.

One of the largest pyramids, the Great Pyramid, contains 2.3 million blocks of stone. Each stone averages 2.5 tons (2.25 metric tons) in weight. Building the Great Pyramid probably required from 10,000

BIOGRAPHY

Cleopatra
(69–30 BC)

After the death of the Egyptian king Ptolemy XII in 51 BC, his daughter, Cleopatra, and her brother became co-rulers of Egypt. From the age of 17, Cleopatra ruled ancient Egypt for more than 20 years. Her reign was during a period of Egyptian history when Egypt was dominated by Rome. Even though she was of Greek descent, Cleopatra was worshipped by ancient Egyptians.

Cleopatra tried to drive out the Romans from Egypt. She feared the Romans would arrest her and take over Egypt. Rather than see them ruling her kingdom, Cleopatra chose to commit suicide. According to tradition, she poisoned herself at the age of 39 with the venom of a deadly snake.

Analyzing What did Cleopatra fear?

to 30,000 workers. They finished the job in about 20 years, and the pyramid still stands thousands of years later.

Egyptian Writing

The ancient Egyptians developed a sophisticated writing system, or hieroglyphics (hy-ruh-GLIH-fiks). This writing system used pictures and symbols that stood for ideas or words. Each symbol represented one or more sounds in the Egyptian language. The Egyptians carved hieroglyphics on their temples and stone monuments. Many of these writings recorded the words and achievements of the pharaohs.

Greek and Arab Civilizations

Because of North Africa's long Mediterranean coastline, the region was open to invaders over the centuries. Those invaders included people from the eastern Mediterranean, Greeks, and Romans. For example, one invader was the Macedonian king Alexander the Great. Alexander founded the city of **Alexandria** in Egypt in 332 BC.

FOCUS ON READING

What details would you use to summarize the ancient Egyptians and their writing?

Ancient Egypt's King Tut

More than 3,300 years ago a 19-year-old named King Tutankhamen ruled ancient Egypt. Since the discovery of Tut's mummy in 1922, many have wondered how he died. In 2005, scientists used modern technology to help find possible clues to King Tut's cause of death.

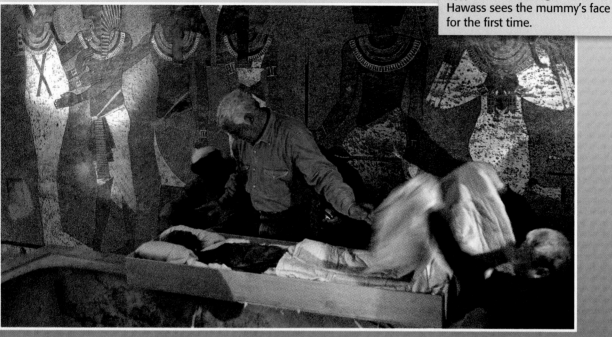

Mummy Unveiled
Inside King Tut's burial chamber, Egyptian archaeologist Zahi Hawass sees the mummy's face for the first time.

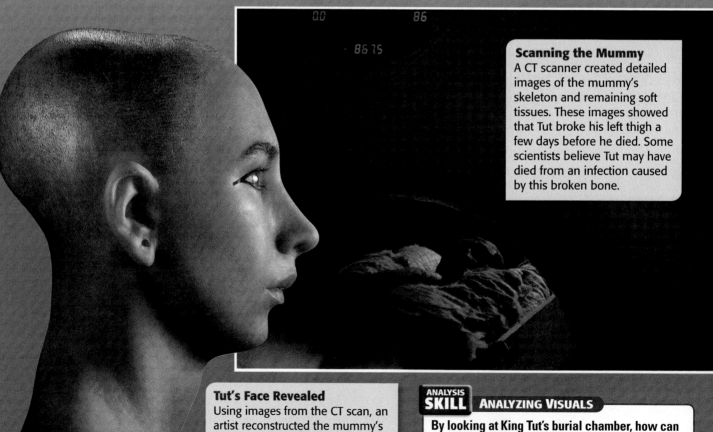

Scanning the Mummy
A CT scanner created detailed images of the mummy's skeleton and remaining soft tissues. These images showed that Tut broke his left thigh a few days before he died. Some scientists believe Tut may have died from an infection caused by this broken bone.

Tut's Face Revealed
Using images from the CT scan, an artist reconstructed the mummy's face with clay and plastic.

ANALYSIS SKILL ANALYZING VISUALS

By looking at King Tut's burial chamber, how can you tell that King Tut was an important person?

Alexandria became an important seaport and trading center. The city was also a great center of learning.

Beginning in the AD 600s, Arab armies from Southwest Asia swept across North Africa. They brought the Arabic language and Islam to the region. Under Muslim rule, North African cities became major centers of learning, trade, and craft making. These cities included Cairo in Egypt and Fès in Morocco.

European Influence

In the 1800s European countries began to take over the region. By 1912 they had authority over all of North Africa. In that year Italy captured Libya. Spain already controlled northern Morocco. France ruled the rest of Morocco as well as Tunisia and Algeria. The British controlled Egypt.

The countries of North Africa gradually gained independence. Egypt gained limited independence in 1922. The British kept military bases there and maintained control of the Suez Canal until 1956. During World War II the region was a major battleground. Libya, Morocco, and Tunisia each won independence in the 1950s.

Algeria was the last North African country to win independence. Many French citizens had moved to the country, and they considered Algeria part of France. Algeria finally won independence in 1962.

Modern North Africa

Since independence, the countries of North Africa have tried to build stronger ties with other Arab countries. Before signing a peace treaty in 1979, Egypt led other Arab countries in several wars against Israel. In 1976 Morocco took over the former Spanish colony of Western Sahara.

READING CHECK **Evaluating** What was one significant event in North Africa's history?

Cultures of North Africa

As you have just read, many of the countries of North Africa share a common history. Likewise, the people of North Africa share many aspects of culture—language, religion, foods, holidays, customs, and arts and literature.

People and Language

Egyptians, Berbers, and Bedouins make up nearly all of Egypt's population. Bedouins are nomadic herders who travel throughout the deserts of Egypt.

Most people in the other countries of North Africa are of mixed Arab and Berber ancestry. The **Berbers** are an ethnic group who are native to North Africa and speak Berber languages. The majority of North Africans speak Arabic.

Most ethnic Europeans left North Africa after the region's countries became independent. However, because of the European influence in the region, some North Africans also speak French, Italian, and English.

Religion

Most North Africans are Muslims who practice the religion of Islam. Islam plays a major role in North African life. For example, North African Muslims stop to pray five times a day. In addition, Fridays are special days when Muslims meet in mosques for prayer. About 6 percent of Egyptians are Christians or practice other religions.

Foods

What kinds of food would you eat on a trip to North Africa? Grains, vegetables, fruits, and nuts are common foods.

You would also notice that most meals in North Africa include couscous (KOOS-koos). This dish is made from wheat and looks like small pellets of pasta.

Couscous is usually steamed over boiling water or soup. Often it is served with vegetables or meat, butter, and olive oil.

Egyptians also enjoy a dish called *fuul*. It is made with fava beans mashed with olive oil, salt, pepper, garlic, and lemons. It is often served with hard-boiled eggs and bread. Many Egyptians eat these foods on holidays and at family gatherings.

FOCUS ON CULTURE

The Berbers

Before the AD 600s when Arabs settled in North Africa, a people called the Berbers lived in the region. The descendants of these ancient peoples live throughout North Africa today—mostly in Morocco and Algeria. Some Berbers are nomadic and live in goat-hair tents. Other Berbers farm crops that include wheat, barley, fruits, and olives. Some also raise cattle, sheep, or goats.

Berber culture is centered on a community made up of different tribes. Once a year, Berber tribes gather at large festivals. At these gatherings Berbers trade goods, and many couples get married in elaborate ceremonies.

Drawing Conclusions How have Berbers kept their culture alive?

Holidays and Customs

Important holidays in North Africa include the birthday of Muhammad, the prophet of Islam. This holiday is marked with lights, parades, and special sweets of honey, nuts, and sugar. During the holy month of Ramadan, Muslims abstain from food and drink during the day.

Gathering at cafes is a custom practiced by many men in North Africa. The cafes are a place where they go to play chess or dominoes. Most women in North Africa socialize only in their homes.

A certain way of greeting each other on the street is another North African custom. People greet each other by shaking hands and then touching their hand to their heart. If they are family or friends, they will kiss each other on the cheek. The number of kisses varies from country to country.

Many North Africans wear traditional clothes, which are long and loosely fitted. Such styles are ideal for the region's hot climate. Many North African women dress according to Muslim tradition. Their clothing covers all of the body except the face and hands.

The Arts and Literature

North Africa has a rich and varied tradition in the arts and literature. Traditional arts include wood carving and weaving. The region is famous for beautiful hand-woven carpets. The women who weave these carpets use bright colors to create complex geometric patterns. Beautifully detailed handpainted tilework is also a major art form in the region.

Other arts in Egypt include its growing movie industry. Egyptian films in Arabic have become popular throughout Southwest Asia and North Africa.

Many North Africans also enjoy popular music based on singing and poetry. The musical scale there has many more notes

Fès, Morocco

At a tannery in the city of Fès, men dye sheepskins in large vats of dyes. The city's craftspeople use yellow-dyed leather to make distinctive leather shoes.

ANALYZING VISUALS What other colors are the sheepskins dyed?

Vats of different colored dyes

Yellow-dyed sheepskins are laid out to dry.

than are common in Western music. As a result of this difference, North African tunes seem to wail or waver. Musicians in Morocco often use instruments such as the three-stringed sintir.

The region has also produced important writers and artists. For example, Egyptian poetry and other writing date back thousands of years. One of Egypt's most famous writers is Naguib Mahfouz. In 1988 he became the first Arab writer to win the Nobel Prize for Literature.

READING CHECK Analyzing What are some important facts about the people and culture of North Africa?

SUMMARY AND PREVIEW In this section you learned about the history and culture of North Africa. Next, you will learn about the region today.

go.hrw.com
Online Quiz
KEYWORD: SK7 HP14

Section 2 Assessment

Reviewing Ideas, Terms, and Places

1. **a. Define** What are hieroglyphics?
   b. Make Inferences What made the city of **Alexandria** important?
   c. Evaluate Why do you think European countries wanted to take over countries in North Africa?
2. **a. Recall** What language do most North Africans speak?
   b. Summarize What is one custom practiced in North Africa?
   c. Elaborate How is Islam a major part of the daily lives of many North Africans?

Critical Thinking

3. **Summarizing** Use your notes to summarize what you learned about the culture of North Africa.

Language	
Religion	
Food	
The Arts	
Literature	

FOCUS ON WRITING

4. **Choosing Details** Which details about North Africa's history and culture will you include in your myth? Write a sentence or two about each detail.

Social Studies Skills

Analyzing a Diagram

Learn

Diagrams are drawings that use lines and labels to explain or illustrate something. Pictorial diagrams show an object in simple form, much like it would look if you were viewing it. Cutaway diagrams, like the one of an Egyptian pyramid below, show the "insides" of an object. These diagrams usually have labels that identify important areas of the diagram.

Practice

Analyze the diagram below, and answer the following questions.

1. What type of diagram is this?

2. What labels in the diagram suggest what this pyramid was used for?

3. Of what materials was the pyramid made?

An Egyptian Pyramid

King's burial chamber

Air shaft

Grand gallery

Tunnel

Apply

Draw a cutaway diagram of your school. Label classrooms, hallways, the cafeteria, and other areas. Use your diagram to answer the following questions.

1. How many stories are in your school?

2. Where is the closest exit located from the classroom you are sitting in now?

3. What are some of the materials your school is made of?

North Africa Today

If YOU lived there...

You live in the colorful city of Marrakesh, Morocco. This is the week of carnival, an exotic celebration where you can stroll among the crowds and see storytellers, musicians, and snake charmers. You may stop at a food stall for a snack and a cup of mint tea. But even in an ordinary week, you can explore the markets and see palaces and gardens with fountains.

Why is Marrakesh an exciting place to live?

BUILDING BACKGROUND Several countries and cultures have influenced modern North Africa. Some countries in North Africa were once colonies of France or Italy. As a result, the region is still linked with events in Europe.

Egypt

With a population of more than 75 million, Egypt is North Africa's most populous country. Egypt's government faces many challenges today. Most Egyptians are poor farmers because Egypt has limited resources and jobs.

Government and Society

Even though Egypt is a republic, its government is heavily influenced by Islamic law. Egypt's government has a constitution and Egyptians elect their government officials. Power is shared between Egypt's president and the prime minister.

Many Egyptians debate over the role of Islam in the country. Some Egyptian Muslims believe Egypt's government, laws, and society should be based on Islamic law. However, some Egyptians worry that such a change in government would mean fewer personal freedoms.

Egyptians are divided over their country's role in the world. Some Egyptians want their government to remain a leader among Arab countries. However, others want their government to focus more on improving their daily life.

What You Will Learn...

Main Ideas

1. Many of Egypt's people are farmers and live along the Nile River.
2. People in the other countries of North Africa are mostly pastoral nomads or farmers, and oil is an important resource in the region.

The Big Idea

Many people of North Africa are farmers, and oil is an important resource.

Key Terms and Places

Cairo, *p. 363*
Maghreb, *p. 364*
souks, *p. 365*
free port, *p. 365*
dictator, *p. 365*

TAKING NOTES Using the chart below, take notes on the governments, economies, and cities in Egypt and the other countries of North Africa.

	Egypt	Other Countries of North Africa
Government		
Economy		
Cities		

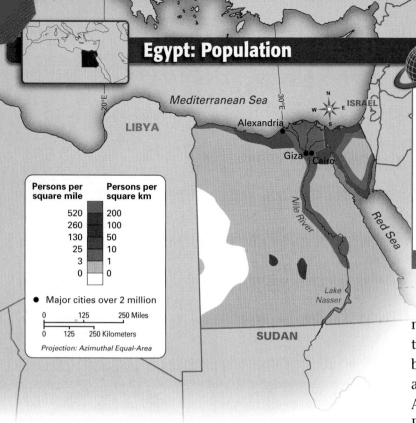

Egypt: Population

Mediterranean Sea

LIBYA

Alexandria

Giza ● ● Cairo

Nile River

Red Sea

ISRAEL

Lake Nasser

SUDAN

Persons per square mile	**Persons per square km**
520	200
260	100
130	50
25	10
3	1
0	0

● Major cities over 2 million

0   125   250 Miles

0   125   250 Kilometers

Projection: Azimuthal Equal-Area

map zone
Geography Skills

Place Most of Egypt's population lives near the Nile River.
1. **Locate** Which cities have more than 2 million people?
2. **Draw Conclusions** Why do some Egyptians live in rural areas instead of cities?

Some supporters of an Islamic government have turned to violence to advance their cause. Attacks on tourists by members of a radical Islamic group in the 1990s and 2000s were particularly worrisome. A loss of tourism would severely hurt Egypt's economy.

Many Egyptians live in severe poverty. Many do not have clean water for cooking or washing. The spread of disease in crowded cities is also a problem. In addition, about half of Egyptians cannot read and write. Still, Egypt's government has made progress. Today Egyptians live longer and are much healthier than they were 50 years ago.

Resources and Economy

Egypt is challenged by its limited resources. For example, the country's only farmland is located in the Nile River Valley and Delta. To keep the land productive, farmers must use more and more fertilizer. In addition, salt water drifting up the Nile from the Mediterranean has brought salts to the surface that are harmful to crops. These problems and a rapidly growing population have forced Egypt to import much of its food.

About 32 percent of Egyptians are farmers, but less than 3 percent of the land is used for farming. Most farming is located

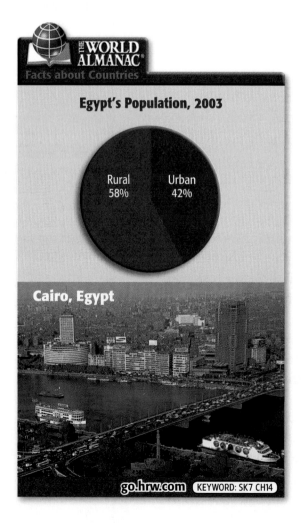

THE WORLD ALMANAC
Facts about Countries

Egypt's Population, 2003

Rural 58%

Urban 42%

Cairo, Egypt

go.hrw.com KEYWORD: SK7 CH14

along the Nile Delta, which is extremely fertile. A warm, sunny climate and water for irrigation make the delta ideal for growing cotton. Farmlands along the Nile River are used for growing vegetables, grain, and fruit.

The Suez Canal is an important part of Egypt's economy. The canal makes about 2 billion dollars a year by requiring tolls from ships that pass through the canal. Thousands of ships use the canal each year to avoid making long trips around Southern Africa. This heavy traffic makes the canal one of the world's busiest waterways.

Egypt's economy depends mostly on agriculture, petroleum exports, and tourism. To provide for its growing population, Egypt is working to expand its industries. Recently, the government has invested in the country's communications and natural gas industries.

Many Egyptians also depend on money sent home by family members working in other countries. Some Egyptians work in other countries because there are not enough jobs in Egypt. Over 7 million Egyptians work in Europe or oil-rich countries in Southwest Asia.

Cities and Rural Life

Most North Africans live in cities along the Mediterranean coast or in villages in the foothills of the Atlas Mountains. Although, in Egypt 99 percent of the population lives in the Nile Valley and Delta. Egypt's capital, **Cairo**, is located in the Nile Delta.

With more than 10 million people, Cairo is the largest urban area in North Africa. The city is crowded, poor, and polluted. Cairo continues to grow as people move into the city from Egypt's rural areas in search of work. For centuries, Cairo's location at the southern end of the Nile Delta helped the city grow. The city also lies along old trading routes.

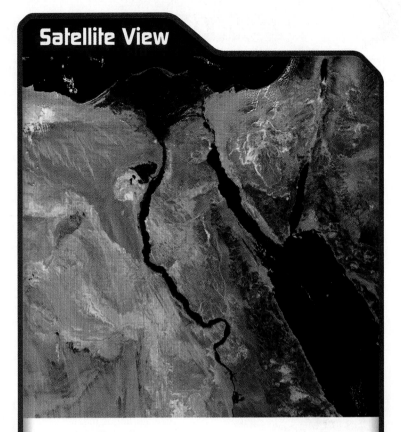

Satellite View

The Nile River

From space, the Nile looks like a river of green. Actually, the areas that appear green in this satellite image are thousands of irrigated fields that line the banks of the river. The river deposits silt along its banks, which makes the land extremely fertile. Farmers also depend on the Nile's waters to irrigate their crops. Without water, they could not farm in the desert.

Notice how the river appears smaller at the bottom of this image. The Aswan High Dam controls the river's flow here, which prevents flooding and provides electricity.

Drawing Conclusions How is the Nile important to Egypt's people?

Today the landscape of Cairo is a mixture of modern buildings, historic mosques, and small, mud-brick houses. However, there is not enough housing in Cairo for its growing population. Many people live in makeshift housing in the slums or boats along the Nile. Communities have even developed in cemeteries, where people convert tombs into bedrooms and kitchens.

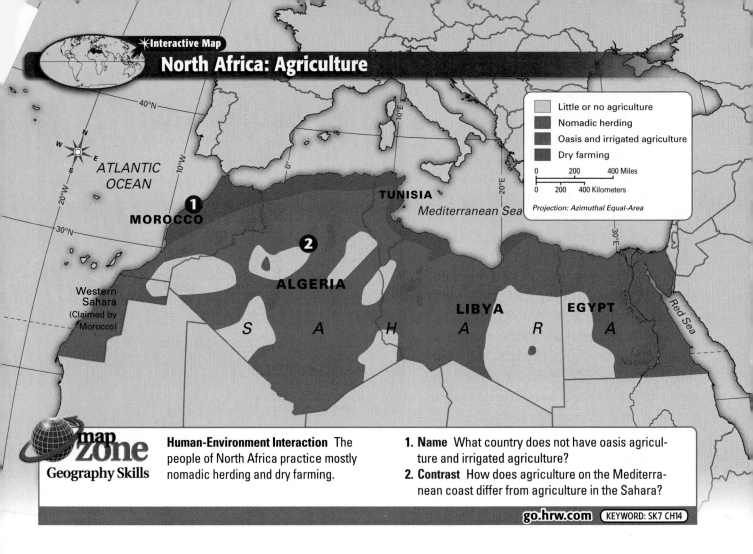

North Africa: Agriculture

Little or no agriculture
Nomadic herding
Oasis and irrigated agriculture
Dry farming

0   200   400 Miles
0   200   400 Kilometers
Projection: Azimuthal Equal-Area

40°N

ATLANTIC OCEAN

W   E
S

TUNISIA

Mediterranean Sea

MOROCCO ①

30°N

②

ALGERIA

Western Sahara (Claimed by Morocco)

S A H A R A

LIBYA   EGYPT

Red Sea

Lake Nasser

map zone
Geography Skills

Human-Environment Interaction The people of North Africa practice mostly nomadic herding and dry farming.

1. **Name** What country does not have oasis agriculture and irrigated agriculture?
2. **Contrast** How does agriculture on the Mediterranean coast differ from agriculture in the Sahara?

go.hrw.com   KEYWORD: SK7 CH14

FOCUS ON READING

Think about what details are important in a summary of Egypt's government, economy, and cities.

Alexandria is Egypt's second-largest city. The city was founded by Alexander the Great. Known in ancient times for its spectacular library, it is now the home to a large university and many industries. Its location on the Mediterranean Sea has made it a major seaport. The home of some 3.5 million people, Alexandria is as poor and crowded as Cairo.

More than half of all Egyptians live in small villages and other rural areas. Most rural Egyptians are farmers called fellahin (fel-uh-HEEN). These farmers own very small plots of land along the Nile River. Some fellahin also work large farms owned by powerful families.

READING CHECK Finding Main Ideas What are some of the challenges Cairo faces today?

Other Countries of North Africa

Western Libya, Tunisia, Algeria, and Morocco are often called the **Maghreb** (MUH-gruhb). This Arabic word means "west" or "the direction of the setting sun." Since most of the Maghreb is covered by the Sahara, cities and farmland are located in narrow bands along the coast.

Government and Economy

A major challenge in North Africa is the conflict over the role of Islam in society. For example, in Algeria some groups want a government based on Islamic principles and laws. In 1992 the government canceled elections that many believed would be won by Islamic groups.

1 Dry farming In Morocco, hungry goats climb an argan nut tree. These trees and other hardy crops thrive without much water.

2 Oasis agriculture In Algeria, oases in the desert provide enough water to irrigate crops and date palms.

Oil, mining, and tourism are important industries for the countries of North Africa. Oil is the most important resource, particularly in Libya and Algeria. Money from oil pays for schools, health care, food, social programs, and military equipment. The region's countries also have large deposits of natural gas, iron ore, and lead. The largest trade partners of Algeria, Libya, and Morocco are European Union members.

Agriculture is a major economic activity in North Africa. About one in six workers in Libya, Tunisia, and Algeria is a farmer. In Morocco, farmers make up about 40 percent of the labor force. North Africa's farmers grow and export wheat, olives, fruits, and nuts. Tourism is also an important economic activity in the region, especially in Morocco and Tunisia.

Cities

Many North African cities have large marketplaces, or **souks**. The souks are located in the old district of a city called the Casbah. These souks sell various goods such as spices, carpets, and copper teapots. The Casbah in Algeria's capital, Algiers, is a maze of winding alleys and tall walls.

Libya and Tunisia's cities and most of its population are found in the coastal areas. Libya is the most urbanized country in the region. More than 86 percent of Libya's 5.7 million people live in cities. The largest cities are Benghazi and the capital, Tripoli. Tunisia's capital and largest city, Tunis, lies on the Mediterranean coast.

Morocco's largest city, Casablanca, has about 3.3 million people. Another Moroccan city, Tangier, overlooks the Strait of Gibraltar. This beautiful city was once a Spanish territory. Today tourists can take a quick ferry ride from Spain across the strait to Tangier, a free port. A **free port** is a city in which almost no taxes are placed on goods sold there.

The Countries Today

In addition to sharing similar economies, the countries of North Africa also share similar challenges. Some countries are dealing with violence, while others are strengthening their trading relationships with the United States and Europe.

Libya Since 1969 Libya has been ruled by a dictator, General Mu'ammar al-Gadhafi. A **dictator** is someone who rules a country with complete power. Gadhafi has supported acts of violence against Israel and its neighbors. As a result, many countries limited their economic relationship with Libya. However, Libya's relations with the West have recently improved. For example, in 2004, the United States lifted trade restrictions with Libya.

Algiers, Algeria

Algeria's capital and major port, Algiers, sits on the Mediterranean Sea.

Algeria Violence between Algeria's government and some Islamic groups claimed thousands of lives in the 1990s. Today, Algeria is trying to recover from the violence and strengthen the country's economy with exports to Europe.

Tunisia Tunisia's government has granted Tunisian women more rights than women in any other North African country. Tunisia has close economic relationships with European countries. Today about two-thirds of Tunisia's imported goods are from European Union countries.

Morocco Morocco is the only North African country with little oil. Today, the country is an important producer and exporter of fertilizer.

READING CHECK **Summarizing** What are some of the challenges these countries face?

SUMMARY AND PREVIEW In this section you learned about North Africa today. In the next chapter you will learn about the region of West Africa.

Section 3 Assessment

Reviewing Ideas, Terms, and Places

1. **a. Define** What is a **souk**?
   b. Draw Conclusions Why is housing scarce in **Cairo**?
   c. Predict In what ways do you think Egypt's government can help solve the country's poverty?
2. **a. Recall** What countries in North Africa make up the **Maghreb**?
   b. Compare and Contrast How is Libya similar to and different from Morocco?
   c. Evaluate How do you think the countries of North Africa can improve their economies?

Critical Thinking

3. **Comparing** Use your notes to compare Egypt with the other countries of North Africa.

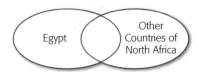

Egypt / Other Countries of North Africa

FOCUS ON WRITING

4. **Taking Notes on North Africa** Are there any characteristics of North Africa today that you might feature in your myth? Write down your ideas.

Geography's Impact
video series
Review the video to answer the closing question:
What are two benefits and two consequences of the Aswan High Dam?

Visual Summary

Use the visual summary below to help you review the main ideas of the chapter.

QUICK FACTS

The Sahara is a major physical feature of North Africa.

One of the world's earliest civilizations thrived on the Nile River in ancient Egypt.

Most major cities in North Africa are located on the Mediterranean Sea.

Reviewing Vocabulary, Terms, and Places

Unscramble each group of letters below to spell a term or place that matches the given definition.

1. **sasoi**—wet, fertile area in a desert where a spring or well provides water

2. **ashraa**—the largest desert in the world that covers most of North Africa

3. **ipmtac**—effect, result

4. **enli virer**—the world's longest river that empties into the Mediterranean Sea in Egypt

5. **oicar**—a city founded more than 1,000 years ago on the Nile and is the capital of Egypt today

6. **uahtroyti**—power; right to rule

7. **tidrotca**—someone who rules a country with complete power

8. **ksuos**—marketplaces

9. **efer tpro**—a city in which almost no taxes are placed on goods sold there

Comprehension and Critical Thinking

SECTION 1 *(Pages 350–353)*

10. **a. Describe** What is the Nile River Valley like? Describe the river and the landscape.

    b. Draw Conclusions How important are oases to people traveling through the Sahara?

    c. Elaborate Why do you think few people live in the Sahara? What role does climate play in where people live? Explain your answer.

SECTION 2 *(Pages 354–359)*

11. **a. Recall** What types of monuments did the ancient Egyptians build?

    b. Make Inferences Why did European countries want to control most of North Africa?

    c. Elaborate Why do you think some groups living in North Africa are nomadic people?

SECTION 3 (*Pages 361–366*)

12. a. Define What is the Maghreb? What physical feature covers this region?

b. Contrast How does Egypt's economy differ from the economies of the other countries of North Africa?

c. Predict In what ways do you think Egypt could improve the lives of its people, who are mostly poor? Explain your answer.

Using the Internet

13. Activity: Exploring the Sahara The Sahara is the largest desert in the world and is covered by great seas of sand dunes. One of the most inhospitable, hostile places on Earth, few people live in the Sahara, and few people even travel through it. Enter the activity keyword to begin your journey through the Sahara. See pictures of the Sahara and learn more about its history and geography. Imagine what it would be like to cross the desert. Then create a PowerPoint presentation or a visual display that summarizes your adventures across the great Sahara.

Social Studies Skills

Analyzing a Diagram Use the diagram of an Egyptian pyramid on this chapter's Social Studies Skills page to answer the following questions.

14. From looking at the title of the diagram, who built this pyramid?

15. How did the pyramid builders get to the king's burial chamber?

16. In what kind of climate is the pyramid located?

17. Summarizing Re-read the paragraphs under Physical Features in Section 1. Create a short summary of each paragraph. Then combine these paragraph summaries into a summary of the whole passage.

18. Writing a Myth Choose one physical feature of North Africa to be the subject of your myth. Then, write two to three paragraphs describing some characteristics of the physical feature. Use your imagination! You might describe the characteristics of the physical feature and how you think ancient peoples would find it mysterious. For example, you might explain why it rarely rains in the Sahara or why the Nile flows to the Mediterranean Sea.

Map Activity

19. North Africa On a separate sheet of paper, match the letters on the map with their correct labels.

Nile River

Atlas Mountains

Cairo

Tripoli

Strait of Gibraltar

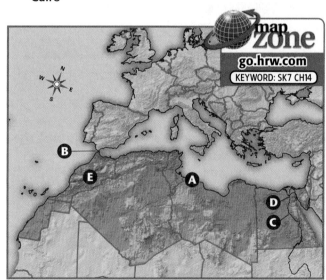

Standardized Test Practice

DIRECTIONS: Read questions 1 through 7 and write the letter of the best response. Question 8 will require a brief essay.

1 **What physical feature covers most of North Africa?**

A the Nile

B the Sahara

C Sinai Peninsula

D Atlas Mountains

2 **The Nile flows through Egypt and empties into the**

A Red Sea.

B Atlantic Ocean.

C Mediterranean Sea.

D Sahara.

3 **The ancient Egyptians built pyramids to bury their**

A relatives.

B pharaohs.

C pets.

D valuable goods.

4 **What language do the majority of North Africans speak?**

A English

B French

C Italian

D Arabic

5 **Most North Africans are**

A Christians.

B Buddhists.

C Muslims.

D Hindus.

North Africa

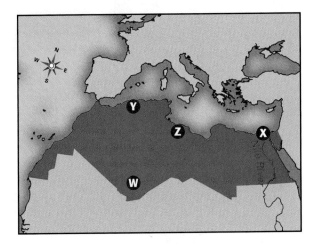

6 **Use the map to answer the following question. Ancient Egyptian civilization thrived in North Africa at the location marked on the map by the letter**

A W.

B Z.

C X.

D Y.

7 **What do ships use to avoid sailing around Southern Africa?**

A the Nile

B the Suez Canal

C the Aswan High Dam

D the Strait of Gibraltar

8 **Extended Response** Look at the physical map in Section 1. Write a short essay describing the physical features of North Africa. Explain why people live only in certain areas of the region.

CHAPTER 15

West Africa

What You Will Learn...

In this chapter you will learn about the 17 countries of West Africa. First, you will learn about the dry plains and major rivers in the region. Then you will learn about West Africa's history and culture as well as what the countries in the region are like today.

SECTION 1
Physical Geography **372**

SECTION 2
History and Culture...................**376**

SECTION 3
West Africa Today.....................**382**

FOCUS ON READING AND SPEAKING

Understanding Comparison-Contrast Comparing and contrasting, or looking for similarities and differences, can help you more fully understand the subject you are studying. As you read, look for ways to compare and contrast the information in your text. **See the lesson, Understanding Comparison-Contrast, on page 682.**

Giving an Oral Description Storytelling is an important part of West Africa's history and culture. Storytellers pass along information to the community about events, places, and people. As you read this chapter, imagine that you are a storyteller. You are going to pass on some information about a person who lives, or has lived, in this region.

History People such as the Dogon cliff dwellers have been living in Mali for hundreds of years.

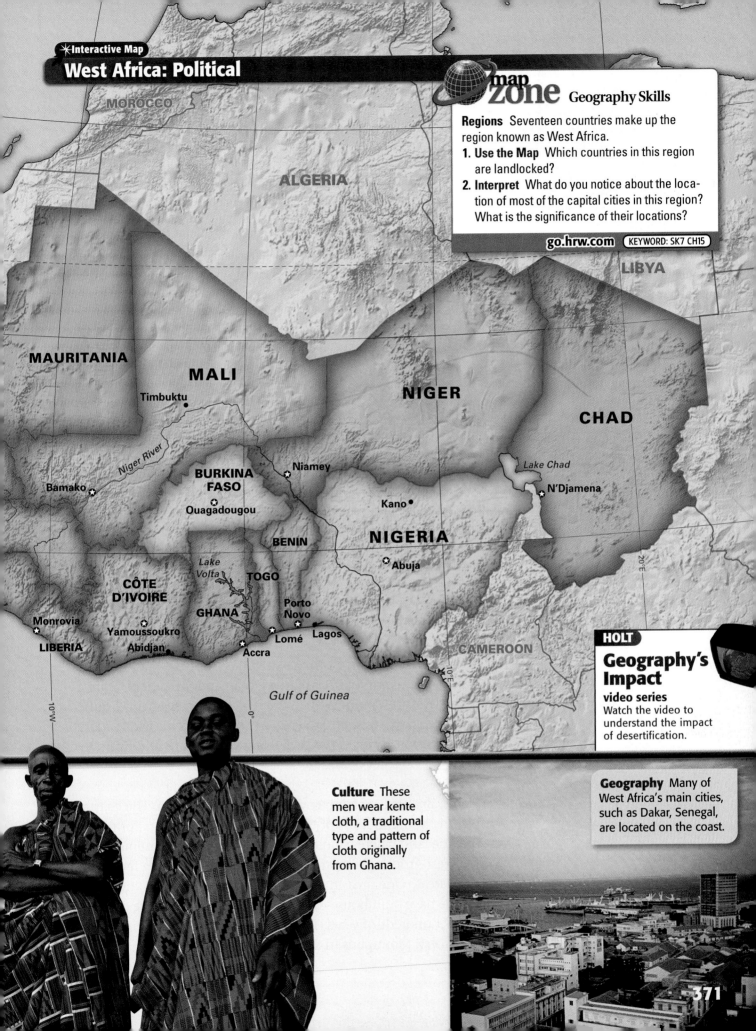

MOROCCO

ALGERIA

LIBYA

MAURITANIA

MALI

Timbuktu

NIGER

CHAD

Niger River

Bamako ★

BURKINA FASO

Niamey ●

Lake Chad

N'Djamena ★

Ouagadougou ★

Kano ●

NIGERIA

BENIN

Abuja ★

CÔTE D'IVOIRE

Lake Volta

TOGO

GHANA

Porto Novo ★

Yamoussoukro ★

Monrovia ★

LIBERIA

Abidjan ●

Lomé ★

Lagos ★

Accra ★

CAMEROON

Gulf of Guinea

10°W

0°

10°E

20°E

map zone — Geography Skills

Regions Seventeen countries make up the region known as West Africa.

1. **Use the Map** Which countries in this region are landlocked?
2. **Interpret** What do you notice about the location of most of the capital cities in this region? What is the significance of their locations?

go.hrw.com KEYWORD: SK7 CH15

HOLT
Geography's Impact
video series
Watch the video to understand the impact of desertification.

Culture These men wear kente cloth, a traditional type and pattern of cloth originally from Ghana.

Geography Many of West Africa's main cities, such as Dakar, Senegal, are located on the coast.

Physical Geography

What You Will Learn...

Main Ideas

1. West Africa's key physical features include plains and the Niger River.
2. West Africa has distinct climate and vegetation zones that go from arid in the north to tropical in the south.
3. West Africa has good agricultural and mineral resources that may one day help the economies in the region.

The Big Idea

West Africa, which is mostly a region of plains, has climates ranging from arid to tropical and has important resources.

Key Terms and Places

Niger River, *p. 373*
zonal, *p. 374*
Sahel, *p. 374*
desertification, *p. 374*
savanna, *p. 374*

 TAKING NOTES As you read, use a chart like the one below to help you organize your notes on the physical geography of West Africa.

Physical features	
Climate and vegetation	
Resources	

If YOU lived there...

Your family grows crops on the banks of the Niger River. Last year, your father let you go with him to sell the crops in a city down the river. This year you get to go with him again. As you paddle your boat, everything looks the same as last year—until suddenly the river appears to grow! It looks as big as the sea, and there are many islands all around. The river wasn't like this last year.

What do you think caused the change in the river?

BUILDING BACKGROUND The Niger River is one of West Africa's most important physical features. It brings precious water to the region's dry plains. Much of the interior of West Africa experiences desertlike conditions, but the region's rivers and lakes help to support life there.

Physical Features

The region we call West Africa stretches from the Sahara in the north to the coasts of the Atlantic Ocean and the Gulf of Guinea in the west and south. While West Africa's climate changes quite a bit from north to south, the region does not have a wide variety of landforms. Its main physical features are plains and rivers.

Plains and Highlands

Plains, flat areas of land, cover most of West Africa. The coastal plain is home to most of the region's cities. The interior plains provide land where people can raise a few crops or animals.

West Africa's plains are vast, interrupted only by a few highland areas. One area in the southwest has plateaus and cliffs. People have built houses directly into the sides of these cliffs for many hundreds of years. The region's only high mountains are the Tibesti Mountains in the northeast.

The Niger River

As you can see on the map below, many rivers flow across West Africa's plains. The most important river is the Niger (NY-juhr). The **Niger River** starts in some low mountains not too far from the Atlantic Ocean. From there, it flows 2,600 miles (4,185 km) into the interior of the region before emptying into the Gulf of Guinea.

The Niger brings life-giving water to West Africa. Many people farm along its banks or fish in its waters. It is also an important transportation route, especially during the rainy season. At that time, the river floods and water flows smoothly over its rapids.

Part of the way along its route the river divides into a network of channels, swamps, and lakes. This watery network is called the inland delta. Although it looks much like the delta where a river flows into the sea, this one is actually hundreds of miles from the coast in Mali.

FOCUS ON READING
The word *although* signals contrast in this paragraph. What is being contrasted?

READING CHECK Summarizing Why is the Niger River important to West Africa?

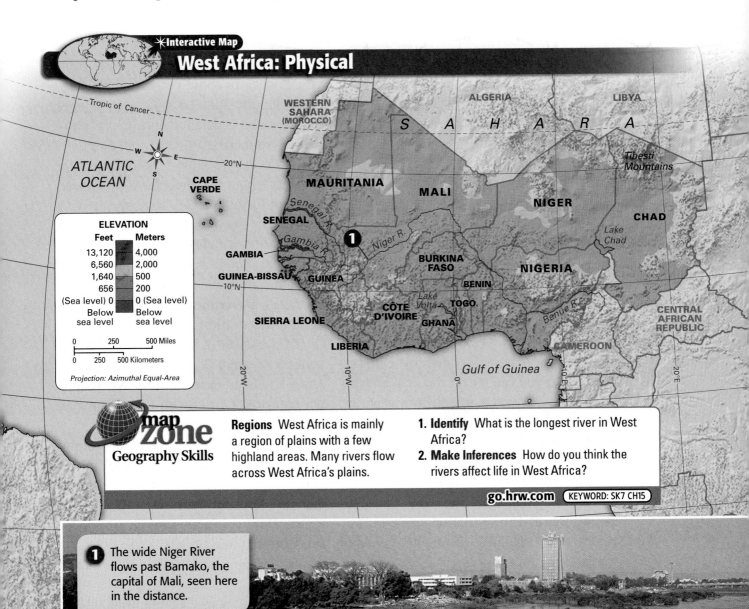

Interactive Map
West Africa: Physical

ATLANTIC OCEAN

Tropic of Cancer

WESTERN SAHARA (MOROCCO)

ALGERIA

LIBYA

S A H A R A

Tibesti Mountains

20°N

CAPE VERDE

MAURITANIA

MALI

NIGER

CHAD

Senegal R.

SENEGAL

Lake Chad

Gambia R.

Niger R.

GAMBIA

BURKINA FASO

NIGERIA

GUINEA-BISSAU

GUINEA

10°N

BENIN

Benue R.

CENTRAL AFRICAN REPUBLIC

SIERRA LEONE

CÔTE D'IVOIRE

Lake Volta

TOGO

GHANA

LIBERIA

CAMEROON

Gulf of Guinea

ELEVATION

Feet	Meters
13,120	4,000
6,560	2,000
1,640	500
656	200
(Sea level) 0	0 (Sea level)
Below sea level	Below sea level

0   250   500 Miles
0   250   500 Kilometers

Projection: Azimuthal Equal-Area

20°W   10°W   0°   10°E   20°E

map Zone
Geography Skills

Regions West Africa is mainly a region of plains with a few highland areas. Many rivers flow across West Africa's plains.

1. **Identify** What is the longest river in West Africa?
2. **Make Inferences** How do you think the rivers affect life in West Africa?

go.hrw.com KEYWORD: SK7 CH15

1 The wide Niger River flows past Bamako, the capital of Mali, seen here in the distance.

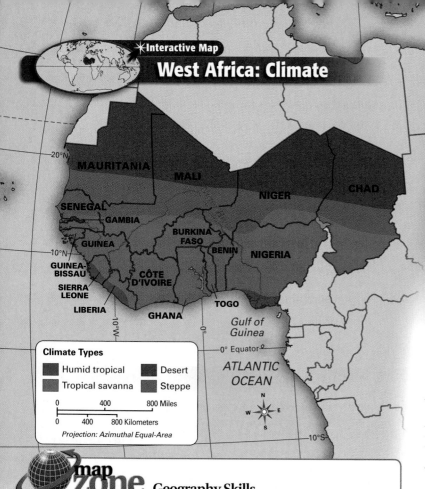

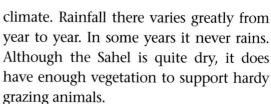

West Africa: Climate

★ Interactive Map

MAURITANIA

MALI

NIGER

CHAD

SENEGAL

GAMBIA

GUINEA

BURKINA FASO

BENIN

NIGERIA

GUINEA-BISSAU

CÔTE D'IVOIRE

SIERRA LEONE

LIBERIA

TOGO

GHANA

Gulf of Guinea

0° Equator

ATLANTIC OCEAN

Climate Types

Humid tropical

Desert

Tropical savanna

Steppe

0   400   800 Miles

0   400   800 Kilometers

Projection: Azimuthal Equal-Area

map zone Geography Skills

Regions Four climate types stretch across West Africa in horizontal bands.

1. Identify What countries have desert climates?

2. Make Inferences What areas do you think get the most rainfall?

go.hrw.com KEYWORD: SK7 CH15

Sahel Vegetation in the semiarid Sahel is limited, but it does support some grazing animals.

Climate and Vegetation

West Africa has four different climate regions. As you can see on the map above, these climate regions stretch from east to west in bands or zones. Because of this, geographers say the region's climates are **zonal**, which means "organized by zone."

The northernmost zone of the region lies within the Sahara, the world's largest desert. Hardly any vegetation grows in the desert, and large areas of this dry climate zone have few or no people.

South of the Sahara is the semiarid **Sahel** (SAH-hel), a strip of land that divides the desert from wetter areas. It has a steppe climate. Rainfall there varies greatly from year to year. In some years it never rains. Although the Sahel is quite dry, it does have enough vegetation to support hardy grazing animals.

However, the Sahel is becoming more like the Sahara. Animals have overgrazed the land in some areas. Also, people have cut down trees for firewood. Without these plants to anchor the soil, wind blows soil away. These conditions, along with drought, are causing desertification in the Sahel. **Desertification** is the spread of desertlike conditions.

To the south of the Sahel is a savanna zone. A **savanna** is an area of tall grasses and scattered trees and shrubs. When rains fall regularly, farmers can do well in this region of West Africa.

The fourth climate zone lies along the coasts of the Atlantic and the Gulf of Guinea. This zone has a humid tropical climate. Plentiful rain supports tropical forests. However, many trees have been cut from these forests to make room for the region's growing populations.

READING CHECK **Categorizing** What are the region's four climate zones?

Savanna Grasses and scattered trees grow on the savanna. This region can be good for farming.

Tropical Forest Thick forests are found along the coasts of West Africa. The tall trees provide homes for many animals.

Resources

West Africa has a variety of resources. These resources include agricultural products, oil, and minerals.

The climate in parts of West Africa is good for agriculture. For example, Ghana is the world's leading producer of cacao, which is used to make chocolate. Coffee, coconuts, and peanuts are also among the region's main exports.

Oil, which is found off the coast of Nigeria, is the region's most valuable resource. Nigeria is a major exporter of oil. West Africa also has mineral riches, such as diamonds, gold, iron ore, and bauxite. Bauxite is the main source of aluminum.

READING CHECK **Summarizing** What are some of the region's resources?

SUMMARY AND PREVIEW West Africa is mostly covered with plains. Across these plains stretch four different climate zones, most of which are dry. In spite of the harsh climate, West Africa has some valuable resources. Next, you will learn about West Africa's history and culture.

go.hrw.com
Online Quiz
KEYWORD: SK7 HP15

Section 1 Assessment

Reviewing Ideas, Terms, and Places

1. **a. Describe** What is the inland delta on the **Niger River** like?
   b. Summarize What is the physical geography of West Africa like?
   c. Elaborate Why do you think most of West Africa's cities are located on the coastal plain?
2. **a. Recall** Why do geographers say West Africa's climates are **zonal**?
   b. Compare and Contrast What is one similarity and one difference between the **Sahel** and the **savanna**?
   c. Evaluate How do you think **desertification** affects people's lives in West Africa?
3. **a. Identify** What is the most valuable resource in West Africa?
   b. Make Inferences Where do you think most of the crops in West Africa are grown?

Critical Thinking

4. **Identifying Cause and Effect**
   Review your notes on climate. Using a graphic organizer like the one here, identify the causes and effects of desertification.

| Causes | → | Desertification | → | Effects |

FOCUS ON SPEAKING

5. **Describing the Physical Geography** The person you will describe will live or have lived in this region. How might the physical geography have affected his or her life?

History and Culture

What You Will Learn...

Main Ideas

1. In West Africa's history, trade made great kingdoms rich, but this greatness declined as Europeans began to control trade routes.
2. The culture of West Africa includes many different ethnic groups, languages, religions, and housing styles.

The Big Idea

Powerful early kingdoms, European slave trade and colonization, and traditions from a mix of ethnic groups have all influenced West African culture.

Key Terms and Places

Timbuktu, *p. 377*
animism, *p. 378*
extended family, *p. 379*

TAKING NOTES As you read, use a graphic organizer like the one below to take notes on West Africa's history and culture.

History	Culture

If YOU lived there...

When you were a small child, your family moved to Lagos, the largest city in Nigeria. You live in a city apartment now, but you still visit your aunts and uncles and cousins in your home village. There are more types of activities in the city, but you also remember that it was fun to have all your family members around.

Do you want to stay in the city or move back to your village? Why?

BUILDING BACKGROUND West African societies are changing as people like this family move to cities where they meet people whose habits and language are strange to them. West Africa has been home to many different ethnic groups throughout its history.

History

Much of what we know about West Africa's early history is based on archaeology. Archaeology is the study of the past based on what people left behind. Oral history—a spoken record of past events—offers other clues.

Merchants from North Africa crossed the Sahara to trade for West African gold and salt.

Great Kingdoms

Ancient artifacts suggest that early trading centers developed into great kingdoms in West Africa. One of the earliest kingdoms was Ghana (GAH-nuh). By controlling the Sahara trade in gold and salt, Ghana became rich and powerful by about 800.

According to legend, Ghana fell to a mighty warrior from a neighboring kingdom in about 1300. Under this leader, the empire of Mali (MAH-lee) replaced Ghana. Mali gained control of the Sahara trade routes. Mali's most famous king, Mansa Musa, used wealth from trade to support artists and scholars. However, invasions caused the decline of Mali by the 1500s.

As Mali declined, the kingdom of Songhai (SAWNG-hy) came to power. With a university, mosques, and more than 100 schools, the Songhai city of **Timbuktu** was a cultural center. By about 1600, however, invasions had weakened this kingdom.

The great West African trade cities also faded when the Sahara trade decreased. Trade decreased partly because Europeans began sailing along the west coast of Africa. They could trade for gold on the coast rather than with the North African traders who carried it through the desert.

The Slave Trade

For a while, both Europeans and Africans profited from trade with each other. However, in the 1500s the demand for labor in Europe's American colonies changed this relationship. European traders met the demand for labor by selling enslaved Africans to colonists.

The slave trade was profitable for these traders, but it devastated West Africa. Many families were broken up when members were kidnapped and enslaved. Africans often died on the voyage to the Americas. By the end of the slave trade in the 1800s, millions of Africans had been enslaved.

Colonial Era and Independence

Even with the end of the slave trade, Europeans wanted access to West Africa's resources. To ensure that access, France, Britain, Germany, and Portugal all claimed colonies in the region in the 1800s.

Some Europeans moved to West Africa to run the colonies. They built schools, roads, and railroads. However, they also created new and difficult problems for the people of West Africa. For example, many West Africans gave up farming and instead earned only low wages working in the new commercial economy.

After World War II, Africans worked for independence. Most of the colonies became independent during the 1950s and 1960s. All were independent by 1974.

FOCUS ON READING
What words in the discussion of the slave trade signal comparison or contrast?

READING CHECK **Summarizing** What impact did Europeans have on West Africa?

Culture

West African societies are very diverse. Their culture reflects three main influences —traditional African cultures, European culture, and Islam.

People and Languages

West Africa's people belong to hundreds of different ethnic groups. In fact, Nigeria alone is made up of more than 250 ethnic groups. The biggest ethnic groups there are Hausa and Fulani, Yoruba, and Igbo. Members of some ethnic groups in West Africa still live in their traditional villages. Other ethnic groups mix with each other in the region's cities.

Because of the way the European colonizers drew political boundaries, country borders sometimes separated members of the same ethnic group. Other borders grouped together peoples that did not get along. As a result, many West Africans are more loyal to their own ethnic groups than they are to their countries.

Because of the huge number of ethnic groups, hundreds of different languages are spoken in West Africa. In some areas, using the colonial languages of French, English, or Portuguese helps people from different groups communicate with each other. Also, West African languages that many people share, such as Fula and Hausa, help with communication in the region.

Religion

Like peoples and languages, many forms of religion exist in West Africa. Traditional religions of West Africa have often been forms of animism. **Animism** is the belief that bodies of water, animals, trees, and other natural objects have spirits. Animists also honor the memories of ancestors.

The two most common religions came from outside the region. They are Islam and Christianity. North African traders brought Islam to West Africa. Europeans introduced Christianity. Today most West Africans of the Sahel practice Islam. Many towns there have mosques built of mud. Christianity is the most common religion south of the Sahel.

Clothing, Families, and Homes

West Africans wear a mix of traditional and modern clothing styles. Some West Africans, particularly in the cities, wear Western-style clothing. Traditional robes, pants, blouses, and skirts are made from colorful cotton fabrics. Women often wear beautiful wrapped headdresses. Because of the warm climate, most clothing is loose.

CONNECTING TO the Arts

Masks

Masks are one of the best-known West African arts. They are traditionally carved out of wood only by skilled and respected men. The colors and shape of a mask have specific meanings. For example, the color white represents the spirit world.

Masks are used in ceremonies to call spirits or to prepare boys and girls for adulthood. Ceremony participants often wear a mask as part of a costume that completely hides the body. The wearer is believed to become what the mask represents.

Drawing Inferences Why would someone want to wear a mask?

A West African Village

These homes are in Burkina Faso. Trees are scarce in the Sahel and savanna so there is little wood for construction.

Women are responsible for painting and decorating the walls of the homes.

These homes are made of a mixture of mud, water, and cow dung.

Rural homes are small and simple. Many homes in the Sahel and savanna zones are circular. Straw or tin roofs sit atop mud, mud-brick, or straw huts. Large extended families often live close together in the same village. An **extended family** includes the father, mother, children, and close relatives in one household.

In urban areas also, members of an extended family may all live together. However, in West Africa's cities you will find modern buildings. People may live in houses or high-rise apartments.

READING CHECK **Generalizing** What are some features of West African culture?

SUMMARY AND PREVIEW Great kingdoms and European colonists once ruled West Africa. These historical influences still affect West Africa's diverse cultures. Next, you will learn about the countries of West Africa today.

Section 2 Assessment

Reviewing Ideas, Terms, and Places

1. **a. Identify** What was the significance of **Timbuktu**?
   b. Explain How did the slave trade affect West Africa?
   c. Evaluate Do you think West Africans mostly appreciated or disliked the European colonizers? Explain your answer.
2. **a. Recall** What do people who believe in **animism** think about natural objects?
   b. Analyze How did European colonizers affect tension between ethnic groups?

Critical Thinking

3. **Sequencing** Look over your notes on the history of West Africa. Then, using a diagram like the one here, put major events in chronological order.

FOCUS ON SPEAKING

4. **Describing History and Culture** What details about West Africa's history might affect the daily life of someone in the area? Many aspects of culture in the region would affect someone in West Africa. What religion would this person practice? How would he or she dress? List some ideas for your description.

Geography and History

The Atlantic Slave Trade

Between 1500 and 1870, British, French, Dutch, Portuguese, and Spanish traders sent millions of enslaved Africans to colonies in the Americas. The highest number of slaves went to British and French colonies in the West Indies. The climate in the colonies was good for growing crops like cotton, tobacco, and sugarcane. These crops required a great deal of labor to grow and process. The colonists relied on enslaved Africans to meet this demand for labor.

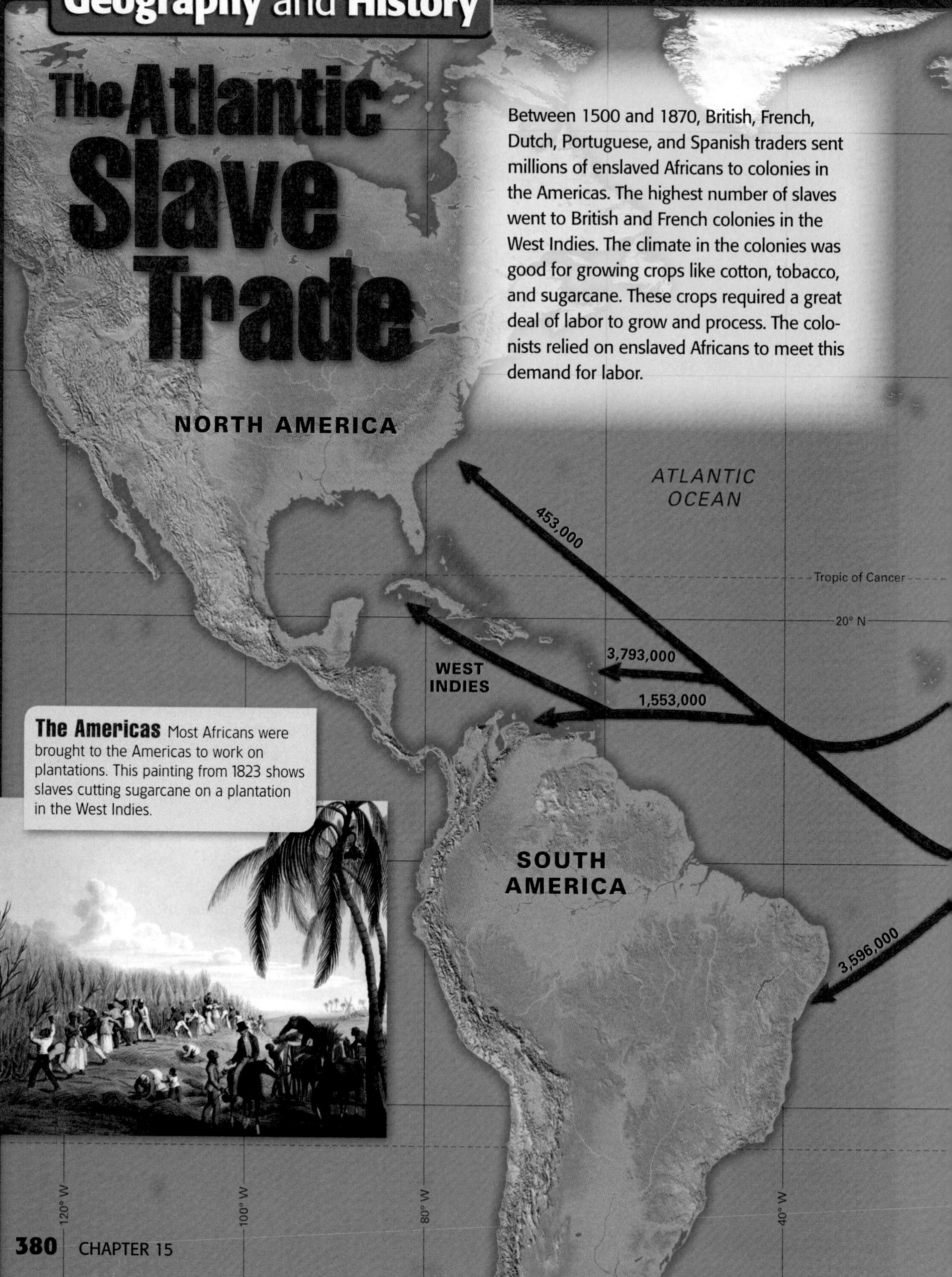

NORTH AMERICA

ATLANTIC OCEAN

453,000

Tropic of Cancer

20° N

3,793,000

WEST INDIES

1,553,000

The Americas Most Africans were brought to the Americas to work on plantations. This painting from 1823 shows slaves cutting sugarcane on a plantation in the West Indies.

SOUTH AMERICA

3,596,000

120° W
100° W
80° W
40° W

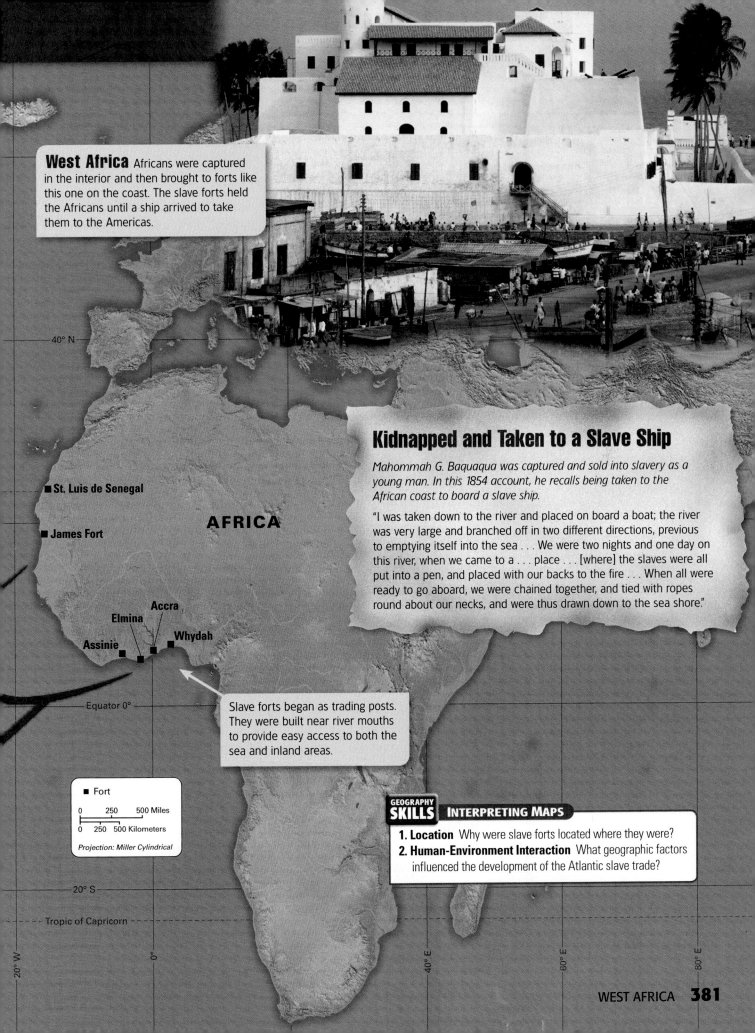

West Africa Africans were captured in the interior and then brought to forts like this one on the coast. The slave forts held the Africans until a ship arrived to take them to the Americas.

40° N

■ St. Luis de Senegal

■ James Fort

AFRICA

Accra

Elmina

Assinie

Whydah

Equator 0°

Kidnapped and Taken to a Slave Ship

Mahommah G. Baquaqua was captured and sold into slavery as a young man. In this 1854 account, he recalls being taken to the African coast to board a slave ship.

"I was taken down to the river and placed on board a boat; the river was very large and branched off in two different directions, previous to emptying itself into the sea . . . We were two nights and one day on this river, when we came to a . . . place . . . [where] the slaves were all put into a pen, and placed with our backs to the fire . . . When all were ready to go aboard, we were chained together, and tied with ropes round about our necks, and were thus drawn down to the sea shore."

Slave forts began as trading posts. They were built near river mouths to provide easy access to both the sea and inland areas.

■ Fort

0   250   500 Miles

0   250   500 Kilometers

Projection: Miller Cylindrical

GEOGRAPHY SKILLS **INTERPRETING MAPS**

1. **Location** Why were slave forts located where they were?
2. **Human-Environment Interaction** What geographic factors influenced the development of the Atlantic slave trade?

20° S

- - Tropic of Capricorn - -

20° W

0°

40° E

60° E

80° E

West Africa Today

If **YOU** lived there...

You live in the Sahel country of Niger, where your family herds cattle. You travel with your animals to find good grazing land for them. In the past few years, however, the desert has been expanding. It is getting harder and harder to find good grass and water for your cattle. You worry about the coming years.

How does this environment affect your life and your future?

BUILDING BACKGROUND The countries of West Africa are very different from one another. Some, such as Niger, have poor soils, little rain, and few resources. Others, such as Nigeria, have good natural resources. None of these countries is wealthy, however.

What You Will Learn...

Main Ideas

1. Nigeria has many different ethnic groups, an oil-based economy, and one of the world's largest cities.
2. Most coastal countries of West Africa have struggling economies and weak or unstable governments.
3. Lack of resources in the Sahel countries is a main challenge to economic development.

The Big Idea

Many countries in West Africa struggle with poor economies and political instability.

Key Terms and Places

secede, *p. 382*
Lagos, *p. 383*
famine, *p. 385*

TAKING NOTES As you read, use a graphic organizer like the one below to take notes on the governments and economies of the different regions of West Africa.

	Government	Economy
Nigeria		
Other Coastal Countries		
Sahel Countries		

Nigeria

Nigeria is the second largest country in West Africa. With more than 130 million people, it has Africa's largest population, its second largest city, and one of the strongest economies.

People and Government

Like many other former colonies, Nigeria has many different ethnic groups within its borders. Conflicts have often taken place among those ethnic groups. In the 1960s one conflict became so serious that one ethnic group, the Igbo, tried to secede from Nigeria. To **secede** means to break away from the main country. This action led to a bloody civil war, which the Igbo eventually lost.

Ethnic and regional conflicts have continued to be an issue in Nigeria. Avoiding conflict was important in choosing a site for a new capital in the 1990s. Leaders chose Abuja (ah-BOO-jah) because it was centrally located in an area of low population density. A low population density meant that there would be fewer people to cause conflicts. Nigeria's government is now a democracy after years of military rule.

Crowding in Lagos

Lagos is a busy seaport and industrial center. Overcrowding leads to problems common in big cities such as traffic jams and poor housing.

ANALYZING VISUALS
What activities are these people participating in?

Africa's Largest Cities

THE WORLD ALMANAC®
Facts about the World

go.hrw.com   KEYWORD: SK7 CH15

Economy

Nigeria has some of Africa's richest natural resources. Major oil fields, the country's most important resource, are located in the Niger River delta and just off the coast. Oil accounts for about 95 percent of the country's export earnings. Income from oil exports has allowed Nigeria to build good roads and railroads for transporting oil. The oil industry is centered around **Lagos** (LAY-gahs). Also the former capital, Lagos is the most populous city in West Africa.

Although Nigeria is rich in resources, many Nigerians are poor. One cause of the poverty there is a high birthrate. Nigeria cannot produce enough food for its growing population. Another cause of Nigeria's poverty is a history of bad government. Corrupt government officials have used their positions to enrich themselves.

READING CHECK **Drawing Inferences** What are some obstacles to progress in Nigeria?

Other Coastal Countries

Several West African countries lie along the Atlantic Ocean and the Gulf of Guinea. Many of these countries have struggling economies and unstable governments.

Senegal and Gambia

Senegal wraps around Gambia. The odd border was created by French and British diplomats during the colonial era. Senegal is larger and richer than Gambia, but the two countries do have many similarities. For example, peanuts are their major crops. Also, tourism is becoming more important in both countries.

FOCUS ON READING
How are Senegal and Gambia similar?

Many people in Senegal and Gambia speak a language called Wolof (WOH-lawf). Griots (GREE-ohz), or storytellers, are important to the Wolof speakers there and to other West Africans.

Guinea, Guinea-Bissau, and Cape Verde

Guinea and its small neighbor, Guinea-Bissau (GI-nee bi-SOW), are poor countries. Guinea's main natural resource is bauxite, which is used to make aluminum. Guinea-Bissau has undeveloped mineral resources.

Cape Verde (VUHRD) is a group of volcanic islands in the Atlantic. It is West Africa's only island country. Once a Portuguese colony, Cape Verde now has one of the most stable democratic governments in Africa. Services such as tourism form the main part of the country's economy.

Liberia and Sierra Leone

Liberia is Africa's oldest republic. Americans founded it in the 1820s as a home for freed slaves. The freed slaves who settled in Liberia and their descendants lived in towns on the coast. They often clashed with Africans already living there. Those Africans were usually poorer and lived in rural areas. In the 1980s these conflicts led to a civil war, which ended in 2003.

Sierra Leone (lee-OHN) also experienced violent civil war, from 1991 to 2002. The fighting wrecked the country's economy, killed thousands of people, and forced millions from their homes.

Now, both Liberia and Sierra Leone are trying to rebuild. They do have natural resources on which to build stronger economies. Liberia exports rubber and iron ore while Sierra Leone exports diamonds.

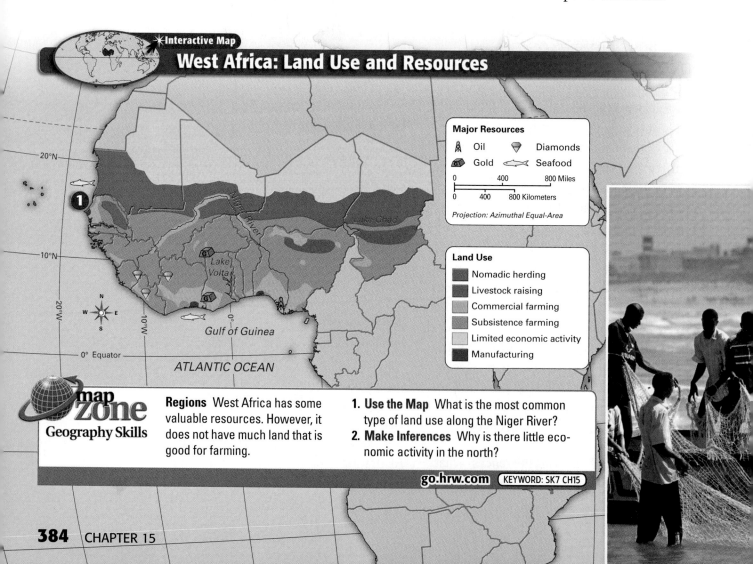

Interactive Map

West Africa: Land Use and Resources

Major Resources

Oil	Diamonds
Gold	Seafood

0   400   800 Miles
0   400   800 Kilometers

Projection: Azimuthal Equal-Area

Land Use

- Nomadic herding
- Livestock raising
- Commercial farming
- Subsistence farming
- Limited economic activity
- Manufacturing

20°N

10°N

20°W

10°W

0° Equator

Niger River

Lake Chad

Lake Volta

Gulf of Guinea

ATLANTIC OCEAN

map zone Geography Skills

Regions West Africa has some valuable resources. However, it does not have much land that is good for farming.

1. **Use the Map** What is the most common type of land use along the Niger River?
2. **Make Inferences** Why is there little economic activity in the north?

go.hrw.com KEYWORD: SK7 CH15

Ghana and Côte d'Ivoire

Ghana is named for an ancient kingdom. Côte d'Ivoire (koht-dee-VWAHR) is a former French colony whose name means "Ivory Coast" in English. Côte d'Ivoire boasts Africa's largest Christian church building.

These two countries have rich natural resources. Gold, timber, and cacao (kuh-KOW) are major products of Ghana. Côte d'Ivoire is a world leader in export of cacao and coffee. However, civil war there has hurt the economy.

Togo and Benin

Unstable governments have troubled Togo and Benin (buh-NEEN) since independence. These two countries have experienced periods of military rule. Their fragile economies have contributed to their unstable and sometimes violent politics.

Both Togo and Benin are poor. The people depend on farming and herding for income. Palm products, cacao, and coffee are the main crops in both countries.

READING CHECK **Generalizing** What are the economies of the coastal countries like?

1 Fishing is an important industry in the coastal countries of West Africa.

Sahel Countries

The Sahel region of West Africa includes some of the poorest and least developed countries in the world. Drought and the expanding desert make feeding the people in these countries difficult.

Mauritania, Niger, and Chad

Most Mauritanians were once nomadic herders. Today the expanding Sahara has driven more than half of the nomads into cities. People in these cities, as well as the rest of the country, are very poor. Only in the far south, near the Senegal River, can people farm. Near the Atlantic Ocean, people fish for a living. Corrupt governments and ethnic tensions between blacks and Arabs add to Mauritania's troubles.

In Niger, only about 3 percent of the land is good for farming. The country's only farmland lies along the Niger River and near the Nigerian border. Farmers there grow staple, or main, food crops, such as millet and sorghum. These grains are cooked like oatmeal.

In the early 2000s, locusts and drought destroyed Niger's crops. The loss of crops caused widespread **famine**, or an extreme shortage of food. International groups provided some aid, but it was impossible to **distribute** food to all who needed it.

Chad has more land for farming than Mauritania or Niger, and conditions there are somewhat better than in the other two countries. In addition to farming, Lake Chad once had a healthy fishing industry and supplied water to several countries. However, drought has evaporated much of the lake's water in the past several years.

The future may hold more promise for Chad. A long civil war finally ended in the 1990s. Also, oil was recently discovered there, and Chad began to export this valuable resource in 2004.

ACADEMIC VOCABULARY

distribute to divide among a group of people

Water well

Dry Croplands

This aerial view shows how, with just a little water, people can farm in the dry Sahel.

Mali and Burkina Faso

The Sahara covers about 40 percent of the land in Mali. The scarce amount of land available for farming makes Mali among the world's poorest countries. The available farmland lies in the southwest, along the Niger River. Most people in Mali fish or farm in this small area along the river. Cotton and gold are Mali's main exports.

Mali's economy does have some bright spots, however. A fairly stable democratic government has begun economic reforms. Also, the ancient cities of Timbuktu and Gao (GOW) continue to attract tourists.

Burkina Faso is also a poor country. It has thin soil and few mineral resources. Few trees remain in or near the capital, Ouagadougou (wah-gah-DOO-goo), because they have been cut for firewood and building material. Jobs in the city are also scarce. To support their families many men try to find work in other countries. Thus, when unrest disrupts work opportunities in other countries, Burkina Faso's economy suffers.

READING CHECK **Summarizing** What are the challenges facing Chad and Burkina Faso?

SUMMARY AND PREVIEW Countries in West Africa struggle with poor economies. In addition, many have faced political instability since independence. In the next chapter, you will learn about how the countries in East Africa face some similar issues.

Section 3 Assessment

Reviewing Ideas, Terms, and Places

1. **a. Recall** Why did the Igbo try to **secede**?
   b. Evaluate What do you think were some benefits and drawbacks to Nigeria's leaders moving the capital from **Lagos** to Abuja?
2. **a. Identify** What is West Africa's only island country?
   b. Compare What are some similarities between Togo and Benin?
   c. Elaborate Why do you think countries with poor economies often have unstable governments?
3. **a. Describe** What caused **famine** in Niger?
   b. Evaluate What do you think is the biggest problem facing the Sahel countries? Explain.

Critical Thinking

4. **Compare and Contrast** Review your notes on the coastal countries and the Sahel countries. Then use a diagram like the one here to compare and contrast the two regions.

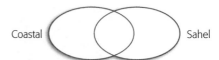

Coastal   Sahel

FOCUS ON SPEAKING

5. **Describing Countries of West Africa** Think about the countries of West Africa. Which one might be a good location for the person you are going to describe? Take some notes about that place.

from
AKÉ: The Years of Childhood

by Wole Soyinka

Merchants from North Africa sometimes traded brass objects in West Africa.

About the Reading *In this excerpt from* Aké: The Years of Childhood, *Nigerian-born Wole Soyinka describes some traders who came to his childhood home in Aké. As a young boy, he was fascinated with the appearance of the exotic goods.*

AS YOU READ Notice the variety of goods the traders brought to the author's house.

It was a strange procedure, one which made little sense to me. ❶ They spread their wares in front of the house and I had to be prised off them. There were brass figures, horses, camels, trays, bowls, ornaments. Human figures spun on a podium, balanced by weights at the end of curved light metal rods. We spun them round and round, yet they never fell off their narrow perch. The smell of fresh leather filled the house as pouffs, handbags, slippers and worked scabbards were unpacked. There were bottles encased in leather, with leather stoppers, . . . scrolls, glass beads, bottles of scent with exotic names—I never forgot, from the first moment I read it on the label—Bint el Sudan, with its picture of a turbanned warrior by a kneeling camel. A veiled maiden offered him a bowl of fruits. They looked unlike anything in the orchard and Essay said they were called dates. ❷ I did not believe him; dates were figures which appeared on a calendar on the wall, so I took it as one of his jokes.

GUIDED READING

WORD HELP

wares goods
prised taken by force
pouff fluffy clothing or accessory
scabbard a case to hold a knife
turbanned wearing a turban, or wrapped cloth, on the head

❶ The author is describing a visit by the Hausa traders who came from northwestern Africa.

❷ Essay is the author's father.

Connecting Literature to Geography

1. **Drawing Inferences** The author describes many unusual things. What descriptions or comments lead you to believe that the trader traveled to Aké from far away?

2. **Analyzing** Think about the way the author described the goods. What senses did the author use as a child to discover the goods the traders brought?

Analyzing a Precipitation Map

Learn

A precipitation map shows how much rain or snow typically falls in a certain area over a year. Studying a precipitation map can help you understand a region's climate.

To read a precipitation map, first look at the legend to see what the different colors mean. Compare the legend to the map to see how much precipitation different areas get.

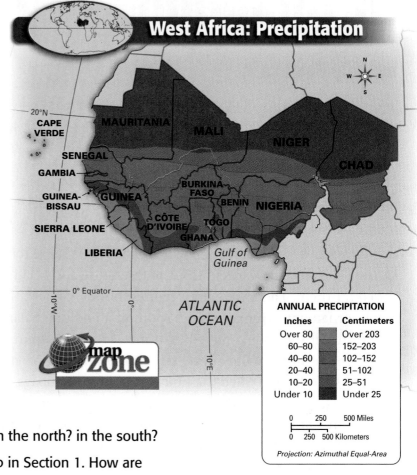

Practice

Use the map on this page to answer the following questions.

1 What countries have areas that get over 80 inches of rain every year?

2 In what part of the region does the least amount of rain fall?

3 What do you think vegetation is like in the north? in the south?

4 Compare this map to the climate map in Section 1. How are the two maps similar?

Apply

Using an atlas or the Internet, find a precipitation map of the United States. Use that map to answer the following questions.

1. What area of the country gets the most precipitation?

2. What area of the country gets the least precipitation?

3. How much annual precipitation does Hawaii get?

Geography's Impact
video series
Review the video to answer the closing question:
What are some of the ways desertification can be slowed, stopped, or even reversed?

Visual Summary

Use the visual summary below to help you review the main ideas of the chapter.

QUICK FACTS

West Africa has four distinct climate zones. In the dry Sahel, people try to farm or herd cattle.

Masks are just one example of traditional African culture that can be seen in West Africa today.

Most West African countries are poor. Fishing is important to the economy along rivers and in coastal areas.

Reviewing Vocabulary, Terms, and Places

For each statement below, write T if it is true and F if it is false. If the statement is false, write the correct term that would make the sentence a true statement.

1. West Africa's climate is described as savanna because it is organized by zone.

2. Animism, a belief that natural objects have spirits, is a traditional religion in West Africa.

3. An extended family is one that includes a mother, father, children, and close relatives in one household.

4. International aid agencies have tried to distribute food in Niger.

5. Timbuktu is the largest city in Nigeria.

6. The Niger River flows through many countries in West Africa and empties into the Gulf of Guinea.

7. The spread of desertlike conditions is famine.

8. Some animals can graze in the Sahel.

Comprehension and Critical Thinking

SECTION 1 *(Pages 372–375)*

9. **a. Identify** What are the four climate zones of West Africa?

   b. Make Inferences What are some problems caused by desertification?

   c. Elaborate West Africa has valuable resources such as gold and diamonds. Why do you think these resources have not made West Africa a rich region?

SECTION 2 *(Pages 376–379)*

10. **a. Recall** What religion do most people in the Sahel practice?

    b. Analyze What role did trade play in the early West African kingdoms and later in West Africa's history?

    c. Elaborate What might be some advantages of living with an extended family?

11. a. Identify Which country in West Africa has an economy based nearly entirely on oil?

b. Compare and Contrast What is one similarity and one difference between the cause of civil war in Nigeria and its cause in Liberia?

c. Predict How might the recent discovery of oil in Chad affect that country in the future?

Using the Internet

12. Activity: Creating a Postcard Come and learn about the mighty baobab tree. This unique tree looks as if it has been plucked from the ground and turned upside down. These trees are known not only for their unique look but also for their great size. Some are so big that a chain of 30 people is needed to surround one tree trunk! Enter the activity keyword to visit Web sites about baobab trees in West Africa. Then create a postcard about this strange wonder of nature.

FOCUS ON READING AND SPEAKING

Understanding Comparison-Contrast *Look over your notes or re-read Section 1. Use the information on climate and vegetation to answer the following questions.*

13. How are the Sahara and the Sahel similar?

14. How are the Sahara and the Sahel different?

15. Compare the Sahel and the savanna zone. How are they similar?

16. Contrast the savanna region and the humid tropical region along the coast. How are these areas different?

17. Giving an Oral Description Read over your notes. Then prepare a brief oral presentation about a day in the life of someone from West Africa. Tell about the land, climate, and vegetation. Describe the culture, including family life. Tell what this person does for a living. Practice your presentation several times before you give it so you can make frequent eye contact with your audience. During your presentation, remember to speak loudly and clearly. Use good descriptive language to interest your audience in your topic.

Social Studies Skills

Analyzing a Precipitation Map *Use the precipitation map in the Social Studies Skills lesson to answer the following questions.*

18. What countries have areas that receive under 10 inches of rain every year?

19. Where in West Africa does the most rain typically fall?

20. How would you describe annual precipitation in Chad?

21. How would you describe annual precipitation in Benin?

Map Activity

22. West Africa On a separate sheet of paper, match the letters on the map with their correct labels.

Niger River   Senegal River

Lagos, Nigeria   Mali

Gulf of Guinea

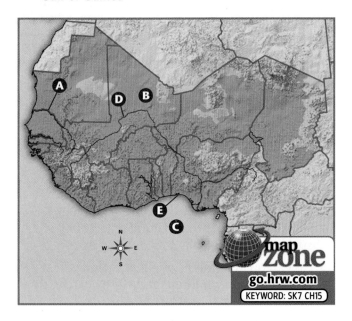

DIRECTIONS: Read questions 1 through 6 and write the letter of the best response. Then read question 7 and write your own well-constructed response.

1 The climate zone just south of the Sahara is called the

   A desert.

   B savanna.

   C Sahel.

   D tropical forest.

2 Which West African country was named for an ancient kingdom in the region?

   A Liberia

   B Nigeria

   C Chad

   D Ghana

3 Which of the following statements about the slave trade is false?

   A European slave traders built schools and railroads in West Africa.

   B European slave traders profited from it.

   C It broke up families in West Africa.

   D Enslaved Africans were sent to the Americas to meet the increased demand for labor there.

4 Which country in West Africa has an economy based on oil?

   A Niger

   B Nigeria

   C Mauritania

   D Mali

5 Which country has one of the most stable democratic governments in Africa?

   A Nigeria

   B Liberia

   C Cape Verde

   D Sierra Leone

West Africa: Population

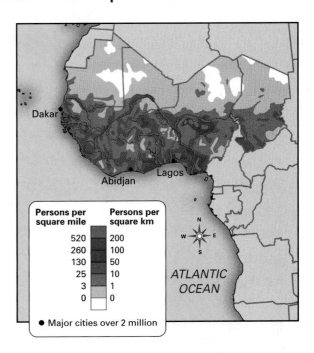

6 Based on the map above, which of the following sentences is true?

   A West Africa has only one city with a population over 2 million.

   B West Africa's highest population density is in the Sahel countries.

   C Most of the region's population is in the south.

   D The region around Dakar has a population density of over 520 people per square mile.

7 Extended Response Compare the map above to the climate map in Section 1. Then write a brief essay explaining factors that affect human settlement in West Africa. One paragraph should explain how the two maps are related. Another paragraph should describe physical factors that influence settlement.

CHAPTER 16

East Africa

○ National capital
● Other cities

0   150   300 Miles
0 150 300 Kilometers

Projection: Lambert Equal-Area

20°N

10°N

0° Equator

10°S

ATLANTIC OCEAN

What You Will Learn...

In this chapter you will learn about the physical geography of East Africa. You will also learn about the region's rich history and culture. Finally, you will study the countries of East Africa today.

SECTION 1
Physical Geography **394**

SECTION 2
History and Culture....................**398**

SECTION 3
East Africa Today......................**402**

FOCUS ON READING AND WRITING

Identifying Supporting Details Supporting details are the facts and examples that provide information to support the main ideas of a chapter, section, or paragraph. At the beginning of each section in this book, there is a list of main ideas. As you read this chapter, look for the details that support each section's main ideas. **See the lesson, Identifying Supporting Details, on page 683.**

Writing a Letter Home Imagine that you are spending your summer vacation visiting the countries of East Africa. You want to write a letter home to a friend in the United States describing the land and its people. As you read this chapter, you will gather information that you can include in your letter.

History This gold artifact is from the early Nubian civilization of northern Sudan.

East Africa: Political

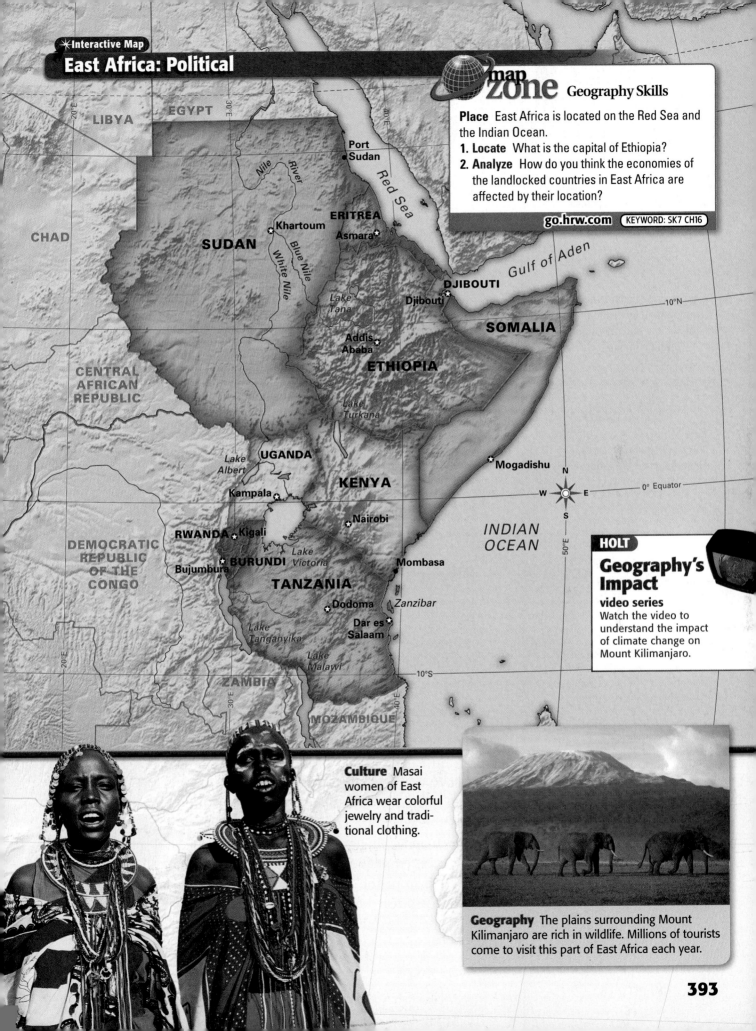

go.hrw.com KEYWORD: SK7 CH16

map zone Geography Skills

Place East Africa is located on the Red Sea and the Indian Ocean.
1. **Locate** What is the capital of Ethiopia?
2. **Analyze** How do you think the economies of the landlocked countries in East Africa are affected by their location?

LIBYA

EGYPT

30°E

20°E

40°E

Port Sudan

Red Sea

Nile River

CHAD

Khartoum

SUDAN

White Nile

Blue Nile

ERITREA

Asmara

Gulf of Aden

10°N

DJIBOUTI

Djibouti

CENTRAL AFRICAN REPUBLIC

Lake Tana

SOMALIA

Addis Ababa

ETHIOPIA

Lake Turkana

Lake Albert

UGANDA

Mogadishu

Kampala

KENYA

N

Nairobi

0° Equator

W   E

RWANDA   Kigali

S

DEMOCRATIC REPUBLIC OF THE CONGO

Lake Victoria

BURUNDI

Bujumbura

Mombasa

INDIAN OCEAN

50°E

HOLT

Geography's Impact

video series
Watch the video to understand the impact of climate change on Mount Kilimanjaro.

TANZANIA

Dodoma

Zanzibar

Lake Tanganyika

Dar es Salaam

Lake Malawi

20°E

30°E

40°E

10°S

ZAMBIA

MOZAMBIQUE

Culture Masai women of East Africa wear colorful jewelry and traditional clothing.

Geography The plains surrounding Mount Kilimanjaro are rich in wildlife. Millions of tourists come to visit this part of East Africa each year.

Physical Geography

If YOU lived there...

You and your friends are planning to hike up Mount Kilimanjaro, near the equator in Tanzania. It is hot in your camp at the base of the mountain. You're wearing shorts and a T-shirt, but your guide tells you to pack a fleece jacket and jeans. You start your climb, and soon you understand this advice. The air is much colder, and there's snow on the nearby peaks.

Why is it cold at the top of the mountain?

BUILDING BACKGROUND The landscapes of East Africa have been shaped by powerful forces. The movement of tectonic plates has stretched the Earth's surface here, creating steep-sided valleys and huge lakes.

Physical Features

East Africa is a region of spectacular landscapes and wildlife. Vast plains and plateaus stretch throughout the region. In the north lie huge deserts and dry grasslands. In the southwest, large lakes dot the plateaus. In the east, sandy beaches and colorful coral reefs run along the coast.

The Rift Valleys

Look at the map on the next page. As you can see, East Africa's rift valleys cut from north to south across the region. **Rift valleys** are places on Earth's surface where the crust stretches until it breaks. Rift valleys form when Earth's tectonic plates move away from each other. This movement causes the land to arch and split along the rift valleys. As the land splits open, volcanoes erupt and deposit layers of rock in the region.

Seen from the air, the **Great Rift Valley** looks like a giant scar. The Great Rift Valley is the largest rift on Earth and is made up of two rifts—the eastern rift and the western rift.

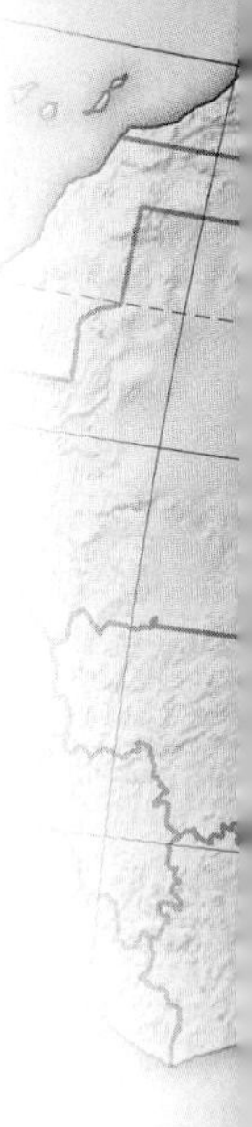

What You Will Learn...

Main Ideas

1. East Africa's physical features range from rift valleys to plains.
2. East Africa's climate is influenced by its location and elevation, and the region's vegetation includes savannas and forests.

The Big Idea

East Africa is a region of diverse physical features, climates, and vegetation.

Key Terms and Places

rift valleys, *p. 394*
Great Rift Valley, *p. 394*
Mount Kilimanjaro, *p. 395*
Serengeti Plain, *p. 395*
Lake Victoria, *p. 396*
droughts, *p. 397*

TAKING NOTES As you read, use the chart below to take notes on East Africa's physical features and climate and vegetation.

Physical Features	
Climate and Vegetation	

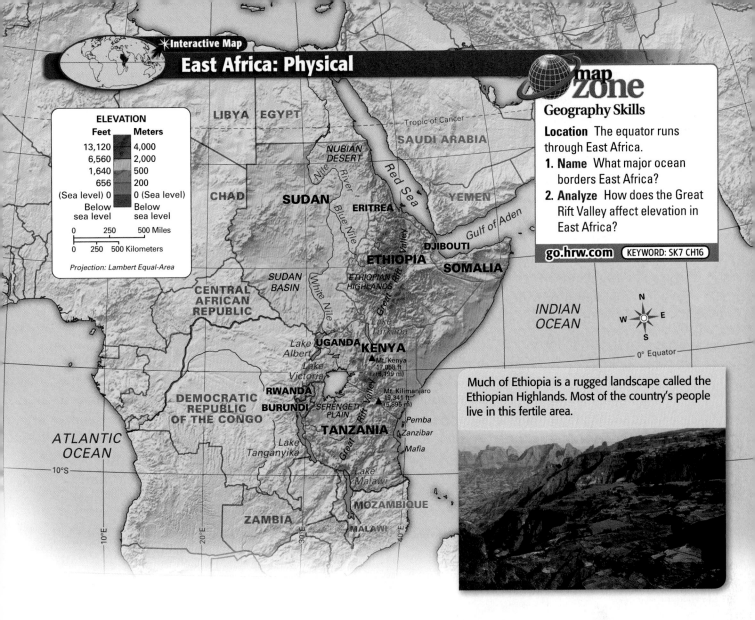

ELEVATION

Feet	Meters
13,120	4,000
6,560	2,000
1,640	500
656	200
(Sea level) 0	0 (Sea level)
Below sea level	Below sea level

0 250 500 Miles
0 250 500 Kilometers

Projection: Lambert Equal-Area

LIBYA EGYPT
Tropic of Cancer
SAUDI ARABIA
NUBIAN DESERT
CHAD SUDAN ERITREA YEMEN
Gulf of Aden
DJIBOUTI
ETHIOPIA SOMALIA
ETHIOPIAN HIGHLANDS
SUDAN BASIN
CENTRAL AFRICAN REPUBLIC
Lake Turkana
White Nile
Lake Albert UGANDA KENYA
Lake Victoria ▲Mt. Kenya 17,058 ft (5,199 m)
DEMOCRATIC REPUBLIC OF THE CONGO
RWANDA ▲Mt. Kilimanjaro 19,341 ft (5,895 m)
BURUNDI
SERENGETI PLAIN *Pemba*
TANZANIA *Zanzibar*
Lake Tanganyika *Mafia*
ATLANTIC OCEAN
Lake Malawi
ZAMBIA MOZAMBIQUE
MALAWI
INDIAN OCEAN
0° Equator
Nile River *Blue Nile* *Red Sea* *Great Rift Valley*

map zone
Geography Skills

Location The equator runs through East Africa.
1. **Name** What major ocean borders East Africa?
2. **Analyze** How does the Great Rift Valley affect elevation in East Africa?

go.hrw.com KEYWORD: SK7 CH16

Much of Ethiopia is a rugged landscape called the Ethiopian Highlands. Most of the country's people live in this fertile area.

The rift walls are usually a series of steep cliffs. These cliffs rise as much as 6,000 feet (2,000 m).

Mountains and Highlands

The landscape of East Africa has many high volcanic mountains. The highest mountain in Africa, **Mount Kilimanjaro** (ki-luh-muhn-JAHR-oh), rises to 19,340 feet (5,895 m). Despite Kilimanjaro's location near the equator, the mountain's peak has long been covered in snow. This much colder climate is caused by Kilimanjaro's high elevation.

Other areas of high elevation in East Africa include the Ethiopian Highlands.

These highlands, which lie mostly in Ethiopia, are very rugged. Deep river valleys cut through this landscape.

Plains

Even though much of East Africa lies at high elevations, some areas are flat. For example, plains stretch as far as the eye can see along the eastern rift in Tanzania and Kenya. Tanzania's **Serengeti Plain** is one of the largest plains. It is here that an abundance of wildlife thrives. The plain's grasses, trees, and water provide nutrition for wildlife that includes elephants, giraffes, lions, and zebras. To protect this wildlife, Tanzania established a national park.

Rivers and Lakes

FOCUS ON READING

What details in this paragraph support this section's second main idea?

East Africa also has a number of rivers and large lakes. The world's longest river, the Nile, begins in East Africa and flows north to the Mediterranean Sea. The Nile is formed by the meeting of the Blue Nile and the White Nile at Khartoum, Sudan. The White Nile is formed by the water that flows into Africa's largest lake, **Lake Victoria**. The Blue Nile is formed from waters that run down from Ethiopia's highlands. As the Nile meanders through Sudan, it provides a narrow, fertile lifeline to farmers in the desert.

The region has a number of great lakes in addition to Lake Victoria. One group of lakes forms a chain in the western rift valleys. There are also lakes along the drier eastern rift valleys. Near the eastern rift, heat from the Earth's interior makes some lakes so hot that no human can swim in them. In addition, some lakes are extremely salty. However, some of these rift lakes provide algae for the region's flamingos.

READING CHECK **Evaluating** What river is the most important in this region? Why?

Climate and Vegetation

When you think of Africa, do you think of it as being a hot or cold place? Most people usually think all of Africa is hot. However, they are mistaken. Some areas of East Africa have a cool climate.

East Africa's location on the equator and differences in elevation influence the climates and types of vegetation in East Africa. For example, areas near the equator receive the greatest amount of rainfall. Areas farther from the equator are much drier and seasonal droughts are common. **Droughts** are periods when little rain falls, and crops are damaged. During a drought, crops and the grasses for cattle die and people begin to starve. Several times in recent decades droughts have affected the people of East Africa.

Further south of the equator the climate changes to tropical savanna. Tall grasses and scattered trees make up the savanna landscape. Here the greatest climate changes occur along the sides of the rift valleys. The rift floors are dry with grasslands and thorn shrubs.

North of the equator, areas of plateaus and mountains have a highland climate and dense forests. Temperatures in the highlands are much cooler than temperatures on the savanna. The highlands experience heavy rainfall because of its high elevation, but the valleys are drier. This mild climate makes farming possible. As a result, most of the region's population lives in the highlands.

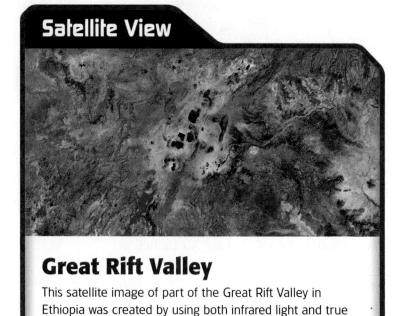

Satellite View

Great Rift Valley

This satellite image of part of the Great Rift Valley in Ethiopia was created by using both infrared light and true color. The bright blue dots are some of the smaller lakes that were created by the rifts. Once active volcanoes, some of these lakes are very deep. Vegetation appears as areas of green. Bare, rocky land appears pink and gray.

Analyzing How were the lakes in the Great Rift Valley created?

Ancient volcanoes surround Uganda's Lake Mutanda. Here villagers rely on the lake's plentiful supply of fish.

Areas east of the highlands and on the Indian Ocean coast are at a much lower elevation. These areas have desert and steppe climates. Vegetation is limited to shrubs and hardy grasses that are adapted to water shortages.

READING CHECK **Categorizing** What are some of East Africa's climate types?

SUMMARY AND PREVIEW In this section you learned about East Africa's rift valleys, mountains, highlands, plains, rivers, and lakes. You also learned that the region's location and elevation affect its climate and vegetation. In the next section you will learn about East Africa's history and culture.

go.hrw.com
Online Quiz
KEYWORD: SK7 HP16

Section 1 Assessment

Reviewing Ideas, Terms, and Places

1. **a. Define** What are **rift valleys**?
   b. Explain Why is there snow on **Mount Kilimanjaro**?
   c. Elaborate What are some unusual characteristics of the lakes in the **Great Rift Valley**?
2. **a. Recall** What is the climate of the highlands in East Africa like?
   b. Draw Conclusions What are some effects of **drought** in the region?
   c. Develop How are the climates of some areas of East Africa affected by elevation?

Critical Thinking

3. **Categorizing** Using your notes and this chart, place details about East Africa's physical features into different categories.

Physical Features			
Rift Valleys	Mountains and Highlands	Plains	Rivers and Lakes

FOCUS ON WRITING

4. **Describing the Physical Geography** Note the physical features of East Africa that you can describe in your letter. How do these features compare to the features where you live?

History and Culture

What You Will Learn...

Main Ideas

1. The history of East Africa is one of religion, trade, and European influence.
2. East Africans speak many different languages and practice several different religions.

The Big Idea

East Africa is a region with a rich history and diverse cultures.

Key Terms and Places

Nubia, *p. 398*
Zanzibar, *p. 399*
imperialism, *p. 399*

 TAKING NOTES As you read, use this diagram to take notes on East Africa's history and culture.

History	Culture

If YOU lived there...

You live on the island of Zanzibar, part of the country of Tanzania. Your hometown has beautiful beaches, historic palaces, and sites associated with the East African slave trade. Although you and your friends learn English in school, you speak the African language of Swahili to each other.

How has your country's history affected your life today?

BUILDING BACKGROUND For almost a century, nearly all the countries of East Africa were controlled by European countries. Before that, however, the region's people had close trade ties with Arabs from Southwest Asia. This Arab influence blended with native African cultures to form a new culture and language.

History

Early civilizations in East Africa were highly developed. Later, Christianity and Islam influenced the lives of many East Africans. Other influences included trade, the arrival of Europeans, ethnic conflict, and independence.

Christianity and Islam

Christian missionaries from Egypt first introduced Christianity to Ethiopia as early as the AD 300s. About 200 years later Christianity spread into **Nubia**, an area of Egypt and Sudan today.

In the early 1200s, a powerful Christian emperor named Lalibela ruled Ethiopia. Lalibela is best known for the 11 rock churches he built during his reign. He claimed that God told him to carve the churches out of the rocky ground. Today, the town where the churches are located is called Lalibela.

By about AD 700, Islam was a major religion in Egypt and other parts of North Africa. Gradually, Muslim Arabs from Egypt

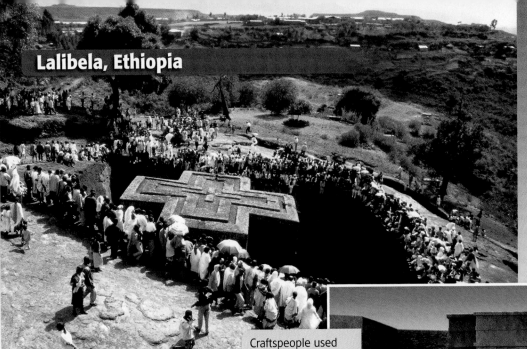

Lalibela, Ethiopia

In the 1200s, highly skilled Ethiopian architects and craftspeople built this Christian church at Lalibela.

ANALYZING VISUALS What Christian symbol does the church resemble?

Workers dug deep trenches to carve out the church.

Craftspeople used special tools to carve windows and doors out of solid rock.

spread into northern Sudan and brought their Islamic faith with them. At the same time, Islam spread to the Indian Ocean coast of what is now Somalia. City-states such as Mogadishu and Mombasa became major Islamic centers and controlled trade on the coast.

The Slave Trade

The slave trade along the Indian Ocean coast dates back more than 1,000 years. East Africans, Arabs, and Europeans all participated in the slave trade in East Africa. They kidnapped Africans, enslaved them, and shipped them to ports throughout Africa and Southwest Asia. Most of these slaves went to Islamic countries. By the early 1500s the Portuguese had begun setting up forts and settlements on the East African coast to support the slave trade.

In the late 1700s the East African island of **Zanzibar** became an international slave-trading center. Later, large plantations with slave labor were set up by Europeans in the interior. They grew crops of cloves and sugarcane.

European Influence and Conflict

Most European nations ended slavery in the early 1800s. They focused instead on trading products such as gold, ivory, and rubber. To get these goods, Europeans believed they needed to dominate regions of Africa rich in these natural resources. Europeans also wanted to expand their empires by establishing more colonies. The British were the most aggressive, and they gained control over much of East Africa.

In the 1880s Britain and other European powers divided up most of Africa. They drew boundaries that separated some ethnic groups. To maintain power over their colonies, Europeans used **imperialism**, a practice that tries to dominate other countries' government, trade, and culture.

Within East Africa, just Kenya was settled by large numbers of Europeans. Under imperialism, colonial rulers usually controlled their countries through African deputies. Many of the deputies were traditional chiefs. These chiefs were loyal to their own peoples, which tended to strengthen ethnic rivalries. Today governments are trying to influence feelings of national identity, but ethnic conflict is still strong in many countries.

In the early 1960s, most East African countries gained independence from European colonizers. Ethiopia was never colonized. Its mountains provided natural protection, and its peoples resisted European colonization.

Independence, however, did not solve all of the problems of the former colonies. In addition, new challenges faced the newly independent countries. For example, some countries experienced ethnic conflicts.

READING CHECK **Evaluating** Why was Ethiopia never a European colony?

FOCUS ON READING

In the second paragraph under Languages, find at least two details to support the main idea in the first sentence.

Culture

Over thousands of years of human settlement, East Africa developed a great diversity of people and ways of life. As a result, East Africans speak many different languages and practice several religions.

Language

East Africa's history of European imperialism influenced language in many countries in the region. For example, French is an official language in Rwanda, Burundi, and Djibouti today. English is the primary language of millions of people in Uganda, Kenya, and Tanzania.

In addition to European languages, many East Africans also speak African languages. Swahili is the most widely spoken African language in the region. As East Africans traded with Arabic speakers from Southwest Asia, Swahili developed. In fact, Swahili comes from the Arabic word meaning "on the coast." Today about 80 million people speak Swahili. Ethiopians speak Amharic, and Somalians speak Somali.

FOCUS ON CULTURE

The Swahili

For more than 1,000 years, a culture unlike any other has thrived along the coast of modern-day Kenya and Tanzania. In the AD 700s, trade contacts between East Africans and Arab traders began. Over time these interactions led to the creation of a unique language and culture known as Swahili.

The Swahili adopted some cultural traits from Arab traders. For example, many East Africans converted to the religion of the Arab traders—Islam. African languages blended with Arabic to form the Swahili language.

Generalizing What effects did Arab traders have on Swahili culture?

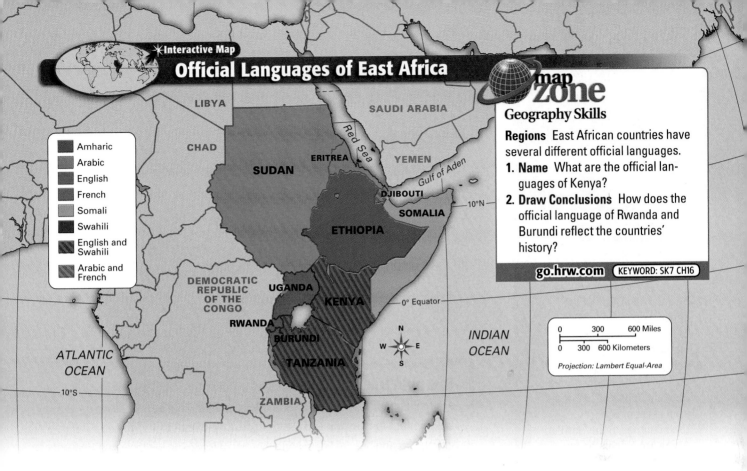

Official Languages of East Africa

map zone

Geography Skills

Regions East African countries have several different official languages.
1. **Name** What are the official languages of Kenya?
2. **Draw Conclusions** How does the official language of Rwanda and Burundi reflect the countries' history?

go.hrw.com KEYWORD: SK7 CH16

Legend:
- Amharic
- Arabic
- English
- French
- Somali
- Swahili
- English and Swahili
- Arabic and French

Map labels: LIBYA, CHAD, SUDAN, ERITREA, SAUDI ARABIA, YEMEN, Red Sea, Gulf of Aden, DJIBOUTI, SOMALIA, ETHIOPIA, DEMOCRATIC REPUBLIC OF THE CONGO, UGANDA, KENYA, RWANDA, BURUNDI, TANZANIA, ZAMBIA, ATLANTIC OCEAN, INDIAN OCEAN, Equator, 10°N, 0° Equator, 10°S

0 300 600 Miles
0 300 600 Kilometers
Projection: Lambert Equal-Area

Religion

Religion is an important aspect of culture for many East Africans. While the religions in East Africa vary greatly, most of them place emphasis on honoring ancestors.

Many East Africans are followers of animist religions. Animists believe the natural world contains spirits. Some people also combine animist worship with religions such as Christianity. Most Christians in East Africa live in Ethiopia. Islam is also practiced in the region. Sudan and Somalia are predominantly Muslim.

READING CHECK **Analyzing** Why might people in East Africa speak a European language?

SUMMARY AND PREVIEW In this section you learned about the history and culture of East Africa. Next, you will learn about the countries of East Africa today.

go.hrw.com
Online Quiz
KEYWORD: SK7 HP16

Section 2 Assessment

Reviewing Ideas, Terms, and Places

1. **a. Define** What is **imperialism**?
   b. Explain What did the emperor Lalibela have architects build in Ethiopia?
   c. Evaluate Why do you think Europeans wanted colonies in East Africa?
2. **a. Identify** What language do most people in Uganda speak?
   b. Explain What do animists believe?
   c. Elaborate How have languages and religion influenced culture in East Africa?

Critical Thinking

3. **Summarizing** Using your notes and this chart, write two sentences that summarize what you learned about the region's languages and religions.

Language	
Religion	

FOCUS ON WRITING

4. **Writing about History and Culture** What aspects of the history and culture of East Africa might interest your friend? Note ideas for your letter.

East Africa Today

If **YOU** lived there...

You are a safari guide in one of Kenya's amazing national parks. Your safari van, filled with tourists, is parked at the edge of the vast savanna. It is early evening, and you are waiting for animals to come to a water hole for a drink. As it grows darker, a huge lion appears and then stalks away on huge paws.

What benefits do tourists bring to your country?

BUILDING BACKGROUND Many of the countries of East Africa are rich in natural resources—including wildlife—but people disagree about the best way to use them. Droughts can make life here difficult. In addition, political and ethnic conflicts have led to unrest and violence in some areas of the region.

What You Will Learn...

Main Ideas

1. National parks are a major source of income for Tanzania and Kenya.
2. Rwanda and Burundi are densely populated rural countries with a history of ethnic conflict.
3. Both Sudan and Uganda have economies based on agriculture, but Sudan has suffered from years of war.
4. The countries of the Horn of Africa are among the poorest in the world.

The Big Idea

East Africa has abundant national parks, but most of the region's countries are poor and recovering from conflicts.

Key Terms and Places

safari, *p. 403*
geothermal energy, *p. 403*
genocide, *p. 405*
Darfur, *p. 405*
Mogadishu, *p. 407*

TAKING NOTES As you read, take notes on East Africa's people and economies. Use a chart like the one below to organize your notes.

	People	Economy
Tanzania and Kenya		
Rwanda and Burundi		
Sudan and Uganda		
Horn of Africa		

Close-up

Serengeti National Park

The Serengeti Plain is home to one of the world's greatest concentrations of wildlife. In Tanzania, part of the plain is a national park. About 100,000 tourists visit the Serengeti each year to view its diverse wildlife.

Huge herds of wildebeest migrate across the Serengeti each year.

ANALYSIS SKILL **ANALYZING VISUALS**

How would you describe the Serengeti landscape?

Tanzania and Kenya Today

The economies of both Tanzania and Kenya rely heavily on tourism and agriculture. However, both countries are among the poorest in the world.

Economy and Resources

Tanzania and Kenya are popular tourist destinations. With more than 2 million tourists visiting each year, tourism is a major source of income for both countries. Today many tourists visit Tanzania and Kenya to go on a safari in the countries' numerous national parks. A **safari** is an overland journey to view African wildlife.

In addition to tourism, Tanzania is particularly rich in gold and diamonds. However, it is still a poor country of mainly subsistence farmers. Poor soils and limited technology have restricted productivity.

In Kenya, much of the land has been set aside as national parkland. Many people would like to farm these lands, but farming would endanger African wildlife. Kenya's economy and tourism industry would likely be **affected** as well.

Kenya's economy relies mostly on agriculture. Mount Kilimanjaro's southern slopes are a rich agricultural region. The rich soils here provide crops of coffee and tea for exports.

Kenya's economy also benefits from another natural resource—geothermal energy. **Geothermal energy** is energy produced from the heat of Earth's interior. This heat—in the form of extremely hot steam—comes up to the surface through cracks in the rift valleys.

ACADEMIC VOCABULARY

affect to change or influence

READING CHECK Finding Main Ideas What activity supports the economies of both Tanzania and Kenya?

Tanzanian guides take visitors on a safari to view Serengeti's wildlife.

Watering holes attract wildlife, which includes flamingos, hippos, and giraffes.

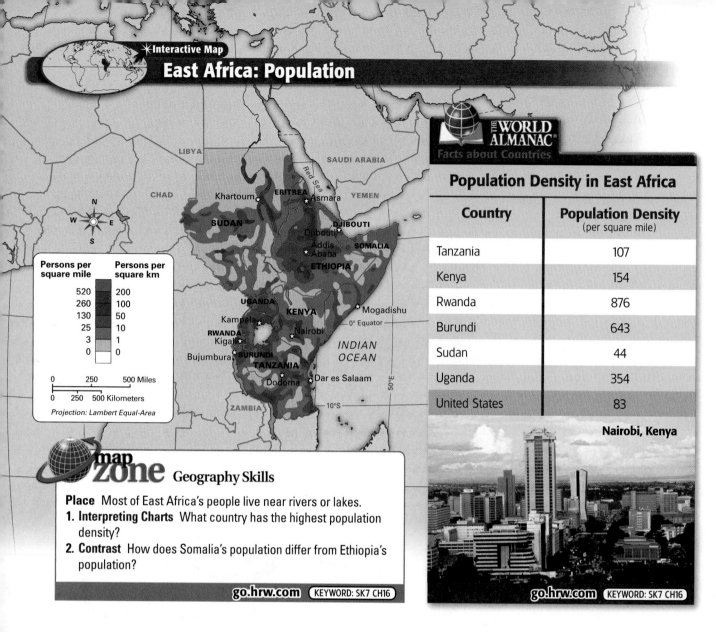

THE WORLD ALMANAC®
Facts about Countries

Population Density in East Africa

Country	Population Density (per square mile)
Tanzania	107
Kenya	154
Rwanda	876
Burundi	643
Sudan	44
Uganda	354
United States	83

Nairobi, Kenya

Persons per square mile / Persons per square km

Persons per square mile	Persons per square km
520	200
260	100
130	50
25	10
3	1
0	0

0   250   500 Miles
0   250   500 Kilometers
Projection: Lambert Equal-Area

map zone Geography Skills

Place Most of East Africa's people live near rivers or lakes.

1. **Interpreting Charts** What country has the highest population density?
2. **Contrast** How does Somalia's population differ from Ethiopia's population?

go.hrw.com   KEYWORD: SK7 CH16

go.hrw.com   KEYWORD: SK7 CH16

Cities

Imagine a large city with businesspeople hurrying to work, colorful outdoor markets, soaring skyscrapers, and beautiful parks. The capitals of Tanzania and Kenya both fit this description of a vibrant, modern African city.

Tanzania's official capital is Dodoma. The Tanzanian government began moving its capital from Dar es Salaam to Dodoma in the mid-1970s. Dar es Salaam, a port city with about 2.3 million people, is located on the Indian Ocean and is Tanzania's business center.

Kenya's capital, Nairobi, also serves as the country's industrial center. In addition, Nairobi is well connected with the rest of East Africa by a network of railways. By rail, Kenya transports tea and other major crops to the major port of Mombasa.

Even though Kenya and Tanzania are peaceful countries, Dar es Salaam and Nairobi have both endured terrorist attacks. In 1998 members of the al Qaeda terrorist group bombed the U.S. embassies in Dar es Salaam and Nairobi. Most of the more than 250 people killed and the thousands injured were Africans.

READING CHECK **Draw Conclusions** Why do you think it would be important for the railroad to link Kenya's cities?

Rwanda and Burundi Today

Rwanda and Burundi are mostly populated by two ethnic groups—the Tutsi and the Hutu. Since gaining independence from Germany, differences between the Tutsi and Hutu ethnic groups have led to conflict in Rwanda and Burundi. These conflicts have roots in the region's history. The colonial borders of Rwanda and Burundi drawn by Europeans often lumped different ethnic groups into one country.

In Rwanda in the 1990s, hatred between the Hutu and the Tutsi led to genocide. A **genocide** is the intentional destruction of a people. The Hutu tried to completely wipe out the Tutsi. Armed bands of Hutu killed hundreds of thousands of Tutsi.

Rwanda and Burundi are two of the most densely populated countries in all of Africa. These two countries are located in fertile highlands and share a history as German colonies. Both countries lack resources and rely on coffee and tea exports for economic earnings.

READING CHECK **Analyzing** What contributed to the region's ethnic conflict?

Sudan and Uganda Today

Sudan is Africa's largest country. It is a mainly agricultural country with few mineral resources. Arab Muslims make up about 40 percent of Sudan's population and have political power. They dominate northern Sudan and the capital, Khartoum.

For decades, Sudan has suffered from religious and ethnic conflict. Muslims and Christians fought a civil war for many years. More recently, a genocide occurred in a region of Sudan called **Darfur**. Ethnic conflict there resulted in tens of thousands of black Sudanese being killed by an Arab militia group. Millions more have fled Darfur and are scattered throughout the region as refugees.

Today Uganda is still recovering from several decades of a military dictatorship. Since 1986 Uganda has become more democratic, but economic progress has been slow. About 80 percent of Uganda's workforce is employed in agriculture, with coffee as the country's major export.

READING CHECK **Summarizing** What ethnic group dominates northern Sudan?

FOCUS ON READING

In the paragraphs under Rwanda and Burundi, what details support the main idea that these countries have a history of ethnic conflict?

Refugees in Sudan

People from Sudan who fled the country's Darfur region receive food and aid at a refugee camp.

More than half of Eritrea's population are Christian and belong to the Ethiopian Orthodox Church.

In Ethiopia, boys help their families herd sheep.

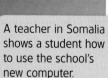

A teacher in Somalia shows a student how to use the school's new computer.

The Horn of Africa

Four East African countries located on the Red Sea and the Indian Ocean are called the Horn of Africa. This area is called the Horn because it resembles the horn of a rhinoceros. The Horn's people, economies, and resources vary by country.

Ethiopia

Unlike the other countries of the Horn of Africa, Ethiopia has never been under foreign rule. The country's mountains have protected the country from invasion.

In addition to providing a natural defense barrier, Ethiopia's rugged mountain slopes and highlands have rich volcanic soil. As a result, agriculture is Ethiopia's chief economic activity. Ethiopia's economy benefits from exports of coffee, livestock, and oilseeds. Many people also herd sheep and cattle.

During the last 30 years Ethiopia has experienced serious droughts. In the 1980s, drought caused the loss of crops and the starvation of several million people. In contrast, Ethiopia has experienced plenty of rainfall in recent years. Farmers are now able to grow their crops.

Most Ethiopians living in the highlands are Christian, while most of the lowland people are Muslim. Many Ethiopians speak Amharic, the country's official language.

Eritrea

In the late 1800s the Italians made present-day Eritrea a colony. In the 1960s it became an Ethiopian province.

After years of war with Ethiopia, Eritrea broke away from Ethiopia in 1993. Since then the economy has slowly improved. The country's Red Sea coastline is lined with spectacular coral reefs, which attract tourists to the country. Most Eritreans are farmers or herders. The country's economy relies largely on cotton exports.

Somalia

Somalia is a country of deserts and dry savannas. Because Somalia is so dry, much of the land is not suitable for farming. As a result, Somalis are nomadic herders. Livestock is the country's main export.

Somalia is less diverse than most other African countries. Most people in the country are members of a single ethnic group, the Somali. In addition, most Somalis are Muslims and speak the same African language, also called Somali.

Somalia has been troubled by violence in the past. In addition, the country has often had no central government of any kind. Different clans have fought over grazing rights and control over port cities such as **Mogadishu**.

In the 1990s Somalis experienced widespread starvation caused by a civil war and a severe drought. The United Nations sent aid and troops to the country. U.S. troops also assisted with this operation.

Djibouti

Djibouti (ji-BOO-tee) is a small, desert country. It lies on the Bab al-Mandab, which is the narrow strait that connects the Red Sea and the Indian Ocean. The strait lies along a major shipping route.

In the 1860s the French took control of Djibouti. It did not gain independence from France until 1977. The French government still contributes economic and military support to the country. As a result, French is one of Djibouti's two official languages. The other is Arabic.

The country's capital and major port is also called Djibouti. The capital serves as a port for landlocked Ethiopia. Since Djibouti has very few resources, the port is a major source of the country's income.

The people of Djibouti include two major ethnic groups—the Issa and the Afar. The Issa are closely related to the people of Somalia. The Afar are related to the people of Ethiopia. Members of both groups are Muslim. In the early 1990s, a civil war between the Afar and Issa broke out. In 2001 the two groups signed a peace treaty, which ended the fighting.

READING CHECK **Generalizing** What do the people of Djibouti have in common with people from other countries in East Africa?

SUMMARY AND PREVIEW The countries of East Africa are poor, but rich in wildlife and resources. Next, you will learn about the region of Central Africa.

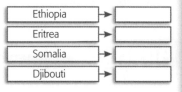

go.hrw.com
Online Quiz
KEYWORD: SK7 HP16

Section 3 Assessment

Reviewing Ideas, Terms, and Places

1. **a. Define** What is **geothermal energy**?
   b. Make Generalizations Why are Kenyans not allowed to farm in national parks?
2. **a. Define** What is **genocide**?
   b. Explain What are the two ethnic groups that make up the population of Rwanda and Burundi?
3. **a. Identify** What is the largest country in Africa?
   b. Analyze Why are millions of Sudanese refugees?
4. **a. Recall** What two major world religions are practiced in Ethiopia?
   b. Analyze How do you think Djibouti's location has helped its economy?

Critical Thinking

5. **Summarize** Draw a chart like this one. Using your notes, summarize in at least two sentences what you learned about each country.

Ethiopia	→	
Eritrea	→	
Somalia	→	
Djibouti	→	

FOCUS ON WRITING

6. **Writing about East Africa Today** Think about what it would be like to travel through the East African countries. What would you want to tell your friend about their people, their governments, their economies? Make a list of the details you would share.

Social Studies Skills

Chart and Graph | Critical Thinking | Geography | Study

Doing Fieldwork and Using Questionnaires

Learn

To a geographer, fieldwork means visiting a place to learn more about it. While there, the geographer might visit major sites or talk to people to learn about their lives. He or she might also distribute a questionnaire.

A questionnaire is a document that asks people to provide information. Geographers use them to find out specific details about the people in an area, such as what languages they speak. Governments and other groups also use questionnaires to learn more about the people they serve.

Practice

The questionnaire to the right is one that might have been created by the government of Kenya. Study it to answer the questions below.

1 What details does the questionnaire ask about people living in the household?

2 How are the questions organized? Why do you think that is?

3 Why would asking for the person's age be an important question?

Republic of Kenya Population and Housing Census
Form A: Information Regarding All Persons

Name
What are the names of all persons who live in this household?
_____

Relationship
What is your relationship to the head of the household? (circle one)

1 Head
2 Spouse
3 Son
4 Daughter
5 Brother/Sister
6 Father/Mother
7 Other relative
8 Non-relative

Age
How old are you?

Tribe Nationality
What is your tribe or nationality?

Religion
What is your religion? (circle one)

1 Catholic
2 Protestant
3 Other Christian
4 Muslim
5 Traditionalist
6 Other Religion
7 No Religion

Birthplace
Where were you born?
_____

Apply

Work with a group of classmates to create a questionnaire about popular music. Think of five questions about popular music that you could ask your fellow students. Try to ask only multiple-choice and yes-or-no questions. These types of questions are easier to study than other questions are. Once you have completed your questionnaire, write a short explanation of what you hope to learn from each question.

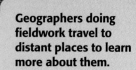

Geographers doing fieldwork travel to distant places to learn more about them.

Geography's Impact
video series
Review the video to answer the closing question:
How might the loss of Mount Kilimanjaro's glaciers affect people living in the area?

Visual Summary

Use the visual summary below to help you review the main ideas of the chapter.

QUICK FACTS

East Africa is a land of spectacular mountains, highlands, and lakes.

East Africa's history includes European imperialism and independence.

Even though many countries in East Africa are very poor, they are trying to improve educational opportunities.

Reviewing Vocabulary, Terms, and Places

Choose one word from each word pair to correctly complete each sentence below.

1. A_____(rift valley/plain) is a place on Earth's surface where the crust stretches until it breaks.

2. The tallest mountain in Africa is_____. (Mount Kilimanjaro/Mount Kenya)

3. _____means to change or influence. (geothermal energy/affect)

4. A_____is an overland journey that is taken to view African wildlife. (drought/safari)

5. _____has experienced serious droughts over the past 30 years. (Ethiopia/Kenya)

6. _____are periods when little rain falls and crops are damaged. (flood/drought)

7. The intentional destruction of a people is called_____. (murder/genocide)

8. _____is a practice which tries to dominate other countries' government, trade, and culture. (imperialism/influence)

9. A genocide committed by an Arab militia occured in the region of Sudan known as _____. (Darfur/Khartoum)

10. _____is an East African island that was an international slave-trading center in the late 1700s. (Madagascar/Zanzibar)

Comprehension and Critical Thinking

SECTION 1 *(Pages 394–397)*

11. a. Identify What is the Great Rift Valley? What is it made of?

b. Draw Conclusions How is the Nile necessary for farming in the desert?

c. Predict How do you think the effects of drought can be avoided in the future?

SECTION 2 *(Pages 398–401)*

12. a. Recall In which East African country did an emperor build 11 rock churches?

   b. Contrast How does the major religion practiced in Ethiopia differ from other religions practiced in East Africa?

   c. Elaborate How did the African language of Swahili develop?

SECTION 3 *(Pages 402–407)*

13. a. Define What is a safari?

   b. Draw Conclusions What economic activity do both Kenya and Tanzania rely on?

   c. Evaluate Why do you think the al Qaeda terrorist group bombed U.S. embassies in Dar es Salaam and Nairobi?

Using the Internet

go.hrw.com
KEYWORD: SK7 CH16

14. Activity: Understanding Cultures In the East African countries you read about in this chapter, there are hundreds of different ethnic groups. Enter the activity keyword. Discover some ethnic groups of East Africa as you visit Web sites about their culture. Then create a graphic organizer or chart that compares East African ethnic groups. It might include comparisons of language, beliefs, traditions, foods, and more.

Social Studies Skills

15. Doing Fieldwork and Using Questionnaires As in Kenya, the United States conducts a census of its population. Research what kinds of questions the U.S. Census asks Americans every 10 years. Go to the official U.S. Census Bureau's Web site at www.census.gov. There you will find examples of the questionnaires used in the 2000 Census. How do you think the questionnaires on the next U.S. Census in 2010 will be the same or different than the questions that were asked on the 2000 Census?

FOCUS ON READING AND WRITING

16. Identifying Supporting Details Look back over the paragraphs under the Culture heading in Section 2. Then make a list of details you find to support the section's main ideas. Make sure you include details about the different languages spoken in East Africa today.

17. Writing a Letter Now that you have information about East Africa, you need to organize it. Think about your audience, a friend at home, and what would feel natural if you had been traveling. Would you organize by topics like physical geography and culture? Or would you organize by country? After you organize your information, write a one-page letter.

Map Activity ✱Interactive

18. East Africa On a separate sheet of paper, match the letters on the map with their correct labels.

Great Rift Valley   Mount Kilimanjaro

Lake Victoria   Nile River

Indian Ocean

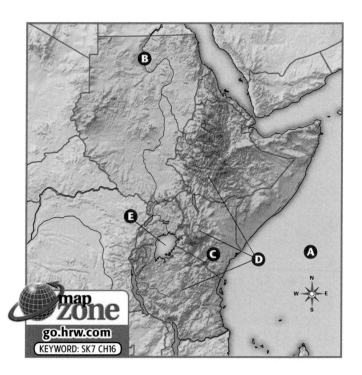
map zone
go.hrw.com
KEYWORD: SK7 CH16

DIRECTIONS: *Read questions 1 through 7 and write the letter of the best response. Then read question 8 and write your own well-constructed response.*

1 **What physical feature of East Africa is covered with snow and ice year-round?**

A Serengeti Plain

B Mount Kilimanjaro

C Great Rift Valley

D Mount Kenya

2 **What is one cause of the cool climate in some areas of East Africa?**

A elevation

B drought

C Indian Ocean

D Great Rift Valley

3 **The Swahili language developed through trade contacts between East Africans and**

A the Chinese.

B Europeans.

C Arabs.

D West Africans.

4 **Tourism is a large part of the economy in**

A Tanzania and Kenya.

B Sudan and Uganda.

C Ethiopia and Eritrea.

D Rwanda and Burundi.

5 **Which East African country used to be a province of Ethiopia?**

A Djibouti

B Somalia

C Eritrea

D Kenya

"Then they were over the first hills and the wildebeeste were trailing up them, and then they were over the mountains with sudden depths of green-rising forest and solid bamboo slopes, and then the heavy forest again, sculptured into peaks and hollows until they crossed, and hills sloped down and then another plain, hot now, and purple brown, bumpy with heat..."

—Ernest Hemingway, "The Snows of Kilimanjaro"

6 **In the passage above, Hemingway describes the view of the Mount Kilimanjaro landscape from a plane. The landscape he describes is filled with**

A rivers.

B lakes.

C deserts.

D forests.

7 **In the passage above, the climate of the plain is described as**

A hot.

B dry.

C wet.

D cold.

8 **Extended Response** Look at the table and map of East Africa's population density in Section 3. Write a paragraph explaining why you think some areas of East Africa are more populated than other areas. Identify at least two reasons.

CHAPTER 17

Central Africa

What You Will Learn...

In this chapter you will learn about the rivers, forests, and resources of Central Africa. This region has been influenced by native traditions and Europeans, and you will read about how these influences have affected Central Africa's culture. Finally, you will learn about the different countries in Central Africa and some of the challenges these countries face.

SECTION 1
Physical Geography **414**

SECTION 2
History and Culture....................**420**

SECTION 3
Central Africa Today....................**424**

FOCUS ON READING AND WRITING

Using Word Parts Many English words have little word parts at the beginning (prefixes) or the end (suffixes) of the word. When you come to an unfamiliar word in your reading, see if you can recognize a prefix or suffix to help you figure out the meaning of the word. **See the lesson, Using Word Parts, on page 684.**

Writing an Acrostic An acrostic is a type of poem in which the first letters of each line spell a word. The lines of the poem describe that word. As you read the chapter, think of a word—maybe a country name or a physical feature—that you would like to describe in your acrostic.

ATLANTIC OCEAN

0° Equator

⊙ National capital
● Other cities

| 0 | 200 | 400 Miles |
| 0 | 200 | 400 Kilometers |

Projection: Lambert Azimuthal Equal-Area

10°S

10°W

0°

Culture Drums and horns are used in traditional music from Central Africa. These musicians are from Cameroon.

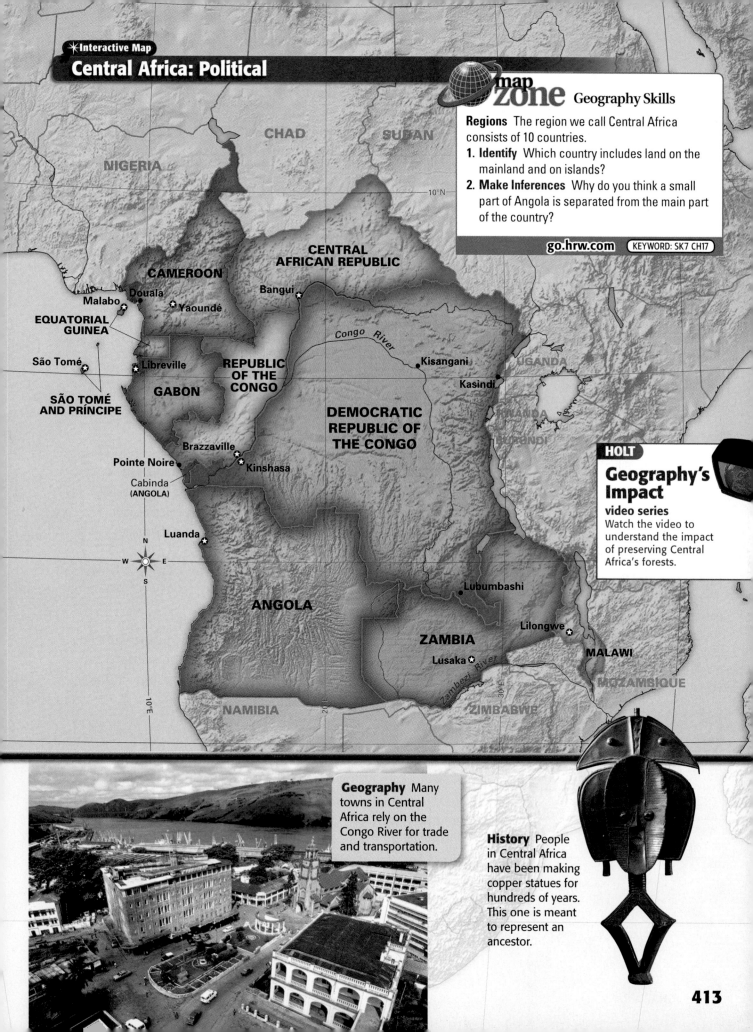

map zone Geography Skills

Regions The region we call Central Africa consists of 10 countries.

1. **Identify** Which country includes land on the mainland and on islands?
2. **Make Inferences** Why do you think a small part of Angola is separated from the main part of the country?

go.hrw.com KEYWORD: SK7 CH17

CHAD

SUDAN

NIGERIA

CENTRAL AFRICAN REPUBLIC

Bangui

CAMEROON

Douala

Malabo

Yaoundé

EQUATORIAL GUINEA

Congo River

Kisangani

São Tomé

Libreville

UGANDA

Kasindi

GABON

REPUBLIC OF THE CONGO

RWANDA

DEMOCRATIC REPUBLIC OF THE CONGO

BURUNDI

SÃO TOMÉ AND PRÍNCIPE

Brazzaville

Pointe Noire

Kinshasa

Cabinda (ANGOLA)

HOLT

Geography's Impact

video series

Watch the video to understand the impact of preserving Central Africa's forests.

N
W E
S

Luanda

Lubumbashi

ANGOLA

Lilongwe

ZAMBIA

MALAWI

Lusaka

Zambezi River

MOZAMBIQUE

NAMIBIA

ZIMBABWE

Geography Many towns in Central Africa rely on the Congo River for trade and transportation.

History People in Central Africa have been making copper statues for hundreds of years. This one is meant to represent an ancestor.

Physical Geography

What You Will Learn...

Main Ideas

1. Central Africa's major physical features include the Congo Basin and plateaus surrounding the basin.
2. Central Africa has a humid tropical climate and dense forest vegetation.
3. Central Africa's resources include forest products and valuable minerals such as diamonds and copper.

The Big Idea

The Congo River, tropical forests, and mineral resources are important features of Central Africa's physical geography.

Key Terms and Places

Congo Basin, *p. 414*
basin, *p. 414*
Congo River, *p. 415*
Zambezi River, *p. 415*
periodic market, *p. 417*
copper belt, *p. 417*

TAKING NOTES As you read, use a chart like the one here to note characteristics of Central Africa's physical geography.

Physical features	
Climate and vegetation	
Resources	

If YOU lived there...

You are on a nature hike with a guide through the forests of the Congo Basin. It has been several hours since you have seen any other people. Sometimes your guide has to cut a path through the thick vegetation, but mostly you try not to disturb any plants or animals. Suddenly, you reach a clearing and see a group of men working hard to load huge tree trunks onto big trucks.

How do you feel about what you see?

BUILDING BACKGROUND Much of Central Africa, particularly in the Congo Basin, is covered with thick, tropical forests. The forests provide valuable resources, but people have different ideas about how the forests should be used. Forests are just one of the many types of landscapes in Central Africa.

Physical Features

Central Africa is bordered by the Atlantic Ocean in the west. In the east, it is bordered by a huge valley called the Western Rift Valley. The land in between has some of the highest mountains and biggest rivers in Africa.

Landforms

You can think of the region as a big soup bowl with a wide rim. Near the middle of the bowl is the **Congo Basin**. In geography, a **basin** is a generally flat region surrounded by higher land such as mountains and plateaus.

Plateaus and low hills surround the Congo Basin. The highest mountains in Central Africa lie farther away from the basin, along the Western Rift Valley. Some of these snowcapped mountains rise to more than 16,700 feet (5,090 m). Two lakes also lie along the rift—Lake Nyasa and Lake Tanganyika (tan-guhn-YEE-kuh). Lake Nyasa is also called Lake Malawi.

Rivers

The huge **Congo River** is fed by hundreds of smaller rivers. They drain the swampy Congo Basin and flow into the river as it runs toward the Atlantic. Many rapids and waterfalls lie along its route, especially near its mouth. These obstacles make it impossible for ships to travel from the interior of Central Africa all the way to the Atlantic. The Congo provides an important transportation route in the interior, however.

In the southern part of the region, the **Zambezi** (zam-BEE-zee) **River** flows eastward toward the Indian Ocean. Many rivers in Angola and Zambia, as well as water from Lake Nyasa, flow into the Zambezi. The Zambezi also has many waterfalls along its route, the most famous of which are the spectacular Victoria Falls.

READING CHECK **Finding Main Ideas** Where is the highest land in Central Africa?

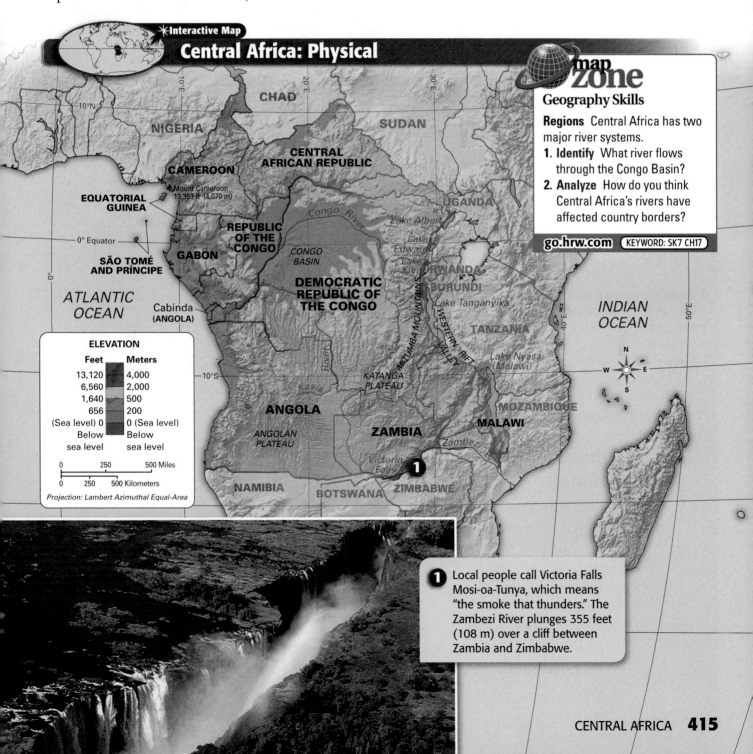

Central Africa: Physical

★Interactive Map

map zone

Geography Skills

Regions Central Africa has two major river systems.
1. **Identify** What river flows through the Congo Basin?
2. **Analyze** How do you think Central Africa's rivers have affected country borders?

go.hrw.com KEYWORD: SK7 CH17

ELEVATION

Feet	Meters
13,120	4,000
6,560	2,000
1,640	500
656	200
(Sea level) 0	0 (Sea level)
Below sea level	Below sea level

0   250   500 Miles
0   250   500 Kilometers
Projection: Lambert Azimuthal Equal-Area

CHAD
NIGERIA
SUDAN
CAMEROON
CENTRAL AFRICAN REPUBLIC
Mount Cameroon 13,353 ft (4,070 m)
EQUATORIAL GUINEA
UGANDA
Congo River
Lake Albert
REPUBLIC OF THE CONGO
0° Equator
Lake Edward
SÃO TOMÉ AND PRÍNCIPE
GABON
CONGO BASIN
Lake Kivu
RWANDA
BURUNDI
ATLANTIC OCEAN
DEMOCRATIC REPUBLIC OF THE CONGO
Lake Tanganyika
Cabinda (ANGOLA)
MITUMBA MOUNTAINS
WESTERN RIFT VALLEY
TANZANIA
INDIAN OCEAN
River
KATANGA PLATEAU
Lake Nyasa (Malawi)
Kasai
ANGOLA
MOZAMBIQUE
ANGOLAN PLATEAU
ZAMBIA
MALAWI
Zambezi
Victoria Falls ①
NAMIBIA
BOTSWANA
ZIMBABWE

① Local people call Victoria Falls Mosi-oa-Tunya, which means "the smoke that thunders." The Zambezi River plunges 355 feet (108 m) over a cliff between Zambia and Zimbabwe.

Central Africa's National Parks

CHAD
NIGERIA
CAMEROON
CENTRAL AFRICAN REPUBLIC
SUDAN
EQUATORIAL GUINEA
UGANDA
Congo River
0° Equator
GABON
SÃO TOMÉ AND PRÍNCIPE
REPUBLIC OF THE CONGO
RWANDA
BURUNDI
ATLANTIC OCEAN
Cabinda (ANGOLA)
DEMOCRATIC REPUBLIC OF THE CONGO
TANZANIA
10°S
ANGOLA
ZAMBIA
MALAWI
Zambezi R.
NAMIBIA
ZIMBABWE
MOZAMBIQUE

National park areas

0   400   800 Miles
0   400   800 Kilometers
Projection: Lambert Azimuthal Equal-Area

National parks in Central Africa protect the habitat of gorillas (above), which are endangered, and okapis (below).

map zone Geography Skills

Human-Environment Interaction People have created national parks in Central Africa to try to protect the region's landscapes and animals.
1. **Use the Map** What Central African countries have national parks in coastal areas?
2. **Explain** Why do people want to protect natural environments?

Climate, Vegetation, and Animals

Central Africa lies along the equator and in the low latitudes. Therefore, the Congo Basin and much of the Atlantic coast have a humid tropical climate. These areas have warm temperatures all year and receive a lot of rainfall.

This climate supports a large, dense tropical forest. The many kinds of tall trees in the forest form a complete canopy. The canopy is the uppermost layer of the trees where the limbs spread out. Canopy leaves block sunlight to the ground below.

Such animals as gorillas, elephants, wild boars, and okapis live in the forest. The okapi is a short-necked relative of the giraffe. However, since little sunlight shines through the canopy, only a few animals live on the forest floor. Some animals, such as birds, monkeys, bats, and snakes, live in the trees. Many insects also live in Central Africa's forest.

The animals in Central Africa's tropical forests, as well as the forests themselves, are in danger. Large areas of forest are being cleared rapidly for farming and logging. Also, people hunt the large animals in the forests to get food. To promote protection of forests and other natural environments, governments have set up national park areas in their countries.

North and south of the Congo Basin are large areas with a tropical savanna climate. Those areas are warm all year, but they have distinct dry and wet seasons. There are grasslands, scattered trees, and shrubs. The high mountains in the east have a highland climate. Dry steppe and even desert climates are found in the far southern part of the region.

READING CHECK **Summarizing** What are the climate and vegetation like in the Congo Basin?

Resources

The tropical environment of Central Africa is good for growing crops. Most people in the region are subsistence farmers. However, many farmers are now beginning to grow crops for sale. Common crops are coffee, bananas, and corn. In rural areas, people trade agricultural and other products in periodic markets. A **periodic market** is an open-air trading market that is set up once or twice a week.

Central Africa is rich in other natural resources as well. The large tropical forest provides timber, while the rivers provide a way to travel and to trade. Dams on the rivers produce hydroelectricity, an important energy resource. Other energy resources in the region include oil, natural gas, and coal.

Central Africa also has many valuable minerals, including copper, uranium, tin, zinc, diamonds, gold, and cobalt. Of these, copper is the most important. Most of Africa's copper is found in an area called the **copper belt**. The copper belt stretches through northern Zambia and southern Democratic Republic of the Congo. However, poor transportation systems and political problems have kept the region's resources from being fully developed.

READING CHECK **Analyzing** Why are Central Africa's rivers an important natural resource?

FOCUS ON READING

What prefix do you recognize in *promote*?

SUMMARY AND PREVIEW Mighty rivers, the tropical forest of the Congo Basin, and mineral resources characterize the physical geography of Central Africa. These landscapes have influenced the region's history. Next, you will read about Central Africa's history and culture.

Section 1 Assessment

go.hrw.com
Online Quiz
KEYWORD: SK7 HP17

Reviewing Ideas, Terms, and Places

1. **a. Describe** What is the **Congo Basin**?
   b. Elaborate How do you think the **Congo River**'s rapids and waterfalls affect the economy of the region?
2. **a. Recall** What part of Central Africa has a highland climate?
   b. Explain Why have governments in the region set up national parks?
   c. Evaluate Is it more important to use the forest's resources or to protect the natural environment? Why?
3. **a. Define** What is a **periodic market**?
   b. Elaborate What kinds of political problems might keep mineral resources from being fully developed?

Critical Thinking

4. **Contrasting** Use your notes and a graphic organizer like this one to list differences between the Congo Basin and the areas surrounding it in Central Africa.

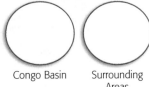

Congo Basin   Surrounding Areas

FOCUS ON WRITING

5. **Describing Physical Geography** What topics in this section might work well in your acrostic? Jot down notes on one or two topics you could feature in your poem.

Mapping Central Africa's Forests

Essential Elements

The World in Spatial Terms
Places and Regions
Physical Systems
Human Systems
Environment and Society
The Uses of Geography

Background Imagine taking a walk along a street in your neighborhood. Your purpose is to see the street in spatial terms and gather information to help you make a map. While you walk, you ask the kinds of questions geographers ask. How many houses, apartment buildings, or businesses are on the street? What kinds of animals or trees do you see? Your walk ends, and you organize your data. Now imagine that you are going to gather data on another walk. This walk will be 2,000 miles long.

A 2,000-Mile Walk In September 1999, an American scientist named Michael Fay began a 465-day, 2,000-mile walk through Central Africa's forests. He and his team followed elephant trails through thick vegetation. They waded through creeks and mucky swamps.

On the walk, Fay gathered data on the number and kinds of animals he saw. He counted elephant dung, chimpanzee nests, leopard tracks, and gorillas. He counted the types of trees and other plants along his

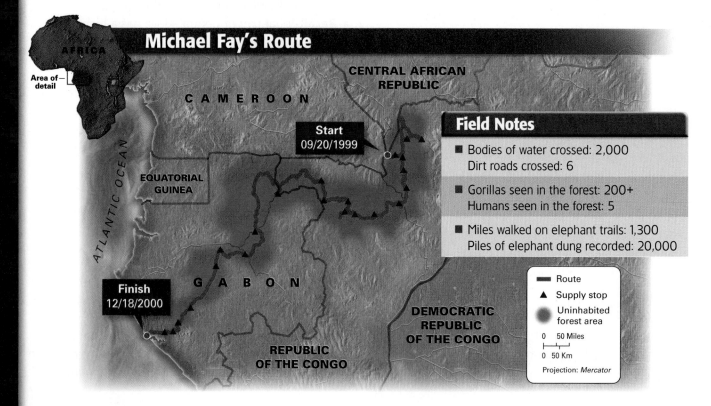

Michael Fay's Route

AFRICA

Area of detail

CAMEROON

CENTRAL AFRICAN REPUBLIC

ATLANTIC OCEAN

EQUATORIAL GUINEA

Start 09/20/1999

GABON

Finish 12/18/2000

REPUBLIC OF THE CONGO

DEMOCRATIC REPUBLIC OF THE CONGO

Field Notes

- Bodies of water crossed: 2,000
  Dirt roads crossed: 6

- Gorillas seen in the forest: 200+
  Humans seen in the forest: 5

- Miles walked on elephant trails: 1,300
  Piles of elephant dung recorded: 20,000

▬ Route
▲ Supply stop
● Uninhabited forest area

0   50 Miles
0   50 Km

Projection: *Mercator*

Michael Fay (above) and his team had to chop their way through thick forest vegetation. In a clearing, they spotted this group of elephants.

route. He also counted human settlements and determined the effect of human activities on the environment.

Fay used a variety of tools to record the data he gathered on his walk. He wrote down what he observed in waterproof notebooks. He shot events and scenes with video and still cameras. To measure the distance he and his team walked each day, he used a tool called a Fieldranger. He also kept track of his exact position in the forest by using a GPS, or global positioning system.

What It Means Michael Fay explained the purpose of his long walk. "The whole idea behind this is to be able to use the data we've collected as a tool." Other geographers can compare Fay's data with their own. Their comparison may help them create more accurate maps. These maps will show where plants, animals, and humans are located in Central Africa's forests.

Fay's data can also help scientists plan the future use of land or resources in a region. For example, Fay has used his data to convince government officials in Gabon to set aside 10 percent of its land to create 13 national parks. The parks will be protected from future logging and farming. They also will preserve many of the plants and animals that Fay and his team observed on their long walk.

Geography for Life Activity

1. Why did Michael Fay walk 2,000 miles?

2. In what practical way has Michael Fay used his data?

3. **Read More about Fay's Walk** Read the three-part article on Michael Fay's walk in *National Geographic* October 2000, March 2001, and August 2001. After you read the article, explain why Fay called his walk a "megatransect."

History and Culture

What You Will Learn...

Main Ideas

1. Great African kingdoms and European colonizers have influenced the history of Central Africa.
2. The culture of Central Africa includes many ethnic groups and languages, but it has also been influenced by European colonization.

The Big Idea

Central Africa's history and culture have been influenced by native traditions and European colonizers.

Key Terms and Places

Kongo Kingdom, *p. 420*
dialects, *p. 422*

TAKING NOTES As you read, use a graphic organizer like the one here to take notes on Central Africa's history and culture.

History	
Culture	

If **YOU** lived there...

You live in Central Africa in the 1300s. Over the past year, many new people have moved to your village. They speak a different language—one that you don't understand. They also have some customs that seem strange to you. But they have begun bringing fancy items such as animal skins and shells to your village. Now your village seems very rich.

How do you feel about these new people?

BUILDING BACKGROUND Different groups of people have influenced Central Africa throughout its history. Whether they came from near or far, and whether they stayed in Central Africa only decades or for more than hundred years, these groups brought their own cultures and customs to the region.

History

Early humans lived in Central Africa many thousands of years ago. However, the descendants of these people have had less impact on the region's history than people from the outside. Tribes from West Africa, and later European colonists, brought their customs to the region and changed the way people lived.

Early History

About 2,000 years ago new peoples began to migrate to Central Africa from West Africa. They eventually formed several kingdoms in Central Africa. Among the most important was the **Kongo Kingdom**. Founded in the 1300s, it was located near the mouth of the Congo River.

The Kongo people established trade routes to western and eastern Africa. Their kingdom grew rich from the trade of animal skins, shells, slaves, and ivory. Ivory is a cream-colored material that comes from elephant tusks.

In the late 1400s, Europeans came to the region. They wanted the region's forest products and other resources such as ivory. They used ivory for fine furniture, jewelry, statues, and piano keys. Europeans also began to trade with some Central African kingdoms for slaves. Over a span of about 300 years, the Europeans took millions of enslaved Africans to their colonies in the Americas.

Some African kingdoms became richer by trading with Europeans. However, all were gradually changed and weakened by European influence. In the late 1800s, European countries divided all of Central Africa into colonies. The colonial powers were France, Belgium, Germany, Spain, the United Kingdom, and Portugal.

These European powers drew colonial borders that ignored the homelands of different ethnic groups. Many different ethnic groups were lumped together in colonies where they had to interact. These groups spoke different languages and had different customs. Their differences caused conflicts, especially after the colonies won independence.

Modern Central Africa

Central African colonies gained their independence from European powers after World War II. Some of the colonies fought bloody wars to win their independence. The last country to become independent was Angola. It won freedom from Portugal in 1975.

Independence did not bring peace to Central Africa, however. Ethnic groups continued to fight one another within the borders of the new countries. Also, the United States and the Soviet Union used Central Africa as a battleground in the Cold War. They supported different allies in small wars throughout Africa. The wars in the region killed many people and caused great damage.

FOCUS ON READING
What prefix do you recognize in *interact*?

READING CHECK **Summarizing** What role did Europeans play in Central Africa's history?

Culture

Today about 100 million people live in Central Africa. These people belong to many different ethnic groups and have different customs.

People and Language

The people of Central Africa speak hundreds of different languages. They also speak different **dialects**, regional varieties of a language. For example, although many Central Africans speak Bantu languages, those languages can be quite different from one another.

The main reason for this variety is the number of ethnic groups. Most ethnic groups have their own language or dialect. Most people in the region speak traditional African languages in their daily lives. However, the official languages of the region are European because of the influence of the colonial powers. For example, French is the official language of the Democratic Republic of the Congo. Portuguese is the language of Angola. English is an official language in Zambia and Malawi.

Religion

Central Africa's colonial history has also influenced religion. Europeans introduced Christianity to the region. Now many people in the former French, Spanish, and Portuguese colonies are Roman Catholic. Protestant Christianity is most common in former British colonies.

Two other religions came to parts of Central Africa from other regions. Influenced by the Muslim countries of the Sahel, the northern part of Central Africa has many Muslims. Zambia is the home of Muslims as well as Hindus.

The Arts

Central Africa's traditional cultures influence the arts of the region. The region is famous for sculpture, carved wooden masks, and beautiful cotton gowns dyed in bright colors.

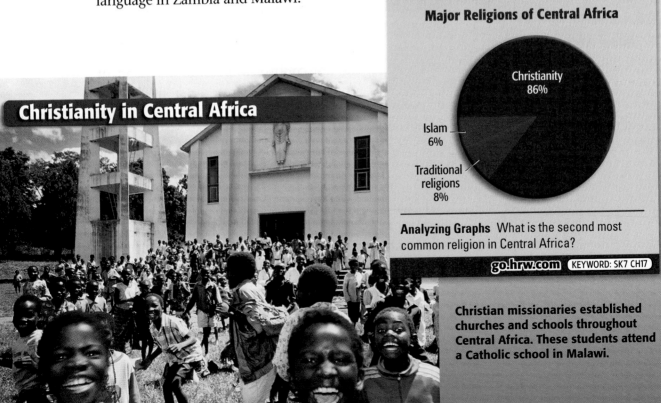

Christianity in Central Africa

THE WORLD ALMANAC
Facts about the World

Major Religions of Central Africa

Christianity 86%
Islam 6%
Traditional religions 8%

Analyzing Graphs What is the second most common religion in Central Africa?

go.hrw.com KEYWORD: SK7 CH17

Christian missionaries established churches and schools throughout Central Africa. These students attend a Catholic school in Malawi.

Bantu Languages

About 2,000 years ago people who spoke Bantu languages migrated out of West Africa. They moved to Central Africa as well as eastern and southern Africa. The Bantu speakers mixed with peoples who already lived in these lands.

The migration of Bantu speakers had important effects on African life. They brought new ways for growing food. They used tools made of iron, which others also began to use. The Bantu speakers also brought their languages. Today many Central Africans speak one or more of the some 500 Bantu languages such as Rundi, Bemba, or Luba.

Drawing Inferences How do you think the number of languages affects communication in the region?

Central Africa also has popular styles of music. The *likembe*, or thumb piano, was invented in the Congo region. Also, a type of dance music called *makossa* originated in Cameroon and has become popular throughout Africa. It can be played with guitars and electric keyboards.

READING CHECK **Generalizing** What are characteristics of culture in Central Africa?

SUMMARY AND PREVIEW Central Africa's history was influenced by great kingdoms that controlled trade and by Europeans, who originally came to the region looking for trade goods. European and traditional African influences have shaped the region's culture. Next, you will learn about the countries of Central Africa and what life is like there today.

Section 2 Assessment

go.hrw.com
Online Quiz
KEYWORD: SK7 HP17

Reviewing Ideas, Terms, and Places

1. **a. Recall** What Central African resource did Europeans value for making jewelry and crafts?
   b. Explain How did the **Kongo Kingdom** become important?
   c. Elaborate How do you think the colonial borders affected Central African countries' fights for independence?
2. **a. Define** What is a **dialect**?
   b. Summarize How did the colonial era affect Central Africa's culture?
   c. Elaborate How might Central Africa's culture be different today if the region had not been colonized by Europeans?

Critical Thinking

3. **Sequencing** Review your notes on Central Africa's history. Using a graphic organizer like this one, put major events in chronological order.

FOCUS ON WRITING

4. **Taking Notes on History and Culture** Your acrostic could describe the region's history and culture as well as physical geography. Take notes on interesting information you might include in your poem.

Central Africa Today

What You Will Learn...

Main Ideas

1. The countries of Central Africa are mostly poor, and many are trying to recover from years of civil war.
2. Challenges to peace, health, and the environment slow economic development in Central Africa.

The Big Idea

War, disease, and environmental problems have made it difficult for the countries of Central Africa to develop stable governments and economies.

Key Terms and Places

Kinshasa, *p. 425*
inflation, *p. 427*
malaria, *p. 428*
malnutrition, *p. 429*

TAKING NOTES As you read, use a graphic organizer like the one here to organize your notes on the countries of Central Africa and the challenges facing those countries.

Countries	Challenges

If YOU lived there...

You are an economic adviser in Zambia. Your country is poor, and most people are farmers. But scientists say Zambia has a lot of copper under ground. With a new copper mine, you could sell valuable copper to other countries. However, the mine would destroy a lot of farmland.

Do you support building the mine? Why or why not?

BUILDING BACKGROUND You have already read about Central Africa's great resources. Many countries in the region have the potential for great wealth. However, several factors throughout history have made it difficult for Central African countries to develop their resources.

Countries of Central Africa

Most of the countries in Central Africa are very poor. After years of colonial rule and then civil war, they are struggling to build stable governments and strong economies.

Democratic Republic of the Congo

The Democratic Republic of the Congo was a Belgian colony until 1960. When the country gained independence, many Belgians left. Few teachers, doctors, and other professionals remained in the former colony. In addition, various ethnic groups fought each other for power. These problems were partly to blame for keeping the new country poor.

A military leader named Joseph Mobutu came to power in 1965. He ruled as a dictator. One way Mobutu used his power was to change the name of the country to Zaire—a name that was traditionally African rather than European. He also changed his own name to Mobutu Sese Seko.

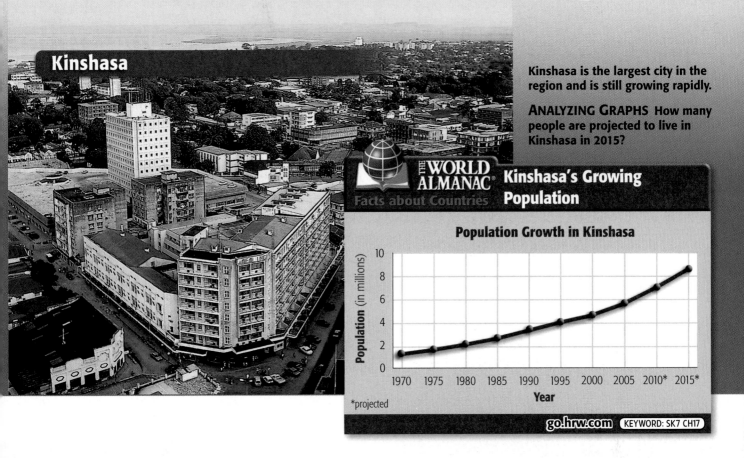

Kinshasa

Kinshasa is the largest city in the region and is still growing rapidly.

ANALYZING GRAPHS How many people are projected to live in Kinshasa in 2015?

THE WORLD ALMANAC Facts about Countries

Kinshasa's Growing Population

Population Growth in Kinshasa

*projected

go.hrw.com KEYWORD: SK7 CH17

During his rule, the government took over foreign-owned industries. It borrowed money from foreign countries to try to expand industry. However, most farmers suffered, and government and business leaders were corrupt. While the economy collapsed, Mobutu became one of the richest people in the world and used violence against people who challenged him.

In 1997, after a civil war, a new government took over. The new government renamed the country the Democratic Republic of the Congo.

The Democratic Republic of the Congo is a treasure chest of minerals that could bring wealth to the country. The south is part of Central Africa's rich copper belt. The country also has gold, diamonds, and cobalt. In addition, the tropical forest provides wood, food, and rubber. However, civil war, bad government, and crime have scared many foreign businesses away. As a result, the country's resources have helped few of its people.

Most people in the Democratic Republic of the Congo are poor. They usually live in rural areas where they must farm and trade for food. Many people are moving to the capital, **Kinshasa**. This crowded city has some modern buildings, but most of the city consists of poor slums.

Central African Republic and Cameroon

North of the Democratic Republic of the Congo is the landlocked country of Central African Republic. Since independence, this country has struggled with military coups, corrupt leaders, and improper elections.

In addition to political instability, the country suffers from a weak economy. Most people there are farmers. Although the country has diamonds and gold, it does not have railroads or ports needed to transport the resources for export. Central African Republic receives some aid from foreign countries, but this is not enough to meet the needs of its people.

FOCUS ON
READING

If *–al* means
"relating to,"
what does
political mean?

Between Central African Republic and the Atlantic Ocean is Cameroon. Unlike most countries in Central Africa, Cameroon is fairly stable. It is a republic. The president is elected and holds most of the power.

Political stability has made economic growth possible. The country has oil reserves and good conditions for farming. Cacao, cotton, and coffee are valuable export crops. A good system of roads and railways helps people transport these goods for export to other countries.

Because of the steady economy, the people of Cameroon have a high standard of living for the region. For example, more people in Cameroon are enrolled in school than in most places in Africa.

Equatorial Guinea and São Tomé and Príncipe

Tiny Equatorial Guinea is divided between the mainland and five islands. The country is a republic. It has held elections, but many have seen the elections as being flawed. These elections have kept the same president ruling the country for more than 25 years. Although the recent discovery of oil has produced economic growth, living conditions for most people are still poor.

The island country of São Tomé and Príncipe has struggled with political instability. In addition, it is a poor country with few resources. It produces much cacao but has to import food. The recent discovery of oil in its waters may help the economy.

Village Architecture

Although Central Africa has several big cities, many people still live in rural villages. Different groups of people have different styles of architecture for their villages. Building materials vary depending on the resources available in the geographic setting.

An extended family lives together in these adobe homes in the mountains of Cameroon.

The strong tropical sun provides power for this hut in Angola.

Gabon and Republic of the Congo

Gabon has had only one president since 1967. For many years, Gabon held no multi-party elections. Gabon's economy provides the highest standard of living in the region. More than half the country's income comes from oil.

Like Gabon, the Republic of the Congo receives much of its income from oil. It also receives income from forest products. Despite these resources, a civil war in the late 1990s hurt the economy.

The Republic of the Congo is mostly urban and growing more so. Many people are moving from villages to cities. The biggest city is the capital, Brazzaville.

A family sits outside their home in Zambia.

How does the construction of the huts help you recognize different climates in Central Africa?

Angola

Angola won independence from Portugal in 1975. The country then plunged into a long civil war. Fighting finally ended in 2002, and the country has been more stable since then. Angola is now a republic with an elected president.

Even with peace, Angola's economy is struggling. For about 85 percent of the population, subsistence farming is the only source of income. Even worse, land mines left over from the civil war endanger the farmers. A high rate of **inflation**, the rise in prices that occurs when currency loses its buying power, has also weakened the economy. Finally, corrupt officials have taken large amounts of money meant for public projects.

Angola does have potential, however. The country has diamonds and oil. The oil is found offshore and in Cabinda. Cabinda is a part of Angola that is separated from the rest of the country by the Democratic Republic of the Congo.

Zambia and Malawi

The southernmost countries in Central Africa are Zambia and Malawi. About 85 percent of Zambia's workers are farmers. Though rich with copper mines, Zambia's economy is growing very slowly. It is hurt by high levels of debt and inflation.

Nearly all of Malawi's people farm for a living. About 75 percent of the people live in villages in rural areas. Aid from foreign countries and religious groups has been important to the economy. However, the country has been slow to build factories and industries. In the future, Malawi will probably have to develop its own industries rather than rely on aid from foreign countries.

READING CHECK Generalizing What are the economies like in Central African countries?

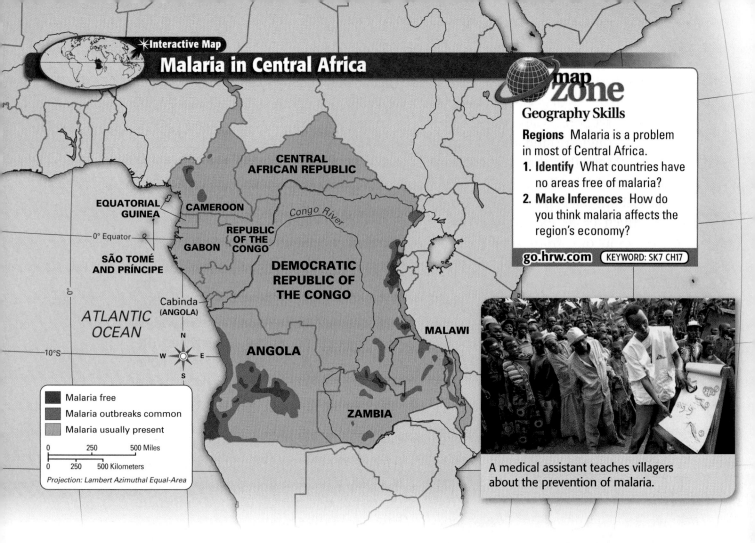

map zone

Geography Skills

Regions Malaria is a problem in most of Central Africa.
1. **Identify** What countries have no areas free of malaria?
2. **Make Inferences** How do you think malaria affects the region's economy?

go.hrw.com KEYWORD: SK7 CH17

CENTRAL AFRICAN REPUBLIC

EQUATORIAL GUINEA

CAMEROON

Congo River

0° Equator

REPUBLIC OF THE CONGO

GABON

SÃO TOMÉ AND PRÍNCIPE

DEMOCRATIC REPUBLIC OF THE CONGO

ATLANTIC OCEAN

Cabinda (ANGOLA)

0°

MALAWI

N W E S

ANGOLA

10°S

ZAMBIA

Malaria free
Malaria outbreaks common
Malaria usually present

0 250 500 Miles
0 250 500 Kilometers
Projection: Lambert Azimuthal Equal-Area

A medical assistant teaches villagers about the prevention of malaria.

Issues and Challenges

As you have read, many of the countries in Central Africa have unstable governments and poor economies. These circumstances have been either the cause or effect of other issues and challenges in the region today.

Ethnic and Regional Conflict

A mix of ethnic groups and competing desires for power has led to civil war in many of the region's countries. Thousands of people have been killed in these wars over the past several years.

ACADEMIC VOCABULARY

implement to put in place

Wars have also contributed to poor economies in the region. The people killed or injured in the fighting can no longer work. In addition, the fighting destroys land and other resources that could be used in more productive ways.

Health

Like war, disease kills many people in the region. **Malaria is a disease spread by mosquitoes that causes fever and pain.** Without treatment it can lead to death. In fact, malaria is by far the most common cause of death in Central Africa. A child there dies from malaria every 30 seconds. On the map above, you can see that this disease is a problem almost everywhere.

International health organizations and some national governments have begun to **implement** strategies to control malaria. These strategies include educating people about the disease and passing out nets treated with insecticide. The nets and medicine are expensive, and not everyone can afford them. However, people who sleep under these nets will be protected from mosquitoes and malaria.

While some countries are beginning to control malaria, another disease is spreading rapidly. HIV, the virus that causes AIDS, is very common in Central Africa. Hundreds of thousands of people die of AIDS each year in Central Africa. There is no cure for HIV infection, and medicines to control it are very expensive. International groups are working hard to find a cure for HIV and to slow the spread of the disease.

Partly because so many people die of disease, Central Africa has a very young population. Almost 50 percent of people living in Central Africa are under age 15. For comparison, only about 20 percent of the people in the United States are under age 15. Although many young people in Central Africa work, they do not contribute to the economy as much as older, more experienced workers do.

Resources and Environment

To help their economies and their people, the countries of Central Africa must begin to develop their natural resources more effectively. Agricultural land is one resource that must be managed more effectively. In some places, partly because of war, food production has actually declined. Also, food production cannot keep up with the demands of the growing population. The results are food shortages and malnutrition. **Malnutrition** is the condition of not getting enough nutrients from food.

The environment is another important resource that must be managed. Some of Central Africa's most important industries are destroying the environment. Lumber companies cut down trees in the tropical forest, threatening the wildlife that lives there. Mining is also harming the environment. Diamonds and copper are mined in huge open pits. This mining process removes large areas of land and destroys the landscape.

Many people in Central Africa and around the world are working hard and spending billions of dollars to improve conditions in the region. National parks have been set up to protect the environment. Projects to provide irrigation and prevent erosion are helping people plant more crops. Central Africa's land and people hold great potential for the future.

READING CHECK Summarizing What are some threats to Central Africa's environment?

SUMMARY AND PREVIEW Countries in Central Africa are trying to build stable governments and strong economies after years of civil war, but challenges slow economic development. Next, you will learn about the places and people of Southern Africa.

Section 3 Assessment

go.hrw.com
Online Quiz
KEYWORD: SK7 HP17

Reviewing Ideas, Terms, and Places

1. a. **Define** What is **inflation**?
   b. **Summarize** What effect did Mobutu Sese Seko's rule have on the Democratic Republic of the Congo?
   c. **Evaluate** Do you think Central African countries would benefit more from a stable government or from a strong economy? Explain your answer.
2. a. **Identify** What causes **malaria**?
   b. **Explain** How are some countries coping with environmental challenges?

Critical Thinking

3. **Evaluating** Look over your notes on Central Africa. Using a graphic organizer like the one here, rank the challenges facing Central Africa. Put the one you see as the biggest challenge first.

Challenges
1.
2.

FOCUS ON WRITING

4. **Describing Countries** You can focus on one country or on the whole region in your acrostic. Review your notes, and then decide whether to select one to write about.

Interpreting a Population Pyramid

Learn

A population pyramid shows the percentages of males and females by age group in a country's population. The pyramids are split into two sides. Each bar on the left shows the percentage of a country's population that is male and of a certain age. The bars on the right show the same information for females.

Population pyramids help us understand population trends in countries. Countries that have large percentages of young people have populations that are growing rapidly. Countries with more older people are growing slowly or not at all.

Practice

Use the population pyramid of Angola to answer the following questions.

❶ What age group is the largest?

❷ What percent of Angola's population is made up of 15- to 19-year-old males?

❸ What does this population pyramid tell you about the population trend in Angola?

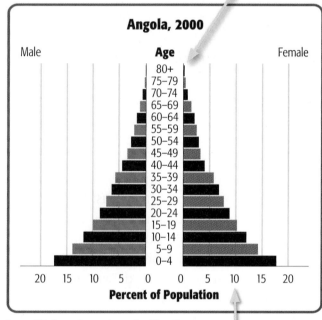

Ages are listed down the middle of the diagram.

Source: U.S. Census Bureau, International Data Base

Percentages are labeled across the bottom of the diagram.

Apply

Do research at the library or on the Internet to find age and population data for the United States. Use that information to answer the following questions.

1. What age group is the largest?

2. Are there more males or females over age 80?

3. How would you describe the shape of the population pyramid?

Geography's Impact
video series
Review the video to answer the closing question:
Why do you think the president of Gabon passed a law establishing new national parks?

Visual Summary

Use the visual summary below to help you review the main ideas of the chapter.

QUICK FACTS

The forests of Central Africa's Congo Basin are home to gorillas and many other kinds of animals.

Ivory attracted Europeans to Central Africa. They left their influence on the region's history and culture.

Countries of Central Africa are looking for ways to solve many of their challenges, such as preventing disease.

Reviewing Vocabulary, Terms, and Places

Using your own paper, complete the sentences below by providing the correct term for each blank.

1. The _____ is a low area near the middle of Central Africa.

2. _____ is a disease spread by mosquitoes that causes fever and aching.

3. People who do not get enough nutrients from their food suffer from _____.

4. To _____ a system is to put it in place.

5. _____ is the rise in prices that occurs when currency loses its buying power.

6. A _____ is a regional variety of a language.

7. The river that flows through Central Africa and into the Atlantic Ocean is the _____.

8. A _____ is an open-air market set up once or twice a week.

9. Much of the copper in Central Africa comes from a region known as the _____.

Comprehension and Critical Thinking

SECTION 1 *(Pages 414–417)*

10. **a. Describe** What are the main landforms in Central Africa?

   b. Make Inferences Why would people in rural areas be more likely to shop at periodic markets than at grocery stores?

   c. Elaborate How does the development of national parks affect people in the region? How does it affect people around the world?

SECTION 2 *(Pages 420–423)*

11. **a. Recall** When did European countries divide Central Africa into colonies?

   b. Analyze What factors besides European colonization influenced where different religions are most common in Central Africa today?

   c. Evaluate What do you think was the most significant influence or effect the Europeans had on Central Africa? Explain your answer.

12. **a. Identify** What are the diseases that affect many people in Central Africa?

    b. Analyze What factors have allowed certain countries like Cameroon and Gabon to have stronger economies than other countries in the region?

    c. Evaluate What are the benefits of foreign aid to Central Africa? What might be some possible drawbacks?

FOCUS ON READING AND WRITING

Using Word Parts *Look at the list of prefixes and suffixes and their meanings below. Then answer the questions that follow.*

mal- (bad)	*-ous* (characterized by)
in- (not)	*-ment* (result, action)
re- (again)	*-ion* (action, condition)

13. Which of the following words means "getting only poor nutrients"?

    a. nutriment   **c.** renutrition

    b. malnutrition   **d.** nutrious

14. Which of the following words means "the condition of being protected"?

    a. reprotect   **c.** protection

    b. protectment   **d.** protectous

15. **Writing Your Acrostic** Your poem will describe Central Africa or a part of it. Choose the place you want to describe and write the letters of that word vertically, with one letter on each line of your paper. For each letter, use your notes to write a descriptive word or phrase that tells about your subject. Make a final copy of your acrostic to share with classmates.

Social Studies Skills

Interpreting a Population Pyramid *Use the population pyramid in the Social Studies Skills lesson to answer the following questions.*

16. What age group is the smallest?

17. How would you describe the current population in Angola?

go.hrw.com
KEYWORD: SK7 CH17

Using the Internet

18. **Activity: Making a Scrapbook** Central Africa is home to many different ethnic groups. Although there are similarities among them, they each have unique characteristics as well. Enter the activity keyword and take a journey to Central Africa. Research some of the many groups that live there. Then create an illustrated scrapbook that documents some of the groups that you have met in your travels. Include information on their towns, ways of life, and environments. You may also want to include maps, souvenirs, and pictures from your journey.

Map Activity   ★Interactive

19. **Central Africa** On a separate sheet of paper, match the letters on the map with their correct labels.

    Congo River   Congo Basin

    Zambezi River   Lake Nyasa

    Angola

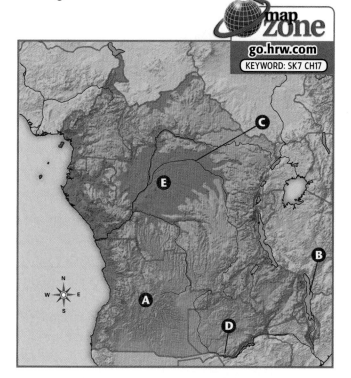

go.hrw.com
KEYWORD: SK7 CH17

DIRECTIONS: *Read questions 1 through 6 and write the letter of the best response. Then read question 7 and write your own well-constructed response.*

1 **What major river flows through Central Africa and into the Atlantic Ocean?**

  A Zambezi River

  B Congo Basin

  C Niger River

  D Congo River

2 **In rural areas, people are most likely to trade goods at a**

  A copper belt.

  B periodic market.

  C supermarket.

  D dialect.

3 **Why did Europeans become interested in Central Africa?**

  A They wanted resources and trade goods.

  B They wanted to teach people European languages.

  C They wanted to divide up ethnic groups.

  D They wanted to destroy African kingdoms.

4 **What disease is spread by mosquitoes and is very common in Central Africa?**

  A malnutrition

  B HIV

  C malaria

  D inflation

5 **Most people in Zambia and Malawi work in**

  A copper mines.

  B the oil industry.

  C cities.

  D farming.

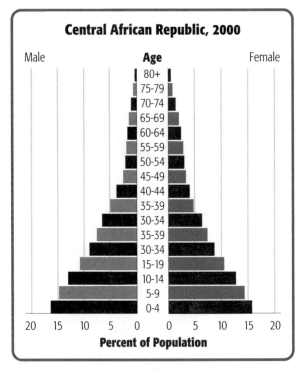

Central African Republic, 2000

Source: U.S. Census Bureau, International Data Base

6 **Based on the graph above, which of the following statements is false?**

  A Females ages 15–19 make up about 10 percent of the population.

  B Males and females ages 0–4 each make up over 15 percent of the population.

  C The population of Central African Republic is growing at a very slow rate.

  D There are more males ages 5–9 than there are females.

7 **Extended Response** Using the graph above and the graph of Population Growth in Kinshasa in Section 3, write a paragraph explaining how Central Africa's population and people's lives in the region are changing.

CHAPTER 18

Southern Africa

What You Will Learn...

In this chapter you will learn about nine countries that are located in the region of Southern Africa—South Africa, Lesotho, Swaziland, Namibia, Botswana, Zimbabwe, Mozambique, Madagascar, and Comoros. You will learn about the region's history, cultures, and economies.

SECTION 1
Physical Geography **436**

SECTION 2
History and Culture....................**440**

SECTION 3
Southern Africa Today **446**

FOCUS ON READING AND VIEWING

Making Generalizations A generalization is a broad, general conclusion drawn from examples, facts, or other information. As you read this chapter, try to make generalizations about the facts and information in the text. Making generalizations will help you understand the meaning of what you are reading. **See the lesson, Making Generalizations, on page 685**.

Viewing a TV News Report You are a journalist covering world news. Your assignment is to create a brief TV news report on something about Southern Africa. As you read this chapter, you will collect information about the region and plan your report. Later you and your classmates will give your TV news reports and evaluate one another's reports.

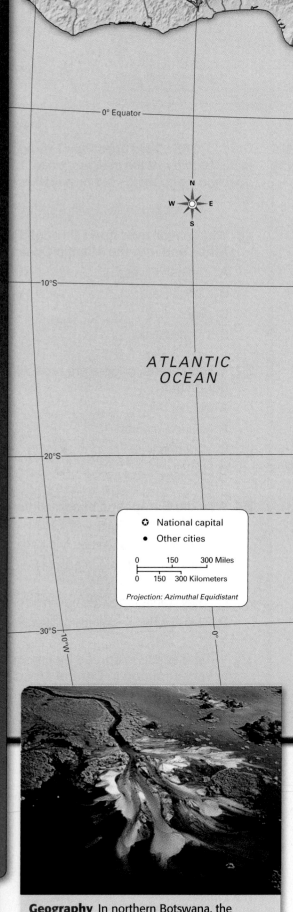

ATLANTIC OCEAN

○ National capital
• Other cities

0   150   300 Miles
0   150   300 Kilometers

Projection: Azimuthal Equidistant

Geography In northern Botswana, the Okavango River forms an enormous inland delta.

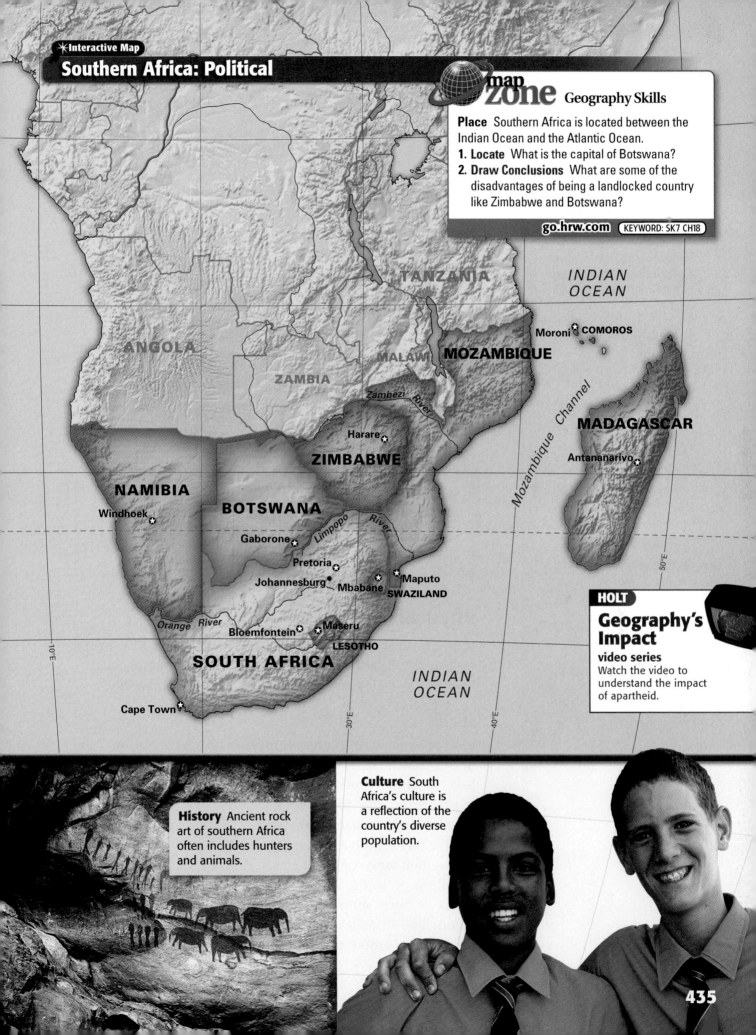

Southern Africa: Political

map **zone** Geography Skills

Place Southern Africa is located between the Indian Ocean and the Atlantic Ocean.
1. **Locate** What is the capital of Botswana?
2. **Draw Conclusions** What are some of the disadvantages of being a landlocked country like Zimbabwe and Botswana?

go.hrw.com KEYWORD: SK7 CH18

INDIAN OCEAN

TANZANIA

ANGOLA

ZAMBIA

MALAWI

MOZAMBIQUE

Moroni COMOROS

Zambezi River

Harare

ZIMBABWE

MADAGASCAR

Antananarivo

Mozambique Channel

NAMIBIA

Windhoek

BOTSWANA

Limpopo River

Gaborone

Pretoria

Johannesburg

Mbabane

Maputo

SWAZILAND

Orange River

Bloemfontein

Maseru

LESOTHO

SOUTH AFRICA

Cape Town

INDIAN OCEAN

10°E   30°E   40°E   50°E

HOLT

Geography's Impact

video series
Watch the video to understand the impact of apartheid.

History Ancient rock art of southern Africa often includes hunters and animals.

Culture South Africa's culture is a reflection of the country's diverse population.

Physical Geography

What You Will Learn...

Main Ideas

1. Southern Africa's main physical feature is a large plateau with plains, rivers, and mountains.
2. The climate and vegetation of Southern Africa is mostly savanna and desert.
3. Southern Africa has valuable mineral resources.

The Big Idea

Southern Africa's physical geography includes a high, mostly dry plateau, grassy plains and rivers, and valuable mineral resources.

Key Terms and Places

escarpment, *p. 436*
veld, *p. 438*
Namib Desert, *p. 438*
pans, *p. 438*

TAKING NOTES As you read, take notes on the physical geography of Southern Africa. Use a chart like this one to organize your notes.

Physical Features	
Climate and Vegetation	
Resources	

If YOU lived there...

You are a member of the San, a people who live in the Kalahari Desert. Your family lives with several others in a group of circular grass huts. You are friends with the other children. Sometimes you help your mom look for eggs or plants to use for carrying water. Your water containers, clothes, carrying bags, and weapons all come from the resources you find in the desert. Next year you will move away to attend school in a town.

How will your life change next year?

BUILDING BACKGROUND Parts of Southern Africa have a desert climate. Little vegetation grows in these areas, but some people do live there. Most of Southern Africa's people live in cooler and wetter areas, such as on the high, grassy plains in the south and east.

Physical Features

Southern Africa has some amazing scenery. On a visit to the region, you might see grassy plains, steamy swamps, mighty rivers, rocky waterfalls, and steep mountains and plateaus.

Plateaus and Mountains

Most of the land in Southern Africa lies on a large plateau. Parts of this plateau reach more than 4,000 feet (1,220 m) above sea level. To form the plateau, the land rises sharply from a narrow coastal plain. The steep face at the edge of a plateau or other raised area is called an **escarpment**.

In eastern South Africa, part of the escarpment is made up of a mountain range called the Drakensberg (DRAH-kuhnz-buhrk). The steep peaks rise as high as 11,425 feet (3,482 m). Farther north, another mountain range, the Inyanga (in-YANG-guh) Mountains, separates Zimbabwe and Mozambique. Southern Africa also has mountains along its western coast.

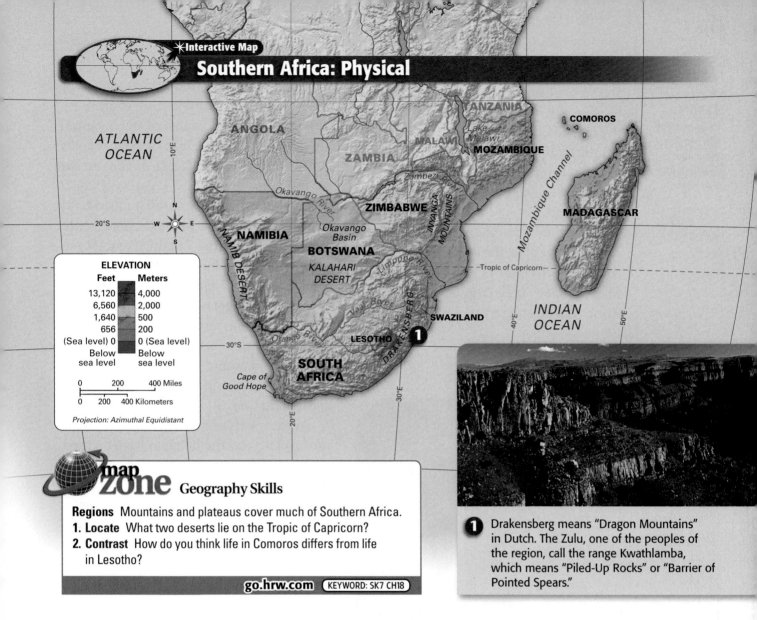

Interactive Map

Southern Africa: Physical

ATLANTIC OCEAN

ANGOLA

ZAMBIA

TANZANIA

MALAWI

Lake Malawi

MOZAMBIQUE

COMOROS

MADAGASCAR

NAMIBIA

Okavango River

Okavango Basin

ZIMBABWE

Zambezi River

INYANGA MOUNTAINS

Mozambique Channel

BOTSWANA

KALAHARI DESERT

Limpopo River

Tropic of Capricorn

INDIAN OCEAN

NAMIB DESERT

Vaal River

SWAZILAND

DRAKENSBERG

Orange River

LESOTHO

SOUTH AFRICA

Cape of Good Hope

ELEVATION

Feet	Meters
13,120	4,000
6,560	2,000
1,640	500
656	200
(Sea level) 0	0 (Sea level)
Below sea level	Below sea level

0   200   400 Miles

0   200   400 Kilometers

Projection: Azimuthal Equidistant

mapZone Geography Skills

Regions Mountains and plateaus cover much of Southern Africa.
1. **Locate** What two deserts lie on the Tropic of Capricorn?
2. **Contrast** How do you think life in Comoros differs from life in Lesotho?

go.hrw.com   KEYWORD: SK7 CH18

1 Drakensberg means "Dragon Mountains" in Dutch. The Zulu, one of the peoples of the region, call the range Kwathlamba, which means "Piled-Up Rocks" or "Barrier of Pointed Spears."

Plains and Rivers

Southern Africa's narrow coastal plain and the wide plateau are covered with grassy plains. These flat plains are home to animals such as lions, leopards, elephants, baboons, and antelope.

Several large rivers cross Southern Africa's plains. The Okavango River flows from Angola into a huge basin in Botswana. This river's water never reaches the ocean. Instead it forms a swampy inland delta that is home to crocodiles, zebras, hippos, and other animals. Many tourists travel to Botswana to see these wild animals in their natural habitat.

The Orange River passes through the rocky Augrabies (oh-KRAH-bees) Falls as it flows to the Atlantic Ocean. When the water in the river is at its highest, the falls are several miles wide. The water tumbles down 19 separate waterfalls. The Limpopo River is another of the region's major rivers. It flows into the Indian Ocean. **Features** such as waterfalls and other obstacles block ships from sailing up these rivers. However, the rivers do allow irrigation for farmland in an otherwise dry area.

READING CHECK **Generalizing** What are Southern Africa's main physical features?

ACADEMIC VOCABULARY

features characteristics

Climate and Vegetation

FOCUS ON READING

What generalization can you make about Southern Africa's climate?

Southern Africa's climates vary from east to west. The wettest place in the region is the east coast of the island of Madagascar. On the mainland, winds carrying moisture blow in from the Indian Ocean. Because the Drakensberg's high elevation causes these winds to blow upward, the eastern slopes of these mountains are rainy.

In contrast to the eastern part of the continent, the west is very dry. From the Atlantic coast, deserts give way to plains with semiarid and steppe climates.

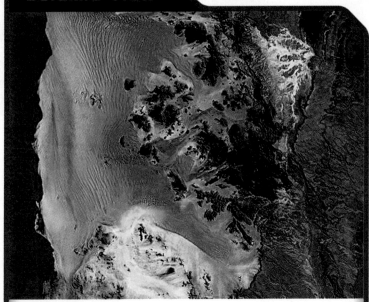

Satellite View

Namib Desert

One of the world's most unusual deserts, the Namib lies on the Atlantic coast in Namibia. As this satellite image shows, the land there is extremely dry. Some of the world's highest sand dunes stretch for miles along the coast.

In spite of its harsh conditions, some insects have adapted to life in the desert. They can survive there because at night a fog rolls in from the ocean. The insects use the fog as a source of water.

Drawing Conclusions How have some insects adapted to living in the Namib Desert?

Savanna and Deserts

A large savanna region covers much of Southern Africa. Shrubs and short trees grow on the grassy plains of the savanna. In South Africa, these open grassland areas are known as the **veld** (VELT). As you can see on the map on the next page, vegetation gets more sparse in the south and west.

The driest place in the region is the **Namib Desert** on the Atlantic coast. Some parts of the Namib get as little as a half an inch (13 mm) of rainfall per year. In this dry area, plants get water from dew and fog rather than from rain.

Another desert, the Kalahari, occupies most of Botswana. Although this desert gets enough rain in the north to support grasses and trees, its sandy plains are mostly covered with scattered shrubs. Ancient streams crossing the Kalahari have drained into low, flat areas, or **pans**. On these flat areas, minerals left behind when the water evaporated form a glittering white layer.

Tropical Forests

Unlike the mainland, Madagascar has lush vegetation and tropical forests. It also has many animals found nowhere else. For example, some 50 species of lemurs, relatives of apes, live only on this island. However, the destruction of Madagascar's forests has endangered many of the island's animals.

READING CHECK **Summarizing** What is the climate and vegetation like in Southern Africa?

Resources

Southern Africa is rich in natural resources. Madagascar's forests provide timber. The region's rivers supply hydroelectricity and water for irrigation. Where rain is plentiful or irrigation is possible, farmers can grow a wide range of crops.

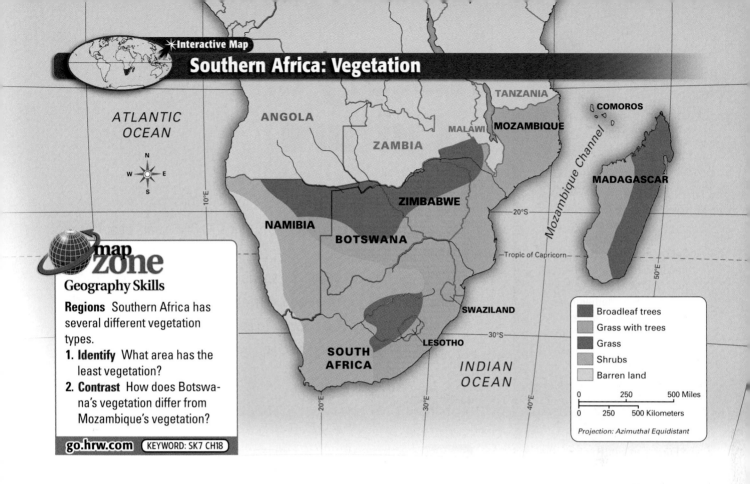

Southern Africa: Vegetation

ATLANTIC
OCEAN

ANGOLA

TANZANIA

COMOROS

MALAWI MOZAMBIQUE

ZAMBIA

Mozambique Channel

MADAGASCAR

ZIMBABWE

20°S

NAMIBIA

BOTSWANA

Tropic of Capricorn

SWAZILAND

30°S

LESOTHO

SOUTH
AFRICA

INDIAN
OCEAN

Broadleaf trees
Grass with trees
Grass
Shrubs
Barren land

0   250   500 Miles
0   250   500 Kilometers

Projection: Azimuthal Equidistant

map zone

Geography Skills

Regions Southern Africa has several different vegetation types.

1. **Identify** What area has the least vegetation?
2. **Contrast** How does Botswana's vegetation differ from Mozambique's vegetation?

go.hrw.com KEYWORD: SK7 CH18

The region's most valuable resources, however, are minerals. Mines in South Africa produce most of the world's gold. In addition, South Africa, Botswana, and Namibia have productive diamond mines. Other mineral resources in Southern Africa include coal, platinum, copper, uranium, and iron ore. Although mining is very important to the economy of the region, the mines can have damaging effects on the surrounding natural environments.

READING CHECK Finding Main Ideas What are the main resources of Southern Africa?

SUMMARY AND PREVIEW Southern Africa is mainly covered with grassy plains and deserts atop a large plateau. Minerals are among the region's main resources. In the next section, you will learn about Southern Africa's history and culture.

Section 1 Assessment

go.hrw.com
Online Quiz
KEYWORD: SK7 HP18

Reviewing Ideas, Terms, and Places

1. **a. Define** What is an **escarpment**?
   b. Elaborate How is the Okavango River different from most other rivers you have studied?
2. **a. Recall** Where in Southern Africa is the driest climate?
   b. Explain What caused minerals to collect in **pans** in the Kalahari Desert?
3. **a. Identify** What are Southern Africa's most valuable resources?
   b. Elaborate How do you think the gold and diamond mines have affected South Africa's economy?

Critical Thinking

4. **Categorizing** Review your notes and use a graphic organizer like this one to sort characteristics by location.

	East	West
Physical Features		
Climate and Vegetation		

FOCUS ON VIEWING

5. **Telling about the Physical Geography** Your TV news report might focus on some part of the geography of Southern Africa. Could you focus on the destruction of the rain forest or life in the desert?

History and Culture

What You Will Learn...

Main Ideas

1. Southern Africa's history began with hunter-gatherers, followed by great empires and European settlements.
2. The cultures of Southern Africa are rich in different languages, religions, customs, and art.

The Big Idea

Native African ethnic groups and European settlements influenced the history and culture of Southern Africa.

Key Terms and Places

Great Zimbabwe, *p. 441*
Cape of Good Hope, *p. 441*
Afrikaners, *p. 442*
Boers, *p. 442*
apartheid, *p. 442*
township, *p. 443*

TAKING NOTES As you read, take notes on the history and culture of Southern Africa. Use the chart below to organize your notes.

History	Culture

If YOU lived there...

You are a hunter living in Southern Africa 10,000 years ago. The animals you hunt include antelope, rhinoceros, and ostrich. A spear is your only weapon. You spend several days following herds of animals until you and several other people are able to surround them. After the hunt, you decide to paint your hunting experience on a rock overhang near where you live.

Why do you paint these images of animals?

BUILDING BACKGROUND Southern Africa's fertile land and its abundance of wildlife have supported different peoples for tens of thousands of years. Hunter-gatherers were the first peoples to thrive in the region. Much later, peoples from West Africa migrated to the region, and then eventually Europeans.

History

As you learned in the previous chapter, Bantu farmers migrated from West Africa to Central Africa as early as 2,000 years ago. These peoples also migrated to Southern Africa at about the same time. Much later, in the 1700s, Europeans arrived on the coast of Southern Africa and forever changed the landscape and ways of life of the people in the region.

Early History

For many centuries the Khoisan peoples lived in Southern Africa. Divided into several ethnic groups, the Khoisan were hunter-gatherers and herders. When the early Bantu peoples migrated from West and Central Africa, they brought new languages and iron tools.

One Bantu group, the Shona, built an empire that reached its height in the 1400s. The Shona Empire included much of what is now the countries of Zimbabwe and Mozambique.

The Shona farmed, raised cattle, and traded gold with other groups on the coast.

The Shona are best known for **Great Zimbabwe**, their stone-walled capital. In fact, the name Zimbabwe is the Shona word for stone-walled towns. The builders of Great Zimbabwe used huge granite boulders and rectangular blocks of stone to build the capital's walls.

Founded in the late 1000s, Great Zimbabwe was a small trading and herding center. In the 1100s, the population grew, and both gold mining and farming grew in importance. Great Zimbabwe may have had 10,000 to 20,000 residents. With these resources, the city eventually became the center of a large trading network.

Trade made Great Zimbabwe's rulers wealthy and powerful. However, in the 1400s the gold trade declined. Deprived of its main source of wealth, Great Zimbabwe weakened. By 1500 it was no longer a capital and trading center.

Archaeologists have found Chinese porcelain and other artifacts from Asia at Great Zimbabwe. These artifacts suggest that the Shona traded widely. In addition to trading with peoples of Asia, the Shona apparently traded with the Swahili. The Swahili were Muslim Africans living along the East Africa coast. In effect, all of these peoples were once connected by an Indian Ocean trade network.

Europeans in Southern Africa

In the late 1400s traders from Portugal explored the Southern African coast on their way to Asia to trade for spices. To get to Asia from Portugal, they had to sail around the southern tip of Africa and then cross the Indian Ocean. The trip was long and difficult, so they set up bases on the Southern African coast. These bases provided the ships with supplies.

The Dutch Other Europeans arrived in Southern Africa after the Portuguese. People from the Netherlands, or the Dutch, were the first Europeans to settle in the region. In 1652 the Dutch set up a trade station at a natural harbor near the **Cape of Good Hope**.

Great Zimbabwe

Highly skilled craftspeople built several stone walls that surrounded the Shona capital of Great Zimbabwe. Today the ruins are a World Heritage Site.

ANALYZING VISUALS Why do you think Great Zimbabwe was made of stone?

The Cape sits at the tip of Africa. The land around the Cape lacked the gold and copper of the interior. However, it had a mild climate, similar to the climate the Dutch were used to back home.

This small colony on the Cape provided supplies to Dutch ships sailing between Dutch colonies in the East Indies and the Netherlands. The Dutch eventually brought in slaves from the region and Southeast Asia to work in the colony.

The Afrikaners and the Boers The people of the colony were very diverse. In addition to the Dutch, other Europeans also settled on the Cape. Dutch, French, and German settlers and their descendants in South Africa were called **Afrikaners**. Over time, a new language called Afrikaans emerged in the Cape colony. This language combined Dutch with Khoisan and Bantu words. German, French, and English also influenced the language's development.

Dutch Settlers

This painting shows a meeting between a Dutch governor and a Khoisan chief.

In the early 1800s, Great Britain took over the area of the Cape. The **Boers**, Afrikaner frontier farmers who had spread out from the original Cape colony, resisted the British. Many Boers packed all their belongings into wagons and soon moved farther east and north.

The Zulu and the British At about the same time, a Bantu-speaking group, the Zulu, became a powerful fighting force in the region. They conquered the surrounding African peoples, creating their own empire. When the Boers moved north of the Cape, they entered Zulu territory. The two sides clashed over control of the land. Eventually the British also wanted Zulu land. After a series of battles, the British defeated the Zulu.

The ending of slavery in the British Empire in the 1830s brought changes to the economy of colonial settlements in the region. Instead of slaves, people traded ivory—the tusks of elephants. Over time, however, hunters wiped out the entire elephant population in some parts of Southern Africa. With ivory in short supply, trade shifted to diamonds and gold, which were discovered in South Africa in the 1860s.

Apartheid

In the early 1900s South Africa's government, which was dominated by white Afrikaners, became increasingly racist. As a result, black South Africans opposed the government. To defend their rights, they formed the African National Congress (ANC) in 1912.

However, the trend toward racial division and inequality continued. South Africa's government set up a policy of separation of races, or **apartheid**, which means "apartness." This policy divided people into four groups: whites, blacks, Coloureds, and Asians.

Music of South Africa

Stomping. Spinning. Swaying. This is the kind of dancing you might see at a performance of South African music. In addition, musicians playing drums, guitars, and traditional flutes provide a rhythmic beat that makes it impossible to stand still.

One of the groups that does it the best are the Mahotella Queens. They are grandmothers who have been singing together for over 40 years. Their songs mix gospel with traditional African music. Performing all over the world, the Queens give unforgettable performances of nonstop singing and dancing.

Drawing Conclusions Why do you think South African music is popular around the world?

Coloureds and Asians were only allowed to live in certain areas. Each African tribe or group was given its own rural "homeland." These homelands generally did not include good farmland, mines, and other natural resources. Those resources were owned by the whites, and blacks had no rights in white areas.

Housing, health care, and schools for blacks were poor compared to those for whites. Schools for Coloureds were poor, but slightly better than the black schools.

During apartheid, many blacks found work in white-owned industries, mines, shops, and farms. Blacks had to live in separate areas called **townships**, which were often crowded clusters of tiny homes. The townships were far from the jobs in the cities and mines.

Independence

Beginning in the 1960s, many colonies gained independence from the European countries that had once colonized them. Some gained independence rather peacefully, but others struggled. For example, the British colonists in Rhodesia fought native Africans for years. Fighting broke out after the colonists declared their own white-dominated republic in 1970. Finally, in 1980, the Africans won independence and renamed their country Zimbabwe.

Independence also did not come easy for other countries. Despite violent resistance, Namibia continued to be ruled by South Africa until 1990. Mozambique was granted independence in 1975 after 10 years of war against Portuguese rule.

READING CHECK **Generalizing** Why did Europeans settle Southern Africa?

Culture

Over time, many groups of people created a diverse culture in Southern Africa. As a result, the region's culture reflects both European and African influence.

Ndebele Village

The Ndebele are one of many ethnic groups in South Africa who have kept their traditional culture alive. Many live in villages of brightly painted houses and courtyards.

ANALYSIS SKILL **ANALYZING VISUALS**

What aspect of Ndebele culture do you see in these two photographs?

People

The people of Southern Africa belong to hundreds of different ethnic groups. Some groups are very large. For example, about 9 million people in South Africa are Zulu. More than 1.2 million of Botswana's 1.6 million people belong to a single ethnic group, the Tswana.

Other ethnic groups are small and usually not native to Africa. For example, about 6 percent of Namibia's population is of German descent. In Madagascar people are a mix of 18 small ethnic groups. These Malagasy groups descended from people who migrated across the Indian Ocean from Indonesia.

Languages

FOCUS ON READING

What generalization can you make about the languages spoken in Southern Africa?

Because people in Southern Africa belong to hundreds of different ethnic groups, they speak many languages. Most of the African languages spoken in Southern Africa are related to one of two language families—Khoisan or Bantu.

The early peoples of Southern Africa spoke different Khoisan languages. Khoisan speakers are known for the "click" sounds they make when they speak. Today, the majority of Khoisan speakers belong to the San ethnic group and live in remote areas of Botswana and Namibia.

Most people in Southern Africa speak one of the more than 200 Bantu languages. For example, most of South Africa's 11 official languages are Bantu.

In countries with European influence, European languages are also spoken. For example, English is the official language of Namibia and Zimbabwe. The official language of Mozambique is Portuguese.

Religion

In addition to language, Europeans brought the religion of Christianity to Southern Africa. As a result, millions of people in Southern Africa are Christians. In Namibia and South Africa the majority of the population is Christian.

Painting houses is traditionally the role of Ndebele women. They use colorful geometric patterns.

Artists in Zimbabwe are known for their beautiful stone sculptures of birds and other animals. Traditional crafts of Botswana include ostrich-eggshell beadwork and woven baskets with complex designs. People there also produce colorful wool rugs.

READING CHECK **Analyzing** Why do you think the people of Southern Africa speak several different languages?

SUMMARY AND PREVIEW Southern Africa's ancient history and later European settlement greatly influenced the region's culture. Next, you will learn about the governments and economies of the region's countries today.

Many people in Southern Africa who are not Christian practice traditional African religions. Some of these people believe that ancestors and the spirits of the dead have divine powers. In Zimbabwe, traditional beliefs and Christianity have been mixed together. About half of the people in Zimbabwe practice a combination of traditional beliefs and Christianity.

Celebrations and Art

Southern Africans celebrate many holidays. On Heritage Day, South Africans celebrate their country's diverse population. Most countries in Southern Africa celebrate their countries' independence day. Many Christian holidays such as Christmas Day are also celebrated throughout the region.

Southern Africa's art reflects its many cultures. For example, South African artists make traditional ethnic designs for items such as clothing, lamps, linens, and other products. Artists in Lesotho are famous for their woven tapestries of daily life.

go.hrw.com
Online Quiz
KEYWORD: SK7 HP18

Section 2 Assessment

Reviewing Ideas, Terms, and Places

1. **a. Define** What was **apartheid**?
   b. Draw Conclusions Why did the Shona capital of **Great Zimbabwe** decline as a trading center?
   c. Elaborate Why do you think the language of Afrikaans developed among the European colonists?
2. **a. Recall** What ethnic group in Southern Africa speaks languages that use click sounds?
   b. Draw Conclusions How do the religions practiced in Southern Africa reflect the region's history?
   c. Evaluate Why do you think Heritage Day is a national holiday in South Africa?

Critical Thinking

3. **Sequencing** Review your notes on the history of Southern Africa. Then organize your information using a time line like the one below. You may add more dates if you need to.

   AD 1000   1990

FOCUS ON VIEWING

4. **Discussing History and Culture** Which information about the history and culture of Southern Africa might make a good TV news report? What visuals would be interesting?

Southern Africa Today

What You Will Learn...

Main Ideas

1. South Africa ended apartheid and now has a stable government and strong economy.
2. Some countries of Southern Africa have good resources and economies, but several are still struggling.
3. Southern African governments are responding to issues and challenges such as drought, disease, and environmental destruction.

The Big Idea

Countries of Southern Africa today are trying to use their governments and resources to improve their economies and deal with challenges.

Key Terms and Places

sanctions, *p. 446*
Cape Town, *p. 448*
enclave, *p. 448*

 TAKING NOTES As you read, take notes on the countries of Southern Africa today and the challenges they face. Use this chart to organize your notes.

South Africa	
Other Countries	
Challenges	

If YOU lived there...

You are an economic adviser in Botswana. In recent years your country has made progress toward improving people's lives, but you think there is room for improvement. One way you plan to help the economy is by promoting tourism. Botswana already has amazing natural landscapes and fascinating animals.

What could your country do to attract more tourists?

BUILDING BACKGROUND Some of the countries in Southern Africa are relatively well off, with plentiful resources and good jobs and transportation systems. Others lack these positive conditions. One of the most successful countries in the region is South Africa.

South Africa

Today South Africa has a stable government and the strongest economy in the region. In addition, many South Africans are enjoying new rights and freedoms. The country has made great progress in resolving the problems of its past, but it still faces many challenges.

End of Apartheid

Ending apartheid, the separation of races, has probably been South Africa's biggest challenge in recent years. Many people around the world objected to the country's apartheid laws. For that reason, they put **sanctions**—economic or political penalties imposed by one country on another to force a change in policy—on South Africa. Some countries banned trade with South Africa. Several companies in the United States and Europe refused to invest their money in South Africa. In addition, many international scientific and sports organizations refused to include South Africans in meetings or competitions.

Celebrating Mandela's Freedom

South Africans in Soweto warmly welcomed Nelson Mandela after he was released from prison in 1990.

BIOGRAPHY

Nelson Mandela
(1918–   )

Because he protested against apartheid, Nelson Mandela was imprisoned for 26 years. In 1990, however, South Africa's President de Klerk released Mandela from prison. Mandela and de Klerk shared the Nobel Peace Prize in 1993. One year later, Mandela became South Africa's first black president. He wrote a new constitution and worked to improve the living conditions of black South Africans.

Summarizing What did Nelson Mandela accomplish when he was South Africa's president?

The sanctions isolated South Africa. As other countries in Southern Africa gained independence, South Africa became even more isolated. Protest within the country increased. In response, the government outlawed the African National Congress (ANC). This group had been formed to protect the rights of black South Africans. Many ANC members were jailed or forced to leave the country.

The antiapartheid protests continued, however. Finally, in the late 1980s South Africa began to move away from the apartheid system. In 1990 the government released its political prisoners, including Nelson Mandela. Mandela was elected president in 1994 after South Africans of all races were given the right to vote.

Today all races have equal rights in South Africa. The country's public schools and universities are open to all people, as are hospitals and transportation. However, economic equality has come more slowly. White South Africans are still wealthier than the vast majority of black South Africans. Still, South Africans now have opportunities for a better future.

Government and Economy

South Africa's government and economy are well positioned to create a better future for the country. South Africa's new government is a republic with an elected president. The country's constitution emphasizes equality and human rights.

In working toward equality, the government is trying to create jobs and better conditions for black workers and farmers. Currently, most of South Africa's wealth and industries are still controlled by whites. However, even some officials who favor reform are afraid to **execute** new policies. They fear that rapid change will weaken the economy. They are also concerned that it might cause educated and wealthy whites to leave the country.

ACADEMIC VOCABULARY
execute to perform, to carry out

South Africa's strong economy may help bring economic opportunities to the entire population. The country has more resources and industry than most African countries. For example, South Africa is the world's largest producer of several valuable minerals—gold, platinum, and chromium. The country is also a major exporter of gold and diamonds.

Large cities in South Africa also contribute to the country's economy. Africa's largest industrial area is located in Johannesburg. In addition, beautiful cities such as **Cape Town** attract many tourists.

READING CHECK **Analyzing** Why and how did South Africa do away with apartheid?

Other Countries of Southern Africa

The eight other countries in the region share some characteristics with South Africa. Some, but not all, have strong economies and stable governments.

Lesotho and Swaziland

These two countries are particularly influenced by South Africa. Lesotho and Swaziland are both enclaves. An **enclave** is a small territory surrounded by foreign territory. Lesotho and Swaziland are both located completely, or almost completely, within South Africa. Swaziland shares part of its border with Mozambique.

Close-up

Cape Town

Founded by the Dutch in 1652, Cape Town is a bustling international port city today. It lies on the South Atlantic Ocean and is home to about 3 million people.

Hiking trails lead to the top of Lion's Head for an amazing view of the city.

ANALYSIS SKILL **ANALYZING VISUALS**

What can you see in this photograph that might reveal why Cape Town is popular with tourists?

Because it is so small, Lesotho has few resources or agricultural land. As a result, it is a poor country. Many of its people work in nearby South Africa. In spite of its poverty, Lesotho has the highest female literacy rate in Africa. Most children, including females, get at least a primary education in free schools run by Christian churches.

Swaziland has some important mineral deposits and industry. Cattle raising and farming are also common there. A good transportation system helps Swaziland to participate in foreign trade.

Lesotho and Swaziland are both kingdoms. Although each country has a king as head of state, each is governed by an elected prime minister and a parliament.

The city's buildings are a mix of modern and Dutch colonial architecture.

People jog, bike, and rollerblade on this trail along the ocean.

Namibia

Namibia gained its independence from South Africa as recently as 1990. Now it is a republic with an elected president and legislature. Its capital, Windhoek, is located in the central highlands.

Very few people live in Namibia's deserts in the east and the west, but these areas are the sites of some of the richest mineral deposits in Africa. Most of the country's income comes from the mining of diamonds, copper, uranium, lead, and zinc. Fishing in the Atlantic Ocean and sheep ranching are also important sources of income. In spite of this strong economy, however, most people are still poor.

Botswana

Botswana is one of Africa's success stories, thanks to mineral resources and a stable democratic government. The main economic activities in Botswana are cattle ranching and diamond mining. Recently, international companies have built factories there, and tourism is increasing. Although unemployment is high, the country has had one of the world's highest rates of economic growth since the 1960s.

FOCUS ON READING
In general, what is Botswana's economy like?

Zimbabwe

Zimbabwe has suffered from a poor economy and political instability. Zimbabwe does not lack resources. It has gold and copper mines as well as productive agriculture and manufacturing. However, high inflation, debts, and war have hurt the economy.

In addition, there is much inequality. Although white residents make up less than 1 percent of the population, they own most of the large farms and ranches. In 2000 the president began a program to take farmland from white farmers and give the land to black residents. This action led many white farmers to leave the country and caused food shortages.

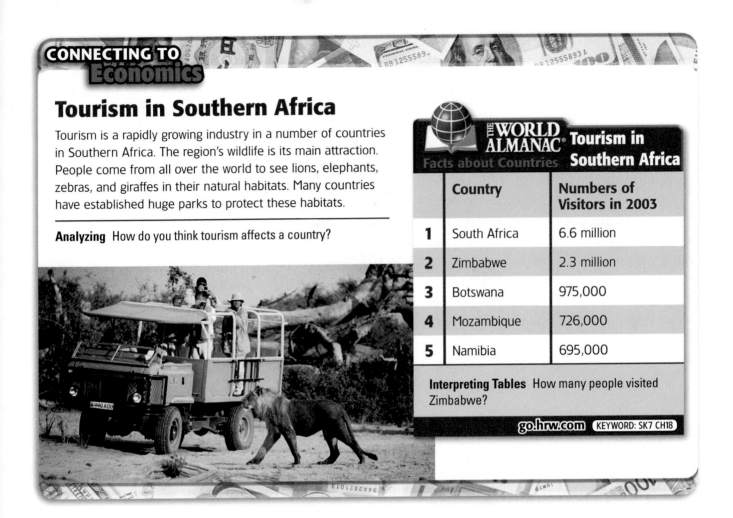

Tourism in Southern Africa

Tourism is a rapidly growing industry in a number of countries in Southern Africa. The region's wildlife is its main attraction. People come from all over the world to see lions, elephants, zebras, and giraffes in their natural habitats. Many countries have established huge parks to protect these habitats.

Analyzing How do you think tourism affects a country?

THE WORLD ALMANAC · Facts about Countries — Tourism in Southern Africa

	Country	Numbers of Visitors in 2003
1	South Africa	6.6 million
2	Zimbabwe	2.3 million
3	Botswana	975,000
4	Mozambique	726,000
5	Namibia	695,000

Interpreting Tables How many people visited Zimbabwe?

go.hrw.com | KEYWORD: SK7 CH18

The attempt at land reform, the poor economy, and violent acts against political opponents have made people in Zimbabwe unhappy with the president. Although he was re-elected in 2002, most people think the election was flawed.

Mozambique

Mozambique is one of the world's poorest countries. The economy has been badly damaged by civil war, but it is improving. Mozambique's ports ship many products from the interior of Africa. Taxes collected on these shipments are an important source of income. Also, plantations grow cashews, cotton, and sugar for export. The country must import more than it exports, however, and it relies on foreign aid.

Madagascar and Comoros

Madagascar was ruled for more than 20 years by a socialist dictator. Today the elected president is working to improve the struggling economy. Most of the country's income comes from exports of coffee, vanilla, sugar, and cloves. Madagascar also has some manufacturing, and the country is popular with tourists who come to see the unique plants and animals.

Comoros is a country made up of four tiny islands. It suffers from a lack of resources and political instability. The government of Comoros is struggling to improve education and promote tourism.

READING CHECK **Contrasting** In what ways are Botswana and Zimbabwe different?

Issues and Challenges

Although conditions in many countries of Southern Africa are better than they are on much of the continent, the region has its own challenges. One of the most serious problems facing Southern Africa is poverty. Terrible droughts often destroy food crops. In addition, many of Southern Africa's people are unemployed.

Disease is another problem. Southern Africa has high numbers of people infected with HIV. The region's governments are trying to educate people to slow the spread of disease.

Another challenge is environmental destruction. For example, in Madagascar, deforestation leads to erosion. There is hope for the future, though. Namibia was the first country in the world to put environmental protection in its constitution. Also, the African Union (AU) works to promote cooperation among African countries. The AU tries to solve problems across the continent.

READING CHECK **Generalizing** What main challenges does Southern Africa face?

Baobab trees line a street in Madagascar.

SUMMARY Southern Africa has valuable mineral resources and landscapes popular with tourists. Some countries have more stable governments and economies than much of Africa. However, the region still faces many challenges.

Section 3 Assessment

Reviewing Ideas, Terms, and Places

1. **a. Describe** What effect did **sanctions** have on South Africa?
   b. Interpret What have been two effects of the end of apartheid?
2. **a. Recall** Which country's president began a program to take farmland from white farmers?
   b. Make Inferences Why might being an **enclave** affect a country's economy?
   c. Rank Besides South Africa, which two countries in the region seem to have the best economies?
3. **a. Describe** How does terrible drought lead to poverty?
   b. Explain How are people in Southern Africa addressing the challenges in the region?

Critical Thinking

4. **Summarizing** Review your notes on South Africa. Then using a graphic organizer like this one, describe what the country has been like at each different period.

Before the 1990s	The 1990s	Today

FOCUS ON VIEWING

5. **Telling about Southern Africa Today** Would you try to include information about all of Southern Africa in your report? Or would you just focus on one country? Take notes on your ideas.

Social Studies Skills

Chart and Graph | Critical Thinking | Geography | Study

Evaluating a Web Site

Learn

The Internet is one of the most valuable tools available for research today. However, not everything that you find on the Internet is useful or accurate. You have to be careful and analyze the sites you use.

A good Web site should be accurate and up-to-date. Before you use a site for research, find out who produced it. The author should be qualified and unbiased. Also, check to see when the site was last updated. If it has not been updated recently, the information it contains may no longer be accurate.

Practice

Study this page taken from a Web site and then answer these questions.

❶ Who do you think produced this Web site? How can you tell?

❷ What kinds of information can you find on this site?

❸ Do you think this would be a good site for research? Why or why not?

A country's official Web site is usually a good source for information.

Check to see how current the articles on the Web site are. Have they been updated regularly?

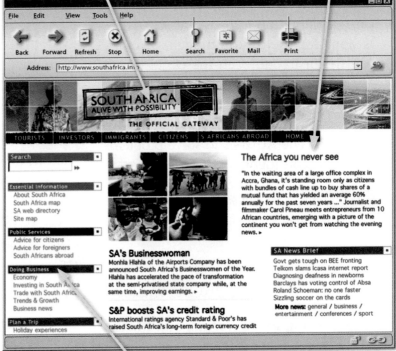

Notice what type of information is present on the Web site. Is the site biased or unbalanced?

Apply

Search the Internet to find a Web page about one of the countries of Southern Africa. Analyze the site and determine whether you think it would be a good site for research. Write a paragraph to explain your decision. Make sure to include the site's URL and the date on which you visited it in your report.

Geography's Impact
video series
Review the video to answer the closing question:
What are some ways South Africans could continue working together?

Visual Summary

Use the visual summary below to help you review the main ideas of the chapter.

QUICK FACTS

The Drakensberg rises to meet a high plateau that dominates the physical geography of Southern Africa.

Traditional African and European cultures mix in Southern Africa. Music and dance are very important.

Countries such as South Africa have strong economies and modern cities. Other countries struggle with poverty.

Reviewing Vocabulary, Terms, and Places

Match the words with their definitions.

1. Great Zimbabwe
2. features
3. Boers
4. apartheid
5. sanctions
6. townships
7. escarpment
8. enclave

a. the steep face at the edge of a plateau or other raised area

b. economic or political penalties imposed by one country on another to force a change in policy

c. a large, stone-walled town built by the Shona

d. Afrikaner frontier farmers in South Africa

e. a small territory surrounded by foreign territory

f. characteristics

g. South Africa's policy of separation of races

h. separate areas with clusters of tiny homes for black South Africans

Comprehension and Critical Thinking

SECTION 1 *(Pages 436–439)*

9. **a. Identify** What are the two main deserts in Southern Africa?

b. Contrast How is the eastern part of Southern Africa different from the western part?

c. Elaborate How do you think the geography of Southern Africa has affected settlement patterns in the region?

SECTION 2 *(Pages 440–445)*

10. **a. Define** Who are the Afrikaners? What country do they live in?

b. Contrast How does the origin of Khoisan languages differ from Bantu languages? What is unusual about Khoisan languages?

c. Elaborate What was life like for non-whites under the policy of apartheid? What rights were blacks, Coloureds, and Asians denied?

11. a. Identify Which countries are enclaves?

b. Analyze In what ways has South Africa changed with the end of apartheid? In what ways has it stayed the same?

c. Evaluate Poverty is the most serious challenge facing Southern Africa. Do you agree or disagree with this statement? Explain your answer.

Using the Internet

go.hrw.com
KEYWORD: SK7 CH18

12. Activity: Researching Apartheid From 1948 until 1994, many people in South Africa were legally discriminated against under the policy known as apartheid. Enter the activity keyword. Imagine that you are a reporter writing an article on the history of apartheid. Using both primary and secondary sources, research who started apartheid, how people struggled against it, and when it finally came to an end. Using that information, create an outline for your article. Be sure to include details from your research that support the main ideas.

FOCUS ON READING AND VIEWING

13. Making Generalizations Re-read the information about South Africa today in Section 3. Based on the specific information you read, make one generalization about the country's economy and one about its resources.

14. Presenting a TV News Report Review your notes and decide on a topic for your report. Next, identify the point you want to make about your topic—your purpose. Your purpose may be to share interesting information—a recently celebrated holiday, for example. Or your purpose may be more serious—perhaps the need to reduce poverty. Decide what images you will show and what you will say to make your point.

Create a script identifying visuals and voice over. Present your report to the class using visuals, just as though you were on the TV news. Listen and watch your classmates' reports. Evaluate their reports based on accuracy of content and visual interest.

Social Studies Skills

15. Analyzing a Web Site Search the Internet to find two Web sites about topics in Southern Africa. One Web site should be one you would consider good to use for research. The other site should be one you do not consider to be a good source of information for research. Write a paragraph comparing and contrasting the two sites. Be sure to explain why one site seems more useful and accurate than the other.

Map Activity ★Interactive

16. Southern Africa On a separate sheet of paper, match the letters on the map with their correct labels.

Cape of Good Hope	Namib Desert
Okavango Basin	Drakensberg
Orange River	

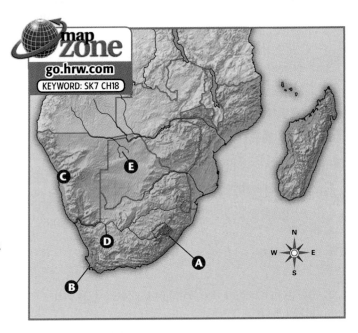

map zone
go.hrw.com
KEYWORD: SK7 CH18

DIRECTIONS: Read questions 1 through 7 and write the letter of the best response. Then read question 8 and write your own well-constructed response.

1 Most of the land in Southern Africa lies on a

A mountain range.

B coastal plain.

C plateau.

D delta.

2 The Dutch first settled in Southern Africa in 1652 near the

A Inyanga Mountains.

B Cape of Good Hope.

C Okavango Basin.

D Namib Desert.

3 Who were the first Europeans to explore the southern coast of Africa?

A Portuguese

B Dutch

C French

D German

4 Which country had a policy called apartheid to separate different races?

A Zimbabwe

B Madagascar

C Namibia

D South Africa

5 Which of the following statements about the end of apartheid is false?

A Sanctions helped bring the end of apartheid.

B Black people and white people now have economic equality.

C Both black people and white people can vote.

D Public schools and universities are open to all people.

Madagascar: Climate

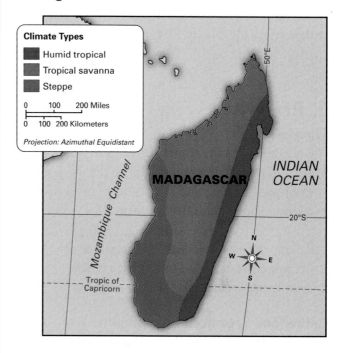

Climate Types
- Humid tropical
- Tropical savanna
- Steppe

0   100   200 Miles

0   100   200 Kilometers

Projection: Azimuthal Equidistant

MADAGASCAR

INDIAN OCEAN

Mozambique Channel

50°E

20°S

Tropic of Capricorn

6 Based on the map above, where would Madagascar's tropical forests likely be located?

A in the east

B in the west

C on the savanna

D in the north

7 Which two countries are enclaves?

A South Africa and Lesotho

B Madagascar and Comoros

C Lesotho and Swaziland

D Zimbabwe and Mozambique

8 **Extended Response** Choose two countries from the table in Section 3 on Tourism in Southern Africa. Think about the information in this chart and what you know about the reources in these two countries. Write a paragraph explaining reasons for the differences and similarities in the number of tourists who visit each country.

Explaining Cause or Effect

"Why did it happen?" "What were the results?" Questions like these help us identify causes and effects. This, in turn, helps us understand the relationships among physical geography, history, and culture.

Assignment

Write a paper about one of these topics:

- causes of economic problems in West Africa
- effects of European colonization in Southern Africa

1. Prewrite

Choose a Topic

- Choose one of the topics above to write about.
- Turn that topic into a big idea, or thesis. For example, "Three main factors cause most of the economic problems in West Africa."

> **TIP** **What Relationships?** Transitional words like *as a result, because, since,* and *so* can help make connections between causes and effects.

Gather and Organize Information

- Depending on the topic you have chosen, identify at least three causes or three effects. Use your textbook, the library, or the Internet.
- Organize causes or effects in their order of importance. To have the most impact on your readers, put the most important cause or effect last.

2. Write

Use a Writer's Framework

A Writer's Framework

Introduction
- Start with an interesting fact or question related to your big idea, or thesis.
- State your big idea and provide background information.

Body
- Write at least one paragraph, including supporting facts and examples, for each cause or effect.
- Organize your causes or effects by order of importance.

Conclusion
- Summarize the causes or effects.
- Restate your big idea.

3. Evaluate and Revise

Review and Improve Your Paper

- Re-read your paper and use the questions below to determine how to make your paper better.
- Make changes to improve your paper.

Evaluation Questions for a Cause and Effect Explanation

1. Do you begin with a fact or question related to your big idea, or thesis?
2. Does your introduction identify your big idea and provide any needed background?
3. Do you have at least one paragraph for each cause or effect?
4. Do you include facts and details to support the connections between causes and effects?
5. Do you explain the causes or effects in order of importance?
6. Do you summarize the causes or effects and restate your big idea?

4. Proofread and Publish

Give Your Explanation the Finishing Touch

- Make sure transitional words and phrases connect causes and effects as clearly as possible.
- Check for capitalization of proper nouns, such as the names of countries and regions.
- Have someone else read your paper.

5. Practice and Apply

Use the steps and strategies outlined in this workshop to write your cause-and-effect paper. Share your paper with other students who wrote on the same topic. Compare your lists of causes or effects.

South and East Asia and the Pacific

Himalayas

The highest mountain range in the world, the Himalayas, separates the Indian Subcontinent from the rest of Asia.

The Outback

About 75 to 80 percent of Australia is covered by the Outback, a dry interior region of ancient rocks and plains.

South and East Asia and the Pacific

Rain Forest

The rich green color of Southeast Asia is caused by tropical rain forests. They are home to rare animals like the orangutan, found only in this region.

Explore the Satellite Image
Towering mountains, dense rain forests, and dry plains are all features of this large region in Asia and the Pacific. What other physical features can you see in this satellite image?

The Satellite's Path

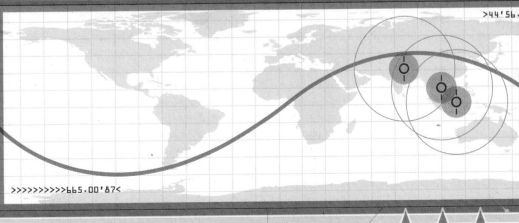

>44'56.0

>>>>>>>>>665.00'87<

+355

+966
+803
+964

567.476.348

456.094.

South and East Asia: Physical

CENTRAL ASIA

MONGOLIA

Altay Mountains

Mongolian Plateau

GOBI DESERT

Tian Shan

Tarim Basin

K2 28,250 ft (8,610 m)

Taklimakan Desert

Kunlun Shan

CHINA

Qinling Shandi

North China Plain

Yellow Sea

NORTH KOREA

SOUTH KOREA

Sea of Japan (East Sea)

Hokkaido

Honshu

JAPAN

Shikoku

Kyushu

Ryukyu Islands

East China Sea

Tropic of Cancer

TAIWAN

PAKISTAN

Plateau of Tibet

Thar Desert

Mount Everest 29,035 ft (8,850 m)

Sichuan Basin

HIMALAYAS

NEPAL

BHUTAN

BANGLADESH

INDIA

Ganges Delta

Deccan Plateau

Eastern Ghats

Western Ghats

Malabar Coast

MYANMAR (BURMA)

LAOS

Hainan

THAILAND

VIETNAM

South China Sea

CAMBODIA

Luzon

PHILIPPINES

Mindanao

PACIFIC OCEAN

Lakshadweep Islands (INDIA)

Bay of Bengal

Andaman Islands (INDIA)

Gulf of Thailand

BRUNEI

New Guinea

MALDIVES

SRI LANKA

Nicobar Islands (INDIA)

Strait of Malacca

MALAYSIA

SINGAPORE

Borneo

Sulawesi (Celebes)

Moluccas

INDIAN OCEAN

INDONESIA

EAST TIMOR

Sumatra

Timor

Java

AUSTRALIA

ELEVATION

Feet		Meters
13,120		4,000
6,560		2,000
1,640		500
656		200
(Sea level) 0		0 (Sea level)
Below sea level		Below sea level

0 250 500 750 Miles

0 250 500 750 Kilometers

Projection: Two-Point Equidistant

map zone
Geography Skills

Regions South and East Asia includes many major rivers, long coastal plains, and large islands.

1. **Identify** What major rivers can you see in China and India?
2. **Make Inferences** How do you think rivers influence where people live in this region?

THE WORLD ALMANAC®
Facts about the World

Geographical Extremes: South and East Asia

Longest River	Chang Jiang (Yangzi River), China: 3,964 miles (6,378 km)
Highest Point	Mount Everest, Nepal/China: 29,035 feet (8,850 m)
Lowest Point	Turpan Depression, China: 505 feet (154 m) below sea level
Highest Recorded Temperature	Tuguegarao, Philippines: 108°F (42°C)
Wettest Place	Mawsynram, India: 467.4 inches (1,187.2 cm) average precipitation per year
Largest Country	China: 3,705,405 square miles (9,596,999 square km)
Smallest Country	Maldives: 116 square miles (300 square km)
Largest Rain Forest	Indonesia: 386,000 square miles (999,740 square km)
Strongest Earthquake	Off the coast of Sumatra, Indonesia, on December 26, 2004: Magnitude 9.0

Mount Everest

go.hrw.com   KEYWORD: SK7 UN4

Size Comparison: The United States and South and East Asia

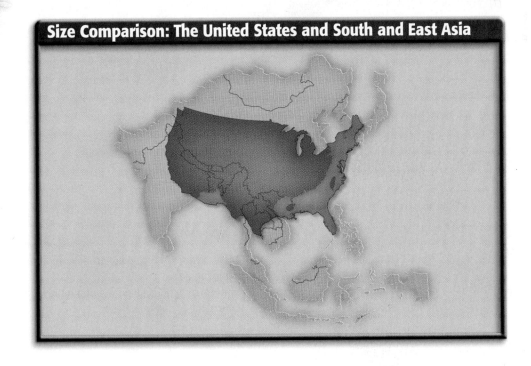

South and East Asia: Political

Map Zone
Geography Skills

Place South and East Asia includes several large countries, many smaller ones, and a number of island countries.

1. **Name** What are the three largest countries in this region?
2. **Analyze** What do you notice about the locations of many capital cities?

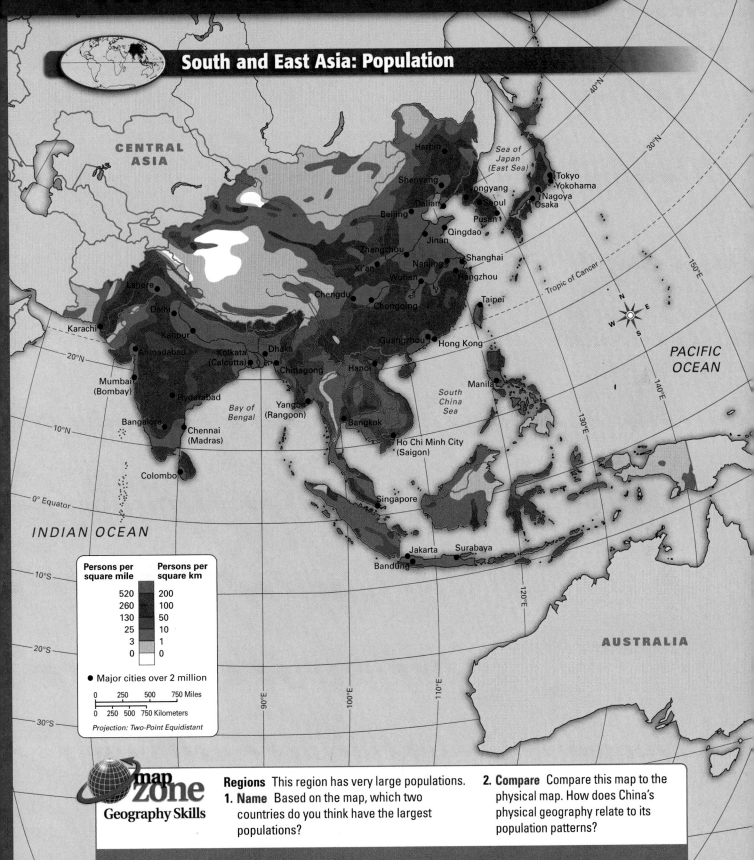

South and East Asia and the Pacific

South and East Asia: Population

CENTRAL ASIA

Harbin

Shenyang

Pyongyang

Dalian

Beijing

Seoul

Pusan

Qingdao

Jinan

Zhengzhou

Xi'an

Nanjing

Shanghai

Wuhan

Hangzhou

Chengdu

Chongqing

Taipei

Lahore

Delhi

Karachi

Kanpur

Guangzhou

Hong Kong

Ahmadabad

Kolkata (Calcutta)

Dhaka

Chittagong

Hanoi

Mumbai (Bombay)

Hyderabad

Bay of Bengal

Yangon (Rangoon)

Manila

Bangalore

Chennai (Madras)

Bangkok

South China Sea

Colombo

Ho Chi Minh City (Saigon)

Singapore

INDIAN OCEAN

Jakarta

Surabaya

Bandung

Sea of Japan (East Sea)

Tokyo

Yokohama

Nagoya

Osaka

Tropic of Cancer

PACIFIC OCEAN

AUSTRALIA

40°N

30°N

150°E

140°E

130°E

120°E

110°E

100°E

90°E

20°N

10°N

0° Equator

10°S

20°S

30°S

Persons per square mile | **Persons per square km**
520 | 200
260 | 100
130 | 50
25 | 10
3 | 1
0 | 0

● Major cities over 2 million

0   250   500   750 Miles
0   250 500 750 Kilometers

Projection: Two-Point Equidistant

map zone
Geography Skills

Regions This region has very large populations.
1. Name Based on the map, which two countries do you think have the largest populations?

2. Compare Compare this map to the physical map. How does China's physical geography relate to its population patterns?

South and East Asia: Climate

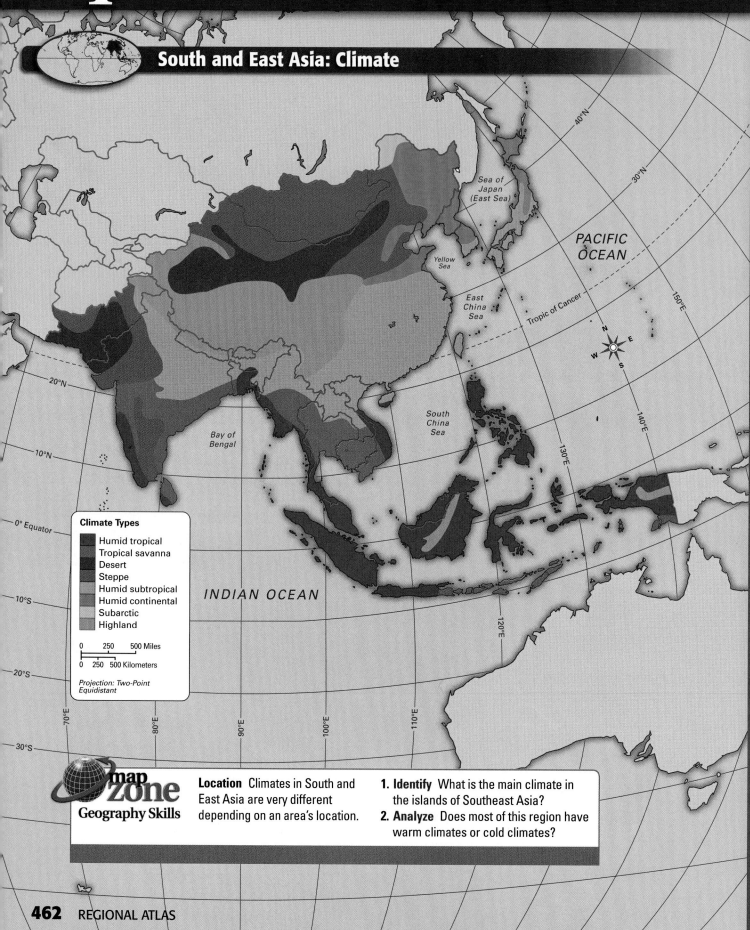

Climate Types
- Humid tropical
- Tropical savanna
- Desert
- Steppe
- Humid subtropical
- Humid continental
- Subarctic
- Highland

0   250   500 Miles
0   250   500 Kilometers

Projection: Two-Point Equidistant

Sea of Japan (East Sea)

PACIFIC OCEAN

Yellow Sea

East China Sea

Tropic of Cancer

South China Sea

Bay of Bengal

INDIAN OCEAN

map zone Geography Skills

Location Climates in South and East Asia are very different depending on an area's location.

1. **Identify** What is the main climate in the islands of Southeast Asia?

2. **Analyze** Does most of this region have warm climates or cold climates?

South and East Asia and the Pacific

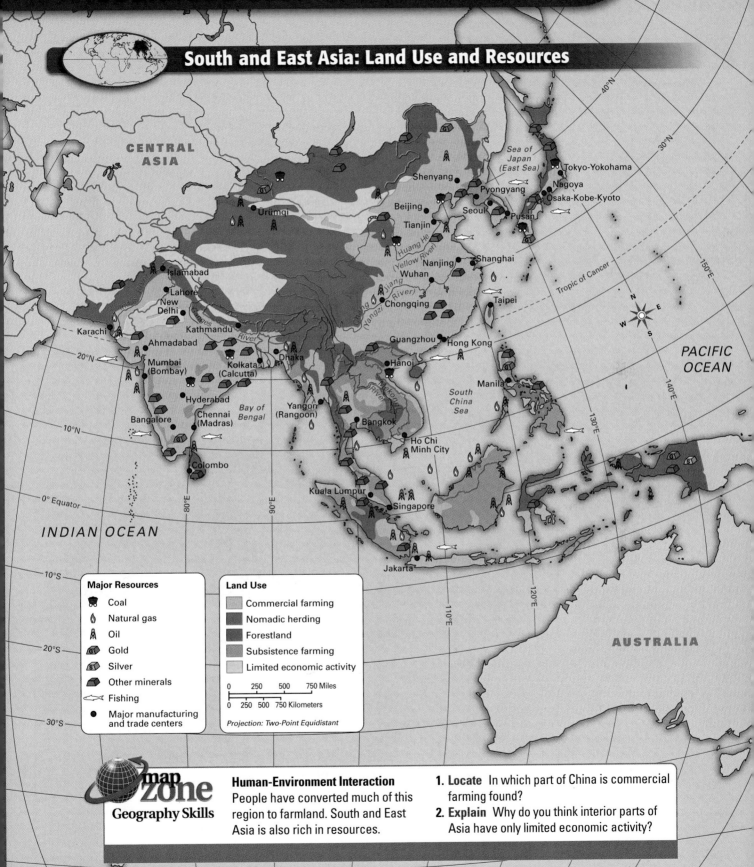

South and East Asia: Land Use and Resources

CENTRAL ASIA

Ürümqi

Shenyang

Pyongyang

Sea of Japan (East Sea)

Tokyo-Yokohama

Nagoya

Osaka-Kobe-Kyoto

Beijing

Seoul

Pusan

Tianjin

Huang He (Yellow River)

Islamabad

Lahore

New Delhi

Nanjing

Shanghai

Wuhan

Karachi

Kathmandu

Ganges River

Chang Jiang (Yangtze River)

Chongqing

Taipei

Tropic of Cancer

Ahmadabad

Dhaka

Guangzhou

Hong Kong

Mumbai (Bombay)

Kolkata (Calcutta)

Hanoi

Mekong River

Manila

PACIFIC OCEAN

Hyderabad

Yangon (Rangoon)

South China Sea

Bangalore

Chennai (Madras)

Bay of Bengal

Bangkok

Ho Chi Minh City

Colombo

INDIAN OCEAN

Kuala Lumpur

Singapore

Jakarta

AUSTRALIA

Major Resources

- Coal
- Natural gas
- Oil
- Gold
- Silver
- Other minerals
- Fishing
- Major manufacturing and trade centers

Land Use

- Commercial farming
- Nomadic herding
- Forestland
- Subsistence farming
- Limited economic activity

0   250   500   750 Miles

0 250 500 750 Kilometers

Projection: Two-Point Equidistant

map zone
Geography Skills

Human-Environment Interaction
People have converted much of this region to farmland. South and East Asia is also rich in resources.

1. **Locate** In which part of China is commercial farming found?
2. **Explain** Why do you think interior parts of Asia have only limited economic activity?

The Pacific World: Physical

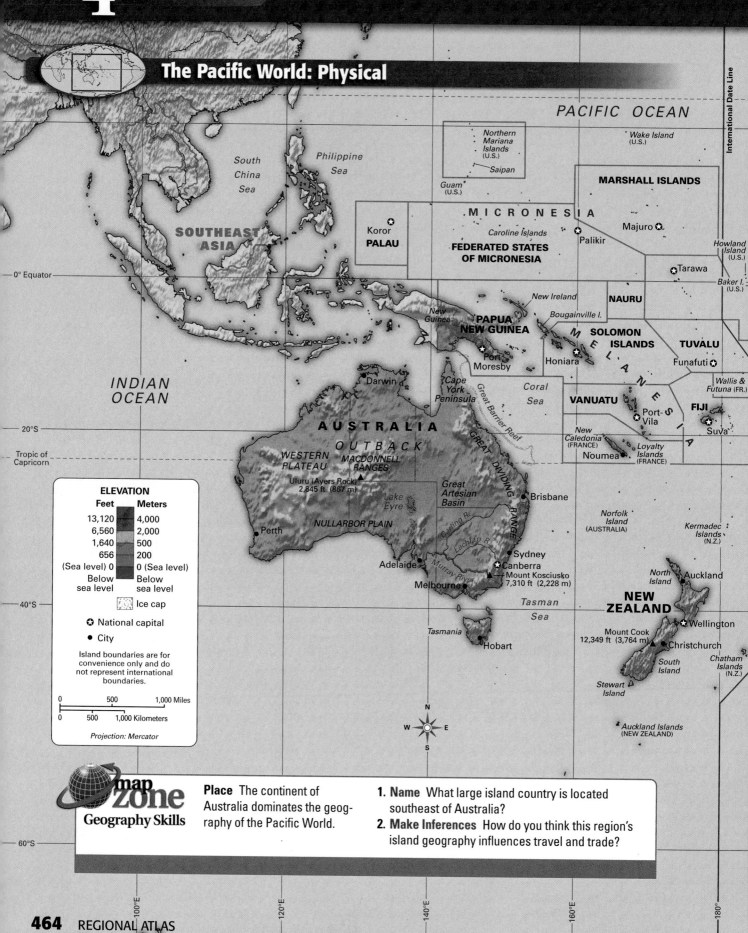

PACIFIC OCEAN

International Date Line

SOUTHEAST ASIA

South China Sea

Philippine Sea

Northern Mariana Islands (U.S.)

Saipan

Wake Island (U.S.)

Guam (U.S.)

MARSHALL ISLANDS

M I C R O N E S I A

Koror

PALAU

Caroline Islands

Palikir

Majuro

FEDERATED STATES OF MICRONESIA

Howland Island (U.S.)

0° Equator

Tarawa

Baker I. (U.S.)

New Ireland

NAURU

New Guinea

Bougainville I.

PAPUA NEW GUINEA

Port Moresby

Honiara

SOLOMON ISLANDS

M E L A N E S I A

TUVALU

Funafuti

Wallis & Futuna (FR.)

INDIAN OCEAN

Darwin

Cape York Peninsula

Great Barrier Reef

Coral Sea

VANUATU

Port-Vila

FIJI

Suva

20°S

AUSTRALIA

OUTBACK

New Caledonia (FRANCE)

Noumea

Loyalty Islands (FRANCE)

Tropic of Capricorn

WESTERN PLATEAU

MACDONNELL RANGES

Uluru (Ayers Rock) 2,845 ft. (867 m)

Lake Eyre

GREAT DIVIDING RANGE

Great Artesian Basin

Brisbane

Norfolk Island (AUSTRALIA)

Kermadec Islands (N.Z.)

ELEVATION

Feet	Meters
13,120	4,000
6,560	2,000
1,640	500
656	200
(Sea level) 0	0 (Sea level)
Below sea level	Below sea level

Ice cap

✪ National capital

● City

Island boundaries are for convenience only and do not represent international boundaries.

| 0 | 500 | 1,000 Miles |
| 0 | 500 | 1,000 Kilometers |

Projection: Mercator

Perth

NULLARBOR PLAIN

Darling R.

Lachlan R.

Murray River

Adelaide

Sydney

Canberra

Mount Kosciusko 7,310 ft (2,228 m)

Melbourne

North Island

Auckland

NEW ZEALAND

40°S

Tasman Sea

Tasmania

Hobart

Mount Cook 12,349 ft (3,764 m)

Wellington

Christchurch

South Island

Chatham Islands (N.Z.)

Stewart Island

N W E S

Auckland Islands (NEW ZEALAND)

map zone
Geography Skills

Place The continent of Australia dominates the geography of the Pacific World.

1. **Name** What large island country is located southeast of Australia?
2. **Make Inferences** How do you think this region's island geography influences travel and trade?

60°S

100°E 120°E 140°E 160°E 180°

South and East Asia and the Pacific

THE WORLD ALMANAC®
Facts about the World

Geographical Extremes: The Pacific World

Highest Point	Mount Wilhelm, Papua New Guinea: 14,793 feet (4,509 m)
Lowest Point	Lake Eyre, Australia: 52 feet (16 m) below sea level
Highest Recorded Temperature	Cloncurry, Australia: 128°F (53°C)
Lowest Recorded Temperature	Charlotte Pass, Australia: –9°F (–22°C)
Wettest Place	Bellenden Ker, Australia: 340 inches (863.6 cm) average precipitation per year
Driest Place	Mulka, Australia: 4.1 inches (10.4 cm) average precipitation per year
Largest Country	Australia: 2,967,908 square miles (7,686,882 square km)
Smallest Country	Nauru: 8 square miles (20.7 square km)
Largest Rain Forest	Papua New Guinea: 130,000 square miles (336,700 square km)
Longest Coral Reef	Great Barrier Reef, Australia: 1,250 miles (2,011 km)

Antarctica

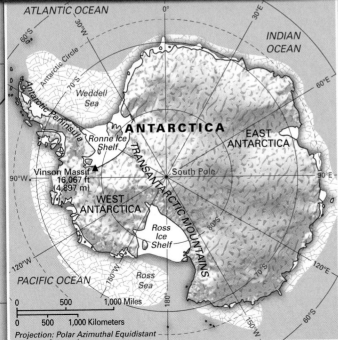

Australia's Great Barrier Reef

go.hrw.com KEYWORD: SK7 UN4

South and East Asia and the Pacific

COUNTRY Capital	FLAG	POPULATION	AREA (sq mi)	PER CAPITA GDP (U.S. $)	LIFE EXPECTANCY AT BIRTH	TVS PER 1,000 PEOPLE
Australia Canberra		20.1 million	2,967,908	$30,700	80.4	716
Bangladesh Dhaka		144.3 million	55,599	$2,000	62.1	7
Bhutan Thimphu		2.2 million	18,147	$1,400	54.4	6
Brunei Bandar Seri Begawan		372,400	2,228	$23,600	74.8	637
Cambodia Phnom Penh		13.6 million	69,900	$2,000	58.9	9
China Beijing		1,306.3 million	3,705,405	$5,600	72.3	291
East Timor Dili		1 million	5,794	$400	65.9	NA
Fiji Suva		893,400	7,054	$5,900	69.5	110
India New Delhi		1,080.3 million	1,269,345	$3,100	64.4	75
Indonesia Jakarta		242 million	741,100	$3,500	69.6	143
Japan Tokyo		127.4 million	145,883	$29,400	81.2	719
Kiribati Tarawa		103,100	313	$800	61.7	23
Laos Vientiane		6.2 million	91,429	$1,900	55.1	10
Malaysia Kuala Lumpur		24 million	127,317	$9,700	72.2	174
Maldives Male		349,100	116	$3,900	64.1	38
United States Washington, D.C.		295.7 million	3,718,710	$40,100	77.7	844

COUNTRY Capital	FLAG	POPULATION	AREA (sq mi)	PER CAPITA GDP (U.S. $)	LIFE EXPECTANCY AT BIRTH	TVS PER 1,000 PEOPLE
Marshall Islands Majuro		59,100	70	$1,600	70.0	NA
Micronesia, Federated States of Palikir		108,100	271	$2,000	69.8	20
Mongolia Ulaanbaatar		2.8 million	604,250	$1,900	64.5	58
Myanmar (Burma) Yangon (Rangoon)		42.7 million	261,970	$1,700	60.7	7
Nauru No official capital		13,000	8	$5,000	62.7	1
Nepal Kathmandu		27.7 million	54,363	$1,500	59.8	6
New Zealand Wellington		4 million	103,738	$23,200	78.7	516
North Korea Pyongyang		22.9 million	46,541	$1,400	71.4	55
Pakistan Islamabad		162.4 million	310,403	$2,200	63.0	105
Palau Koror		20,300	177	$9,000	70.1	98
Papua New Guinea Port Moresby		5.5 million	178,703	$2,200	64.9	13
Philippines Manila		87.9 million	115,831	$5,000	69.9	110
Samoa Apia		177,300	1,137	$5,600	70.7	56
Singapore Singapore		4.4 million	267	27,800	81.6	341
Solomon Islands Honiara		538,000	10,985	$1,700	72.7	16
United States Washington, D.C.		295.7 million	3,718,710	$40,100	77.7	844

COUNTRY Capital	FLAG	POPULATION	AREA (sq mi)	PER CAPITA GDP (U.S. $)	LIFE EXPECTANCY AT BIRTH	TVS PER 1,000 PEOPLE
South Korea Seoul		48.4 million	38,023	$19,200	76.9	364
Sri Lanka Colombo		20.1 million	25,332	$4,000	73.2	102
Taiwan Taipei		22.9 million	13,892	$25,300	77.3	327
Thailand Bangkok		65.4 million	198,456	$8,100	72.0	274
Tonga Nuku'alofa		112,400	289	$2,300	69.5	61
Tuvalu Funafuti		11,600	10	$1,100	68.0	9
Vanuatu Port-Vila		205,800	4,710	$2,900	62.5	12
Vietnam Hanoi		83.5 million	127,244	$2,700	70.6	184
United States Washington, D.C.		295.7 million	3,718,710	$40,100	77.7	844

Palm trees along the coast of Tonga

ANALYSIS SKILL · ANALYZING TABLES

1. Which five countries in this region have the highest per capita GDPs? How do they compare to the per capita GDP of the United States?
2. Compare the life expectancy and number of TVs per 1,000 people in Japan and Kiribati. What might this comparison indicate about life in the two countries?

Population Giants

World's Largest Populations

Country	Population
China	1.3 billion
India	1.08 billion
United States	295.7 million
Indonesia	242 million
Brazil	186.1 million
Pakistan	162.4 million
Russia	144.3 million
Bangladesh	143.4 million
Nigeria	128.8 million
Japan	127.4 million

■ Asian Countries
■ Other Countries

Of the ten countries with the largest populations, six are located in South and East Asia.

Percent of World Population

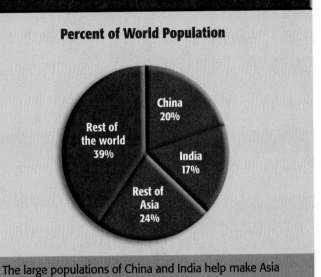

China 20%
India 17%
Rest of Asia 24%
Rest of the world 39%

The large populations of China and India help make Asia home to more than 60 percent of the world's people.

Economic Powers

Japan

- World's third-largest economy
- $538.8 billion in exports
- Per capita GDP of $29,400
- Major exports: transportation equipment, cars, semiconductors, electronics

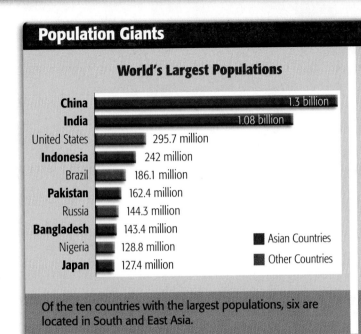

Japan is one of the most technologically advanced countries and is a leading producer of hi-tech goods.

China

- World's second-largest economy
- $583.1 billion in exports
- GDP growth rate of 9.1%
- Major exports: machinery and electronics, clothing, plastics, furniture, toys

China is an emerging economic powerhouse with a huge population and a fast growing economy.

ANALYSIS SKILL **ANALYZING VISUALS**

1. Which two countries have the largest populations?
2. What kinds of exports help make Japan and China economic powers?

History of Ancient India

2300 BC–AD 500

What You Will Learn...

In this chapter you will learn about the ancient civilization of India, the birthplace of two major world religions—Hinduism and Buddhism. You will also learn about the early civilizations and powerful empires that developed in India.

SECTION 1
Early Indian Civilizations **472**

SECTION 2
Origins of Hinduism **478**

SECTION 3
Origins of Buddhism **484**

SECTION 4
Indian Empires **490**

SECTION 5
Indian Achievements **495**

FOCUS ON READING AND WRITING

Sequencing When you read, it is important to keep track of the sequence, or order, in which events happen. Look for dates and other clues to help you figure out the proper sequence. **See the lesson, Sequencing, on page 686.**

Creating a Poster Ancient India was the home of amazing cities, strong empires, and influential religions. As you read this chapter, think about how you could illustrate one aspect of Indian culture in a poster.

0   250   500 Miles
0   250   500 Kilometers

Projection: Albers Equal-Area

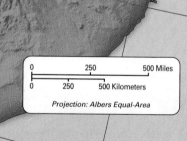

Early India The first civilization in India, the Harappans, were skilled builders and artists.

Ancient India, 2300 BC–AD 500

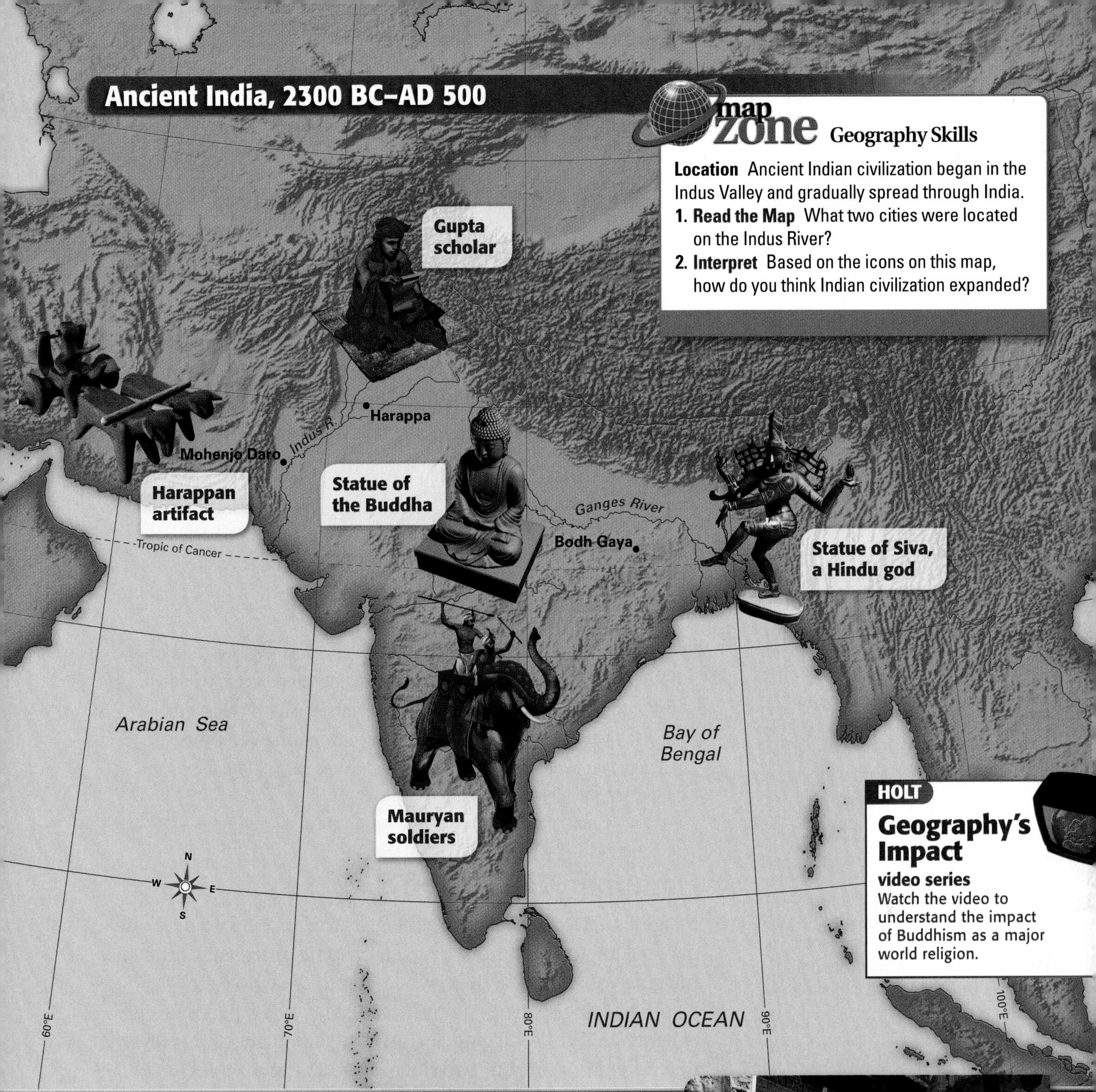

map zone Geography Skills

Location Ancient Indian civilization began in the Indus Valley and gradually spread through India.
1. **Read the Map** What two cities were located on the Indus River?
2. **Interpret** Based on the icons on this map, how do you think Indian civilization expanded?

Gupta scholar

Harappa

Mohenjo Daro · Indus R.

Harappan artifact

Statue of the Buddha

Ganges River

Bodh Gaya

Statue of Siva, a Hindu god

Tropic of Cancer

Arabian Sea

Bay of Bengal

Mauryan soldiers

INDIAN OCEAN

60°E 70°E 80°E 90°E 100°E

HOLT

Geography's Impact
video series
Watch the video to understand the impact of Buddhism as a major world religion.

Hinduism A major world religion, Hinduism, developed in India. In this photo, Hindus bathe in the sacred river Ganges.

Buddhism India was also the birthplace of another religion, Buddhism. Buddhist temples like this one at Ajanta are found all over India.

Early Indian Civilizations

What You Will Learn...

Main Ideas

1. Located on the Indus River, the Harappan civilization also had contact with people far from India.
2. Harappan achievements included a writing system, city planning, and art.
3. The Aryan invasion changed India's civilization.

The Big Idea

Indian civilization developed on the Indus River.

Key Terms and Places

Indus River, *p. 472*
Harappa, *p. 473*
Mohenjo Daro, *p. 473*
Sanskrit, *p. 477*

TAKING NOTES As you read this section, take notes on India's two earliest civilizations, the Harappans and Aryans. Record what you find in a graphic organizer like this one.

Ancient Indian Civilizations	
Harappan Civilization	Aryan Civilization

If YOU lived there...

You are a trader in the huge city of Mohenjo Daro. Your business is booming, as traders come to the city from all over Asia. With your new wealth, you have bought a huge house with a rooftop terrace and even indoor plumbing! This morning, however, you heard that invaders are headed toward the city. People are telling you that you should flee for your safety.

What will you miss most about life in the city?

BUILDING BACKGROUND India was home to one of the world's first civilizations. Like other early civilizations, the one in India grew up in a river valley. As archaeologists discovered, however, the society that eventually developed in India was very different from the ones that developed elsewhere.

Harappan Civilization

Imagine that you are an archaeologist. You are out in a field one day looking for a few pots, a tablet, or some other small artifact. Imagine your surprise, then, when you find a whole city!

Archaeologists working in India in the 1920s had that very experience. While digging for artifacts along the **Indus River**, they found not one but two huge cities. The archaeologists had thought people had lived along the Indus long ago, but they had no idea that an advanced civilization had existed there.

India's First Civilization

Historians now call the civilization that developed along the Indus River the Harappan (huh-RA-puhn) civilization. The name comes from the modern city of Harappa (huh-RA-puh), Pakistan. It was near this city that the ruins of the ancient civilization were first discovered. From studying these ruins, archaeologists think that the civilization thrived between 2300 and 1700 BC.

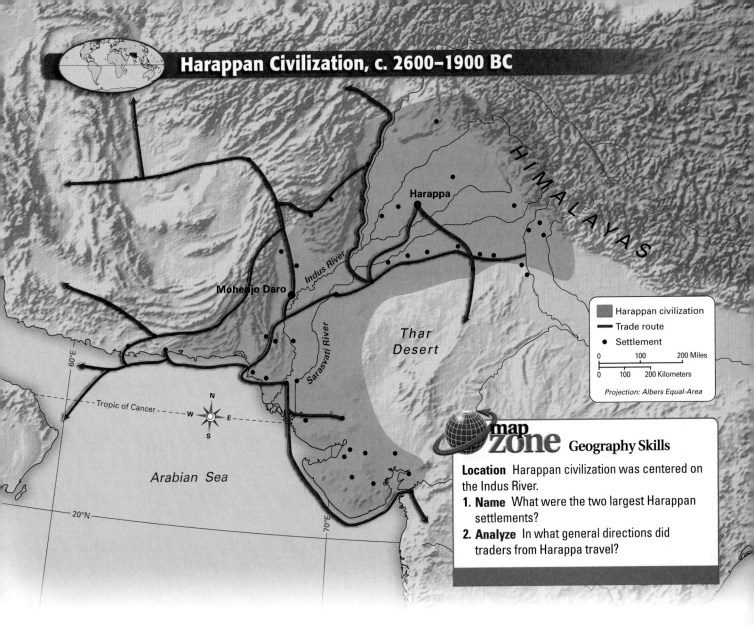

Harappan Civilization, c. 2600–1900 BC

Harappa

Mohenjo Daro

Indus River

Saraswati River

Thar Desert

HIMALAYAS

Arabian Sea

Tropic of Cancer

20°N

60°E

70°E

Harappan civilization
Trade route
• Settlement

0   100   200 Miles
0   100   200 Kilometers

Projection: Albers Equal-Area

map zone Geography Skills

Location Harappan civilization was centered on the Indus River.
1. **Name** What were the two largest Harappan settlements?
2. **Analyze** In what general directions did traders from Harappa travel?

The Harappan civilization controlled large areas on both sides of the Indus River. As you can see on the map, settlements were scattered over a huge area. Most of these settlements lay next to rivers. The largest settlements were two cities, **Harappa** and **Mohenjo Daro** (mo-HEN-joh DAR-oh).

Like most other ancient societies, the Harappan civilization was dependent on agriculture. Farmers in the Indus Valley grew a variety of crops—from wheat and barley to dates and vegetables—to feed both themselves and city dwellers. They used irrigation canals to bring water from the Indus and other rivers to their fields.

Contact with Other Cultures

Although the Harappan civilization was centered on the Indus, its influence reached far beyond that area. In fact, archaeologists have found evidence that the Harappans had contact with people as far away as southern India and Mesopotamia.

Most of this contact with other cultures was in the form of trade. The Harappans traded to obtain raw materials. They then used these materials to make products such as pottery, stamps and seals, and statues.

READING CHECK **Finding Main Ideas** Where was the Harappan civilization located?

Harappan Achievements

Historians do not know much about the Harappan civilization. They think the Harappans had kings and strong central governments, but they are not sure. They also know little about Harappan religion.

Although we do not know much about how the Harappans lived, we do know that they made great achievements in many fields. Everything we know about these achievements comes from artifacts.

Writing System

The ancient Harappans developed India's first writing system. However, scholars have not yet learned to read this language. Archaeologists have found many examples of Harappan writing, but none of them is more than a few words long. This lack of long passages has made translating the language difficult. Because we cannot read what they wrote, we rely on other clues to study Harappan society.

Close-up

Life in Mohenjo Daro

Mohenjo Daro was one of the two major cities of the Harappan civilization. Located next to the Indus River in what is now Pakistan, the city probably covered one square mile. The people who lived in the city enjoyed some of the most advanced comforts of their time, including indoor plumbing.

Harappan merchants used a standard set of weights to measure goods such as precious stones.

City Planning

Most of what we have learned about the Harappans has come from studying their cities, especially Harappa and Mohenjo Daro. The two cities lay on the Indus more than 300 miles apart, but they appear to have been remarkably similar.

Both Harappa and Mohenjo Daro were well-planned cities. A close examination of their ruins shows that the Harappans were careful planners and skilled engineers.

Harappa and Mohenjo Daro were built with defense in mind. Each city stood near a towering fortress. From these fortresses, defenders could look down on the cities' carefully laid out brick streets. These streets crossed at right angles and were lined with storehouses, workshops, market stalls, and houses. Using their engineering skills, the Harappans built extensive sewer systems to keep their streets from flooding. They also installed plumbing in many buildings.

Next to the city was a huge citadel, or fortress, to guard against invasions.

The houses of Mohenjo Daro had flat roofs. People climbed to their roofs to take advantage of cooling breezes.

The city's streets were paved and well drained. They met at right angles, creating a grid pattern.

ANALYSIS SKILL **ANALYZING VISUALS**

What in this picture suggests that Mohenjo Daro was a well-planned city?

Artistic Achievements

In Harappan cities, archaeologists have found many artifacts that show that the Harappans were skilled artisans. For example, they have found sturdy pottery vessels, jewelry, and ivory objects.

Some of these ancient artifacts have helped historians draw conclusions about Harappan society. For example, they found a statue that shows two animals pulling a cart. Based on this statue, they conclude that the Harappans built and used wheeled vehicles. Likewise, a statue of a man with elaborate clothes and jewelry suggests that Harappan society had an upper class.

Harappan civilization ended by the early 1700s BC, but no one is sure why. Perhaps invaders destroyed the cities or natural disasters, like floods or earthquakes, caused the civilization to collapse.

FOCUS ON READING

In what order did the Aryans settle lands in India?

READING CHECK **Analyzing** Why do we not know much about Harappan civilization?

Harappan Art

Like other ancient peoples, the Harappans made small seals like the one below that were used to stamp goods. They also used clay pots like the one at right decorated with a goat.

Aryan Migration

Not long after the Harappan civilization crumbled, a new group arrived in the Indus Valley. These people were called the Aryans (AIR-ee-uhnz). They were originally from the area around the Caspian Sea in Central Asia. Over time, however, they became the dominant group in India.

Arrival and Spread

The Aryans first arrived in India in the 2000s BC. Historians and archaeologists believe that the Aryans crossed into India through mountain passes in the northwest. Over many centuries, they spread east and south into central India. From there they moved even farther east into the Ganges River Valley.

Much of what we know about Aryan society comes from religious writings known as the Vedas (VAY-duhs). These are collections of poems, hymns, myths, and rituals that were written by Aryan priests. You will read more about the Vedas later in this chapter.

Government and Society

As nomads, the Aryans took along their herds of animals as they moved. But over time, they settled in villages and began to farm. Unlike the Harappans, they did not build big cities.

The Aryan political system was also different from the Harappan system. The Aryans lived in small communities, based mostly on family ties. No single ruling authority existed. Instead, each group had its own leader, often a skilled warrior.

Aryan villages were governed by rajas (RAH-juhz). A raja was a leader who ruled a village and the land around it. Villagers farmed some of this land for the raja. They used other sections as pastures for their cows, horses, sheep, and goats.

Although many rajas were related, they didn't always get along. Sometimes rajas joined forces before fighting a common enemy. Other times, however, rajas went to war against each other. In fact, Aryan groups fought each other nearly as often as they fought outsiders.

Language

The first Aryan settlers did not read or write. Because of this, they had to memorize the poems and hymns that were important in their culture, such as the Vedas. If people forgot these poems and hymns, the works would be lost forever.

The language in which these Aryan poems and hymns were composed was **Sanskrit, the most important language of ancient India.** At first, Sanskrit was only a spoken language. Eventually, however, people figured out how to write it down so they could keep records. These Sanskrit records are a major source of information about Aryan society. Sanskrit is no longer spoken today, but it is the root of many modern South Asian languages.

READING CHECK **Identifying** What source provides much of the information we have about the Aryans?

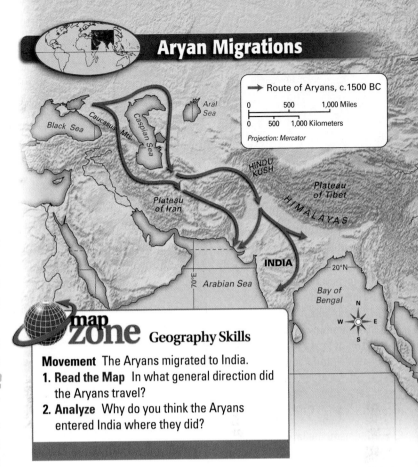

Aryan Migrations

→ Route of Aryans, c.1500 BC

0   500   1,000 Miles
0   500   1,000 Kilometers
Projection: Mercator

Black Sea · Caucasus Mts. · Caspian Sea · Aral Sea · HINDU KUSH · Plateau of Tibet · Plateau of Iran · HIMALAYAS · INDIA · Arabian Sea · Bay of Bengal · 70°E · 20°N

map zone Geography Skills

Movement The Aryans migrated to India.
1. **Read the Map** In what general direction did the Aryans travel?
2. **Analyze** Why do you think the Aryans entered India where they did?

SUMMARY AND PREVIEW The earliest civilizations in India were centered in the Indus Valley. In the next section, you will learn about a new religion that developed in the Indus Valley after the Aryans settled there—Hinduism.

Section 1 Assessment

Reviewing Ideas, Terms, and Places

1. **a. Recall** Where did the Harappan civilization develop?
   b. Explain Why did the Harappans make contact with people far from India?
2. **a. Identify** What was **Mohenjo Daro**?
   b. Analyze What is one reason that scholars do not completely understand some important parts of Harappan society?
3. **a. Identify** Who were the Aryans?
   b. Contrast How was Aryan society different from Harappan society?

Critical Thinking

4. **Summarizing** Using your notes, list the major achievements of India's first two civilizations. Record your conclusions in a diagram like this one.

Early Indian Achievements

| Harappan society |
| Aryan society |

FOCUS ON WRITING

5. **Illustrating Geography and Early Civilizations** This section described two possible topics for your poster: geography and early civilizations. Which of them is more interesting to you? Write down some ideas for a poster about that topic.

Origins of Hinduism

What You Will Learn...

Main Ideas

1. Indian society divided into distinct groups.
2. The Aryans formed a religion known as Brahmanism.
3. Hinduism developed out of Brahmanism and influences from other cultures.
4. The Jains reacted to Hinduism by breaking away.

The Big Idea

Hinduism, the largest religion in India, developed out of ancient Indian beliefs and practices.

Key Terms

caste system, *p. 479*
reincarnation, *p. 481*
karma, *p. 482*
nonviolence, *p. 483*

TAKING NOTES As you read, take notes on Hinduism using a diagram like the one below. Pay attention to its origins, teachings, and other religions that developed alongside it.

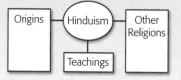

If **YOU** lived there...

Your family are skillful weavers who make beautiful cotton cloth. You belong to the class in Aryan society who are traders, farmers, and craftspeople. Often the raja of your town leads the warriors into battle. You admire their bravery but know you can never be one of them. To be an Aryan warrior, you must be born into that noble class. Instead, you have your own duty to carry out.

How do you feel about remaining a weaver?

BUILDING BACKGROUND As the Aryans migrated into India, they developed a strict system of social classes. As the Aryans' influence spread through India, so did their class system. Before long, this class system was a key part of Indian society.

Indian Society Divides

As Aryan society became more complex, their society became divided into groups. These groups were largely organized by people's occupations. Strict rules developed about how people of different groups could interact. As time passed, these rules became stricter and became central to Indian society.

The *Varnas*

According to the Vedas, there were four main *varnas*, or social divisions, in Aryan society. These *varnas* were

- Brahmins (BRAH-muhns), or priests,
- Kshatriyas (KSHA-tree-uhs), or rulers and warriors,
- Vaisyas (VYSH-yuhs), or farmers, craftspeople, and traders, and
- Sudras (SOO-drahs), or laborers and non-Aryans.

The Brahmins were seen as the highest ranking because they performed rituals for the gods. This gave the Brahmins great influence over the other *varnas*.

The Caste System

As the rules of interaction between *varnas* got stricter, the Aryan social order became more complex. In time, each of the *varnas* in Aryan society was further divided into many castes, or groups. This **caste system** divided Indian society into groups based on a person's birth, wealth, or occupation. At one time, some 3,000 separate castes existed in India.

The caste to which a person belonged determined his or her place in society. However, this ordering was by no means permanent. Over time, individual castes gained or lost favor in society as caste members gained wealth or power. On rare occasions, people could change caste.

People in the lowest class, the Sudra castes, had hard lives. But there was a group of people who were even worse off, a group who didn't belong to any caste at all. They were called untouchables because people from the castes were not supposed to have contact with them. The only jobs open to them were unpleasant ones, such as tanning animal hides and disposing of dead animals. As a result, they were seen as unclean and were the outcasts of society.

Caste Rules

To keep their classes distinct, the Aryans developed sutras, or guides, which listed the rules for the caste system. For example, people could not marry someone from a different class. It was even forbidden for people from one class to eat with people from another. People who broke the caste rules could be banned from their homes and their castes, which would make them untouchables. Because of these strict rules, people spent almost all of their time with others in their same class.

READING CHECK **Drawing Inferences** How did a person become a member of a caste?

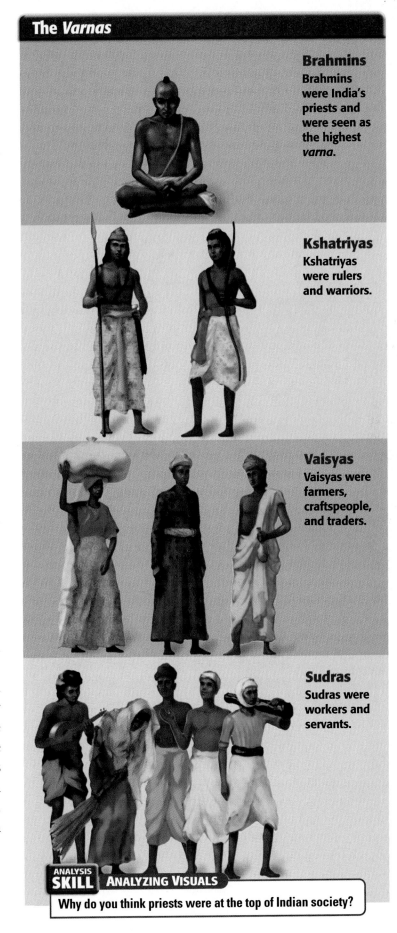

The *Varnas*

Brahmins
Brahmins were India's priests and were seen as the highest *varna*.

Kshatriyas
Kshatriyas were rulers and warriors.

Vaisyas
Vaisyas were farmers, craftspeople, and traders.

Sudras
Sudras were workers and servants.

ANALYSIS SKILL **ANALYZING VISUALS**
Why do you think priests were at the top of Indian society?

Hindu Gods and Beliefs

Hindus believe in many gods, but they believe that all the gods are aspects of a single universal spirit called Brahman. Three aspects of Brahman are particularly important in Hinduism—Brahma, Siva, and Vishnu.

Major Beliefs of Hinduism

- A universal spirit called Brahman created the universe and everything in it. Everything in the world is just a part of Brahman.

- Every person has a soul or atman that will eventually join with Brahman.

- People's souls are reincarnated many times before they can join with Brahman.

- A person's karma affects how he or she will be reincarnated.

The god Brahma represents the creator aspect of Brahman. His four faces symbolize the four Vedas.

Brahmanism

Religion had been an important part of Aryan life even before the Aryans moved to India. Eventually in India, religion took on even more meaning. Because Aryan priests were called Brahmins, their religion is often called Brahmanism.

The Vedas

FOCUS ON READING

Which were written first, the Vedas or the Vedic texts?

Aryan religion was based on the Vedas. There are four Vedas, each containing sacred hymns and poems. The oldest of the Vedas, the *Rigveda*, was probably written before 1000 BC. It includes hymns of praise to many gods. This passage, for example, is the opening of a hymn praising Indra, a god of the sky and war.

> "The one who is first and possessed of wisdom when born; the god who strove to protect the gods with strength; the one before whose force the two worlds were afraid because of the greatness of his virility [power]: he, O people, is Indra."
>
> –from the *Rigveda*, in *Reading about the World, Volume I*, edited by Paul Brians, et al

Later Vedic Texts

Over the centuries, Aryan Brahmins wrote down their thoughts about the Vedas. In time these thoughts were compiled into collections called Vedic texts.

One collection of Vedic texts describes Aryan religious rituals. For example, it describes how to perform sacrifices. Priests prepared animals, food, or drinks to be sacrificed in a fire. The Aryans believed that the fire would carry these offerings to the gods.

A second collection of Vedic texts describes secret rituals that only certain people could perform. In fact, the rituals were so secret that they had to be done in the forest, far from other people.

The final group of Vedic texts are the Upanishads (oo-PAHN-ee-shads), most of which were written by about 600 BC. These writings are reflections on the Vedas by religious students and teachers.

READING CHECK **Finding Main Ideas** What are the Vedic texts?

Siva, the destroyer aspect of Brahman, is usually shown with four arms and three eyes. Here he is shown dancing on the back of a demon he has defeated.

Vishnu is the preserver aspect of Brahman. In his four arms, he carries a conch shell, a mace, and a discus, symbols of his power and greatness.

Hinduism Develops

The Vedas, the Upanishads, and the other Vedic texts remained the basis of Indian religion for centuries. Eventually, though, the ideas of these sacred texts began to blend with ideas from other cultures. People from Persia and other kingdoms in Central Asia, for example, brought their ideas to India. In time, this blending of ideas created a religion called Hinduism, the largest religion in India today.

Hindu Beliefs

The Hindus believe in many gods. Among them are three major gods: Brahma the Creator, Siva the Destroyer, and Vishnu the Preserver. At the same time, however, Hindus believe that each god is part of a single universal spirit called Brahman. They believe that Brahman created the world and preserves it. Gods such as Brahma, Siva, and Vishnu are different aspects of Brahman. In fact, Hindus believe that everything in the world is part of Brahman.

Life and Rebirth

According to Hindu teachings, everyone has a soul, or atman. This soul holds the person's personality, those qualities that make a person who he or she is. Hindus believe that a person's ultimate goal should be to reunite that soul with Brahman, the universal spirit.

Hindus believe that their souls will eventually join Brahman because the world we live in is an illusion. Brahman is the only reality. The Upanishads taught that people must try to see through the illusion of the world. Since it is hard to see through illusions, it can take several lifetimes. That is why Hindus believe that souls are born and reborn many times, each time in a new body. This process of rebirth is called reincarnation.

THE IMPACT TODAY

More than 800 million people in India practice Hinduism today.

Hinduism and the Caste System

According to the traditional Hindu view of reincarnation, a person who has died is reborn in a new physical form.

The type of form depends upon his or her **karma, the effects that good or bad actions have on a person's soul.** Evil actions during one's life will build bad karma. A person with bad karma will be reborn into a lower caste or as a lesser creature, such as a pig or an ant.

In contrast, good actions build good karma. People with good karma are born into a higher caste in their next lives. In time, good karma will bring salvation, or freedom from life's worries and the cycle of rebirth. This salvation is called *moksha*.

Hinduism taught that each person had a duty to accept his or her place in the world without complaint. This is called obeying one's dharma. People could build good karma by fulfilling the duties required of their specific caste. Through reincarnation, Hinduism offered rewards to those who lived good lives. Even untouchables could be reborn into a higher caste.

Hinduism was popular at all levels of Hindu society, through all four *varnas*. By teaching people to accept their places in life, Hinduism helped preserve the caste system in India.

Hinduism and Women

Early Hinduism taught that both men and women could gain salvation. However, Hinduism also taught that women were inferior to men. Women were not allowed to study the Vedas or other sacred texts.

Over the centuries, Hindu women have gained more rights. This change has been the result of efforts by influential Hindu leaders like Mohandas Gandhi, who led the movement for Indian independence. As a result, many of the restrictions once placed on Hindu women have been lifted.

READING CHECK **Summarizing** What factors determined how a person would be reborn?

FOCUS ON CULTURE

The Sacred Ganges

Hindus believe that there are many sacred places in India. Making a pilgrimage to one of these places, they believe, will help improve their karma and increase their chance for salvation. The most sacred of all the pilgrimage sites in India is the Ganges River in the northeast.

Known to Hindus as Mother Ganga, the Ganges flows out of the Himalayas. In traditional Hindu teachings, however, the river flows from the feet of Vishnu and over the head of Siva before it makes its way across the land. Through this contact with the gods, the river's water is made holy. Hindus believe that bathing in the Ganges will purify them and remove some of their bad karma.

Although the entire Ganges is considered sacred, a few cities along its path are seen as especially holy. At these sites, pilgrims gather to bathe and celebrate Hindu festivals. Steps lead down from the cities right to the edge of the water so people can more easily reach the river.

Summarizing Why is the Ganges a pilgrimage site?

Jains React to Hinduism

Although Hinduism was widely followed in India, not everyone agreed with its beliefs. Some unsatisfied people and groups looked for new religious ideas. One such group was the Jains (JYNZ), believers in a religion called Jainism (JY-ni-zuhm).

Jainism was based on the teachings of a man named Mahariva. Born into the Kshatriya *varna* around 599 BC, he was unhappy with the control of religion by the Brahmins, whom he thought put too much emphasis on rituals. Mahariva gave up his life of luxury, became a monk, and established the principles of Jainism.

The Jains try to live by four principles: injure no life, tell the truth, do not steal, and own no property. In their efforts not to injure anyone or anything, the Jains practice **nonviolence, or the avoidance of violent actions**. The Sanskrit word for this nonviolence is *ahimsa* (uh-HIM-sah). Many Hindus also practice *ahimsa*.

The Jains' emphasis on nonviolence comes from their belief that everything is alive and part of the cycle of rebirth. Jains are very serious about not injuring or killing any creature—humans, animals, insects, or plants. They do not believe in animal sacrifice, such as the ones the ancient Brahmins performed. Because they do not want to hurt any living creatures, Jains are vegetarians. They do not eat any food that comes from animals.

READING CHECK **Identifying Points of View**
Why do Jains avoid eating meat?

SUMMARY AND PREVIEW You have learned about two religions that grew in ancient India—Hinduism and Jainism. In Section 3, you will learn about a third religion that began there—Buddhism.

These Jain women are wearing masks to make sure they don't accidentally inhale and kill insects.

go.hrw.com
Online Quiz
KEYWORD: SK7 HP19

Section 2 Assessment

Reviewing Ideas, Terms, and Places

1. **a. Identify** What is the **caste system**?
   b. Explain Why did strict caste rules develop?
2. **a. Identify** What does the *Rigveda* include?
   b. Analyze What role did sacrifice play in Aryan society?
3. **a. Define** What is **karma**?
   b. Sequence How did Brahmanism develop into Hinduism?
   c. Elaborate How does Hinduism reinforce followers' willingness to remain within their castes?
4. **a. Recall** What are the four main teachings of Jainism?
   b. Predict How do you think the idea of **nonviolence** affected the daily lives of Jains in ancient India?

Critical Thinking

5. **Analyzing Causes** Draw a graphic organizer like this one. Using your notes, explain how Hinduism developed from Brahmanism and how Jainism developed from Hinduism.

Brahmanism → Hinduism → Jainism

FOCUS ON WRITING

6. **Illustrating Hinduism** Now you have another possible topic for your poster. How might you illustrate a complex religion like Hinduism? What pictures would work?

Origins of Buddhism

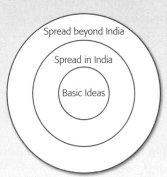
If YOU lived there...

You are a trader traveling in northern India in about 520 BC. As you pass through a town, you see a crowd of people sitting silently in the shade of a huge tree. A man sitting at the foot of the tree is speaking about how one ought to live. His words are like nothing you have heard from the Hindu priests.

Will you stay to listen? Why or why not?

BUILDING BACKGROUND The Jains were not the only ones to break from Hinduism. In the 500s BC a young Indian prince attracted many people to his teachings about how people should live.

Siddhartha's Search for Wisdom

In the late 500s BC a restless young man, dissatisfied with the teachings of Hinduism, began to ask his own questions about life and religious matters. In time, he found answers. These answers attracted many followers, and the young man's ideas became the foundation of a major new religion in India.

The Quest for Answers

The restless young man was Siddhartha Gautama (si-DAHR-tuh GAU-tuh-muh). Born around 563 BC in northern India near the Himalayas, Siddhartha was a prince who grew up in luxury. Born a Kshatriya, a member of the warrior class, Siddhartha never had to struggle with the problems that many people of his time faced. However, Siddhartha was not satisfied. He felt something was missing in his life.

Siddhartha looked around him and saw how hard most people had to work and how much they suffered. He saw people grieving for lost loved ones and wondered why there was so much pain in the world. As a result, Siddhartha began to ask questions about the meaning of human life.

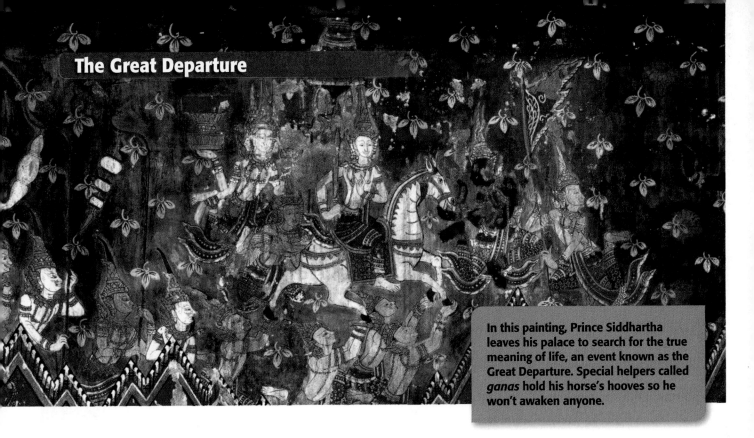

The Great Departure

In this painting, Prince Siddhartha leaves his palace to search for the true meaning of life, an event known as the Great Departure. Special helpers called *ganas* hold his horse's hooves so he won't awaken anyone.

Before Siddhartha reached age 30, he left his home and family to look for answers. His journey took him to many regions in India. Wherever he traveled, he had discussions with priests and people known for their wisdom. Yet no one could give convincing answers to Siddhartha's questions.

The Buddha Finds Enlightenment

Siddhartha did not give up. Instead, he became even more determined to find the answers he was seeking. For several years, he wandered in search of answers.

Siddhartha wanted to free his mind from daily concerns. For a while, he did not even wash himself. He also started **fasting**, or going without food. He devoted much of his time to **meditation**, the focusing of the mind on spiritual ideas.

According to legend, Siddhartha spent six years wandering throughout India. He eventually came to a place near the town of Gaya, close to the Ganges River. There, he sat down under a tree and meditated.

After seven weeks of deep meditation, he suddenly had the answers that he had been looking for. He had realized that human suffering comes from three things:

- wanting what we like but do not have,
- wanting to keep what we like and already have, and
- not wanting what we dislike but have.

Siddhartha spent seven more weeks meditating under the tree, which his followers later named the Tree of Wisdom. He then described his new ideas to five of his former companions. His followers later called this talk the First Sermon.

Siddhartha Gautama was about 35 years old when he found enlightenment under the tree. From that point on, he would be called the Buddha (BOO-duh), or the "Enlightened One." The Buddha spent the rest of his life traveling across northern India and teaching people his ideas.

READING CHECK **Summarizing** What did the Buddha conclude about the cause of suffering?

FOCUS ON READING
What steps did the Buddha take in his search for enlightenment?

Teachings of Buddhism

As he traveled, the Buddha gained many followers. Many of these followers were merchants and artisans, but he even taught a few kings. These followers were the first believers in Buddhism, the religion based on the teachings of the Buddha.

The Buddha was raised Hindu, and many of his teachings reflected Hindu ideas. Like Hindus, he believed that people should act morally and treat others well. In one of his sermons, he said

" Let a man overcome anger by love. Let him overcome the greedy by liberality [giving], the liar by truth. This is called progress in the discipline [training] of the Blessed. "
–The Buddha, quoted in *The History of Nations: India*

Four Noble Truths

At the heart of the Buddha's teachings were four guiding principles. These became known as the Four Noble Truths:

1. Suffering and unhappiness are a part of human life. No one can escape sorrow.

2. Suffering comes from our desires for pleasure and material goods. People cause their own misery because they want things they cannot have.

3. People can overcome their desires and ignorance and reach **nirvana**, a state of perfect peace. Reaching nirvana would free a person's soul from suffering and from the need for further reincarnation.

4. People can overcome ignorance and desire by following an eight-fold path that leads to wisdom, enlightenment, and salvation.

The chart on the next page shows the steps in the Eightfold Path. The Buddha believed that this path was a middle way between human desires and denying oneself any pleasure. He believed that people should overcome their desire for material goods. They should, however, be reasonable, and not starve their bodies or cause themselves unnecessary pain.

This giant statue of the Buddha is just south of the town of Gaya in Bodh Gaya, India—the place where Buddhists believe Siddhartha reached enlightenment.

The Eightfold Path

1. **Right Thought**
Believe in the nature of existence as suffering and in the Four Noble Truths.

2. **Right Intent**
Incline toward goodness and kindness.

3. **Right Speech**
Avoid lies and gossip.

4. **Right Action**
Don't steal from or harm others.

5. **Right Livelihood**
Reject work that hurts others.

6. **Right Effort**
Prevent evil and do good.

7. **Right Mindfulness**
Control your feelings and thoughts.

8. **Right Concentration**
Practice proper meditation.

Challenging Hindu Ideas

Some of the Buddha's teachings challenged traditional Hindu ideas. For example, the Buddha rejected many of the ideas contained in the Vedas, such as animal sacrifice. He told people that they did not have to follow these texts.

The Buddha challenged the authority of the Hindu priests, the Brahmins. He did not believe that they or their rituals were necessary for enlightenment. Instead, he taught that it was the responsibility of each person to work for his or her own salvation. Priests could not help them. However, the Buddha did not reject the Hindu teaching of reincarnation. He taught that people who failed to reach nirvana would have to be reborn time and time again until they achieved it.

The Buddha was opposed to the caste system. He didn't think that people should be confined to a particular place in society. He taught that every person who followed the Eightfold Path properly would reach nirvana. It didn't matter what *varna* or caste they had belonged to in life as long as they lived the way they should.

The Buddha's opposition to the caste system won him the support of the masses. Many herders, farmers, artisans, and untouchables liked hearing that their low social rank would not be a barrier to their enlightenment. Unlike Hinduism, Buddhism made them feel that they had the power to change their lives.

The Buddha also gained followers among the higher classes. Many rich and powerful Indians welcomed his ideas about avoiding extreme behavior while seeking salvation. By the time of his death around 483 BC, the Buddha's influence was spreading rapidly throughout India.

READING CHECK **Comparing** How did Buddha's teachings agree with Hinduism?

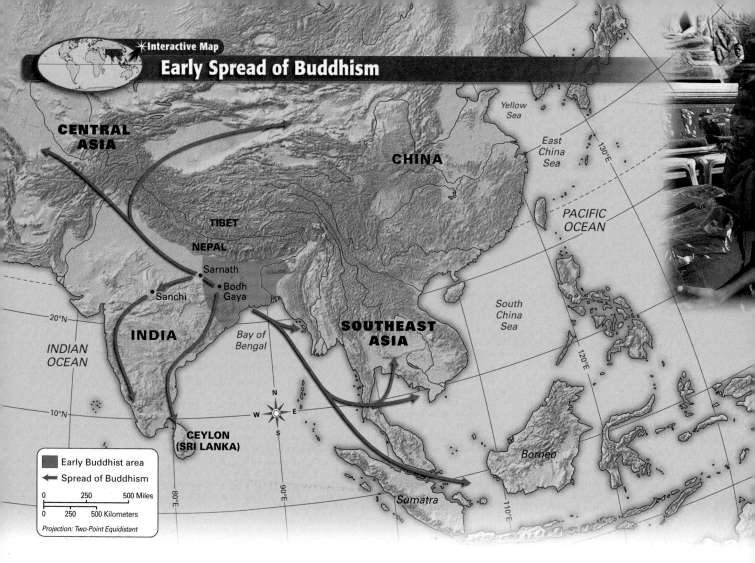

Early Spread of Buddhism

CENTRAL
ASIA

CHINA

Yellow
Sea

East
China
Sea

130°E

TIBET

PACIFIC
OCEAN

NEPAL

Sarnath

• Bodh
Gaya

• Sanchi

South
China
Sea

20°N

INDIA

Bay of
Bengal

SOUTHEAST
ASIA

120°E

INDIAN
OCEAN

N

W   E

S

10°N

CEYLON
(SRI LANKA)

Borneo

Early Buddhist area

Spread of Buddhism

0   250   500 Miles

0   250   500 Kilometers

Projection: Two-Point Equidistant

80°E

90°E

Sumatra

110°E

Buddhism Spreads

Buddhism continued to attract followers after the Buddha's death. After spreading through India, the religion began to spread to other areas as well.

Buddhism Spreads in India

According to Buddhist tradition, 500 of the Buddha's followers gathered together shortly after he died. They wanted to make sure that the Buddha's teachings were remembered correctly.

In the years after this council, the Buddha's followers spread his teachings throughout India. The ideas spread very quickly, because Buddhist teachings were popular and easy to understand. Within 200 years of the Buddha's death, Buddhism had spread through most of India.

Buddhism Spreads beyond India

The spread of Buddhism increased after one of the most powerful kings in India, Asoka, became Buddhist in the 200s BC. Once he converted, he built Buddhist temples and schools throughout India. More importantly, though, he worked to spread Buddhism into areas outside of India. You will learn more about Asoka and his accomplishments in the next section.

Asoka sent Buddhist **missionaries**, or people who work to spread their religious beliefs, to other kingdoms in Asia. One group of these missionaries sailed to the island of Sri Lanka around 251 BC. Others followed trade routes east to what is now Myanmar and to other parts of Southeast Asia. Missionaries also went north to areas near the Himalayas.

Young Buddhist students carry gifts in Sri Lanka, one of the many places outside of India where Buddhism spread.

map zone Geography Skills

Movement After the Buddha died, his teachings were carried through much of Asia.

1. **Identify** Buddhism spread to what island south of India?
2. **Interpret** What physical feature kept Buddhist missionaries from moving directly into China?

go.hrw.com (KEYWORD: SK7 CH19)

Missionaries also introduced Buddhism to lands west of India. They founded Buddhist communities in Central Asia and Persia. They even taught about Buddhism as far away as Syria and Egypt.

Buddhism continued to grow over the centuries. Eventually, it spread via the Silk Road into China, then Korea and Japan. Through their work, missionaries taught Buddhism to millions of people.

A Split within Buddhism

Even as Buddhism spread through Asia, however, it began to change. Not all Buddhists could agree on their beliefs and practices. Eventually, disagreements between Buddhists led to a split within the religion. Two major branches of Buddhism were created—Theravada and Mahayana.

Members of the Theravada branch tried to follow the Buddha's teachings exactly as he had stated them. Mahayana Buddhists, though, believed that other people could interpret the Buddha's teachings to help people reach nirvana. Both branches have millions of believers today, but Mahayana is by far the larger branch.

READING CHECK **Sequencing** How did the Buddha's teachings spread out of India?

SUMMARY AND PREVIEW Buddhism, one of India's major religions, grew more popular once it was adopted by rulers of India's great empires. You will learn more about those empires in the next section.

Section 3 Assessment

go.hrw.com
Online Quiz
KEYWORD: SK7 HP19

Reviewing Ideas, Terms, and Places

1. a. **Identify** Who was the Buddha, and what does the term *Buddha* mean?
   b. **Summarize** How did Siddhartha Gautama free his mind and clarify his thinking as he searched for wisdom?
2. a. **Identify** What is **nirvana**?
   b. **Contrast** How are Buddhist teachings different from Hindu teachings?
   c. **Elaborate** Why do Buddhists believe that following the Eightfold Path leads to a better life?
3. a. **Describe** Into what lands did Buddhism spread?
   b. **Summarize** What role did **missionaries** play in spreading Buddhism?

Critical Thinking

4. **Finding Main Ideas** Draw a diagram like this one. Use it and your notes to identify and describe Buddhism's Four Noble Truths. Write a sentence explaining how the Truths are central to Buddhism.

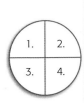

FOCUS ON WRITING

5. **Considering Indian Religions** Look back over what you've just read and your notes about Hinduism. Perhaps you will want to focus your poster on ancient India's two major religions. Think about how you could design a poster around this theme.

Indian Empires

If YOU lived there...

You are a merchant in India in about 240 BC. You travel from town to town on your donkey, carrying bolts of colorful cloth. In the heat of summer, you are grateful for the banyan trees along the road. They shelter you from the blazing sun. You stop at wells for cool drinks of water and rest houses for a break in your journey. You know these are all the work of your king, Asoka.

How do you feel about your king?

What You Will Learn...

Main Ideas

1. The Mauryan Empire unified most of India.
2. Gupta rulers promoted Hinduism in their empire.

The Big Idea

The Mauryas and the Guptas built great empires in India.

Key Terms

mercenaries, *p. 490*
edicts, *p. 491*

TAKING NOTES As you read, take notes about the rise and fall of ancient India's two greatest empires. Record your notes in a chart like the one shown here.

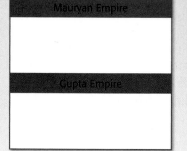

Mauryan Empire

Gupta Empire

BUILDING BACKGROUND For centuries after the Aryan invasion, India was divided into small states. Each state had its own ruler and laws. Then, in the 300s BC, a foreign general, Alexander the Great, took over and unified part of northwestern India. Inspired by his example, a strong leader soon united India for the first time.

Mauryan Empire Unifies India

In the 320s BC a military leader named Candragupta Maurya (kuhn-druh-GOOP-tuh MOUR-yuh) rose to power in northern India. Using an army of **mercenaries**, or hired soldiers, he seized control of the entire northern part of India. By doing so, he founded the Mauryan Empire. Mauryan rule lasted for about 150 years.

The Mauryan Empire

Candragupta Maurya ruled his empire with the help of a complex government. It included a network of spies and a huge army of some 600,000 soldiers. The army also had thousands of war elephants and thousands of chariots. In return for the army's protection, farmers paid a heavy tax to the government.

In 301 BC Candragupta decided to become a Jainist monk. To do so, he had to give up his throne. He passed the throne to his son, who continued to expand the empire. Before long, the Mauryas ruled all of northern India and much of central India as well.

Asoka

Around 270 BC Candragupta's grandson Asoka (uh-SOH-kuh) became king. Asoka was a strong ruler, the strongest of all the Mauryan emperors. He extended Mauryan rule over most of India. In conquering other kingdoms, Asoka made his own empire both stronger and richer.

For many years, Asoka watched his armies fight bloody battles against other peoples. A few years into his rule, however, Asoka converted to Buddhism. When he did, he swore that he would not launch any more wars of conquest.

After converting to Buddhism, Asoka had the time and resources to improve the lives of his people. He had wells dug and roads built throughout the empire. Along these roads, workers planted shade trees, built rest houses for travelers, and raised large stone pillars carved with Buddhist edicts, or laws. Asoka also encouraged the spread of Buddhism in India and the rest of Asia. As you read in the previous section, he sent missionaries to lands all over Asia.

Asoka died in 233 BC, and the empire began to fall apart soon afterward. His sons fought for power, and invaders threatened the empire. In 184 BC the last Mauryan king was killed by one of his generals. India divided into smaller states once again.

FOCUS ON READING

What were some key events in Asoka's life? In what order did they occur?

READING CHECK Finding Main Ideas How did the Mauryans gain control of most of India?

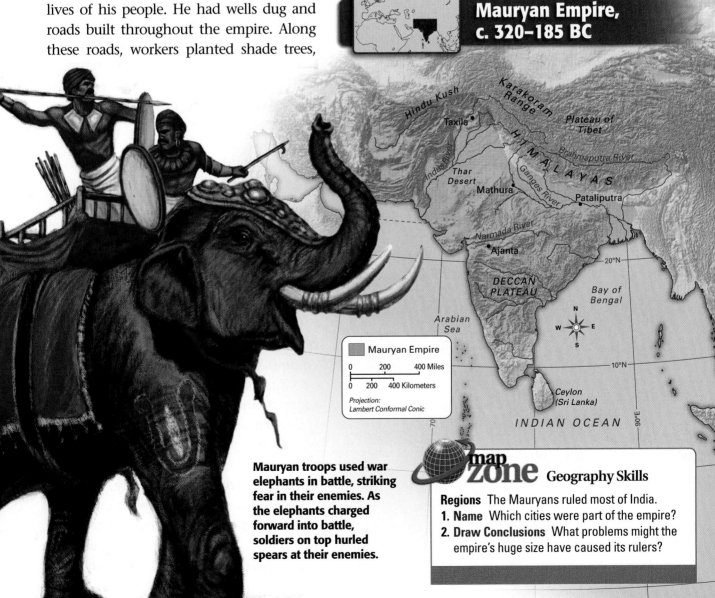

Mauryan Empire, c. 320–185 BC

Mauryan troops used war elephants in battle, striking fear in their enemies. As the elephants charged forward into battle, soldiers on top hurled spears at their enemies.

map zone Geography Skills

Regions The Mauryans ruled most of India.
1. **Name** Which cities were part of the empire?
2. **Draw Conclusions** What problems might the empire's huge size have caused its rulers?

Gupta Rulers Promote Hinduism

After the collapse of the Mauryan Empire, India remained divided for about 500 years. During that period, Buddhism continued to prosper and spread in India, and so the popularity of Hinduism declined.

A New Hindu Empire

ACADEMIC VOCABULARY

establish to set up or create

Eventually, however, a new dynasty was **established** in India. It was the Gupta (GOOP-tuh) dynasty, which took over India around AD 320. Under the Guptas, India was once again united, and it once again became prosperous.

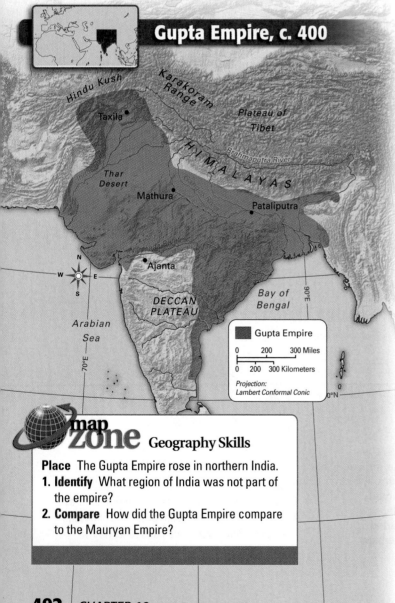

Gupta Empire, c. 400

Hindu Kush
Karakoram Range
Taxila
Plateau of Tibet
Brahmaputra River
HIMALAYAS
Thar Desert
Mathura
Pataliputra
Ajanta
DECCAN PLATEAU
Bay of Bengal
Arabian Sea

Gupta Empire

0   200   300 Miles
0   200   300 Kilometers
Projection: Lambert Conformal Conic

map zone Geography Skills

Place The Gupta Empire rose in northern India.
1. **Identify** What region of India was not part of the empire?
2. **Compare** How did the Gupta Empire compare to the Mauryan Empire?

The first Gupta emperor was Candra Gupta I. Although their names are similar, he was not related to Candragupta Maurya. From his base in northern India, Candra Gupta's armies invaded and conquered neighboring lands. Eventually, he brought much of the northern part of India under his control.

Indian civilization flourished under the Gupta rulers. These rulers were Hindu, so Hinduism became India's dominant religion. Gupta kings built many Hindu temples, some of which became models for later Indian architecture. They also promoted a revival of Hindu writings and worship practices.

Although they were Hindus, the Gupta rulers also supported the religious beliefs of Buddhism and Jainism. They promoted Buddhist art and built Buddhist temples. They also established a university at Nalanda that became one of Asia's greatest centers for Buddhist studies.

Gupta Society

In 375 Emperor Candra Gupta II took the throne in India. Gupta society reached its high point during his rule. Under Candra Gupta II, the empire continued to grow, eventually stretching all the way across northern India. At the same time, the empire's economy strengthened, and so people prospered. They created fine works of art and literature. Outsiders admired the empire's wealth and beauty.

Gupta kings believed the strict social order of the Hindu caste system would strengthen their rule. They also thought it would keep the empire stable. As a result, the Guptas considered the caste system an important part of Indian society.

This was not good news for women, whose roles were limited by caste rules. Brahmins taught that a woman's role was to marry and have children. Women couldn't

even choose their own husbands. Parents arranged all marriages. Once married, wives had few rights. They were expected to serve their husbands. Widows had an even lower social status than other women.

Gupta rule remained strong in India until the late 400s. At that time the Huns, a group from Central Asia, invaded India from the northwest. Their fierce attacks drained the Gupta Empire of its power and wealth. As the Hun armies marched farther into India, the Guptas lost hope.

By the middle of the 500s, Gupta rule had ended, and India had divided into small kingdoms yet again.

READING CHECK **Summarizing** What was the Gupta dynasty's position on religion?

SUMMARY AND PREVIEW The Mauryans and Guptas united much of India in their empires. Next, you will learn about their many achievements.

go.hrw.com
Online Quiz
KEYWORD: SK7 HP19

Section 4 Assessment

Reviewing Ideas, Terms, and Places

1. **a. Identify** Who created the Mauryan Empire?
   b. Summarize What happened after Asoka became a Buddhist?
   c. Elaborate Why do you think many people consider Asoka the greatest of all Mauryan rulers?
2. **a. Recall** What religion did most of the Gupta rulers belong to?
   b. Compare and Contrast How were the rulers Candragupta Maurya and Candra Gupta I alike, and how were they different?
   c. Evaluate Do you think the Gupta enforcement of caste rules was a good idea? Why or why not?

Critical Thinking

3. **Categorizing** Draw a chart like this one. Fill it with facts about India's rulers.

Ruler	Dynasty	Accomplishments

FOCUS ON WRITING

4. **Comparing Indian Empires** Another possible topic for your poster would be a comparison of the Mauryan and Gupta empires. Jot down ideas on what you could show in such a comparison.

Asoka

How can one decision change a man's entire life?

When did he live? before 230 BC

Where did he live? Asoka's empire included much of northern and central India.

What did he do? After fighting many bloody wars to expand his empire, Asoka gave up violence and converted to Buddhism.

Why is he important? Asoka is one of the most respected rulers in Indian history and one of the most important figures in the history of Buddhism. As a devout Buddhist, Asoka worked for years to spread the Buddha's teachings. In addition to sending missionaries around Asia, he had huge columns carved with Buddhist teachings raised all over India. Largely through his efforts, Buddhism became one of Asia's main religions.

Generalizing How did Asoka's life change after he became Buddhist?

KEY EVENTS

- **c. 270 BC** Asoka becomes the Mauryan emperor.

- **c. 261 BC** Asoka's empire reaches its greatest size.

- **c. 261 BC** Asoka becomes a Buddhist.

- **c. 251 BC** Asoka begins to send Buddhist missionaries to other parts of Asia.

This Buddhist shrine, located in Sanchi, India, was built by Asoka.

Indian Achievements

If YOU lived there...

You are a traveler in western India in the 300s. You are visiting a cave temple that is carved into a mountain cliff. Inside the cave it is cool and quiet. Huge columns rise all around you. You don't feel you're alone, for the walls and ceilings are covered with paintings. They are filled with lively scenes and figures. In the center is a large statue with calm, peaceful features.

How does this cave make you feel?

BUILDING BACKGROUND The Mauryan and Gupta empires united most of India politically. During these empires, Indian artists, writers, scholars, and scientists made great advances. Some of their works are still studied and admired today.

Religious Art

The Indians of the Mauryan and Gupta periods created great works of art, many of them religious. Many of their paintings and sculptures illustrated either Hindu or Buddhist teachings. Magnificent temples—both Hindu and Buddhist—were built all around India. They remain some of the most beautiful buildings in the world today.

Temples

Early Hindu temples were small, stone structures. They had flat roofs and contained only one or two rooms. In the Gupta period, though, temple architecture became more complex. Gupta temples were topped by huge towers and were covered with carvings of the god worshipped inside.

Buddhist temples of the Gupta period are also impressive. Some Buddhists carved entire temples out of mountainsides. The most famous such temples are at Ajanta and Ellora. Builders filled the caves there with beautiful paintings and sculpture.

What You Will Learn...

Main Ideas

1. Indian artists created great works of religious art.
2. Sanskrit literature flourished during the Gupta period.
3. The Indians made scientific advances in metalworking, medicine, and other sciences.

The Big Idea

The people of ancient India made great contributions to the arts and sciences.

Key Terms

metallurgy, *p. 498*
alloys, *p. 498*
Hindu-Arabic numerals, *p. 498*
inoculation, *p. 498*
astronomy, *p. 499*

TAKING NOTES As you read, look for information on achievements of ancient India. Take notes about these achievements in a chart.

Type of Achievement	Details about Achievements
Religious Art	
Sanskrit Literature	
Scientific Advances	

This Hindu temple is covered with finely detailed carvings and decorations. Many individual sculptures are images of major Hindu gods, like the statue of Vishnu above.

Another type of Buddhist temple was the stupa. Stupas had domed roofs and were built to house sacred items from the life of the Buddha. Many of them were covered with detailed carvings.

Paintings and Sculpture

The Gupta period also saw the creation of countless works of art, both paintings and statues. Painting was a greatly respected profession, and India was home to many skilled artists. However, we don't know the names of many artists from this period. Instead, we know the names of many rich and powerful members of Gupta society who paid artists to create works of beauty and religious and social significance.

Most Indian paintings of the Gupta period are clear and colorful. Some of them show graceful Indians wearing fine jewelry and stylish clothes. Such paintings offer us a glimpse of the Indians' daily and ceremonial lives.

Artists from both of India's major religions, Hinduism and Buddhism, drew on their beliefs to create their works. As a result, many of the finest paintings of ancient India are found in temples. Hindu painters drew hundreds of gods on temple walls and entrances. Buddhists covered the walls and ceilings of temples with scenes from the life of the Buddha.

Indian sculptors also created great works. Many of their statues were made for Buddhist cave temples. In addition to the temples' intricately carved columns, sculptors carved statues of kings and the Buddha. Some of these statues tower over the cave entrances. Hindu temples also featured impressive statues of their gods. In fact, the walls of some temples, such as the one pictured above, were completely covered with carvings and images.

READING CHECK Summarizing How did religion influence ancient Indian art?

Sanskrit Literature

As you read earlier, Sanskrit was the main language of the ancient Aryans. During the Mauryan and Gupta periods, many works of Sanskrit literature were created. These works were later translated into many other languages.

Religious Epics

The greatest of these Sanskrit writings are two religious epics, the *Mahabharata* (muh-HAH-BAH-ruh-tuh) and the *Ramayana* (rah-MAH-yuh-nuh). Still popular in India, the *Mahabharata* is one of the longest literary works ever written. It is a story about a struggle between two families for control of a kingdom. Included within the story are long passages about Hindu beliefs. The most famous is called the *Bhagavad Gita* (BUG-uh-vuhd GEE-tah).

The *Ramayana*, another great epic, tells about a prince named Rama. In truth, the prince was the god Vishnu in human form. He had become human so he could rid the world of demons. He also had to rescue his wife, a princess named Sita. For centuries, characters from the *Ramayana* have been seen as models for how Indians should behave. For example, Rama is seen as the ideal ruler and his relationship with Sita as the ideal marriage.

Other Works

Writers in the Gupta period also created plays, poetry, and other types of literature. One famous writer of this time was Kalidasa (kahl-ee-DAHS-uh). His work was so brilliant that Candra Gupta II hired him to write plays for the royal court.

Sometime before 500, Indian writers also produced a book of stories called the *Panchatantra* (PUHN-chuh-TAHN-truh). The stories in this collection were intended to teach lessons. They praise people for cleverness and quick thinking. Each story ends with a message about winning friends, losing property, waging war, or some other idea. For example, the message below warns listeners to think about what they are doing before they act.

> *"The good and bad of given schemes
> Wise thought must first reveal:
> The stupid heron saw his chicks
> Provide a mongoose meal."*
> –from the *Panchatantra*, translated
> by Arthur William Ryder

Eventually, translations of this popular collection spread throughout the world. It became popular in countries even as far away as Europe.

READING CHECK **Categorizing** What types of literature did writers of ancient India create?

In this illustration of the *Ramayana*, the monkey king orders the monkey general Hanuman to find Sita. Hanuman helped Rama defeat the demons and win back Sita. Many Indians view him as a model of devotion and loyalty.

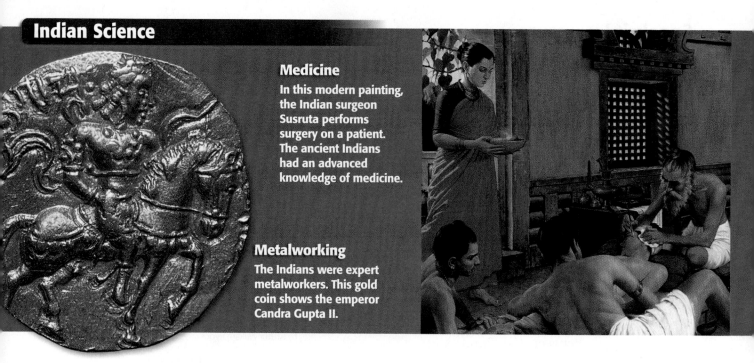

Indian Science

Medicine
In this modern painting, the Indian surgeon Susruta performs surgery on a patient. The ancient Indians had an advanced knowledge of medicine.

Metalworking
The Indians were expert metalworkers. This gold coin shows the emperor Candra Gupta II.

Scientific Advances

Indian achievements were not limited to art, architecture, and literature. Indian scholars also made important advances in metalworking, math, and the sciences.

Metalworking

The ancient Indians were pioneers of **metallurgy** (MET-uhl-uhr-jee), the science of working with metals. Their knowledge allowed them to create high-quality tools and weapons. The Indians also knew **processes** for mixing metals to create **alloys, mixtures of two or more metals**. Alloys are sometimes stronger or easier to work with than pure metals.

Metalworkers made their strongest products out of iron. Indian iron was very hard and pure. These features made the iron a valuable trade item.

During the Gupta dynasty, metalworkers built the famous Iron Pillar near Delhi. Unlike most iron, which rusts easily, the pillar is very resistant to rust. The tall column still attracts crowds of visitors. Scholars study this column even today to learn the Indians' secrets.

Mathematics and Other Sciences

Gupta scholars also made advances in math and science. In fact, they were among the most skilled mathematicians of their day. They developed many of the elements of our modern math system. The very numbers we use today are called **Hindu-Arabic numerals** because they were created by Indian scholars and brought to Europe by Arabs. The Indians were also the first people to create the zero. Although it may seem like a small thing, modern math wouldn't be possible without the zero.

The ancient Indians were also very skilled in the medical sciences. As early as the AD 100s, doctors were writing their knowledge down in textbooks. Among the skills these books describe is how to make medicines from plants and minerals.

Besides curing people with medicines, Indian doctors knew how to protect them against diseases. They used **inoculation** (i-nah-kyuh-LAY-shuhn), the practice of injecting a person with a small dose of a virus to help him or her build a defense to a disease. By fighting off this small dose, the body learns to protect itself.

ACADEMIC VOCABULARY
process a series of steps by which a task is accomplished

THE IMPACT TODAY
People still get inoculations against many diseases.

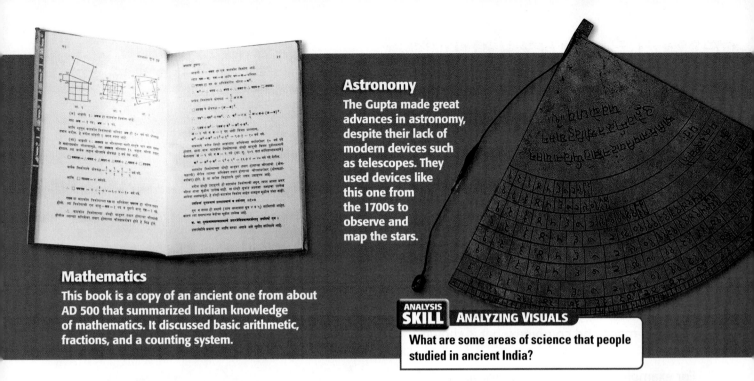

Astronomy

The Gupta made great advances in astronomy, despite their lack of modern devices such as telescopes. They used devices like this one from the 1700s to observe and map the stars.

Mathematics

This book is a copy of an ancient one from about AD 500 that summarized Indian knowledge of mathematics. It discussed basic arithmetic, fractions, and a counting system.

ANALYSIS SKILL **ANALYZING VISUALS**

What are some areas of science that people studied in ancient India?

For people who were injured, Indian doctors could perform surgery. Surgeons repaired broken bones, treated wounds, removed infected tonsils, reconstructed broken noses, and even reattached torn earlobes! If they could find no other cure for an illness, doctors would cast magic spells to help people recover.

Indian interest in **astronomy**, the study of stars and planets, dates back to early times as well. Indian astronomers knew of seven of the nine planets in our solar system. They knew that the sun was a star and that the planets revolved around it. They also knew that the earth was a sphere and that it spun on its axis. In addition, they could predict eclipses of the sun and the moon.

READING CHECK Finding Main Ideas What were two Indian achievements in mathematics?

SUMMARY AND PREVIEW From a group of cities on the Indus River, India grew into a major empire whose people made great achievements. In the next chapter, you'll read about another civilization that experienced similar growth—China.

go.hrw.com
Online Quiz
KEYWORD: SK7 HP19

Section 5 Assessment

Reviewing Ideas, Terms, and Places

1. **a. Describe** What did Hindu temples of the Gupta period look like?
   b. Analyze How can you tell that Indian artists were well respected?
   c. Evaluate Why do you think Hindu and Buddhist temples contained great works of art?
2. **a. Identify** What is the *Bhagavad Gita*?
   b. Explain Why were the stories of the *Panchatantra* written?
   c. Elaborate Why do you think people are still interested in ancient Sanskrit epics today?
3. **a. Define** What is **metallurgy**?
   b. Explain Why do we call the numbers we use today **Hindu-Arabic numerals**?

Critical Thinking

4. **Categorizing** Draw a chart like this one. Identify the scientific advances that fall into each category below.

Metallurgy	Math	Medicine	Astronomy

FOCUS ON WRITING

5. **Highlighting Indian Achievements** List the Indian achievements you could include on a poster. Consider these topics as well as your topic ideas from earlier sections in this chapter. Choose one topic for your poster.

Social Studies Skills

Chart and Graph | Critical Thinking | Geography | Study

Comparing Maps

Learn

Maps are a necessary tool in the study of both history and geography. Sometimes, however, a map does not contain all the information you need. In those cases, you may have to compare two or more maps and combine what is shown on each.

For example, if you look at a physical map of India you can see what landforms are in a region. You can then look at a population map to see how many people live in that region. From this comparison, you can conclude how the region's landforms affect its population distribution.

Practice

Compare the two maps on this page to answer the following questions.

1. What was the northeastern boundary of the Gupta Empire? What is the physical landscape like there?

2. What region of India was never part of the Gupta Empire? Based on the physical map, what might have been one reason for this?

Apply

Choose two maps from this chapter or two maps from the Atlas in this book. Study the two maps and then write three questions that someone could answer by comparing them. Remember that the questions should have people look at both maps to determine the correct answers.

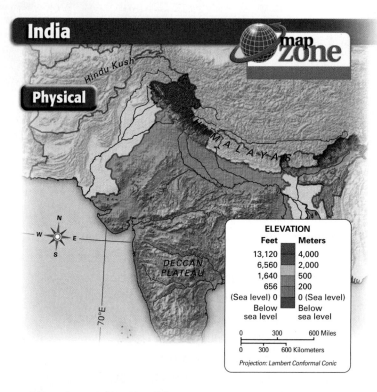

India

Physical

ELEVATION

Feet	Meters
13,120	4,000
6,560	2,000
1,640	500
656	200
(Sea level) 0	0 (Sea level)
Below sea level	Below sea level

0   300   600 Miles
0   300   600 Kilometers

Projection: Lambert Conformal Conic

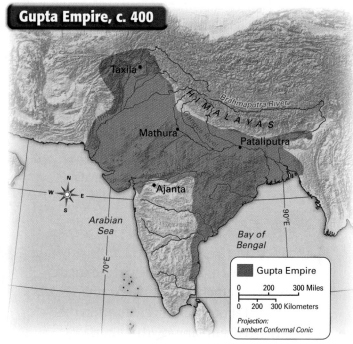

Gupta Empire, c. 400

Gupta Empire

0   200   300 Miles
0   200   300 Kilometers

Projection: Lambert Conformal Conic

Chapter Review

Geography's Impact
video series
Review the video to answer the closing question:
Do you think enlightenment is an achievable goal in today's world? Why or why not?

Visual Summary

Use the visual summary below to help you review the main ideas of the chapter.

QUICK FACTS

The Harappan civilization began in the Indus River Valley.

Hinduism and Buddhism both developed in India.

Indians made great advances in art, literature, science, and other fields.

Reviewing Vocabulary, Terms, and Places

Fill in the blanks with the correct term or name from this chapter.

1. _____ are hired soldiers.

2. A _____ is a division of people into groups based on birth, wealth, or occupation.

3. Hindus believe in _____, the belief that they will be reborn many times after death.

4. Harappa and _____ were the largest cities of the Harappan civilization.

5. The focusing of the mind on spiritual things is called _____.

6. People who work to spread their religious beliefs are called _____.

7. People who practice _____ use only peaceful ways to achieve change.

8. Indian civilization first developed in the valley of the _____.

9. A mixture of metals is called an _____.

Comprehension and Critical Thinking

SECTION 1 *(Pages 472–477)*

10. **a. Describe** What caused floods on the Indus River, and what was the result of those floods?

   b. Contrast How was Aryan culture different from Harappan culture?

   c. Elaborate In what ways was Harappan society an advanced civilization?

SECTION 2 *(Pages 478–483)*

11. **a. Identify** Who were the Brahmins, and what role did they play in Aryan society?

   b. Analyze How do Hindus believe karma affects reincarnation?

   c. Elaborate Hinduism has been called both a polytheistic religion—one that worships many gods—and a monotheistic religion—one that worships only one god. Why do you think this is so?

SECTION 3 *(Pages 484–489)*

12. a. Describe What did the Buddha say caused human suffering?

b. Analyze How did Buddhism grow and change after the Buddha died?

c. Elaborate Why did the Buddha's teachings about nirvana appeal to many people of lower castes?

SECTION 4 *(Pages 490–493)*

13. a. Identify What was Candragupta Maurya's greatest accomplishment?

b. Compare and Contrast What was one similarity between the Mauryans and the Guptas? What was one difference between them?

c. Predict How might Indian history have been different if Asoka had not become a Buddhist?

SECTION 5 *(Pages 495–499)*

14. a. Describe What kinds of religious art did the ancient Indians create?

b. Make Inferences Why do you think religious discussions are included in the *Mahabharata?*

c. Evaluate Which of the ancient Indians' achievements do you think is most impressive? Why?

Using the Internet

go.hrw.com
KEYWORD: SK7 CH19

15. Activity: Making a Brochure In this chapter, you learned about India's early history. That history was largely shaped by India's geography. Enter the activity keyword to research the geography and civilizations of India, taking notes as you go along. Finally, use the interactive brochure template to present what you have found.

Social Studies Skills

Comparing Maps *Study the physical and population maps of South and East Asia in the Atlas. Then answer the following questions.*

16. Along what river in northeastern India is the population density very high?

17. Why do you think fewer people live in far northwestern India than in the northeast?

18. Sequencing Arrange the following list of events in the order in which they happened. Then write a brief paragraph describing the events, using clue words such as *then* and *later* to show the proper sequence.
- The Gupta Empire is created.
- Harappan civilization begins.
- The Aryans invade India.
- The Mauryan Empire is formed.

19. Designing Your Poster Now that you have a topic for your poster, it's time to create it. On a large sheet of paper or poster board, write a title that identifies your topic. Then draw pictures, maps, or diagrams that illustrate it. Next to each picture, write a short caption to identify what the picture, map, or diagram shows.

Map Activity Interactive

20. Ancient India On a separate sheet of paper, match the letters on the map with their correct labels.

Mohenjo Daro	Indus River
Harappa	Ganges River
Bodh Gaya	

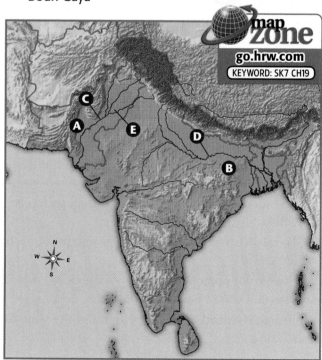

map Zone
go.hrw.com
KEYWORD: SK7 CH19

DIRECTIONS: Read questions 1 through 7 and write the letter of the best response. Then read question 8 and write your own well-constructed response.

1 The earliest civilizations in India developed along which river?

A Indus

B Ganges

C Brahmaputra

D Krishna

2 The people of which *varna* in early India had the hardest lives?

A Brahmins

B Kshatriyas

C Sudras

D Vaisyas

3 What is the *main* goal of people who follow Buddhism as it was taught by the Buddha?

A wealth

B rebirth

C missionary work

D reaching nirvana

4 Which Indian ruler greatly enlarged his empire before giving up violence and promoting the spread of Buddhism?

A Candragupta Maurya

B Asoka

C Buddha

D Mahariva

5 Early India's contributions to world civilization included

A developing the world's first calendar.

B creating what is now called algebra.

C inventing the plow and the wheel.

D introducing zero to the number system.

*"*From anger comes confusion;
from confusion memory lapses;
from broken memory understanding is lost;
from loss of understanding, he is ruined.

But a man of inner strength
whose senses experience objects
without attraction and hatred,
in self control, finds serenity.*"*

–from the *Bhagavad Gita*,
translated by Barbara Stoler Miller

6 Read the psssage above. According to this passage, what will a person find if he has inner strength and self control?

A confusion

B loss of understanding

C serenity

D anger

7 According to Hindu teachings, the universal spirit of which everything in the world is part is called

A Vishnu.

B Brahman.

C Siva.

D Buddha.

8 **Extended Response** The passage from the *Bhagavad Gita* above is advice about how a person can reach nirvana. As you have learned, nirvana is an important concept in both Hinduism and Buddhism. Write a brief paragraph explaining what nirvana is and how Hindus and Buddhists believe one can reach it.

History of Ancient China

1600 BC–AD 1450

Mongols

CENTRAL ASIA

Traders on the Silk Road

SOUTH ASIA

0	250	500 Miles
0	250	500 Kilometers

Projection: Two-Point Equidistant

What You Will Learn...

In this chapter you will learn about the history and culture of ancient China. China was one of the world's early centers of civilization. You will also study the powerful dynasties that arose to rule China and reshape Chinese culture.

SECTION 1
Early China............................**506**

SECTION 2
The Han Dynasty.......................**510**

SECTION 3
The Sui, Tang, and Song Dynasties..**518**

SECTION 4
Confucianism and Government......**524**

SECTION 5
The Yuan and Ming Dynasties.......**528**

FOCUS ON READING AND WRITING

Understanding Chronological Order When you read about history, it is important to keep track of the order in which events happened. You can often use words in the text to help figure this order out. **See the lesson, Understanding Chronological Order, on page 687.**

Writing a Magazine Article You are a freelance writer who has been asked to write a magazine article about the achievements of the anicent Chinese. As you read this chapter, you will collect information. Then you will use the information to write your magazine article.

Early China The first dynasties to rule China left behind artifacts such as this clay figure of a soldier.

Ancient China, 1600 BC–AD 1450

ASIA
The Great Wall
Beijing
Xi'an (Chang'an)
Kaifeng
Qin dynasty emperor
Porcelain vase
Sea of Japan (East Sea)
Yellow Sea
East China Sea
South China Sea
PACIFIC OCEAN
SOUTHEAST ASIA
Tropic of Cancer
40°N
30°N
20°N
110°E
120°E
130°E
N
E
S
W

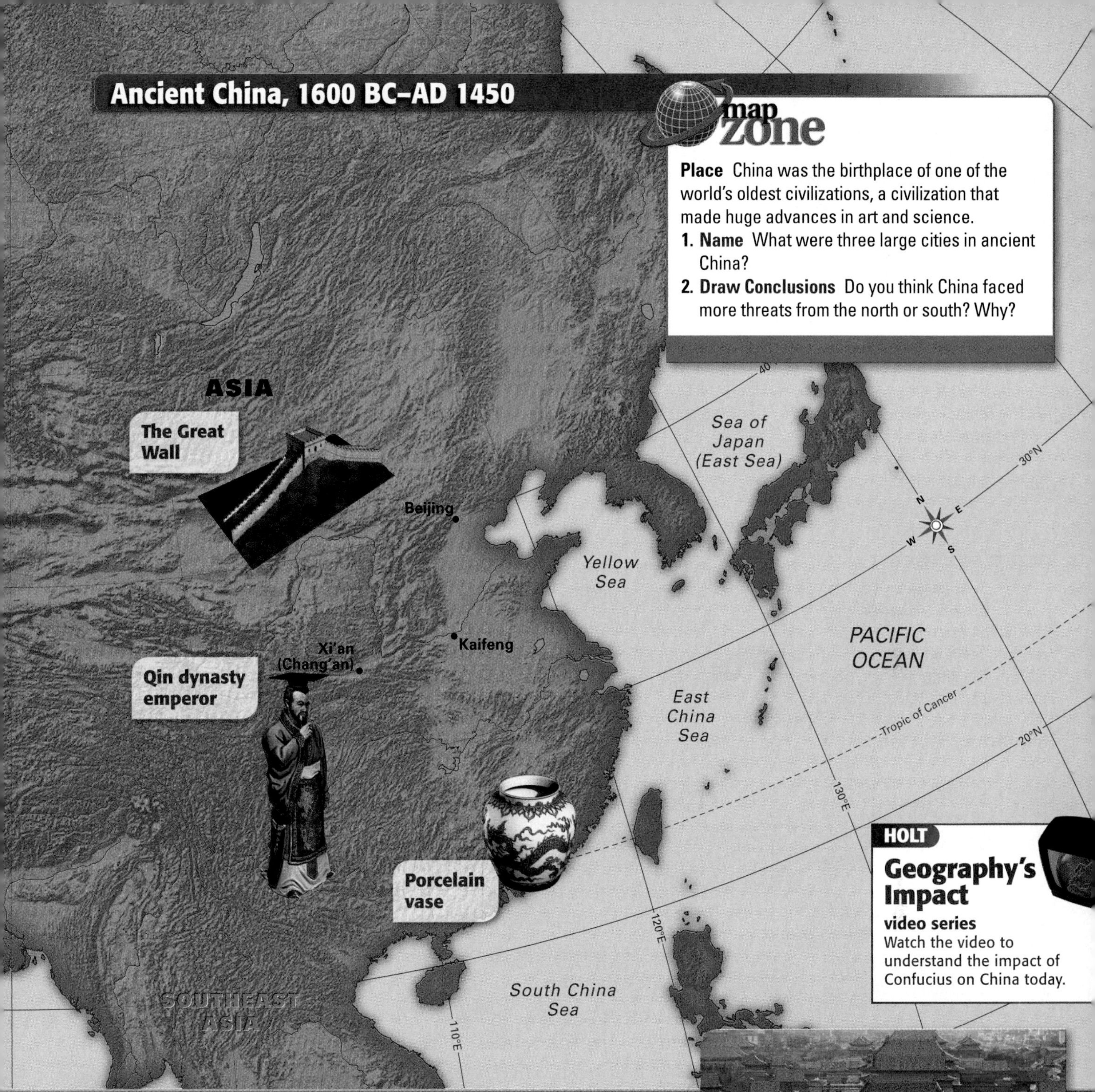

map zone

Place China was the birthplace of one of the world's oldest civilizations, a civilization that made huge advances in art and science.

1. **Name** What were three large cities in ancient China?
2. **Draw Conclusions** Do you think China faced more threats from the north or south? Why?

HOLT

Geography's Impact

video series

Watch the video to understand the impact of Confucius on China today.

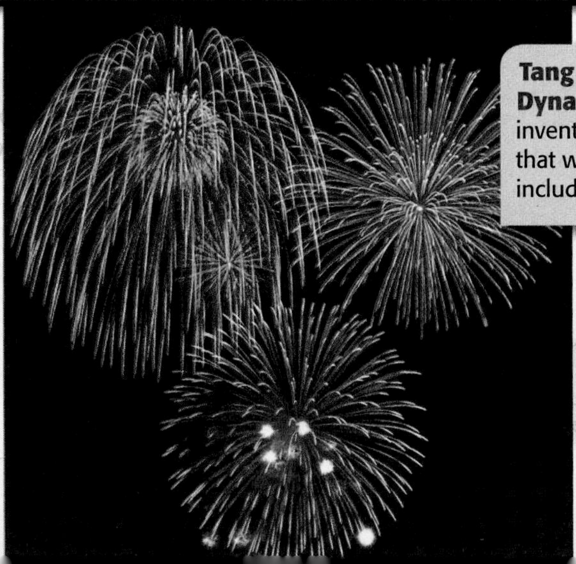

Tang and Song Dynasties The Chinese invented many items that we still use today, including fireworks.

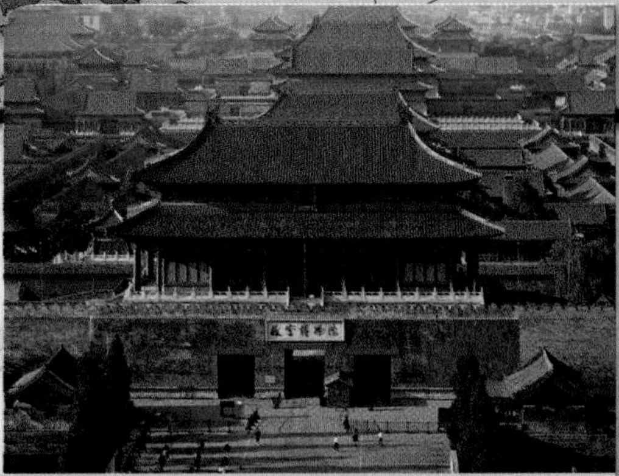

Yuan and Ming Dynasties During the Yuan and Ming dynasties, Beijing became China's largest city and a center of Chinese culture.

Early China

Main Ideas

1. Chinese civilization began along two rivers.
2. The Shang dynasty was the first known dynasty to rule China.
3. The Zhou and Qin dynasties changed Chinese society and made great advances.

The Big Idea

Early Chinese history was shaped by three dynasties—the Shang, the Zhou, and the Qin.

Key Terms and Places

Chang Jiang, *p. 506*
Huang He, *p. 506*
mandate of heaven, *p. 508*
Xi'an, *p. 509*
Great Wall, *p. 509*

TAKING NOTES Draw a chart like the one below. As you read this section, fill in the chart with details about each period in China's early history.

Beginnings	Shang dynasty	Zhou dynasty	Qin dynasty

If YOU lived there...

You are the ruler of China, and hundreds of thousands of people look to you for protection. For many years, your country has lived in peace. Large cities have grown up, and traders travel freely from place to place. Now, however, a new threat looms. Invaders from the north are threatening China's borders. Frightened by the ferocity of these invaders, the people turn to you for help.

What will you do to protect your people?

BUILDING BACKGROUND As in India, people in China first settled near rivers. Two rivers were particularly important in early China—the Huang He and the Chang Jiang. Along these rivers, people began to farm, cities grew up, and China's government was born. The head of that government was an emperor, the ruler of all China.

Chinese Civilization Begins

As early as 7000 BC people had begun to farm in China. They grew rice in the middle **Chang Jiang** Valley. North, along the **Huang He**, the land was better for growing cereals such as millet and wheat. At the same time, people domesticated animals such as pigs and sheep. Supported by these sources of food, China's population grew. Villages appeared along the rivers.

Some of the villages along the Huang He grew into large towns. Walls surrounded these towns to defend them against floods and hostile neighbors. In towns like these, the Chinese left many artifacts, such as arrowheads, fishhooks, tools, and pottery. Some village sites even contained pieces of cloth.

Over time, Chinese culture became more advanced. After 3000 BC people began to use potter's wheels to make many types of pottery. They also learned to dig water wells. As populations grew, villages spread out over larger areas in both northern and southeastern China.

READING CHECK Analyzing When and where did China's earliest civilizations develop?

Early Dynasties of China

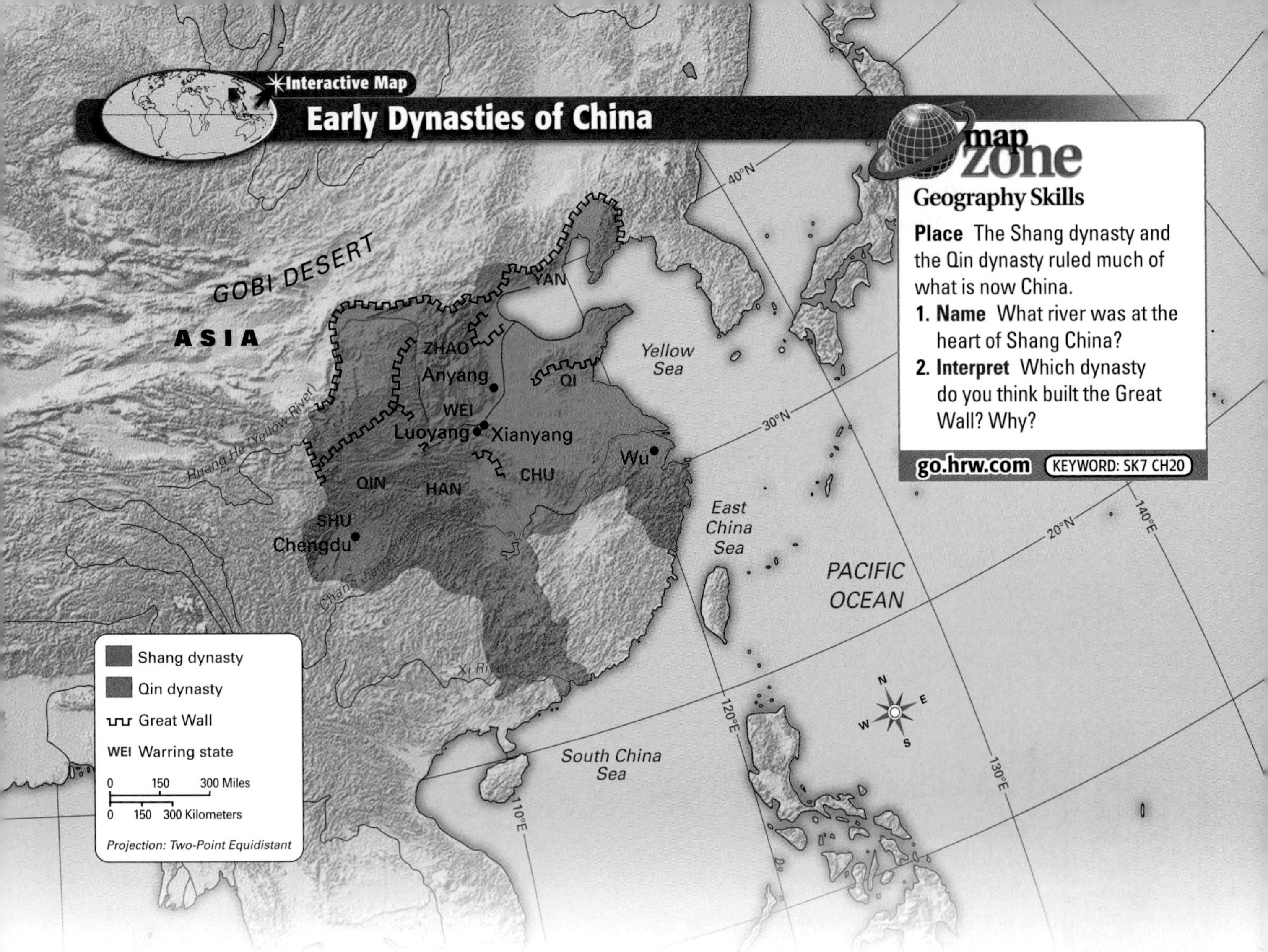
Interactive Map
Early Dynasties of China
GOBI DESERT
ASIA
YAN
ZHAO
Anyang
QI
WEI
Luoyang
Xianyang
CHU
Wu
QIN
HAN
SHU
Chengdu
Huang He (Yellow River)
Xi River
Yellow Sea
East China Sea
PACIFIC OCEAN
South China Sea
40°N
30°N
20°N
110°E
120°E
130°E
140°E
N
E
S
W
Shang dynasty
Qin dynasty
Great Wall
WEI Warring state
0 150 300 Miles
0 150 300 Kilometers
Projection: Two-Point Equidistant

map zone

Geography Skills

Place The Shang dynasty and the Qin dynasty ruled much of what is now China.

1. **Name** What river was at the heart of Shang China?
2. **Interpret** Which dynasty do you think built the Great Wall? Why?

go.hrw.com KEYWORD: SK7 CH20

Shang Dynasty

As time passed, dynasties, or families, of strong rulers began to take power in China. The first dynasty for which we have clear evidence is the Shang, which was firmly established by the 1500s BC. Strongest in the Huang He Valley, the Shang ruled a broad area of northern China, as you can see on the map. Shang emperors ruled in China until the 1100s BC.

The Shang made many advances, including China's first writing system. This system used more than 2,000 symbols to express words or ideas. Although the system has gone through changes over the years, the Chinese symbols used today are based on those of the Shang period.

Many examples of Shang writing that we have found were on cattle bones and turtle shells. Priests had carved questions about the future on bones or shells, which were then heated, causing them to crack. The priests believed they could "read" these cracks to predict the future.

In addition to writing, the Shang also made other achievements. Artisans made beautiful bronze containers for cooking and religious ceremonies. They also made axes, knives, and ornaments from jade. Soldiers developed war chariots, powerful bows, and bronze armor. The Shang also invented a calendar based on the cycles of the moon.

READING CHECK **Summarizing** What were two Shang achievements?

Zhou and Qin Dynasties

The Shang dynasty was only the first of many dynasties described in Chinese records. After the Shang lost power, other dynasties rose up to take control of China. Two of those dynasties were the Zhou (JOH) and the Qin (CHIN).

Zhou Dynasty

FOCUS ON READING

Which dynasty ruled earlier, the Zhou or the Qin?

In the 1100s, the Shang rulers of China were overthrown in a rebellion. In their place, the rebels from the western part of China took power. This event marked the beginning of the Zhou dynasty. This dynasty lasted longer than any other in Chinese history. Zhou rulers held power in China until 771 BC.

The Zhou claimed that they had been chosen by heaven to rule China. They believed that no one could rule without heaven's permission. This idea that heaven chose China's ruler and gave him or her power was called the **mandate of heaven**.

Under the Zhou, a new political order formed in China. The emperor was at the top of society. Everything in China belonged to him, and everyone had to be loyal to him.

Emperors gave land to people in exchange for loyalty or military service. Those people who received this land became lords. Below the lords were peasants, or farmers who owned little land. In addition to growing their own food, peasants had to grow food for lords.

Warring States Period

The Zhou political system broke down as lords grew less loyal to the emperors. When invaders attacked the capital in 771 BC, many lords would not fight. As a result, the emperor was overthrown. China broke apart into many kingdoms that fought each other. This time of disorder in China is called the Warring States period.

BIOGRAPHY

Emperor Shi Huangdi
(c. 259–210 BC)

Shi Huangdi was a powerful emperor and a very strict one. He demanded that everyone in China believe the same things he did. To prevent people from having other ideas, he ordered all books that did not agree with his beliefs burned. When a group of scholars protested the burning of these books, Shi Huangdi had them buried alive. These actions led many Chinese people to resent the emperor. As a result, they were eager to bring the Qin dynasty to an end.

In 1974 archaeologists found the tomb of Emperor Shi Huangdi near Xi'an and made an amazing discovery. Buried close to the emperor was an army of more than 6,000 life-size terra-cotta, or clay, soldiers. They were designed to be with Shi Huangdi in the afterlife. In other nearby chambers of the tomb there were another 1,400 clay figures of cavalry and chariots.

Qin Dynasty

The Warring States period came to an end when one state became strong enough to defeat all its rivals. That state was called Qin. In 221 BC, a king from Qin managed to unify all of China under his control and name himself emperor.

As emperor, the king took a new name. He called himself Shi Huangdi (SHEE hwahng-dee), a name that means "first emperor." Shi Huangdi was a very strict ruler, but he was an effective ruler as well. He expanded the size of China both to the north and to the south, as the map at the beginning of this section shows.

Shi Huangdi greatly changed Chinese politics. Unlike the Zhou rulers, he refused to share his power with anyone. Lords who had enjoyed many rights before now lost those rights. In addition, he ordered thousands of noble families to move to his capital, now called **Xi'an** (SHEE-AHN). He thought nobles that he kept nearby would be less likely to rebel against him.

The Qin dynasty did not last long. While Shi Huangdi lived, he was strong enough to keep China unified. The rulers who followed him, however, were not as strong. In fact, China began to break apart within a few years of Shi Huangdi's death. Rebellions began all around China, and the country fell into civil war.

Qin Achievements

Although the Qin did not rule for long, they saw great advances in China. As emperor, Shi Huangdi worked to make sure that people all over China acted and thought the same way. He created a system of laws that would apply equally to people in all parts of China. He also set up a new system of money. Before, people in each region had used local currencies. He also created a uniform system of writing that eliminated minor differences between regions.

The Qin's best known achievements, though, were in building. Under the Qin, the Chinese built a huge network of roads and canals. These roads and canals linked distant parts of the empire to make travel and trade easier.

To protect China from invasion, Shi Huangdi built the **Great Wall**, a barrier that linked earlier walls that stood near China's northern border. Building the wall took years of labor from hundreds of thousands of workers. Later dynasties added to the wall, parts of which still stand today.

SUMMARY AND PREVIEW Early Chinese history was shaped by the Shang, Zhou, and Qin dynasties. Next, you will read about another strong dynasty, the Han.

go.hrw.com
Online Quiz
KEYWORD: SK7 HP20

Section 1 Assessment

Reviewing Ideas, Terms, and Places

1. **a. Identify** On what rivers did Chinese civilization begin?
   b. Analyze What advances did the early Chinese make?
2. **a. Describe** What area did the Shang rule?
   b. Evaluate What do you think was the Shang dynasty's most important achievement? Why?
3. **a. Define** What is the **mandate of heaven**?
   b. Generalize How did Shi Huangdi change China?

Critical Thinking

4. **Analyzing** Draw a chart like the one shown here. Using your notes, write details about the achievements and political system of China's early dynasties.

	Achievements	Political System
Shang		
Zhou		
Qin		

FOCUS ON WRITING

5. **Identifying Advances** The Shang, Zhou, and Qin made some of the greatest advances in Chinese history. Which of these will you mention in your magazine article? Write down some notes.

The Han Dynasty

If YOU lived there...

You are a young Chinese student from a poor family. Your family has worked hard to give you a good education so that you can get a government job and have a great future. Your friends laugh at you. They say that only boys from wealthy families win the good jobs. They think it is better to join the army.

Will you take the exam or join the army? Why?

Han Dynasty Government

When the Qin dynasty collapsed, many groups fought for power. After years of fighting, an army led by Liu Bang (lee-oo bang) won control. Liu Bang became the first emperor of the Han dynasty, which lasted more than 400 years.

The Rise of a New Dynasty

Liu Bang, a peasant, was able to become emperor in large part because of the Chinese belief in the mandate of heaven. He was the first common person to become emperor. He earned people's loyalty and trust. In addition, he was well liked by both soldiers and peasants, which helped him keep control.

What You Will Learn...

Main Ideas

1. Han dynasty government was largely based on the ideas of Confucius.
2. Han China supported and strengthened family life.
3. The Han made many achievements in art, literature, and learning.

The Big Idea

The Han dynasty created a new form of government that valued family, art, and learning.

Key Terms

sundial, *p. 514*
seismograph, *p. 514*
acupuncture, *p. 515*

TAKING NOTES As you read, take notes on Han government, family life, and achievements. Use a diagram like the one here to help you organize your notes.

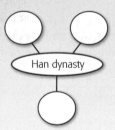

Han dynasty

Time Line

The Han Dynasty

205 BC
The Han dynasty begins.

AD 220
The Han dynasty falls.

200 BC

BC 1 AD

AD 200

140 BC
Wudi becomes emperor and tries to strengthen China's government.

AD 25
The Han move their capital east to Luoyang.

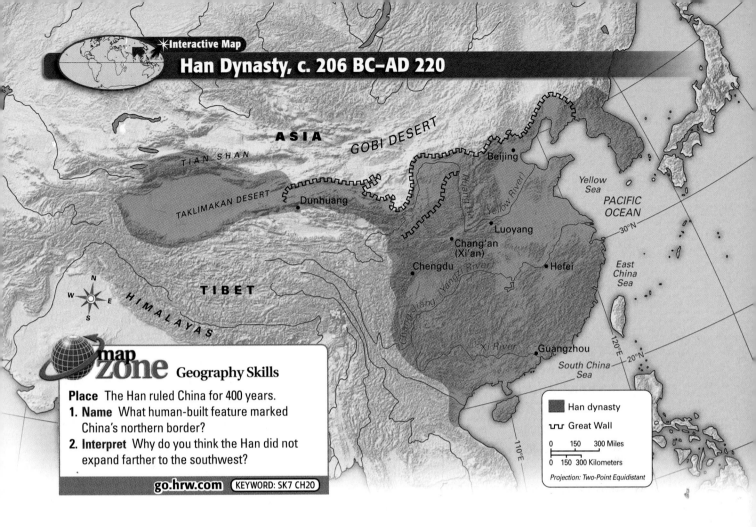

ASIA
GOBI DESERT
TIAN SHAN
TAKLIMAKAN DESERT
•Dunhuang
Beijing
Yellow River
Huang He
Yellow Sea
PACIFIC OCEAN
•Luoyang
Chang'an (Xi'an)
•Chengdu
Chang Jiang (Yangzi River)
•Hefei
East China Sea
TIBET
HIMALAYAS
N W E S
Xi River
•Guangzhou
South China Sea
30°N
20°N
120°E
110°E

map zone Geography Skills

Place The Han ruled China for 400 years.
1. **Name** What human-built feature marked China's northern border?
2. **Interpret** Why do you think the Han did not expand farther to the southwest?

go.hrw.com KEYWORD: SK7 CH20

Han dynasty
⊔⊔⊔ Great Wall

0   150   300 Miles
0   150 300 Kilometers

Projection: Two-Point Equidistant

Liu Bang's rule was different from the strict government of the Qin. He wanted to free people from harsh government policies. He lowered taxes for farmers and made punishments less severe. He gave large blocks of land to his supporters.

In addition to setting new policies, Liu Bang changed the way government worked. He set up a government structure that built on the foundation begun by the Qin. He also relied on educated officials to help him rule.

Wudi Creates a New Government

In 140 BC Emperor Wudi (WOO-dee) took the throne. He wanted to create a stronger government. To do that, he took land from the lords, raised taxes, and put the supply of grain under government control. He also made Confucianism China's official government philosophy.

Confucianism is a philosophy based on the teachings of a man named Confucius. It emphasizes the importance of ethics and moral values, such as respect for elders and loyalty toward family members. Under the Han, government officials were expected to practice Confucianism. Wudi even began a university to teach Confucian ideas.

Studying Confucianism could also get a person a good job in China. If a person passed an exam on Confucian teachings, he could get a position working for the government. Not just anyone could take the test, though. The exams were only open to people who had been recommended for government service already. As a result, wealthy or influential families continued to control the government.

READING CHECK **Analyzing** How was the Han government based on the ideas of Confucius?

FOCUS ON READING

Who ruled first, Liu Bang or Wudi?

Family Life

The Han period was a time of great social change in China. Class structure became more rigid. The family once again became important within Chinese society.

Social Classes

Based on the Confucian system, people were divided into four classes. The upper class was made up of the emperor, his court, and scholars who held government positions. The second class, the largest, was made up of the peasants. Next were artisans who produced items for daily life and some luxury goods. Merchants were the lowest class because they did not actually produce anything. They only bought and sold what others made. The military was not a class in the Confucian system. Still, joining the army offered men a chance to rise in social status because the military was considered part of the government.

This Han artifact is an oil lamp held by a servant.

Lives of Rich and Poor

The classes only divided people into social rank. They did not indicate wealth or power. For instance, even though peasants made up the second highest class, they were poor. Many merchants, on the other hand, were wealthy and powerful despite being in the lowest class.

People's lifestyles varied according to wealth. The emperor and his court lived in a large palace. Less important officials lived in multilevel houses built around courtyards. Many of these wealthy families owned large estates and employed laborers to work the land. Some families even hired private armies to defend their estates.

The wealthy filled their homes with expensive decorations. These included paintings, pottery, bronze lamps, and jade figures. Rich families hired musicians for entertainment. Even the tombs of dead family members were filled with beautiful, expensive objects.

Most people in Han China, however, did not live so comfortably. Nearly 60 million people lived in China during the Han dynasty, and about 90 percent of them were peasants who lived in the countryside. Peasants put in long, tiring days working the land. Whether it was in the millet fields of the north or in the rice paddies of the south, the work was hard. In the winter, peasants were forced to work on building projects for the government. Heavy taxes and bad weather forced many farmers to sell their land and work for rich landowners. By the last years of the Han dynasty, only a few farmers were independent.

Chinese peasants lived simple lives. They wore plain clothing made of fiber from a native plant. The main foods they ate were cooked grains like barley. Most peasants lived in small villages. Their small, wood-framed houses had walls made of mud or stamped earth.

The Revival of the Family

Since Confucianism was the government's official philosophy during Wudi's reign, Confucian teachings about the family were also honored. Children were taught from birth to respect their elders. Disobeying one's parents was a crime. Even emperors had a duty to respect their parents.

Confucius had taught that the father was the head of the family. Within the family, the father had absolute power. The Han taught that it was a woman's duty to obey her husband, and children had to obey their father.

Han officials believed that if the family was strong and people obeyed the father, then they would also obey the emperor. Since the Han rewarded strong family ties and respect for elders, some men even gained government jobs based on the respect they showed their parents.

Children were encouraged to serve their parents. They were also expected to honor dead parents with ceremonies and offerings. All members of a family were expected to care for family burial sites.

Chinese parents valued boys more highly than girls. This was because sons carried on the family line and took care of their parents when they were old. On the other hand, daughters became part of their husband's family. According to a Chinese proverb, "Raising daughters is like raising children for another family." Some women, however, still gained power. They could gain influence in their sons' families. An older widow could even become the head of the family.

READING CHECK Identifying Cause and Effect
Why did the family take on such importance during the Han dynasty?

Han Achievements

During the Han dynasty, the Chinese made many advances in art and learning. Some of these advances are shown here.

Science

This is a model of an ancient Chinese seismograph. When an earthquake struck, a lever inside caused a ball to drop from a dragon's mouth into a toad's mouth, indicating the direction from which the earthquake had come.

Han Achievements

Han rule was a time of great achievements. Art and literature thrived, and inventors developed many useful devices.

Art and Literature

The Chinese of the Han period produced many works of art. They became experts at figure painting—a style of painting that includes portraits of people. Portraits often showed religious figures and Confucian scholars. Han artists also painted realistic scenes from everyday life. Their creations covered the walls of palaces and tombs.

In literature, Han China is known for its poetry. Poets developed new styles of verse, including the *fu* style, which was the most popular. *Fu* poets combined prose and poetry to create long literary works. Another style, called *shi*, featured short lines of verse that could be sung. Many Han rulers hired poets known for the beauty of their verse.

ACADEMIC VOCABULARY

innovation a new idea, method, or device

Han writers also produced important works of history. One historian by the name of Sima Qian wrote a complete history of all the dynasties through the early Han. His format and style became the model for later historical writings.

Inventions and Advances

The Han Chinese invented one item that we use every day—paper. They made it by grinding plant fibers, such as mulberry bark and hemp, into a paste. Then they let it dry in sheets. Chinese scholars produced books by pasting several pieces of paper together into a long sheet. Then they rolled the sheet into a scroll.

The Han also made other **innovations** in science. These included the sundial and the seismograph. A **sundial** is a device that uses the position of shadows cast by the sun to tell the time of day. It was an early type of clock. A **seismograph** is a device that measures the strength of earthquakes. Han emperors were very interested in knowing

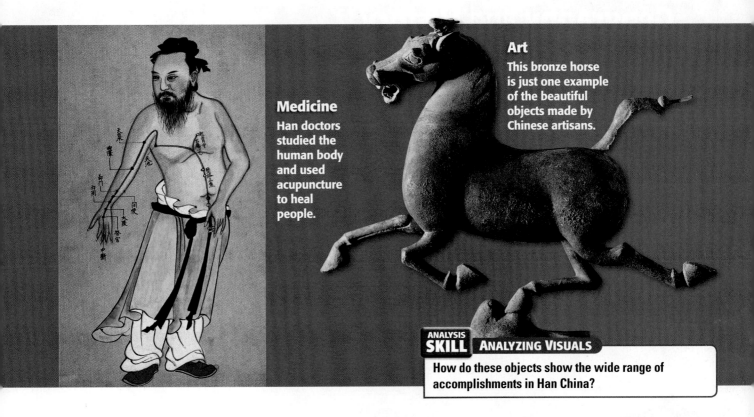

Medicine
Han doctors studied the human body and used acupuncture to heal people.

Art
This bronze horse is just one example of the beautiful objects made by Chinese artisans.

ANALYSIS SKILL **ANALYZING VISUALS**

How do these objects show the wide range of accomplishments in Han China?

about the movements of the Earth. They believed that earthquakes were signs of future evil events.

Another Han innovation, acupuncture (AK-yoo-punk-cher), improved medicine. **Acupuncture** is the practice of inserting fine needles through the skin at specific points to cure disease or relieve pain. Many Han inventions in science and medicine are still used today.

READING CHECK **Categorizing** What advances did the Chinese make during the Han period?

SUMMARY AND PREVIEW Rulers of the Han dynasty based their government on Confucianism, which strengthened family bonds in China. In addition, art and learning thrived under Han rule. In the next section you will learn about two dynasties that also made great advances, the Tang and the Song.

go.hrw.com
Online Quiz
KEYWORD: SK7 HP20

Section 2 Assessment

Reviewing Ideas, Terms, and People
1. **a. Identify** What is Confucianism? How did it affect the government during the Han dynasty?
   b. Summarize How did Emperor Wudi create a strong central government?
   c. Evaluate Do you think that an exam system is the best way to make sure that people are fairly chosen for government jobs? Why or why not?
2. **a. Describe** What was the son's role in the family?
   b. Contrast How did living conditions for the wealthy differ from those of the peasants during the Han dynasty?
3. **Identify** What device did the Chinese invent to measure the strength of earthquakes?

Critical Thinking
4. **Analyzing** Use your notes to complete this diagram about how Confucianism influenced Han government and family.

Government
↑
Confucianism
↓
Family

FOCUS ON WRITING

5. **Analyzing the Han Dynasty** The Han dynasty was one of the most influential in all of Chinese history. How will you describe the dynasty's many achievements in your article? Make a list of achievements you want to include.

The Silk Road

The Silk Road was a long trade route that stretched across the heart of Asia. Along this route, an active trade developed between China and Southwest Asia by about 100 BC. By AD 100, the Silk Road connected Han China in the east with the Roman Empire in the west.

The main goods traded along the Silk Road were luxury goods—ones that were small, light, and expensive. These included goods like silk, spices, and gold. Because they were small and valuable, merchants could carry these goods long distances and still sell them for a large profit. As a result, people in both the east and the west were able to buy luxury goods that were unavailable at home.

GAUL

SPAIN

EUROPE

Aral Sea

Rome

ROMAN EMPIRE

Black Sea

Caspian Sea

Byzantium

Merv

Carthage

GREECE

Asia Minor

Mediterranean Sea

Antioch

Ecbatana

Ctesiphon

Babylon

PERSIA

Alexandria

Petra

Persepolis

AFRICA

Goods from the West Roman merchants like this man grew rich from Silk Road trade. Merchants in the west traded goods like those you see here—wool, amber, and gold.

Aden

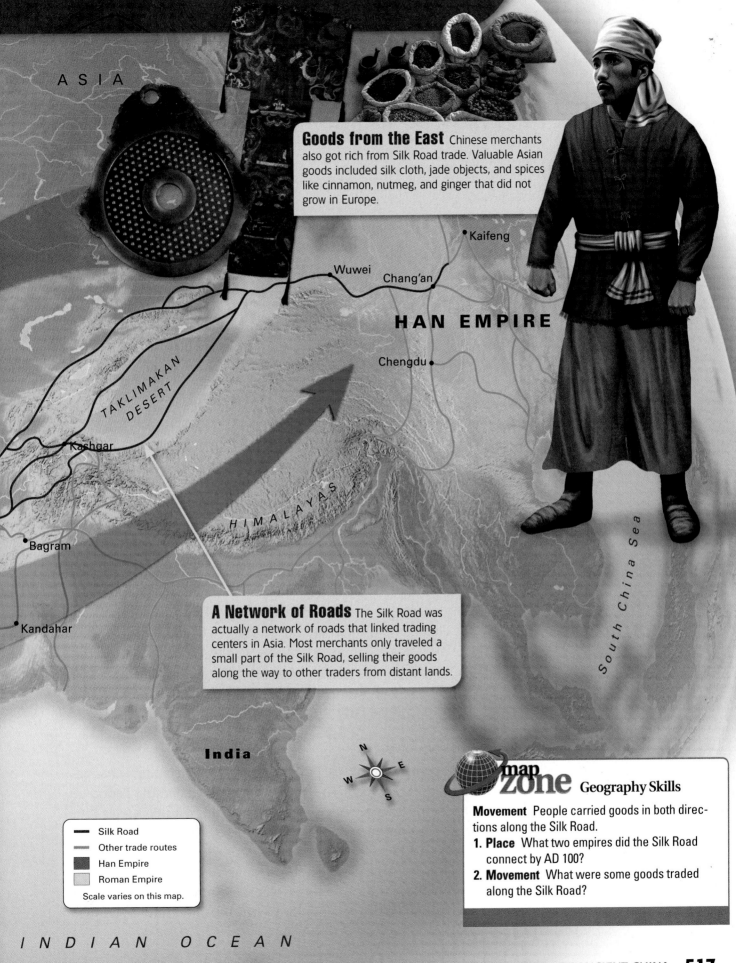

ASIA

Goods from the East Chinese merchants also got rich from Silk Road trade. Valuable Asian goods included silk cloth, jade objects, and spices like cinnamon, nutmeg, and ginger that did not grow in Europe.

• Kaifeng

• Wuwei   Chang'an

HAN EMPIRE

Chengdu •

TAKLIMAKAN DESERT

• Kashgar

HIMALAYAS

• Bagram

A Network of Roads The Silk Road was actually a network of roads that linked trading centers in Asia. Most merchants only traveled a small part of the Silk Road, selling their goods along the way to other traders from distant lands.

• Kandahar

South China Sea

India

N W E S

	Silk Road
	Other trade routes
■	Han Empire
■	Roman Empire
	Scale varies on this map.

map zone Geography Skills

Movement People carried goods in both directions along the Silk Road.
1. **Place** What two empires did the Silk Road connect by AD 100?
2. **Movement** What were some goods traded along the Silk Road?

INDIAN OCEAN

The Sui, Tang, and Song Dynasties

What You Will Learn...

Main Ideas

1. After the Han dynasty, China fell into disorder but was reunified by new dynasties.
2. Cities and trade grew during the Tang and Song dynasties.
3. The Tang and Song dynasties produced fine arts and inventions.

The Big Idea

The Tang and Song dynasties were periods of economic, cultural, and technological accomplishments.

Key Terms and Places

Grand Canal, p. 518
Kaifeng, p. 520
porcelain, p. 521
woodblock printing, p. 522
gunpowder, p. 522
compass, p. 522

 As you read, look for information about accomplishments of the Tang and Song dynasties. Keep track of these accomplishments in a chart like this one.

Tang dynasty	Song dynasty

If YOU lived there...

It is the year 1270. You are a rich merchant in a Chinese city of about a million people. The city around you fills your senses. You see people in colorful clothes among beautiful buildings. Glittering objects lure you into busy shops. You hear people talking—discussing business, gossiping, laughing at jokes. You smell delicious food cooking at a restaurant down the street.

How do you feel about your city?

BUILDING BACKGROUND The Tang and Song dynasties were periods of great wealth and progress. Changes in farming formed the basis for other advances in Chinese civilization.

Disorder and Reunification

When the Han dynasty collapsed, China split into several rival kingdoms, each ruled by military leaders. Historians sometimes call the time of disorder that followed the collapse of the Han the Period of Disunion. It lasted from 220 to 589.

War was common during the Period of Disunion. At the same time, however, Chinese culture spread. New groups moved into China from nearby areas. Over time, many of these groups adopted Chinese customs and became Chinese themselves.

Sui Dynasty

Finally, after centuries of political confusion and cultural change, China was reunified. The man who finally ended the Period of Disunion was a northern ruler named Yang Jian (YANG jee-EN). In 589, he conquered the south, unified China, and created the Sui (SWAY) dynasty.

The Sui dynasty did not last long, only from 589 to 618. During that time, however, its leaders restored order and began the **Grand Canal**, a canal linking northern and southern China.

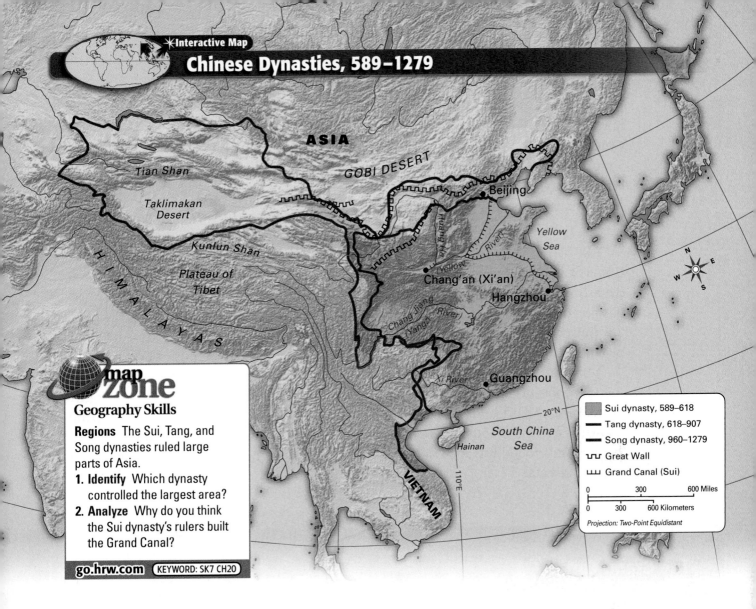

ASIA

Tian Shan

GOBI DESERT

Taklimakan
Desert

Beijing

Kunlun Shan

Yellow
Sea

HIMALAYAS

Plateau of
Tibet

Huang He (Yellow River)

Chang'an (Xi'an)

Hangzhou

Chang Jiang (Yangzi River)

Xi River

Guangzhou

20°N

South China
Sea

Hainan

110°E

VIETNAM

map zone

Geography Skills

Regions The Sui, Tang, and Song dynasties ruled large parts of Asia.

1. **Identify** Which dynasty controlled the largest area?
2. **Analyze** Why do you think the Sui dynasty's rulers built the Grand Canal?

go.hrw.com KEYWORD: SK7 CH20

	Sui dynasty, 589–618
	Tang dynasty, 618–907
	Song dynasty, 960–1279
	Great Wall
	Grand Canal (Sui)

0   300   600 Miles
0   300   600 Kilometers
Projection: Two-Point Equidistant

Tang Dynasty

The Sui dynasty was followed by the Tang, which would rule for nearly 300 years. As you can see on the map, China grew under the Tang dynasty to include much of eastern and central Asia.

Historians view the Tang dynasty as a golden age. Tang rulers conquered many lands, reformed the military, and created law codes. The Tang period also saw great advances in art. Some of China's finest poets, for example, lived during this time.

The Tang dynasty also included the only woman to rule China—Empress Wu. Her methods were sometimes vicious, but she was intelligent and talented.

Song Dynasty

After the Tang dynasty fell, China entered another brief period of chaos and disorder, with separate kingdoms competing for power. As a result, this period in China's history is called the Five Dynasties and Ten Kingdoms. The disorder only lasted 53 years, though, from 907 to 960.

In 960, China was again reunified, this time by the Song dynasty. Like the Tang, the Song ruled for about 300 years, until 1279. Also like the Tang, the Song dynasty was a time of great achievements.

READING CHECK Finding Main Ideas What dynasties restored order to China?

FOCUS ON READING

What dynasty followed the Tang?

Cities and Trade

Throughout the Tang and Song dynasties, much of the food grown on China's farms flowed into the growing cities and towns. China's cities were crowded, busy places. Shopkeepers, government officials, doctors, artisans, entertainers, religious leaders, and artists made them lively places as well.

City Life

China's capital and largest city of the Tang dynasty was Chang'an (chahng-AHN), a huge, bustling trade center now called Xi'an. With a population of more than a million, it was by far the largest city in the world.

Chang'an, like other trading cities, had a mix of people from many cultures—China, Korea, Persia, Arabia, and Europe. It was also known as a religious and philosophical center, not just for Buddhists and Daoists but for Asian Christians as well.

Cities continued to grow under the Song. Several cities, including the Song capital, **Kaifeng** (KY-fuhng), had about a million people. A dozen more cities had populations of close to half a million.

Trade in China and Beyond

Trade grew along with Chinese cities. This trade, combined with China's agricultural base, made China richer than ever before.

Much trade took place within China itself. Traders used the country's rivers to ship goods on barges and ships.

The Grand Canal, a series of waterways that linked major cities, carried a huge amount of trade goods, especially farm products. Construction on the canal had begun during the Sui dynasty. During the Tang dynasty, it was improved and expanded. The Grand Canal allowed the Chinese to move goods and crops from distant agricultural areas into cities.

The Grand Canal

Beijing
Huang He (Yellow River)
Yellow Sea
Chang'an
Zhenjiang
Chang Jiang (Yangzi River)
Hangzhou
East China Sea

☲ Grand Canal (Sui)

The Chinese also carried on trade with other lands and peoples. During the Tang dynasty, most foreign trade was over land routes leading to India and Southwest Asia, though Chinese traders also went to Korea and Japan in the east. The Chinese exported many goods, including tea, rice, spices, and jade. However, one export was especially important—silk. So valuable was silk that the Chinese kept the method of making it secret. In exchange for their exports, the Chinese imported different foods, plants, wool, glass, and precious metals like gold and silver.

During the Song dynasty, sea trade became more important. China opened its Pacific ports to foreign traders. The sea-trade routes connected China to many other countries. During this time, the Chinese also developed another valuable product—a thin, beautiful type of pottery called **porcelain**.

China's Grand Canal (left) is the world's longest human-made waterway. It was built largely to transport rice and other foods from the south to feed China's cities and armies in the north. Barges like the one above crowd the Grand Canal, which is still an important transportation link in China.

All of this trade helped create a strong economy. As a result, merchants became important members of Chinese society during the Song dynasty. Also as a result of the growth of trade and wealth, the Song invented the world's first system of paper money in the 900s.

READING CHECK **Summarizing** How far did China's trade routes extend?

Arts and Inventions

While China grew rich economically, its cultural riches also increased. In literature, art, and science, China made huge advances.

Artists and Poets

The artists and writers of the Tang dynasty were some of China's greatest. Wu Daozi (DOW-tzee) painted murals that celebrated Buddhism and nature. Li Bo and Du Fu wrote poems that readers still enjoy for their beauty. This poem by Li Bo expresses the homesickness that one feels late at night:

" Before my bed
there is bright moonlight
So that it seems
like frost on the ground:
Lifting my head
I watch the bright moon,
Lowering my head
I dream that I'm home. "
–Li Bo, *Quiet Night Thoughts*

THE IMPACT TODAY

Porcelain became so popular in the West that it became known as chinaware, or just china.

Also noted for its literature, the Song period produced Li Qingzhao (ching-ZHOW), perhaps China's greatest female poet. She once said that the purpose of her poetry was to capture a single moment in time.

Artists of both the Tang and Song dynasties made exquisite objects in clay. Tang figurines of horses clearly show the animals' strength. Song artists made porcelain items covered in a pale green glaze called celadon (SEL-uh-duhn).

Chinese Inventions

Paper
Invented during the Han dynasty around 105, paper was one of the greatest of all Chinese inventions. It gave the Chinese a cheap and easy way of keeping records and made printing possible.

Porcelain
Porcelain was first made during the Tang dynasty, but it wasn't perfected for many centuries. Chinese artists were famous for their work with this fragile material.

Woodblock printing
The Chinese invented printing during the Tang dynasty, centuries before it was known in Europe. Printers could copy drawings or texts quickly, much faster than they could be copied by hand.

Gunpowder
Invented during the late Tang or early Song dynasty, gunpowder was used to make fireworks and signals. The Chinese did not generally use it as a weapon.

Movable type
Inventors of the Song dynasty created movable type, which made printing much faster. Carved letters could be rearranged and reused to print many different messages.

Magnetic compass
Invented no later than the Han period, the compass was greatly improved by the Tang. The new compass allowed sailors and merchants to travel vast distances.

Paper money
The world's first paper money was invented by the Song. Lighter and easier to handle than coins, paper money helped the Chinese manage their growing wealth.

Important Inventions

The Tang and Song dynasties produced some of the most remarkable—and most important—inventions in human history. Some of these inventions influenced events around the world.

According to legend, a man named Cai Lun invented paper in the year 105 during the Han dynasty. A later Tang invention built on this achievement— **woodblock printing, a form of printing in which an entire page is carved into a block of wood.** The printer applies ink to the block and presses paper against the block to create a printed page. The world's first known printed book was printed in this way in China in 868.

Another invention of the Tang dynasty was gunpowder. **Gunpowder is a mixture of powders used in guns and explosives.** It was originally used only in fireworks, but it was later used to make small bombs and rockets. Eventually, gunpowder was used to make explosives, firearms, and cannons. Gunpowder dramatically altered how wars were fought and, in doing so, changed the course of human history.

One of the most useful achievements of Tang China was the perfection of the magnetic **compass. This instrument, which uses Earth's magnetic field to show direction,** revolutionized travel. A compass made it possible to find direction more accurately than ever before. The perfection of the compass had far-reaching effects. Explorers the world over used the compass to travel vast distances. Both trading ships and warships also came to rely on the compass for their navigation. Thus, the compass has been a key factor in some of the most important sailing voyages in history.

The Song dynasty also produced many important inventions. Under the Song, the Chinese invented movable type. Movable type is a set of letters or characters that are

CONNECTING TO Economics

The Paper Trail

The dollar bill in your pocket may be crisp and new, but paper money has been around a long time. Paper money was printed for the first time in China in the AD 900s and was in use for about 700 years, through the Ming dynasty, when the bill shown here was printed. However, so much money was printed that it lost value. The Chinese stopped using paper money for centuries. Its use caught on in Europe, though, and eventually became common. Most countries now issue paper money.

Drawing Conclusions How would life be different today without paper money?

---

used to print books. Unlike the blocks used in block printing, movable type can be rearranged and reused to create new lines of text and different pages.

The Song dynasty also introduced the concept of paper money. People were used to buying goods and services with bulky coins made of metals such as bronze, gold, and silver. Paper money was far lighter and easier to use. As trade increased and many people in China grew rich, paper money became more popular.

READING CHECK Finding Main Ideas What were some important inventions of the Tang and Song dynasties?

SUMMARY AND PREVIEW The Tang and Song dynasties were periods of great advancement. Many great artists and writers lived during these periods. Tang and Song inventions also had dramatic effects on world history. Next, you will learn about major changes in China's government during the Song dynasty.

go.hrw.com
Online Quiz
KEYWORD: SK7 HP20

Section 3 Assessment

Reviewing Ideas, Terms, and People

1. **a. Recall** What was the Period of Disunion? What dynasty brought an end to that period?
   b. Explain How did China change during the Tang dynasty?
2. **a. Describe** What were the capital cities of Tang and Song China like?
   b. Draw Conclusions How did geography affect trade in China?
3. **a. Identify** Who was Li Bo?
   b. Draw Conclusions How may the inventions of paper money and **woodblock printing** have been linked?
   c. Rank Which Tang or Song invention do you think was most important? Defend your answer.

Critical Thinking

4. **Categorizing** Copy the chart at right. Use it to organize your notes on the Tang and Song into categories.

	Tang dynasty	Song dynasty
Cities		
Trade		
Art		
Inventions		

FOCUS ON WRITING

5. **Identifying Achievements** Which achievements and inventions of the Tang and Song dynasties seem most important or most interesting? Make a list for later use.

Confucianism and Government

What You Will Learn...

Main Ideas

1. Confucianism, based on Confucius's teachings about proper behavior, dramatically influenced the Song system of government.
2. Scholar-officials ran China's government during the Song dynasty.

The Big Idea

Confucian thought influenced the Song government.

Key Terms

bureaucracy, *p. 526*
civil service, *p. 526*
scholar-official, *p. 526*

TAKING NOTES As you read, use a diagram like this one to note details about Confucianism and the Song government.

Confucianism   Song government

If **YOU** lived there...

You are a student in China in 1184. Night has fallen, but you cannot sleep. Tomorrow you have a test. You know it will be the most important test of your entire life. You have studied for it, not for days or weeks or even months—but for *years*. As you toss and turn, you think about how your entire life will be determined by how well you do on this one test.

How could a single test be so important?

BUILDING BACKGROUND The Song dynasty ruled China from 960 to 1279. This was a time of improvements in agriculture, growing cities, extensive trade, and the development of art and inventions. It was also a time of major changes in Chinese government.

Confucianism

The dominant philosophy in China, Confucianism is based on the teachings of Confucius. He lived more than 1,000 years before the Song dynasty. His ideas, though, had a dramatic effect on the Song system of government.

Confucian Ideas

Confucius's teachings focused on ethics, or proper behavior, for individuals and governments. He said that people should conduct their lives according to two basic principles. These principles were *ren*, or concern for others, and *li*, or appropriate behavior. Confucius argued that society would **function** best if everyone followed *ren* and *li*.

Confucius thought that everyone had a proper role to play in society. Order was maintained when people knew their place and behaved appropriately. For example, Confucius said that young people should obey their elders and that subjects should obey their rulers.

Influence of Confucianism

After his death, Confucius's ideas were spread by his followers, but they were not widely accepted. In fact, the Qin dynasty officially suppressed Confucian ideas and teachings. By the time of the Han dynasty, Confucianism had again come into favor, and Confucianism became the official state philosophy.

During the Period of Disunion, which followed the Han dynasty, Confucianism was overshadowed by Buddhism as the major tradition in China. Many Chinese people had turned to Buddhism for peace and comfort during those troubled times. In doing so, they largely turned away from Confucian ideas and outlooks.

Later, during the Sui and early Tang dynasties, Buddhism was very influential. Unlike Confucianism, which focused on ethical behavior, Buddhism stressed a more spiritual outlook that promised escape from suffering. As Buddhism became more popular in China, Confucianism lost some of its influence.

ACADEMIC VOCABULARY
function work or perform

In addition to ethics, Confucianism stressed the importance of a good education. This painting, created in the Song period, shows Confucian scholars during the Period of Disunion sorting scrolls containing classic Confucian texts.

Civil Service Exams

This painting from the 1600s shows civil servants writing essays for China's emperor. Difficult exams were designed to make sure that government officials were chosen by ability—not by wealth or family connections.

Difficult Exams

- Students had to memorize entire Confucian texts.

- To pass the most difficult tests, students might study for more than 20 years!

- Some exams lasted up to 72 hours, and students were locked in private rooms while taking them.

- Some dishonest students cheated by copying Confucius's works on the inside of their clothes, paying bribes to the test graders, or paying someone else to take the test for them.

- To prevent cheating, exam halls were often locked and guarded.

Neo-Confucianism

Late in the Tang dynasty, many Chinese historians and scholars again became interested in the teachings of Confucius. Their interest was sparked by their desire to improve Chinese government and society.

During and after the Song dynasty, a new philosophy called Neo-Confucianism developed. The term *neo* means "new." Based on Confucianism, Neo-Confucianism was similar to the older philosophy in that it taught proper behavior. However, it also emphasized spiritual matters. For example, Neo-Confucian scholars discussed such issues as what made human beings do bad things even if their basic nature was good.

Neo-Confucianism became much more influential under the Song. Its influence grew even more later. In fact, the ideas of Neo-Confucianism became official government teachings after the Song dynasty.

ACADEMIC VOCABULARY

incentive
something that leads people to follow a certain course of action

READING CHECK Contrasting How did Neo-Confucianism differ from Confucianism?

Scholar-Officials

The Song dynasty took another major step that affected China for centuries. They improved the system by which people went to work for the government. These workers formed a large **bureaucracy,** or a body of unelected government officials. They joined the bureaucracy by passing civil service examinations. **Civil service** means service as a government official.

To become a civil servant, a person had to pass a series of written examinations. The examinations tested students' grasp of Confucianism and related ideas.

Because the tests were so difficult, students spent years preparing for them. Only a very small fraction of the people who took the tests would reach the top level and be appointed to a position in the government. However, candidates for the civil service examinations had a strong **incentive** for studying hard. Passing the tests meant life as a **scholar-official**—an educated member of the government.

Scholar-Officials

First rising to prominence under the Song, scholar-officials remained important in China for centuries. These scholar-officials, for example, lived during the Qing dynasty, which ruled from the mid-1600s to the early 1900s. Their typical responsibilities might include running government offices; maintaining roads, irrigation systems, and other public works; updating and keeping official records; or collecting taxes.

Scholar-officials were elite members of society. They performed many important jobs in the government and were widely admired for their knowledge and ethics. Their benefits included considerable respect and reduced penalties for breaking the law. Many also became wealthy from gifts given by people seeking their aid.

The civil service examination system helped ensure that talented, intelligent people became scholar-officials. The civil service system was a major factor in the stability of the Song government.

READING CHECK Analyzing How did the Song dynasty change China's government?

SUMMARY AND PREVIEW During the Song period, Confucian ideas helped shape China's government. In the next section, you will read about the two dynasties that followed the Song—the Yuan and the Ming.

go.hrw.com
Online Quiz
KEYWORD: SK7 HP20

Section 4 Assessment

Reviewing Ideas, Terms, and People

1. **a. Identify** What two principles did Confucius believe people should follow?
   b. Explain What was Neo-Confucianism?
   c. Elaborate Why do you think Neo-Confucianism appealed to many people?

2. **a. Define** What was a **scholar-official**?
   b. Explain Why would people want to become scholar-officials?
   c. Evaluate Do you think **civil service** examinations were a good way to choose government officials? Why or why not?

Critical Thinking

3. **Sequencing** Review your notes to see how Confucianism led to Neo-Confucianism and Neo-Confucianism led to government bureaucracy. Use a graphic organizer like the one here.

Confucianism → Neo-Confucianism → Government bureaucracy

FOCUS ON WRITING

4. **Gathering Ideas about Confucianism and Government** Think about what you might say about Confucianism in your article. Also, decide whether to include any of the Song dynasty's achievements in government.

The Yuan and Ming Dynasties

What You Will Learn...

Main Ideas

1. The Mongol Empire included China, and the Mongols ruled China as the Yuan dynasty.
2. The Ming dynasty was a time of stability and prosperity.
3. The Ming brought great changes in government and relations with other countries.

The Big Idea

The Chinese were ruled by foreigners during the Yuan dynasty, but they threw off Mongol rule and prospered during the Ming dynasty.

Key Terms and Places

Beijing, *p. 530*
Forbidden City, *p. 532*
isolationism, *p. 534*

TAKING NOTES As you read, use a chart like this one to keep track of important details about the Yuan and Ming dynasties.

	Yuan	Ming
Government		
Religion		
Trade		
Building		
Foreign Relations		

If **YOU** lived there...

You are a farmer in northern China in 1212. As you pull weeds from a wheat field, you hear a sound like thunder. Looking toward the sound, you see hundreds—no, *thousands*—of warriors on horses on the horizon, riding straight toward you. You are frozen with fear. Only one thought fills your mind—the Mongols are coming.

What can you do to save yourself?

BUILDING BACKGROUND Throughout its history, northern China had been attacked over and over by nomadic peoples. During the Song dynasty these attacks became more frequent and threatening.

The Mongol Empire

Among the nomadic peoples who attacked the Chinese were the Mongols. For centuries, the Mongols had lived as tribes in the vast plains north of China. Then in 1206, a strong leader, or khan, united them. His name was Temüjin. When he became leader, though, he was given a new title: "Universal Ruler," or Genghis Khan (JENG-guhs KAHN).

The Mongol Conquest

Genghis Khan organized the Mongols into a powerful army and led them on bloody expeditions of conquest. The brutality of the Mongol attacks terrorized people throughout much of Asia and Eastern Europe. Genghis Khan and his army killed all of the men, women, and children in countless cities and villages. Within 20 years, he ruled a large part of Asia.

Genghis Khan then turned his attention to China. He first led his armies into northern China in 1211. They fought their way south, wrecking whole towns and ruining farmland. By the time of Genghis Khan's death in 1227, all of northern China was under Mongol control.

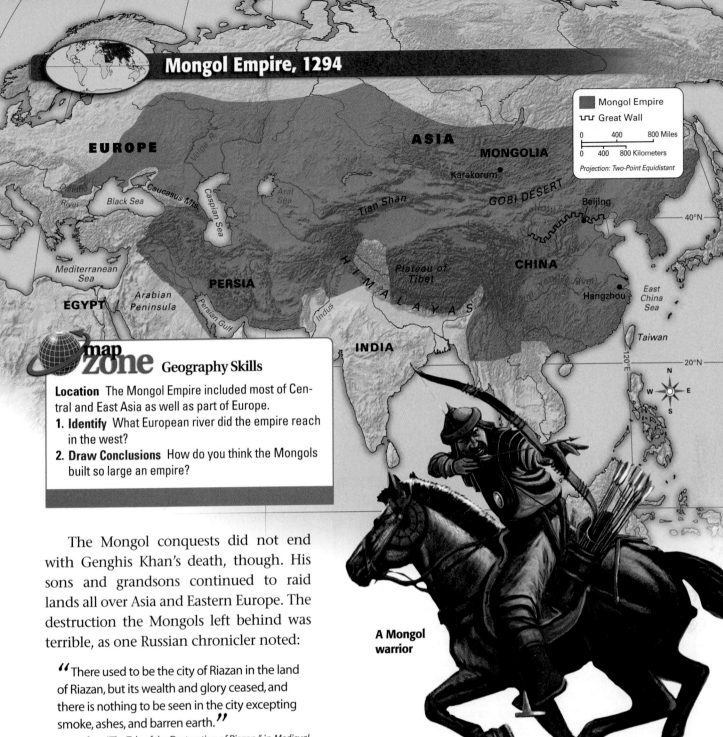

Mongol Empire, 1294

EUROPE

ASIA

MONGOLIA

Karakorum●

GOBI DESERT

Beijing●

CHINA

Hangzhou●

Volga R.

Danube River

Black Sea

Caucasus Mts.

Caspian Sea

Aral Sea

Tian Shan

Mediterranean Sea

PERSIA

Arabian Peninsula

Persian Gulf

EGYPT

Indus

H I M A L A Y A S

Plateau of Tibet

Yangzi River

East China Sea

INDIA

Taiwan

40°N

20°N

120°E

| Mongol Empire |
| Great Wall |

0   400   800 Miles
0   400   800 Kilometers
Projection: Two-Point Equidistant

map zone Geography Skills

Location The Mongol Empire included most of Central and East Asia as well as part of Europe.

1. **Identify** What European river did the empire reach in the west?
2. **Draw Conclusions** How do you think the Mongols built so large an empire?

The Mongol conquests did not end with Genghis Khan's death, though. His sons and grandsons continued to raid lands all over Asia and Eastern Europe. The destruction the Mongols left behind was terrible, as one Russian chronicler noted:

"There used to be the city of Riazan in the land of Riazan, but its wealth and glory ceased, and there is nothing to be seen in the city excepting smoke, ashes, and barren earth."

–from "The Tale of the Destruction of Riazan," in *Medieval Russia's Epics, Chronicles, and Tales*, edited by Serge Zenkovsky

In 1260 Genghis Khan's grandson Kublai Khan (KOO-bluh KAHN) became ruler of the Mongol Empire. He completed the conquest of China and in 1279 declared himself emperor of China. This began the Yuan dynasty, a period that some people also call the Mongol Ascendancy. For the first time in its long history, foreigners ruled all of China.

A Mongol warrior

Life in Yuan China

Kublai Khan and the Mongol rulers he led belonged to a different ethnic group than the Chinese did. They spoke a different language, worshipped different gods, wore different clothing, and had different customs. The Chinese resented being ruled by these foreigners, whom they saw as rude and uncivilized.

However, Kublai Khan did not force the Chinese to accept Mongol ways of life. Some Mongols even adopted aspects of the Chinese culture, such as Confucianism. Still, the Mongols made sure to keep control of the Chinese. They prohibited Confucian scholars from gaining too much power in the government, for example. The Mongols also placed heavy taxes on the Chinese.

Much of the tax money the Mongols collected went to pay for vast public-works projects. These projects required the labor of many Chinese people. The Yuan added to the Grand Canal and built new roads and palaces. Workers also improved the roads used by China's postal system. In addition, the Yuan emperors built a new capital, Dadu, near modern **Beijing**.

Primary Source

BOOK
A Chinese City

In this passage Marco Polo describes his visit to Hangzhou (HAHNG-JOH), a city in southeastern China.

❝Inside the city there is a Lake . . . and all round it are erected [built] beautiful palaces and mansions, of the richest and most exquisite [finest] structure that you can imagine . . . In the middle of the Lake are two Islands, on each of which stands a rich, beautiful and spacious edifice [building], furnished in such style as to seem fit for the palace of an Emperor. And when any one of the citizens desired to hold a marriage feast, or to give any other entertainment, it used to be done at one of these palaces. And everything would be found there ready to order, such as silver plate, trenchers [platters], and dishes, napkins and table-cloths, and whatever else was needful. The King made this provision for the gratification [enjoyment] of his people, and the place was open to every one who desired to give an entertainment.❞

–Marco Polo, from *Description of the World*

ANALYSIS SKILL ANALYZING PRIMARY SOURCES

From this description, what impression might Europeans have of Hangzhou?

Mongol soldiers were sent throughout China to keep the peace as well as to keep a close watch on the Chinese. The soldiers' presence kept overland trade routes safe for merchants. Sea trade between China, India, and Southeast Asia continued, too. The Mongol emperors also welcomed foreign traders at Chinese ports. Some of these traders received special privileges.

Part of what we know about life in the Yuan dynasty comes from one such trader, an Italian merchant named Marco Polo. Between 1271 and 1295 he traveled in and around China. Polo was highly respected by the Mongols and even served in Kublai Khan's court. When Polo returned to Europe, he wrote of his travels. Polo's descriptions of China fascinated many Europeans. His book sparked much European interest in China.

The End of the Yuan Dynasty

Despite their vast empire, the Mongols were not content with their lands. They decided to invade Japan. A Mongol army sailed to Japan in 1274 and 1281. The campaigns, however, were disastrous. Violent storms and fierce defenders destroyed most of the Mongol force.

The failed campaigns against Japan weakened the Mongol military. The huge, expensive public-works projects had already weakened the economy. These weaknesses, combined with Chinese resentment, made China ripe for rebellion.

In the 1300s many Chinese groups rebelled against the Yuan dynasty. In 1368 a former monk named Zhu Yuanzhang (JOO yoo-ahn-JAHNG) took charge of a rebel army. He led this army in a final victory over the Mongols. China was once again ruled by the Chinese.

READING CHECK Finding Main Ideas How did the Mongols come to rule China?

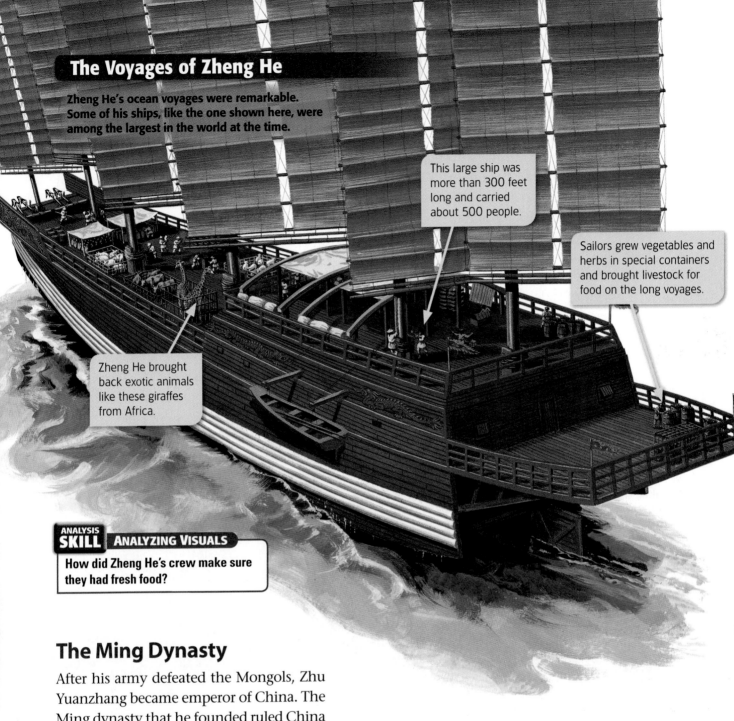

The Voyages of Zheng He

Zheng He's ocean voyages were remarkable. Some of his ships, like the one shown here, were among the largest in the world at the time.

This large ship was more than 300 feet long and carried about 500 people.

Sailors grew vegetables and herbs in special containers and brought livestock for food on the long voyages.

Zheng He brought back exotic animals like these giraffes from Africa.

ANALYSIS SKILL **ANALYZING VISUALS**

How did Zheng He's crew make sure they had fresh food?

The Ming Dynasty

After his army defeated the Mongols, Zhu Yuanzhang became emperor of China. The Ming dynasty that he founded ruled China from 1368 to 1644—nearly 300 years. Ming China proved to be one of the most stable and prosperous times in Chinese history. The Ming expanded China's fame overseas and sponsored incredible building projects across China.

Great Sea Voyages

During the Ming dynasty, the Chinese improved their ships and their sailing skills. The greatest sailor of the period was

Zheng He (juhng HUH). Between 1405 and 1433, he led seven grand voyages to places around Asia. Zheng He's fleets were huge. One included more than 60 ships and 25,000 sailors. Some of the ships were gigantic too, perhaps more than 300 feet long. That is longer than a football field!

In the course of his voyages Zheng He sailed his fleet throughout the Indian Ocean. He sailed as far west as the Persian Gulf and the easternmost coast of Africa.

Everywhere his ships landed, Zheng He presented leaders with beautiful gifts from China. He boasted about his country and encouraged foreign leaders to send gifts to China's emperor. From one voyage, Zheng He returned to China with representatives of some 30 nations, sent by their leaders to honor the emperor. He also brought goods and stories back to China.

Zheng He's voyages rank among the most impressive in the history of seafaring. Although they did not lead to the creation of new trade routes or the exploration of new lands, they served as a clear sign of China's power.

Great Building Projects

The Ming were also known for their grand building projects. Many of these projects were designed to impress both the Chinese people and their enemies to the north.

In Beijing, for example, the Ming emperors built the **Forbidden City**, a huge palace complex that included hundreds of imperial residences, temples, and other government buildings. Within them were some 9,000 rooms. The name Forbidden City came from the fact that the common people were not even allowed to enter the complex. For centuries, this city within a city was a symbol of China's glory.

The Forbidden City

The Forbidden City is not actually a city. It's a huge complex of almost 1,000 buildings in the heart of China's capital. The Forbidden City was built for the emperor, his family, his court, and his servants, and ordinary people were forbidden from entering.

The Forbidden City's main buildings were built of wood and featured gold-colored tile roofs that could only be used for the emperor's buildings.

The crowds of government and military officials who gathered to watch ceremonies were carefully lined up according to their ranks.

Sometimes, the emperor was carried on a special seat called a palanquin as his officers lined the route.

Ming rulers also directed the restoration of the famous Great Wall of China. Large numbers of soldiers and peasants worked to rebuild fallen portions of walls, connect existing walls, and build new ones. The result was a construction feat unmatched in history. The wall was more than 2,000 miles long. It would reach from San Diego to New York! The wall was about 25 feet high and, at the top, 12 feet wide. Protected by the wall—and the soldiers who stood guard along it—the Chinese people felt safe from invasions by the northern tribes.

READING CHECK **Generalizing** In what ways did the Ming dynasty strengthen China?

China under the Ming

During the Ming dynasty, Chinese society began to change. This change was largely due to the efforts of the Ming emperors. Having expelled the Mongols, the Ming emperors worked to eliminate all foreign influences from Chinese society. As a result, China's government and relations with other countries changed dramatically.

The Hall of Supreme Harmony is the largest building in the Forbidden City. Grand celebrations for important holidays, like the emperor's birthday and the New Year, were held there.

ANALYSIS SKILL **ANALYZING VISUALS**

How did the Forbidden City show the power and importance of the emperor?

Government

When the Ming took over China, they adopted many government programs that had been created by the Tang and the Song. However, the Ming emperors were much more powerful than the Tang and Song emperors had been. They abolished the offices of some powerful officials and took a larger role in running the government themselves. These emperors fiercely protected their power, and they punished anyone whom they saw as challenging their authority.

ACADEMIC VOCABULARY
consequences effects of a particular event or events

Despite their personal power, though, the Ming did not disband the civil service system. Because he personally oversaw the entire government, the emperor needed officials to keep his affairs organized.

The Ming also used examinations to appoint censors. These officials were sent all over China to investigate the behavior of local leaders and to judge the quality of schools and other institutions. Censors had existed for many years in China, but under the Ming emperors their power and influence grew.

Relations with Other Countries

In the 1430s a new Ming emperor made Zheng He return to China and dismantle his fleet. At the same time, he banned foreign trade. China entered a period of isolationism. **Isolationism** is a policy of avoiding contact with other countries.

In the end, this isolationism had great **consequences** for China. By the late 1800s the Western world had made huge leaps in technological progress. Westerners were able to take power in some parts of China. Partly due to its isolation and lack of progress, China was too weak to stop them. Gradually, China's glory faded.

READING CHECK Identifying Cause and Effect How did isolationism affect China?

SUMMARY AND PREVIEW In the last two chapters, you have learned about the long histories of India and China. Next, you will learn what the two countries are like today and how their pasts influence the present.

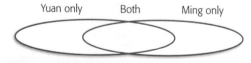
go.hrw.com
Online Quiz
KEYWORD: SK7 HP20

Section 5 Assessment

Reviewing Ideas, Terms, and People

1. **a. Identify** Who was Genghis Khan?
   b. Explain How did the Mongols gain control of China?
   c. Evaluate Judge this statement: "The Mongols should never have tried to invade Japan."
2. **a. Identify** Who was Zheng He, and what did he do?
   b. Analyze What impression do you think the Forbidden City had on the residents of Beijing?
   c. Develop How may the Great Wall have both helped and hurt China?
3. **a. Define** What is **isolationism**?
   b. Explain How did the Ming change China?
   c. Develop How might a policy of isolationism have both advantages and disadvantages?

Critical Thinking

4. **Comparing and Contrasting** Draw a diagram like this one. Use your notes to see how the Yuan and Ming dynasties were alike and different.

   Yuan only   Both   Ming only

FOCUS ON WRITING

5. **Identifying Achievements of the Later Dynasties** Make a list of the achievements of the Yuan and Ming dynasties. Then look back over all your notes and rate the achievements or inventions. Which four or five do you think are most important?

Kublai Khan

How did a Mongol nomad settle down to rule a vast empire?

When did he live? 1215–1294

Where did he live? Kublai Khan came from Mongolia but spent much of his life in China. His capital, Dadu, was near the modern city of Beijing.

What did he do? Kublai Khan completed the conquest of China that Genghis Khan had begun. He ruled China as the emperor of the Yuan dynasty.

Why is he important? The lands Kublai Khan ruled made up one of the largest empires in world history. It stretched from the Pacific Ocean to Eastern Europe. As China's ruler, Kublai Khan welcomed foreign visitors, including the Italian merchant Marco Polo and the Arab historian Ibn Battutah. The stories these two men told helped create interest in China and its products among Westerners.

Generalizing How did Kublai Khan's actions help change people's views of China?

This painting from the 1200s shows Kublai Khan hunting on horseback.

Making Economic Choices

Learn

Economic choices are a part of geography. World leaders must make economic choices every day. For example, a country's president might face a choice about whether to spend government money on improving defense, education, or health care.

You also have to make economic choices in your own life. For example, you might have to decide whether to go to a movie with a friend or buy a CD. You cannot afford to do both, so you must make a choice.

Making economic choices involves sacrifices, or trade-offs. If you choose to spend your money on a movie, the trade-offs are the other things you want but cannot buy. By considering trade-offs, you can make better economic choices.

Practice

Imagine that you are in the school band. The band has enough money to make one major purchase this year. As the diagram below shows, the band can spend the money on new musical instruments, new uniforms, or a band trip. The band decides to buy new instruments.

❶ Based on the diagram below, what are the trade-offs of the band's choice?

❷ What would have been the trade-offs if the band had voted to spend the money on a trip instead?

❸ How do you think creating a diagram like the one below might have helped the band make its economic choice?

New Instruments (instead of using old, worn-out ones)

New Uniforms (instead of playing in school clothes)

Band Trip (instead of not taking a trip this year)

Choice: New Instruments

Apply

1. Describe an example of an economic choice you might face that has three possible trade-offs.

2. For each possible economic choice, identify what the trade-offs are if you make that choice.

3. What final choice will you make? Why?

4. How did considering trade-offs help you make your choice?

Chapter Review

Geography's Impact
video series
Review the video to answer the closing question:
Do you agree with Confucius's ideas concerning family? Why or why not?

Visual Summary

Use the visual summary below to help you review the main ideas of the chapter.

QUICK FACTS

The Shang, Qin, and Han dynasties ruled China and made many advances that were built on later.

Under the Tang and Song dynasties, Confucianism was an important element of Chinese government.

The Mongols invaded China and ruled it as the Yuan dynasty.

The powerful Ming dynasty strengthened China and expanded trade, but then China became isolated.

Reviewing Vocabulary, Terms, and Places

Match the words or names with their definitions or descriptions.

a. gunpowder
b. scholar-official
c. mandate of heaven
d. bureaucracy
e. seismograph
f. porcelain
g. Great Wall
h. isolationism
i. incentive

1. a device to measure the strength of earthquakes
2. something that leads people to follow a certain course of action
3. body of unelected government officials
4. thin, beautiful pottery
5. educated government worker
6. policy of avoiding contact with other countries
7. a barrier along China's northern border
8. a mixture of powders used in explosives
9. the idea that heaven chose who should rule

Comprehension and Critical Thinking

SECTION 1 *(Pages 506–509)*

10. **a. Identify** What was the first known dynasty to rule China? What did it achieve?

   b. Analyze Why did the Qin dynasty not last long after Shi Huangdi's death?

   c. Evaluate Do you think Shi Huangdi was a good ruler for China? Why or why not?

SECTION 2 *(Pages 510–515)*

11. **a. Define** What is Confucianism? How did it affect Han society?

   b. Analyze What was life like for peasants in the Han period?

   c. Elaborate What inventions show that the Han studied nature?

SECTION 3 *(Pages 518–523)*

12. a. Describe What did Wu Daozi, Li Bo, Du Fu, and Li Qingzhao contribute to Chinese culture?

b. Analyze How did the Tang rulers change China's government?

c. Evaluate Which Chinese invention has had a greater effect on world history—the magnetic compass or gunpowder? Why do you think so?

SECTION 4 *(Pages 524–527)*

13. a. Define How did Confucianism change in and after the Song dynasty?

b. Make Inferences Why do you think the civil service examination system was created?

c. Elaborate Why were China's civil service examinations so difficult?

SECTION 5 *(Pages 528–534)*

14. a. Describe How did the Mongols create their huge empire? What areas were included in it?

b. Draw Conclusions How did Marco Polo and Zheng He help shape ideas about China?

c. Elaborate Why do you think the Ming spent so much time and money on the Great Wall?

Using the Internet

go.hrw.com
KEYWORD: SK7 CH20

15. Activity: Creating a Mural The Tang and Song periods saw many agricultural, technological, and commercial developments. New irrigation techniques, movable type, and gunpowder were a few of them. Enter the activity keyword and learn more about such developments. Imagine that a city official has hired you to create a mural showing all of the great things the Chinese developed during the Tang and Song dynasties. Create a large mural that depicts as many advances as possible.

Social Studies Skills

Making Economic Choices *You have enough money to buy one of the following items: shoes, a DVD, or a book.*

16. What are the trade-offs if you buy the DVD?

17. What are the trade-offs if you buy the book?

18. Understanding Chronological Order Arrange the following list of events in the order in which they happened. Then write a brief paragraph describing the events, using clue words such as *then* and *later* to show the proper sequence.

- The Han dynasty rules China.
- The Shang dynasty takes power.
- Mongol armies invade China.
- The Ming dynasty takes control.

19. Writing Your Magazine Article Now that you have identified the achievements or inventions that you want to write about, begin your article. Open with a sentence that states your main idea. Include a paragraph of two or three sentences about each invention or achievement. Describe each achievement or invention and explain why it was so important. End your article with a sentence or two that summarize China's importance to the world.

Map Activity ✳Interactive

20. Ancient China On a separate sheet of paper, match the letters on the map with their correct labels.

Chang'an	Beijing	Huang He
Kaifeng	Chang Jiang	

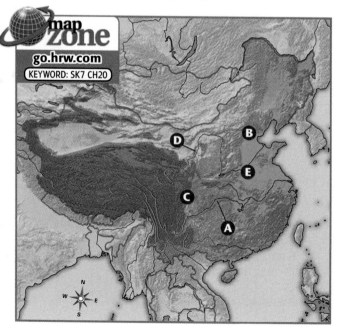

map zone
go.hrw.com
KEYWORD: SK7 CH20

DIRECTIONS: Read questions 1 through 7 and write the letter of the best response. Then read question 8 and write your own well-constructed response.

1 Who was the Chinese admiral who sailed all around Asia during the Ming dynasty?

A Li Bo

B Genghis Khan

C Zhu Yuanzhang

D Zheng He

2 Trade and other contact with peoples far from China stopped under which dynasty?

A Ming

B Yuan

C Song

D Sui

3 Which of the following was one way that Confucianism influenced China?

A emphasis on family and family values

B expansion of manufacturing and trade

C increase in taxes

D elimination of the government

4 Which of the following was an achievement of the Shang dynasty?

A invention of fireworks

B building of the Grand Canal

C creation of a writing system

D construction of the Forbidden City

5 This object displays Chinese expertise at working with

A woodblocks.

B gunpowder.

C cotton fibers.

D porcelain.

6 Emperor Shi Huangdi had laborers work on a structure that Ming rulers improved. What was that structure?

A the Great Wall

B the Great Tomb

C the Forbidden City

D the Temple of Buddha

7 The ruler who completed the Mongol conquest of China was named

A Shi Huangdi.

B Du Fu.

C Kublai Khan.

D Confucius.

8 Extended Response The ancient Chinese made advances in many fields. Some of their advances still affect life today. Choose one of these advances and write a short paragraph that explains how it still influences people's lives. Be sure to give examples to support your main idea.

The Indian Subcontinent

What You Will Learn...

In this chapter you will learn about the physical geography of the Indian Subcontinent. You will also discover the history and culture of the region. Finally, you will learn about the countries of the Indian Subcontinent today.

SECTION 1
Physical Geography **542**

SECTION 2
History and Culture of India.........**546**

SECTION 3
India Today **552**

SECTION 4
India's Neighbors **556**

FOCUS ON READING AND VIEWING

Visualizing As you read, try to visualize the people, places, or events that the text describes. Visualizing, or creating mental images, helps you to better understand and remember the information that is presented. Use your senses to imagine how things look, sound, smell, and feel. **See the lesson, Visualizing, on page 688.**

Presenting and Viewing a Travelogue You are journeying through the Indian Subcontinent, noting the sights and sounds of this beautiful and bustling region of the world. As you read this chapter you will gather details about this region. Then you will create an oral presentation of a travelogue, or traveler's journal. After you present your travelogue, you will watch and listen as your classmates present their travelogues.

IRAQ

IRAN

30°N

☸ National capital
● Major city
--- Disputed boundary

0   150   300 Miles
0   150   300 Kilometers

Projection: Albers Equal-Area

SAUDI ARABIA

20°N

OMAN

YEMEN   50°E

map Zone
Geography Skills

Place The Indian Subcontinent is a large landmass in South Asia.
1. **Locate** What bodies of water border the Indian Subcontinent?
2. **Analyze** What separates the Indian Subcontinent from the rest of Asia?

go.hrw.com **KEYWORD: SK7 CH21**

Culture The people of the subcontinent represent the many cultures and religions of the region.

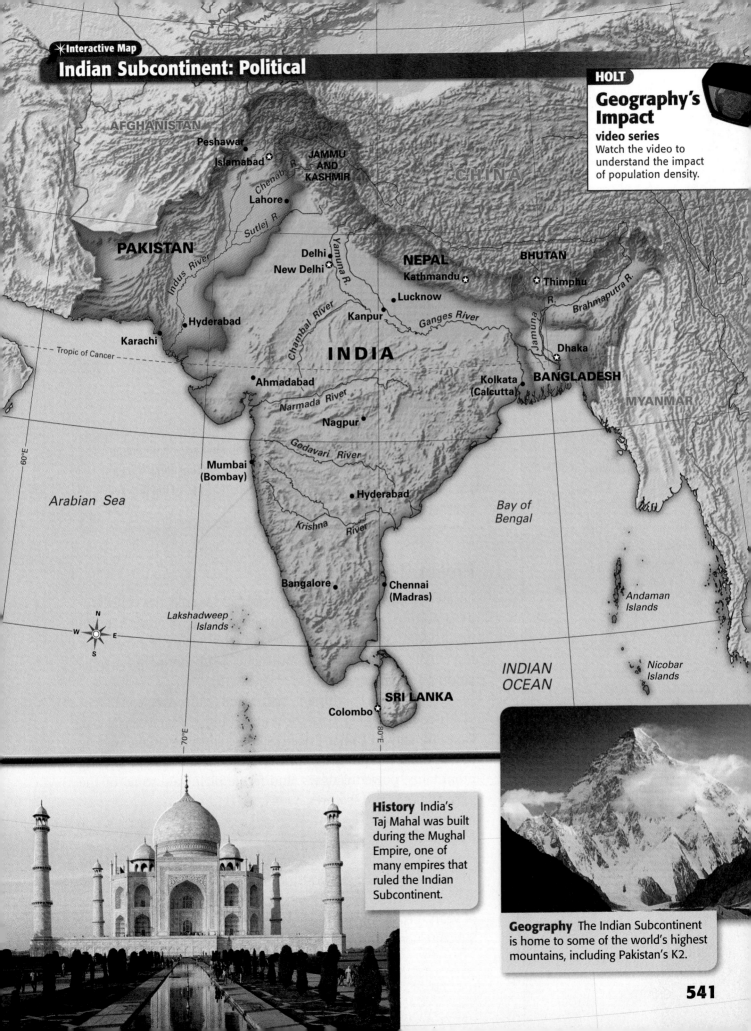

HOLT

Geography's Impact
video series
Watch the video to understand the impact of population density.

AFGHANISTAN

Peshawar

Islamabad

JAMMU AND KASHMIR

Chenab R.

CHINA

Lahore

Sutlej R.

PAKISTAN

Indus River

Delhi

New Delhi

Yamuna R.

NEPAL

Kathmandu

BHUTAN

Thimphu

Lucknow

Kanpur

Ganges River

Jamuna R.

Brahmaputra R.

Hyderabad

Chambal River

Karachi

Tropic of Cancer

INDIA

Dhaka

BANGLADESH

Ahmadabad

Narmada River

Kolkata (Calcutta)

MYANMAR

Nagpur

Godavari River

Mumbai (Bombay)

Hyderabad

Arabian Sea

Krishna River

Bay of Bengal

60°E

Andaman Islands

Bangalore

Chennai (Madras)

N
W   E
S

Lakshadweep Islands

INDIAN OCEAN

Nicobar Islands

SRI LANKA

Colombo

70°E

80°E

History India's Taj Mahal was built during the Mughal Empire, one of many empires that ruled the Indian Subcontinent.

Geography The Indian Subcontinent is home to some of the world's highest mountains, including Pakistan's K2.

Physical Geography

What You Will Learn...

Main Ideas

1. Towering mountains, large rivers, and broad plains are the key physical features of the Indian Subcontinent.
2. The Indian Subcontinent has a great variety of climate regions and resources.

The Big Idea

The physical geography of the Indian Subcontinent features unique physical features and a variety of climates and resources.

Key Terms and Places

subcontinent, *p. 542*
Mount Everest, *p. 543*
Ganges River, *p. 543*
delta, *p. 543*
Indus River, *p. 544*
monsoons, *p. 545*

TAKING NOTES As you read, take notes on the physical features, climates, and resources of the Indian Subcontinent. Use a diagram like the one below to organize your notes.

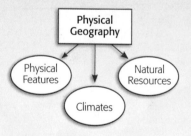

If YOU lived there...

You live in a small farming village in central India. Every year your father talks about the summer monsoons, winds that can bring heavy rains to the region. You know that too much rain can cause floods that may threaten your house and family. Too little rain could cause your crops to fail.

How do you feel about the monsoons?

BUILDING BACKGROUND Weather in the Indian Subcontinent, a region in southern Asia, is greatly affected by monsoon winds. Monsoons are just one of the many unique features of the physical geography of the Indian Subcontinent.

Physical Features

Locate Asia on a map of the world. Notice that the southernmost portion of Asia creates a triangular wedge of land that dips into the Indian Ocean. The piece of land jutting out from the rest of Asia is the Indian Subcontinent. A **subcontinent** is a large landmass that is smaller than a continent.

The Indian Subcontinent, also called South Asia, consists of seven countries—Bangladesh, Bhutan, India, Maldives, Nepal, Pakistan, and Sri Lanka. Together these countries make up one of the most unique geographic regions in the world. Soaring mountains, powerful rivers, and fertile plains are some of the region's dominant features.

Mountains

Huge mountain ranges separate the Indian Subcontinent from the rest of Asia. The rugged Hindu Kush mountains in the northwest divide the subcontinent from Central Asia. For thousands of years, peoples from Asia and Europe have entered the Indian Subcontinent through mountain passes in the Hindu Kush.

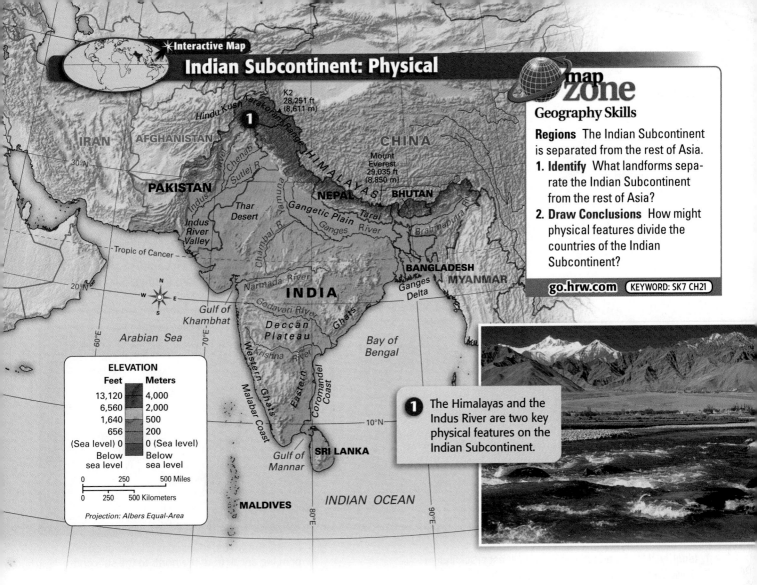

Indian Subcontinent: Physical

K2
28,251 ft
(8,611 m)

Hindu Kush
Karakoram Range
HIMALAYAS

IRAN
AFGHANISTAN
CHINA
30°N

Mount
Everest
29,035 ft
(8,850 m)

PAKISTAN
NEPAL
BHUTAN

Indus River
Chenab R.
Sutlej R.
Thar Desert
Gangetic Plain
Tarai
Ganges River
Yamuna R.
Brahmaputra R.

Indus
River
Valley

Tropic of Cancer

BANGLADESH
MYANMAR

20°N

Narmada River
Chambal R.
INDIA
Ganges
Delta

N
W E
S

Gulf of
Khambhat
Godavari River
Deccan
Plateau
Ghats
Bay of
Bengal

Arabian Sea
Krishna River
Western Ghats
Eastern Ghats
60°E
70°E

Malabar Coast
Coromandel Coast

10°N

ELEVATION

Feet	Meters
13,120	4,000
6,560	2,000
1,640	500
656	200
(Sea level) 0	0 (Sea level)
Below sea level	Below sea level

Gulf of
Mannar
SRI LANKA

0   250   500 Miles
0   250   500 Kilometers

Projection: Albers Equal-Area

MALDIVES
INDIAN OCEAN
80°E
90°E

1 The Himalayas and the Indus River are two key physical features on the Indian Subcontinent.

Two smaller mountain ranges stretch down India's coasts. The Eastern and Western Ghats (GAWTS) are low mountains that separate India's east and west coasts from the country's interior.

Perhaps the most impressive physical features in the subcontinent, however, are the Himalayas. These enormous mountains stretch about 1,500 miles (2,415 km) along the northern border of the Indian Subcontinent. Formed by the collision of two massive tectonic plates, the Himalayas are home to the world's highest mountains. On the border between Nepal and China is **Mount Everest**, the highest mountain on the planet. It measures some 29,035 feet (8,850 m). K2 in northern Pakistan is the world's second highest peak.

Rivers and Plains

Deep in the Himalayas are the sources of some of Asia's mightiest rivers. Two major river systems—the Ganges (GAN-jeez) and the Indus—originate in the Himalayas. Each carries massive amounts of water from the mountains' melting snow and glaciers. For thousands of years, these rivers have flooded the surrounding land, leaving rich soil deposits and fertile plains.

India's most important river is the Ganges. The **Ganges River** flows across northern India and into Bangladesh. There, the Ganges joins with other rivers and creates a huge delta. A **delta** is a landform at the mouth of a river created by sediment deposits. Along the length of the Ganges is a vast area of rich soil and fertile farmland.

FOCUS ON READING
What words in this paragraph help you to visualize the information?

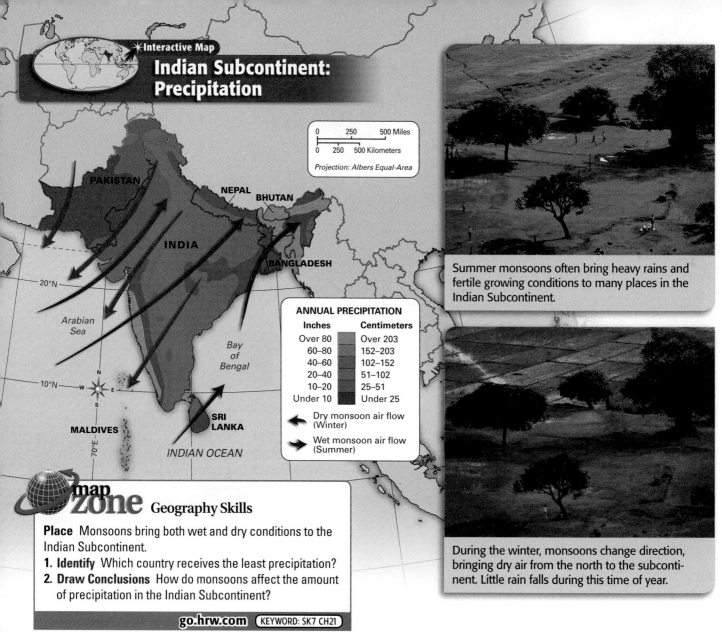

0   250   500 Miles
0   250   500 Kilometers
Projection: Albers Equal-Area

PAKISTAN

NEPAL
BHUTAN

INDIA

BANGLADESH

20°N

Arabian
Sea

Bay
of
Bengal

ANNUAL PRECIPITATION

Inches		Centimeters
Over 80		Over 203
60–80		152–203
40–60		102–152
20–40		51–102
10–20		25–51
Under 10		Under 25

10°N

Dry monsoon air flow
(Winter)

Wet monsoon air flow
(Summer)

MALDIVES

SRI
LANKA

INDIAN OCEAN

map
zone Geography Skills

Place Monsoons bring both wet and dry conditions to the Indian Subcontinent.
1. **Identify** Which country receives the least precipitation?
2. **Draw Conclusions** How do monsoons affect the amount of precipitation in the Indian Subcontinent?

go.hrw.com KEYWORD: SK7 CH21

Summer monsoons often bring heavy rains and fertile growing conditions to many places in the Indian Subcontinent.

During the winter, monsoons change direction, bringing dry air from the north to the subcontinent. Little rain falls during this time of year.

Known as the Ganges Plain, this region is India's farming heartland.

Likewise, Pakistan's **Indus River** also creates a fertile plain known as the Indus River Valley. This valley was once home to the earliest Indian civilizations. Today, it is Pakistan's most densely populated region.

Other Features

Other geographic features are scattered throughout the subcontinent. South of the Ganges Plain, for example, is a large, hilly plateau called the Deccan. East of the Indus Valley is the Thar (TAHR), or Great Indian Desert. Marked by rolling sand dunes, parts of this desert receive as little as 4 inches (100 mm) of rain per year. Still another geographic region is the Tarai (tuh-RY) in southern Nepal. It has fertile farmland and tropical jungles.

READING CHECK **Summarizing** What are the physical features of the Indian Subcontinent?

Climates and Resources

Just as the physical features of the Indian Subcontinent differ, so do its climates and resources. A variety of climates and natural resources exist throughout the region.

Climate Regions

From the Himalayas' snow-covered peaks to the dry Thar Desert, the climates of the Indian Subcontinent differ widely. In the Himalayas, a highland climate brings cool temperatures to much of Nepal and Bhutan. The plains south of the Himalayas have a humid subtropical climate. Hot, humid summers with plenty of rainfall are common in this important farming region.

Tropical climates dominate much of the subcontinent. The tropical savanna climate in central India and Sri Lanka keeps temperatures there warm all year long. This region experiences wet and dry seasons during the year. A humid tropical climate brings warm temperatures and heavy rains to parts of southwest India, Sri Lanka, Maldives, and Bangladesh.

The remainder of the subcontinent has dry climates. Desert and steppe climates extend throughout southern and western India and most of Pakistan.

Monsoons have a huge influence on the weather and climates in the subcontinent. **Monsoons** are seasonal winds that bring either moist or dry air to an area. From June to October, summer monsoons bring moist air up from the Indian Ocean, causing heavy rains. Flooding often accompanies these summer monsoons. In 2005, for example, the city of Mumbai (Bombay), India received some 37 inches (94 cm) of rain in just 24 hours. However, in winter the monsoons change direction, bringing dry air from the north. Because of this, little rain falls from November to January.

Natural Resources

A wide variety of resources are found on the Indian Subcontinent. Agricultural and mineral resources are the most plentiful.

Perhaps the most important resource is the region's fertile soil. Farms produce many different crops, such as tea, rice, nuts, and jute, a plant used for making rope. Timber and livestock are also key resources in the subcontinent, particularly in Nepal and Bhutan.

The Indian Subcontinent also has an abundance of mineral resources. Large deposits of iron ore and coal are found in India. Pakistan has natural gas reserves, while Sri Lankans mine many gemstones.

READING CHECK **Summarizing** What climates and resources are located in this region?

SUMMARY AND PREVIEW In this section you learned about the wide variety of physical features, climates, and resources in the Indian Subcontinent. Next, you will learn about the rich history and culture of this unique region.

go.hrw.com
Online Quiz
KEYWORD: SK7 HP21

Section 1 Assessment

Reviewing Ideas, Terms, and Places

1. **a. Define** What is a **subcontinent**?
   b. Make Inferences Why do you think the **Indus River** Valley is so heavily populated?
   c. Rank Which physical features in the Indian Subcontinent would you most want to visit? Why?
2. **a. Identify** What natural resources are found in the Indian Subcontinent?
   b. Analyze What are some of the benefits and drawbacks of **monsoons**?

Critical Thinking

3. **Drawing Inferences** Draw a chart like the one shown here. Using your notes, write a sentence explaining how each aspect affects life on the Indian Subcontinent.

	Effect on Life
Physical Features	
Climates	
Natural Resources	

FOCUS ON VIEWING

4. **Telling about Physical Geography** What information and images of the region's physical geography might you include in your travelogue? Jot down some ideas.

History and Culture of India

Main Ideas

1. Advanced civilizations and powerful empires shaped the early history of India.
2. Powerful empires controlled India for hundreds of years.
3. Independence from Great Britain led to the division of India into several countries.
4. Religion and the caste system are two important parts of Indian culture.

The Big Idea

Ancient civilizations and powerful empires have shaped the history and culture of India.

Key Terms and Places

Delhi, *p. 548*
colony, *p. 548*
partition, *p. 549*
Hinduism, *p. 550*
Buddhism, *p. 550*
caste system, *p. 550*

TAKING NOTES As you read, take notes on the history and culture of India. Use a diagram like this one to organize your notes.

History

Culture

If YOU lived there...

You live in New Delhi, India's capital city. Museums in your city display artifacts from some of India's oldest civilizations. People can visit beautiful buildings built by powerful empires. Statues and parades celebrate your country's independence.

How does your city celebrate India's history?

BUILDING BACKGROUND The Indian Subcontinent has a rich and interesting history. Ancient civilizations, powerful empires, rule by foreigners, and the struggle for independence have shaped not only the history, but also the culture of India and its neighbors.

Early Civilizations and Empires

India, the largest country on the Indian Subcontinent, is one of the world's oldest civilizations. Early civilizations and empires greatly influenced the history of the Indian Subcontinent.

India's History

Ancient Civilizations

- Around 2300 BC the Harappan civilization begins in the Indus River Valley.
- The Aryans, invaders from Central Asia, enter India around 1500 BC.
- Aryan culture helps shape the languages, religion, and caste system of India.

Harappan artifact

Early Empires

- By 233 BC the Mauryan Empire controls most of the Indian Subcontinent.
- Emperor Asoka helps spread Buddhism in India.
- Indian trade and culture flourish during the Gupta Empire.

Mauryan troops atop a war elephant

Ancient Civilizations

The first urban civilization in the Indian Subcontinent was centered around the Indus River Valley in present-day Pakistan. We call this ancient Indian civilization the Harappan (huh-RA-puhn) civilization after one of its main cities. Historians believe that the Harappan civilization flourished between 2300 and 1700 BC. By about 1700 BC, however, this civilization began to decline. No one is certain what led to its decline. Perhaps invaders or natural disasters destroyed the Harappan civilization.

Not long after the Harappan civilization ended, a new group rose to power. Around 1500 BC the Aryans (AIR-ee-uhnz), a group of people from Central Asia, entered the Indian Subcontinent. Powerful warriors, the Aryans eventually conquered and settled the fertile plains along the Indus and Ganges rivers.

The Aryans greatly **influenced** Indian culture. Their language, called Sanskrit, served as the basis for several languages in South Asia. For example, Hindi, the official language of India, is related to Sanskrit. As the Aryans settled in India, they mixed with Indian groups already living there. Their religious beliefs and customs mixed as well, forming the beginnings of India's social system and Hindu religion.

Early Empires

Over time, powerful kingdoms began to emerge in northern India. One kingdom, the Mauryan Empire, dominated the region by about 320 BC. Strong Mauryan rulers raised huge armies and conquered almost the entire subcontinent. Asoka (uh-SOH-kuh), one of the greatest Mauryan emperors, helped expand the empire and improve trade. Asoka also encouraged the acceptance of other religions. After his death, however, the empire slowly crumbled. Power struggles and invasions destroyed the Mauryan Empire.

After the fall of the Mauryan Empire, India split into many small kingdoms. Eventually, however, a strong new empire rose to power. In the AD 300s, the Gupta (GOOP-tuh) Empire united much of northern India. Under Gupta rulers, trade and culture thrived. Scholars made important advances in math, medicine, and astronomy. Indian mathematicians, for example, first introduced the concept of zero.

Gradually, the Gupta Empire also declined. Attacks by invaders from Asia weakened the empire. By about 550, India was once again divided.

READING CHECK **Summarizing** How did early civilizations and empires influence India?

The Mughal Empire

■ Babur establishes the Mughal Empire in northern India in 1526.

■ Indian trade, culture, and religion thrive under the rule of Akbar the Great.

■ By 1700 the Mughal Empire rules almost all of the Indian Subcontinent.

The first Mughal emperor, Babur

The British Empire

■ The British East India Company establishes trade in northern India in the early 1600s.

■ Indian troops trigger a massive revolt against the East India Company.

■ The British government takes direct control of India in 1858.

■ India and Pakistan gain independence in 1947.

Indian troop in the British Army

Powerful Empires

Powerful empires controlled India for much of its history. First the Mughal Empire and then the British Empire ruled India for hundreds of years.

The Mughal Empire

In the late 600s Muslim armies began launching raids into India. Some Muslims tried to take over Indian kingdoms. Turkish Muslims, for example, established a powerful kingdom at **Delhi** in northern India. In the 1500s a new group of Muslim invaders swept into the subcontinent. Led by the great warrior Babur (BAH-boohr), they conquered much of India. In 1526 Babur established the Mughal (MOO-guhl) Empire.

Babur's grandson, Akbar, was one of India's greatest rulers. Under Akbar's rule, trade flourished. Demand for Indian goods like spices and tea grew. The Mughal Empire grew rich from trade.

Akbar and other Mughal rulers also promoted culture. Although the Mughals were Muslim, most Indians continued to practice Hinduism. Akbar's policy of religious tolerance, or acceptance, encouraged peace throughout his empire. Architecture also thrived in the Mughal Empire. One of India's most spectacular buildings, the Taj Mahal, was built during Mughal rule.

The British Empire

The Mughals were not the only powerful empire in India. As early as the 1500s Europeans had tried to control parts of India. One European country, England, rose to power as the Mughal Empire declined.

The English presence in India began in the 1600s. At the time, European demand for Indian goods, such as cotton and sugar, was very high. Mughal rulers granted the East India Company, a British trading company, valuable trading rights.

At first, the East India Company controlled small trading posts. However, the British presence in India gradually grew. The East India Company expanded its territory and its power. By the mid-1800s the company controlled more than half of the Indian Subcontinent. India had become a British **colony, a territory inhabited and controlled by people from a foreign land.**

British rule angered and frightened many Indians. The East India Company controlled India with the help of an army made up mostly of Indian troops commanded by British officers. In 1857 Indian troops revolted, triggering violence all across India. The British government crushed the rebellion and took control of India away from the East India Company. With that, the British government began to rule India directly.

READING CHECK **Analyzing** How did powerful empires affect Indian history?

Independence and Division

By the late 1800s many Indians had begun to question British rule. Upset by their position as second-class citizens, a group of Indians created the Indian National Congress. Their goal was to gain more rights and opportunities.

As more and more Indians became dissatisfied with British rule, they began to demand independence. Mohandas Gandhi was the most important leader of this Indian independence movement. During the 1920s and 1930s his strategy of nonviolent protest convinced millions of Indians to support independence.

Finally, Great Britain agreed to make India independent. However, tensions between the Hindu and Muslim communities caused a crisis. Fearing they would have little say in the new government, India's Muslims called for a separate nation.

To avoid a civil war, the British government agreed to the **partition**, or division, of India. In 1947 two independent countries were formed. India was mostly Hindu. Pakistan, which included the area that is now Bangladesh, was mostly Muslim. As a result, some 10 million people rushed to cross the border. Muslims and Hindus wanted to live in the country where their religion held a majority.

Soon after India and Pakistan won their independence, other countries in the region gradually did too. Sri Lanka and Maldives gained their independence from Great Britain. In 1971, after a bloody civil war that killed almost 1 million people, East Pakistan broke away to form the country of Bangladesh.

READING CHECK **Identifying Cause and Effect** What were the effects of Indian independence from Great Britain?

FOCUS ON READING

As you read this paragraph, visualize the events described. Draw a rough sketch to depict what you imagine.

The Partition of India

A massive wave of migration took place after the partition of India and Pakistan. Millions of Hindus and Muslims crowded onto trains that would take them to their new homelands in India and Pakistan.

Indian Culture

As you might imagine, the rich and unique history of the Indian Subcontinent has created an equally unique culture. Two aspects of that culture are religion and a strict social class system.

Religion

Religion has played a very important role in Indian history. In fact, India is the birthplace of several major religions, including Hinduism and Buddhism.

Hinduism One of the world's oldest religions is **Hinduism, the dominant religion of India.** According to Hindu beliefs, everything in the universe is part of a single spirit called Brahman. Hindus believe that their ultimate goal is to reunite their souls with that spirit. Hinduism teaches that souls are reincarnated, or reborn, many times before they join with Brahman.

Buddhism Another Indian religion is Buddhism, which began in northern India in the late 500s BC. **Buddhism is a religion based on the teachings of Siddhartha Gautama—the Buddha.** According to the Buddha's teachings, people can rise above their desire for material goods and reach nirvana. Nirvana is a state of perfect peace in which suffering and reincarnation end.

Caste System

Thousands of years ago, the Aryans organized Indian society into a unique social class system known as the caste system. The **caste system** divided Indian society into groups based on a person's birth or occupation.

The caste system features four main classes, or castes, originally based on occupations. Below these four castes are the Dalits, members of India's lowest class. Many rules guided interaction between the classes. For example, people from different castes were not allowed to eat together.

READING CHECK **Analyzing** How do religion and the caste system influence Indian culture?

SUMMARY AND PREVIEW In this section you learned about the rich history and culture of the Indian Subcontinent. Next, you will learn about important issues that affect India today.

Section 2 Assessment

Reviewing Ideas, Terms, and Places

1. **a. Identify** What different peoples ruled India?
   b. Analyze How did these early civilizations and empires influence Indian culture?
2. **a. Describe** What were some accomplishments of the Mughal Empire?
   b. Predict How might Indian history have been different if the British had not ruled India?
3. **a. Recall** Who was the leader of India's independence movement?
   b. Explain What led to the **partition** of India?
4. **a. Define** What is the **caste system**?
   b. Elaborate Why do you think India is home to some of the world's oldest religions?

Critical Thinking

5. **Summarizing** Use your notes and a diagram like the one here to write a sentence summarizing each aspect of Indian history and culture.

Early History | Foreign Rule
Self-Rule | Culture

FOCUS ON VIEWING

6. **Discussing History and Culture** Which details about India's history and culture will you use? How will you explain and illustrate them?

Analyzing a Line Graph

Learn

Line graphs are drawings that display information in a clear, visual form. People often use line graphs to track changes over time. For example, you may want to see how clothing prices change from year to year. Line graphs also provide an easy way to see patterns, like increases or decreases, that emerge over time. Use the following guidelines to analyze a line graph.

- Read the title. The title will tell you about the subject of the line graph.
- Examine the labels. Note the type of information in the graph, the time period, and the units of measure.
- Analyze the information. Be sure to look for patterns that emerge over time.

Practice

Examine the line graph carefully, then answer the questions below.

❶ What is the subject of this line graph?

❷ What units of measure are used? What period of time does the line graph reflect?

❸ What pattern does the line graph indicate? How can you tell?

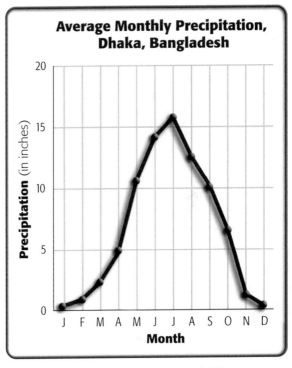

Average Monthly Precipitation, Dhaka, Bangladesh

Source: *National Geographic Atlas of the World, Seventh Edition*

Apply

Create a line graph that tracks your grades in a particular class. Start by organizing your grades by the date of the assignment. Then plot your grades on a line graph. Be sure to use labels and a title to identify the subject and information presented in your line graph. Finally, identify any patterns that you see in the line graph.

SECTION 3

India Today

What You Will Learn...

Main Ideas

1. Daily life in India is centered around cities, villages, and religion.
2. Today India faces many challenges, including a growing population and economic development.

The Big Idea

India today features a blend of modern and traditional culture, a rapidly growing population, and a developing economy.

Key Terms and Places

Mumbai (Bombay), *p. 552*
Kolkata (Calcutta), *p. 552*
urbanization, *p. 554*
green revolution, *p. 555*

TAKING NOTES As you read, take notes on India today. Use a graphic organizer like the one below to organize your notes on the daily life and challenges in India.

Daily Life	Challenges

If YOU lived there...

You live in Mumbai, India's largest city. A major port, Mumbai is home to many industries, such as textiles and electronics. Museums and theaters offer entertainment. Every year, thousands of people flock to Mumbai in search of jobs or to enroll in its universities. The streets are crowded, and pollution is often heavy.

Do you enjoy living in Mumbai? Why or why not?

BUILDING BACKGROUND India has undergone many changes since gaining its independence from Great Britain. Cities have grown dramatically, new businesses and industries have developed, and the population has exploded. India today faces many challenges.

Daily Life in India

More than 1 billion people live in India today. This huge population represents modern India's many different ethnic groups, religions, and lifestyles. Despite these many differences, city life, village life, and religion all help unite the people of India.

Cities

Millions of Indians live in large, bustling cities. In fact, India's two largest cities, **Mumbai (Bombay)** and **Kolkata (Calcutta)**, are among the world's most populous cities. Many people in Indian cities work in factories and offices. Some cities, like Bangalore and Mumbai, are home to universities, research centers, and high-tech businesses. Most city-dwellers, however, struggle to earn a living. Many people live in shacks made of scraps of wood or metal. They often have no plumbing and little clean water.

Villages

Most Indians still live in rural areas. Hundreds of thousands of villages are home to more than 70 percent of India's population. Most villagers work as farmers and live with an extended family in simple homes. Only recently have paved roads and electricity reached many Indian villages.

Religion

In both cities and villages, religion plays a key role in Indian daily life. While most Indians practice Hinduism, many people follow several other religions such as Islam, Buddhism, and traditional religions. In addition, millions of Indians practice two native religions, Sikhism and Jainism.

Religious celebrations are an important part of Indian life today. One of India's most popular festivals is Diwali, the festival of lights. Diwali celebrates Hindu, Sikh, and Jain beliefs.

READING CHECK Contrasting How does life in Indian cities and villages differ?

Close-up

Diwali: The Festival of Lights

Diwali, or the festival of lights, is one of the most important celebrations in India. A variety of activities on each of the five days of Diwali celebrate Hindu, Sikh, and Jain beliefs.

Beautiful firework displays are common during Diwali.

Elaborate chalk designs, called rangolis, often decorate floors and walls.

Diwali is a time to spend with friends and family. Cards and small gifts, such as sweets and candles, are often exchanged.

Small oil lamps, or *diyas*, decorate homes inside and out.

ANALYSIS SKILL **ANALYZING VISUALS**

What elements of Indian daily life do you see in the illustration?

India's Challenges

India has undergone drastic changes since gaining independence. Today the country faces several challenges, such as dealing with a growing population and managing its economic development.

Population

Its more than 1 billion people make India the world's second most populous country. Only China has a larger population. India's population has grown rapidly, doubling since 1947. This huge population growth places a strain on India's environment and many of its resources, including food, housing, and schools.

India's cities are particularly affected by the growing population. As the country's population has grown, urbanization has taken place. **Urbanization is the increase in the percentage of people who live in cities.** Many millions of people have moved to India's cities in search of jobs.

Government and Economy

Since India gained independence, its leaders have strengthened the government and economy. Today India is the world's

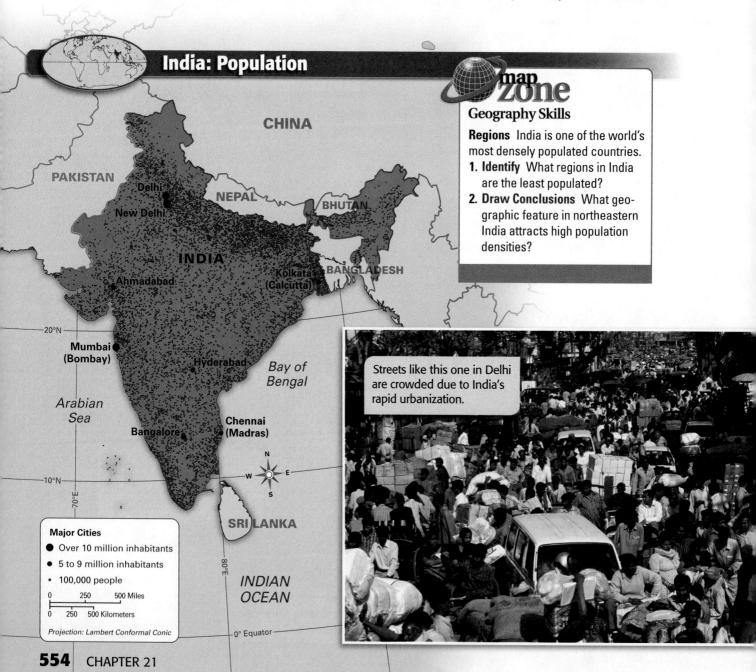

India: Population

CHINA

PAKISTAN

Delhi

New Delhi

NEPAL

BHUTAN

INDIA

Ahmadabad

Kolkata (Calcutta)

BANGLADESH

20°N

Mumbai (Bombay)

Hyderabad

Bay of Bengal

Arabian Sea

Bangalore

Chennai (Madras)

70°E

10°N

SRI LANKA

80°E

INDIAN OCEAN

0° Equator

Major Cities
- ● Over 10 million inhabitants
- ● 5 to 9 million inhabitants
- · 100,000 people

0   250   500 Miles
0 250 500 Kilometers

Projection: Lambert Conformal Conic

map zone

Geography Skills

Regions India is one of the world's most densely populated countries.
1. **Identify** What regions in India are the least populated?
2. **Draw Conclusions** What geographic feature in northeastern India attracts high population densities?

Streets like this one in Delhi are crowded due to India's rapid urbanization.

largest democracy and one of the strongest nations in Asia. The greatest challenges facing India's government are providing for a growing population and resolving conflicts with its neighbor, Pakistan.

India's gross domestic product (GDP) places it among the world's top 5 industrial countries. However, its per capita, or per person, GDP is only $3,100. As a result, millions of Indians live in poverty.

India's government has taken steps to reduce poverty. In the 1960s and 1970s the **green revolution**, a program that encouraged farmers to adopt modern agricultural methods, helped farmers produce more food. Recently, the government has succeeded in attracting many high-tech businesses to India.

READING CHECK **Finding Main Ideas** What are India's government and economy like?

SUMMARY AND PREVIEW India today faces many challenges as it continues to modernize. Next, you will learn about India's neighbors on the subcontinent.

CONNECTING TO Economics

Bollywood

One of India's largest industries is its moviemaking industry. Much of India's film industry is located in Mumbai (Bombay). Many people refer to the industry as Bollywood—a combination of Bombay and Hollywood. Bollywood produces more films every year than any other country. In 2003, for example, India produced 1,100 films—almost twice the number of films produced in the United States. In recent years, Bollywood films have become increasingly popular outside of India—particularly in the United Kingdom and the United States.

Drawing Conclusions How might the film industry affect India's economy?

---

go.hrw.com
Online Quiz
KEYWORD: SK7 HP21

Section 3 Assessment

Reviewing Ideas, Terms, and Places
1. **a. Identify** What different religions are practiced in India today?
   b. Compare and Contrast In what ways are Indian cities similar to cities in the United States? How are they different from U.S. cities?
   c. Elaborate Why do you think that a majority of Indians live in villages?
2. **a. Recall** What is **urbanization**? What is one cause of urbanization?
   b. Make Inferences How did the **green revolution** affect India's economy?
   c. Predict What effects might India's growing population have on its resources and environment in the future?

Critical Thinking
3. **Finding Main Ideas** Using your notes and the web diagram, write the main idea for each element of India today.

Cities and Villages — Population — India Today — Government — Economy

FOCUS ON VIEWING

4. **Telling about India Today** You will need some images, or pictures, for your travelogue. What images can you use to tell about India today?

THE INDIAN SUBCONTINENT **555**

India's Neighbors

What You Will Learn...

Main Ideas

1. Many different ethnic groups and religions influence the culture of India's neighbors.
2. Rapid population growth, ethnic conflicts, and environmental threats are major challenges to the region today.

The Big Idea

Despite cultural differences, the countries that border India share similar challenges.

Key Terms and Places

Sherpas, *p. 556*
Kashmir, *p. 557*
Dhaka, *p. 558*
Kathmandu, *p. 558*

TAKING NOTES As you read, take notes on the culture and issues in the countries that border India. Organize your notes in a diagram like the one below.

If YOU lived there...

You live in the mountainous country of Bhutan. For many years Bhutan's leaders kept the country isolated from outsiders. Recently, they have begun to allow more tourists to enter the country. Some of your neighbors believe that tourism will greatly benefit the country. Others think it could harm the environment.

How do you feel about tourism in Bhutan?

BUILDING BACKGROUND After years of isolation or control by Great Britain, the 1900s brought great changes to the countries on the Indian Subcontinent. Today these countries face rapid population growth and economic and environmental concerns.

Culture

Five countries—Pakistan, Bangladesh, Nepal, Bhutan, and Sri Lanka—share the subcontinent with India. Though they are neighbors, these countries have significantly different cultures.

People The cultures of the countries that border India reflect the customs of many ethnic groups. For example, the Sherpas, an ethnic group from the mountains of Nepal, often serve as guides through the Himalayas. Members of Bhutan's largest ethnic group originally came from Tibet, a region in southern China. Many of Sri Lanka's Tamil (TA-muhl) people came from India to work the country's huge plantations.

Religion As you can see on the map on the next page, a variety of religions exist on the Indian Subcontinent. Most countries, like India, have one major religion. In Pakistan and Bangladesh, for example, most people practice Islam and small portions of the population follow Hinduism, Christianity, and tribal religions. In Nepal, the dominant religion is Hinduism, although Buddhism is practiced in some parts of the country. Buddhism dominates both Bhutan and Sri Lanka.

READING CHECK **Contrasting** In what ways are the cultures of this region different?

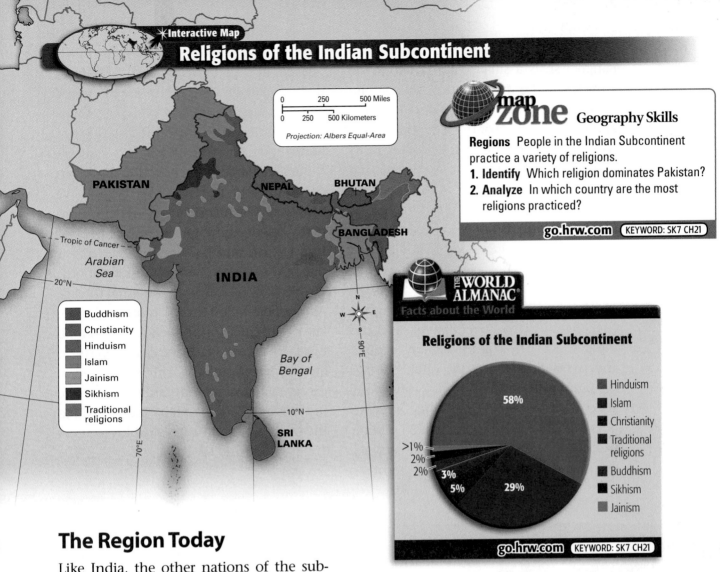

Interactive Map
Religions of the Indian Subcontinent

0   250   500 Miles
0   250   500 Kilometers
Projection: Albers Equal-Area

map zone Geography Skills

Regions People in the Indian Subcontinent practice a variety of religions.
1. **Identify** Which religion dominates Pakistan?
2. **Analyze** In which country are the most religions practiced?

go.hrw.com | KEYWORD: SK7 CH21

PAKISTAN   NEPAL   BHUTAN
                        BANGLADESH

Tropic of Cancer
Arabian Sea   INDIA
20°N

Buddhism
Christianity
Hinduism
Islam
Jainism
Sikhism
Traditional religions

Bay of Bengal

10°N

SRI LANKA

70°E   90°E

THE WORLD ALMANAC®
Facts about the World

Religions of the Indian Subcontinent

58%
>1%
2%
2%   3%
5%   29%

Hinduism
Islam
Christianity
Traditional religions
Buddhism
Sikhism
Jainism

go.hrw.com | KEYWORD: SK7 CH21

The Region Today

Like India, the other nations of the subcontinent face a variety of challenges. Two of the greatest challenges are population growth and poverty.

Pakistan

One of the greatest challenges Pakistan faces is the lack of government stability. Since its creation in 1947, Pakistan has suffered from rebellions and assassinations of government leaders. Without government stability, Pakistan has been unable to adequately address other problems.

Another challenge in Pakistan is rapid population growth. The country's government struggles to manage resources and to reduce poverty. Some people fear that Pakistan's large population and high levels of poverty could cause even more instability.

Relations with India are another important issue in Pakistan today. Since the partition in 1947, the two countries have clashed over the territory of **Kashmir**. Both India and Pakistan claim control of the region. Today Pakistan controls western Kashmir while India controls the east. Armed troops from both countries guard a "line of control" that divides Kashmir.

Since 2001 Pakistan has aided the United States in its war on terrorism. Pakistan's military has arrested hundreds of terrorists and provided information about suspected terrorists. Despite this crackdown, however, many people believe that there are still terrorists within Pakistan's borders.

FOCUS ON READING
Visualize the information described in this paragraph. Describe what mental images you see.

Bangladesh

Bangladesh is a small country about the same size as the state of Wisconsin. Despite its small size, Bangladesh's population is almost half the size of the U.S. population. As a result, it is one of the world's most densely populated countries with some 2,734 people per square mile (1,055 per square km). The capital and largest city, **Dhaka** (DA-kuh), is home to over 11.5 million people. Overcrowding is not limited to urban areas, however. Rural areas are also densely populated.

Flooding is one of Bangladesh's biggest challenges. Many <u>circumstances</u> cause these floods. The country's many streams and rivers flood annually, often damaging farms and homes. Summer monsoons also cause flooding. For example, massive flooding in 2004 left more than 25 million people homeless. It also destroyed schools, farms, and roads throughout the country.

Nepal

The small kingdom of Nepal also faces many challenges today. Its population is growing rapidly. In fact, the population has more than doubled in the last 30 years. **Kathmandu** (kat-man-DOO), the nation's capital and largest city, is troubled by overcrowding and poverty. Thousands have moved to Kathmandu in search of jobs and better opportunities. As a result of population growth and poor resources, Nepal is one of the world's least-developed nations.

Nepal also faces environmental threats. As the population grows, more and more land is needed to grow enough food. To meet this need, farmers clear forests to create more farmland. This deforestation causes soil erosion and harms the wildlife in the region. Nepal's many tourists add to the problem as they use valuable resources and leave behind trash.

Bhutan

Bhutan is a small mountain kingdom that lies in the Himalayas between India and China. Because of the rugged mountains, Bhutan has been isolated throughout much of its history. This isolation limited outside influences until the 1900s, when Bhutan's king established ties first with Great Britain and later with India. By the mid-1900s Bhutan had ended its long isolation. Efforts to modernize Bhutan resulted in the construction of new roads, schools, and hospitals.

Today Bhutan continues to develop economically. Most Bhutanese earn a living as farmers, growing rice, potatoes, and

Nepal

Many of Nepal's people live in the rugged Himalayas and earn a living herding animals.

corn. Some raise livestock like yaks, pigs, and horses. Another important industry is tourism. The government, however, limits the number of visitors to Bhutan to protect Bhutan's environment and way of life.

Sri Lanka

Sri Lanka is a large island country located some 20 miles (32 km) off India's southeast coast. As a result of its close location, India has greatly influenced Sri Lanka. In fact, Sri Lanka's two largest ethnic groups—the Tamil and the Sinhalese (sing-guh-LEEZ)—are descended from Indian settlers.

Conflicts between the Sinhalese and the Tamil divide Sri Lanka today. The Tamil minority has fought for years to create a separate state. Despite a 2002 cease-fire, violence between the two sides continues to disrupt the island nation.

Parts of Sri Lanka were devastated by the 2004 tsunami in the Indian Ocean. Thousands of Sri Lankans were killed, and more than 500,000 people were left homeless. The tsunami also damaged Sri Lanka's fishing and agricultural industries, which are still struggling to rebuild.

READING CHECK Summarizing What key issues affect India's neighbors today?

Sri Lanka

These women are picking tea on one of Sri Lanka's many tea plantations.

SUMMARY AND PREVIEW You have learned about the important challenges that face India's neighbors on the subcontinent. In the next chapter, you will learn about the physical geography, history, and culture of China, Mongolia, and Taiwan.

go.hrw.com
Online Quiz
KEYWORD: SK7 HP21

Section 4 Assessment

Reviewing Ideas, Terms, and Places

1. **a. Identify** What are the major religions of the Indian Subcontinent?
   b. Summarize What cultural differences exist among India's neighbors?
   c. Elaborate Why do you think there are so many different religions in this region?
2. **a. Identify** What is the capital of Nepal?
   b. Compare and Contrast In what ways are the countries of this region similar and different?
   c. Predict How might conflict over **Kashmir** cause problems in the future?

Critical Thinking

3. **Solving Problems** Using your notes and a chart like the one here, identify one challenge facing each of India's neighbors. Then develop a solution for each challenge.

Challenges	Solutions

FOCUS ON VIEWING

4. **Telling about India's Neighbors** Your travels include voyages to India's neighbors. Include important or intriguing details and images in your travelogue.

A Pakistani bride on her wedding day

GUIDED READING

WORD HELP

chadrs cloths worn by women as a head cover

henna a reddish dye made from a shrub; often used to decorate the hands and feet

cacophony a combination of loud sounds

curry a dish prepared in a highly spiced sauce

lapis a stone with a rich, deep blue color

❶ At a *mahendi* celebration women gather to prepare the bride for her wedding day.

❷ To line the eyes means to darken the rims of the eyelids with black kohl, an eyeliner.

from
Shabanu: Daughter of the Wind

by Suzanne Fisher Staples

About the Reading *In* Shabanu, *writer Suzanne Fisher Staples writes about the life of Shabanu, a young girl who is part of a nomadic desert culture in Pakistan. In this passage, Shabanu and her family prepare for the wedding of her older sister.*

AS YOU READ Look for details about the customs and traditions of Shabanu and her people.

Two days before the wedding, Bibi Lal . . . heads a procession of women to our house for the *mahendi* celebration ❶ . . . Bibi Lal looks like a giant white lily among her cousins and nieces, who carry baskets of sweets atop their flower-colored *chadrs*. They sing and dance through the fields, across the canal, to our settlement at the edge of the desert.

Sakina carries a wooden box containing henna. The *mahendi* women, Hindus from a village deep in the desert who will paint our hands and feet, walk behind her. Musicians and a happy cacophony of horns, pipes, and cymbals drift around them.

Mama, the servant girl, and I have prepared a curry of chicken, dishes of spiced vegetables, sweet rice, and several kinds of bread to add to the food that the women of Murad's family bring . . . Sharma has washed and brushed my hair. I wear a new pink tunic. She lines my eyes and rubs the brilliant lapis powder into my lids. ❷

Connecting Literature to Geography

1. Describing How did the women prepare for the upcoming wedding? What was the *mahendi* celebration like?

2. Interpreting Women come to Shabanu's house from a distant village to paint the girls' hands and feet with henna. Why do you think this custom is important to the women and girls? What might it symbolize?

CHAPTER **21** **Chapter Review**

Geography's Impact
video series
Review the video to answer the closing question:
How might population density affect a country?

Visual Summary

Use the visual summary below to help you review the main ideas of the chapter.

QUICK FACTS

Towering mountains and powerful monsoons characterize the physical geography of the Indian Subcontinent.

India's Taj Mahal represents the subcontinent's rich history and culture.

The nations that border India face many economic, political, and environmental challenges today.

Reviewing Vocabulary, Terms, and Places

Choose one word from each word pair to correctly complete each sentence below.

1. _____ often bring heavy rains to the Indian Subcontinent in summer. **(Monsoons/Ghats)**

2. The most popular religion in India today is _____. **(Buddhism/Hinduism/Islam)**

3. A _____ is a condition that influences an event or activity. **(feature/circumstance)**

4. _____ are an ethnic group from the mountains of Nepal. **(Tamil/Sherpas)**

5. The highest peak in the Indian Subcontinent and the world is _____. **(Mount Everest/K2)**

6. India's _____ system divides society based on a person's birth, wealth, and job. **(caste/colonial)**

7. Pakistan is located on the Indian _____, a large landmass. **(Peninsula/Subcontinent)**

Comprehension and Critical Thinking

SECTION 1 *(Pages 542–545)*

8. **a. Recall** What is a delta?

   b. Draw Conclusions Why are rivers important to the people of the Indian Subcontinent?

   c. Evaluate Do you think monsoons have a positive or negative effect on India? Why?

SECTION 2 *(Pages 546–550)*

9. **a. Describe** What was the partition of India? When and why did it take place?

   b. Compare and Contrast In what ways were Mughal and British rule of India similar and different?

   c. Evaluate In your opinion, was partitioning India a good decision? Why or why not?

SECTION 3 *(Pages 552–555)*

10. **a. Identify** What program introduced modern agricultural methods to India?

SECTION 3 (continued)

b. Analyze How has population growth affected India's economy?

c. Elaborate If you lived in India, would you prefer to live in a city or a village? Why?

SECTION 4 (Pages 556–559)

11. a. Identify What countries share the subcontinent with India?

b. Analyze How was Sri Lanka affected by the 2004 tsunami?

c. Predict How might conflict between India and Pakistan lead to problems in the future?

Social Studies Skills

Analyzing Line Graphs *Use the line graph to help you answer the questions that follow.*

Indian Film Production, 1960–2000

Source: *Bollywood, India's Film Industry*

12. What is the subject of the line graph?

13. What general pattern or trend does the line graph indicate?

Using the Internet

go.hrw.com KEYWORD: SK7 CH21

14. Touring India Pack your bags and experience India! It's a country where you can climb towering mountains, journey across vast deserts, and even hike through rain forests. Enter the activity keyword and discover the regions of India. Then make an illustrated travel brochure that features some of the regions you have explored.

FOCUS ON READING AND VIEWING

15. Visualizing Read the literature selection *Shabanu: Daughter of the Wind.* As you read, visualize the scenes the author describes. Then make a list of words from the passage that help you create a mental image of the events. Lastly, draw a rough sketch of your mental image of the *mahendi* celebration.

16. Creating and Viewing a Travelogue Use your notes to create a one- to two-minute script describing your travels in the Indian Subcontinent. Identify and collect the images you need to illustrate your talk. Present your oral travelogue to the class, giving an exciting view of the region. Observe as others present their travelogues. How is each travelogue unique? How are they similar?

Map Activity ✳Interactive

17. The Indian Subcontinent On a separate sheet of paper, match the letters on the map with their correct labels.

Deccan	Mount Everest
Himalayas	Mumbai (Bombay)
Indus River	New Delhi
Kashmir	Sri Lanka

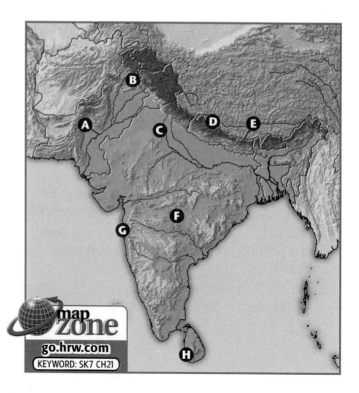

map zone go.hrw.com KEYWORD: SK7 CH21

DIRECTIONS: *Read questions 1 through 7 and write the letter of the best response. Then read question 8 and write your own well-constructed response.*

1 Which of the following is the *oldest* Indian civilization?

A Aryan

B Harappan

C Mughal

D Pakistani

2 Which of the following is a cause of India's rapid urbanization?

A People have moved away from cities to escape overcrowding and poverty.

B People have left villages to avoid rural warfare.

C People have left India in search of land.

D People have moved to cities in search of jobs.

3 Isolationism, Buddhism, and monarchy are all associated with which country?

A Bhutan

B India

C Nepal

D Sri Lanka

4 The majority of Indians today live

A in the Indus River Valley.

B on the coast.

C in cities.

D in villages.

5 The division of Indian society is known as

A the caste system.

B Diwali.

C Hinduism.

D the partition of India.

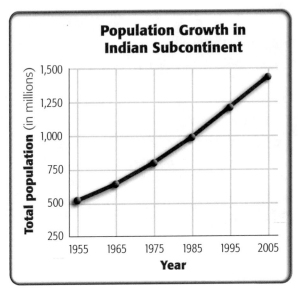

Source: United Nations Population Division

6 Based on the line graph above, what was the approximate population of South Asia in 1985?

A 500,000,000

B 760,000,000

C 1,000,000,000

D 1,400,000,000

7 These seasonal winds bring both wet and dry conditions to much of the Indian Subcontinent.

A hurricanes

B monsoons

C tsunamis

D typhoons

8 **Extended Response** Using information from the map in Section 3 titled India: Population, write a paragraph describing the settlement patterns in India today.

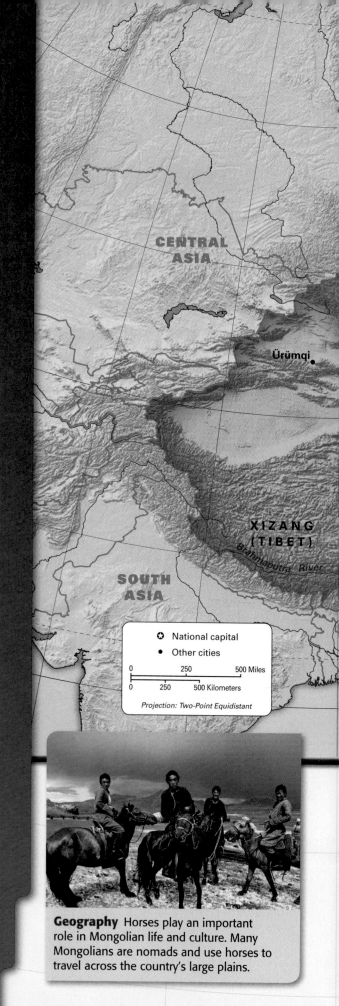

CHAPTER 22
China, Mongolia, and Taiwan

What You Will Learn...

In this chapter you will learn about the physical features, climate, and resources of China, Mongolia, and Taiwan. You will also study the histories of these countries, how different influences have shaped their cultures, and what life is like in these regions today.

SECTION 1
Physical Geography 766

SECTION 2
History and Culture of China 570

SECTION 3
China Today 577

SECTION 4
Mongolia and Taiwan 582

FOCUS ON READING AND WRITING

Identifying Implied Main Ideas The main idea in a piece of writing is sometimes stated directly. Other times, you must figure out the main idea. As you read, look for key details or ideas to help you identify the implied main ideas. **See the lesson, Identifying Implied Main Ideas, on page 689.**

Writing a Legend Since ancient times, people have passed along legends. These stories often tell about supernatural people or events from the past. Read the chapter. Then write your own legend describing the supernatural creation of a physical feature in this region.

Geography Horses play an important role in Mongolian life and culture. Many Mongolians are nomads and use horses to travel across the country's large plains.

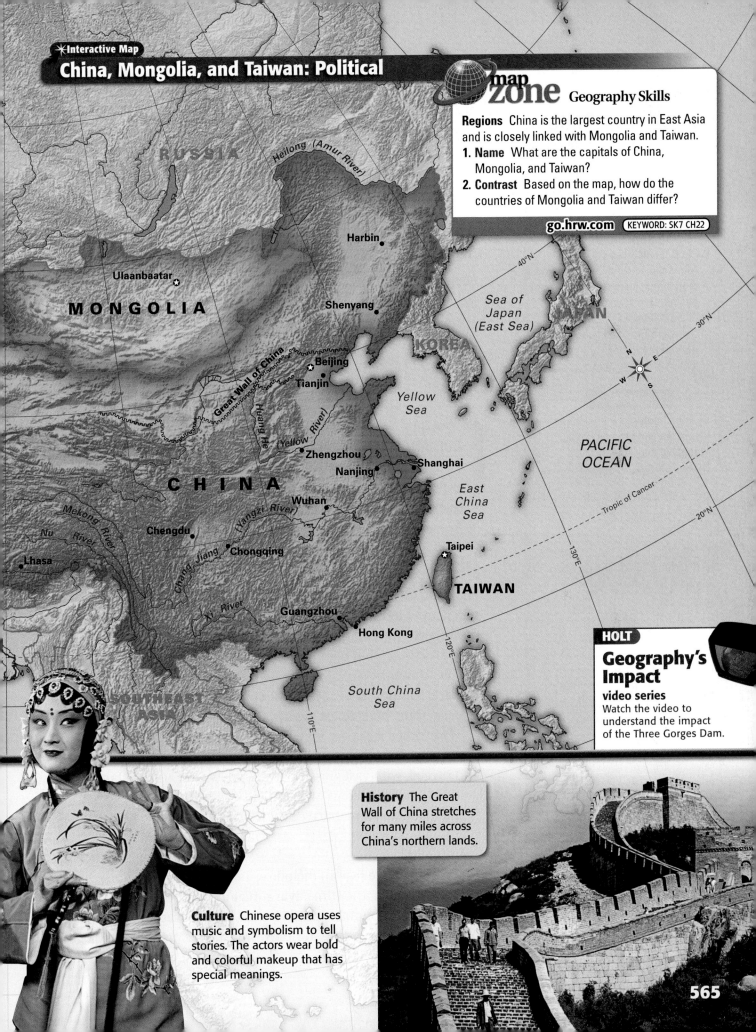

map zone Geography Skills

Regions China is the largest country in East Asia and is closely linked with Mongolia and Taiwan.
1. **Name** What are the capitals of China, Mongolia, and Taiwan?
2. **Contrast** Based on the map, how do the countries of Mongolia and Taiwan differ?

go.hrw.com   KEYWORD: SK7 CH22

RUSSIA

Heilong (Amur River)

Harbin

Ulaanbaatar

MONGOLIA

Shenyang

KOREA

Sea of Japan (East Sea)

JAPAN

Great Wall of China

Beijing

Tianjin

Huang He (Yellow River)

Yellow Sea

PACIFIC OCEAN

Zhengzhou

Nanjing

Shanghai

CHINA

Wuhan

East China Sea

Mekong River

Nu River

Chengdu

Chang Jiang (Yangzi River)

Chongqing

Tropic of Cancer

Taipei

Lhasa

Xi River

Guangzhou

Hong Kong

TAIWAN

SOUTHEAST ASIA

South China Sea

40°N

30°N

20°N

130°E

120°E

110°E

HOLT

Geography's Impact
video series
Watch the video to understand the impact of the Three Gorges Dam.

History The Great Wall of China stretches for many miles across China's northern lands.

Culture Chinese opera uses music and symbolism to tell stories. The actors wear bold and colorful makeup that has special meanings.

Physical Geography

What You Will Learn...

Main Ideas

1. Physical features of China, Mongolia, and Taiwan include mountains, plateaus and basins, plains, and rivers.
2. China, Mongolia, and Taiwan have a range of climates and natural resources.

The Big Idea

Physical features, climate, and resources vary across China, Mongolia, and Taiwan.

Key Terms and Places

Himalayas, *p. 566*
Plateau of Tibet, *p. 567*
Gobi, *p. 567*
North China Plain, *p. 568*
Huang He, *p. 568*
loess, *p. 568*
Chang Jiang, *p. 568*

TAKING NOTES As you read, use a chart like the one below to take notes on the physical features, climate, and resources of China, Mongolia, and Taiwan.

Mountains	
Other Landforms	
Rivers	
Climate and Resources	

If YOU lived there...

You are a young filmmaker who lives in Guangzhou, a port city in southern China. You are preparing to make a documentary film about the Huang He, one of China's great rivers. To make your film, you will follow the river across northern China. Your journey will take you from the Himalayas to the coast of the Yellow Sea.

What do you expect to see on your travels?

BUILDING BACKGROUND China, Mongolia, and Taiwan make up a large part of East Asia. They include a range of physical features and climates—dry plateaus, rugged mountains, fertile plains. This physical geography has greatly influenced life in each country.

Physical Features

Have you seen the view from the top of the world? At 29,035 feet (8,850 m), Mount Everest in the **Himalayas** is the world's highest mountain. From atop Everest, look east. Through misty clouds, icy peaks stretch out before you, fading to land far below. This is China. About the size of the United States, China has a range of physical features. They include not only the world's tallest peaks but also some of its driest deserts and longest rivers.

Two other areas are closely linked to China. To the north lies Mongolia (mahn-GOHL-yuh). This landlocked country is dry and rugged, with vast grasslands and desert. In contrast, Taiwan (TY-WAHN), off the coast of mainland China, is a green tropical island. Look at the map to see the whole region's landforms.

Mountains

Much of the large region, including Taiwan, is mountainous. In southwest China, the Himalayas run along the border. They are Earth's tallest mountain range. Locate on the map the region's other ranges. As a tip, the Chinese word *shan* means "mountain."

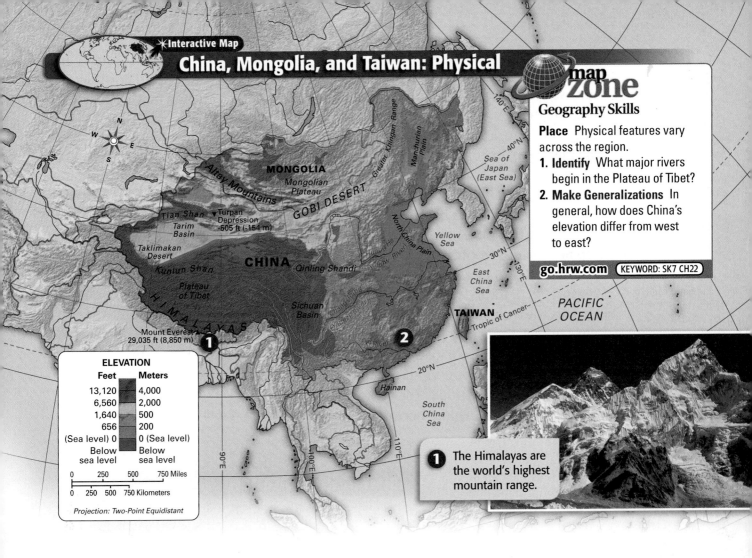

map zone

Geography Skills

Place Physical features vary across the region.

1. **Identify** What major rivers begin in the Plateau of Tibet?
2. **Make Generalizations** In general, how does China's elevation differ from west to east?

go.hrw.com KEYWORD: SK7 CH22

ELEVATION

Feet	Meters
13,120	4,000
6,560	2,000
1,640	500
656	200
(Sea level) 0	0 (Sea level)
Below sea level	Below sea level

0   250   500   750 Miles

0   250   500   750 Kilometers

Projection: Two-Point Equidistant

1 The Himalayas are the world's highest mountain range.

Other Landforms

Many of the mountain ranges are separated by plateaus, basins, and deserts. In southwest China, the **Plateau of Tibet** lies north of the Himalayas. The world's highest plateau, it is called the Roof of the World.

Moving north, we find a low, dry area. A large part of this area is the Taklimakan (tah-kluh-muh-KAHN) Desert, a barren land of sand dunes and blinding sandstorms.

In fact, sandstorms are so common that the desert's Turkish name, Taklimakan, has come to mean "Enter and you will not come out." To the northeast, the Turpan (toohr-PAHN) Depression is China's lowest point, at 505 feet (154 m) below sea level.

Continuing northeast, in Mongolia we find the **Gobi**. This harsh area of gravel and rock is the world's coldest desert. Temperatures can drop to below –40°F (–40°C).

2 Hills that are called karst towers line the Li River in southeast China. These dramatic hills formed over time as rainwater eroded limestone.

In east China, the land levels out into low plains and river valleys. These fertile plains, such as the **North China Plain**, are China's main population centers and farmlands. On Taiwan, a plain on the west coast is the island's main population center.

Rivers

FOCUS ON READING
Which details help you identify the main idea of the paragraph to the right?

In China, two great rivers run west to east. The **Huang He** (HWAHNG HEE), or the Yellow River, flows across northern China. Along its course, this river picks up large amounts of **loess** (LES), or fertile, yellowish soil. The soil colors the river and gives it its name.

In summer, the Huang He often floods. The floods spread layers of loess, enriching the soil for farming. However, such floods have killed millions of people. For this reason, the river is called China's Sorrow.

The mighty **Chang** (CHAHNG) **Jiang**, or the Yangzi (YAHNG-zee) River, flows across central China. It is Asia's longest river and a major transportation route.

READING CHECK **Summarizing** What are the main physical features found in this region?

Climate and Resources

Climate varies widely across the region. The tropical southeast is warm to hot, and monsoons bring heavy rains in summer. In addition, typhoons can strike the southeast coast in summer and fall. Similar to hurricanes, these violent storms bring high winds and rain. As we move to the northeast, the climate is drier and colder. Winter temperatures can drop below 0°F (–18°C).

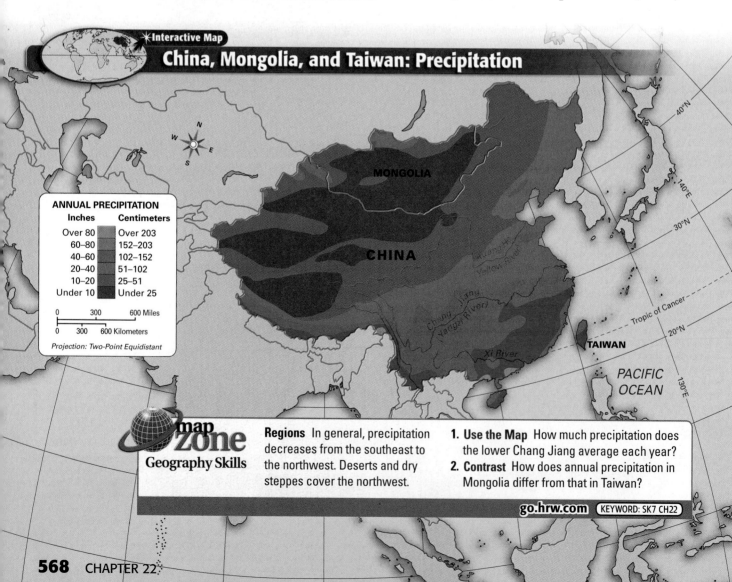

★Interactive Map

China, Mongolia, and Taiwan: Precipitation

ANNUAL PRECIPITATION

Inches	Centimeters
Over 80	Over 203
60–80	152–203
40–60	102–152
20–40	51–102
10–20	25–51
Under 10	Under 25

0   300   600 Miles
0   300   600 Kilometers

Projection: Two-Point Equidistant

MONGOLIA

CHINA

Huang He
Yellow River

Chang Jiang
Yangzi River

Xi River

TAIWAN

Tropic of Cancer

PACIFIC OCEAN

40°N
30°N
20°N
140°E
130°E

map zone
Geography Skills

Regions In general, precipitation decreases from the southeast to the northwest. Deserts and dry steppes cover the northwest.

1. **Use the Map** How much precipitation does the lower Chang Jiang average each year?
2. **Contrast** How does annual precipitation in Mongolia differ from that in Taiwan?

go.hrw.com **KEYWORD: SK7 CH22**

Flooding in China

China's rivers and lakes often flood during the summer rainy season. The satellite images here show Lake Dongting Hu in southern China. The lake appears blue, and the land appears red. Soon after the Before image was taken, heavy rains led to flooding. The After image shows the results. Compare the two images to see the extent of the flood, which killed more than 3,000 people and destroyed some 5 million homes.

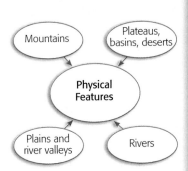

Before

After

For comparison, these arrows are pointing to the same place in each image.

Drawing Inferences Why might people continue to live in areas that often flood?

In the north and west, the climate is mainly dry. Temperatures vary across the area and can get both very hot and cold.

Like the climate, the region's natural resources cover a wide range. China has a wealth of natural resources. The country is rich in mineral resources and is a leading producer of coal, lead, tin, and tungsten. China produces many other minerals and metals as well. China's forestland and farmland are also valuable resources.

Mongolia's natural resources include minerals such as coal, iron, and tin as well as livestock. Taiwan's major natural resource is its farmland. Important crops include sugarcane, tea, and bananas.

READING CHECK **Contrasting** Which of these three countries has the most natural resources?

SUMMARY AND PREVIEW As you have read, China, Mongolia, and Taiwan have a range of physical features, climate, and resources. Next, you will read about the history and culture of China.

go.hrw.com
Online Quiz
KEYWORD: SK7 HP22

Section 1 Assessment

Reviewing Ideas, Terms, and Places

1. **a. Identify** What two major rivers run through China?
   b. Explain How does the **Huang He** both benefit and hurt China's people?
   c. Elaborate Why do you think many people in China live on the **North China Plain**?
2. **a. Define** What is a typhoon?
   b. Contrast What are some differences between the climates of southeast and northwest China?
   c. Rate Based on the different climates in this region, which part of the region would you prefer to live in? Why?

Critical Thinking

3. **Categorizing** Look back over your notes for this section. Then use a chart like the one shown here to organize, identify, and describe the main physical features of China, Mongolia, and Taiwan.

   Mountains → Physical Features ← Plateaus, basins, deserts
   Plains and river valleys → Physical Features ← Rivers

FOCUS ON WRITING

4. **Writing about Physical Geography** Note the main physical features of this region. Consider which feature you might want to explain in your legend. Features to consider include mountains, plateaus, and deserts.

History and Culture of China

What You Will Learn...

Main Ideas

1. Family lines of emperors ruled China for most of its early history.
2. In China's modern history, revolution and civil war led to a Communist government.
3. China has the world's most people and a rich culture shaped by ancient traditions.

The Big Idea

Ruled by dynasties in its early history, China is a Communist country with an enormous population and ancient traditions.

Key Terms

dynasty, *p. 571*
dialect, *p. 573*
Daoism, *p. 574*
Confucianism, *p. 574*
pagodas, *p. 575*

TAKING NOTES As you read, use a chart like the one below to take notes on China's history, people, and culture.

China		
History	People	Culture

If **YOU** lived there...

Your parents own a small farm in the Chinese countryside in the mid-1950s. China's new leaders are making changes, however. They are taking people's farms and combining them to create large government-run farms. Your family and neighbors will now work a large farm together. China's leaders will tell you what to grow and pay you based on how much the farm produces.

How do you feel about these changes?

BUILDING BACKGROUND In 1949 China established a strong central government. This new government changed many familiar patterns of life. For much of its history, though, China had been ruled by family lines of emperors. During this period, China developed one of the world's most advanced civilizations.

China's Early Dynasties

Dynasties ruled China for some 3,500 years. The major achievements of the early dynasties are shown here.

Shang, c. 1500–1050 BC

- First recorded Chinese dynasty
- Strongest in the Huang He valley
- Developed China's first writing system, a calendar, and chopsticks
- Skilled at bronze casting

Shang bronze tigress container

Zhou, c. 1050–400 BC

- Longest-lasting Chinese dynasty
- Expanded China but declined into a period of disorder
- Influenced by the new teachings of Confucianism, Daoism, and Legalism
- Began using iron tools and plows

Confucius, a Zhou thinker

China's Early History

When we enjoy the colorful fireworks on the Fourth of July, we can thank the early Chinese people. They invented fireworks. China's early civilization was one of the most advanced in the world. Its many achievements include the magnetic compass, gunpowder, paper, printing, and silk.

Today China can boast a civilization some 4,000 years old, older than any other. Understanding this long history is central to understanding China and its people.

China's Dynasties

For much of its history, China was ruled by dynasties. A **dynasty** is a series of rulers from the same family line. The rulers of China's dynasties were called emperors. Over time, many dynasties rose and fell in China. Between some dynasties, periods of chaos occurred as kingdoms or warlords fought for power. At other times, invaders came in and took control. Through it all, Chinese culture endured and evolved.

One of the most important dynasties is the Qin (CHIN), or Ch'in. It was the first dynasty to unite China under one empire. The greatest Qin ruler was Shi Huangdi (SHEE hwahng-dee). He ordered the building of much of the Great Wall of China.

Made to keep out invaders, the wall linked many older walls in northern China. In addition, Shi Huangdi had thousands of terra-cotta, or clay, warriors made to guard his tomb. These life-size warriors, each of which is unique, are skillful works of art.

The last dynasty in China was the Qing (CHING). Invaders called the Manchu ruled this dynasty starting in 1644. In time, outside influences would help lead to its end.

Outside Influences in China

Throughout history, China often limited contact with the outside world. The Chinese saw their culture as superior and had little use for foreigners. The tall mountains, deserts, and seas around China further limited contact and isolated the region.

Yet, other people increasingly wanted Chinese goods such as silk and tea. To gain access to the goods, some European powers forced China to open up trade in the 1800s. Europeans took control of parts of the country as well. These actions angered many Chinese, some of whom blamed the emperor. At the same time, increased contact with the West exposed the Chinese to new ideas.

READING CHECK **Drawing Conclusions** How did geography affect China's early history?

Qin, c. 221–206 BC

- First unified Chinese empire
- Strong central government with strict laws
- Created standardized money and writing systems
- Built a network of roads and canals and much of the Great Wall

Qin life-size terra-cotta warrior

Han, c. 206 BC–AD 220

- Based government on Confucianism
- Began trading over Silk Road
- Spread of Buddhism from India
- Invented paper, sundial, and acupuncture

Han bronze oil lamp

China's Modern History

As foreign influences increased, China's people grew unhappy with imperial rule. This unhappiness sparked a revolution.

Revolution and Civil War

In 1911, rebels forced out China's last emperor. They then formed a republic, a political system in which voters elect their leaders. Power struggles continued, however. In time, two rival groups emerged— the Nationalists, led by Chiang Kai-shek (chang ky-SHEK), and the Communists, led by Mao Zedong (MOW ZUH-DOOHNG).

The two groups fought a violent civil war. That war ended in October 1949 with the Communists as victors. They founded a new government, the People's Republic of China. The Nationalists fled to Taiwan, where they founded the Republic of China.

FOCUS ON READING

What is the main idea of the second paragraph under Population and Settlement on the next page?

Communist China under Mao

Mao, the Communists' leader, became the head of China's new government. In a Communist system, the government owns most businesses and land and controls all areas of life. China's new Communist government began by taking over control of the economy. The government seized all private farms and organized them into large, state-run farms. It also took over all businesses and factories.

While some changes improved life, others did not. On one hand, women gained more rights and were able to work. On the other hand, the government limited freedoms and imprisoned people who criticized it. In addition, many economic programs were unsuccessful, and some were outright disasters. In the early 1960s, for example, poor planning and drought led to a famine that killed millions.

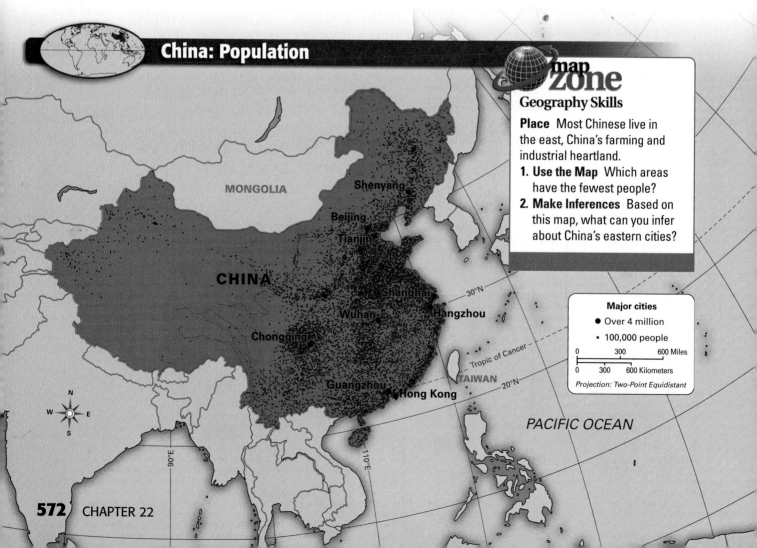

China: Population

map Zone

Geography Skills

Place Most Chinese live in the east, China's farming and industrial heartland.

1. **Use the Map** Which areas have the fewest people?
2. **Make Inferences** Based on this map, what can you infer about China's eastern cities?

Major cities
- Over 4 million
- 100,000 people

0   300   600 Miles
0   300   600 Kilometers

Projection: Two-Point Equidistant

MONGOLIA
Shenyang
Beijing
Tianjin
CHINA
Shanghai
Wuhan   Hangzhou
Chongqing
30°N
Guangzhou
Hong Kong
TAIWAN
20°N
Tropic of Cancer
PACIFIC OCEAN

Communist China Since Mao

Mao died in 1976, and Deng Xiaoping (DUHNG SHOW-PING) soon rose to power. Deng admitted the government had made mistakes. He then worked to modernize and improve China's economy. He allowed some private businesses and encouraged countries to invest in China. As a result, the economy began growing rapidly. Leaders after Deng continued economic reforms.

READING CHECK **Summarizing** How did communism change life in China?

China's People and Culture

One of China's best known features is its people—all 1.3 billion of them. China has the world's largest population. More people live there than in all of Europe, Russia, and the United States combined.

Population and Settlement

As the map shows, this huge population is not evenly spread out. Only 10 percent of the people live in the west, while the rest are jam-packed into the east. In fact, more people live in the Manchurian and North China Plains than in the United States!

Meanwhile, China's population continues to grow—by about 7.5 million each year. China's officials have worked to slow this growth. Officials have urged people to delay having children and have tried to limit each couple to one child. These actions have succeeded in slowing China's population growth.

Ethnic Groups and Language

Of China's millions of people, 92 percent identify their ancestry as Han Chinese. These people share the same culture and traditions. Many Han speak Mandarin, China's official language. Others speak a **dialect**, a regional version of a language.

Ethnic Groups

The majority of Chinese are Han. However, China includes 55 other ethnic groups. Most of these people live in western and southern China.

Han This woman and other Han make up about 92 percent of China's population. They share the same culture and traditions.

Hui This Hui man is from Gansu province, in central China. Most Hui are Sunni Muslims.

Zhuang This Zhuang man is from Guizhou, in southern China. The Zhuang are China's largest minority group.

Some 55 other ethnic groups make up the remaining 8 percent of China's population. Most of these minority groups live in western and southern China, where they have their own distinct cultures.

Religion, Values, and Beliefs

ACADEMIC VOCABULARY

values ideas that people hold dear and try to live by

elements parts of a whole

Ancient religions, **values**, and beliefs shape life for China's many people, even though the Communist government discourages the practice of religion. China's two main belief systems are Daoism (DOW-i-zuhm) and Buddhism. **Daoism** stresses living simply and in harmony with nature. It takes its name from the word *Dao*, which means "the way."

Buddhism came to China from India about AD 100. This religion is based on the teachings of Siddhartha Gautama—the Buddha, who lived from 563 to 483 BC. Buddhists believe moral behavior, kindness, and meditation can lead to peace.

Many Chinese blend **elements** of Daoism and Buddhism with **Confucianism**, a philosophy based on the ideas and teachings of Confucius. This philosophy stresses the importance of family, moral values, and respect for one's elders.

Other major religions in China include Christianity and Islam. Ancestor worship and fortune-telling are popular among the Chinese as well.

Close-up

Beijing's National Day

China celebrates National Day on October 1 with huge parades in Tiananmen Square. This square is one of the world's largest public gathering places. The space is needed because parades can include more than 500,000 participants.

Beijing
CHINA
PACIFIC OCEAN

The Gate of Heavenly Peace displays Mao Zedong's portrait above the entrance.

The parades include couples married on National Day, a popular time to wed.

A military parade of soldiers, tanks, and other equipment shows China's power.

The Arts and Popular Culture

China has a rich artistic tradition. Chinese crafts include items made of bronze, jade, ivory, silk, or wood. Chinese porcelain, which the ancient Chinese developed, is highly prized for its quality and beauty.

Traditional Chinese painting is done on silk or fine paper and reflects a focus on balance and harmony with nature. Popular subjects are landscapes, such as scenes of rugged mountains, trees, and lakes.

Chinese art often includes calligraphy, or decorative writing. Chinese writing uses symbols, or characters, instead of letters. This writing makes beautiful art, and some paintings feature just Chinese calligraphy.

In literature, the Chinese are known for their beautiful poetry. The Chinese highly value poetry, and poems appear on paintings and in novels and plays.

In theater, traditional Chinese opera is popular. These operas tell stories through spoken words, music, and dance. Actors wear elaborate costumes and makeup that have special meanings.

Traditional Chinese architecture features wooden buildings on stone bases. Large tiled roofs curve upward at the edge. Also common are **pagodas**, Buddhist temples that have multi-storied towers with an upward curving roof at each floor. Many cities are a mix of traditional and modern.

ANALYSIS SKILL **ANALYZING VISUALS**

Why might China's government include so many different groups in the National Day parades?

The Chinese believe dragon dances bring good fortune to important events.

世界人民大团结万岁

Lion dances are performed to spread good blessings to the community.

Chinese Martial Arts

Can you imagine getting up each day at 5 AM and exercising for 12 hours or more? Chinese teenagers who attend martial arts schools do just that. Many of the schools' instructors are Buddhist monks trained in the Chinese martial art of kung fu. These instructors teach their students self-defense techniques as well as the importance of hard work, discipline, and respect for one's elders. These values are important in Chinese culture and religion.

Starting as early as age 6, students memorize up to several hundred martial arts movements. These movements include different kicks, jumps, and punches. Some students dream of one day using their martial arts skills to star in a Chinese or American action movie.

Drawing Conclusions Why do you think discipline, hard work, and respect might be important for learning martial arts?

Popular culture includes many activities. Popular sports are martial arts and table tennis. A popular game is mah-jongg, played with small tiles. People also enjoy karaoke clubs, where they sing to music.

READING CHECK **Evaluating** Which aspect of Chinese culture most interests you? Why?

SUMMARY AND PREVIEW After centuries of imperial rule under dynasties, China became a Communist country. China has a rich and ancient culture and is the world's most populous country. In the next section you will read about China's economy, government, and cities.

Section 2 Assessment

go.hrw.com
Online Quiz
KEYWORD: SK7 HP22

Reviewing Ideas, Terms, and Places

1. **a. Define** What is a **dynasty**?
   b. Summarize How did outside influences affect China's early history?
2. **a. Recall** Which two groups fought for power during China's civil war, and which group won?
   b. Contrast How did China's economy under Mao differ from China's economy since his death?
3. **a. Recall** What are some popular pastimes in China today?
   b. Explain What are China's population problems, and how is China addressing them?
   c. Elaborate How are Buddhism, **Confucianism**, and **Daoism** important in Chinese culture?

Critical Thinking

4. **Sequencing** Look back over your notes and then create a chart like this one. List the main events in China's history in the order in which they occurred. Add or remove boxes as necessary.

FOCUS ON WRITING

5. **Collecting Information about China's History and Culture** Note historical or cultural details that you might want to include in your legend. For example, you might include some aspect of Chinese beliefs or artistic traditions in your legend.

China Today

If YOU lived there...

For many years your parents have been farmers, growing tea plants. Since the government began allowing private businesses, your parents have been selling tea in the market as well. With the money they have made, they are considering opening a tea shop.

What do you think your parents should do?

BUILDING BACKGROUND When a Communist government took over China in 1949, it began strictly controlling all areas of life. Over time, China's government has loosened control of the economy. Control over politics and other areas of life remains strict, however.

China's Economy

Think ahead to the day you start working. Would you rather choose your career or have the government choose it for you? The first situation describes a market economy, which we have in the United States. In this type of economy, people can choose their careers, decide what to make or sell, and keep the profits they earn. The second situation describes a **command economy,** an economic system in which the government owns all the businesses and makes all decisions, such as where people work. Communist China used to have a command economy. Then in the 1970s, China began allowing aspects of a market economy.

Farmers near Yunnan, in southern China, use traditional methods to work rice paddies.

What You Will Learn...

Main Ideas

1. China's booming economy is based on agriculture, but industry is growing rapidly.
2. China's government controls many aspects of life and limits political freedom.
3. China is mainly rural, but urban areas are growing.
4. China's environment faces a number of serious problems.

The Big Idea

China's economy and cities are growing rapidly, but the Chinese have little political freedom and many environmental problems.

Key Terms and Places

command economy, *p. 577*
Beijing, *p. 578*
Tibet, *p. 579*
Shanghai, *p. 580*
Hong Kong, *p. 580*

TAKING NOTES As you read, use a chart like the one below to take notes on China today.

577

China developed a mixed economy because it had major economic problems. For example, the production of goods had fallen. In response, the government closed many state-run factories and began allowing privately owned businesses. In addition, the government created special economic zones where foreign business-people could own companies. A mixed economic approach has helped China's economy boom. Today China has the world's second largest economy.

Agriculture and Industry

More Chinese work in farming than in any other economic activity. The country is a leading producer of several crops, such as rice, wheat, corn, and potatoes. China's main farmlands are in the eastern plains and river valleys. To the north, wheat is the main crop. To the south, rice is.

Only about 10 percent of China's land is good for farming. So how does China produce so much food? More than half of all Chinese workers are farmers. This large labor force can work the land at high levels. In addition, farmers cut terraces into hillsides to make the most use of the land.

Although China is mainly agricultural, industry is growing rapidly. Today China produces everything from satellites and chemicals to clothing and toys. Moreover, industry and manufacturing are now the most profitable part of China's economy.

Results of Economic Growth

Economic growth has improved wages and living standards in China. Almost all homes now have electricity, even in rural areas. More and more Chinese can afford goods such as TVs, computers, and even cars. At the same time, many rural Chinese remain poor, and unemployment is high.

READING CHECK Summarizing How has China changed its economy in recent times?

China's Government

More economic freedom in China has not led to more political freedom. The Communist government tightly controls most areas of life. For example, the government controls newspapers and Internet access, which helps to restrict the flow of information and ideas.

In addition, China harshly punishes people who oppose the government. In 1989 more than 100,000 pro-democracy protestors gathered in Tiananmen Square in **Beijing**, China's capital. The protestors were demanding more political rights and freedoms. The Chinese government tried to get the protestors to leave the square. When they refused, the government used troops and tanks to make them leave. Hundreds of protestors were killed, and many more were injured or imprisoned.

China's Urban Growth

China has taken harsh actions against ethnic rebellions as well. As an example, since 1950 China has controlled the Buddhist region of **Tibet**, in southwest China. When the Tibetans rebelled in 1959, the Chinese quickly crushed the revolt. The Dalai Lama (dah-ly LAH-muh), Tibet's Buddhist leader, had to flee to India. China then cracked down on Tibetans' rights.

Because of actions such as these, many other countries have accused China of not respecting human rights. Some of these countries have considered limiting or stopping trade with China. For example, some U.S. politicians want our government to limit trade with China until it shows more respect for human rights.

READING CHECK **Analyzing** What adjectives might you use to describe China's government?

Rural and Urban China

China is a land in the midst of change. Although its countryside remains set in the past, China's cities are growing rapidly and rushing headlong toward the future.

Rural China

Most of China's people live in small, rural villages. Farmers work the fields using the same methods they have used for decades. In small shops and along the streets, sellers cook food and offer goods. Although some villagers' standards of living are improving, the modern world often seems far away.

Urban China

Many people are leaving China's villages for its booming cities, however. The graph below shows how China's urban population is expected to rise in the future.

FOCUS ON READING
What is the implied main idea of the text under Rural China?

China's Projected Urban Population

— Urban population
— Rural population

go.hrw.com KEYWORD: SK7 CH22

Many of China's rapidly growing cities are severely crowded, as can be seen in this Shanghai shopping area. Overcrowding is expected to worsen as China's cities continue to grow.

INTERPRETING GRAPHS About when is China's urban population expected to be larger than its rural population?

CHINA, MONGOLIA, AND TAIWAN **579**

China's Environmental Challenges

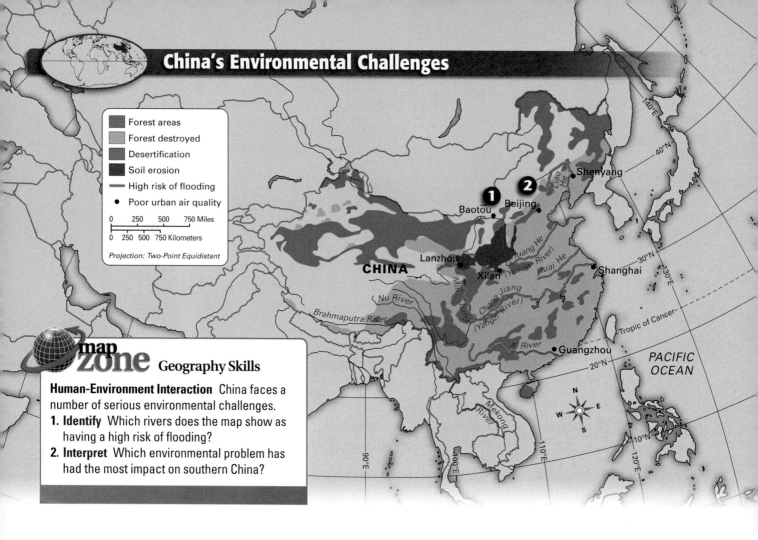

Legend:
- Forest areas
- Forest destroyed
- Desertification
- Soil erosion
- High risk of flooding
- Poor urban air quality

0   250   500   750 Miles
0 250 500 750 Kilometers

Projection: Two-Point Equidistant

CHINA

Cities and features labeled: Baotou, Beijing, Shenyang, Lanzhou, Xi'an, Shanghai, Guangzhou, Liao He, Huang He (Yellow River), Huai He, Chang Jiang (Yangzi River), Min Jiang, Nu River, Brahmaputra River, Xi River, Mekong River, PACIFIC OCEAN, Tropic of Cancer

map zone Geography Skills

Human-Environment Interaction China faces a number of serious environmental challenges.
1. **Identify** Which rivers does the map show as having a high risk of flooding?
2. **Interpret** Which environmental problem has had the most impact on southern China?

China's growing economy has led to its rapid city growth. Look at the population map in Section 2 and find the cities with more than 4 million people. Most are on the coast or along major rivers. These areas have benefited from growing industry and trade. Places that were rice fields not long ago are now bustling urban centers with skyscrapers, factories, and highways.

China's largest city is **Shanghai**, with some 13 million people. Located where the Chang Jiang meets the East China Sea, it is China's leading seaport and an industrial and commercial center. The city is also known for its European feel and nightlife.

China's second-largest city is its capital, Beijing. Also known as Peking, this historic city has many beautiful palaces and temples. A mix of the old and new, Beijing is China's political and cultural center.

In central Beijing, large walls hide the golden-roofed palaces of the Forbidden City, former home of China's emperors. Once off-limits to all but the emperor's household, the city is now a museum open to the public. Nearby, Tiananmen Square is the site of many parades and other public events. Government buildings and museums line this immense square.

In southern China, **Hong Kong** and Macao (muh-KOW) are major port cities and centers of trade and tourism. Both cities were European colonies until recently. The United Kingdom returned Hong Kong to China in 1997, and Portugal returned Macao in 1999. The two modern, crowded cities provide a mix of cultures.

READING CHECK **Contrasting** In what ways might rural life differ from city life in China?

1 Residents of Baotou, in north-central China, wear masks to keep from inhaling harmful particles in the city's polluted air.

2 These children are planting trees to help create new forestland north of Beijing.

China's Environment

China's economic and urban growth has created serious environmental problems. A major problem is pollution. The country's rising number of cars and factories pollute the air and water. At the same time, China burns coal for much of its electricity, which further pollutes the air.

Another serious problem is the loss of forestland and farmland. For centuries the Chinese cut down trees without replanting more. In addition, many of China's expanding cities are in its best farmlands.

The Chinese are working to address such problems. For example, China hopes to lessen pollution by using more hydroelectric power, electricity produced from dams. China is currently building the Three Gorges Dam on the Chang Jiang.

This dam is set to be finished in 2009. When completed, it will be the world's largest dam and generate as much power as 15 coal-burning power plants. On the other hand, the dam will drown hundreds of towns and huge amounts of farmland. Millions of people will have to move, and plant and animal habitats will be harmed.

READING CHECK **Finding Main Ideas** What are some of China's environmental problems?

SUMMARY AND PREVIEW China's economy and cities are growing rapidly, but its government restricts political freedom and faces environmental problems. In the next section you will learn about Mongolia and Taiwan.

go.hrw.com
Online Quiz
KEYWORD: SK7 HP22

Section 3 Assessment

Reviewing Ideas, Terms, and Places

1. **a. Define** What is a **command economy**?
   b. Identify Cause and Effect What changes have helped lead to China's rapid economic growth?
2. **a. Describe** In what ways does China's government restrict freedom?
   b. Evaluate What is your opinion of China's handling of the 1989 demonstration at Tiananmen Square?
3. **a. Identify** What is China's largest city and leading port?
   b. Compare How are **Hong Kong** and Macao similar?
4. **a. Recall** What are China's environmental problems?
   b. Evaluate Do you think China should build the Three Gorges Dam? Why or why not?

Critical Thinking

5. **Categorizing** Create a table like the one shown to organize the challenges that China faces today.

Challenges Facing China		
Economic	Political	Environmental

FOCUS ON WRITING

6. **Collecting Information about China Today** Note any details about China's current economy, government, cities, or environment that you might include in your legend.

Mongolia and Taiwan

If YOU lived there...

Like many Mongolians, you have loved horses since you were a small child. You live in an apartment in the city of Ulaanbaatar, however. Some of your family are talking about leaving the city and becoming nomadic herders like your ancestors were. You think you might like being able to ride horses more. You're not sure you would like living in a tent, though, especially in winter.

Do you want to move back to the land?

BUILDING BACKGROUND While Mongolia is a rugged land where some people still live as nomads, Taiwan is a modern and highly industrialized island. The two regions do have a few things in common, however. Mongolia and Taiwan are both neighbors of China, both are becoming more urban, and both are democracies.

Mongolia

A wild and rugged land, Mongolia is home to the Mongol people. They have a proud and fascinating history. This history includes conquests and empires and a culture that prizes horses.

Mongolia's History

Today when people discuss the world's leading countries, they do not mention Mongolia. However, 700 years ago Mongolia was perhaps the greatest power in the world. Led by the ruler Genghis Khan, the Mongols conquered much of Asia, including China. Later Mongol leaders continued the conquests. They built the greatest empire the world had seen at the time.

The Mongol Empire reached its height in the late 1200s. During that time, the empire stretched from Europe's Danube River in the west to the Pacific Ocean in the east. As time passed, however, the Mongol Empire declined. In the late 1600s China conquered Mongolia and ruled it for more than 200 years.

Nomadic Life in Mongolia

Some Mongolians are nomads, who live in tents called gers. Inside, gers are furnished mainly with rugs. Different areas of the gers are used for specific purposes. For example, the back is used for an altar.

ANALYZING VISUALS What do you think it is like to live in a ger?

Gers have wooden, painted doors. The doors always face south because the wind usually blows from the northeast.

With Russia's help, Mongolia declared independence from China in 1911. Soon Communists gained control and in 1924 formed the Mongolian People's Republic. Meanwhile, Russia had become part of the Soviet Union, a large Communist country north of Mongolia. The Soviet Union strongly influenced Mongolia and gave it large amounts of economic aid. This aid ended, however, after the Soviet Union collapsed in 1991. Since then, Mongolians have struggled to build a democratic government and a free-market economy.

Mongolia's Culture

In spite of years of Communist rule, the Mongolian way of life remains fairly traditional. Nearly half of Mongolia's people live as nomads. They herd livestock across Mongolia's vast grasslands and make their homes in **gers** (GUHRZ). These are large, circular, felt tents that are easy to put up, take down, and move.

Since many Mongols live as herders, horses play a major **role** in Mongolian life. As a result, Mongolian culture highly prizes horse skills, and Mongolian children often learn to ride when they are quite young.

Mongolia Today

Mongolia is sparsely populated. Slightly larger than Alaska, it has about 2.7 million people. More than a quarter of them live in **Ulaanbaatar** (oo-lahn-BAH-tawr), the capital and only large city. Mongolia's other cities are quite small. However, Mongolia's urban population is slowly growing.

The country's main industries include textiles, carpets, coal, copper, and oil. The city of Ulaanbaatar is the main industrial and commercial center. Mongolia produces little food other than from livestock, however, and faces food and water shortages.

READING CHECK **Summarizing** What are some features of Mongolian culture?

ACADEMIC VOCABULARY

role part or function

FOCUS ON READING

Read the second paragraph under Mongolia Today. Determine the topic of each sentence. What is the implied main idea?

Taiwan

When Portuguese sailors visited the island of Taiwan in the late 1500s, they called it *Ilha Formosa*, or "beautiful island." For many years, Westerners called Taiwan by the name Formosa. Today the loveliness of Taiwan's green mountains and waterfalls competes with its modern, crowded cities.

Taiwan's History

The Chinese began settling Taiwan in the 600s. At different times in history, both China and Japan have controlled Taiwan. In 1949, though, the Chinese Nationalists took over Taiwan. Led by Chiang Kai-shek, the Nationalists were fleeing the Communists, who had taken control of China's mainland. The Chinese Nationalist Party ruled Taiwan under martial law, or military rule, for 38 years. Today Taiwan's government is a multiparty democracy.

As the chart below explains, tensions remain between China and Taiwan. The Chinese government claims that Taiwan is a rebel part of China. In contrast, Taiwan's government claims to be the true government of China. For all practical purposes, though, Taiwan functions as an independent country.

Taiwan's Culture

Taiwan's history is reflected in its culture. Its population is about 85 percent native Taiwanese. These people are descendants of Chinese people who migrated to Taiwan largely in the 1700s and 1800s. As a result, Chinese ways dominate Taiwan's culture.

Other influences have shaped Taiwan's culture as well. Because Japan once ruled Taiwan, Japanese culture can be seen in some Taiwanese buildings and foods. More recently, European and American practices and customs are becoming noticeable in Taiwan, particularly in larger cities.

Taiwan Today

Taiwan is a modern country with a population of about 23 million. These people live on an island about the size of Delaware and Maryland combined. Because much of Taiwan is mountainous, most people live on the island's western coastal plain. This region is home to Taiwan's main cities.

The two largest cities are **Taipei** (TY-PAY) and **Kao-hsiung** (KOW-SHYOOHNG). Taipei, the capital, is Taiwan's main financial center. Because it has grown so quickly, it faces serious overcrowding and environmental problems. Kao-hsiung is a center of heavy industry and Taiwan's main seaport.

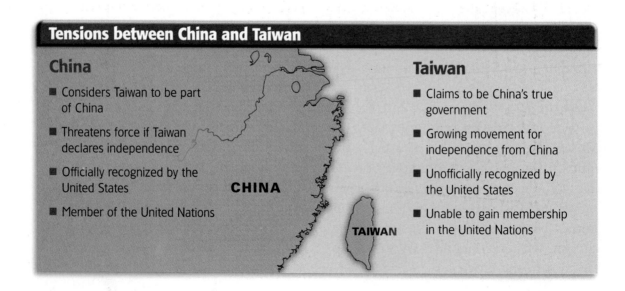

Tensions between China and Taiwan

China
- Considers Taiwan to be part of China
- Threatens force if Taiwan declares independence
- Officially recognized by the United States
- Member of the United Nations

CHINA

Taiwan
- Claims to be China's true government
- Growing movement for independence from China
- Unofficially recognized by the United States
- Unable to gain membership in the United Nations

TAIWAN

Taipei

Taipei, Taiwan's capital, is a bustling city of more than 2 million people. The tall tower in the photo is the Taipei 101, which is 101 stories tall.

Taiwan is one of Asia's richest and most industrialized countries. It is a leader in the production and export of computers and sports equipment. Taiwan's farmers grow many crops as well, such as sugarcane.

READING CHECK **Contrasting** How does Taiwan's economy differ from Mongolia's?

SUMMARY AND PREVIEW Mongolia and Taiwan are smaller countries bordering China. Mongolia is a wild land with a nomadic people who prize horses. In contrast, Taiwan is a modern and industrialized island. In the next chapter, you will learn about Japan and the Koreas.

Section 4 Assessment

Reviewing Ideas, Terms, and Places

1. **a. Define** What are **gers**, and what are their roles in Mongolia's culture?
   b. Make Inferences Why might many Mongolians be proud of their country's history?
   c. Elaborate Why does Mongolia's culture prize horses?
2. **a. Recall** Why is **Taipei** an important Taiwanese city, and what problems does the city face?
   b. Summarize What is the significance of Chiang Kai-shek in Taiwan's history?
   c. Evaluate Would you rather live in Taiwan or Mongolia? Provide information about each place to explain your answer.

Critical Thinking

3. **Comparing and Contrasting** Create a Venn diagram like the one shown. Use your notes and compare and contrast the histories, cultures, and societies of Mongolia and Taiwan.

Mongolia   Taiwan

FOCUS ON WRITING

4. **Collecting Information about Mongolia and Taiwan** Consider Mongolia and Taiwan as settings for your legend. For example, your legend might explain the creation of the Gobi, a large desert located partly in Mongolia.

CHINA, MONGOLIA, AND TAIWAN **585**

Social Studies Skills

Analyzing Points of View

Learn

Geography involves issues, situations where people disagree. The way people look at an issue is their point of view. To analyze points of view, use these tips:

- Consider a person's background. Think about where the person lives, what the person does, and what his or her beliefs and attitudes are.

- Look for emotional language, such as name calling or biased terms. Emotional language often reveals a person's point of view.

- Look at the evidence, or facts and statistics, to see what point of view they support.

- Put it all together to identify the point of view.

Practice

Read the passage below about a law forbidding any part of China to declare independence. Then answer the questions that follow.

❶ What is China's point of view about Taiwan?

❷ What is Taiwan's point of view about China?

New Law Angers Taiwan

Taiwan's government has warned that China's new anti-secession [anti-independence] law . . . will have a "serious impact" on security in the region. . .

Taiwan officials were quick to call the measure a "war bill," coming as China boosts its military spending by 13 percent to $30 billion. . .

But Chinese Premier Wen Jiabao said the new legislation [law] was not a "war bill" and warned outsiders not to get involved . . . "It is not targeted at the people of Taiwan, nor is it a war bill," Wen said at a news conference.

Source: *CNN International*, March 14, 2005

Consider background—China considers Taiwan a rebel province. Taiwan has a growing movement for independence.

Look for emotional language—The phrase "war bill" appeals to the emotions. People have strong feelings about war.

Look at the evidence—The information about military spending is evidence supporting one point of view.

Put it all together to identify each point of view.

Apply

1. In the passage above, how does each side's background affect its point of view?

2. Which point of view does the evidence about China's military spending support?

Chapter Review

Geography's Impact
video series
Review the video to answer the closing question:
If you were involved in the decision to build the Three Gorges Dam, would you support it or vote against it? Why?

Visual Summary

Use the visual summary below to help you review the main ideas of the chapter.

QUICK FACTS

China is a large Communist country with a rich culture. Both its economy and population are growing rapidly.

Mongolia lies to the north of China. It is a harsh, wild land. Many Mongolians are nomads who herd livestock.

Taiwan is an island off the southern coast of China. It is a modern and industrialized region.

Reviewing Vocabulary, Terms, and Places

Match the words or places below with their definitions or descriptions.

1. command economy

2. North China Plain

3. pagodas

4. gers

5. Tibet

6. dialect

7. Himalayas

8. Taipei

a. Buddhist region in southwest China

b. world's highest mountain range

c. regional version of a language

d. capital city of Taiwan

e. system in which the government owns most businesses and makes most economic decisions

f. fertile and highly populated region in eastern China

g. circular, felt tents in which Mongol nomads live

h. Buddhist temples with multiple stories

Comprehension and Critical Thinking

SECTION 1 *(Pages 566–569)*

9. a. Recall What physical features separate many of the mountain ranges in this region?

b. Explain What is the Huang He called in English, and how did the river get its name?

c. Elaborate What major physical features might a traveler see during a trip from the Himalayas, in southwestern China, to Beijing, in northeastern China?

SECTION 2 *(Pages 570–576)*

10. a. Identify Who is Mao Zedong, and why is he significant in China's history?

b. Summarize What are some of China's artistic traditions, and how have they contributed to world culture?

c. Predict What future challenges do you think China might face if its population continues to grow at its current rate?

SECTION 3 *(Pages 577–581)*

11. a. Recall What do more than half of China's workers do for a living?

b. Summarize What elements of free enterprise does China's command economy now include?

c. Evaluate What is your opinion about China's treatment of Tibet?

SECTION 4 *(Pages 582–585)*

12. a. Identify What is the capital of Mongolia?

b. Analyze How is Taiwan's history reflected in the island's culture today?

c. Predict Do you think China and Taiwan can resolve their disagreements? Why or why not?

Using the Internet

go.hrw.com
KEYWORD: SK7 CH22

13. Activity: Touring China's Great Wall The construction of the Great Wall of China began more than 2,000 years ago. The wall was built over time to keep out invaders and to protect China's people. Enter the activity keyword to explore this wonder of the world. Take notes on the wall's history, myths and legends, and other interesting facts. Then make a brochure about your virtual visit to the Great Wall of China.

Social Studies Skills

Analyzing Points of View *Read the following passage from this chapter. Then answer the questions below.*

*"*In 1989 more than 100,000 pro-democracy protestors gathered in Tiananmen Square in Beijing, China's capital. The protestors were demanding more political rights and freedoms. The Chinese government tried to get the protestors to leave the square. When they refused, the government used troops and tanks to make them leave. Hundreds of protestors were injured or killed.*"*

14. What was the point of view of the protestors toward China's government?

15. What was the point of view of China's government toward the protestors?

---

16. Identifying Implied Main Ideas Read the first paragraph under the heading Revolution and Civil War in Section 2. What is the implied main idea of this paragraph? What words and phrases help signal the implied main idea?

17. Writing a Legend Choose one physical feature and decide how you will explain its creation. Then review your notes and choose characters, events, and settings for your legend. Your legend should be two to three paragraphs. It should include (a) a beginning; (b) a middle that includes a climax, or high point of the story; and (c) a conclusion, or end. Remember, legends tell about extraordinary events, so you should use your imagination and creativity.

Map Activity ★Interactive

18. China, Mongolia, and Taiwan On a separate sheet of paper, match the letters on the map with their correct labels below.

Beijing, China	Hong Kong, China
Chang Jiang	Huang He
Great Wall of China	Taipei, Taiwan
Himalayas	Ulaanbaatar, Mongolia

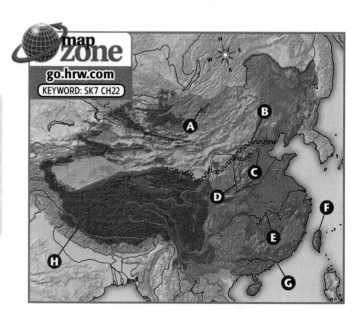

map zone
go.hrw.com
KEYWORD: SK7 CH22

DIRECTIONS: Read questions 1 through 7 and write the letter of the best response. Then read question 8 and write your own well-constructed response.

1 **What is the world's highest mountain range?**

   A Himalayas

   B Kunlun Shan

   C Tian Shan

   D Qinling Shandi

2 **Why is China's Qin dynasty significant?**

   A first recorded dynasty in China

   B longest-lasting dynasty in China

   C first dynasty to unify China

   D first dynasty to practice Buddhism

3 **In which area do most people in China live?**

   A west

   B east

   C south

   D north

4 **Which of these challenges faces China?**

   A slow population growth

   B a weak economy

   C lack of urban growth

   D air and water pollution

5 **Which phrase *best* describes Taiwan?**

   A a nomadic culture that prizes horses

   B modern and industrialized cities

   C strict government and few political freedoms

   D mainly rural and agricultural

6 **Who was a great ruler in Mongolian history?**

   A Genghis Khan

   B Chiang Kai-shek

   C Mao Zedong

   D Shi Huangdi

China, Mongolia, and Taiwan: Precipitation

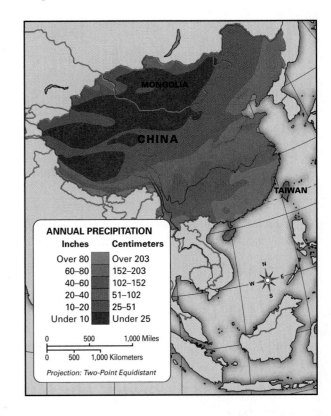

ANNUAL PRECIPITATION

Inches	Centimeters
Over 80	Over 203
60–80	152–203
40–60	102–152
20–40	51–102
10–20	25–51
Under 10	Under 25

0   500   1,000 Miles

0   500   1,000 Kilometers

Projection: Two-Point Equidistant

7 **Based on the map, which statement best describes precipitation across this region?**

   A increases from east to west

   B decreases from north to south

   C decreases from the southeast to the northwest

   D increases from the southeast to the northwest

8 **Extended Response** Look at the map titled China's Environmental Challenges in Section 3. Write two to three paragraphs explaining why the Chinese government should take action to address environmental problems. Make certain to include a description of the ways in which each problem affects China.

CHAPTER 23

Japan and the Koreas

What You Will Learn...

In this chapter you will learn about three countries—Japan, South Korea, and North Korea. Although the three share some physical features and have intertwined histories, they are all very different today. Both Japan and South Korea are democratic countries with prosperous economies, while North Korea is a Communist dictatorship faced with economic hardships.

SECTION 1
Physical Geography 592

SECTION 2
History and Culture.................597

SECTION 3
Japan Today 602

SECTION 4
The Koreas Today 608

FOCUS ON READING AND WRITING

Understanding Fact and Opinion A fact is a statement that can be proved true. An opinion is someone's belief about something. When you read a textbook, you need to recognize the difference between facts and opinions. **See the lesson, Understanding Fact and Opinion, on page 690.**

Composing a Five-Line Poem For centuries, Japanese poets have written five-line poems called tankas. Most tanka poems describe a single image or emotion in very few words. After you read this chapter, you will choose one image of Japan or the Koreas to use as the subject of your own five-line poem.

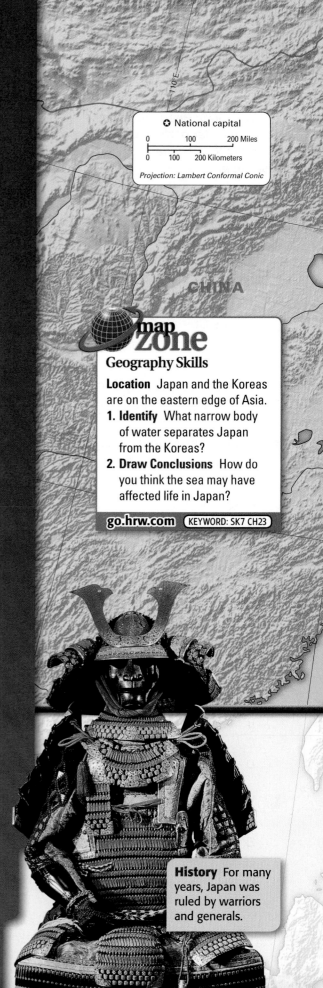

❂ National capital

| 0 | 100 | 200 Miles |
| 0 | 100 | 200 Kilometers |

Projection: Lambert Conformal Conic

CHINA

map Zone

Geography Skills

Location Japan and the Koreas are on the eastern edge of Asia.
1. **Identify** What narrow body of water separates Japan from the Koreas?
2. **Draw Conclusions** How do you think the sea may have affected life in Japan?

go.hrw.com KEYWORD: SK7 CH23

History For many years, Japan was ruled by warriors and generals.

RUSSIA

Sea of Okhotsk

Sapporo

Tumen R.

Yalu River

NORTH KOREA

Pyongyang

Nampo

Sea of Japan (East Sea)

Seoul

Inchon

SOUTH KOREA

Taegu

Yellow Sea

Pusan

Korea Strait

Hiroshima

JAPAN

Kyoto

Osaka

Nagoya

Tokyo

Yokohama

PACIFIC OCEAN

Nagasaki

East China Sea

30°N

Philippine Sea

HOLT

Geography's Impact
video series
Watch the video to understand the impact of natural hazards.

Geography Mount Fuji, a common symbol of Japan, is one of the thousands of mountains found in the region.

Culture Under Kim Il Sung, North Korea became a Communist country.

591

SECTION 1

Physical Geography

What You Will Learn...

Main Ideas

1. The main physical features of Japan and the Koreas are rugged mountains.
2. The climates and resources of Japan and the Koreas vary from north to south.

The Big Idea

Japan and Korea are both rugged, mountainous areas surrounded by water.

Key Terms and Places

Fuji, *p. 593*
Korean Peninsula, *p. 593*
tsunamis, *p. 594*
fishery, *p. 595*

TAKING NOTES Draw a table like the one below. As you read, take notes about the physical geography of Japan in one column and about the Korean Peninsula in the other column.

Physical Geography	
Japan	Korean Peninsula

If YOU lived there...

You are a passenger on a very fast train zipping its way across the countryside. If you look out the window to your right, you can see the distant sparkle of sunlight on the ocean. If you look to the left, you see rocky, rugged mountains. Suddenly the train leaves the mountains, and you see hundreds of trees covered in delicate pink flowers. Rising above the trees is a single snowcapped volcano.

How does this scenery make you feel?

BUILDING BACKGROUND The train described above is one of the many that cross the islands of Japan every day. Japan's mountains, trees, and water features give the islands a unique character. Not far away, the Korean Peninsula also has a distinctive landscape.

Physical Features

Japan, North Korea, and South Korea are on the eastern edge of the Asian continent, just east of China. Separated from each other only by a narrow strait, Japan and the Koreas share many common landscape features.

Physical Features of Japan

Japan is an island country. It is made up of four large islands and more than 3,000 smaller islands. These islands are arranged in a long chain more than 1,500 miles (2,400 km) long. This is about the same length as the eastern coast of the United States, from southern Florida to northern Maine. All together, however, Japan's land area is slightly smaller than the state of California.

About 95 percent of Japan's land area is made up of four large islands. From north to south, these major islands are Hokkaido (hoh-KY-doh), Honshu (HAWN-shoo), Shikoku (shee-KOH-koo), and Kyushu (KYOO-shoo). Together they are called the home islands. Most of Japan's people live there.

Rugged, tree-covered mountains are a common sight in Japan. In fact, mountains cover some 75 percent of the country. For the most part, Japan's mountains are very steep and rocky. As a result, the country's largest mountain range, the Japanese Alps, is popular with climbers and skiers.

Japan's highest mountain, **Fuji**, is not part of the Alps. In fact, it is not part of any mountain range. A volcano, Mount Fuji rises high above a relatively flat area in eastern Honshu. The mountain's cone-shaped peak has become a symbol of Japan. In addition, many Japanese consider Fuji a sacred place. As a result, many shrines have been built at its foot and summit.

Physical Features of Korea

Jutting south from the Asian mainland, the **Korean Peninsula** includes both North Korea and South Korea. Like the islands of Japan, much of the peninsula is covered with rugged mountains. These mountains form long ranges that run along Korea's eastern coast. The peninsula's highest mountains are in the north.

Unlike Japan, Korea also has some large plains. These plains are found mainly along the peninsula's western coast and in river valleys. Korea also has more rivers than Japan does. Most of these rivers flow westward across the peninsula and pour into the Yellow Sea.

FOCUS ON READING
Are these sentences facts or opinions? How can you tell?

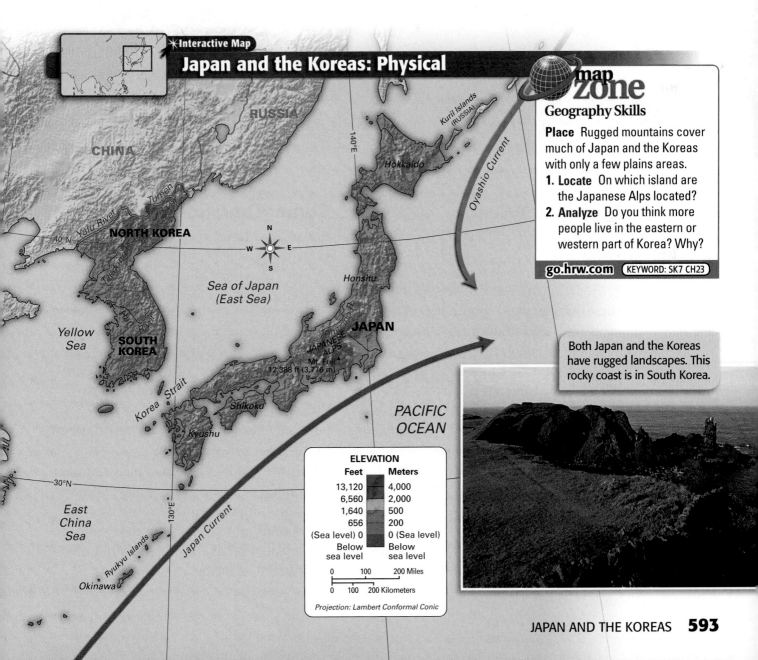

★Interactive Map

Japan and the Koreas: Physical

map zone

Geography Skills

Place Rugged mountains cover much of Japan and the Koreas with only a few plains areas.

1. **Locate** On which island are the Japanese Alps located?
2. **Analyze** Do you think more people live in the eastern or western part of Korea? Why?

go.hrw.com KEYWORD: SK7 CH23

Both Japan and the Koreas have rugged landscapes. This rocky coast is in South Korea.

ELEVATION

Feet		Meters
13,120		4,000
6,560		2,000
1,640		500
656		200
(Sea level) 0		0 (Sea level)
Below sea level		Below sea level

0   100   200 Miles
0   100   200 Kilometers

Projection: Lambert Conformal Conic

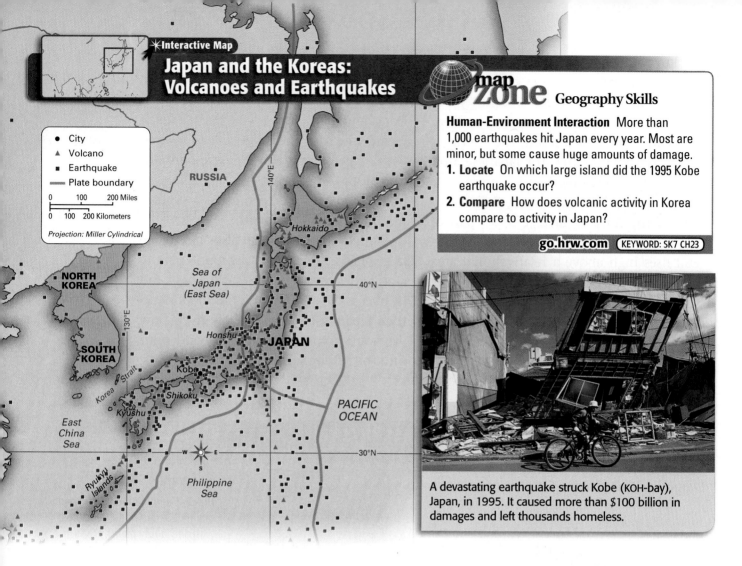

Japan and the Koreas: Volcanoes and Earthquakes

map **zone** Geography Skills

Human-Environment Interaction More than 1,000 earthquakes hit Japan every year. Most are minor, but some cause huge amounts of damage.

1. Locate On which large island did the 1995 Kobe earthquake occur?

2. Compare How does volcanic activity in Korea compare to activity in Japan?

go.hrw.com KEYWORD: SK7 CH23

- City
- Volcano
- Earthquake
- Plate boundary

0 100 200 Miles
0 100 200 Kilometers

Projection: Miller Cylindrical

RUSSIA

140°E

Hokkaido

NORTH KOREA

Sea of Japan (East Sea)

40°N

130°E

Honshu

JAPAN

SOUTH KOREA

Kobe

Korea Strait

Shikoku

Kyushu

East China Sea

PACIFIC OCEAN

30°N

Ryukyu Islands

Philippine Sea

A devastating earthquake struck Kobe (KOH-bay), Japan, in 1995. It caused more than $100 billion in damages and left thousands homeless.

Natural Disasters

Because of its location, Japan is subject to many sorts of natural disasters. Among these disasters are volcanic eruptions and earthquakes. As you can see on the map, these disasters are common in Japan. They can cause huge amounts of damage in the country. In addition, large underwater earthquakes sometimes cause destructive waves called **tsunamis** (sooh-NAH-mees).

Korea does not have many volcanoes or earthquakes. From time to time, though, huge storms called typhoons sweep over the peninsula from the Pacific. These storms cause great damage in both the Korean Peninsula and Japan.

READING CHECK Contrasting How are the physical features of Japan and Korea different?

Climate and Resources

Just as Japan and the Koreas have many similar physical features, they also have similar climates. The resources found in each country, however, differ greatly.

Climate

The climates of Japan and the Koreas vary from north to south. The northern parts of the region have a humid continental climate. This means that summers are cool, but winters are long and cold. In addition, the area has a short growing season.

To the south, the region has a humid subtropical climate with mild winters and hot, humid summers. These areas see heavy rains and typhoons in the summer. Some places receive up to 80 inches (200 cm) of rain each year.

Resources

Resources are not evenly distributed among Japan and the Koreas. Neither Japan nor South Korea, for example, is very rich in mineral resources. North Korea, on the other hand, has large deposits of coal, iron, and other minerals.

Although most of the region does not have many mineral resources, it does have other resources. For example, the people of the Koreas have used their land's features to generate electricity. The peninsula's rocky terrain and rapidly flowing rivers make it an excellent location for creating hydroelectric power.

In addition, Japan has one of the world's strongest fishing economies. The islands lie near one of the world's most productive fisheries. A **fishery** is a place where lots of fish and other seafood can be caught. Swift ocean currents near Japan carry countless fish to the islands. Fishers then use huge nets to catch the fish and bring them to Japan's many bustling fish markets. These fish markets are among the busiest in the world.

READING CHECK **Analyzing** What are some resources found in Japan and the Koreas?

This fish market in Tokyo, Japan, is the busiest in the world. People gather here every morning to buy freshly caught fish.

SUMMARY AND PREVIEW The islands of Japan and the Korean Peninsula share many common features. In the next section, you will see how the people of Japan and Korea also share some similar customs and how their histories have been intertwined for centuries.

Section 1 Assessment

Reviewing Ideas, Terms, and Places

1. **a. Identify** What types of landforms cover Japan and the **Korean Peninsula**?
   b. Compare and Contrast How are the physical features of Japan and Korea similar? How are they different?
   c. Predict How do you think natural disasters affect life in Japan and Korea?
2. **a. Describe** What kind of climate is found in the northern parts of the region? What kind of climate is found in the southern parts?
   b. Draw Conclusions Why are **fisheries** important to Japan's economy?

Critical Thinking

3. **Categorizing** Draw a chart like this one. In each row, describe the region's landforms, climate, and resources.

	Japan	Korean Peninsula
Landforms		
Climate		
Resources		

FOCUS ON WRITING

4. **Thinking about Nature** Many Japanese poems deal with nature—the beauty of a flower, for example. What could you write about the region's physical environment in your poem?

Using a Topographic Map

Learn

Topographic maps show elevation, or the height of land above sea level. They do so with contour lines, lines that connect points on the map that have equal elevation. Every point on a contour line has the same elevation. In most cases, everything inside that line has a higher elevation. Everything outside the line is lower. Each contour line is labeled to show the elevation it indicates.

An area that has lots of contour lines is more rugged than an area with few contour lines. The distance between contour lines shows how steep an area is. If the lines are very close together, then the area has a steep slope. If the lines are farther apart, then the area has a much gentler incline. Other symbols on the map show features such as rivers and roads.

Practice

Use the topographic map on this page to answer the following questions.

1. Is Awaji Island more rugged in the south or the north? How can you tell?

2. Does the land get higher or lower as you travel west from Yura?

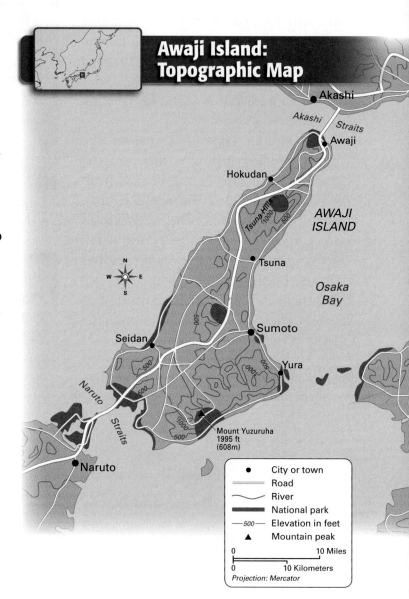

Awaji Island: Topographic Map

Akashi

Akashi Straits

Awaji

Hokudan

Tsuna Hills 1000 500

AWAJI ISLAND

Tsuna

Osaka Bay

500

Seidan

1000

Sumoto

Yura

Naruto Straits

500

1000

500

Naruto

Mount Yuzuruha 1995 ft (608m)

- • City or town
- ═══ Road
- ～ River
- ▬ National park
- —500— Elevation in feet
- ▲ Mountain peak

0 ———— 10 Miles
0 ———— 10 Kilometers
Projection: Mercator

Apply

Search the Internet or look in a local library to find a topographic map of your area. Study the map to find three major landmarks and write down their elevations. Then write two statements about the information you can see on the map.

History and Culture

If YOU lived there...

You live in Kyoto, one of the most beautiful cities in Japan. Your class is visiting a museum to see an amazing demonstration by a sword maker. You all stare in amazement as he hammers red-hot metal into a curved sword, then plunges it into cold water. He tells you that his family has been making swords for 300 years.

What kind of craft would you like to know?

BUILDING BACKGROUND Even though Japan is an industrial nation, the Japanese still respect and admire traditional arts and crafts and the people who make them, such as this sword maker. In fact, traditions continue to shape life in Japan and the Koreas today.

History

Both Japan and the Koreas have very long histories. Early in these histories, their cultures were intertwined. As time passed, though, Japan and the Koreas developed very differently.

Early History

Early in their histories, both Japan and the Koreas were influenced by China. Since the Korean Peninsula borders China, and Japan lies just across the sea, elements of Chinese culture seeped into both places.

Among the elements of Chinese culture that influenced Japan and Korea was Buddhism. Scholars and missionaries first brought Buddhism into Korea. From there, visitors carried it to Japan. Before long, Buddhism was the main religion in both countries.

Japanese Buddha statue

What You Will Learn...

Main Ideas

1. The early histories of Japan and Korea were closely linked, but the countries developed very differently.
2. Japanese culture blends traditional customs with modern innovations.
3. Though they share a common culture, life is very different in North and South Korea.

The Big Idea

History and tradition are very important to the people of Japan and the Koreas.

Key Terms and Places

Kyoto, *p. 598*
shoguns, *p. 598*
samurai, *p. 598*
kimonos, *p. 600*
kimchi, *p. 601*

TAKING NOTES Draw two boxes like the ones shown here. In one box, take notes about the history and culture of Japan. In the other, take notes about the history and culture of the Koreas.

History and Culture	
Japan	The Koreas

Emperors, Shoguns, and Samurai

The first central government in Japan was based on China's government. For many centuries, emperors ruled in Japan just as they did in China. The imperial capital at Heian, now called **Kyoto**, was a center of art, literature, and learning. At times, some of Japan's emperors were more concerned with art than with running the country. Eventually, their power slipped away.

As the emperors' power faded, Japan fell under the control of military leaders called shoguns. Powerful generals, the shoguns ruled Japan in the emperor's name. Only one shogun could hold power at a time.

Serving under the shogun were armies of samurai, or highly trained warriors. They were fierce in battle and devoted to their leaders. As a result, the samurai were very respected in Japanese society. With their support, the shoguns continued to rule Japan well into the 1800s.

BIOGRAPHY

Hirohito
(1901–1989)

Hirohito was Japan's emperor for most of the 1900s. As such, he led the country through periods of great crisis and change. He was emperor when Japan launched wars against China and Russia in the 1930s. He was also in power in 1945 when the United States bombed Hiroshima and Nagasaki. After World War II ended, Hirohito led Japan through changes in its government and economy. Many of these changes affected Hirohito personally. For example, he gave up much of the power he had once held as emperor in favor of a democratic government.

Drawing Conclusions Why might a ruler give up much of his power?

Later Japan

Not everyone was happy with the rule of the shoguns. In 1868 a group of samurai overthrew the shogun and gave power back to the emperor.

When World War II began, Japan allied itself with Germany and Italy. It wanted to build an empire in Southeast Asia and the Pacific. The Japanese drew the United States into the war in 1941 when they bombed the naval base at Pearl Harbor, Hawaii. After many years of fighting, the Americans took drastic measures to end the war. They dropped devastating atomic bombs on two Japanese cities, Hiroshima and Nagasaki. Shocked by these terrible weapons, the Japanese surrendered.

Korea

Although Japan was influenced by China, it remained independent. The Koreas, on the other hand, did not. For centuries, the Koreas were ruled by China. Later, the Japanese invaded the Korean Peninsula. They were harsh rulers, and the Korean people grew to resent the Japanese.

After World War II, Korea was taken away from Japan and once again made independent. Rather than forming one country, though, the Koreans formed two. Aided by the Soviet Union, North Korea created a Communist government. In South Korea, the United States helped build a democratic government.

In 1950 North Korea invaded South Korea, starting the Korean War. The North Koreans wanted to unify all of Korea under a Communist government. With the aid of many other countries, including the United States, the South Koreans drove the invaders back. The Korean War was costly, and its effects linger in the Koreas today.

READING CHECK Analyzing How did the Koreas change after the Korean War?

A Buddhist Temple

Buddhism has been a major religion in Japan for centuries. This temple in Nara was built around the year 600.

Japanese Culture

Japan's culture reflects the country's long and varied history. For example, some elements of the culture reflect the influence of the Chinese, while others are native to Japan. Since World War II, Western ideas and innovations have also helped shape Japanese life.

Language

Nearly everyone in Japan speaks Japanese. The Japanese language is complicated and can be difficult for other people to learn. This difficulty stems in large part from the Japanese writing system. Japanese writing uses two different types of characters. Some characters, called kanji, represent whole words. There are about 2,000 kanji characters in common use today. Other characters, called kana, stand for parts of words. Most texts written in Japanese use both kanji and kana characters.

Religion

Religion can also be complicated in Japan. Most people who live there blend elements of two religions—Shinto and Buddhism.

Unlike Buddhism, which was brought to Japan from Korea, Shinto is native to the islands. According to Shinto teachings, nature spirits called *kami* (KAH-mee) live in the world. Shintoists believe everything in nature—the sun, the moon, trees, rocks, waterfalls, and animals—has *kami*. They also believe that some *kami* help people live and keep them from harm. As a result, they build shrines to the *kami* and perform ceremonies to ask for their blessings.

Buddhists have also built shrines and temples all over Japan. Some temples, like the one pictured above, are very old. They date back to the earliest days of Buddhism in Japan. People visit these temples to seek peace and enlightenment. The search for enlightenment is Buddhists' main goal.

Traditional Dress

These Korean dancers are wearing traditional costumes to perform a fan dance. Most of the time, people in both South Korea and North Korea wear Western-style clothes.

Customs and Traditions

Japan's history lives on in its customs and traditions. For example, many Japanese wear traditional robes called kimonos on special occasions, just as samurai did long ago. Most of the time, though, people in Japan wear Western-style clothing.

Traditional forms of art are also still popular in Japan. Among these art forms are two types of drama, Noh and Kabuki. Noh plays use music and dance to tell a story. Actors do not move much and wear masks, using their gestures to convey their tale. Kabuki actors, on the other hand, are much more active. Kabuki plays tell stories, but they often teach lessons about duty and other **abstract** ideas as well.

READING CHECK Summarizing How did Japan's history affect its culture today?

ACADEMIC VOCABULARY

abstract expressing a quality or idea without reference to an actual thing

Korean Culture

Like Japan's, Korea's culture reflects the peninsula's long history. Traditional ways of life influence how people act and think.

Language and Religion

People in both North Korea and South Korea speak Korean. Unlike Japanese, Korean is written with an alphabet. People combine letters to form words, rather than using symbols to represent entire words or syllables as in Japanese.

In the past, most people in Korea were Buddhists and Confucianists. Recently, though, Christianity has also become widespread. About one-fourth of South Korea's people are Christian. North Korea, like many Communist countries, discourages people from practicing any religion.

Kimonos are the traditional clothing style in Japan. Both men and women wear kimonos for special occasions, such as weddings.

Customs and Traditions

Like the Japanese, the people of Korea have kept many ancient traditions alive. Many Korean foods, for example, have been part of the Korean diet for centuries.

One example of a long-lasting Korean food is **kimchi, a dish made from pickled cabbage and various spices.** First created in the 1100s, kimchi is still served at many Korean meals. In fact, many people think of it as Korea's national dish.

Traditional art forms have also remained popular in parts of the Koreas. This is true especially in North Korea. Since World War II, the Communist government of North Korea has encouraged people to retain many of their old customs and traditions. The Communists think that Korean culture is the best in the world and do everything they can to preserve it.

In South Korea, though, some customs have been lost. Many people there have moved into cities and come into contact with people from other countries. As a result, people have adopted new ways of life. Many of these ways are combinations of old and new ideas. For example, Korean art today combines traditional themes such as nature with modern forms, like film.

READING CHECK **Contrasting** How are North and South Korea's cultures different?

SUMMARY AND PREVIEW In this section, you learned that the cultures of Japan and the Koreas have been shaped by the countries' histories. In the next section you will see how traditional cultures continue to influnce life in Japan today.

go.hrw.com
Online Quiz
KEYWORD: SK7 HP23

Section 2 Assessment

Reviewing Ideas, Terms, and Places

1. a. **Define** Who were **shoguns**?
   b. **Elaborate** How did World War II affect life in Japan?
2. a. **Identify** What is one traditional style of clothing in Japan? What do people wear most of the time?
   b. **Elaborate** How does Japan's religion reflect its history?
3. a. **Recall** What is **kimchi**? Why is it important in Korea?
   b. **Explain** What has led to many of the differences between modern culture in North and South Korea?

Critical Thinking

4. **Analyzing** Draw a diagram like this one. Using your notes, list two features of Japanese culture in the left box and of Korean culture in the right box. Below each box, write a sentence about how each country's culture reflects its history.

Japanese Culture	Korean Culture

FOCUS ON WRITING

5. **Analyzing Cultures** Traditions and customs are central to life in Japan and the Koreas. How can you reflect this importance in your poem?

Japan Today

What You Will Learn...

Main Ideas

1. Since World War II, Japan has developed a democratic government and one of the world's strongest economies.
2. A shortage of open space shapes daily life in Japan.
3. Crowding, competition, and pollution are among Japan's main issues and challenges.

The Big Idea

Japan has overcome many challenges to become one of the most highly developed countries in Asia.

Key Terms and Places

Diet, *p. 602*
Tokyo, *p. 602*
work ethic, *p. 603*
trade surplus, *p. 603*
tariff, *p. 603*
Osaka, *p. 606*

TAKING NOTES Draw a diagram like the one below. As you read, take notes about Japan's government, economy, and daily life in the appropriate ovals.

If YOU lived there...

You and your family live in a small apartment in the crowded city of Tokyo. Every day you and your friends crowd into jammed subway trains to travel to school. Since your work in school is very hard and demanding, you really look forward to weekends. You especially like to visit mountain parks where there are flowering trees, quiet gardens, and ancient shrines.

Do you like your life in Tokyo? Why or why not?

BUILDING BACKGROUND Although Japan has become an economic powerhouse, it is still a small country in area. Its cities have become more and more crowded with high-rise office and apartment buildings. Most people live in these cities today, though many feel a special fondness for natural areas like mountains and lakes.

Government and Economy

Do you own any products made by Sony? Have you seen ads for vehicles made by Honda, Toyota, or Mitsubishi? Chances are good that you have. These companies are some of the most successful in the world, and all of them are Japanese.

Since World War II, Japan's government and economy have changed dramatically. Japan was once an imperial state that was shut off from the rest of the world. Today Japan is a democracy with one of the world's strongest economies.

Government

Since the end of World War II, Japan's government has been a constitutional monarchy headed by an emperor. Although the emperor is officially the head of state, he has little power. His main role is to act as a symbol of Japan and of the Japanese people. In his place, power rests in an elected legislature called the **Diet** and in an elected prime minister. From the capital city of **Tokyo**, the Diet and the prime minister make the laws that govern life in Japan today.

Economy

Today Japan is an economic powerhouse. However, this was not always the case. Until the 1950s, Japan's economy was not that strong. Within a few decades, though, the economy grew tremendously.

The most successful area of Japan's economy is manufacturing. Japanese companies are known for making high-quality products, especially cars and electronics. Japanese companies are among the world's leading manufacturers of televisions, DVD players, CD players, and other electronic items. The methods that companies use to make these products are also celebrated. Many Japanese companies are leaders in new technology and ideas.

Reasons for Success Many factors have contributed to Japan's economic success. One factor is the government. It works closely with business leaders to control production and plan for the future.

Japan's workforce also contributed to its success. Japan has well-educated, highly trained workers. As a result, its companies tend to be both efficient and productive. Most workers in Japan also have a strong work ethic. A **work ethic** is the belief that work in itself is worthwhile. Because of their work ethic, most Japanese work hard and are loyal to their companies. As a result, the companies are successful.

Trade Japan's economy depends on trade. In fact, many products manufactured in the country are intended to be sold outside of Japan. Many of these goods are sent to China and the United States. The United States is Japan's major trading partner.

Japan's trade has been so successful that it has built up a huge trade surplus. A **trade surplus** exists when a country exports more goods than it imports. Because of this surplus, many Japanese companies have become very wealthy.

CONNECTING TO Technology

Building Small

The Japanese are known as masters of technology. Companies use this technology in many ways to create new products and improve existing ones. One way many Japanese companies have sought to improve their products—especially personal electronics products—is by making them smaller.

Since Sony released the first personal stereo system in the late 1970s, making small products has been a major business in Japan. Now shoppers can buy tiny radios, video games, cell phones, and cameras. Some of these products are smaller than the palm of your hand.

Generalizing Why might people want to buy small versions of products?

Japan is able to export more than it imports in part because of high tariffs. A **tariff** is a fee that a country charges on imports or exports. For many years, Japan's government has placed high tariffs on goods brought into the country. This makes imported goods more expensive, and so people buy Japanese goods rather than imported ones.

Resources Although its economy is based on manufacturing, Japan has few natural resources. As a result, the country must import raw materials. In addition, Japan has little arable land. Farms cannot grow enough food for the country's growing population. Instead, the Japanese have to buy food from other countries, including China and the United States.

READING CHECK Summarizing What have the Japanese done to build their economy?

Daily Life

Japan is a densely populated country. Slightly smaller than California, it has nearly four times as many people! Most of these people live in crowded cities such as the capital, Tokyo.

Life in Tokyo

FOCUS ON READING

How can you tell that the statements in this paragraph are facts?

Besides serving as the national capital, Tokyo is the center of Japan's banking and communication industries. As a result, the city is busy, noisy, and very crowded. Nearly 30 million people live in a relatively small area. Because Tokyo is so densely populated, land is scarce. As a result, Tokyo's real estate prices are among the highest in the world. Some people save up for years to buy homes in Tokyo. They earn money by putting money in savings accounts or by investing in stocks and bonds.

Because space is so limited in Tokyo, people have found creative ways to adapt. Buildings tend to be fairly tall and narrow so that they take less land area. People also use space under ground. For example, shops and restaurants can be found below the streets in subway stations. Another way the Japanese have found to save space is the capsule hotel. Guests in these hotels—mostly traveling businesspeople—crawl into tiny sleeping chambers rather than having rooms with beds.

Many people work in Tokyo but live outside the city. So many people commute to and from Tokyo that trains are very crowded. During peak travel times, commuters are crammed into train cars.

Tokyo is not all about work, though. During their leisure time, people can visit Tokyo's many parks, museums, and stores. They can also take short trips to local amusement parks, baseball stadiums, or other attractions. Among these attractions are a huge indoor beach and a ski resort filled with artificial snow.

Close-up
Life in Tokyo

Home to some 30 million people, Tokyo is one of the world's busiest cities. This illustration shows what a typical day in Tokyo is like.

Small Shinto shrines can be found even in the heart of busy Tokyo.

During peak travel times, Tokyo's trains are so crowded that people need to be pushed aboard.

Gardens planted on the roofs of buildings help keep Tokyo's temperature down.

To save space, the Japanese build capsule hotels. Each guest sleeps in his or her own tiny chamber.

Bustling shopping centers can be found below many of Tokyo's streets.

ANALYSIS
SKILL **ANALYZING VISUALS**

Based on this image, how does life in Tokyo compare to daily life where you are?

Life in Other Cities

Most of Japan's other cities, like Tokyo, are crowded and busy. Many of them serve as centers of industry or transportation.

The second largest city in Japan, **Osaka**, is located in western Honshu. In Osaka—as in Tokyo and other cities—tall, modern skyscrapers stand next to tiny Shinto temples. Another major city is Kyoto. Once Japan's capital, Kyoto is full of historic buildings.

Transportation between Cities

ACADEMIC VOCABULARY

efficient
productive and not wasteful

To connect cities that lie far apart, the Japanese have built a network of rail lines. Some of these lines carry very fast trains called *Shinkansen*, or bullet trains. They can reach speeds of more than 160 miles per hour (250 kph). Japan's train system is very **efficient**. Trains nearly always leave on time and are almost never late.

Rural Life

Not everyone in Japan lives in cities. Some people live in the country in small villages. The people in these villages own or work on farms.

Relatively little of Japan's land is arable, or suitable for farming. Much of the land is too rocky or steep to grow crops on. As a result, most farms are small. The average Japanese farm is only about 2.5 acres (1 hectare). In contrast, the average farm in the United States is 175 times that size.

Because their farms are so small and Japan imports so much of its food, many farmers cannot make a living from their crops. As a result, many people have left rural areas to find jobs in cities.

READING CHECK Finding Main Ideas What are Japanese cities like?

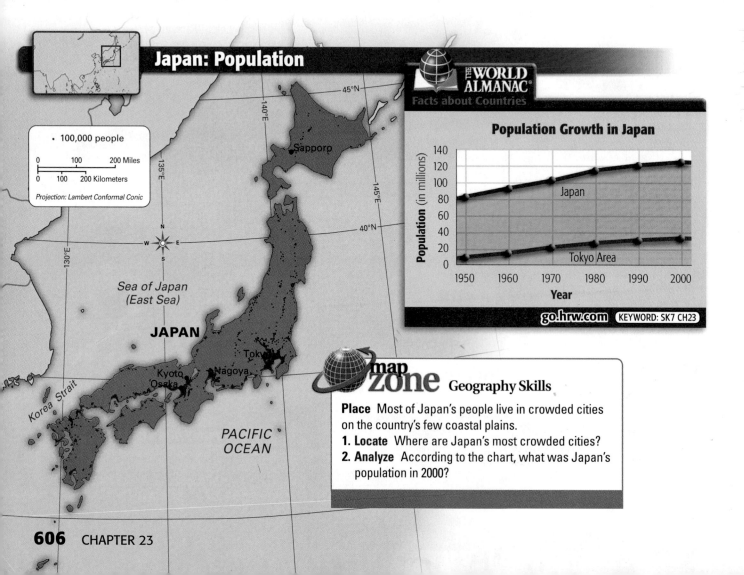

Japan: Population

THE WORLD ALMANAC®
Facts about Countries

- • 100,000 people

0   100   200 Miles
0   100   200 Kilometers

Projection: Lambert Conformal Conic

Sapporo

Sea of Japan
(East Sea)

JAPAN

Kyoto
Osaka
Nagoya
Tokyo

Korea Strait

PACIFIC
OCEAN

45°N
140°E
135°E
145°E
40°N
130°E

Population Growth in Japan

Population (in millions)

| 140 |
| 120 |
| 100 |
| 80 |
| 60 |
| 40 |
| 20 |
| 0 |

Japan

Tokyo Area

1950   1960   1970   1980   1990   2000

Year

go.hrw.com   KEYWORD: SK7 CH23

map zone Geography Skills

Place Most of Japan's people live in crowded cities on the country's few coastal plains.
1. **Locate** Where are Japan's most crowded cities?
2. **Analyze** According to the chart, what was Japan's population in 2000?

Issues and Challenges

Many people consider Japan one of the world's most successful countries. In recent years, however, a few issues have arisen that present challenges for Japan's future.

One of these issues is Japan's lack of space. As cities grow, crowding has become a serious issue. To make space, some people have begun to construct taller buildings. Such buildings have to be carefully planned, though, to withstand earthquakes.

Japan also faces economic challenges. For many years, it had the only strong economy in East Asia. Recently, however, other countries have challenged Japan's economic dominance. Competition from China and South Korea has begun taking business from some Japanese companies.

Pollution has also become a problem in Japan. In 1997 officials from more than 150 countries met in Japan to discuss the pollution problem. They signed the Kyoto Protocol, an agreement to cut down on pollution and improve air quality.

READING CHECK **Finding Main Ideas** What are three issues facing Japan?

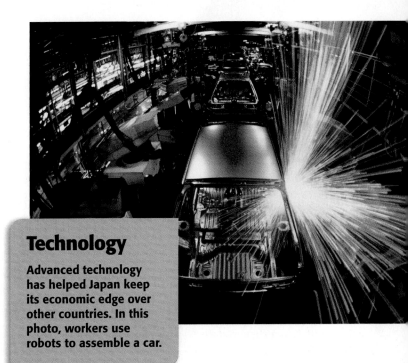

Technology
Advanced technology has helped Japan keep its economic edge over other countries. In this photo, workers use robots to assemble a car.

SUMMARY AND PREVIEW Since World War II, Japan has created a democratic government and a strong, highly technological economy. In the next section, you will learn about changes that have occurred in South Korea and North Korea in the same time period.

Section 3 Assessment

go.hrw.com
Online Quiz
KEYWORD: SK7 HP23

Reviewing Ideas, Terms, and Places
1. **a. Identify** What are some goods made in Japan?
   b. Explain How has Japan's government changed since World War II?
   c. Elaborate Why do you think **work ethic** is so important to the Japanese economy?
2. **a. Describe** How have people tried to save space in Japanese cities?
   b. Evaluate Do you think you would like living in **Tokyo**? Why or why not?
3. **a. Identify** What is one issue that crowding has caused for Japan?
   b. Analyze How are other countries presenting challenges to Japan's economy?

Critical Thinking
4. **Analyzing** Draw a graphic organizer like the one shown here. In one circle, write two sentences about city life in Japan. In another, write two sentences about rural life. In the third, write two sentences about issues facing the Japanese.

City Life   Rural Life   Issues

FOCUS ON WRITING

5. **Thinking about Japan** What image, or picture, of life in Japan could you write about in your poem? List two or three ideas. Then decide which is the most promising idea for your poem.

The Koreas Today

If **YOU** lived there...

You live in Inchon, one of South Korea's largest cities. Sometimes your grandparents tell you about the other family members who still live in North Korea. You have never met them, of course, and your grandparents have not seen them since they were children, more than 50 years ago. After hearing stories about these family members, you are curious about their lives.

Would you like to visit North Korea?

BUILDING BACKGROUND A peace treaty ended the Korean War in 1953, but it left the Korean Peninsula divided into two very different countries. The conflict separated families from their relatives on the other side of the zone that divides South Korea from North Korea. Since then, the countries have developed in very different ways.

South Korea Today

Japan's closest neighbor is both a major economic rival and a key trading partner. That neighbor is South Korea. Like Japan, South Korea is a democratic country with a strong economy. Unlike Japan, South Korea shares a border with a potentially hostile neighbor—North Korea.

Government and Economy

The official name of South Korea is the Republic of Korea. As the name suggests, South Korea's government is a republic. It is headed by a president and an assembly elected by the people, much like the United States is. In fact, the United States helped create South Korea's government after World War II.

The United States also helped make South Korea's economy one of the strongest in East Asia. In addition, Korean business leaders and government officials have worked together to ensure that the economy stays strong. In recent years, South Korea has become a major manufacturing country, exporting goods to places all around the world.

What You Will Learn...

Main Ideas

1. The people of South Korea today have freedom and economic opportunities.
2. The people of North Korea today have little freedom or economic opportunity.
3. Some people in both South and North Korea support the idea of Korean reunification.

The Big Idea

Though they share a common history and culture, the two Koreas have very different governments and economies.

Key Terms and Places

Seoul, *p. 609*
demilitarized zone, *p. 609*
Pyongyang, *p. 611*

TAKING NOTES In your notebook, draw two boxes like the ones shown here. As you read, take notes about South Korea in the left box and notes about North Korea in the right. Note similarities between the countries below the boxes.

South Korea	North Korea

The Demilitarized Zone

NORTH
KOREA

Demilitarized
Zone (DMZ)

SOUTH
KOREA

The demilitarized zone separates the countries of South and North Korea. It is about 2.5 miles (4 km) wide and 150 miles (240 km) long. Armed guards patrol both sides of the zone.

Daily Life

Like Japan, South Korea is very densely populated. The capital city, **Seoul** (SOHL), is one of the most densely populated cities in the world. It has more than 40,000 people per square mile (15,000/sq km).

Although parts of South Korea are densely populated, very few people live in the mountainous interior. Most people live near the coast. A coastal plain in western South Korea is the most crowded part of the country.

In South Korea's cities, most people live in small apartments. Because space is scarce, housing is expensive. Also, cities sometimes suffer from pollution from the many factories, cars, and coal-fired heating systems found there. In some cities, industrial waste has also polluted the water.

Outside the cities, many South Koreans still follow traditional ways of life. Most of them are farmers who grow rice, beans, and cabbage they can use to make kimchi. They usually live on small farms.

Issues and Challenges

Government policies and international politics have led to some challenges for South Korea. Although South Korea has a successful economy, some people feel that it is corrupt. For many years, four families have controlled much of the country's industry. Some members of these families have used their wealth and power as a tool for making themselves even wealthier. The government hopes to put an end to this corruption through reform programs.

A bigger challenge to South Korea is its relationship with North Korea. Since the end of the Korean War in the 1950s, the two countries have been separated. Between them is a **demilitarized zone**, an empty buffer zone created to keep the two countries from fighting. Although troops are not allowed in the demilitarized zone, guards patrol both sides.

READING CHECK Summarizing What issues face South Korea today?

North Korea Today

The official name of North Korea is the Democratic People's Republic of Korea. Its name, however, is misleading. North Korea is neither a democracy nor a republic. It is a totalitarian state, and the Communist Party controls both the government and the economy.

Government and Economy

The Communist government of North Korea has existed since World War II. It was created by a man named Kim Il Sung. He ruled the country from 1948 until his death in 1994. During this time, he created many **policies** that are still in effect today.

Kim ruled North Korea as a dictator. According to North Korea's constitution, most power rests in an elected legislature.

ACADEMIC
VOCABULARY
policy rule, course of action

In truth, though, the legislature never had much power. Advised by members of the Communist Party, Kim ruled alone.

When Kim Il Sung died in 1994, his son Kim Jong Il took over. Like his father, the younger Kim rules as a dictator. He was elected by the North Korean legislature. The people had no say in his election.

As a Communist country, North Korea has a command economy. This means that the government plans the economy and decides what is produced. It also owns all land and controls access to jobs.

Unlike Japan and South Korea, North Korea is rich in mineral resources. With these resources, factories in North Korea make machinery and military supplies. However, most factories use out-of-date technology. As a result, North Korea is much poorer than Japan and South Korea.

Life in Korea

610

Because it is so rocky, very little of North Korea's land can be farmed. The farmland that does exist is owned by the government. It is farmed by cooperatives—large groups of farmers who work the land together. These cooperatives are not able to grow enough food for the country. As a result, the government has to import food. This can be a difficult task because North Korea's relations with most other countries are strained.

Daily Life

Like Japan and South Korea, North Korea is largely an urban society. Most people live in cities. The largest city is the capital, **Pyongyang** (PYUHNG-YAHNG), in the west. Like Tokyo and Seoul, Pyongyang is very crowded. More than 3 million people live in the city.

The differences between life in South Korea and North Korea can be seen in their capitals. Seoul, South Korea (shown to the left), is a busy, modern city and a major commercial center. In comparison, North Korea's capital, Pyongyang (shown above), has little traffic or commercial development.

ANALYZING VISUALS What do these photos suggest about life in Seoul and Pyongyang?

Life in Pyongyang is very different from life in Tokyo or Seoul. For example, few people in Pyongyang own private cars. The North Korean government allows only top Communist officials to own cars. Most residents have to use buses or the subway to get around. At night, many streets are dark because of electricity shortages.

The people of North Korea have fewer rights than the people of Japan or South Korea. For example, the government controls individual speech as well as the press. Because the government feels that religion conflicts with many Communist ideas, it also discourages people from practicing any religions.

Issues and Challenges

Why does North Korea, which is rich in resources, have shortages of electricity and food? These problems are due in part to choices the government has made. For years, North Korea had ties mostly with other Communist countries. Since the breakup of the Soviet Union, North Korea has been largely isolated from the rest of the world. It has closed its markets to foreign goods, which means that other countries cannot sell their goods there. At the same time, North Korea lacks the technology to take advantage of its resources. As a result, many people suffer and resources go unused.

In addition, many countries worry about North Korea's ability to make and use nuclear weapons. In 2002 North Korea's government announced that it had enough materials to build six nuclear bombs. This announcement troubled world leaders, who feared what North Korea might do with atomic weapons. Negotiations are underway to resolve the issue.

READING CHECK Generalizing What is North Korea's relationship with the world?

Young people at a political rally express support for reunification. The flag in the background shows a united Korea.

The governments of both South Korea and North Korea have also expressed their support for reunification. Leaders from the two countries met in 2000 for the first time since the Korean War. As part of their meeting, they discussed ways to improve relations and communication between the two countries. For example, they agreed to build a road through the demilitarized zone to connect the two Koreas.

The chief obstacle to the reunification of Korea is the question of government. South Koreans want a unified Korea to be a democracy. North Korean leaders, on the other hand, have insisted that Korea should be Communist. Until this issue is resolved, the countries will remain separate.

READING CHECK **Summarizing** What issues stand in the way of Korean reunification?

Korean Reunification

FOCUS ON READING

What opinion do many Koreans hold toward reunification?

For years, people from both South and North Korea have called for their countries to be reunited. Because the two Koreas share a common history and culture, these people believe they should be one country. As time has passed, more and more people have voiced support for reunification.

SUMMARY AND PREVIEW In this chapter you learned about the history, cultures, and people of Japan and the Koreas. In the next chapter, you will examine a region that lies farther south, a region called Southeast Asia.

go.hrw.com
Online Quiz
KEYWORD: SK7 HP23

Section 4 Assessment

Reviewing Ideas, Terms, and Places

1. **a. Define** What is the **demilitarized zone**? Why does it exist?
   b. Summarize What factors have helped South Korea develop a strong economy?
2. **a. Identify** What is the capital of North Korea? What is life like there?
   b. Contrast How is North Korea's government different from South Korea's?
3. **a. Recall** Why do many Koreans support the idea of reunification?
   b. Evaluate If you lived in North or South Korea, do you think you would support the reunification of the countries? Why or why not?

Critical Thinking

4. **Analyze** Draw a diagram like the one below. In the left box, write three statements about South Korea. In the right box, write three statements about North Korea. In the oval, list one factor that supports reunification and one that hinders it.

South Korea — Reunification — North Korea

FOCUS ON WRITING

5. **Considering Korea** As you read about the Koreas, did you think of an image, or picture, that would work in a poem? List your ideas.

Chapter Review

Geography's Impact
video series
Review the video to answer the closing question:
How has Japan's location on the Ring of Fire made it so prone to natural hazards?

Visual Summary

Use the visual summary below to help you review the main ideas of the chapter.

QUICK FACTS

Japan and the Korean Peninsula have rugged landscapes that are largely covered by mountains.

Japan has one of the world's strongest economies, due in large part to its superior technology.

Since World War II, life in democratic South Korea has been very different from life in Communist North Korea.

Reviewing Vocabulary, Terms, and Places

Imagine these terms from the chapter are correct answers to items in a crossword puzzle. Write the clues for the answers.

1. Tokyo
2. abstract
3. trade surplus
4. tariff
5. kimono
6. efficient
7. work ethic
8. Seoul
9. fishery
10. Pyongyang
11. kimchi
12. policy

Comprehension and Critical Thinking

SECTION 1 *(Pages 592–595)*

13. **a. Identify** What physical feature covers most of Japan and the Korean Peninsula? What is one famous example of this landform?

b. Draw Conclusions Fish and seafood are very important in the Japanese diet. Why do you think this is so?

c. Predict How do you think earthquakes and typhoons would affect your life if you lived in Japan?

SECTION 2 *(Pages 597–601)*

14. **a. Identify** Who were the shoguns? What role did they play in Japanese history?

b. Explain What caused the Korean War? What happened as a result of the war?

c. Elaborate How have the histories of Japan and Korea affected their cultures?

SECTION 3 *(Pages 602–607)*

15. **a. Recall** What is the most important aspect of Japan's economy?

b. Make Inferences Why is Tokyo such a busy and crowded city?

c. Develop How might Japan try to address the problem of crowding in its cities?

SECTION 4 *(Pages 608–612)*

16. a. Recall What type of government does South Korea have? What type of government does North Korea have?

b. Contrast How is South Korea's economy different from North Korea's? Which has been more successful?

c. Predict Do you think the reunification of the Koreas will happen in the near future? Why or why not?

Social Studies Skills

Using a Topographic Map *Use the topographic map in this chapter's Social Studies Skills lesson to answer the following questions.*

17. What elevations do the contour lines on this map show?

18. Where are the highest points on Awaji Island located? How can you tell?

19. Is the city of Sumoto located more or less than 500 feet above sea level?

FOCUS ON READING AND WRITING

Understanding Fact and Opinion *Decide whether each of the following statements is a fact or an opinion.*

20. Japan would be a great place to live.

21. Japan is an island country.

22. North Korea should give up Communism.

23. The Koreas should reunify.

Writing Your Five-Line Poem *Use your notes and the instructions below to create your poem.*

24. Review your notes and decide on a topic to write about. Remember that your poem should describe one image or picture—an object, a place, etc.—from Japanese or Korean culture.

　　The first three lines of your poem should describe the object or place you have chosen. The last two should express how it makes you feel. Try to use the traditional Tanka syllable count in your poem: five syllables in lines 1 and 3; seven syllables in lines 2, 4, and 5. Remember that your poem does not have to rhyme.

Using the Internet

25. Activity: Comparing Schools Have you ever wondered what it would be like to live in Japan? How would your life be different if you lived there? What would you learn about in school? What would you eat for lunch? Enter the activity keyword. Learn about Japan and about Japanese culture by exploring the Web links provided. Look for ways that your life would be the same and different if you grew up in Japan. Then record what you learned in a chart or a graphic organizer.

Map Activity ★Interactive

26. Japan and the Koreas On a separate sheet of paper, match the letters on the map with their correct labels.

North Korea	Tokyo, Japan
South Korea	Hokkaido
Korea Strait	Sea of Japan (East Sea)

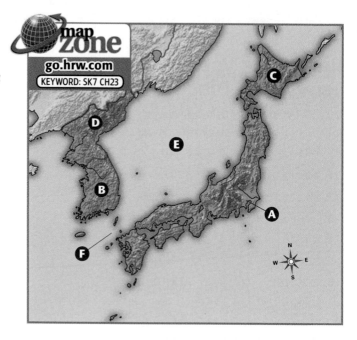

Standardized Test Practice

DIRECTIONS: Read questions 1 through 7 and write the letter of the best response. Then read question 8 and write your own well-constructed response.

1 Tokyo, Osaka, and Kyoto are all cities in

A South Korea.

B North Korea.

C Japan.

D Honshu.

2 North Korea and South Korea are separated by the

A Korea Strait.

B demilitarized zone.

C Sea of Japan.

D Japanese Alps.

3 Kim Il Sung and Kim Jong Il are famous leaders from which country?

A South Korea

B North Korea

C Japan

D Honshu

4 The country that had the strongest influence on early Korea and Japan was

A Russia.

B India.

C China.

D the United States.

5 What is one reason that many people in Korea support reunification?

A They want a socialist government.

B They want to have the largest country in the world.

C They are afraid of an attack from Japan.

D They believe that all Koreans share a common history and culture.

Japan and the Koreas

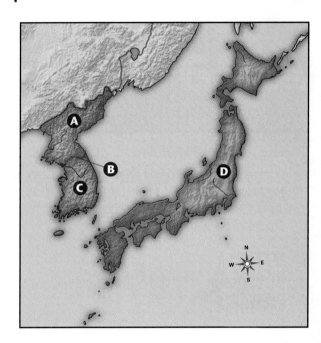

6 On the map above, which letter appears in an area led by a Communist government?

A A

B B

C C

D D

7 Which country has the strongest economy in the region?

A South Korea

B North Korea

C Japan

D Honshu

8 **Extended Response** Write a brief paragraph explaining why the coastal plains of Japan and Korea are so crowded. You may wish to refer to the map above as you prepare your answer.

CHAPTER 24

Southeast Asia

What You Will Learn...

In this chapter you will learn about the physical features, climate, and natural resources of Southeast Asia. You will also examine the histories and cultures of the countries in this region and explore what life is like there today.

SECTION 1
Physical Geography **618**

SECTION 2
History and Culture **624**

SECTION 3
Mainland Southeast Asia Today **629**

SECTION 4
Island Southeast Asia Today **634**

FOCUS ON READING AND SPEAKING

Using Context Clues—Definitions As you read, you may run across words you do not know. You can often figure out the meaning of an unknown word by using context clues. One type of context clue is a definition of a word, or a restatement of its meaning. To use these clues, look at the words and sentences around the unknown word—its context. **See the lesson, Using Context Clues—Definitions, on page 691.**

Presenting an Interview With a partner, you will role-play a journalist interviewing a regional expert on Southeast Asia. First, read about the region. Then with your partner create a question-and-answer interview script about the region to present to your classmates.

Geography Boats lie along the shore of Phi Phi Don Island in Thailand. The island's beauty makes it a popular vacation spot.

Southeast Asia: Political

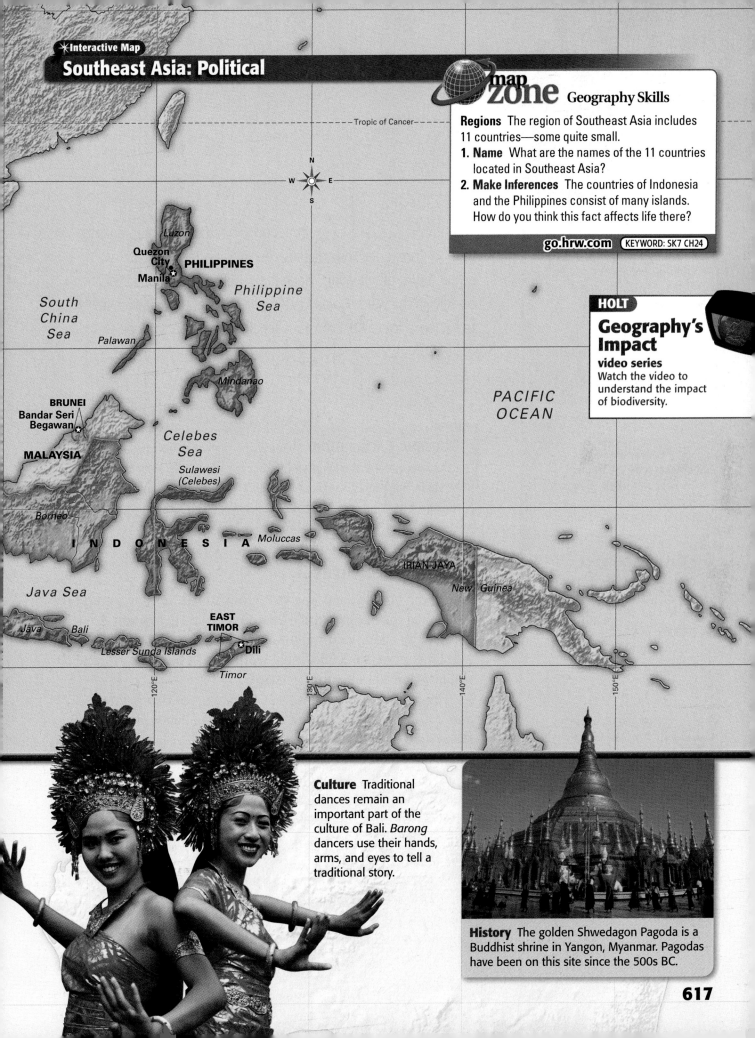

map zone Geography Skills

Regions The region of Southeast Asia includes 11 countries—some quite small.

1. **Name** What are the names of the 11 countries located in Southeast Asia?
2. **Make Inferences** The countries of Indonesia and the Philippines consist of many islands. How do you think this fact affects life there?

go.hrw.com KEYWORD: SK7 CH24

HOLT

Geography's Impact
video series
Watch the video to understand the impact of biodiversity.

Tropic of Cancer

N
W — E
S

Luzon

Quezon City
Manila
PHILIPPINES

Philippine Sea

South China Sea

Palawan

Mindanao

PACIFIC OCEAN

BRUNEI
Bandar Seri Begawan

Celebes Sea

MALAYSIA

Sulawesi (Celebes)

Borneo

I N D O N E S I A   Moluccas

IRIAN JAYA

New Guinea

Java Sea

Java   Bali

EAST TIMOR
Dili

Lesser Sunda Islands

Timor

120°E   130°E   140°E   150°E

Culture Traditional dances remain an important part of the culture of Bali. *Barong* dancers use their hands, arms, and eyes to tell a traditional story.

History The golden Shwedagon Pagoda is a Buddhist shrine in Yangon, Myanmar. Pagodas have been on this site since the 500s BC.

617

Physical Geography

What You Will Learn...

Main Ideas

1. Southeast Asia's physical features include peninsulas, islands, rivers, and many seas, straits, and gulfs.
2. The tropical climate of Southeast Asia supports a wide range of plants and animals.
3. Southeast Asia is rich in natural resources such as wood, rubber, and fossil fuels.

The Big Idea

Southeast Asia is a tropical region of peninsulas, islands, and waterways with diverse plants, animals, and resources.

Key Terms and Places

Indochina Peninsula, *p. 618*
Malay Peninsula, *p. 618*
Malay Archipelago, *p. 618*
archipelago, *p. 618*
New Guinea, *p. 619*
Borneo, *p. 619*
Mekong River, *p. 619*

TAKING NOTES As you read, use a chart like this one to help you take notes on the physical geography of Southeast Asia.

Physical Features	
Climate, Plants, Animals	
Natural Resources	

If YOU lived there...

Your family lives on a houseboat on a branch of the great Mekong River in Cambodia. You catch fish in cages under the boat. Your home is part of a floating village of houseboats and houses built on stilts in the water. Boats loaded with fruits and vegetables travel from house to house. Even your school is on a nearby boat.

How does water shape life in your village?

BUILDING BACKGROUND Waterways, such as rivers, canals, seas, and oceans, are important to life in Southeast Asia. Waterways are both "highways" and sources of food. Where rivers empty into the sea, they form deltas, areas of rich soil good for farming.

Physical Features

Where can you find a flower that grows up to 3 feet across and smells like rotting garbage? How about a lizard that can grow up to 10 feet long and weigh up to 300 pounds? These amazing sights as well as some of the world's most beautiful tropical paradises are all in Southeast Asia.

The region of Southeast Asia is made up of two peninsulas and two large island groups. The **Indochina Peninsula** and the **Malay** (muh-LAY) **Peninsula** extend from the Asian mainland. We call this part of the region Mainland Southeast Asia. The two island groups are the Philippines and the **Malay Archipelago**. An **archipelago** (ahr-kuh-PE-luh-goh) is a large group of islands. We call this part of the region Island Southeast Asia.

Landforms

In Mainland Southeast Asia, rugged mountains fan out across the countries of Myanmar (MYAHN-mahr), Thailand (TY-land), Laos (LOWS), and Vietnam (vee-ET-NAHM). Between these mountains are low plateaus and river floodplains.

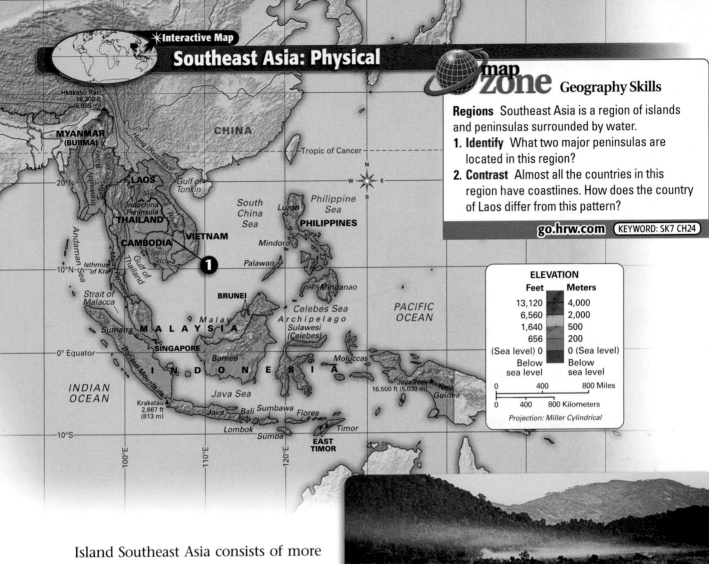

Southeast Asia: Physical

map zone Geography Skills

Regions Southeast Asia is a region of islands and peninsulas surrounded by water.
1. **Identify** What two major peninsulas are located in this region?
2. **Contrast** Almost all the countries in this region have coastlines. How does the country of Laos differ from this pattern?

go.hrw.com **KEYWORD: SK7 CH24**

ELEVATION

Feet		Meters
13,120		4,000
6,560		2,000
1,640		500
656		200
(Sea level) 0		0 (Sea level)
Below sea level		Below sea level

0 400 800 Miles
0 400 800 Kilometers
Projection: Miller Cylindrical

① Mist hovers over the Mekong River as it flows through the forested mountains of northern Thailand.

Island Southeast Asia consists of more than 20,000 islands, some of them among the world's largest. **New Guinea** is Earth's second largest island, and **Borneo** its third largest. Many of the area's larger islands have high mountains. A few peaks are high enough to have snow and glaciers.

Island Southeast Asia is a part of the Ring of Fire as well. As a result, earthquakes and volcanic eruptions often rock the area. When such events occur underwater, they can cause tsunamis, or giant series of waves. In 2004 a tsunami in the Indian Ocean killed hundreds of thousands of people, many in Southeast Asia.

Bodies of Water

Water is a central part of Southeast Asia. Look at the map to identify the many seas, straits, and gulfs in this region.

In addition, several major rivers drain the mainland's peninsulas. Of these rivers, the mighty **Mekong** (MAY-KAWNG) **River** is the most important. The mainland's fertile river valleys and deltas support farming and are home to many people.

READING CHECK Finding Main Ideas What are Southeast Asia's major physical features?

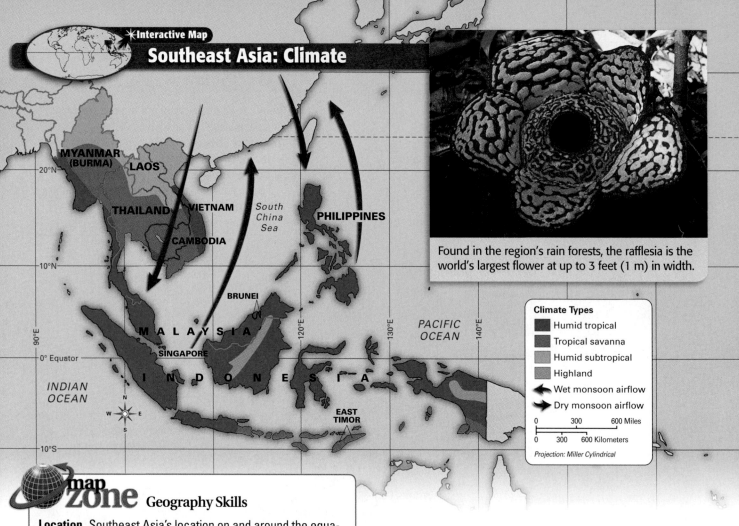

MYANMAR (BURMA)

LAOS

THAILAND   VIETNAM   South China Sea   PHILIPPINES

CAMBODIA

BRUNEI

M A L A Y S I A

SINGAPORE

I N D O N E S I A

PACIFIC OCEAN

INDIAN OCEAN

0° Equator

EAST TIMOR

Found in the region's rain forests, the rafflesia is the world's largest flower at up to 3 feet (1 m) in width.

Climate Types

- ◼ Humid tropical
- ◼ Tropical savanna
- ◼ Humid subtropical
- ◼ Highland
- ← Wet monsoon airflow
- → Dry monsoon airflow

0   300   600 Miles
0   300   600 Kilometers
Projection: Miller Cylindrical

map zone Geography Skills

Location Southeast Asia's location on and around the equator affects the region's climate.

1. **Identify** What is the main climate found in Indonesia, Malaysia, and the Philippines?
2. **Interpret** Based on the map, how do monsoons affect the climate of this region?

go.hrw.com   KEYWORD: SK7 CH24

Climate, Plants, and Animals

Southeast Asia lies in the tropics, the area on and around the equator. Temperatures are warm to hot year-round, but become cooler to the north and in the mountains.

Much of the mainland has a tropical savanna climate. Seasonal <u>monsoon</u> winds from the oceans bring heavy rain in summer and drier weather in winter. Severe flooding is common during wet seasons. This climate supports savannas—areas of tall grasses and scattered trees and shrubs.

FOCUS ON READING

What context clues help you figure out the definition of *monsoon*?

The islands and the Malay Peninsula mainly have a humid tropical climate. This climate is hot, muggy, and rainy all year. Showers or storms occur almost daily. In addition, huge storms called typhoons can bring heavy rains and powerful winds.

The humid tropical climate's heat and heavy rainfall support tropical rain forests. These lush forests are home to a huge number of different plants and animals. About 40,000 kinds of flowering plants grow in Indonesia alone. These plants include the rafflesia, the world's largest flower. Measuring up to 3 feet (1 m) across, this flower produces a horrible, rotting stink.

Rain forest animals include elephants, monkeys, tigers, and many types of birds. Some species are found nowhere else. They include orangutans and Komodo dragons, lizards that can grow 10 feet (3 m) long.

Orangutans live in the rain forests of Borneo and Sumatra. Deforestation has seriously reduced their habitat.

Natural Resources

Southeast Asia has a number of valuable natural resources. The region's hot, wet climate and rich soils make farming highly productive. Rice is a major crop, and others include coconuts, coffee, sugarcane, palm oil, and spices. Some countries, such as Indonesia and Malaysia (muh-LAY-zhuh), also have large rubber tree plantations.

The region's seas provide fisheries, and its tropical rain forests provide valuable hardwoods and medicines. The region also has many minerals and fossil fuels, including tin, iron ore, natural gas, and oil. For example, the island of Borneo sits atop an oil field.

READING CHECK **Summarizing** What are the region's major natural resources?

Many of these plants and animals are endangered because of loss of habitat. People are clearing the tropical rain forests for farming, wood, and mining. These actions threaten the area's future diversity.

READING CHECK **Analyzing** How does climate contribute to the region's diversity of life?

SUMMARY AND PREVIEW Southeast Asia is a tropical region of peninsulas, islands, and waterways with diverse life and rich resources. Next, you will read about the region's history and culture.

Section 1 Assessment

Reviewing Ideas, Terms, and Places

1. **a. Define** What is an **archipelago**?
   b. Compare and Contrast How do the physical features of Mainland Southeast Asia compare and contrast to those of Island Southeast Asia?
2. **a. Recall** What type of forest occurs in the region?
   b. Summarize What is the climate like across much of Southeast Asia?
   c. Predict What do you think might happen to the region's wildlife if the tropical rain forests continue to be destroyed?
3. **a. Identify** Which countries in the region are major producers of rubber?
   b. Analyze How does the region's climate contribute to its natural resources?

Critical Thinking

4. **Summarizing** Draw a chart like this one. Use your notes to provide information about the climate, plants, and animals in Southeast Asia. In the left-hand box, also note how climate shapes life in the region.

| Climate of Southeast Asia | → | Plants |
| | → | Animals |

FOCUS ON SPEAKING

5. **Writing Questions about the Region's Physical Geography** Note information about the region's physical features, climate, plants, animals, and natural resources. Write two questions and answers for your interview. For example, you might ask a question about the region's tropical rain forests.

Tsunami!

Essential Elements

The World in Spatial Terms
Places and Regions
Physical Systems
Human Systems
Environment and Society
The Uses of Geography

Background "Huge Waves Hit Japan." This event is a tsunami (soo-NAH-mee), a series of giant sea waves. Records of deadly tsunamis go back 3,000 years. Some places, such as Japan, have been hit time and again.

Tsunamis occur when an earthquake, volcanic eruption, or other event causes seawater to move in huge waves. The majority of tsunamis occur in the Pacific Ocean because of the region's many earthquakes.

Warning systems help alert people to tsunamis. The Pacific Tsunami Warning Center monitors tsunamis in the Pacific Ocean. Sensors on the ocean floor and buoys on the water's surface help detect earthquakes and measure waves. When a tsunami threatens, radio, TV, and sirens alert the public.

Indian Ocean Catastrophe

On December 26, 2004, a massive earthquake erupted below the Indian Ocean. The earthquake launched a monster tsunami. Within half an hour, walls of water up to 65 feet high came barreling ashore in Indonesia. The water swept away boats, buildings, and people. Meanwhile, the tsunami kept traveling in ever-widening rings across the ocean. The waves eventually wiped out coastal communities in a dozen countries. Some 300,000 people eventually died.

At the time, the Indian Ocean did not have a tsunami warning system. Tsunamis are rare in that part of the world. As a result, many countries there had been unwilling to invest in a warning system.

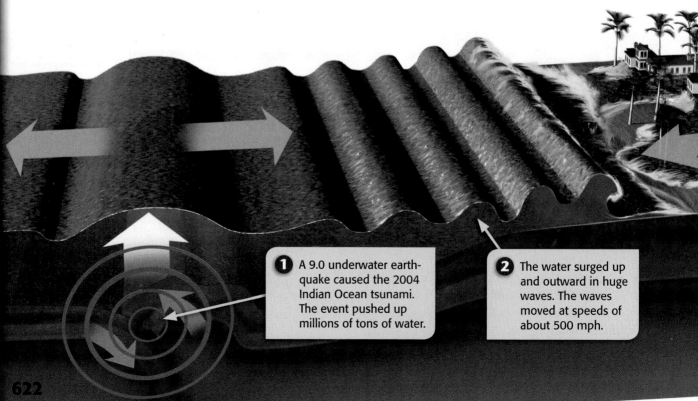

1 A 9.0 underwater earthquake caused the 2004 Indian Ocean tsunami. The event pushed up millions of tons of water.

2 The water surged up and outward in huge waves. The waves moved at speeds of about 500 mph.

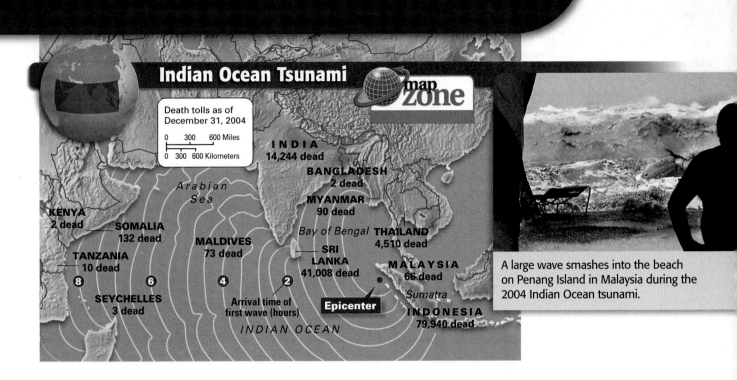

Indian Ocean Tsunami

Death tolls as of December 31, 2004

0   300   600 Miles
0 300 600 Kilometers

KENYA 2 dead

Arabian Sea

INDIA 14,244 dead

BANGLADESH 2 dead

MYANMAR 90 dead

SOMALIA 132 dead

MALDIVES 73 dead

Bay of Bengal

THAILAND 4,510 dead

TANZANIA 10 dead

SRI LANKA 41,008 dead

MALAYSIA 66 dead

⑧   ⑥   ④   ②

SEYCHELLES 3 dead

Arrival time of first wave (hours)

Epicenter

Sumatra

INDONESIA 79,940 dead

INDIAN OCEAN

A large wave smashes into the beach on Penang Island in Malaysia during the 2004 Indian Ocean tsunami.

In 2004 these countries paid a terrible price for their decision. As the map shows, the 2004 tsunami hit countries from South Asia to East Africa. Most people had no warning of the tsunami. In addition, many people did not know how to protect themselves. Instead of heading to high ground, some people went to the beach for a closer look. Many died when later waves hit.

Tilly Smith, a 10-year-old on vacation in Thailand, was one of the few who understood the danger. Two weeks earlier, her geography teacher had discussed tsunamis. As the water began surging, Smith warned her family and other tourists to flee. Her geographic knowledge saved their lives.

What It Means No one can prevent tsunamis. Yet, by studying geography, we can prepare for these disasters and help protect lives and property. The United Nations is now working to create a global tsunami warning system. People are also trying to plant more mangroves along coastlines. These bushy swamp trees provide a natural barrier against high waves.

③ When they strike, tsunamis often look like a rapidly rising tide or swell of water. The water then rushes far inland and back out.

Geography for Life Activity

1. What steps are being taken to avoid another disaster such as the Indian Ocean tsunami in 2004?

2. About 75 percent of tsunami warnings since 1948 were false alarms. What might be the risks and benefits of early warnings to move people out of harm's way?

3. **Creating a Survival Guide** Create a tsunami survival guide. List the dos and don'ts for this emergency.

History and Culture

What You Will Learn...

Main Ideas

1. Southeast Asia's early history includes empires, colonial rule, and independence.
2. The modern history of Southeast Asia involves struggles with war and communism.
3. Southeast Asia's culture reflects its Chinese, Indian, and European heritage.

The Big Idea

People, ideas, and traditions from China, India, Europe, and elsewhere have shaped Southeast Asia's history and culture.

Key Terms and Places

Timor, *p. 625*
domino theory, *p. 626*
wats, *p. 626*

TAKING NOTES As you read, use a chart like the one here to help you take notes on the history and culture of Southeast Asia.

History	Culture

If **YOU** lived there...

You and your friends are strolling through the market in Jakarta, Indonesia, looking for a snack. You have many choices—tents along the street, carts called gerobak, and vendors on bicycles all sell food. You might choose satay, strips of chicken or lamb grilled on a stick. Or you might pick one of many rice dishes. For dessert, you can buy fruit or order an ice cream cone.

What do you like about living in Jakarta?

BUILDING BACKGROUND Colonial rule helped shape Southeast Asia's history and culture—including foods. Throughout the region you can see not only a blend of different Asian influences but also a blend of American, Dutch, French, and Spanish influences.

Early History

Southeast Asia lies south of China and east of India, and both countries have played a strong role in the region's history. Over time, many people from China and India settled in Southeast Asia. As settlements grew, trade developed with China and India.

Early Civilization

The region's most advanced early civilization was the Khmer (kuh-MER). From the AD 800s to the mid-1200s the Khmer controlled a large empire in what is now Cambodia. The remains of Angkor Wat, a huge temple complex the Khmer built in the 1100s, reflect their advanced civilization and Hindu religion.

In the 1200s the Thai (TY) from southern China settled in the Khmer area. Around the same time, Buddhism, introduced earlier from India and Sri Lanka, began replacing Hinduism in the region.

Colonial Rule and Independence

As in many parts of the world, European powers started colonizing Southeast Asia during the 1500s. Led by Portugal, European powers came to the region in search of spices and other trade goods.

In 1521 explorer Ferdinand Magellan reached the Philippines and claimed the islands for Spain. The Spaniards who followed came to colonize, trade, and spread Roman Catholicism. This religion remains the main faith in the Philippines today.

In the 1600s and 1700s Dutch traders drove the Portuguese out of much of the region. Portugal kept only the small island of **Timor**. The Dutch gained control of the tea and spice trade on what became the Dutch East Indies, now Indonesia.

In the 1800s the British and French set up colonies with plantations, railroads, and mines. Many people from China and India came to work in the colonies. The British and French spread Christianity as well.

In 1898 the United States entered the region when it won the Philippines from Spain after the Spanish-American War. By the early 1900s, colonial powers ruled most of the region, as the map on the next page shows. Only Siam (sy-AM), now Thailand, was never colonized, although it lost land.

In World War II (1939–1945), Japan invaded and occupied most of Southeast Asia. After Japan lost the war, the United States gave the Philippines independence. Soon, other people in the region began to fight for their independence.

One of the bloodiest wars for independence was in French Indochina. In 1954 the French left. Indochina then split into the independent countries of Cambodia, Laos, and Vietnam. By 1970, most of Southeast Asia had thrown off colonial rule.

READING CHECK **Identifying Cause and Effect** What reasons led other countries to set up colonies across most of Southeast Asia?

Angkor Wat

The stone towers of Angkor Wat rise from the rain forest in what is now Cambodia. To the Khmer, Angkor Wat symbolized the center of the universe.

ANALYZING VISUALS
What conclusions can you draw about the Khmer based on Angkor Wat?

The towers represent the peaks of Mount Meru, the home of the gods in Hindu mythology.

The pathway represents the rainbow bridge, the link between the gods and people in Hindu mythology.

Angkor Wat was built in the mid-1100s as a Hindu temple. It later became a Buddhist temple.

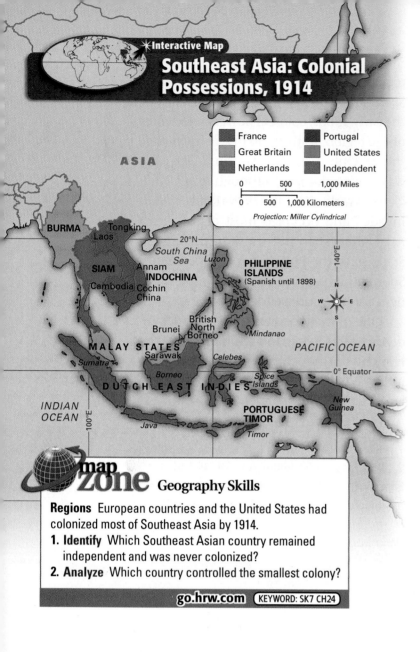

Southeast Asia: Colonial Possessions, 1914

ASIA

France	Portugal
Great Britain	United States
Netherlands	Independent

0   500   1,000 Miles

0   500   1,000 Kilometers

Projection: Miller Cylindrical

BURMA

Tongking

Laos

20°N

South China Sea

Luzon

SIAM

Annam

INDOCHINA

PHILIPPINE ISLANDS
(Spanish until 1898)

Cambodia

Cochin China

140°E

N W E S

Brunei

British North Borneo

Mindanao

PACIFIC OCEAN

MALAY STATES

Sarawak

Sumatra

Celebes

Borneo

Spice Islands

0° Equator

DUTCH EAST INDIES

INDIAN OCEAN

100°E

Java

PORTUGUESE TIMOR

New Guinea

Timor

map zone Geography Skills

Regions European countries and the United States had colonized most of Southeast Asia by 1914.
1. **Identify** Which Southeast Asian country remained independent and was never colonized?
2. **Analyze** Which country controlled the smallest colony?

go.hrw.com KEYWORD: SK7 CH24

Modern History

FOCUS ON READING

How does the context help explain the meaning of the term *oust* in the paragraph to the right?

The move toward independence was not easy. In Vietnam, the fighting to <u>oust</u> the French left the country divided into North and South Vietnam. A civil war then broke out in the South. To defend South Vietnam from Communist forces in that war, the United States sent in troops in the 1960s.

The United States based its decision to send troops on one **criterion**—the potential spread of communism. According to the **domino theory**, if one country fell to communism, other countries nearby would follow like falling dominoes.

ACADEMIC VOCABULARY

criterion rule or standard for defining

Years of war caused millions of deaths and terrible destruction. In the end, North and South Vietnam reunited as one Communist country. As the Communists took over, about 1 million refugees fled South Vietnam. Many went to the United States.

Civil wars also raged in Cambodia and Laos. In 1975 Communist forces took over both countries. The government in Cambodia was brutal, causing the deaths of more than 1 million people there. Then in 1978 Vietnam helped to overthrow Cambodia's government. This event sparked further fighting, which continued off and on until the mid-1990s. The United Nations then helped Cambodia achieve peace.

READING CHECK **Summarizing** What are some key events in the region's modern history?

Culture

The many groups that influenced Southeast Asia's history also shaped its culture. This diverse culture blends native, Chinese, Indian, and European ways of life.

People and Languages

The countries in Southeast Asia have many ethnic groups. As an example, Indonesia has more than 300 ethnic groups. Most of the countries have one main ethnic group plus many smaller ethnic groups.

Not surprisingly, many languages are spoken in Southeast Asia. These languages include native languages and dialects as well as Chinese and European languages.

Religions

The main religions in Southeast Asia are Buddhism, Christianity, Hinduism, and Islam. Buddhism is the main faith on the mainland. This area features many beautiful **wats**, Buddhist temples that also serve as monasteries.

Islam is the main religion in Malaysia, Brunei, and Indonesia. In fact, Indonesia has more Muslims than any other country. In the Philippines, most people are Roman Catholic. Hinduism is practiced in Indian communities and on the island of Bali.

Customs

Customs differ widely across the region, but some similarities exist. For example, religion often shapes life, and people celebrate many religious festivals. Some people continue to practice traditional customs, such as dances and music. These customs are especially popular in rural areas. In addition, many people wear traditional clothing, such as sarongs, strips of cloth worn wrapped around the body.

READING CHECK **Generalizing** How has Southeast Asia's history influenced its culture?

SUMMARY AND PREVIEW Southeast Asia has a long history that has helped shape its diverse culture. Next, you will read about Mainland Southeast Asia.

Thai Teenage Buddhist Monks

Would you be willing to serve as a monk for a few months? In Thailand, many Buddhist boys and young men serve as monks for a short period. This period might last from one week to a few months. These temporary monks follow the lifestyle of actual Buddhist monks, shaving their heads, wearing robes, and maintaining a life of simplicity. During their stay, the teenage monks learn about Buddhism and practice meditation. Some Thai teens decide to become Buddhist monks permanently. This decision is considered a great honor for their families.

Summarizing What are some of the things that Thai boys and young men do while serving as Buddhist monks?

go.hrw.com
Online Quiz
KEYWORD: SK7 HP24

Section 2 Assessment

Reviewing Ideas, Terms, and Places

1. **a. Describe** What was the significance of the Khmer Empire?
   b. Identify Cause and Effect What was the result of the war for independence in French Indochina?
   c. Elaborate How did European colonization shape Southeast Asia's history?
2. **a. Define** What was the **domino theory**?
   b. Summarize What role has communism played in Southeast Asia's modern history?
3. **a. Define** What is a **wat**?
   b. Contrast How does religion in the mainland and island countries differ?
   c. Elaborate How has the history of Southeast Asia shaped the region's culture?

Critical Thinking

4. **Sequencing** Copy the time line shown below. Using your notes, identify on the time line the important people, periods, events, and years in Southeast Asia's history.

800s   2000

FOCUS ON SPEAKING

5. **Writing Questions about History and Culture** What interesting questions could you ask about the history and culture of Southeast Asia? Write two questions and their answers to add to your notes.

Social Studies Skills

Chart and Graph | Critical Thinking | Geography | Study

Analyzing Visuals

Learn

Geographers get information from many sources. These sources include not only text and data but also visuals, such as diagrams and photographs. Use these tips to analyze visuals:

- **Identify the subject.** Read the title and caption, if available. If not, look at the content of the image. What does it show? Where is it located?

- **Analyze the content.** What is the purpose of the image? What information is in the image? What conclusions can you draw from this information? Write your conclusions in your notes.

- **Summarize your analysis.** Write a summary of the information in the visual and of the conclusions you can draw from it.

Practice

Analyze the photograph at right. Then answer the following questions.

❶ What is the title of the photograph?

❷ Where is this scene, and what is happening?

❸ What conclusions can you draw from the information in the photograph?

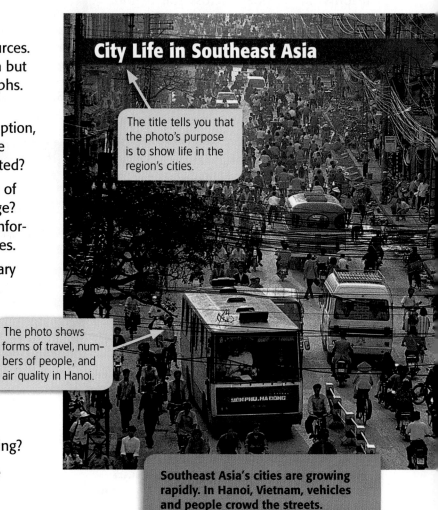

City Life in Southeast Asia

The title tells you that the photo's purpose is to show life in the region's cities.

The photo shows forms of travel, numbers of people, and air quality in Hanoi.

Southeast Asia's cities are growing rapidly. In Hanoi, Vietnam, vehicles and people crowd the streets.

Apply

Analyze the images of the rubber tree plantation in Section 4. Then answer the following questions.

1. What is the purpose of the two photos?

2. What do the photos show about rubber tree farming?

3. Based on the information in the photos, what conclusions can you draw about rubber tree farming in particular and about agriculture in Island Southeast Asia in general?

Mainland Southeast Asia Today

If **YOU** lived there...

You live in Vietnam, where your family works on a collective state-run farm. On the side, your family also sells vegetables. Now your older brother wants to start his own business—a bicycle repair shop. The Communist government allows this, but your parents think it is safer for him to keep working on the farm.

What do you think your brother should do?

BUILDING BACKGROUND After decades of war and hardship, most countries in Mainland Southeast Asia are moving forward. Even those countries with Communist governments, such as Vietnam and Laos, are working to develop freer and stronger economies.

The Area Today

Look at the map at the start of the chapter and identify the countries of Mainland Southeast Asia. These countries include Myanmar, Thailand, Cambodia, Laos, and Vietnam.

War, harsh governments, and other problems have slowed progress in most of Mainland Southeast Asia. However, the area's countries have rich resources and are working to improve their futures. For example, as of 2005 all the countries of Southeast Asia except East Timor had joined the Association of Southeast Asian Nations (ASEAN). This organization promotes political, economic, and social cooperation throughout the region.

Rural Life

Mainland Southeast Asia is largely rural. Most people are farmers who live in small villages and work long hours in the fields. Most farm work is done by hand or using traditional methods. Farmers grow rice, the region's main crop, on fertile slopes along rivers and on terraced shelves of land. The wet, tropical climate enables farmers to grow two or three crops each year.

What You Will Learn...

Main Ideas

1. The area today is largely rural and agricultural, but cities are growing rapidly.
2. Myanmar is poor with a harsh military government, while Thailand is a democracy with a strong economy.
3. The countries of Indochina are poor and struggling to rebuild after years of war.

The Big Idea

Many of the farming countries in Mainland Southeast Asia are poor but are working to improve their economies.

Key Terms and Places

Yangon, *p. 630*
human rights, *p. 630*
Bangkok, *p. 630*
klongs, *p. 630*
Phnom Penh, *p. 633*
Hanoi, *p. 633*

TAKING NOTES As you read, use a chart like this one to take notes on each country in Mainland Southeast Asia. Add a row for each country.

| Country | → | Description |

Most rural people live in the area's fertile river valleys and deltas, which have the best farmland. A delta is an area of fertile land around the mouth of a river. A few people live in remote villages in the rugged, forested mountains. These areas have poor soils that make farming difficult. Many of the people who live there belong to small ethnic groups known as hill peoples.

Urban Life

Although most people live in rural areas, Mainland Southeast Asia has several large cities. Most are growing rapidly as people move to them for work. Rapid growth has led to crowding and pollution. People, bicycles, scooters, cars, and buses clog city streets. Smog hangs in the still air. Growing cities also mix the old and new. Skyscrapers tower over huts, and cars zip past pedicabs, taxicabs that are pedaled like bikes.

FOCUS ON READING
What words in the paragraph to the right tell you the definition of *pedicabs*?

READING CHECK Finding Main Ideas Where do most people in Mainland Southeast Asia live?

BIOGRAPHY

Aung San Suu Kyi
(1945–)

Aung San Suu Kyi has dedicated herself to making life better in her native Myanmar. Suu Kyi is the best-known opponent of the country's harsh military government. Her party, the National League for Democracy (NLD), won control of the country's parliament in 1990. The military government refused to give up power, however. The government then placed Aung San Suu Kyi and other NLD members under house arrest. For her efforts to bring democracy to Myanmar, Suu Kyi received the Nobel Peace Prize in 1991. As of 2005 she remained under house arrest. However, Suu Kyi continues to fight for democratic reform and free elections in Myanmar. Her efforts have led the United States and some Asian countries to press Myanmar's government to change.

Identifying Points of View What does Aung San Suu Kyi hope to achieve through her efforts in Myanmar?

Myanmar and Thailand

Myanmar and Thailand form the northwestern part of Mainland Southeast Asia. While Myanmar is poor, Thailand boasts the area's strongest economy.

Myanmar

Myanmar lies south of China on the Bay of Bengal. Also known as Burma, the country gained independence from Great Britain in 1948. The capital is **Yangon**, or Rangoon. It is a major port on the Andaman Sea.

Most of the people in Myanmar are Burmese. Many live in small farming villages in houses built on stilts. Buddhism is the main religion, and village life often centers around a local Buddhist monastery.

Life is difficult in Myanmar because a harsh military government rules the country. The government abuses **human rights**, rights that all people deserve such as rights to equality and justice. A Burmese woman, Aung San Suu Kyi (awng sahn soo chee), has led a movement for more democracy and rights. She and others have been jailed and harassed for their actions.

Myanmar's poor human-rights record has isolated the country and hurt its economy. Some countries, such as the United States, will no longer trade with Myanmar. Despite rich natural resources—such as oil, timber, metals, jade, and gems—Myanmar and most of its people remain poor.

Thailand

To the southwest of Malaysia is Thailand, once known as Siam. The capital and largest city is **Bangkok**. Modern and crowded, it lies near the mouth of the Chao Phraya (chow PRY-uh) River. Bangkok is known for its many spectacular palaces and Buddhist wats. The city is also famous for its **klongs, or canals**. Klongs are used for transportation and trade, and to drain floodwater.

A Bangkok Canal

Sick of crowded roads? In Bangkok, you can use a network of canals, called klongs, to travel through parts of the city. Water taxis and boats transport people and goods. At floating markets, vendors sell fish, fruit, and other foods to locals and tourists.

To move around the klongs, people use narrow, shallow boats and poles.

ประกาศ
ห้ามเรือทุกชนิดใช้เครื่องยนต์
ตั้งแต่เวลา 08.00-12.00 น.
ฝ่าฝืนมีโทษปรับไม่เกิน 1,000 บาท
สภ.อ.ดำเนินสะดวก

Many sellers wear bamboo hats with wide brims to block the sun and rain.

Small bananas, called finger bananas, are displayed on green banana leaves.

Vendors sell both cooked food and raw produce, such as the pomelos shown here.

ANALYSIS SKILL **ANALYZING VISUALS**

What advantages do you think klongs provide to both travelers and people selling goods?

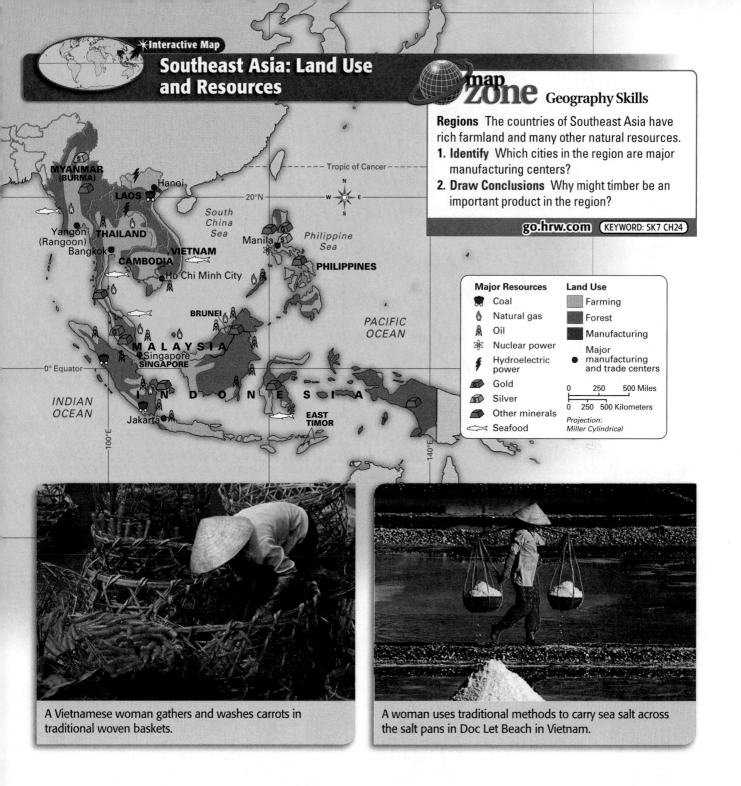

Southeast Asia: Land Use and Resources

Interactive Map

map zone Geography Skills

Regions The countries of Southeast Asia have rich farmland and many other natural resources.
1. **Identify** Which cities in the region are major manufacturing centers?
2. **Draw Conclusions** Why might timber be an important product in the region?

go.hrw.com KEYWORD: SK7 CH24

Major Resources
- Coal
- Natural gas
- Oil
- Nuclear power
- Hydroelectric power
- Gold
- Silver
- Other minerals
- Seafood

Land Use
- Farming
- Forest
- Manufacturing
- Major manufacturing and trade centers

0   250   500 Miles
0   250   500 Kilometers
Projection: Miller Cylindrical

MYANMAR (BURMA) · Hanoi · LAOS · Tropic of Cancer · 20°N · South China Sea · Philippine Sea · Yangon (Rangoon) · THAILAND · Manila · PHILIPPINES · Bangkok · VIETNAM · CAMBODIA · Ho Chi Minh City · BRUNEI · PACIFIC OCEAN · MALAYSIA · Singapore · SINGAPORE · 0° Equator · INDONESIA · INDIAN OCEAN · Jakarta · EAST TIMOR · 100°E · 140°E

A Vietnamese woman gathers and washes carrots in traditional woven baskets.

A woman uses traditional methods to carry sea salt across the salt pans in Doc Let Beach in Vietnam.

Thailand is a constitutional monarchy. A monarch, or king, serves as a ceremonial head of state. A prime minister and elected legislature hold the real power, however.

A democratically elected government and rich resources have helped Thailand's economy to grow. Industry, farming, fishing, mining, and tourism fuel this growth.

Farms produce rice, pineapples, and rubber. Factories produce computers, textiles, and electronics. Magnificent Buddhist wats and unspoiled beaches draw tourists.

READING CHECK Comparing and Contrasting What are some similarities and differences between Myanmar and Thailand?

The Countries of Indochina

The former countries of French Indochina lie to the east and south of Thailand. They are struggling to overcome decades of war.

Cambodia

Cambodia lies to the northeast of the Gulf of Thailand. **Phnom Penh** (puh-NAWM pen) is the capital and chief city. Located in the Mekong River valley, it is a center of trade.

Some 20 years of war, terror, and devastation in Cambodia finally ended in the early 1990s. Today the country has a stable, elected government similar to Thailand's. Years of conflict left their mark, however. Although farming has improved, the country has little industry. In addition, many land mines remain hidden in the land.

Laos

Laos is landlocked with rugged mountains. Poor and undeveloped, it has few roads, no railroads, and limited electricity.

The Communist government of Laos has been increasing economic freedom in hopes of improving the economy. Even so, Laos remains the area's poorest country.

The economy is based on farming, but good farmland is limited. Most people are subsistence farmers, meaning they grow just enough food for their families.

Vietnam

Like Laos, Vietnam is rugged and mountainous. The capital, **Hanoi**, is located in the north in the Hong (Red) River delta. The largest city, Ho Chi Minh City, is in the south in the Mekong delta.

Vietnam's Communist government has been allowing more economic freedom and private business. The changes have helped the economy grow. Most people still farm, but industry and services are expanding. Fishing and mining are also important.

READING CHECK **Evaluating** How would you rate the economies of these three countries?

SUMMARY AND PREVIEW The mainland countries are rural and agricultural with fast-growing cities. Most of the countries are poor despite rich resources. Next, you will read about Island Southeast Asia.

Section 3 Assessment

Reviewing Ideas, Terms, and Places

1. **a. Recall** In what areas do most people in Mainland Southeast Asia live?
   b. Identify Cause and Effect How has rapid growth affected the area's cities?
2. **a. Define** What are **klongs**, and in what ways are they used?
   b. Contrast How does Thailand's economy differ from Myanmar's economy?
   c. Predict How might Myanmar's economy change if the country had a government that respected **human rights**? Explain your answer.
3. **a. Identify** What is the area's poorest country?
   b. Summarize What issues and challenges face Cambodia, Laos, and Vietnam?

Critical Thinking

4. **Categorizing** Draw a chart like the one shown. Use your notes to provide information for each category in the chart.

FOCUS ON SPEAKING

5. **Writing Questions about Mainland Southeast Asia Today** Write one question about each country covered in this section. Your questions might highlight differences among the countries or focus on similarities across the area.

Island Southeast Asia Today

What You Will Learn...

Main Ideas

1. The area today has rich resources and growing cities but faces challenges.
2. Malaysia and its neighbors have strong economies but differ in many ways.
3. Indonesia is big and diverse with a growing economy, and East Timor is small and poor.
4. The Philippines has less ethnic diversity, and its economy is improving.

The Big Idea

The countries of Island Southeast Asia range from wealthy and urban to poor and rural.

Key Terms and Places

kampong, *p. 635*
Jakarta, *p. 635*
Kuala Lumpur, *p. 635*
free ports, *p. 636*
sultan, *p. 637*
Java, *p. 637*
Manila, *p. 638*

TAKING NOTES As you read, use a chart like this one to take notes on each country in Island Southeast Asia. Add a row for each country.

Country → Description

If YOU lived there...

You live in Canada but are visiting your cousins in Singapore. You start to cross the street in the middle of a block, but your cousin quickly stops you. "You have to pay a big fine if you do that!" he says. Singapore has many strict laws and strong punishments, he explains. These laws are meant to make the city safe.

What do you think about Singapore's laws?

BUILDING BACKGROUND Singapore and the other countries of Island Southeast Asia present many contrasts. You can quickly go from skyscrapers to rice paddies to tropical rain forests. Many ethnic groups may live in one country, which can lead to unrest.

The Area Today

Island Southeast Asia lies at a crossroads between major oceans and continents. The area's six countries are Malaysia, Singapore, Brunei (brooh-NY), Indonesia, East Timor, and the Philippines.

The future for these countries could be bright. They have the potential for wealth and good standards of living, such as rich resources and a large, skilled labor force. In addition, most of the countries have growing economies and belong to ASEAN. This organization promotes cooperation in Southeast Asia.

Island Southeast Asia faces challenges, however. First, violent ethnic conflicts have hurt progress in some countries. Second, many people live in poverty, while a few leaders and business-people control much of the money. Third, the area has many environmental problems, such as pollution.

Rural and Urban Life

Many people in Island Southeast Asia live in rural areas, where they farm or fish. As on the mainland, rice is the main crop. Others include coffee, spices, sugarcane, tea, and tropical fruit.

Rubber is a major crop as well, and Indonesia and Malaysia are the world's largest producers of natural rubber. Seafood is the area's main source of protein.

As on the mainland, many people in Island Southeast Asia are leaving rural villages to move to cities for work. The largest cities, the major capitals, are modern and crowded. Common problems in these cities include smog and heavy traffic. Some cities also have large slums.

In Malaysia, Indonesia, and other parts of the area, many people live in kampongs. A **kampong** is a village or city district with traditional houses built on stilts. The stilts protect the houses from flooding, which is common in the area. The term *kampong* also refers to the slums around the area's cities such as **Jakarta**, Indonesia's capital.

READING CHECK **Summarizing** Why could the future be bright for Island Southeast Asia?

Malaysia and Its Neighbors

Malaysia and its much smaller neighbors, Singapore and Brunei, were all once British colonies. Today all three countries are independent and differ in many ways.

Malaysia

Malaysia consists of two parts. One is on the southern end of the Malay Peninsula. The other is on northern Borneo. Most of the country's people live on the peninsula. **Kuala Lumpur** (KWAH-luh LOOHM-poohr), Malaysia's capital, is there as well. The capital is a cultural and economic center.

Malaysia is ethnically diverse. The Malays are the main ethnic group, but many Chinese and other groups live in Malaysia as well. As a result, the country has many languages and religions. Bahasa Malay is the main language, and Islam and Buddhism are the main religions.

Rubber Tree Plantations

Southeast Asia's tropical climate is well suited to rubber trees. At left, a man taps, or cuts, a rubber tree at a Malaysia plantation. A milky liquid drains from the cut into a cup, as shown above. The liquid dries to form a rubbery material.

ANALYZING VISUALS What do you think it is like to work on a rubber tree plantation?

Singapore

INTERVIEW
Lee Kuan Yew on Singapore

Lee Kuan Yew was Singapore's prime minister from 1959 to 1990. He remade the tiny country into an economic power. In a 1994 interview, Lee discussed Singapore's strict laws.

" *The expansion of the right of the individual to behave or misbehave as he pleases has come at the expense of orderly society. In the East the main object is to have a well-ordered society so that everybody can have maximum enjoyment of his freedoms. This freedom can exist only in an ordered state.*"

—from "A Conversation with Lee Kuan Yew"

ANALYSIS SKILL **ANALYZING PRIMARY SOURCES**

Do you agree with Lee that freedom for all can exist only in a society with strict order? Why or why not?

Singapore

A populous country, Singapore is squeezed onto a tiny island at the tip of the Malay Peninsula. The island lies on a major shipping route. This location has helped make Singapore a rich country.

Today Singapore is one of the world's busiest **free ports**, ports that place few if any taxes on goods. It is also an industrial center, and many foreign banks and high-tech firms have located offices there.

Singapore sparkles as the gem of Southeast Asia. The country is modern, wealthy, orderly, and clean. Crime rates are low.

How has Singapore achieved such success? The government has worked hard to clean up slums and improve housing. In addition, laws are extremely strict. To provide **concrete** examples, fines for littering are stiff, and people caught with illegal drugs can be executed. Moreover, the government strictly controls politics and the media. Certain movies are banned, as are satellite dishes. Recently, however, Singapore has loosened up some restrictions.

ACADEMIC VOCABULARY
concrete
specific, real

Malaysia is a constitutional monarchy. The king's duties are largely ceremonial, and local rulers take turns being king. A prime minister and elected legislature hold the real power.

Malaysia's economy is one of the stronger in the area. Well-educated workers and rich resources help drive this economy. The country produces and exports natural rubber, palm oil, electronics, oil, and timber.

Brunei

The tiny country of Brunei is on the island of Borneo, which it shares with Malaysia and Indonesia. A **sultan**, the supreme ruler of a Muslim country, governs Brunei.

The country has grown wealthy from large oil and gas deposits. Because of this wealth, Brunei's citizens do not pay income tax and receive free health care and other benefits. Brunei's oil will run out around 2020, however. As a result, the government is developing other areas of the economy.

READING CHECK **Contrasting** How do Malaysia, Singapore, and Brunei differ?

Indonesia and East Timor

Indonesia is the largest of the island countries. East Timor, once part of Indonesia, is one of the area's smallest countries.

Indonesia

Indonesia has several claims to fame. It is the world's largest archipelago, with some 13,500 islands. It has the fourth-largest population of any country as well as the largest Muslim population. Indonesia is extremely diverse as well, as you have read. It has more than 300 ethnic groups who speak more than 250 languages.

Indonesia's main island is **Java**. The capital, Jakarta, is there, as are more than half of Indonesia's people. For this reason, Java is extremely crowded. To reduce the crowding, the government has been moving people to less-populated islands. Many people on those islands dislike that policy.

Indonesia's rich resources have helped its economy to grow. The main resources include rubber, oil and gas, and timber. The country also has good farmland for rice and other crops. Factories turn out clothing and electronics. Islands such as Bali draw thousands of tourists each year.

At the same time, problems have hurt Indonesia's economy. Many of the people are poor, and unemployment is high. In some areas, ethnic and religious conflicts have led to fighting and terrorism.

East Timor

East Timor is located on the small island of Timor. In 1999 East Timor declared independence from Indonesia. The island then plunged into violence. East Timor only gained its independence after the United Nations sent in troops to restore peace. Years of fighting have left East Timor one of the region's poorest countries. Most people farm, and coffee is the main export.

READING CHECK **Generalizing** How has violence affected Indonesia and East Timor?

FOCUS ON READING
How does the highlighted text help you understand the meaning of *sultan?*

Rice Farming

Terraced rice paddies, such as these in Quezon in the Philippines, are common throughout Southeast Asia.

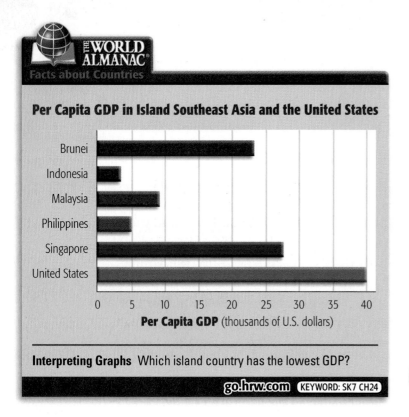

Per Capita GDP in Island Southeast Asia and the United States

Per Capita GDP (thousands of U.S. dollars)

Interpreting Graphs Which island country has the lowest GDP?

go.hrw.com (KEYWORD: SK7 CH24)

The Philippines has many resources to fuel economic growth. Natural resources include copper and other metals, oil, and tropical wood. Farmers grow coconuts, sugarcane, rice, and corn. Factories produce and export clothing and electronics.

Although the economy of the Philippines has improved in recent years, a wide gap still exists between the rich and the poor. A few Filipinos are wealthy. Most, however, are poor farmers who do not own the land they work.

The Philippines has experienced religious conflict as well. Although the country is mainly Roman Catholic, some areas are largely Muslim and want independence.

READING CHECK **Contrasting** How does the Philippines differ from much of the area?

The Philippines

The Philippines includes more than 7,000 islands. The largest and most populated is Luzon, which includes the capital, **Manila**. The Philippines has less ethnic diversity than the other island countries. Almost all Filipinos are ethnic Malays.

SUMMARY AND PREVIEW You have read that Island Southeast Asia has many contrasts. While some countries are wealthy, others are poor. While some countries are modern and urban, others are more traditional and rural. In the next chapter you will read about the Pacific World.

Section 4 Assessment

go.hrw.com
Online Quiz
KEYWORD: SK7 HP24

Reviewing Ideas, Terms, and Places

1. **a. Identify** What problems does the area face?
   b. Compare How does urban life compare between the island and mainland countries?
2. **a. Define** What is a **sultan**?
   b. Explain How have Singapore and Brunei become rich countries?
3. **a. Recall** What island is **Jakarta** located on?
   b. Sequence What series of events led to East Timor's independence?
4. **a. Identify** What are the capital city and the main island in the Philippines?
   b. Analyze Why is the Philippines' economic improvement not benefiting many of its people?

Critical Thinking

5. **Categorizing** Draw a chart like the one shown. Use your notes to provide information for each category in the chart.

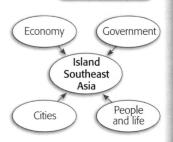

FOCUS ON SPEAKING

6. **Writing Questions about Island Southeast Asia Today** Write one question about each country covered in this section. Your questions might highlight differences among the countries or focus on similarities across the area.

Geography's Impact
video series
Review the video to answer the closing question:
Why do you think it is important to preserve the environment of the Malay Archipelago?

Visual Summary

Use the visual summary below to help you review the main ideas of the chapter.

QUICK FACTS

Southeast Asia is a tropical region of peninsulas, islands, and water. Its history includes empires and colonization.

Mainland Southeast Asia is rural with growing cities. Many of the countries are poor but have rich resources.

Island Southeast Asia is more urban, and some countries are wealthy. Many of the area's people are poor, though.

Reviewing Vocabulary, Terms, and Places

For each group of terms below, write a sentence that shows how all the terms in the group are related.

1. archipelagos
   Indonesia
   Philippines

2. Aung San Suu Kyi
   human rights
   Myanmar

3. Bangkok
   klongs

4. Indochina
   domino theory

5. Jakarta
   kampongs

6. Singapore
   free port

7. Brunei
   sultan

Comprehension and Critical Thinking

SECTION 1 *(Pages 618–621)*

8. **a. Identify** What are the two peninsulas and the two archipelagos that make up the region of Southeast Asia?

   b. Compare and Contrast In what ways are the main climate of Mainland Southeast Asia and of Island Southeast Asia similar and different?

   c. Develop What different needs should people weigh when considering how best to protect the region's tropical rain forests?

SECTION 2 *(Pages 624–627)*

9. **a. Recall** What theory led the U.S. military to become involved in Southeast Asia?

   b. Identify Cause and Effect Why are so many languages spoken in Southeast Asia?

   c. Predict How do you think Southeast Asia might be different today if Europeans had never explored and colonized the area?

SECTION 3 (Pages 629–633)

10. a. Describe Where do most people live and work in Mainland Southeast Asia?

b. Summarize What factors have slowed economic progress in Mainland Southeast Asia?

c. Develop What actions might Myanmar take to try to improve its economy?

SECTION 4 (Pages 634–638)

11. a. Identify Which two countries in Island Southeast Asia have wealthy economies?

b. Compare What are some ways in which Indonesia and the Philippines are similar?

c. Elaborate How has ethnic diversity affected the countries of Island Southeast Asia?

Using the Internet

go.hrw.com
KEYWORD: SK7 CH24

12. Activity: Writing a Report on Rain Forests The tropical rain forests of Indonesia are home to a rich diversity of life. Enter the activity keyword to research these rain forests. Then write a short report that summarizes the threats they face.

FOCUS ON READING AND SPEAKING

Using Context Clues—Definitions *Add a phrase or sentence to provide a definition for the underlined word.*

13. In Thailand, many young men serve for short periods in Buddhist <u>monasteries</u>.

14. Much of the <u>cultivated</u> land in Southeast Asia is used to grow rice.

Presenting an Interview *Use your interview notes to complete the activity below.*

15. Working with your partner, choose the five best questions and answers for your interview. Write a brief introduction and conclusion for the journalist to present at the start and end of the interview. Decide who will play the journalist and who will play the expert. Practice the interview until it sounds natural and then present it to your class.

Social Studies Skills

Analyzing Visuals *Turn to Section 3 and analyze the large photograph of a Bangkok canal. Then answer the following questions about the photograph.*

16. What are the title and location of the photo?

17. How do the captions help you understand the information in the photograph?

18. What types of activities are taking place in the photograph?

19. Based on the information in the photo, what conclusions can you draw about the use of canals in the city of Bangkok?

Map Activity ★Interactive

20. Southeast Asia On a separate sheet of paper, match the letters on the map with their correct labels below.

Bangkok, Thailand	Jakarta, Indonesia
Borneo	Malay Peninsula
Hanoi, Vietnam	Manila, Philippines
Indochina Peninsula	Singapore

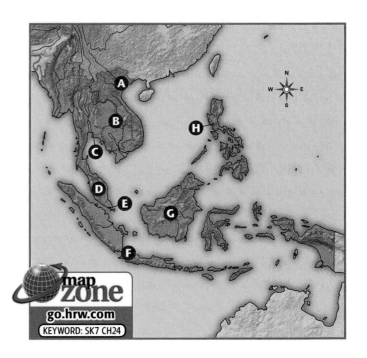

map zone
go.hrw.com
KEYWORD: SK7 CH24

DIRECTIONS: Read questions 1 through 7 and write the letter of the best response. Then read question 8 and write your own well-constructed response.

1 The two peninsulas in Southeast Asia are the Indochina Peninsula and the

A Burma Peninsula.

B Malay Peninsula.

C Philippine Peninsula.

D Thai Peninsula.

2 What is the largest island in this region?

A Bali

B Borneo

C Java

D New Guinea

3 Which early advanced society in Southeast Asia was located in what is now Cambodia?

A Burmese

B Khmer

C Malays

D Thais

4 Which country in Mainland Southeast Asia has a harsh military government?

A Cambodia

B Laos

C Myanmar

D Thailand

5 What interesting feature of Bangkok helps people get around the city?

A kampongs

B klongs

C sultans

D wats

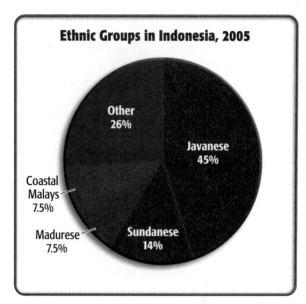

Ethnic Groups in Indonesia, 2005

Other 26%

Javanese 45%

Coastal Malays 7.5%

Madurese 7.5%

Sundanese 14%

Source: Central Intelligence Agency, *The World Factbook 2005*

6 Based on the circle graph above, which of the following was the largest ethnic group in Indonesia in 2005?

A Coastal Malays

B Javanese

C Madurese

D Sundanese

7 Which small country in Island Southeast Asia has become wealthy from oil?

A Brunei

B East Timor

C Indonesia

D Philippines

8 **Extended Response** Examine the Section 1 map titled Southeast Asia: Climate. Based on the information in the map and in the text, write two or three paragraphs explaining how climate affects life in Southeast Asia. Consider plant life, animal life, and how people live and work.

The Pacific World

What You Will Learn...

In this chapter you will learn about the vast world located in the Pacific Ocean. You will study the geography, history, and culture of Australia and New Zealand. You will also discover one of the most unique places in the world—the Pacific Islands. Finally, you will examine the immense and isolated continent of Antarctica.

SECTION 1
Australia and New Zealand **644**

SECTION 2
The Pacific Islands **653**

SECTION 3
Antarctica.............................**658**

FOCUS ON READING AND WRITING

Drawing Conclusions When you read, you discover new information. However, to find out what that information means, you have to draw conclusions. Conclusions are judgments we make as we combine new information with what we already know. As you read about the Pacific world, draw conclusions about the information you come across. **See the lesson, Drawing Conclusions, on page 692.**

Creating a Brochure You work for an advertising agency, and your assignment is to create a brochure about the natural resources in the Pacific world. Your goal is to encourage people to invest money in developing the local economies. As you read this chapter, collect information to use in your brochure.

Geography From Uluru in the dry Australian Outback to freezing Antarctica, the Pacific world is a land of great geographic variety.

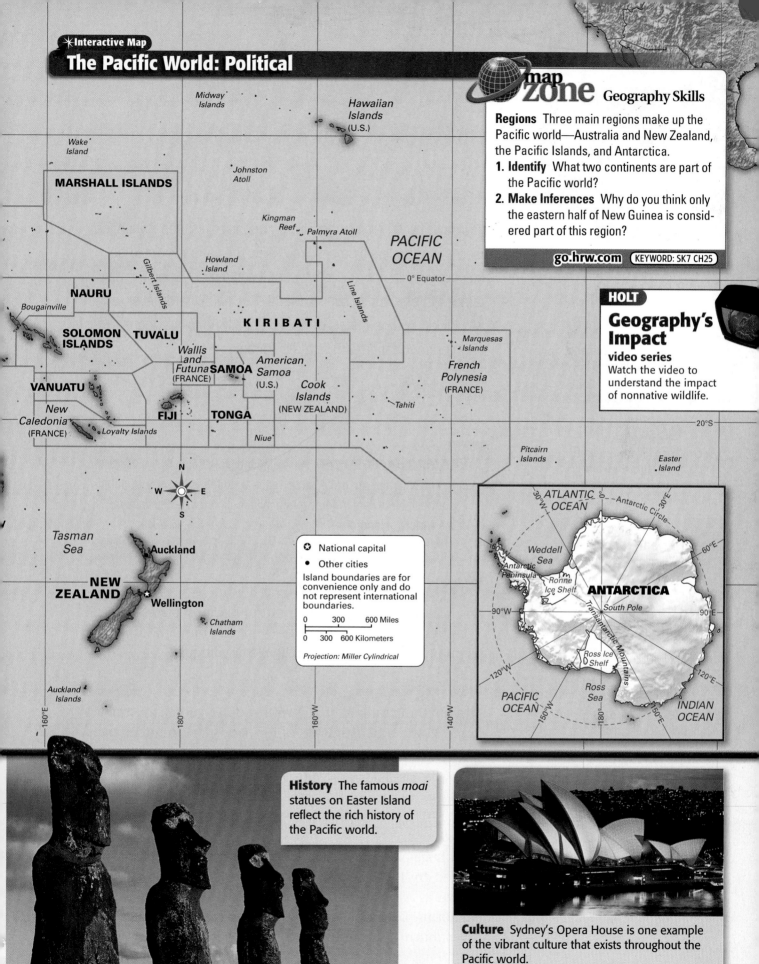

Midway Islands

Hawaiian Islands (U.S.)

Wake Island

MARSHALL ISLANDS

Johnston Atoll

Kingman Reef

Palmyra Atoll

PACIFIC OCEAN

0° Equator

Gilbert Islands

Howland Island

NAURU

Line Islands

Bougainville

K I R I B A T I

Marquesas Islands

SOLOMON ISLANDS

TUVALU

Wallis and Futuna (FRANCE)

SAMOA

American Samoa (U.S.)

Cook Islands (NEW ZEALAND)

French Polynesia (FRANCE)

VANUATU

Tahiti

New Caledonia (FRANCE)

FIJI

TONGA

Niue

Loyalty Islands

Pitcairn Islands

20°S

Easter Island

Tasman Sea

Auckland

N
W   E
S

NEW ZEALAND

Wellington

Chatham Islands

Auckland Islands

● National capital
• Other cities

Island boundaries are for convenience only and do not represent international boundaries.

0   300   600 Miles
0   300   600 Kilometers

Projection: Miller Cylindrical

160°E   180°   160°W   140°W

map zone Geography Skills

Regions Three main regions make up the Pacific world—Australia and New Zealand, the Pacific Islands, and Antarctica.
1. **Identify** What two continents are part of the Pacific world?
2. **Make Inferences** Why do you think only the eastern half of New Guinea is considered part of this region?

go.hrw.com   KEYWORD: SK7 CH25

HOLT
Geography's Impact
video series
Watch the video to understand the impact of nonnative wildlife.

ATLANTIC OCEAN

30°W   Antarctic Circle   30°E

Weddell Sea

60°E

Antarctic Peninsula

Ronne Ice Shelf

ANTARCTICA

90°W   South Pole   90°E

Transantarctic Mountains

120°W

Ross Ice Shelf

120°E

PACIFIC OCEAN

150°W   180°   150°E

Ross Sea

INDIAN OCEAN

History The famous *moai* statues on Easter Island reflect the rich history of the Pacific world.

Culture Sydney's Opera House is one example of the vibrant culture that exists throughout the Pacific world.

Australia and New Zealand

What You Will Learn...

Main Ideas

1. The physical geography of Australia and New Zealand is diverse and unusual.
2. Native peoples and British settlers shaped the history of Australia and New Zealand.
3. Australia and New Zealand today are wealthy and culturally diverse countries.

The Big Idea

Australia and New Zealand share a similar history and culture but have unique natural environments.

Key Terms and Places

Great Barrier Reef, *p. 645*
coral reef, *p. 645*
Aborigines, *p. 647*
Maori, *p. 647*
Outback, *p. 648*

TAKING NOTES As you read, take notes on Australia and New Zealand's physical geography, history, and situation today. Organize your notes in a chart like this one.

	Australia	New Zealand
Physical Geography		
History		
The Region Today		

If YOU lived there...

You have just taken a summer job working at a sheep station, or ranch, in Australia's Outback. You knew the Outback would be hot, but you did not realize how hot it could get! During the day, temperatures climb to over 100°F (40°C), and it hardly ever rains. In addition, you have learned that there are no towns nearby. Your only communication with home is by radio.

How will you adapt to living in the Outback?

BUILDING BACKGROUND Australia and New Zealand are very different. Much of Australia, such as the Outback, is hot, dry, and flat. In contrast, New Zealand has much milder climates, fertile valleys, and a variety of landforms.

Physical Geography

Australia and New Zealand are quite unlike most places on Earth. The physical features, variety of climates, unusual wildlife, and plentiful resources make the region truly unique.

Physical Features

The physical features of the region differ widely. Australia is home to wide, flat stretches of dry land. On the other hand, New Zealand features beautiful green hills and tall mountains.

Australia Similar to an island, Australia is surrounded by water. However, due to its immense size—almost 3 million square miles (7.7 million square km)—geographers consider Australia a continent.

A huge plateau covers the western half of Australia. Mostly flat and dry, this plateau is home to Uluru, a rock formation also known as Ayers Rock. Uluru is one of Australia's best-known landforms. Low mountains, valleys, and a major river system cover much of Eastern Australia. Fertile plains lie along the

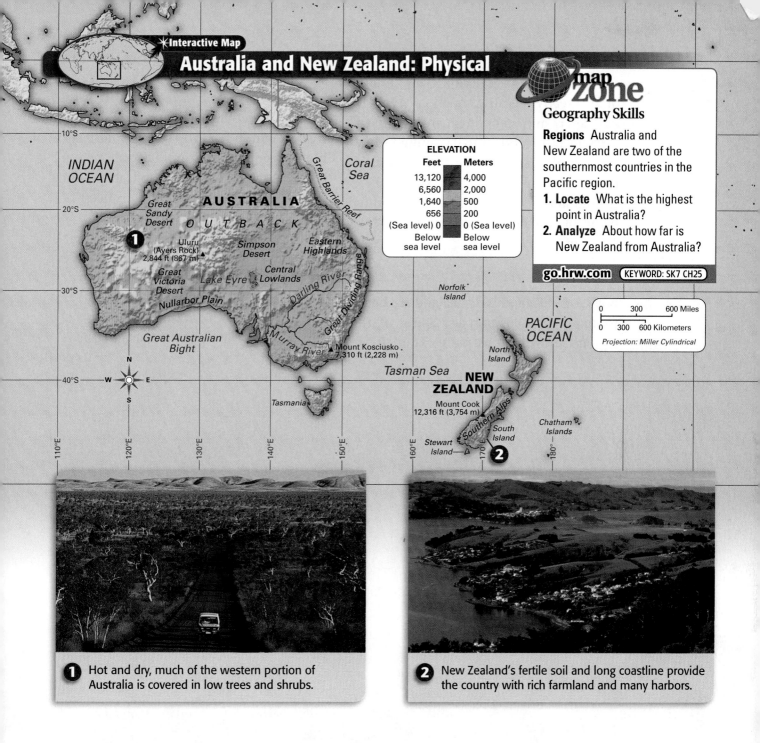

map zone

Geography Skills

Regions Australia and New Zealand are two of the southernmost countries in the Pacific region.

1. **Locate** What is the highest point in Australia?
2. **Analyze** About how far is New Zealand from Australia?

go.hrw.com KEYWORD: SK7 CH25

ELEVATION

Feet	Meters
13,120	4,000
6,560	2,000
1,640	500
656	200
(Sea level) 0	0 (Sea level)
Below sea level	Below sea level

0   300   600 Miles
0   300   600 Kilometers
Projection: Miller Cylindrical

INDIAN OCEAN

AUSTRALIA

OUTBACK

Great Sandy Desert

Uluru (Ayers Rock) 2,844 ft (867 m)

Great Victoria Desert

Simpson Desert

Lake Eyre

Central Lowlands

Eastern Highlands

Nullarbor Plain

Great Australian Bight

Murray River

Darling River

Great Dividing Range

Mount Kosciusko 7,310 ft (2,228 m)

Great Barrier Reef

Coral Sea

Norfolk Island

PACIFIC OCEAN

Tasman Sea

Tasmania

NEW ZEALAND

North Island

Mount Cook 12,316 ft (3,754 m)

Southern Alps

Stewart Island

South Island

Chatham Islands

Caption 1: Hot and dry, much of the western portion of Australia is covered in low trees and shrubs.

Caption 2: New Zealand's fertile soil and long coastline provide the country with rich farmland and many harbors.

coasts. Off Australia's northeastern coast is the **Great Barrier Reef**, the world's largest coral reef. A **coral reef** is a collection of rocky material found in shallow, tropical waters. The Great Barrier Reef is home to an incredible variety of marine animals.

New Zealand New Zealand, located some 1,000 miles southeast of Australia, includes two main islands, North Island and South Island. North Island is covered by hills and coastal plains. It is also home to volcanoes, geysers, and hot springs. One of the key features on South Island is a large mountain range called the Southern Alps. Thick forests, deep lakes, and even glaciers are found in the Southern Alps. The rest of the island is covered by fertile hills and rich plains. Fjords, or narrow inlets of the sea, create many natural harbors along the coasts of both islands.

Climates

The climates of Australia and New Zealand differ greatly. Because much of Australia has desert and steppe climates, temperatures are warm and rainfall is limited. However, along the coasts the climate is more temperate. Unlike Australia, New Zealand is mild and wet. A marine climate brings plentiful rainfall and mild temperatures to much of the country.

Wildlife and Resources

Both Australia and New Zealand are home to many unique animals. Some of the region's most famous native animals are Australia's kangaroo and koala and New Zealand's kiwi, a flightless bird.

Australia is rich in resources. It is the world's top producer of bauxite and lead as well as diamonds and opals. Australia is also home to energy resources like coal, natural gas, and oil. Despite poor soil, farms and ranches raise wheat, cotton, and sheep.

Unlike Australia, New Zealand has a great deal of fertile land but few mineral resources. New Zealand's main resources are wool, timber, and gold.

READING CHECK **Contrasting** How does the physical geography of the two countries differ?

Close-up

Maori Culture

The Maori, the descendants of New Zealand's earliest settlers, lived in small settlements throughout the islands. Their rich culture and traditions are still alive in New Zealand today.

Beautifully decorated storehouses served as a sign of a village's wealth and power. They often held weapons, tools, and foods.

The *moko*, or tattoos, of Maori warriors were symbols of a warrior's bravery. They also helped intimidate the enemy during battle.

The Maori used elaborately carved war canoes to launch attacks on their enemies.

History

Despite their many geographic differences, Australia and New Zealand share a similar history. Both countries were originally inhabited by settlers from other parts of the Pacific. Later, both Australia and New Zealand were colonized by the British.

Early Settlers

The first settlers in Australia likely migrated there from Southeast Asia at least 40,000 years ago. These settlers, the **Aborigines** (a-buh-RIJ-uh-nees), were the first humans to live in Australia. Early Aborigines were nomads who gathered various plants and hunted animals with boomerangs and spears. Nature played an important role in the religion of the early Aborigines, who believed that it was their duty to preserve the land.

New Zealand's first settlers came from other Pacific islands more recently, about 1,200 years ago. The descendants of these early settlers, the **Maori** (MOWR-ee), settled throughout New Zealand. Like Australia's Aborigines, the Maori were fishers and hunters. Unlike the Aborigines, however, the Maori also used farming to survive.

The Arrival of Europeans

European explorers first sighted Australia and New Zealand in the 1600s. It wasn't until later, however, that Europeans began to explore the region. In 1769 British explorer James Cook explored the main islands of New Zealand. The following year, Cook landed on the east coast of Australia and claimed the land for Britain.

Within 20 years of Cook's claim, the British began settling in Australia. Many of the first to arrive were British prisoners, but other settlers came, too. As the settlers built farms and ranches, they took over the Aborigines' lands. Many Aborigines died of diseases introduced by the Europeans.

In New Zealand, large numbers of British settlers started to arrive in the early 1800s. After the British signed a treaty with the Maori in 1840, New Zealand became a part of the British Empire. However, tensions between the Maori and British settlers led to a series of wars over land.

Australia and New Zealand both gained their independence in the early 1900s. Today the two countries are members of the British Commonwealth of Nations and are close allies of the United Kingdom.

FOCUS ON READING
What conclusions can you draw about why European settlers were attracted to Australia?

Maori life centered around a village meetinghouse, where important gatherings like weddings and funerals were held.

ANALYSIS SKILL **ANALYZING VISUALS**
Based on the illustration, what elements were important in Maori culture?

READING CHECK Finding Main Ideas How did early settlers influence the region?

THE PACIFIC WORLD **647**

Australian Sports

Outdoor sports are tremendously popular in sunny Australia. Some of Australia's most popular activities include water sports, such as swimming, surfing, and water polo. In recent years, many Australians have dominated the swimming competition at the summer Olympic Games.

Australia's national sport is cricket, a game played with a bat and ball. Cricket was first introduced to Australia by British settlers. Other popular sports with British roots are rugby and Australian Rules football. These two sports allow players to kick, carry, or pass the ball with their hands or feet. Every year hundreds of thousands of Australians attend professional rugby matches like the one in the photo below.

Drawing Conclusions Why do you think outdoor sports are so popular in Australia?

Australia and New Zealand Today

Despite their isolation from other nations, Australia and New Zealand today are rich and well-developed. Their governments, economies, and people make them among the world's most successful countries.

Government

As former British colonies, the British style of government has influenced both Australia and New Zealand. As a result, both countries have similar governments.

For example, the British monarch is the head of state in both Australia and New Zealand. Both countries are parliamentary democracies, a type of government in which citizens elect members to represent them in a parliament. Each country has a prime minister. The prime minister, along with Parliament, runs the government.

The governments of Australia and New Zealand have many features in common with the U.S. government. For example, Australia has a federal system like that of the United States. In this system, a central government shares power with the states. Australia's Parliament, similar to the U.S. Congress, consists of two houses—a House of Representatives and a Senate. A Bill of Rights also protects the individual rights of New Zealand's citizens.

Economy

Australia and New Zealand are both rich, economically developed countries. Agriculture is a major part of their economies. The two countries are among the world's top producers of wool. In fact, Australia regularly supplies about one-quarter of the wool used in clothing. Both countries also export meat and dairy products.

Australia and New Zealand also have other important industries. Mining is one of Australia's main industries. Companies mine bauxite, gold, and uranium throughout the **Outback**, Australia's interior. Other industries include steel, heavy machines, and computers. New Zealand has also become more industrialized in recent years. Factories turn out processed food, clothing, and paper products. Banking, insurance, and tourism are also important industries.

People

Today Australia and New Zealand have diverse populations. Most Australians and New Zealanders are of British ancestry. In recent years, however, peoples from around the world have migrated to the region. For example, since the 1970s Asians and Pacific Islanders have settled in Australia and New Zealand in growing numbers.

Native Maori and Aborigines make up only a small percentage of New Zealand's and Australia's populations. One challenge facing both countries today is improving the economic and political status of the those populations. Many of the region's Maori and Aborigines trail the rest of the population in terms of education, land ownership, and employment.

Most Australians and New Zealanders live in urban areas. About 85 percent of Australia's population lives in large cities along the coasts. Sydney and Melbourne, Australia's two largest cities, are home to almost 8 million people. Rural areas like the Outback, on the other hand, have less than 15 percent of the population. In New Zealand, a majority of the population lives on the North Island. There, large cities like Auckland are common.

READING CHECK **Summarizing** What are the economic strengths of these countries?

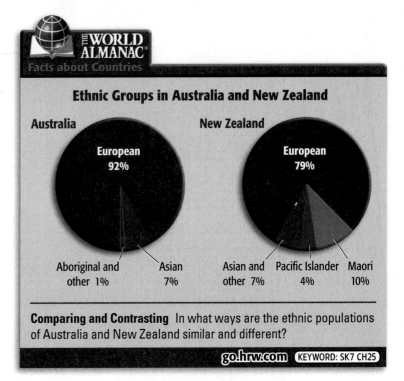

THE WORLD ALMANAC®
Facts about Countries

Ethnic Groups in Australia and New Zealand

Australia

European 92%

Aboriginal and other 1%   Asian 7%

New Zealand

European 79%

Asian and other 7%   Pacific Islander 4%   Maori 10%

Comparing and Contrasting In what ways are the ethnic populations of Australia and New Zealand similar and different?

go.hrw.com [KEYWORD: SK7 CH25]

SUMMARY AND PREVIEW Despite their geographical differences, Australia and New Zealand have much in common. The two countries share a similar history, culture, and economy. In the next section you will learn about another region in the Pacific world—the Pacific Islands.

go.hrw.com
Online Quiz
KEYWORD: SK7 HP25

Section 1 Assessment

Reviewing Ideas, Terms, and Places

1. **a. Identify** What is the **Great Barrier Reef**? Where is it located?
   b. Elaborate Given its harsh climate, why do you think so many people have settled in Australia?
2. **a. Describe** Who are the **Maori**? From where did they originate?
   b. Draw Conclusions How might the **Aborigines'** relationship with nature have differed from that of other peoples?
3. **a. Recall** Where do most Australians and New Zealanders live?
   b. Compare and Contrast How are the governments of Australia and New Zealand similar to and different from that of the United States?

Critical Thinking

4. **Comparing and Contrasting** Use your notes and a diagram like the one here to compare and contrast the geography, history, and culture of Australia and New Zealand.

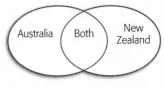

Australia   Both   New Zealand

FOCUS ON WRITING

5. **Describing Australia and New Zealand** What natural resources do these two countries produce? Make a list of the ones people might want to invest in. What illustrations could you include?

Settling the Pacific

For years scholars have puzzled over a mystery in the Pacific. How exactly did humans reach the thousands of islands scattered throughout the Pacific world? While we don't know all the details, evidence suggests that people from Southeast Asia originally settled the Pacific Islands. Over tens of thousands of years these people and their descendants slowly migrated throughout the Pacific. Thanks to expert canoe-building and navigational skills, these settlers reached islands many thousands of miles apart.

Many long-distance voyages were likely made on double canoes built with two hulls carved from huge logs.

Stick Charts Stick charts like this one from the Marshall Islands were probably used to navigate from island to island. The sticks show the direction of waves and ocean currents. Shells indicate the location of islands and other landmarks.

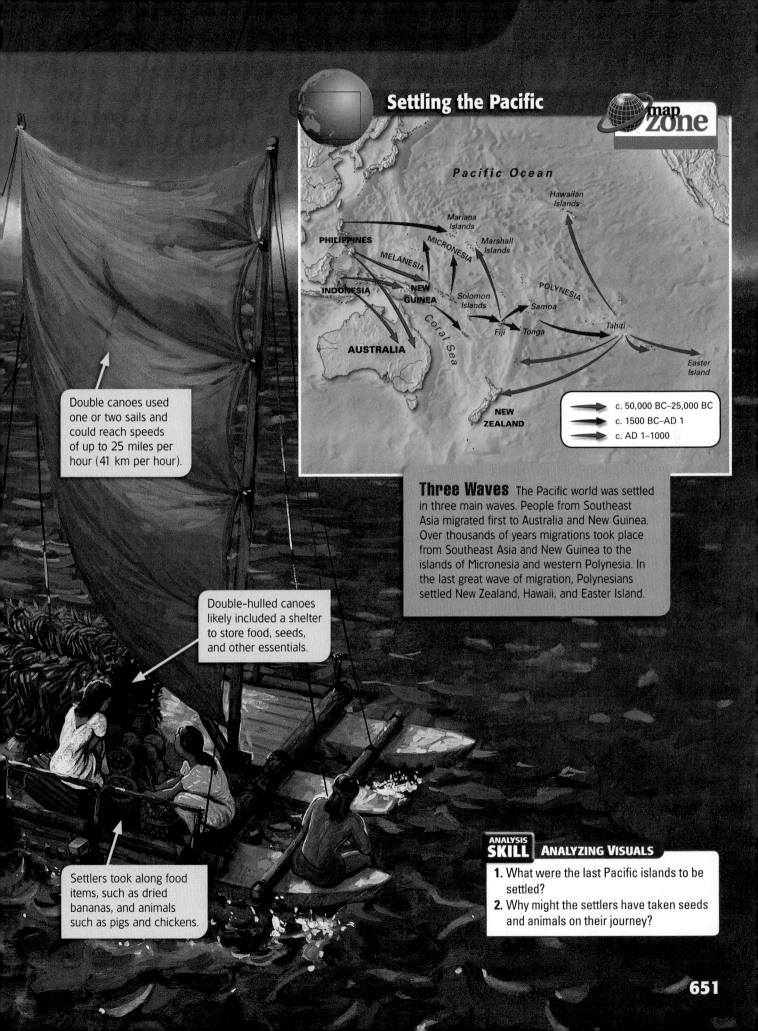

Settling the Pacific

map zone

Pacific Ocean

Hawaiian Islands

Mariana Islands

PHILIPPINES

MICRONESIA

Marshall Islands

MELANESIA

NEW GUINEA

INDONESIA

Solomon Islands

POLYNESIA

Samoa

Coral Sea

Fiji   Tonga

Tahiti

AUSTRALIA

Easter Island

NEW ZEALAND

→	c. 50,000 BC–25,000 BC
→	c. 1500 BC–AD 1
→	c. AD 1–1000

Double canoes used one or two sails and could reach speeds of up to 25 miles per hour (41 km per hour).

Double-hulled canoes likely included a shelter to store food, seeds, and other essentials.

Three Waves The Pacific world was settled in three main waves. People from Southeast Asia migrated first to Australia and New Guinea. Over thousands of years migrations took place from Southeast Asia and New Guinea to the islands of Micronesia and western Polynesia. In the last great wave of migration, Polynesians settled New Zealand, Hawaii, and Easter Island.

Settlers took along food items, such as dried bananas, and animals such as pigs and chickens.

ANALYSIS SKILL **ANALYZING VISUALS**

1. What were the last Pacific islands to be settled?
2. Why might the settlers have taken seeds and animals on their journey?

Locating Information

Learn

Your teacher has asked you to find information about New Zealand's Maori. Where should you go? What should you do? The best place to start your search for information is in the library. The chart at right includes some library resources you may find helpful.

Practice

Determine which of the sources described here you would most likely use to locate the information in the questions that follow.

1 Which different sources could you use to find information about Maori culture?

2 In which source would you most likely find maps of Maori migration routes to New Zealand?

3 Where might you look to find videos about Maori art and music?

4 Which resource would be best for locating information about the current population of Maori in New Zealand?

Library Resources	
almanac	a collection of current statistics and general information usually published annually
atlas	a collection of maps and charts
electronic database	a collection of information you can access and search by computer
encyclopedias	books or computer software with short articles on a variety of subjects, usually arranged in alphabetical order
magazine and newspaper indexes	listings of recent and past articles from newspapers and magazines
online catalog	a computerized listing of books, videos, and other library resources; you search for resources by title, author, keyword, or subject
World Wide Web	a collection of information on the Internet; if you use a Web site, be sure to carefully examine its reliability, or trustworthiness

Apply

Use resources from a local library to answer the questions below.

1. About when did the Maori first settle in New Zealand?

2. What different subtopics can you find on the Maori in the library catalog?

3. Write a list of important facts about the Maori.

The Pacific Islands

If YOU lived there...

You live on a small island in the South Pacific. For many years, the people on your island have made their living by fishing. Now, however, a European company has expressed interest in building an airport and a luxury hotel on your island. It hopes that tourists will be drawn by the island's dazzling beaches and tropical climate. The company's leaders want your permission before they build.

Will you give them permission? Why or why not?

BUILDING BACKGROUND Thousands of islands are scattered across the Pacific Ocean. Many of these islands are tiny and have few mineral resources. Among the resources they do have are pleasant climates and scenic landscapes. As a result, many Pacific islands have become popular tourist destinations.

Physical Geography

The Pacific Ocean covers more than one-third of Earth's surface. Scattered throughout this ocean are thousands of islands with similar physical features, climates, and resources.

Island Regions

We divide the Pacific Islands into three regions—Micronesia, Melanesia, and Polynesia—based on their culture and geography. **Micronesia**, which means "tiny islands," is located just east of the Philippines. Some 2,000 small islands make up this region. South of Micronesia is **Melanesia**, which stretches from New Guinea in the west to Fiji in the east. Melanesia is the most heavily populated Pacific Island region. The largest region is **Polynesia**, which means "many islands." Among Polynesia's many islands are Tonga, Samoa, and the Hawaiian Islands.

Physical Features

The Pacific Islands differ greatly. Some islands, like New Guinea (GI-nee), cover thousands of square miles. Other islands are tiny. For example, Nauru covers only 8 square miles (21 square km).

What You Will Learn...

Main Ideas

1. Unique physical features, tropical climates, and limited resources shape the physical geography of the Pacific Islands.
2. Native customs and contact with the western world have influenced the history and culture of the Pacific Islands.
3. Pacific Islanders today are working to improve their economies and protect the environment.

The Big Idea

The Pacific islands have tropical climates, rich cultures, and unique challenges.

Key Terms and Places

Micronesia, *p. 653*
Melanesia, *p. 653*
Polynesia, *p. 653*
atoll, *p. 654*
territory, *p. 655*

TAKING NOTES As you read, use a diagram like this one to take notes on the Pacific Islands.

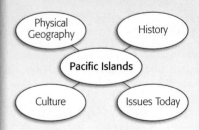

Geographers classify the islands of the Pacific as either high islands or low islands. High islands tend to be mountainous and rocky. Most high islands are volcanic islands. They were formed when volcanic mountains grew from the ocean floor and reached the surface. The islands of Tahiti and Hawaii in Polynesia are examples of high islands. Other high islands, such as New Guinea, are formed from continental rock rather than volcanoes. For example, the country of Papua (PA-pyooh-wuh) New Guinea, located on the eastern half of the island of New Guinea, has rocky mountains that rise above 13,000 feet (3,960 m).

Low islands are typically much smaller than high islands. Most barely rise above sea level. Many low islands are atolls. An atoll is a small, ring-shaped coral island that surrounds a lagoon. Wake Island, west of the Hawaiian Islands, is an example of an atoll. Wake Island rises only 21 feet (6.4 m) above sea level and covers only 2.5 square miles (6.5 square km).

High and Low Islands

Many high islands, like the island of Hawaii, often have mountainous terrain, rich soils, and dense rain forests. Many low islands, like this small island in the Society Islands chain, are formed from coral reefs. Because most low islands have poor soils, agriculture is limited.

Climate and Resources

All but two of the Pacific Island countries lie in the tropics. As a result, most islands have a humid tropical climate. Rain falls all year and temperatures are warm. Tropical savanna climates with rainy and dry seasons exist in a few places, such as New Caledonia. The mountains of New Guinea are home to a cool highland climate.

Resources in the Pacific Islands vary widely. Most low islands have thin soils and little vegetation. They have few trees other than the coconut palm. In addition, low islands have few mineral or energy resources. Partly because of these conditions, low islands have small populations.

In contrast to low islands, the Pacific's high islands have many natural resources. Volcanic soils provide fertile farmland and dense forests. Farms produce crops such as coffee, cocoa, bananas, and sugarcane. Some high islands also have many mineral resources. Papua New Guinea, for example, exports gold, copper, and oil.

READING CHECK Contrasting How do the Pacific's low islands differ from high islands?

The Formation of an Atoll

The Pacific Islands are home to many atolls, or small coral islands that surround shallow lagoons. Coral reefs are formed from the skeletons of many tiny sea animals. When a coral reef forms on the edges of a volcanic island, it often forms a barrier reef around the island.

As the volcanic island sinks, the coral remains. Sand and other debris gradually collects on the reef's surface, raising the land above sea level. Eventually, all that remains is an atoll.

Sequencing Describe the process in which atolls form.

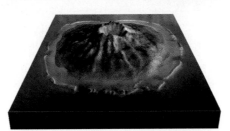

Coral reefs will sometimes form along the edges of a volcanic island, creating a ring around the island.

As the island sinks into the ocean floor, the coral reef grows upward and forms an offshore barrier reef.

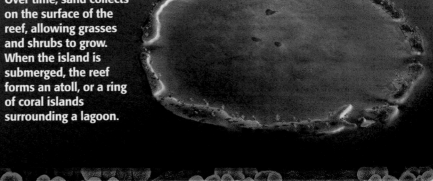

Over time, sand collects on the surface of the reef, allowing grasses and shrubs to grow. When the island is submerged, the reef forms an atoll, or a ring of coral islands surrounding a lagoon.

History and Culture

The Pacific Islands were one of the last places settled by humans. Because of their isolation from other civilizations, the islands have a unique history and culture.

Early History

Scholars believe that people began settling the Pacific Islands at least 35,000 years ago. The large islands of Melanesia were the first to be settled. Over time, people spread to the islands of Micronesia and Polynesia.

Europeans first encountered the Pacific Islands in the 1500s. Two centuries later, British captain James Cook explored all the main Pacific Island regions. By the late 1800s European powers such as Spain, Great Britain, and France controlled most of the Pacific Islands.

Modern History

By the early 1900s, other countries were entering the Pacific as well. In 1898 the United States defeated Spain in the Spanish-American War. As a result, Guam became a U.S. territory. A **territory** is an area that is under the authority of another government. Japan also expanded its empire into the Pacific Ocean in the early 1900s. In World War II, the Pacific Islands were the scene of many tough battles between Allied and Japanese forces. After Japan's defeat in 1945, the United Nations placed some islands under the control of the United States and other Allies.

In the last half of the 1900s many Pacific Islands gained their independence. However, several countries—including the United States, France, and New Zealand—still have territories in the Pacific Islands.

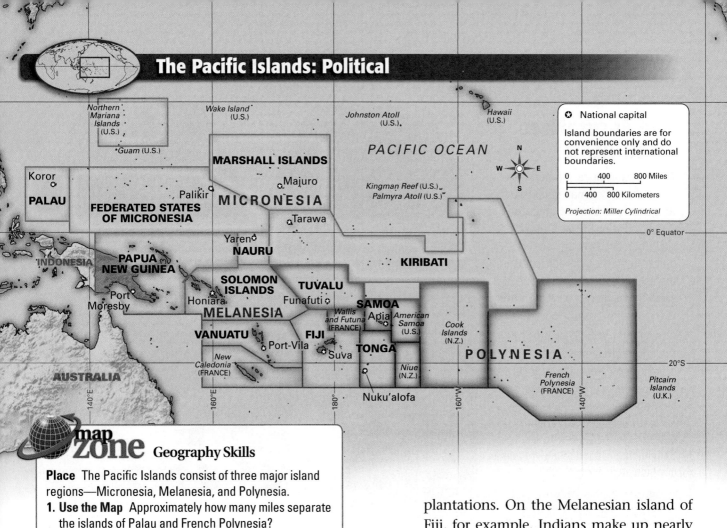

The Pacific Islands: Political

Northern Mariana Islands (U.S.)

Guam (U.S.)

Wake Island (U.S.)

Johnston Atoll (U.S.)

Hawaii (U.S.)

PACIFIC OCEAN

MARSHALL ISLANDS

Majuro

Kingman Reef (U.S.)
Palmyra Atoll (U.S.)

Koror

PALAU

Palikir

MICRONESIA

FEDERATED STATES OF MICRONESIA

Tarawa

Yaren

NAURU

KIRIBATI

0° Equator

INDONESIA

PAPUA NEW GUINEA

Port Moresby

SOLOMON ISLANDS

Honiara

MELANESIA

TUVALU

Funafuti

Wallis and Futuna (FRANCE)

SAMOA

Apia

American Samoa (U.S.)

Cook Islands (N.Z.)

VANUATU

Port-Vila

FIJI

Suva

TONGA

POLYNESIA

New Caledonia (FRANCE)

Niue (N.Z.)

French Polynesia (FRANCE)

Pitcairn Islands (U.K.)

20°S

AUSTRALIA

Nuku'alofa

140°E 160°E 180° 160°W 140°W

National capital

Island boundaries are for convenience only and do not represent international boundaries.

0   400   800 Miles
0   400   800 Kilometers

Projection: Miller Cylindrical

map zone Geography Skills

Place The Pacific Islands consist of three major island regions—Micronesia, Melanesia, and Polynesia.
1. **Use the Map** Approximately how many miles separate the islands of Palau and French Polynesia?
2. **Contrast** Based on the map, how do the Melanesian islands differ from those of French Polynesia?

Culture

A variety of cultures thrive throughout the Pacific Islands. Some culture traits, such as fishing, are common throughout the entire region. Others are only found on a specific island or island chain.

FOCUS ON READING

What can you conclude about the influence other cultures have had on the Pacific Islands?

People Close to 9 million people live in the Pacific Islands today. Most Pacific Islanders are descendants of the region's original settlers. However, the population of the Pacific Islands also includes large numbers of ethnic Europeans and Asians, particularly Indians and Chinese. Many ethnic Asians are descended from people brought to the islands to work on colonial plantations. On the Melanesian island of Fiji, for example, Indians make up nearly half of the population.

Before the arrival of Europeans, the people of the Pacific Islands practiced hundreds of different religions. Today most Pacific Islanders are Christian. In Melanesia, however, some people continue to practice traditional local religions.

Traditions Although modern culture exists throughout the Pacific Islands, many people continue to practice traditional customs. In parts of Polynesia, for example, people still construct their homes from bamboo and palm leaves. Many Pacific Islanders today continue to live in ancient villages, practice customary art styles, and hold ceremonies that feature traditional costumes and dances.

READING CHECK **Making Inferences** In what ways have the Pacific Islands been influenced by contact with westerners?

The Pacific Islands Today

Many people imagine sunny beaches and tourists when they think of the Pacific Islands today. Despite the region's healthy tourism industry, however, Pacific Island countries face important challenges.

The countries of the Pacific Islands have developing economies. Fishing, tourism, and agriculture are key industries. Some countries, particularly Papua New Guinea, export minerals and timber. The region's isolation from other countries, however, hinders its ability to trade.

The environment is an important concern in the Pacific Islands. The Pacific Islands were used for nuclear testing grounds from the 1940s to the 1990s. Many people fear that one **effect** of these tests may be health problems for people in the region. Global warming also concerns Pacific Islanders. Some researchers believe that rising temperatures may cause polar ice to melt. The rise in ocean levels would threaten low-lying Pacific Islands.

READING CHECK **Summarizing** What are some challenges Pacific Islanders face today?

Villagers on Tanna Island in Vanuatu perform a traditional dance.

SUMMARY AND PREVIEW The Pacific Islands are one of the most isolated regions in the world. As a result, unique cultures and challenges exist in the region. In the next section you will learn about another isolated part of the globe—Antarctica.

Section 2 Assessment

go.hrw.com
Online Quiz
KEYWORD: SK7 HP25

Reviewing Ideas, Terms, and Places

1. **a. Describe** Into what regions are the Pacific Islands divided?
   b. Draw Conclusions Why might high islands have larger populations than low islands?
2. **a. Define** What is a **territory**?
   b. Make Inferences Why did other countries seek to control the Pacific Islands?
   c. Elaborate Why do you think that many Pacific Islanders continue to practice traditional customs?
3. **a. Recall** What economic resources are available to the Pacific Islands?
   b. Predict How might the Pacific Islands be affected by global warming in the future?

Critical Thinking

4. **Finding Main Ideas** Draw a chart like the one shown. Using your notes, identify the main idea of each topic and write a sentence for each.

Physical Geography	History	Culture	Issues Today

FOCUS ON WRITING

5. **Telling about the Resources of the Pacific** Add to your list by noting the natural resources of the Pacific Islands. Which resources will you describe in your brochure? How might you describe them?

Antarctica

If **YOU** lived there...

You are a scientist working at a research laboratory in Antarctica. One day you receive an e-mail message from a friend. She wants to open a company that will lead public tours through Antarctica so people can see its spectacular icy landscapes and wildlife. Some of your fellow scientists think that tours are a good idea, while others think that they could ruin the local environment.

What will you tell your friend?

BUILDING BACKGROUND Antarctica, the continent surrounding the South Pole, has no permanent residents. The only people there are scientists who research the frozen land. For many years, people around the world have debated the best way to use this frozen land.

Physical Geography

In the southernmost part of the world is the continent of Antarctica. This frozen land is very different from any other place on Earth.

The Land

Ice covers about 98 percent of Antarctica's 5.4 million square miles (14 million square km). This ice sheet contains more than 90 percent of the world's ice. On average the ice sheet is more than 1 mile (1.6 km) thick.

Penguins live in the icy waters around Antarctica, a continent almost completely covered in ice.

The weight of Antarctica's ice sheet causes ice to flow slowly off the continent. As the ice reaches the coast, it forms a ledge over the surrounding seas. This ledge of ice that extends over the water is called an **ice shelf**. Antarctica's ice shelves are huge. In fact, the Ross Ice Shelf, Antarctica's largest, is about the size of France.

Sometimes parts of the ice shelf break off into the surrounding water. Floating masses of ice that have broken off a glacier are **icebergs**. When one iceberg recently formed, it was approximately the size of the country of Luxembourg.

In western Antarctica, the **Antarctic Peninsula** extends north of the Antarctic Circle. As a result, temperatures there are often warmer than in other parts of the continent.

Climate and Resources

Most of Antarctica's interior is dominated by a freezing ice-cap climate. Temperatures can drop below −120°F (−84°C), and very little precipitation falls. As a result, much of Antarctica is considered a **polar desert**, a high-latitude region that receives very little precipitation. The precipitation that does fall does not melt due to the cold temperatures. Instead, it remains as ice.

Because of Antarctica's high latitude, the continent is in almost total darkness during winter months. Seas clog with ice as a result of the extreme temperatures. In the summer, the sun shines around the clock and temperatures rise to near freezing.

Plant life only survives in the ice-free tundra areas. Insects are the frozen land's only land animals. Penguins, seals, and whales live in Antarctica's waters. Antarctica has many mineral resources, including iron ore, gold, copper, and coal.

READING CHECK **Summarizing** What are the physical features and resources of Antarctica?

Primary Source

BOOK
Crossing Antarctica

In 1989 a six-person team set off to cross Antarctica on foot. The 3,700-mile journey took seven months to complete. Team member Will Steger describes his first view of the continent.

Now, flying over the iceberg-laden Weddell Sea, the biggest adventure of my life was about to begin . . .

To the south I could barely pick out the peaks of mountains, mountains I knew jutted three thousand feet into the air. They lined the peninsula's coast for hundreds of miles. Leading up to them was a two-mile-wide sheet of snow and ice, preceded by the blue of the sea. It was a picture of purity, similar to many I had seen in the picture books . . .

—from *Crossing Antarctica*, by Will Steger and John Bowermaster

ANALYSIS SKILL **ANALYZING PRIMARY SOURCES**

What physical features does the author notice on his trip over Antarctica?

Antarctic Exploration

BIOGRAPHY

Sir Ernest Shackleton
(1874–1922)

Irish-born Ernest Shackleton was one of several early explorers of Antarctica. Shackleton led a British expedition in 1907–1909 that climbed Mt. Erebus, an active volcano, discovered the Beardmore Glacier, and came within 97 miles of the South Pole—the farthest south anyone had ever been.

In the early 1900s several expeditions set out to find the South Pole. The first to reach the pole were members of a Norwegian expedition led by Roald Amundsen. In this photo a member of the Norwegian expedition poses with his team of dogs near the flag that marks the South Pole.

Early Explorations

The discovery of Antarctica is a fairly recent one. Although explorers long believed there was a southern continent, it was not until 1775 that James Cook first sighted the Antarctic Peninsula. In the 1800s explorers first investigated Antarctica. One **motive** of many explorers was to discover the South Pole and other new lands. In 1911 a team of Norwegian explorers became the first people to reach the South Pole.

Since then, several countries—including the United States, Australia, and Chile—have claimed parts of Antarctica. In 1959 the international Antarctic Treaty was signed to preserve the continent "for science and peace." This treaty banned military activity in Antarctica and set aside the entire continent for research.

READING CHECK **Making Inferences** Why do you think Antarctica is set aside for research?

Antarctica Today

Today Antarctica is the only continent without a permanent human population. Scientists use the continent to conduct research and to monitor the environment.

Scientific Research

While they are conducting research in Antarctica, researchers live in bases, or stations. Several countries, including the United States, the United Kingdom, and Russia, have bases in Antarctica.

Antarctic research covers a wide range of topics. Some scientists concentrate on the continent's plant and animal life. Others examine weather conditions. One group of researchers is studying Earth's ozone layer. The **ozone layer** is a layer of Earth's atmosphere that protects living things from the harmful effects of the sun's ultraviolet rays. Scientists have found a thinning in the ozone layer above Antarctica.

Environmental Threats

Many people today are concerned about Antarctica's environment. Over the years, researchers and tourists have left behind trash and sewage, polluting the environment. Oil spills have damaged surrounding seas. In addition, companies have hoped to exploit Antarctica's valuable resources.

Some people fear that any mining of the resources in Antarctica will result in more environmental problems. To prevent this, a new international agreement was reached in 1991. This agreement forbids most activities that do not have a scientific purpose. It bans mining and drilling and limits tourism.

READING CHECK **Finding Main Ideas** What are some issues that affect Antarctica today?

SUMMARY In this section, you have learned about Antarctica's unusual physical geography and harsh climates. Despite the difficulty of living in such harsh conditions, Antarctica remains an important place for scientific research.

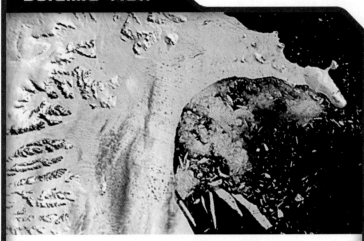

Satellite View

Antarctica's Ice Shelves

Antarctica is home to many large ice shelves. An ice shelf is a piece of a glacier that extends over the surrounding seas. In recent years, scientists have become concerned that rising temperatures on the planet are causing the rapid disintegration of some of Antarctica's ice shelves. This satellite image from 2002 shows the breakup of a huge portion of Antarctica's Larsen B Ice Shelf, located on the Antarctic Peninsula. The breakup of this ice shelf released some 720 billion tons of ice into the Weddell Sea.

Identifying Cause and Effect What do scientists believe has led to growing disintegration of Antarctica's ice shelves?

---

go.hrw.com
Online Quiz
KEYWORD: SK7 HP25

Section 3 Assessment

Reviewing Ideas, Terms, and Places

1. **a. Define** What are **ice shelves** and **icebergs**?
   b. Contrast How does Antarctica differ from most other continents?
   c. Elaborate What aspects of Antarctica's physical geography would you most like to see? Why?
2. **a. Identify** What was the Antarctic Treaty of 1959?
   b. Predict What might have happened if countries had not agreed to preserve Antarctica for research?
3. **a. Recall** What is Antarctica used for today?
   b. Analyze How has Antarctic research benefited science?
   c. Elaborate Do you agree with bans on tourism and mining in Antarctica? Why or why not?

Critical Thinking

4. **Summarizing** Draw a diagram like the one here. Use your notes to list three facts about each aspect of Antarctica's physical geography.

Physical Geography	→	
Early Explorers	→	
Antarctica Today	→	

FOCUS ON WRITING

5. **Describing Antarctica** In your notebook, describe the natural resources of Antarctica. Decide which to include in your brochure. What illustrations might you include?

Adelie penguins jump off an iceberg near Antarctica.

from
Antarctic Journal:
Four Months at the Bottom of the World

by Jennifer Owings Dewey

About the Reading *In her book,* Antarctic Journal: Four Months at the Bottom of the World, *writer and artist Jennifer Owings Dewey describes her four-month visit to Antarctica as a visiting researcher at Palmer research station.*

AS YOU READ Identify what the icebergs mean to the author.

Coming back we see icebergs drifting south out of the Weddell Sea. ❶ The bergs originate hundreds of miles away and ride ocean currents.

We sail close, but not too close, for beneath the waves is where the bulk of an iceberg is.

Seawater splashes up on iceberg shores shaped by years of wave action. Sunlight strikes gleaming ramparts that shine with rainbow colors. Erosion works at the ice, creating caves and hollows, coves and inlets.

Penguins and seals hitch rides on icebergs. Gulls and other seabirds rest on high points.

One iceberg collides in slow motion with another. The smaller one topples, rolls, and heaves like a dying rhinoceros, emerald seawater mixed with spray drenching its surfaces.

I yearn to ride an iceberg like a penguin or a gull, touching its frozen sides, drifting slowly on the waves. I draw them, but I can't capture their splendour.

GUIDED READING

WORD HELP

rampart an embankment made of earth used for protection

❶ The Weddell Sea is a part of the Southern Ocean, which borders the western part of Antarctica.

Connecting Literature to Geography

1. **Describe** What details in the passage tell you that the icebergs are in motion?
2. **Compare and Contrast** The author was clearly moved by the beauty of the place she was observing. Think of a place you have seen in your community that made a lasting impression on your senses. Tell how the author's description of icebergs is like your own experience. Then explain how the two experiences are different.

Chapter Review

Geography's Impact
video series
Review the video to answer the closing question:
What are some ways to help prevent or limit the spread of nonnative species?

Visual Summary

Use the visual summary below to help you review the main ideas of the chapter.

QUICK FACTS

Australia and New Zealand
Despite different physical features, Australia and New Zealand have much in common.

The Pacific Islands
The Pacific Islands are home to tropical climates and beautiful beaches.

Antarctica
Antarctica's unique environment makes it an important site for scientific research.

Reviewing Vocabulary, Terms, and Places

Choose the letter of the answer that best completes each statement below.

1. The original inhabitants of Australia are the
   a. Aborigines   **c.** Papuans
   b. Maori   **d.** Polynesians

2. A floating mass of ice that has broken off a glacier is a(n)
   a. atoll   **c.** iceberg
   b. coral reef   **d.** polar desert

3. Located off the northeast coast of Australia, this is the world's largest coral reef
   a. Australian Reef   **c.** Kiwi Reef
   b. Great Barrier Reef   **d.** Reef of the Coral Sea

4. The result of an action or decision is a(n)
   a. agreement   **c.** motive
   b. effect   **d.** purpose

Comprehension and Critical Thinking

SECTION 1 *(Pages 644–649)*

5. **a. Describe** What is the physical geography of Australia like?

   b. Compare and Contrast In what ways are the countries of Australia and New Zealand similar and different?

   c. Elaborate Why do you think the economies of Australia and New Zealand are so strong?

SECTION 2 *(Pages 653–657)*

6. **a. Identify** What two types of islands are commonly found in the Pacific Ocean? How are they different?

   b. Analyze How were the islands of the Pacific Ocean originally settled?

   c. Elaborate Many Pacific Islands are very isolated from other societies. Would you want to live in these isolated communities? Why or why not?

SECTION 3 *(Pages 658–661)*

7. a. Describe What types of wildlife are found in and around Antarctica?

b. Draw Conclusions Why do you think many of the world's countries supported setting aside Antarctica for scientific research?

c. Predict What effects might the thinning of the ozone layer have on Antarctica?

Social Studies Skills

Locating Information *Use your knowledge about locating information to answer the questions below.*

8. Where might you look to find information about recent weather statistics in Antarctica?

9. What types of sources might you use to find books about early explorations of Antarctica?

10. What sources could you use to find electronic resources about Antarctica?

FOCUS ON READING AND WRITING

Drawing Conclusions *Read the paragraph below, then answer the questions that follow.*

> Getting an education can be a challenge in the Outback, where students are spread across vast distances. Basic education comes from the School of the Air, which broadcasts classes via radio. Other lessons are conducted by videotape or mail. The use of the Internet, e-mail, and videoconferencing has been used in recent years.

11. What did you know about education and Australia before you read this paragraph?

12. What conclusion(s) can you draw about education in Australia's Outback?

Writing a Brochure *Use your notes and the directions below to create a brochure.*

13. Divide your brochure into sections—one on Australia and New Zealand, one on the Pacific Islands, and one on Antarctica. In each section, identify the important resources and try to convince the reader to invest in them. Use illustrations to support the points you want to make. Finally, design a cover page for your brochure.

Using the Internet

 **go.hrw.com** KEYWORD: SK7 CH25

14. Activity: Creating a Display The original people of Australia, the Aborigines, came from Southeast Asia at least 40,000 years ago. Today many Aborigines work to keep their languages, religion, and customs alive. Enter the activity keyword. Take notes about Aborigine culture as you view the Web sites. Then create a visual display or multimedia presentation featuring the interesting things you discover about their dreamtime, art, music, and more.

Map Activity ✳ Interactive

15. The Pacific World On a separate sheet of paper, match the letters on the map with their correct labels.

Great Barrier Reef	Pacific Ocean
Melbourne	Papua New Guinea
North Island	Perth
Outback	Sydney

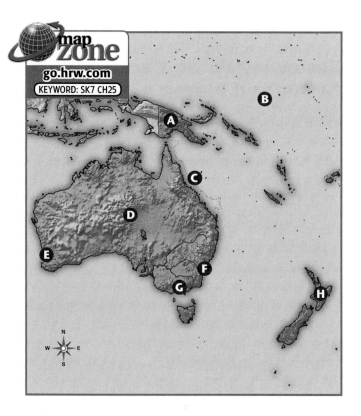

DIRECTIONS: Read questions 1 through 7 and write the letter of the best response. Then read question 8 and write your own well-constructed response.

1 **What is the world's only country that is also a continent?**

A Australia

B Micronesia

C New Zealand

D Polynesia

2 **What physical feature lies off Australia's northeastern coast?**

A Nullarbor Plain

B Central Lowlands

C Great Barrier Reef

D Outback

3 **The descendants of the first people to live in New Zealand are called the**

A Aborigines.

B Maori.

C goa.

D kiwi.

4 **A ring-shaped island surrounding a lagoon is called**

A a high island.

B a low island.

C an atoll.

D a territory.

5 **The only people who live in Antarctica are**

A scientists.

B tourists.

C miners.

D government officials.

Australia and New Zealand: Climate

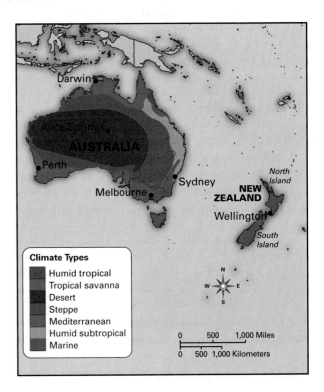

6 **According to the map above, what city has a desert climate?**

A Alice Springs

B Melbourne

C Perth

D Wellington

7 **According to the map, New Zealand has what kind of climate?**

A humid tropical

B steppe

C tropical savanna

D marine

8 **Extended Response** The vast majority of Australia's population lives along the coast, especially the east and southeast coasts. Study the map above. Then write a brief paragraph explaining why you think this is true.

Persuasion

Persuasion is about convincing others to act or believe in a certain way. Just as you use persuasion to convince your friends to see a certain movie, people use persuasion to convince others to help them solve the world's problems.

Assignment

Write a persuasive paper about an issue faced by the people of Asia and the Pacific. Choose an issue related to the natural environment or culture of the area.

1. Prewrite

Choose an Issue

- Choose an issue to write about. For example, you might choose the danger of tsunamis or the role of governments.
- Create a statement of opinion. For example, you might say, "Countries in this region must create a warning system for tsunamis."

Gather and Organize Information

- Search your textbook, the library, or the Internet for evidence that supports your opinion.
- Identify at least two reasons to support your opinion. Find facts, examples, and expert opinions to support each reason.

> **TIP** **That's a Reason** Convince your readers by presenting reasons to support your opinion. For example, one reason to create a warning system for tsunamis is to save lives.

2. Write

Use a Writer's Framework

A Writer's Framework

Introduction
- Start with a fact or question related to the issue you will discuss.
- Clearly state your opinion in a sentence.

Body
- Write one paragraph for each reason. Begin with the least important reason and end with the most important.
- Include facts, examples and expert opinions as support.

Conclusion
- Restate your opinion and summarize your reasons.

3. Evaluate and Revise

Review and Improve Your Paper

- As you review your paper, use the questions below to evaluate it.
- Make changes to improve your paper.

Evaluation Questions for a Persuasive Essay

1. Do you begin with an interesting fact or question related to the issue?
2. Does your introduction clearly state your opinion and provide any necessary background information?
3. Do you discuss your reasons from least to most important?
4. Do you provide facts, examples, or expert opinions to support each of your reasons?
5. Does your conclusion restate your opinion and summarize your reasons?

4. Proofread and Publish

Give Your Paper the Finishing Touch

- Make sure you have correctly spelled and capitalized all names of people or places.
- Check for correct comma usage when presenting a list of reasons or evidence.
- Decide how to share your paper. For example, could you publish it in a school paper or in a classroom collection of essays?

5. Practice and Apply

Use the steps and strategies outlined in this workshop to write your persuasive essay. Share your opinion with others to see whether they find your opinion convincing.

References

Reading Social Studies 668

Economics Handbook 693

Facts about the World 698

Atlas 702

Gazetteer 722

Biographical Dictionary 730

English and Spanish Glossary 733

Index 745

Credits and Acknowledgments 766

Using Prior Knowledge

FOCUS ON READING

When you put together a puzzle, you search for pieces that are missing to complete the picture. As you read, you do the same thing when you use prior knowledge. You take what you already know about a subject and then add the information you are reading to create a full picture. The example below shows how using prior knowledge about computer mapping helped one reader fill in the pieces about how geographers use computer mapping.

In the past, maps were always drawn by hand. Many were not very accurate. Today, though, most maps are made using computers and satellite images. Through advances in mapmaking, we can make accurate maps on almost any scale, from the whole world to a single neighborhood, and keep them up to date.

From Section 3, The Branches of Geography

Computer Mapping	
What I know before reading	What else I learned
• My dad uses the computer to get a map for trips. • I can find maps on the Internet of states and countries.	• Maps have not always been very accurate. • Computers help make new kinds of maps that are more than just cities and roads. • These computer maps are an important part of geography.

YOU TRY IT!

Draw a chart like the one above. Think about what you know about satellite images and list this prior knowledge in the left column of your chart. Then read the passage below. Once you have read it, add what you learned about satellite images to the right column.

Much of the information gathered by these satellites is in the form of images. Geographers can study these images to see what an area looks like from above Earth. Satellites also collect information that we cannot see from the planet's surface. The information gathered by satellites helps geographers make accurate maps.

From Section 1, Studying Geography

Using Word Parts

FOCUS ON READING

Many English words are made up of several word parts: roots, prefixes, and suffixes. A root is the base of the word and carries the main meaning. A prefix is a letter or syllable added to the beginning of a root. A suffix is a letter or syllable added to the end to create new words. When you come across a new word, you can sometimes figure out the meaning by looking at its parts. Below are some common word parts and their meanings.

Common Roots		
Word Root	**Meaning**	**Sample Words**
-graph-	write, writing	autograph, biography
-vid-, -vis-	see	videotape, visible

Common Prefixes		
Prefix	**Meaning**	**Sample Words**
geo-	earth	geology
inter-	between, among	interpersonal, intercom
in-	not	ineffective
re-	again	restate, rebuild

Common Suffixes		
Suffix	**Meaning**	**Sample Words**
-ible	capable of	visible, responsible
-less	without	penniless, hopeless
-ment	result, action	commitment
-al	relating to	directional
-tion	the act or condition of	rotation, selection

YOU TRY IT!

Read the following words. First separate any prefixes or suffixes and identify the word's root. Use the chart above to define the root, the prefix, or the suffix. Then write a definition for each word.

geography	visualize	movement
seasonal	reshaping	interact
regardless	separation	invisible

Understanding Cause and Effect

FOCUS ON READING

Learning to identify causes and effects can help you understand geography. A **cause** is something that makes another thing happen. An **effect** is the result of something else that happened. A cause may have several effects, and an effect may have several causes. In addition, as you can see in the example below, causes and effects may occur in a chain. Then, each effect in turn becomes the cause for another event.

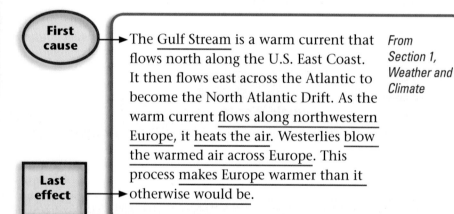

First cause → The Gulf Stream is a warm current that flows north along the U.S. East Coast. It then flows east across the Atlantic to become the North Atlantic Drift. As the warm current flows along northwestern Europe, it heats the air. Westerlies blow the warmed air across Europe. This process makes Europe warmer than it **Last effect** → otherwise would be.

From Section 1, Weather and Climate

Cause
Gulf Stream

↓

Effect
Warm water flows along the coast of northwest Europe.

↓

Effect
Warm water raises temperature of the air above.

↓

Effect
Winds blow warm air across Europe.

↓

Effect
Warm winds make Europe warmer.

YOU TRY IT!

Read the following sentences, and then use a graphic organizer like the one below right to analyze the cause and effects. Create as many boxes as you need to list the causes and effects.

Mountains also create wet and dry areas. . . A mountain forces wind blowing against it to rise. As it rises, the air cools and precipitation falls as rain or snow. Thus, the side of the mountain facing the wind is often green and lush. However, little moisture remains for the other side. This effect creates a rain shadow.

From Section 1, Weather and Climate

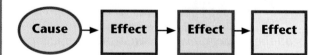

Cause → **Effect** → **Effect** → **Effect**

Understanding Main Ideas

FOCUS ON READING

Main ideas are like the hub of a wheel. The hub holds the wheel together, and everything circles around it. In a paragraph, the main idea holds the paragraph together and all the facts and details revolve around it. The main idea is usually stated clearly in a topic sentence, which may come at the beginning or end of a paragraph. Topic sentences always summarize the most important idea of a paragraph.

To find the main idea, ask yourself what one point is holding the paragraph together. See how the main idea in the following example holds all the details from the paragraph together.

A single country may also include more than one culture region within its borders. Mexico is one of many countries that is made up of different culture regions. People in northern Mexico and southern Mexico, for example, have different culture traits. The culture of northern Mexico tends to be more modern, while traditional culture remains strong in southern Mexico.

From Section 1, Culture

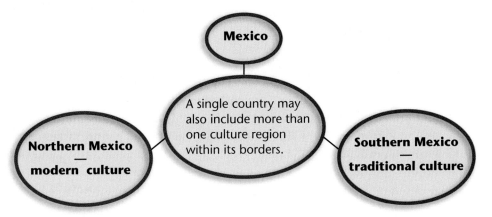

YOU TRY IT!

Read the following paragraph, and then use a graphic organizer like the one above to identify the main idea. Create as many circles as you need to list the supporting facts and details.

At the same time, the United States is influenced by global culture. Martial arts movies from Asia attract large audiences in the United States. Radio stations in the United States play music by African, Latin American, and European musicians. We even adopt many foreign words, like *sushi* and *plaza*, into English.

From Section 4, Global Connections

Paraphrasing

FOCUS ON READING

When you paraphrase, you explain someone else's idea in your own words. When you put an idea in your own words, you will understand it better and remember it longer. To paraphrase a passage, first read it carefully. Make sure you understand the main ideas. Then, using your own words, restate what the writer is saying. Keep the ideas in the same order and focus on using your own, familiar vocabulary. Your sentences may be shorter and simpler, but they should match the ideas in the text. Below is an example of a paraphrased passage.

Original Text	Paraphrase
Priests, people who performed or led religious ceremonies, had great status in Sumer. People relied on them to help gain the gods' favor. Priests interpreted the wishes of the gods and made offerings to them. These offerings were made in temples, special buildings where priests performed their religious ceremonies.	Priests hold the religious services, so people respect them. People want the priests to help them get on the gods' good side. Priests do this by explaining what the gods want and by making offerings. They make offerings in a special building where they lead services.

From Section 2, The Rise of Sumer

To paraphrase:
- Understand the ideas.
- Use your own words.
- Keep the same order.
- Make it sound like you.
- Keep it about the same length.

YOU TRY IT!

Read the following passage, and then write a paraphrase using the steps described above.

Irrigation increased the amount of food farmers were able to grow. In fact, farmers could produce a food surplus, or more than they needed. Farmers also used irrigation to water grazing areas for cattle and sheep. As a result, Mesopotamians ate a variety of foods. Fish, meat, wheat, barley, and dates were plentiful.

From Section 1, Geography of the Fertile Crescent

READING SOCIAL STUDIES

Understanding Implied Main Ideas

FOCUS ON READING

Do you ever "read between the lines" when people say things? You understand what people mean even when they don't come right out and say it. You can do the same thing with writing. Writers don't always state the main idea directly, but you can find clues to the main idea in the details. To understand an implied main idea, first read the text carefully and think about the topic. Next, look at the facts and details and ask yourself what the paragraph is saying. Then create a statement that sums up the main idea. Notice the way this process works with the paragraph below.

As a young man Jesus lived in the town of Nazareth and probably studied with Joseph to become a carpenter. Like many young Jewish men of the time, Jesus also studied the laws and teachings of Judaism. By the time he was about 30, Jesus had begun to travel and teach.

From Section 2, Origins of Christianity

1. What is the topic?
Jesus as a young man

2. What are the facts and details?
• lived in Nazareth
• studied to be a carpenter
• learned about Judaism

3. What is the main idea?
Jesus lived the typical life of a young Jewish man.

YOU TRY IT!

Read the following sentences. Notice the main idea is not stated. Using the three steps described above, develop a statement that expresses the main idea of the paragraph.

Justinian was stopped from leaving by his wife, Theodora. She convinced Justinian to stay in the city. Smart and powerful, Theodora helped her husband rule effectively. With her advice, he found a way to end the riots. Justinian's soldiers killed all the rioters—some 30,000 people—and saved the emperor's throne.

From Section 3, The Byzantine Empire

Sequencing

READING SOCIAL STUDIES

FOCUS ON READING

When you read about how countries and cultures developed, it is necessary to understand the sequence, or order, of events. Writers sometimes signal the sequence by using words or phrases such as *first, before, then, later, soon, next,* or *finally.* Sometimes writers use dates instead to indicate the order of events. Developing a sequence chain is a way to help you mark and understand the order of events. Clue words indicating sequence are underlined in the example.

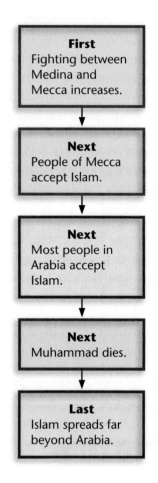

As the Muslim community in Medina grew stronger, other Arab tribes began to accept Islam. Conflict with the Meccans, however, increased. <u>In 630, after several years</u> of fighting, the people of Mecca gave in. They accepted Islam as their religion.

<u>Before long,</u> most people in Arabia had accepted Muhammad as their spiritual and political leader and become Muslims. Muhammad died <u>in 632</u>, but the religion he taught would <u>soon</u> spread far beyond the Arabian Peninsula.

From Section 1, Origins of Islam

YOU TRY IT!

Read the following passage. Create a graphic organizer like the one above to list the events in sequence. Use as many boxes as you need to create a full sequence chain.

In 711 a combined Arab and Berber army invaded Spain and quickly conquered it. Next, the army moved into what is now France, but it was stopped by a Christian army near the city of Tours. Despite this defeat, Muslims called Moors ruled parts of Spain for the next 700 years.

From Section 3, Muslim Empires

Setting a Purpose

FOCUS ON READING

When you start on a trip, you have a purpose or a destination in mind. When you read, you should also have a purpose in mind before you start. This purpose keeps you focused and moving toward your goal. To decide on a purpose, look over the headings, pictures, and study tips before you read. Then ask yourself a question that can guide your reading. See how a heading suggested a purpose for the passage below.

Notice Headings, Pictures or Tips
Here's a heading about teenagers and a picture.

Set a Purpose
I wonder who these teenagers are and what they're doing for peace. I'll read to find out.

Ask Questions
What's so important about these teenagers?

Israeli Teens for Peace

Peace between Israeli Jews and Palestinian Arabs has not been easy in the past. Moreover, some believe peace in the region might be impossible ever to accomplish. But don't tell that to a group of 200 Jewish and Arab teenagers who are making a difference in Israel. These teens belong to an organization called Seeds of Peace. To learn more about each other's culture and thus understand each other better, these teens meet regularly.

From Section 3, Israel

YOU TRY IT!

Read the following introduction to the section on Israel. Ask yourself questions that can set a purpose for your reading. Following the steps given above, develop a purpose for reading about Israel. State this purpose in one to two sentences.

Do you know that Israel is often referred to as the Holy Land? Some people call Israel the Holy Land because it is home to sacred sites for three of the world's major religions—Judaism, Christianity, and Islam. Throughout the region's history, these three groups have fought over their right to the land.

From Section 3, Israel

Re-Reading

FOCUS ON READING

When you read about other countries, you will come across some information that is completely new to you. Sometimes it can seem difficult to keep all the people, places, dates, and events straight. Re-reading can help you absorb new information and understand the main facts of a passage. Follow these three steps in re-reading. First, read the whole passage. Look over the passage and identify the main details you need to focus on. Then re-read the passage slowly. As you read, make sure you understand the details by restating the details silently. If necessary, go back and re-read until you have the details firmly in your mind. Here's how this process works with the following passage.

The Persian Empire was later conquered by several Muslim empires. Muslims converted the Persians to Islam, but most people retained their Persian culture. They built beautiful mosques with colorful tiles and large domes.

From Section 4, Iran

1. Read the passage.

2. Identify the main details to focus on.
Persian Empire, Muslims, culture

3. Re-read and restate the details silently.
The Persian Empire was first. Then it was conquered by Muslims. Persian and Muslim cultures blended. Mosques show the region's culture.

YOU TRY IT!

Read the following sentences. Then, following the three steps above, write down the main details to focus on. After you re-read the paragraph, write down the information restated in your own words to show that you understood what you read.

The Tigris and Euphrates rivers flow across a low, flat plain in Iraq. They join together before they reach the Persian Gulf. The Tigris and Euphrates are what are known as exotic rivers, or rivers that begin in humid regions and then flow through dry areas. The rivers create a narrow fertile area, which in ancient times was called Mesopotamia, or the "land between rivers."

From Section 1, Physical Geography

Using Context Clues

FOCUS ON READING

One way to figure out the meaning of an unfamiliar word or term is by finding clues in its context, the words or sentences surrounding it. A common context clue is a restatement. Restatements simply define the new word using ordinary words you already know. Notice how the following passage uses a restatement to define nomads.

> For centuries, Central Asians have made a living by raising horses, cattle, sheep, and goats. Many herders live as <u>nomads</u>, people who move often from place to place. The nomads move their herds from mountain pastures in the summer to lowland pastures in the winter.
>
> *From Section 2, History and Culture*

Restatement:
people who move often from place to place

YOU TRY IT!

Read the following sentences and identify the restatement for each underlined term.

> Many other people in Kyrgyzstan are farmers. Fertile soils there allow a mix of irrigated crops and <u>dryland farming</u>, or farming that relies on rainfall instead of irrigation.
>
> Even with the decline in agricultural production, Tajikistan still relies on cotton farming for much of its income. However, only 5 to 6 percent of the country's land is <u>arable</u>, or suitable for growing crops.
>
> *From Section 3, Central Asia Today*

Categorizing

FOCUS ON READING

When you sort things into groups of similar items, you are categorizing. Think of folding laundry. First you might sort into different piles: towels, socks, and T-shirts. The piles—or categories—help you manage the laundry because towels go to a different place than socks. When you read, categorizing helps you to manage the information by identifying the main types, or groups, of information. Then you can more easily see the individual facts and details in each group. Notice how the information in the paragraph below has been sorted into three main groups.

The subjects of Egyptian paintings vary widely. Some of the paintings show important historical events, such as the crowning of a new king or the founding of a temple. Others show major religious rituals. Still other paintings show scenes from everyday life, such as farming or hunting.

From Section 4, Egyptian Achievements

Subjects of Egyptian Paintings		
Category 1: Important historical events	**Category 2:** Major religious rituals	**Category 3:** Everyday life

YOU TRY IT!

Read the following sentences. Then use a graphic organizer like the one above to categorize the natural barriers in ancient Egypt. Create as many categories as you need.

In addition to a stable food supply, Egypt's location offered another advantage. It had natural barriers, which made it hard to invade Egypt. To the west, the desert was too big and harsh to cross. To the north, the Mediterranean Sea kept many enemies away. To the east, more desert and the Red Sea provided protection. Finally, to the south, cataracts in the Nile made it difficult for invaders to sail into Egypt that way.

From Section 1, Geography and Early Egypt

Asking Questions

FOCUS ON READING

When newspaper reporters want to get to the heart of a story, they ask certain questions: who, what, when, where, why, and how. When you are reading a textbook, you can use these same questions to get to the heart of the information you are reading. Notice how asking and answering questions about the passage below gets at the important information.

When?
751 BC

Where?
Egypt

What?
Conquered
Upper Egypt

Why?
To gain more
power

As Kush was growing stronger, Egypt was losing power. A series of weak pharaohs left Egypt open to attack. In the 700s BC a Kushite king, Kashta, took advantage of Egypt's weakness. Kashta attacked Eqypt. By about 751 BC he had conquered Upper Egypt.

From Section 1, Kush and Egypt

Who?
Kashta

How?
Attacking
during reign
of weak
pharaoh

YOU TRY IT!

Read the following passage carefully. Review the information in the passage by asking and answering the questions below.

By the AD 300s, Kush had lost much of its wealth and military might. Seeing that the Kushites were weak, the king of Aksum sent an army to conquer his former trade rival. In AD 350, the army of Aksum's King Ezana destroyed Meroë and took over the kingdom of Kush.

From Section 2, Later Kush

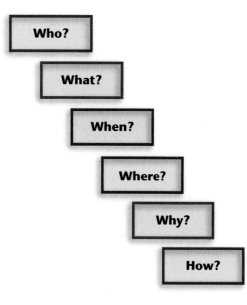

Who?

What?

When?

Where?

Why?

How?

Understanding Cause and Effect

FOCUS ON READING

To understand a country's history, you should look for cause and effect chains. A cause makes something happen, and an effect is what happens as a result of a cause. The effect can then become a cause and create another effect. Notice how the events below create a cause-and-effect chain.

> As the trade in gold and salt increased, Ghana's rulers gained power. Over time, their military strength grew as well. With their armies they began to take control of this trade from the merchants who had once controlled it. Merchants from the north and south met to exchange goods in Ghana. As a result of their control of trade routes, the rulers of Ghana became wealthy.
>
> *From Section 1, Empire of Ghana*

| **First Cause** Increase in gold and salt trade | → | **Effect** Ghana's rulers became powerful | → | **Effect** More military strength | → | **Effect** Took control of trade routes | → | **Final Effect** Rulers of Ghana became wealthy |

YOU TRY IT!

Read the following sentences, and then use a graphic organizer like the one above to analyze causes and effects. Create as many boxes as you need to list the causes and effects.

> When Mansa Musa died, his son Maghan took the throne. Maghan was a weak ruler. When raiders from the southeast poured into Mali, he couldn't stop them. The raiders set fire to Timbuktu's great schools and mosques. Mali never fully recovered from this terrible blow. The empire continued to weaken and decline.
>
> *From Section 2, Mali and Songhai*

Summarizing

FOCUS ON READING

Learning about countries means understanding a lot of information. Summarizing is one way to help you handle large amounts of information. A summary is a short restatement of the most important ideas in a text. The example below shows three steps you can use to write a summary. First underline important details. Then write a short summary of each paragraph. Finally, combine these paragraph summaries into a short summary of the whole passage.

With more than 10 million people, Cairo is the largest urban area in North Africa. The city is crowded, poor, and polluted. Cairo continues to grow as people move into the city from Egypt's rural areas in search of work. For centuries, Cairo's location at the southern end of the Nile Delta helped the city grow. The city also lies along old trading routes.

Today the landscape of Cairo is a mixture of modern buildings, historic mosques, and small, mud-brick houses. However, there is not enough housing in Cairo for its growing population. Many people live in makeshift housing in the slums or boats along the Nile. Communities have even developed in cemeteries, where people convert tombs into bedrooms and kitchens.

From Section 3, North Africa Today

Summary of Paragraph 1
The crowded city of Cairo is North Africa's largest city and continues to grow.

Summary of Paragraph 2
Without enough housing, people in Cairo live in slums, boats and cemeteries.

Combined Summary
Cairo is North Africa's largest city, and it is so crowded that people live in houses, boats and cemeteries.

YOU TRY IT!

Read the following paragraphs. First, write a summary for each paragraph and then write a combined summary of the whole passage.

Even though Egypt is a republic, its government is heavily influenced by Islamic law. Egypt's government has a constitution and Egyptians elect their government officials. Power is shared between Egypt's president and the prime minister.

Many Egyptians debate over the role of Islam in the country. Some Egyptian Muslims believe Egypt's government, laws, and society should be based on Islamic law. However, some Egyptians worry that such a change in government would mean fewer personal freedoms.

From Section 3, North Africa Today

Understanding Comparison-Contrast

FOCUS ON READING

Comparing shows how things are alike. Contrasting shows how things are different. You can understand comparison-contrast by learning to recognize clue words and points of comparison. Clue words let you know whether to look for similarities or differences. Points of comparison are the main topics that are being compared or contrasted. Notice how the passage below compares and contrasts life in rural and urban areas.

> Highlighted words are points of comparison.

Rural homes are small and simple. Many homes in the Sahel and savanna zones are circular. Straw or tin roofs sit atop mud, mud-brick, or straw huts. Large extended families often live close together in the same village . . .

In urban areas, also, members of an extended family may all live together. However, in West Africa's cities you will find modern buildings. People may live in houses or high rise apartments.

From Section 2, History and Culture

> Underlined words are clue words.

Clue Words	
Comparison	**Contrast**
share, similar, like, also, both, in addition, besides	however, while, unlike, different, but, although

YOU TRY IT!

Read the following passage about Liberia and Sierra Leone. Use a diagram like the one here to compare and contrast the two countries.

Now, both Liberia and Sierra Leone are trying to rebuild. They do have natural resources on which to build stronger economies. Liberia has rubber and iron ore while Sierra Leone exports diamonds.

From Section 3, West Africa Today

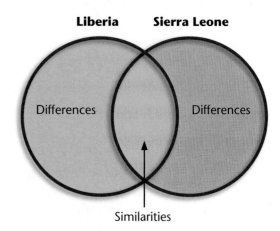

Liberia   Sierra Leone

Differences   Differences

Similarities

Identifying Supporting Details

FOCUS ON READING

Why believe what you read? One reason is because of details that support or prove the main idea. These details might be facts, statistics, examples, or definitions. In the example below, notice what kind of proof or supporting details help you believe the main idea.

> The landscape of East Africa has many high volcanic mountains. The highest mountain in Africa, Mount Kilimanjaro, rises to 19,340 feet. Despite Kilimanjaro's location near the equator, the mountain's peak has long been covered in snow. This much colder climate is caused by Kilimanjaro's high elevation.
>
> *From Section 1, Physical Geography*

Main Idea
The landscape of East Africa has many high volcanic mountains.

Supporting Details			
Example	**Statistic**	**Fact**	**Fact**
The highest mountain is Mount Kilimanjaro.	Mount Kilimanjaro is 19,340 feet high.	Kilimanjaro has a cold climate.	It is near the equator, but its peak is snow covered.

YOU TRY IT!

Read the following sentences, and then use a graphic organizer like the one above to identify the supporting details.

> Somalia is less diverse than most other African countries. Most people in the country are members of a single ethnic group, the Somali. In addition, most Somalis are Muslims and speak the same African language, also called Somali.
>
> *From Section 3, East Africa Today*

Using Word Parts

FOCUS ON READING

Many English words are made up of several word parts: roots, prefixes, and suffixes. A root is the main part of the word. A prefix is a letter or syllable added before the root. A suffix is a word part added after the root. Knowing the meanings of common prefixes and suffixes may help you figure out unfamiliar words. Below are some common prefixes and suffixes along with their meanings and examples of their use.

Common Prefixes		
Prefix	**Meaning**	**Sample Words**
en-	to cause to be	enforce
in-	not	ineffective
inter-	between, among	interpersonal
mal-	bad	malfunction
pro-	for, in front	proclaim
re-	again	rerun, rebuild

Common Suffixes		
Suffix	**Meaning**	**Sample Words**
-al	relating to	directional
-dom	state, condition	freedom
-ion	action, condition	rotation, selection
-ous	characterized by	victorious
-ment	result, action	development, entertainment

YOU TRY IT!

Read the following words from the chapter. Underline any prefixes or suffixes. Use the chart above to find the meaning of the prefix or the suffix. Then come up with a definition for each word.

> independence, instability, malnutrition,
> interact, re-elected, enslaved,
> industrial, endanger, migration

Making Generalizations

FOCUS ON READING

As you read about different people and cultures, you probably notice many similarities. Seeing those similarities may lead you to make a generalization. A generalization is a statement that applies to many different situations or people even though it is based on a few specific situations or people. In the following example, a generalization is made from combining new information with information from personal experience. Sometimes you might also make a generalization from reading about several new situations, even though you don't have personal experience with the situation.

> Several large rivers cross Southern Africa's plains. The Okavango River flows from Angola into a huge basin in Botswana. This river's water never reaches the ocean. Instead it forms a swampy inland delta that is home to crocodiles, zebras, hippos, and other animals. Many tourists travel to Botswana to see these wild animals in their natural habitat.
>
> *From Section 1, Physical Geography*

1. What you read:
Tourists will travel a long way to see wild animals.

2. What you know from personal experience:
My family loves to see wild animals in the zoo.

Generalization:
Many people enjoy seeing wild animals in person.

YOU TRY IT!

Read the following sentences about four South African countries. Then make a generalization from these four situations about political instability and a country's economy.

> Zimbabwe has suffered from a poor economy and political instability.
>
> Mozambique is one of the world's poorest countries. The economy has been badly damaged by civil war, but it is improving.
>
> Comoros suffers from a lack of resources and political instability. The government of Comoros is struggling to improve education and promote tourism.
>
> Madagascar was ruled for more than 20 years by a socialist dictator. Today the elected president is working to improve the struggling economy.
>
> *From Section 3, Southern Africa Today*

Sequencing

READING SOCIAL STUDIES

FOCUS ON READING

Have you ever used written instructions to put together an item you bought? If so, you know that the steps in the directions need to be followed in order. The instructions probably included words like *first, next,* and *then* to help you figure out what order you needed to do the steps in. The same kinds of words can help you when you read a history book. Words such as *first, then, later, next,* and *finally* can help you figure out the sequence, or order, in which events occurred. Read the passage below, noting the underlined clue words. Notice how they indicate the order of the events listed in the sequence chain at right.

> Not long <u>after</u> the Harappan civilization crumbled, a new group arrived in the Indus Valley. These people were called the Aryans. They were <u>originally</u> from the area around the Caspian Sea in Central Asia. <u>Over time,</u> however, they became the dominant group in India.
>
> *From Section 1, Early Indian Civilizations*

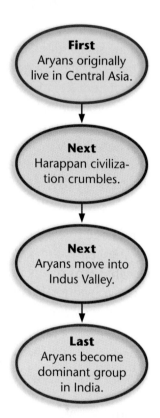

YOU TRY IT!

Read the following passage. Look for clue words to help you figure out the order of the events described in it. Then make a sequence chain like the one above to show that order.

> For many years, Asoka watched his armies fight bloody battles against other peoples. A few years into his rule, however, Asoka converted to Buddhism. When he did, he swore that he would not launch any more wars of conquest. After converting to Buddhism, Asoka had the time and resources to improve the lives of his people.
>
> *From Section 4, Indian Empires*

Understanding Chronological Order

FOCUS ON READING

When you read a paragraph in a history text, you can usually use clue words to help you keep track of the order of events. When you read a longer section of text that includes many paragraphs, though, you may need more clues. One of the best clues you can use in this case is dates. Each of the sentences below includes at least one date. Notice how those dates were used to create a time line that lists events in chronological, or time, order.

As early as 7000 BC people had begun to farm in China.

After 3000 BC people began to use potter's wheels to make many types of pottery.

The first dynasty for which we have clear evidence is the Shang, which was firmly established by the 1500s BC.

Shang emperors ruled in China until the 1100s BC.

From Section 1, Early China

7000 BC
People begin farming in China.

3000 BC
People begin using potter's wheels.

1500s BC
Shang dynasty rules China.

1100s BC
Shang lose power.

5000 BC ———————— 1 BC

YOU TRY IT!

Read the following sentences. Use the dates in the sentences to create a time line listing events in chronological order.

The Ming dynasty that he founded ruled China from 1368 to 1644.

Genghis Khan led his armies into northern China in 1211.

Between 1405 and 1433, Zheng He led seven grand voyages to places around Asia.

In the 1300s many Chinese groups rebelled against the Yuan dynasty.

From Section 5, The Yuan and Ming Dynasties

Visualizing

FOCUS ON READING

Maybe you have heard the saying "a picture is worth a thousand words." That means a picture can show in a small space what might take many words to describe. Visualizing, or creating mental pictures, can help you see and remember what you read. When you read, try to imagine what a snapshot of the images in the passage might look like. First, form the background or setting in your mind. Then keep adding specific details that can help you picture the rest of the information.

Form the background picture:
I see the shape of the Indian subcontinent.

Add more specific details:
I see two large crowds of people moving toward the diagonal line and the number 10,000,000.

To avoid a civil war, the British agreed to the partition, or division, of India. In 1947 two independent countries were formed. India was mostly Hindu. Pakistan, which included the area that is now Bangladesh, was mostly Muslim. As a result, some 10 million people rushed to cross the border. Muslims and Hindus wanted to live in the country where their religion held a majority.

From Section 2, History and Culture of India

Add specific details:
I see a huge diagonal line near the top left dividing the country into India and Pakistan.

Add more specific details:
I see two large arrows. The arrow pointing left says, "This way to Pakistan for Muslims." The arrow pointing right says, "This way to India for Hindus."

YOU TRY IT!

Read the following sentences. Then, using the process explained above, describe the images you see.

Flooding is one of Bangladesh's biggest challenges. Many circumstances cause these floods. The country's many streams and rivers flood annually, often damaging farms and homes. Summer monsoons also cause flooding. For example, massive flooding in 2004 left more than 25 million people homeless. It also destroyed schools, farms, and roads throughout the country.

From Section 4, India's Neighbors

Identifying Implied Main Ideas

FOCUS ON READING

Main ideas are often stated in a paragraph's topic sentence. When the main idea is not stated directly, however, you can find it by looking at the details in the paragraph. First, read the text carefully and think about the topic. Next, look at the facts and details and ask yourself what details are repeated. What points do those details make? Then create a statement that sums up the main idea. Examine how this process works for the paragraph below.

> Yet, other people increasingly wanted Chinese goods such as silk and tea. To gain access to the goods, some European powers forced China to open up trade in the 1800s. Europeans took over parts of the country as well. These actions angered many Chinese, some of whom blamed the emperor. At the same time, increased contact with the West exposed the Chinese to new ideas.
>
> *From Section 2, History and Culture of China*

YOU TRY IT!

Read the following sentences. Then use the steps listed to the right to develop a statement that expresses the main idea of the paragraph.

> Only about 10 percent of China's land is good for farming. So how does China produce so much food? More than half of all Chinese workers are farmers. This large labor force can work the land at high levels. In addition, farmers cut terraces into hillsides to make the most use of the land.
>
> *From Section 3, China Today*

1. What is the topic?
China's contact with Europe

↓

2. What are the facts and details?
• Europe forced China to trade.

• Europeans took over some parts of China.

• Some Chinese blamed the emperor.

• The Chinese heard new ideas.

↓

3. What details are repeated?
Increased contact with Europe

↓

4. What is the main idea?
Increased contact with Europe had both positive and negative effects on China.

READING SOCIAL STUDIES

Understanding Fact and Opinion

FOCUS ON READING

When you read, it is important to distinguish facts from opinions. A fact is a statement that can be proved or disproved. An opinion is a personal belief or attitude, so it cannot be proved true or false. When you are reading a social studies text, you want to read only facts, not the author's opinions. To determine whether a sentence is a fact or an opinion, ask if it can be proved using outside sources. If it can, the sentence is a fact. The following pairs of statements show the difference between facts and opinions.

Fact: Hirohito was Japan's emperor for most of the 1900s. *(This fact can be proved through research.)*

Opinion: I believe Hirohito was Japan's best emperor. *(The word* best *signifies that this is the writer's judgment, or opinion.)*

Fact: One example of a long-lasting Korean food is kimchi, a dish made from pickled cabbage and various spices. *(The ingredients in kimchi can be checked for accuracy.)*

Opinion: Kimchi is a delicious dish made from pickled cabbage and various spices. *(No one can prove kimchi is delicious, because it is a matter of personal taste.)*

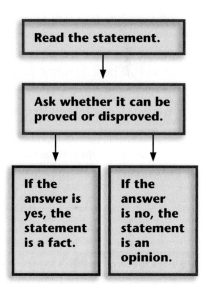

Read the statement.

↓

Ask whether it can be proved or disproved.

If the answer is yes, the statement is a fact.

If the answer is no, the statement is an opinion.

YOU TRY IT!

Read the following sentences and identify each as a fact or an opinion.

1. The second largest city in Japan, Osaka, is located in western Honshu.

2. In Osaka—as in Tokyo and other cities—tall modern skyscrapers stand next to tiny Shinto temples.

3. Osaka is more beautiful than Tokyo.

4. Another major city is Kyoto, which all tourists should visit.

5. Once Japan's capital, Kyoto is full of historic buildings.

6. Tokyo has the country's best restaurants.

Using Context Clues—Definitions

FOCUS ON READING

One way to figure out the meaning of an unfamiliar word or term is by finding clues in its context, the words or sentences surrounding the word or term. A common context clue is a restatement. Restatements are simply a definition of the new word using ordinary words you already know. Notice how the following passage uses a restatement to define archipelago. Some context clues are not as complete or obvious. Notice how the following passage provides a description that is a partial definition of peninsula.

The region of Southeast Asia is made up of two *peninsulas* and two large island groups. The Indochina *Peninsula* and the Malay (muh-LAY) *Peninsula* extend from the Asian mainland. . . The two island groups are the Philippines and the Malay Archipelago. An *archipelago* (ahr-kuh-PE-luh-goh) is a large group of islands.

From Section 1, Physical Geography

Peninsula: land that extends from a mainland out into water

Archipelago: large group of islands

YOU TRY IT!

Read the following passages and identify the meaning of the italicized words by using definitions, or restatements, in context.

The many groups that influenced Southeast Asia's history also shaped its culture. This *diverse* culture blends native, Chinese, Indian, and European ways of life.

From Section 2, History and Culture

The economy is based on farming, but good farmland is limited. Most people are *subsistence farmers*, meaning they grow just enough food for their families.

From Section 3, Mainland Southeast Asia Today

READING SOCIAL STUDIES

Drawing Conclusions

FOCUS ON READING

You have probably heard the phrase, "Put two and two together." When people say that, they don't mean "2 + 2 = 4." They mean, "put the information together." When you put together information you already know with information you have read, you can draw a conclusion. To make a conclusion, read the passage carefully. Then think about what you already know about the topic. Put the two together to draw a conclusion.

Off Australia's northeastern coast is the Great Barrier Reef, the world's largest coral reef. A coral reef is a collection of rocky material found in wshallow, tropical waters. The Great Barrier Reef is home to an incredible variety of marine animals.

From Section 1, Australia and New Zealand

| **Information gathered from the passage:** Australia has the world's largest coral reef. | **+** | **What you already know:** I know many people like to snorkel or scuba dive at coral reefs. | **=** | **Add the information up to reach your conclusion:** Australia probably has many tourists who come to visit the Great Barrier Reef. |

YOU TRY IT!

Read the following paragraphs. Think about what you know about living in a very cold climate. Then use the process described above to draw a conclusion about the following passage.

Because of Antarctica's high latitude, the continent is in almost total darkness during winter months. Seas clog with ice as a result of the extreme temperatures. In the summer, the sun shines around the clock and temperatures rise to near freezing.

Plant life only survives in the ice-free tundra areas. Insects are the frozen land's only land animals. Penguins, seals, and whales live in Antarctica's waters.

From Section 3, Antarctica

Economics Handbook

What Is Economics?

Economics may sound dull, but it touches almost every part of your life. Here are some examples of the kinds of economic choices you may have made yourself:

- Which pair of shoes to buy—the ones on sale or the ones you really like, which cost much more
- Whether to continue saving your money for the DVD player you want or use some of it now to go to a movie
- Whether to give some money to a fundraiser for a new park or to housing for the homeless

As these examples show, we can think of economics as a study of choices. These choices are the ones people make to satisfy their needs or their desires.

Glossary of Economic Terms

Here are some of the words we use to talk about economics:

ECONOMIC SYSTEMS

Countries have developed different economic systems to help them make choices, such as what goods and services to produce, how to produce them, and for whom to produce them. The most common economic systems in the world are market economies and mixed economies.

capitalism See market economy.

command economy an economic system in which the central government makes all economic decisions, such as in the countries of Cuba and North Korea

communism a political system in which the government owns all property and runs a command economy

free enterprise a system in which businesses operate with little government involvement, such as in a country with a market economy

market economy an economic system based on private ownership, free trade, and competition; the government has little to say about what, how, or for whom goods and services are produced; examples include Germany and the United States

mixed economy an economy that is a combination of command, market, and traditional economies

traditional economy an economy in which production is based on customs and tradition, and in which people often grow their own food, make their own goods, and use barter to trade

THE ECONOMY AND MONEY

People, businesses, and countries obtain the items they need and want through economic activities such as producing, selling, and buying goods or services. Countries differ in the amount of economic activity that they have and in the strength of their economies.

consumer a person who buys goods or services for personal use

consumer good a finished product sold to consumers for personal or home use

corporation a business in which a group of owners share in the profits and losses

currency paper or coins that a country uses for its money supply

demand the amount of goods and services that consumers are willing and able to buy at a given time

depression a severe drop in overall business activity over a long period of time

developed countries countries with strong economies and a high quality of life; often have high per capita GDPs and high levels of industrialization and technology

developing countries countries with less productive economies and a lower quality of life; often have less industrialization and technology

economic development the level of a country's economic activity, growth, and quality of life

economy the structure of economic life in a country

goods objects or materials that humans can purchase to satisfy their wants and needs

gross domestic product (GDP) total market value of all goods and services produced in a country in a given year; *per capita GDP* is the average value of goods and services produced per person in a country in a given year

industrialization the process of using machinery for all major forms of production

inflation an increase in overall prices

investment the purchase of something with the expectation that it will gain in value; usually property, stocks, etc.

money any item, usually coins or paper currency, that is used in payment for goods or services

producer a person or group that makes goods or provides services to satisfy consumers' wants and needs

productivity the amount of goods or services that a worker or workers can produce within a given amount of time

profit the gain or excess made by selling goods or services over their costs

purchasing power the amount of income that people have available to spend on goods and services

services any activities that are performed for a fee

standard of living how well people are living; determined by the amount of goods and services they can afford

stock a share of ownership in a corporation

supply the amount of goods and services that are available at a given time

INTERNATIONAL TRADE

Countries trade with each other to obtain resources, goods, and services. Growing global trade has helped lead to the development of a global economy.

balance of trade the difference between the value of a country's exports and imports

barter the exchange of one good or service for another

black market the illegal buying and selling of goods, often at high prices

comparative advantage the ability of a company or country to produce something at a lower cost than other companies or countries

competition rivalry between businesses selling similar goods or services; a condition that often leads to lower prices or improved products

e-commerce the electronic trading of goods and services, such as over the Internet

exports goods or services that a country sells and sends to other countries

free trade trade among nations that is not affected by financial or legal barriers; trade without barriers

imports goods or services that a country brings in or purchases from another country

interdependence a relationship between countries in which they rely on one another for resources, goods, or services

market the trade of goods and services

market clearing price the price of a good or service at which supply equals demand

one-crop economy an economy that is dominated by the production of a single product

opportunity cost the value of the next-best alternative that is sacrificed when choosing to consume or produce another good or service

scarcity a condition of limited resources and unlimited wants by people

specialization a focus on only one or two aspects of production in order to produce a product more quickly and cheaply; for example, one worker washes the wheels of the car, another cleans the interior, and another washes the body

trade barriers financial or legal limitations to trade; prevention of free trade

trade-offs the goods or services sacrificed in order to consume or produce another good or service

underground economy illegal economic activities and unreported legal economic activities

PERSONAL ECONOMICS

Individuals make personal choices in how they manage and use their money to satisfy their needs and desires. Individuals have the choice to spend, save, or invest their money.

budget a plan listing the expenses and income of an individual or organization

credit a system that allows consumers to pay for goods and services over time

debt an amount of money that is owed

financial institutions businesses that keep and invest people's money and loan money to people; include banks or credit unions

income a gain of money that comes typically from labor or capital

interest the money that a borrower pays to a lender in return for a loan

loan money given on the condition that it will be paid back, often with interest

savings money or income that is not used to purchase goods or services

tax a required payment to a local, state, or national government; different kinds of taxes include sales taxes, income taxes, and property taxes

wage the payment a worker receives for his or her labor

RESOURCES

People and businesses need resources—such as land, labor, and money—to produce goods and services.

capital generally refers to wealth, in particular wealth that can be used to finance the production of goods or services

human capital sometimes used to refer to human skills and education that affect the production of goods and services in a company or country

labor force all people who are legally old enough to work and are either working or looking for work

natural resource any material in nature that people use and value

nonrenewable resource a resource that cannot be replaced naturally, such as coal or petroleum

raw material a natural resource used to make a product or good

renewable resource a resource that Earth replaces naturally, such as water, soil, and trees

ORGANIZATIONS

Countries have formed many organizations to promote economic cooperation, growth, and trade. These organizations are important in today's global economy.

European Union (EU) an organization that promotes political and economic cooperation in Europe

International Monetary Fund (IMF) a UN agency that promotes cooperation in international trade and that works to maintain stability in the exchange of countries' currencies

Organization of Economic Cooperation and Development (OECD) an organization of countries that promotes democracy and market economies

United Nations (UN) an organization of countries that promotes peace and security around the globe

World Bank a UN agency that provides loans to countries for development and recovery

World Trade Organization (WTO) an international organization dealing with trade between nations

Economic Handbook Review

Reviewing Vocabulary and Terms

On a separate sheet of paper, fill in the blanks in the following sentences:

ECONOMIC SYSTEMS

1. **A.** Businesses are able to operate with little government involvement in a _____ system.
   B. In a _____, a central government makes all economic decisions.
   C. _____ is a political system in which the government owns all property and runs a command economy.
   D. Economies that combine parts of command, market, or traditional economies are called _____.
   E. _____ is another name for a market economy, which is based on private ownership, free trade, and competition.

THE ECONOMY AND MONEY

2. **A.** _____ are objects or materials that people can buy to satisfy their needs and wants.
   B. A _____ is any activity that is performed for a fee.
   C. A person who buys goods or services is a _____, and a person or group that makes goods or provides services is a _____.
   D. The amount of goods and services that consumers are willing and able to buy at any given time is known as _____.
   E. The total value of all the goods and services produced in the United States in one year is its _____.

INTERNATIONAL TRADE

3. A. If we have an unlimited demand for a natural resource, such as oil, and there is only so much oil in the ground, we have a condition called _____.

B. Goods or services that a country sells to other countries are _____.

C. Rivalry between producers that provide the same good or service is called _____.

D. If a country is able to produce a good or service at a lower cost than other countries, it is said to have a _____.

E. Trade among nations that is not limited by legal or economic barriers is called _____.

PERSONAL ECONOMICS

4. A. A _____ is a required payment to a local, state, or national government that is used to support public services such as education, road construction, and government aid.

B. The money we do not spend on goods or services is our _____.

C. You can use _____ to pay for goods and services over time.

D. The payment that a worker receives for his or her labor is called a _____.

E. Individuals and companies use _____ to plan and manage their expenses and income.

Activities

1. With a partner, compare prices in two grocery stores. Create a chart showing the price of five items in the two stores. Also, figure the average price of the items in each store. How do you think the fact that the stores are near each other affects prices? How might prices be different if one store went out of business? How might the prices be different or similar if the United States had a command economy? Present what you have learned about prices and competition to your class.

2. With a group, choose five countries from a unit region to research. Look up the per capita GDP and the life expectancy rates for each of these countries in the regional atlas. Then use your textbook, go to your library, or use the Internet to research the literacy rate and the number of TVs per 1,000 people for each of these countries. Organize this information in a five-column chart like the one shown here. Study the information to see if you can find any patterns. Write a brief paragraph explaining what you have learned about the five countries.

Region				
Country	Per Capita GDP (U.S. $)	Life Expectancy at Birth	Literacy Rate	TVs per 1,000 People

3. Work with a partner to identify some of the many types of currency used in either Africa or Asia. Then imagine that you are the owners of a business in the United States. You have created a new product that you want to sell in the continent you selected, but people there do not use the same currency as you do. To sell your product, you will need to be able to exchange one type of currency for another. Search the Internet or look in a newspaper to find a list of currency exchange rates. For example, if your product sells for 1,000 dollars, what should the cost be in South African rand? In Indian rupees? In Chinese yuan? In Japanese yen?

RESOURCES

5. **A.** Diamonds and gold are examples of _____, which are any materials in nature that people use and value.

   B. The _____ consists of all people who are legally able to work and are working or looking for work.

   C. Wealth that can be used to finance the production of goods and services is called _____.

   D. Oil is an example of a _____, which is a resource that cannot be replaced naturally.

   E. Water and trees are examples of _____, resources that Earth replaces naturally.

ORGANIZATIONS

6. **A.** Many European countries have joined the _____ to help promote political and economic cooperation across Europe.

   B. The _____ consists of many agencies that promote peace and security around the world.

   C. The _____ is a UN agency that provides loans to countries to help them develop their economies.

   D. The _____ is a UN agency that helps protect the stability of countries' currencies.

   E. Many democratic countries promote market economies through the _____.

4. With three or four partners, create a skit that illustrates one of the following basic economic concepts: scarcity and limited resources, supply and demand, or opportunity costs and trade-offs. For example, a skit might illustrate supply and demand by showing how the high demand for the best seats at a concert increases the ticket prices for those seats. Write a script for your skit that includes an introduction stating which economic concept you are illustrating. Each member of your group must participate in the skit. Then practice the skit and perform it for the class.

5. Conduct research to find the following information for each country in the chart below: main trading partners, exports, imports, industrial products, agricultural products, and resources. Organize the information into a second chart. Then use the information in the two charts to write a one-page report explaining how international trade, specialization, and available natural resources affects each country's per capita GDP and standard of living.

THE WORLD ALMANAC
Facts about Countries — Southwest and Central Asia

COUNTRY Capital	FLAG	POPULATION	AREA (sq mi)	PER CAPITA GDP (U.S. $)	LIFE EXPECTANCY AT BIRTH	TVS PER 1,000 PEOPLE
Afghanistan Kabul		29.9 million	250,001	$800	42.9	14
Iraq Baghdad		26.1 million	168,754	$3,500	68.7	82
Kazakhstan Astana		15.2 million	1,049,155	$7,800	66.6	240
Kuwait Kuwait City		2.3 million	6,880	$21,300	77.0	480
Saudi Arabia Riyadh		26.4 million	756,985	$12,000	75.5	263
United States Washington, D.C.		295.7 million	3,718,710	$40,100	77.7	844

The Physical World

Inside the Earth

Earth's interior has several different layers. Deep inside the planet is the core. The inner core is solid, and the outer core is liquid. Above the core is the mantle, which is mostly solid rock with a molten layer on top. The surface layer of Earth includes the crust, which is made up of rocks and soil. Finally, the atmosphere extends from the crust into space. It supports much of the life on Earth.

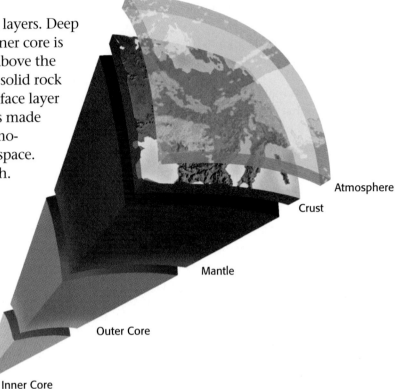

Atmosphere

Crust

Mantle

Outer Core

Inner Core

Tectonic Plates

Earth's crust is divided into huge pieces called tectonic plates, which fit together like a puzzle. As these plates slowly move, they collide and break apart, forming surface features like mountains, ocean basins, and ocean trenches.

Earth Facts	
Age:	4.6 billion years
Mass:	5,974,000,000,000,000,000,000 metric tons
Distance around the equator:	24,902 miles (40,067 km)
Distance around the poles:	24,860 miles (40,000 km)
Distance from the sun:	about 93 million miles (150 million km)
Earth's speed around the sun:	18.5 miles a second (29.8 km a second)
Percent of Earth's surface covered by water:	71%
What makes Earth unique:	large amounts of liquid water, tectonic activity, and life

The Continents

Geographers identify seven large landmasses, or continents, on Earth. Most of these continents are almost completely surrounded by water. Europe and Asia, however, are not. They share a long land boundary.

The world's continents are very different. For example, much of Australia is dry and rocky, while Antarctica is cold and icy. The information below highlights some key facts about each continent.

North America

- Percent of Earth's land: 16.5%
- Percent of Earth's population: 5.1%
- Lowest point: Death Valley, 282 feet (86 m) below sea level

Europe

- Percent of Earth's land: 6.7%
- Percent of Earth's population: 11.5%
- People per square mile: 187

South America

- Percent of Earth's land: 12%
- Percent of Earth's population: 8.6%
- Longest mountains: Andes, 4,500 miles (7,240 km)

Africa

- Percent of Earth's land: 20.2%
- Percent of Earth's population: 13.6%
- Longest river: Nile River, 4,160 miles (6,693 km)

Australia

- Percent of Earth's land: 5.2%
- Percent of Earth's population: 0.3%
- Oldest rocks: 3.7 billion years

Asia

- Percent of Earth's land: 30%
- Percent of Earth's population: 60.7%
- Highest point: Mount Everest, 29,035 feet (8,850 m)

Antarctica

- Percent of Earth's land: 8.9%
- Percent of Earth's population: 0%
- Coldest place: Plateau Station, -56.7°C (-70.1°F) average temperature

The Human World

World Population

More than 6 billion people live in the world today, and that number is growing quickly. Some people predict the world's population will reach 9 billion by 2050. As our population grows, it is also becoming more urban. Soon, as many people will live in cities and in towns as live in rural areas.

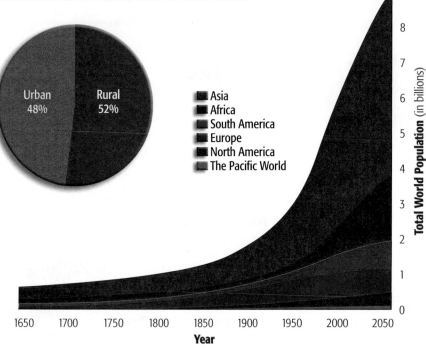

Urban 48%
Rural 52%

- Asia
- Africa
- South America
- Europe
- North America
- The Pacific World

Total World Population (in billions)

Year

As the world's population grows, people are moving to already large cities such as Shanghai (above) and Hong Kong (right) in China.

Geographers divide the world into developed and less developed regions. In general, developed countries are wealthier and more urban, have lower population growth rates and higher life expectancies. As you can imagine, life is very different in developed and less developed regions.

Developed and Less Developed Countries

	Population	Rate of Natural Increase	Life Expectancy	Percent Urban	Per Capita GNP (U.S. $)
Developed Countries	1.2 billion	0.1%	76	76%	$23,690
Less Developed Countries	5.2 billion	1.5%	65	41%	$3,850
The World	6.4 billion	1.3%	67	48%	$7,590

World Religions

A large percentage of the world's people follow one of several major world religions. Christianity is the largest religion. About 33 percent of the world's people are Christian. Islam is the second-largest religion with about 20 percent. It is also the fastest-growing religion. Hinduism and Buddhism are also major world religions.

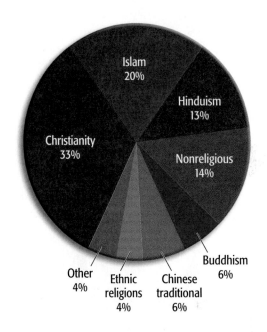

Islam 20%

Hinduism 13%

Christianity 33%

Nonreligious 14%

Buddhism 6%

Other 4%

Ethnic religions 4%

Chinese traditional 6%

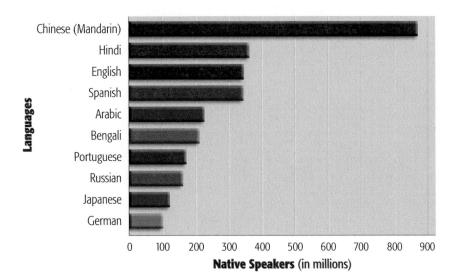

Native Speakers (in millions)

Languages

World Languages

Although several thousand languages are spoken today, a handful of major languages have the largest numbers of native speakers. Chinese (Mandarin) is spoken by nearly one in six people. Hindi, English, Spanish, and Arabic are next, with native speakers all over the world.

United States: Physical

PACIFIC
OCEAN

Strait of
Juan de Fuca

Puget
Sound

Mount Rainier
14,410 ft
(4,392 m)

COAST RANGES

CASCADE RANGE

Columbia River

Columbia
Plateau

Franklin D.
Roosevelt Lake

Bitterroot

Salmon
River Mts.

Sawtooth
Mts.

Snake

River

Cape
Mendocino

Goose
Lake

Shasta
Lake

Pyramid
Lake

GREAT

BASIN

Great
Salt
Lake

Uinta
Mts.

R O C K Y

Milk River

Missouri River

Fort Peck
Lake

Yellowstone River

Bighorn Mts.

Powder

River

Lake
Sakakawea

Lake
Oahe

G R E A T

Black
Hills

Cheyenne

River

White River

Niobrara River

James River

I N T E R I

San Francisco Bay

Sacramento River

Central
Valley

SIERRA NEVADA

San Joaquin River

Coast Ranges

Monterey
Bay

Mount Whitney
14,494 ft
(4,419 m)

Death Valley

Mojave
Desert

Salton
Sea

Imperial
Valley

Channel
Islands

Grand
Teton

Gannett Peak
13,804 ft
(4,207 m)

Wind River Range

Front Range

Mount Elbert
14,433 ft
(4,400 m)

M O U N T A I N S

North Platte River

South Platte River

Platte River

Republican River

Smoky Hill River

P L A I N S

COLORADO

PLATEAU

Grand
Canyon

Painted Desert

Colorado River

Pikes Peak
14,110 ft
(4,301 m)

San Luis
Valley

Sangre De Cristo Mts.

DIVIDE

CONTINENTAL

Lake
Mead

Gila River

Sonoran
Desert

Gulf of
California

Canadian

River

Pecos River

Amistad
Reservoir

Rio Grande

Colorado River

MEXICO

Padre
Island

Nueces

River

45°N

40°N

35°N

30°N

125°W

120°W

To understand the relative locations of Alaska and
Hawaii, as well as the vast distances separating them
from the rest of the United States, see the world map.

Kauai
Niihau
Oahu
Molokai
Lanai
Maui
Kahoolawe

HAWAII

PACIFIC
OCEAN

Mauna Kea
13,796 ft
(4,206 m)

Hawaii

22°N

19°N

0   75   150 Miles
0   75   150 Kilometers
Projection: Mercator

160°W

155°W

ARCTIC OCEAN

RUSSIA

Bering
Strait

Arctic Circle

St. Lawrence
Island

St. Matthew
Island

Nunivak
Island

Kuskokwim River

Yukon River

BROOKS RANGE

Porcupine River

ALASKA RANGE

Mount McKinley
20,320 ft
(6,194 m)

CANADA

Gulf of Alaska

Kodiak Island

Alexander
Archipelago

55°N

Bering Sea

Attu Island

ALEUTIAN ISLANDS

PACIFIC
OCEAN

55°N

50°N

170°E

180°

170°W

160°W

150°W

140°W

0   250   500 Miles
0   250   500 Kilometers
Projection: Albers Equal Area

CANADA

Red River

Mesabi Range

Isle Royale

Lake Superior

Minnesota River

Mississippi River

Wisconsin River

Lake Michigan

Lake Huron

Lake Ontario

Lake Erie

St. Lawrence River

St. Lawrence Seaway

St. John River

Penobscot River

Longfellow Mts.

White Mts.

Green Mts.

Connecticut River

Adirondack Mts.

Lake Champlain

Catskill Mts.

Cape Cod

Long Island Sound

Long Island

ALLEGHENY PLATEAU

APPALACHIAN MOUNTAINS

Allegheny R.

Susquehanna River

Delaware River

Delaware Bay

Chesapeake Bay

40°N

70°W

ATLANTIC OCEAN

Des Moines River

Missouri River

Illinois River

Wabash River

Scioto River

Ohio River

Kansas R.

P L A I N S

Monongahela River

Kanawha River

James River

Roanoke River

Pamlico Sound

Cape Hatteras

35°N

Lake of the Ozarks

OZARK PLATEAU

Keystone Lake

White River

Lake Barkley

Cumberland River

Cumberland Plateau

Great Smoky Mts.

BLUE RIDGE MOUNTAINS

P I E D M O N T

Arkansas River

Kentucky Lake

Eufaula Lake

Ouachita Mts.

Lake Texoma

Tennessee River

Tombigbee River

Coosa River

Oconee River

Savannah River

Altamaha River

Sea Islands

ELEVATION

Feet	Meters
13,120	4,000
6,560	2,000
1,640	500
656	200
(Sea level) 0	0 (Sea level)
Below sea level	Below sea level

0   100   200 Miles

0   100   200 Kilometers

Projection: Albers Equal Area

Trinity River

Saline River

Red River

Brazos River

Toledo Bend Reservoir

Mississippi River

Pearl River

Alabama R.

Chattahoochee River

C O A S T A L   P L A I N

G U L F

Chandeleur Islands

Mississippi Delta

Okefenokee Swamp

FLORIDA PENINSULA

Cape Canaveral

N
W   E
S

80°W

Gulf of Mexico

Lake Okeechobee

The Everglades

Cape Sable

Florida Keys

BAHAMAS

25°N

75°W

Straits of Florida

85°W

95°W

90°W

ATLAS   **703**

ATLAS

United States: Political

ATLAS

CANADA

MINNESOTA
Grand Forks
Fargo
Duluth
Marquette
Sault Ste. Marie
Lake Superior
Superior

MAINE
Augusta
Burlington
Montpelier
Portland
Concord
Manchester
VT
NH
Lake Champlain
Hudson R.
Connecticut R.

WISCONSIN
Minneapolis
St. Paul
Green Bay
Madison
Milwaukee
MICHIGAN
Lake Michigan
Lake Huron
Grand Rapids
Saginaw
Lansing
Detroit
Ann Arbor

Rochester
Syracuse
Albany
Springfield
Buffalo
NEW YORK
Boston
Worcester
Providence
MA
CT
RI
Hartford
New Haven
Bridgeport
Jersey City
Yonkers
New York City
Newark
Long Island Sound
Long Island
Cape Cod

Sioux Falls
Sioux City
IOWA
Cedar Rapids
Davenport
Des Moines
Rockford
Chicago
Gary
South Bend
Fort Wayne
Peoria
Springfield
INDIANA
Indianapolis
OHIO
Columbus
Dayton
Cincinnati
Toledo
Cleveland
Akron
Youngstown
Lake Erie
Lake Ontario

Omaha
Lincoln
MISSOURI
Kansas City
Kansas City
St. Louis
East St. Louis
ILLINOIS
Illinois River
Mississippi River
Missouri River

PENNSYLVANIA
Pittsburgh
Harrisburg
Allentown
Philadelphia
Camden
Trenton
NJ
Atlantic City
DE
Dover
MD
Baltimore
Annapolis
Washington D.C.
Delaware Bay
Susquehanna River

Topeka
Wichita
Keystone Lake
Tulsa
Eufaula Lake
Lake Texoma
Lake of the Ozarks
Jefferson City
Springfield

Louisville
Evansville
Frankfort
Lexington
KENTUCKY
Ohio River
Lake Barkley
Nashville
Kentucky Lake
TENNESSEE
Chattanooga
Knoxville
Memphis
Kentucky Lake

WEST VIRGINIA
Charleston
VIRGINIA
Richmond
Newport News
Norfolk
Virginia Beach
Chesapeake Bay

ATLANTIC OCEAN

40°N
70°W
35°N
Cape Hatteras

Greensboro
Durham
Raleigh
Winston-Salem
Asheville
Charlotte
NORTH CAROLINA
Greenville
SOUTH CAROLINA
Columbia
Charleston
Savannah River

ARKANSAS
Fayetteville
Little Rock
Pine Bluff
Huntsville
MISSISSIPPI
Vicksburg
Jackson
Meridian
ALABAMA
Birmingham
Montgomery
Columbus
GEORGIA
Atlanta
Macon
Savannah
Sea Islands

LOUISIANA
Dallas
Waco
Shreveport
Beaumont
Houston
Galveston
Toledo Bend Reservoir
Red River
Baton Rouge
New Orleans
Biloxi
Mobile
Pensacola
Chandeleur Islands
Gulf of Mexico

Tallahassee
Gainesville
Jacksonville
FLORIDA
Orlando
Tampa
St. Petersburg
Lake Okeechobee
Fort Myers
Fort Lauderdale
Miami
Cape Canaveral
Cape Sable
Florida Keys
Straits of Florida

BAHAMAS

National capital
State capitals
Other cities

0   100   200 Miles
0   100   200 Kilometers

Projection: Albers Equal Area

N
W   E
S

95°W
90°W
85°W
80°W
75°W
30°N
25°N

Chattahoochee R.

ATLAS

World: Physical

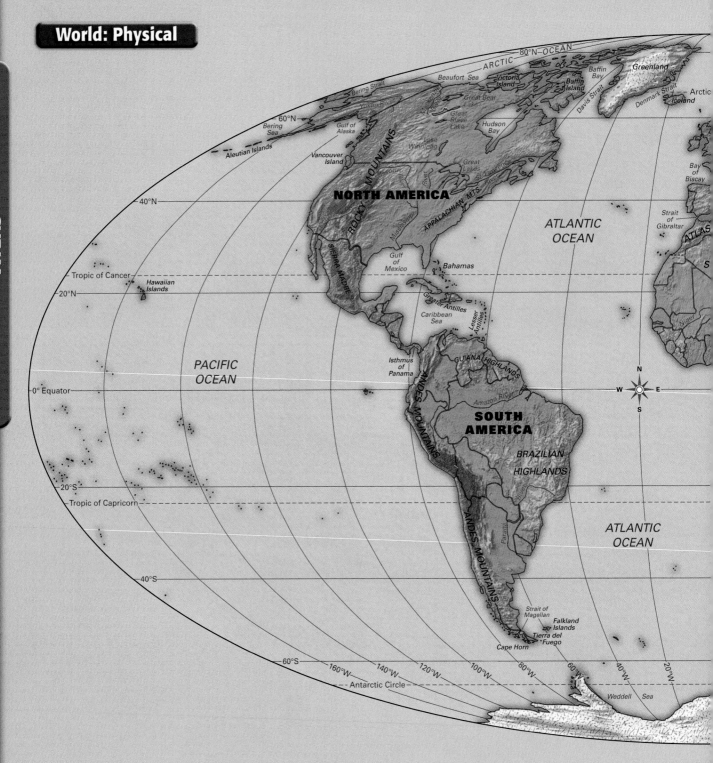

ARCTIC 80°N OCEAN
Beaufort Sea
Victoria Island
Baffin Bay
Greenland
Baffin Island
Davis Strait
Denmark Strait
Iceland
Arctic
Bering Strait
Yukon River
60°N
Bering Sea
Gulf of Alaska
Great Bear Lake
Great Slave Lake
Hudson Bay
Aleutian Islands
Vancouver Island
Winnipeg
Bay of Biscay
ROCKY MOUNTAINS
Great Lakes
St. Lawrence
40°N
NORTH AMERICA
APPALACHIAN MTS.
ATLANTIC OCEAN
Strait of Gibraltar
ATLAS
SIERRA MADRE
Mississippi
Gulf of Mexico
Tropic of Cancer
Hawaiian Islands
Bahamas
S
20°N
Greater Antilles
Niger
Caribbean Sea
Lesser Antilles
PACIFIC OCEAN
Isthmus of Panama
GUIANA HIGHLANDS
N
0° Equator
Amazon River
W   E
S
SOUTH AMERICA
ANDES MOUNTAINS
BRAZILIAN HIGHLANDS
20°S
Tropic of Capricorn
Paraná
ATLANTIC OCEAN
ANDES MOUNTAINS
40°S
Strait of Magellan
Falkland Islands
Tierra del Fuego
Cape Horn
60°S   160°W   140°W   120°W   100°W   80°W   60°W   40°W   20°W
Antarctic Circle
Weddell Sea

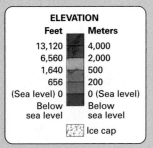

ELEVATION

Feet	Meters
13,120	4,000
6,560	2,000
1,640	500
656	200
(Sea level) 0	0 (Sea level)
Below sea level	Below sea level

Ice cap

0   500   1,000   1,500   2,000 Miles
0   1,000   2,000 Kilometers

Projection: Mollweide

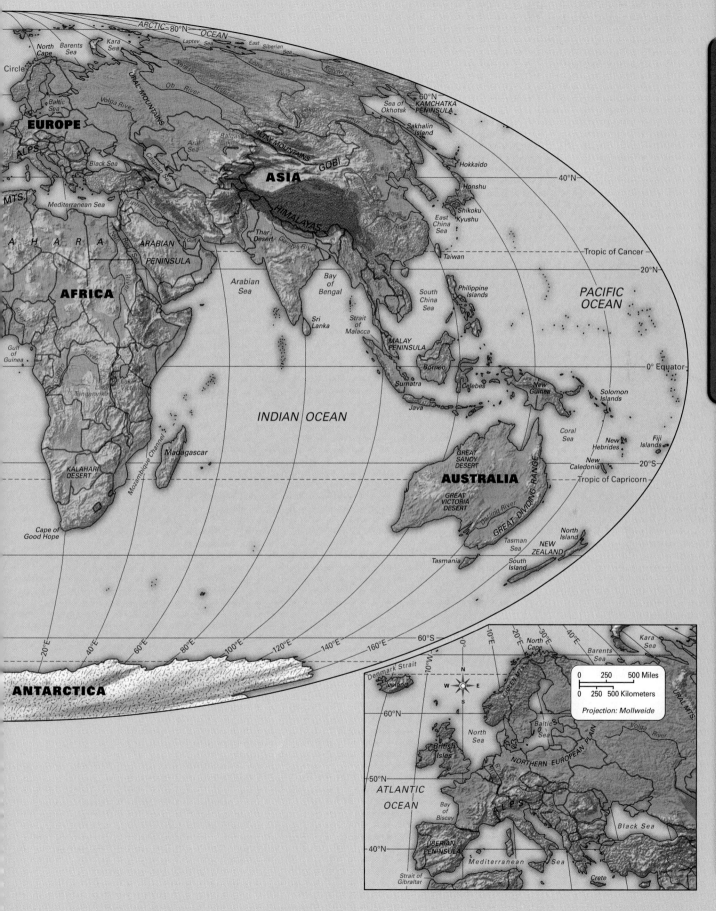

ARCTIC—80°N—OCEAN
North Cape
Barents Sea
Kara Sea
Laptev Sea
East Siberian Sea
Baltic Sea
EUROPE
Volga River
URAL MOUNTAINS
Ob River
Yenisei River
Lena River
60°N
Sea of Okhotsk
KAMCHATKA PENINSULA
Sakhalin Island
ALPS
Black Sea
Aral Sea
Caspian Sea
Balqas
ALTAY MOUNTAINS
ASIA
GOBI
Hokkaido
40°N
Honshu
MTS.
Mediterranean Sea
Euphrates River
Tigris River
Persian Gulf
HIMALAYAS
Yangtze River
Shikoku
Kyushu
East China Sea
SAHARA
ARABIAN PENINSULA
Thar Desert
Ganges River
Tropic of Cancer
Taiwan
20°N
Red Sea
Arabian Sea
Bay of Bengal
South China Sea
Philippine Islands
PACIFIC OCEAN
AFRICA
Gulf of Guinea
Niger River
Congo River
Sri Lanka
Strait of Malacca
MALAY PENINSULA
Borneo
0° Equator
Lake Tanganyika
Lake Victoria
Sumatra
Celebes
New Guinea
Solomon Islands
Java
INDIAN OCEAN
Coral Sea
Fiji Islands
Madagascar
Mozambique Channel
GREAT SANDY DESERT
New Hebrides
New Caledonia
20°S
KALAHARI DESERT
AUSTRALIA
GREAT VICTORIA DESERT
Tropic of Capricorn
GREAT DIVIDING RANGE
Darling River
Cape of Good Hope
Tasman Sea
NEW ZEALAND
North Island
20°E 40°E 60°E 80°E 100°E 120°E 140°E 160°E
Tasmania
South Island
60°S
ANTARCTICA

Denmark Strait
Iceland
North Cape
Barents Sea
Kara Sea
60°N
SCANDINAVIAN PENINSULA
North Sea
Baltic Sea
URAL MTS.
N
W   E
S
Volga River
British Isles
NORTHERN EUROPEAN PLAIN
50°N
ATLANTIC OCEAN
Bay of Biscay
ALPS
Black Sea
40°N
IBERIAN PENINSULA
Mediterranean   Sea
Strait of Gibraltar
Crete

0   250   500 Miles
0   250   500 Kilometers
Projection: Mollweide

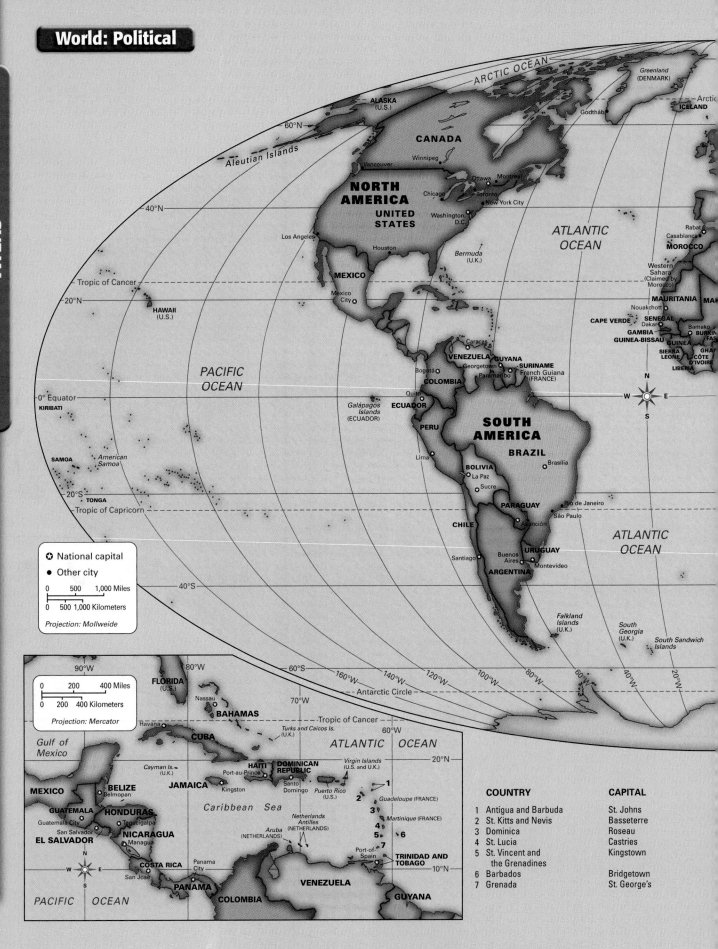

ATLAS

ARCTIC OCEAN
Greenland
(DENMARK)
Godthåb
ICELAND
Arctic

ALASKA
(U.S.)
60°N
CANADA
Winnipeg
Vancouver
NORTH
AMERICA
Ottawa   Montreal
Chicago
Toronto
40°N
New York City
UNITED
STATES
Washington,
D.C.
ATLANTIC
OCEAN
Rabat
Casablanca
MOROCCO
Los Angeles
Houston
Bermuda
(U.K.)
Western
Sahara
(Claimed by
Morocco)
Tropic of Cancer
MEXICO
20°N
MAURITANIA   MA
HAWAII
(U.S.)
Mexico
City
Nouakchott
CAPE VERDE   SENEGAL
Dakar
BURKIN
Caracas
GAMBIA
GUINEA-BISSAU
Bamako
BURKIN
FAS
PACIFIC
OCEAN
VENEZUELA   GUYANA
Georgetown
SIERRA
LEONE
GUINEA
GHA
CÔTE
D'IVOIRE
Bogotá
COLOMBIA
Paramaribo
SURINAME
French Guiana
(FRANCE)
LIBERIA
0° Equator
Quito
ECUADOR
KIRIBATI
Galápagos
Islands
(ECUADOR)
N
W   E
S
PERU
SOUTH
AMERICA
BRAZIL
SAMOA
American
Samoa
Lima
Brasília
BOLIVIA
La Paz
Sucre
20°S
TONGA
Rio de Janeiro
Tropic of Capricorn
PARAGUAY
São Paulo
CHILE
Asunción
ATLANTIC
OCEAN
URUGUAY
Buenos
Aires
Santiago
Montevideo
ARGENTINA
National capital
Other city
0   500   1,000 Miles
0   500 1,000 Kilometers
Projection: Mollweide
40°S
Falkland
Islands
(U.K.)
South
Georgia
(U.K.)
South Sandwich
Islands
60°S
160°W
140°W
120°W
100°W
80°W
60°W
40°W
20°W
Antarctic Circle

90°W   80°W
FLORIDA
(U.S.)
70°W
Tropic of Cancer
60°W
0   200   400 Miles
0   200   400 Kilometers
Projection: Mercator
Nassau
BAHAMAS
20°N
Gulf of
Mexico
Havana
Turks and Caicos Is.
(U.K.)
ATLANTIC   OCEAN
CUBA
Cayman Is.
(U.K.)
HAITI   DOMINICAN
REPUBLIC
Virgin Islands
(U.S. and U.K.)
MEXICO
BELIZE
Belmopan
JAMAICA
Port-au-Prince
Kingston
Santo
Domingo
Puerto Rico
(U.S.)
1
Guadeloupe (FRANCE)
2
GUATEMALA
HONDURAS
Caribbean   Sea
3
Martinique (FRANCE)
Guatemala City
Tegucigalpa
Netherlands
Antilles
(NETHERLANDS)
4
San Salvador
NICARAGUA
Aruba
(NETHERLANDS)
5   6
EL SALVADOR
Managua
7
Port-of-
Spain
N
W   E
S
COSTA RICA
Panama
City
TRINIDAD AND
TOBAGO
San Jose
10°N
PACIFIC   OCEAN
PANAMA
VENEZUELA
COLOMBIA
GUYANA

COUNTRY	CAPITAL
1 Antigua and Barbuda	St. Johns
2 St. Kitts and Nevis	Basseterre
3 Dominica	Roseau
4 St. Lucia	Castries
5 St. Vincent and the Grenadines	Kingstown
6 Barbados	Bridgetown
7 Grenada	St. George's

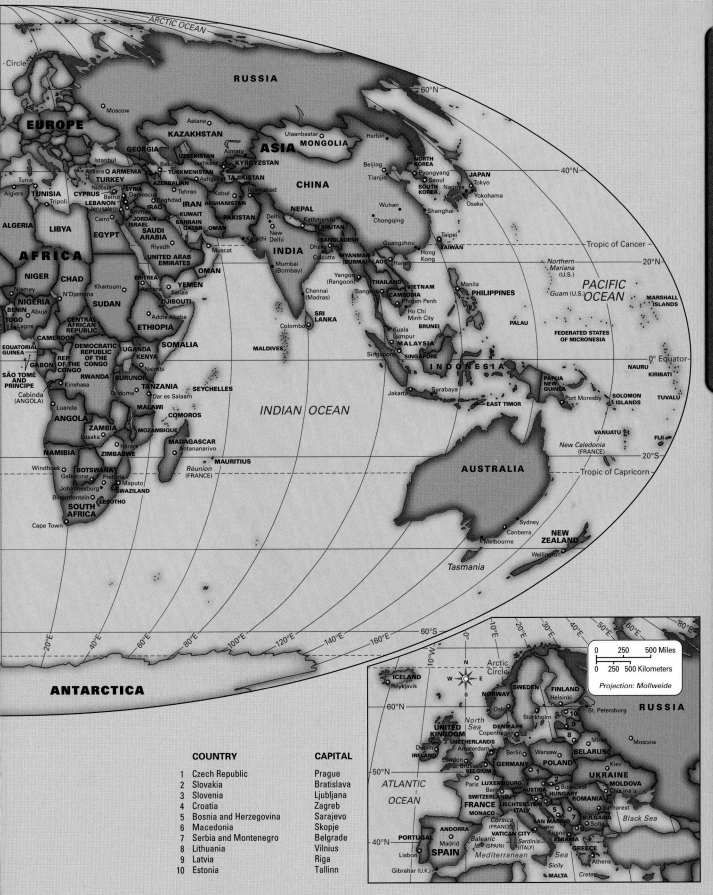

ARCTIC OCEAN

Circle

RUSSIA

60°N

EUROPE

Moscow

KAZAKHSTAN

Astana

ASIA

Ulaanbaatar

MONGOLIA

Harbin

40°N

GEORGIA

UZBEKISTAN

Almaty

KYRGYZSTAN

Beijing

NORTH
KOREA

JAPAN

Istanbul

Baku

Tashkent

Pyongyang

Tokyo

Ankara

ARMENIA

TURKMENISTAN

TAJIKISTAN

CHINA

SOUTH
KOREA

Seoul

Nagoya

Tunis

TURKEY

Nicosia

AZERBAIJAN

Ashgabat

Tianjin

Yokohama

Algiers

CYPRUS

SYRIA

Beirut

Tehran

Kabul

Islamabad

Wuhan

Osaka

TUNISIA

LEBANON

Damascus

IRAN

AFGHANISTAN

Chongqing

Shanghai

Tripoli

Baghdad

IRAQ

KUWAIT

PAKISTAN

NEPAL

Kathmandu

ALGERIA

Jerusalem

Amman

ISRAEL

JORDAN

BAHRAIN

QATAR

OMAN

Delhi

New
Delhi

BHUTAN

Taipei

Tropic of Cancer

LIBYA

EGYPT

Cairo

SAUDI
ARABIA

Riyadh

Muscat

INDIA

BANGLADESH

Dhaka

Guangzhou

Hong
Kong

TAIWAN

20°N

Northern
Mariana
(U.S.)

PACIFIC
OCEAN

AFRICA

UNITED ARAB
EMIRATES

OMAN

Mumbai
(Bombay)

Calcutta

MYANMAR
(BURMA)

LAOS

Hanoi

NIGER

CHAD

Khartoum

ERITREA

Asmara

YEMEN

Sanaa

Chennai
(Madras)

Yangon
(Rangoon)

THAILAND

VIETNAM

Bangkok

CAMBODIA

Manila

PHILIPPINES

Guam (U.S.)

MARSHALL
ISLANDS

Niamey

N'Djamena

DJIBOUTI

SRI
LANKA

Phnom Penh

Ho Chi
Minh City

PALAU

NIGERIA

Addis Ababa

Colombo

BRUNEI

BENIN

Abuja

SUDAN

ETHIOPIA

SOMALIA

Kuala
Lumpur

FEDERATED STATES
OF MICRONESIA

TOGO

Lagos

MALDIVES

MALAYSIA

Lagos

CENTRAL
AFRICAN
REPUBLIC

UGANDA

KENYA

Singapore

SINGAPORE

0° Equator

CAMEROON

EQUATORIAL
GUINEA

GABON

DEMOCRATIC
REPUBLIC
OF THE
CONGO

REP.
OF THE
CONGO

RWANDA

Nairobi

I N D O N E S I A

NAURU

KIRIBATI

SÃO TOMÉ
AND
PRINCIPE

BURUNDI

Dodoma

TANZANIA

SEYCHELLES

Jakarta

Surabaya

PAPUA
NEW
GUINEA

SOLOMON
ISLANDS

TUVALU

Kinshasa

Cabinda
(ANGOLA)

Luanda

MALAWI

COMOROS

Port Moresby

EAST TIMOR

INDIAN OCEAN

ANGOLA

ZAMBIA

MOZAMBIQUE

MADAGASCAR

Antananarivo

VANUATU

FIJI

Lusaka

Harare

MAURITIUS

New Caledonia
(FRANCE)

20°S

NAMIBIA

ZIMBABWE

Réunion
(FRANCE)

AUSTRALIA

Tropic of Capricorn

Windhoek

BOTSWANA

Gaborone

Pretoria

Johannesburg

Maputo

SWAZILAND

Bloemfontein

LESOTHO

SOUTH
AFRICA

Sydney

Canberra

NEW
ZEALAND

Cape Town

Melbourne

Wellington

Tasmania

20°E

40°E

60°E

80°E

100°E

120°E

140°E

160°E

60°S

ANTARCTICA

	COUNTRY	CAPITAL
1	Czech Republic	Prague
2	Slovakia	Bratislava
3	Slovenia	Ljubljana
4	Croatia	Zagreb
5	Bosnia and Herzegovina	Sarajevo
6	Macedonia	Skopje
7	Serbia and Montenegro	Belgrade
8	Lithuania	Vilnius
9	Latvia	Riga
10	Estonia	Tallinn

0   250   500 Miles

0   250   500 Kilometers

Projection: Mollweide

ICELAND

Reykjavik

Arctic
Circle

NORWAY

SWEDEN

FINLAND

RUSSIA

Helsinki

60°N

Oslo

Stockholm

St. Petersburg

10

UNITED
KINGDOM

North
Sea

DENMARK

Copenhagen

9

8

Minsk

Moscow

Dublin

NETHERLANDS

Amsterdam

Berlin

Warsaw

BELARUS

IRELAND

London

GERMANY

POLAND

Kiev

50°N

Brussels

BELGIUM

1

UKRAINE

ATLANTIC
OCEAN

Paris

LUXEMBOURG

2

Budapest

MOLDOVA

Bern

AUSTRIA

HUNGARY

Chişinău

FRANCE

SWITZERLAND

LIECHTENSTEIN

3

ROMANIA

MONACO

4

ITALY

SAN MARINO

7

Bucharest

Corsica
(FRANCE)

5

BULGARIA

Black Sea

PORTUGAL

ANDORRA

VATICAN CITY

Rome

6

Sofia

Madrid

Balearic
Isl.
(SPAIN)

Sardinia
(ITALY)

Tirane

ALBANIA

GREECE

40°N

Lisbon

SPAIN

Mediterranean

Sicily

Sea

Athens

Gibraltar (U.K.)

MALTA

Crete

North America: Physical

ATLAS

ASIA

ARCTIC OCEAN

+North Pole

POLAR ICE PACK

EUROPE

Arctic Circle

Queen Elizabeth Islands

BROOKS RANGE

Greenland

Denmark Strait

St. Lawrence Island

Bering Sea

Bering Strait

Nunivak Island

ALASKA RANGE

Mt. McKinley 20,320 ft (6,194 m)

Beaufort Sea

Banks Island

Victoria Island

Great Bear Lake

Baffin Bay

Baffin Island

Davis Strait

Cape Farewell

Kodiak Island

Gulf of Alaska

YUKON PLATEAU

Great Slave Lake

Southampton Island

Hudson Strait

Labrador Sea

Alexander Archipelago

Lake Athabasca

Coats Island

Mansel Island

Queen Charlotte Islands

Vancouver Island

Mackenzie River

Lake Winnipeg

C A N A D I A N   S H I E L D

Hudson Bay

Anticosti Island

Newfoundland

PACIFIC OCEAN

Mount Rainier 14,410 ft (4,392 m)

COAST RANGES

Saskatchewan River

Nelson River

Prince Edward Island

Gulf of St. Lawrence

Cape Breton Island

Cape Mendocino

Missouri River

L. Superior

L. Huron

Lake Ontario

ATLANTIC OCEAN

SIERRA NEVADA

GREAT BASIN

BLACK HILLS

L. Michigan

Lake Erie

APPALACHIAN MOUNTAINS

Cape Cod

Long Island

DEATH VALLEY

Platte River

INTERIOR PLAINS

OZARK PLATEAU

Ohio River

PIEDMONT

Cape Hatteras

Bermuda

CENTRAL VALLEY

Mount Whitney 14,494 ft (4,418 m)

COLORADO PLATEAU

Colorado River

Arkansas River

Mississippi River

Red River

Tennessee River

ATLANTIC COASTAL PLAIN

Guadalupe Island

BAJA CALIFORNIA

Brazos River

Rio Grande

GULF COASTAL PLAIN

FLORIDA PENINSULA

Cape Canaveral

Tropic of Cancer

SIERRA MADRE OCCIDENTAL

Gulf of California

SIERRA MADRE ORIENTAL

Gulf of Mexico

Florida Keys

Straits of Florida

Bahamas

Cuba

Greater Antilles

Hispaniola

Puerto Rico

Lesser Antilles

Jamaica

Caribbean Sea

Trinidad

Popocatépetl 17,887 ft (5,452 m)

YUCATÁN PENINSULA

SIERRA MADRE DEL SUR

Lake Nicaragua

CENTRAL AMERICA

ISTHMUS OF PANAMA

SOUTH AMERICA

ELEVATION

Feet	Meters
13,120	4,000
6,560	2,000
1,640	500
656	200
(Sea level) 0	0 (Sea level)
Below sea level	Below sea level

Ice cap

0   300   600 Miles

0   300   600 Kilometers

Projection: Azimuthal Equal Area

0° Equator

North America: Political

ASIA

ARCTIC OCEAN

North Pole

10°E

Bering Strait

St. Lawrence Island

Bering Sea

Nunivak Island

160°E

170°E

180°

170°W

160°W

150°W

140°W

130°W

120°W

111°W

100°W

90°W

80°N

70°N

60°N

50°N

40°N

30°N

20°N

10°N

Point Barrow

Beaufort Sea

Banks Island

Victoria Island

Great Bear Lake

ALASKA (U.S.)

Anchorage

Kodiak Island

Gulf of Alaska

Alexander Archipelago

Juneau

Queen Charlotte Islands

Vancouver Island

Queen Elizabeth Islands

Ellesmere Island

Baffin Bay

Baffin Island

Southampton Island

Coats Island

Mansel Island

Hudson Strait

Hudson Bay

Great Slave Lake

Greenland (DENMARK)

Denmark Strait

Arctic Circle

ICELAND

10°W

0°

20°W

30°W

Cape Farewell

Davis Strait

Labrador Sea

50°N

40°W

Anticosti Island

Newfoundland

St. Pierre and Miquelon (FRANCE)

Cape Breton Island

Prince Edward Island

Gulf of St. Lawrence

PACIFIC OCEAN

Edmonton

Calgary

Vancouver

Seattle

Portland

CANADA

Lake Winnipeg

Winnipeg

Lake Superior

Lake Huron

Lake Michigan

Lake Ontario

Lake Erie

Ottawa

Toronto

Montreal

Quebec

Boston

Cape Cod

New York City

Philadelphia

Baltimore

Washington, D.C.

Norfolk

ATLANTIC OCEAN

50°W

40°N

30°N

San Francisco

San Jose

Los Angeles

San Diego

Tijuana

Great Salt Lake

Salt Lake City

Denver

Kansas City

St. Louis

Minneapolis

Milwaukee

Detroit

Chicago

Cleveland

Columbus

Indianapolis

UNITED STATES

Memphis

Phoenix

Dallas

Austin

San Antonio

Houston

New Orleans

Atlanta

Birmingham

Jacksonville

Bermuda (U.K.)

Tropic of Cancer

Gulf of California

Monterrey

Gulf of Mexico

MEXICO

Guadalajara

Mexico City

Puebla

Mérida

Miami

Florida Keys

Nassau

BAHAMAS

Havana

Straits of Florida

CUBA

Cayman Is. (U.K.)

Kingston

JAMAICA

HAITI

Port-au-Prince

Santo Domingo

DOMINICAN REPUBLIC

Turks and Caicos Islands (U.K.)

Puerto Rico (U.S.)

San Juan

Virgin Is. (U.S. & U.K.)

ST. KITTS & NEVIS

ANTIGUA & BARBUDA

Guadeloupe (FRANCE)

DOMINICA

Martinique (FRANCE)

ST. LUCIA

BARBADOS

ST. VINCENT AND THE GRENADINES

GRENADA

Netherlands Antilles (NETHERLANDS)

Aruba (NETHERLANDS)

Caribbean Sea

TRINIDAD AND TOBAGO

Belmopan

BELIZE

GUATEMALA

Guatemala City

San Salvador

EL SALVADOR

HONDURAS

Tegucigalpa

NICARAGUA

Managua

San José

COSTA RICA

Panama Canal

PANAMA

Panama City

SOUTH AMERICA

Equator

0°

130°W

120°W

111°W

100°W

90°W

Legend

⭐ National capital

● Other city

0   300   600 Miles

0   300   600 Kilometers

Projection: Azimuthal Equal-Area

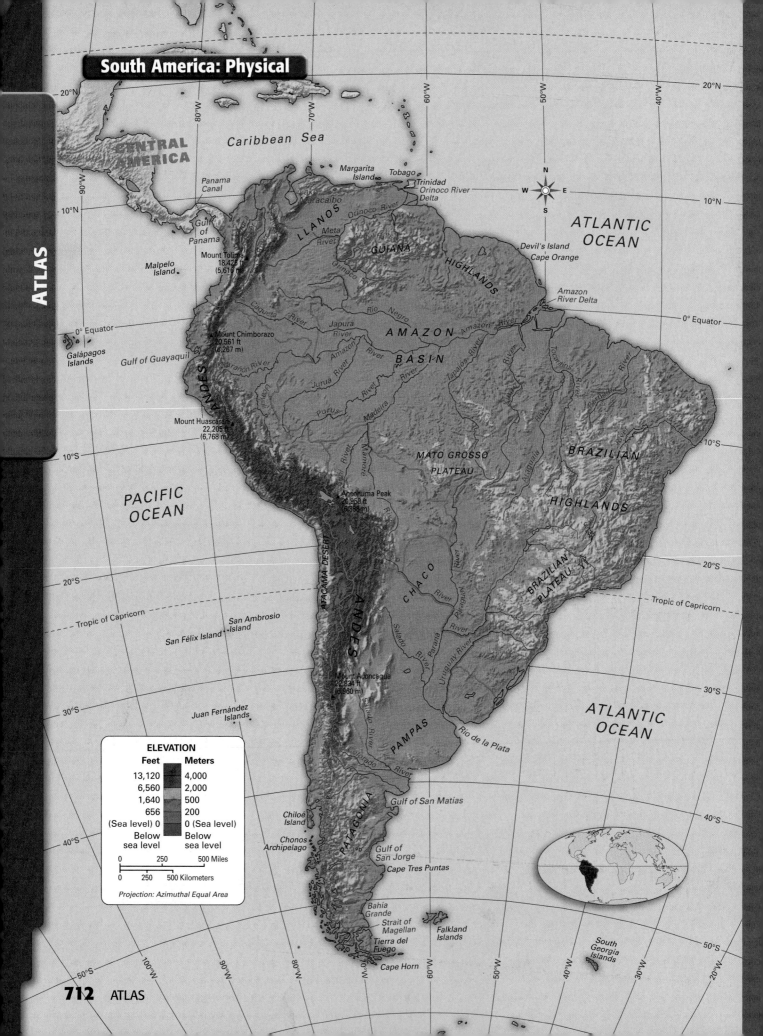

South America: Physical

Caribbean Sea

CENTRAL AMERICA

Panama Canal

Gulf of Panama

Malpelo Island

Mount Tolima
18,425 ft
(5,616 m)

Mount Chimborazo
20,561 ft
(6,267 m)

Gulf of Guayaquil

Galápagos Islands

Mount Huascarán
22,205 ft
(6,768 m)

Lake Maracaibo

Margarita Island   Tobago

Trinidad
Orinoco River Delta

LLANOS

Meta River

Orinoco River

Angel Falls

GUIANA

HIGHLANDS

Devil's Island
Cape Orange

Amazon River Delta

Rio Negro

Caquetá River

Japurá River

Orinoco River

AMAZON

BASIN

Amazon River

Jurua River

Purus River

Madeira River

Marañón River

Ucayali River

Amazon River

Tapajós River

Xingu River

Tocantins River

Araguaia River

Parnaíba River

MATO GROSSO PLATEAU

BRAZILIAN

HIGHLANDS

Ancohuma Peak
20,958 ft
(6,388 m)

ATACAMA DESERT

ANDES

CHACO

Paraguay River

Araguaia River

BRAZILIAN PLATEAU

PACIFIC OCEAN

ATLANTIC OCEAN

San Ambrosio Island

San Félix Island

Juan Fernández Islands

Mount Aconcagua
22,834 ft
(6,960 m)

Salado River

Paraná River

Uruguay River

Rio de la Plata

PAMPAS

Colorado River

Chiloé Island

Chonos Archipelago

PATAGONIA

Gulf of San Matías

Gulf of San Jorge

Cape Tres Puntas

Bahía Grande

Strait of Magellan

Tierra del Fuego

Cape Horn

Falkland Islands

South Georgia Islands

ATLANTIC OCEAN

Tropic of Capricorn

20°N

10°N

0° Equator

10°S

20°S

Tropic of Capricorn

30°S

40°S

50°S

ELEVATION

Feet		Meters
13,120		4,000
6,560		2,000
1,640		500
656		200
(Sea level) 0		0 (Sea level)
Below sea level		Below sea level

0   250   500 Miles

0   250   500 Kilometers

Projection: Azimuthal Equal Area

South America: Political

CENTRAL AMERICA

Caribbean Sea

Barranquilla
Cartagena
Caracas
VENEZUELA
Lake Maracaibo

Medellín
Bogotá
COLOMBIA
Cali

Malpelo Island (COLOMBIA)

Quito
ECUADOR
Guayaquil

Galápagos Islands (ECUADOR)

PERU

Trujillo

Callao Lima

PACIFIC OCEAN

Arequipa

Lake Titicaca
La Paz
Lake Poopó
BOLIVIA
Sucre

Georgetown
Paramaribo
GUYANA
Cayenne
SURINAME
French Guiana (FRANCE)

ATLANTIC OCEAN

Belém

BRAZIL

Recife

Brasília

Salvador

Belo Horizonte

Campinas
São Paulo
PARAGUAY
Asunción

Rio de Janeiro

Curitiba

CHILE

Córdoba

Rosario
Valparaíso
Santiago
Buenos Aires

URUGUAY
Montevideo

Pôrto Alegre

ATLANTIC OCEAN

ARGENTINA

San Ambrosio Island (CHILE)
San Félix Island (CHILE)

Juan Fernández Islands (CHILE)

Tropic of Capricorn

Tropic of Capricorn

Strait of Magellan

Falkland Islands (U.K.)

Tierra del Fuego

South Georgia Island (U.K.)

Key:
⊛ National capital
● Other city

0   250   500 Miles
0   250   500 Kilometers

Projection: Azimuthal Equal-Area

20°N   10°N   0° Equator   10°S   20°S   30°S   40°S   50°S

ASIA

Europe: Physical

ELEVATION

Feet	Meters
13,120	4,000
6,560	2,000
1,640	500
656	200
(Sea level) 0	0 (Sea level)
Below sea level	Below sea level

Ice cap

300 Miles
300 Kilometers
0 150
0 150

Projection: Azimuthal Equal Area

ARCTIC OCEAN

Barents Sea

White Sea

KOLA PENINSULA

URAL MOUNTAINS

Pechora River

Dvina River

North Dvina River

NORTHERN EUROPEAN PLAIN

Caspian Sea

CAUCASUS MTS.

Mt. Elbrus 18,510 ft (5,642 m)

Sea of Azov

CRIMEAN PENINSULA

Black Sea

Volga River

Don River

Dnipro

Lake Ladoga

Lake Onega

Rybinsk Reservoir

Gulf of Finland

BALTIC PLAINS

Baltic Sea

Gulf of Bothnia

KJÖLEN MOUNTAINS

Lake Vänern

Lake Vättern

Kattegat

Skagerrak

Norwegian Sea

North Cape

Vistula River

Oder River

Dniester River

Nistru River

Prut River

CARPATHIAN MTS.

TRANSYLVANIAN ALPS

BALKAN PENINSULA

Danube River

Sea of Marmara

Aegean Sea

Rhodes

Crete

DINARIC ALPS

Adriatic Sea

APENNINES

Tyrrhenian Sea

Sicily

Malta

Corsica

Sardinia

Balearic Islands

Mediterranean Sea

ALPS

Mont Blanc 15,781 ft (4,810 m)

Rhine River

Elbe River

Seine River

Loire River

Garonne River

PYRENEES

IBERIAN PENINSULA

Strait of Gibraltar

Cape Finisterre

Bay of Biscay

ATLANTIC OCEAN

English Channel

Thames River

PENNINES

Irish Sea

Hebrides

British Isles

Orkney Islands

Shetland Islands

Faeroe Islands

Iceland

North Sea

Arctic Circle

SOUTHWEST ASIA

AFRICA

N E S W

Europe: Political

National capital ✪
Other city ●

300 Miles
0 150 300 Kilometers
0 150

Projection: Azimuthal Equal-Area

ASIA

URAL MOUNTAINS

RUSSIA

Nizhny Novgorod

Moscow ✪

Caspian Sea

SOUTHWEST ASIA

Black Sea

Barents Sea

White Sea

St. Petersburg

North Cape

ARCTIC OCEAN

FINLAND
Helsinki ✪

Gulf of Bothnia

ESTONIA
Tallinn

LATVIA
Riga

LITHUANIA
Vilnius

RUSSIA

Minsk ✪

BELARUS

Kiev ✪

UKRAINE

MOLDOVA
Chisinau ✪

ROMANIA
Bucharest ✪

Belgrade
SERBIA AND MONTENEGRO

BULGARIA
Sofia ✪

Skopje ✪
MACEDONIA

GREECE
Athens ✪

Aegean Sea

Rhodes

Crete

SWEDEN

NORWAY
Oslo ✪

Stockholm ✪

Göteborg

Baltic Sea

POLAND
Warsaw ✪

Krakow ●

CZECH REPUBLIC
Prague ✪

SLOVAKIA
Bratislava ✪

Budapest ✪
HUNGARY

Zagreb ✪
CROATIA

BOSNIA AND HERZEGOVINA
Sarajevo ✪

Adriatic Sea

Tirana ✪
ALBANIA

Sea

Bergen

North Sea

DENMARK
Copenhagen ✪

Hamburg ●

GERMANY
Berlin ✪

Dresden ●

Cologne ● Bonn ●

Munich ●

Vienna ✪
AUSTRIA

LIECHTENSTEIN
Vaduz ✪

SLOVENIA
Ljubljana ✪

Lake Geneva

Bern ✪
SWITZERLAND

Milan ●

SAN MARINO ✪

San Marino

MONACO
Monaco ✪

ITALY
Rome ✪

VATICAN CITY ✪

Naples ●

Sicily

MALTA
Valletta ✪

THE NETHERLANDS
Amsterdam ✪

BELGIUM
Brussels ✪

LUXEMBOURG ✪
Luxembourg

Paris ✪
FRANCE

Lyon ●

Marseille ●

Corsica (FRANCE)

Sardinia (ITALY)

Mediterranean Sea

AFRICA

SCOTLAND
Edinburgh ●

Liverpool ●

UNITED KINGDOM

ENGLAND
London ✪

WALES

NORTHERN IRELAND
Belfast ●

IRELAND
Dublin ✪

British Isles

Channel Islands (U.K.)

English Channel

Bay of Biscay

PYRENEES

ANDORRA
Andorra la Vella ✪

Barcelona ●

Balearic Islands (SPAIN)

SPAIN
Madrid ✪

Valencia ●

Seville ●

Gibraltar (U.K.)

Strait of Gibraltar

PORTUGAL
Lisbon ✪

ATLANTIC OCEAN

ICELAND
Reykjavík ✪

Arctic Circle

Faeroe Islands (DENMARK)

Shetland Islands

ATLAS

70°N

60°N

50°N

40°N

30°W

20°W

10°W

0°

10°E

20°E

30°E

40°E

50°E

70°N

60°N

50°N

40°N

Arctic Circle

Asia: Physical

ELEVATION

Feet	Meters
13,120	4,000
6,560	2,000
1,640	500
656	200
(Sea level) 0	0 (Sea level)
Below sea level	Below sea level

Ice cap

750 Miles
250 500 750 Kilometers

Projection: Two-Point Equidistant

PACIFIC OCEAN

AUSTRALIA

EUROPE

AFRICA

INDIAN OCEAN

S I B E R I A

North Pole

Arctic Circle

Aleutian Islands

KAMCHATKA PENINSULA

Bering Sea

CENTRAL RANGE

KOLYMA MTS.

CHERSKIY RANGE

VERKHOYANSKY RANGE

STANOVOY MOUNTAINS

Sea of Okhotsk

Sakhalin Island

Kuril Islands

Hokkaido

Honshu

Shikoku

Kyushu

Sea of Japan (East Sea)

Korea Strait

Okinawa

Ryukyu Islands

East China Sea

Yellow Sea

BOHAI HILLS

NORTH CHINA PLAIN

QIN LING

Taiwan

Hainan

South China Sea

Luzon Strait

Luzon

Philippines

Mindanao

Celebes Sea

Celebes

Borneo

Java Sea

Java

Bangka

Sumatra

MALAY PENINSULA

Mentawai Islands

Andaman Sea

Nicobar Islands

Andaman Islands

Bay of Bengal

Sri Lanka

Maldives

Lakshadweep Islands

Arabian Sea

Socotra Island

Gulf of Aden

Red Sea

SINAI PENINSULA

Mediterranean Sea

Cyprus

Black Sea

Bosporus

CAUCASUS MTS.

Mount Ararat 16,945 ft (5,165 m)

ANATOLIAN PLATEAU

SYRIAN DESERT

AN-NAFUD

Euphrates River

Tigris River

Persian Gulf

ZAGROS MTS.

Gulf of Oman

RUB' AL KHALI

Caspian Sea

URAL MOUNTAINS

Ob River

WEST SIBERIAN PLAIN

Irtysh River

Ishim River

Tobol River

CENTRAL SIBERIAN PLATEAU

Yenisey River

Tunguska River

Lena River

Aldan River

YABLONOVY RANGE

SAYAN MOUNTAINS

ALTAY MOUNTAINS

MONGOLIAN PLATEAU

GREATER KHINGAN RANGE

G O B I

Huang He River

TAYMYR PENINSULA

North Land

Kara Sea

Novaya Zemlya

Laptev Sea

New Siberian Islands

Wrangel Island

Barents Sea

Franz Josef Land

KAZAKH UPLANDS

Balqash Lake

TIAN SHAN

TARIM BASIN

TAKLIMAKAN DESERT

KUNLUN MOUNTAINS

PLATEAU OF TIBET

Mount Everest 29,035 ft (8,850 m)

H I M A L A Y A S

INDO-GANGETIC PLAIN

Ganges River

Brahmaputra River

Indus River

Sutlej River

THAR DESERT

DECCAN PLATEAU

EASTERN GHATS

WESTERN GHATS

HINDU KUSH

KYZYL KUM

KARA KUM

Amu Darya

Syr Darya

Aral Sea

Ural River

USTYURT PLATEAU

TURKMEN LOWLAND

GREAT SALT DESERT

INDOCHINA PENINSULA

Mekong River

Chao Phraya River

Gulf of Thailand

Gulf of Tonkin

Arafura Sea

Banda Sea

Molucca Sea

New Guinea

MACKENZIE MOUNTAINS

Equator

Tropic of Cancer

10°N 20°N 30°N 40°N 50°N 60°N 70°N 80°N

60°N 70°N

10°S

10°E 20°E 30°E 40°E 50°E 60°E 70°E 80°E 90°E 100°E 110°E 120°E 130°E 140°E 150°E

170°W 180° 170°E 160°E 150°E 140°E 130°E 120°E 110°E

Asia: Political

Legend:
- ✪ National capitals
- • Other cities

750 Miles
0   250   500   750 Kilometers

Projection: Two-Point Equidistant

North Pole

Arctic Circle

RUSSIA

URAL MOUNTAINS

EUROPE

RUSSIA

Moscow

Yekaterinburg

Chelyabinsk

Omsk

Novosibirsk

Astana

Yakutsk

Irkutsk

Lake Baykal

Ulaanbaatar

MONGOLIA

KAZAKHSTAN

Aral Sea

Lake Balkhash

Almaty

Bishkek

KYRGYZSTAN

UZBEKISTAN

Tashkent

TAJIKISTAN

Dushanbe

TURKMENISTAN

Ashgabat

Caspian Sea

GEORGIA

Tbilisi

ARMENIA

Yerevan

Baku

AZERBAIJAN

TURKEY

Ankara

Istanbul

Izmir

Black Sea

CYPRUS

Nicosia

LEBANON

Beirut

ISRAEL

Tel Aviv

Jerusalem

Mediterranean Sea

SYRIA

Damascus

Amman

JORDAN

Mosul

Baghdad

IRAQ

Basra

Kuwait City

KUWAIT

Mecca

Jidda

Red Sea

SAUDI ARABIA

Riyadh

Medina

YEMEN

Sanaa

Gulf of Aden

AFRICA

IRAN

Tehran

Shiraz

Persian Gulf

BAHRAIN

Manama

QATAR

Doha

Abu Dhabi

UNITED ARAB EMIRATES

OMAN

Masqat (Muscat)

Socotra (YEMEN)

Arabian Sea

AFGHANISTAN

Kabul

PAKISTAN

Lahore

Karachi

Islamabad

INDIA

New Delhi

Delhi

Jaipur

Ahmadabad

Mumbai (Bombay)

Lakshadweep Islands (INDIA)

Bangalore

Chennai (Madras)

MALDIVES

Male

Colombo

SRI LANKA

NEPAL

Kathmandu

BHUTAN

Thimphu

BANGLADESH

Dhaka

Kolkata (Calcutta)

Bay of Bengal

MYANMAR (BURMA)

Yangon (Rangoon)

Mandalay

Andaman Islands (INDIA)

Nicobar Islands (INDIA)

Andaman Sea

CHINA

Chengdu

Chongqing

Wuhan

Nanjing

Beijing

Shanghai

Yellow Sea

Qingdao

Fushun

Harbin

NORTH KOREA

Pyongyang

SOUTH KOREA

Seoul

Pusan

JAPAN

Tokyo

Yokohama

Osaka

Hiroshima

Nagasaki

Sapporo

Vladivostok

Sea of Okhotsk

Sakhalin Island (RUSSIA)

Kuril Islands (RUSSIA)

Aleutian Islands

Bering Sea

East China Sea

TAIWAN

Taipei

Hong Kong

Macao

Guangzhou

VIETNAM

Hanoi

Ho Chi Minh City

LAOS

Vientiane

THAILAND

Bangkok

CAMBODIA

Phnom Penh

Gulf of Thailand

Hainan (CHINA)

South China Sea

PHILIPPINES

Manila

Luzon Strait

Celebes Sea

BRUNEI

Bandar Seri Begawan

MALAYSIA

Kuala Lumpur

SINGAPORE

Singapore

Medan

INDONESIA

Jakarta

Bandung

Java Sea

Ujung Pandang

Surabaya

EAST TIMOR

Dili

Arafura Sea

AUSTRALIA

New Guinea

PACIFIC OCEAN

INDIAN OCEAN

Tropic of Cancer

Equator

ATLAS   717

Africa: Physical

EUROPE

SOUTHWEST ASIA

Azores

Strait of Gibraltar

Madeira Islands

Canary Islands

Mediterranean Sea

Gulf of Sidra

Suez Canal

Persian Gulf

Tropic of Cancer

Cape Blanc

ATLAS MOUNTAINS

S A H A R A

LIBYAN DESERT

QATTARA DEPRESSION

Cape Verde Islands

EL DJOUF

AHAGGAR MOUNTAINS

AIR MTS.

TIBESTI MOUNTAINS

NUBIAN DESERT

Lake Nasser

Red Sea

Cape Verde

Niger River

S A H E L

S U D A N

CHAD BASIN

Lake Chad

FOUTA DJALLON

Lake Volta

Benue River

SUDAN BASIN

ETHIOPIAN HIGHLANDS

HORN OF AFRICA

Gulf of Aden

SOMALI PENINSULA

Cape Palmas

Gulf of Guinea

ADAMAWA MTS.

CONGO BASIN

Congo River

Lake Albert

Lake Edward

Mount Kenya 17,058 ft (5,199 m)

Equator

Cape Lopez

SERENGETI PLAIN

MASAI STEPPE

Mount Kilimanjaro 19,340 ft (5,895 m)

INDIAN OCEAN

Zanzibar

Ascension

ATLANTIC OCEAN

MPUMBA MOUNTAINS

WESTERN RIFT VALLEY

EASTERN RIFT VALLEY

Lake Tanganyika

Seychelles

Cape Delgado

Comoro Islands

KALAHARI BASIN

Okavango Delta

Victoria Falls

Lake Kariba

Madagascar

Mauritius

NAMIB DESERT

KALAHARI DESERT

Mozambique Channel

Réunion

Tropic of Capricorn

Orange River

GREAT KARROO

DRAKENSBERG MOUNTAINS

Cape of Good Hope

ELEVATION

Feet		Meters
13,120		4,000
6,560		2,000
1,640		500
656		200
(Sea level) 0		0 (Sea level)
Below sea level		Below sea level

0   250   500 Miles

0   250   500 Kilometers

Projection: Azimuthal Equal-Area

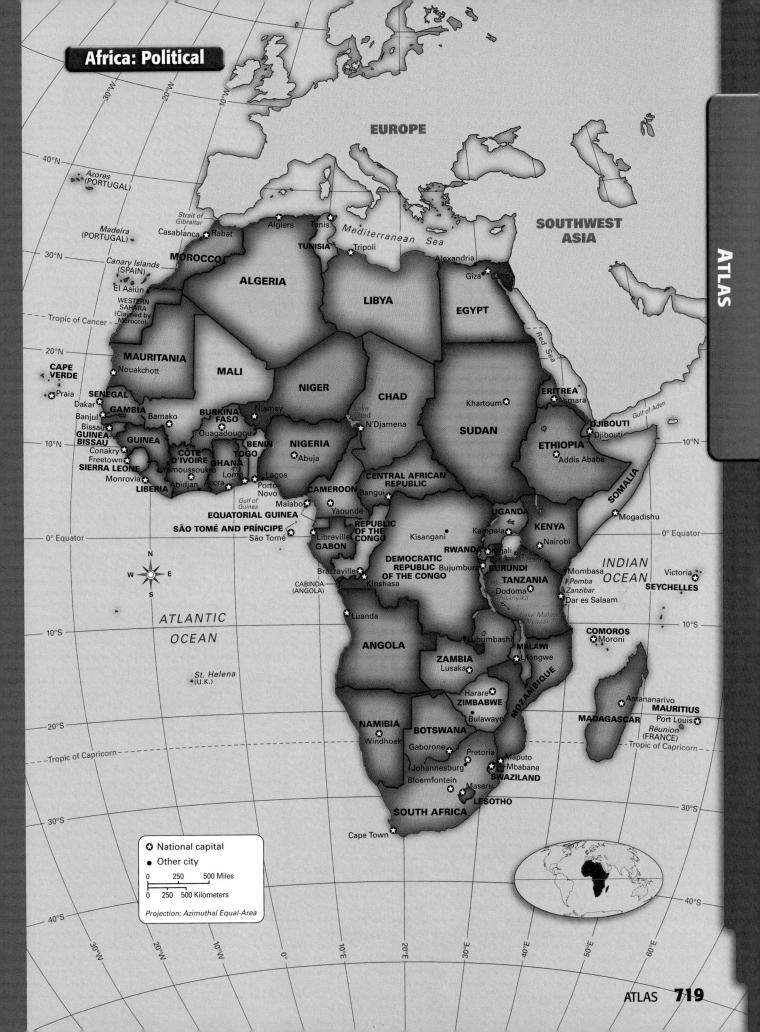

Africa: Political

EUROPE

SOUTHWEST ASIA

Azores (PORTUGAL)

Madeira (PORTUGAL)

Strait of Gibraltar

Casablanca • Rabat ⊛

Algiers ⊛

Tunis ⊛

TUNISIA

Tripoli ⊛

Mediterranean Sea

Alexandria •

Canary Islands (SPAIN)

MOROCCO

30°N

40°N

ALGERIA

LIBYA

EGYPT

Giza • Cairo ⊛

El Aaiún ⊛

WESTERN SAHARA (Claimed by Morocco)

Tropic of Cancer

Red Sea

20°N

MAURITANIA

Nouakchott ⊛

MALI

NIGER

CHAD

Khartoum ⊛

ERITREA

Asmara ⊛

Gulf of Aden

CAPE VERDE

• Praia ⊛

SENEGAL

Dakar •

Niamey ⊛

DJIBOUTI

Djibouti ⊛

GAMBIA

Banjul ⊛

Bamako •

BURKINA FASO

N'Djamena ⊛

SUDAN

ETHIOPIA

Addis Ababa ⊛

10°N

Bissau ⊛

GUINEA-BISSAU

GUINEA

Ouagadougou ⊛

BENIN

NIGERIA

Abuja ⊛

CENTRAL AFRICAN REPUBLIC

Conakry ⊛

Freetown ⊛

SIERRA LEONE

CÔTE D'IVOIRE

Yamoussoukro ⊛

GHANA

TOGO

Lomé ⊛

Lagos •

Bangui ⊛

SOMALIA

Monrovia ⊛

LIBERIA

Abidjan •

Accra ⊛

Porto Novo ⊛

CAMEROON

Yaoundé ⊛

Malabo ⊛

EQUATORIAL GUINEA

Gulf of Guinea

UGANDA

Kampala ⊛

KENYA

Mogadishu ⊛

0° Equator

SÃO TOMÉ AND PRÍNCIPE

São Tomé ⊛

Libreville ⊛

GABON

REPUBLIC OF THE CONGO

Kisangani •

RWANDA

Kigali ⊛

Nairobi •

INDIAN OCEAN

Victoria ⊛

SEYCHELLES

Brazzaville ⊛

DEMOCRATIC REPUBLIC OF THE CONGO

Bujumbura ⊛

BURUNDI

Mombasa •

Pemba

0° Equator

CABINDA (ANGOLA)

Kinshasa •

TANZANIA

Dodoma ⊛

Zanzibar •

Dar es Salaam •

Lake Tanganyika

Luanda ⊛

ATLANTIC OCEAN

10°S

COMOROS

Moroni ⊛

Lubumbashi •

Lake Malawi (Nyasa)

ANGOLA

MALAWI

Lilongwe ⊛

St. Helena (U.K.)

ZAMBIA

Lusaka ⊛

MOZAMBIQUE

Antananarivo ⊛

Harare ⊛

ZIMBABWE

MAURITIUS

Port Louis ⊛

MADAGASCAR

Réunion (FRANCE)

Tropic of Capricorn

Bulawayo •

NAMIBIA

BOTSWANA

Windhoek ⊛

Gaborone ⊛

Pretoria ⊛

Maputo ⊛

Mbabane ⊛

Johannesburg •

SWAZILAND

Bloemfontein •

Maseru ⊛

LESOTHO

SOUTH AFRICA

30°S

Cape Town •

N W E S

10°S

20°S

Legend

- ⊛ National capital
- • Other city

0   250   500 Miles

0   250   500 Kilometers

Projection: Azimuthal Equal-Area

The Pacific: Political

ASIA

NORTH AMERICA

National capital
Other city

1,000 Miles
1,000 Kilometers
500
500
0
0

Projection: Azimuthal Equal-Area

NORTH PACIFIC OCEAN

Tropic of Cancer

30°N

15°N

0° Equator

15°S

Tropic of Capricorn

Easter Island (CHILE)

Pitcairn (U.K.) Ducie Island
Pitcairn Island

Marquesas Islands (FRANCE)

Tuamotu Archipelago (FRANCE)

Rapa Island (FRANCE)

French Polynesia

P O L Y N E S I A

Society Islands (FRANCE) Tahiti (FRANCE) Tubuai Islands (FRANCE)

Papeete

Starbuck Island

Kingman Reef (U.S.) Palmyra Island (U.S.)
Washington Fanning Island
Island

Jarvis I. (U.S.)

K I R I B A T I

Manihiki Cook Islands (NEW ZEALAND)
Rarotonga Island

Howland I. (U.S.) Phoenix Islands

Baker I. (U.S.)

McKean I.

Gardner

Tokelau (N.Z.)

American Samoa
SAMOA Pago Pago
Apia

Niue (N.Z.)
TONGA Nuku'alofa

SOUTH PACIFIC OCEAN

Midway Island (U.S.)

Johnston Island (U.S.)

Hawaiian Islands

Hawaii (U.S.)

International Date Line

180°

Wallis & Futuna (FR.) Suva
TUVALU Funafuti FIJI
M E L A N E S I A

Kermadec Islands (N.Z.)

Chatham Islands (N.Z.)

Wellington
Auckland Christchurch
North Island

Bounty Islands (N.Z.)

NEW ZEALAND
South Island

Auckland Islands (NEW ZEALAND)

Wake Island (U.S.)

MARSHALL ISLANDS

Eniwetok I. Kwajalein Island

Tarawa
Gilbert Islands

Majuro

Palikir

M I C R O N E S I A

NAURU

SOLOMON ISLANDS

Espiritu Santo
VANUATU Malekula I.
Port Vila

New Caledonia (FRANCE)
Noumea

Loyalty Islands (FRANCE)

Norfolk Island (AUSTRALIA)

Bonin Islands (JAPAN)

Volcano Islands (JAPAN)

Northern Marianas (U.S.)

Truk Is.

FEDERATED STATES OF MICRONESIA

Guam Agana (U.S.)

PALAU
Koror

Bismarck Archipelago

PAPUA NEW GUINEA
Port Moresby

Honiara Guadalcanal I.

Coral Sea

New Guinea

Philippine Sea

South China Sea

Arafura Sea

Timor Sea

Darwin

A U S T R A L I A

Brisbane

Sydney
Canberra

Melbourne

Hobart

Tasman Sea

Adelaide

Perth

INDIAN OCEAN

Christmas Island (AUSTRALIA)

135°E 150°E 165°E 180° 165°W 150°W 135°W 120°W

30°N 15°N 0° 15°S 30°S 45°S

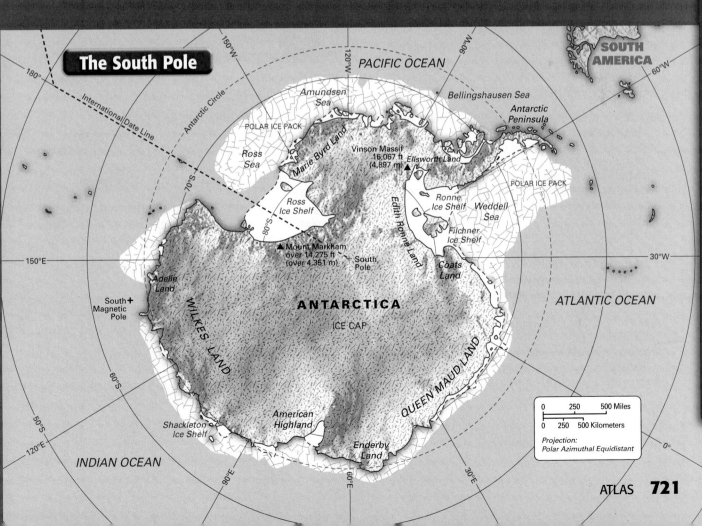

The North Pole

0   200   400 Miles
0   200   400 Kilometers

Projection:
Polar Azimuthal Equidistant

EUROPE

Barents
Sea

Kara
Sea

Norwegian
Sea

Laptev
Sea

ARCTIC
OCEAN

Greenland
Sea

Arctic Circle

30°W

ASIA

150°E

North
Pole

Greenland
(DENMARK)

ATLANTIC
OCEAN

POLAR ICE PACK

International Date Line

North
Magnetic
Pole

Baffin
Bay

60°W

60°N

Beaufort
Sea

Bering Sea

NORTH
AMERICA

50°N

180°

The South Pole

SOUTH
AMERICA

60°W

International Date Line

PACIFIC OCEAN

Antarctic Circle

120°W

90°W

180°

Amundsen
Sea

Bellingshausen Sea

Antarctic
Peninsula

POLAR ICE PACK

Ross
Sea

Marie Byrd Land

Vinson Massif
16,067 ft
(4,897 m)

Ellsworth Land

POLAR ICE PACK

150°E

70°S

Ross
Ice Shelf

Edith Ronne Land

Ronne
Ice Shelf

Weddell
Sea

80°S

Mount Markham
over 14,275 ft
(over 4,351 m)

South
Pole

Filchner
Ice Shelf

30°W

Adelie
Land

South
Magnetic
Pole

WILKES LAND

ANTARCTICA

ICE CAP

Coats
Land

QUEEN MAUD LAND

ATLANTIC OCEAN

60°S

50°S

120°E

Shackleton
Ice Shelf

American
Highland

Enderby
Land

90°E

60°E

30°E

0°

INDIAN OCEAN

0   250   500 Miles
0   250   500 Kilometers

Projection:
Polar Azimuthal Equidistant

ATLAS

Gazetteer

A

Abuja (ah-BOO-jah) (9°N, 7°E) the capital of Nigeria (p. 371)

Accra (6°N, 0°) the capital of Ghana (p. 371)

Addis Ababa (AH-dis AH-bah-bah) (9°N, 39°E) the capital of Ethiopia (p. 393)

Adelaide (35°S, 139°E) a city in southern Australia (p. 642)

Afghanistan a landlocked country in Central Asia (p. 243)

Africa the second-largest continent; surrounded by the Atlantic Ocean, Indian Ocean, and Mediterranean Sea (p. 718)

Akkad a city in Mesopotamia, now in modern Iraq; it was the center of the Akkadian empire (p. 123)

Alexandria an ancient city in Egypt built by Alexander the Great (p. 355)

Algeria a country in North Africa between Morocco and Libya (p. 349)

Algiers (37°N, 3°E) the capital of Algeria (p. 349)

Almaty (ahl-mah-TUH) (43°N, 77°E) the former capital of Kazakhstan (p. 254)

Amazon River a major river in South America (p. 712)

Amman (32°N, 36°E) the capital of Jordan (p. 213)

Amu Darya (uh-MOH duhr-YAH) a river in Central Asia that flows along Afghanistan's border with Tajikistan, Uzbekistan, and Turkmenistan to the Aral Sea (p. 245)

Anatolia (a-nuh-TOH-lee-uh) a mountainous region in Southwest Asia forming most of Turkey; also referred to as Asia Minor (p. 196)

Andes a mountain range along the west coast of South America (p. 712)

Angola a country in Central Africa that borders the Atlantic Ocean (p. 413)

Ankara (40°N, 33°E) the capital and second-largest city of Turkey (p. 202)

An Nafud (ahn nah-FOOD) a large desert in northern Saudi Arabia; known for its giant sand dunes (p. 222)

Antananarivo (19°S, 48°E) the capital of Madagascar (p. 435)

Antarctica a continent around the South Pole (p. 643)

Antarctic Circle the line of latitude located at 66.5° south of the equator; parallel beyond which no sunlight shines on the June solstice (p. 721)

Antarctic Peninsula a large peninsula in Antarctica (p. 643)

Apia (14°S, 172°W) the capital of Western Samoa (p. 656)

Arabia or Arabian Peninsula the world's largest peninsula; located in Southwest Asia (p. 219)

Arabian Sea a large arm of the Indian Ocean between India and Arabia (p. 541)

Aral Sea an inland sea in Central Asia fed by the Syr Darya and Amu Darya rivers; it has been steadily shrinking (p. 258)

Arctic Circle the line of latitude located at 66.5° north of the equator; parallel beyond which no sunlight shines on the December solstice (p. 721)

Arctic Ocean the ocean north of the Arctic Circle; the world's fourth-largest ocean (p. 721)

Ashgabat (38°N, 58°E) the capital of Turkmenistan (p. 243)

Asia Minor a large peninsula in Southwest Asia, between the Black Sea and the Mediterranean Sea, forming most of Turkey (p. 117)

Asia the world's largest continent; located between Europe and the Pacific Ocean (p. 716)

Asmara (15°N, 39°E) the capital of Eritrea (p. 393)

Astana (51°N, 72°E) the capital of Kazakhstan (p. 254)

Athens (38°N, 24°E) an ancient city and the modern capital of Greece (p. 157)

Atlantic Ocean the ocean between the continents of North and South America and the continents of Europe and Africa; the world's second-largest ocean (p. 706)

Atlas Mountains a high mountain range in northwestern Africa (p. 351)

Australia a country and continent in the Pacific (p. 642)

B

Babylon an ancient city in Mesopotamia on the Euphrates River; it was the capital of the Babylonian Empire (p. 134)

Baghdad (33°N, 44°E) the capital of Iraq (p. 219)

Bahrain a small country on the Persian Gulf (p. 219)

Bamako (BAH-mah-koh) (13°N, 8°W) the capital of Mali (p. 371)

Bandar Seri Begawan (5°N, 115°E) capital of Brunei (p. 617)

Bangkok (14°N, 100°E) the capital of Thailand (p. 616)

Bangladesh a country in South Asia (p. 541)

Bangui (bahn-GEE) (4°N, 19°E) the capital of the Central African Republic (p. 413)

Banjul (BAHN-jool) (13°N, 17°W) the capital of Gambia (p. 370)

Bay of Bengal a large bay of the Indian Ocean between India and Southeast Asia (p. 541)

Beijing (40°N, 116°E) the capital of China (p. 565)

Beirut (34°N, 36°E) the capital of Lebanon (p. 212)

Benin a country in West Africa between Togo and Nigeria (p. 371)

Bethlehem (BETH-li-hem) (32°N, 35°E) a town in Judea; traditionally regarded as the birthplace of Jesus (p. 153)

Bhutan a country in South Asia north of India (p. 541)

Bishkek (43°N, 75°E) the capital of Kyrgyzstan (p. 243)

Bissau (bis-OW) (12°N, 16°W) the capital of Guinea-Bissau (p. 370)

Bloemfontein (BLOOM-fahn-tayn) (29°S, 26°E) the judicial capital of South Africa (p. 435)

Blue Mosque (41°N, 29°E) a mosque in Istanbul built in the 1600s; it is known for its beautiful blue-colored interior tiles (p. 188)

Borneo the world's third-largest island; located in Southeast Asia (p. 619)

Bosporus (bahs-puh-ruhs) a narrow strait in Turkey that connects the Mediterranean Sea with the Black Sea (p. 197)

Botswana a country in Southern Africa between Namibia and Zimbabwe (p. 435)

Brazzaville (4°S, 15°E) the capital of the Republic of the Congo (p. 413)

Brunei (brooh-NY) a country in Southeast Asia on the northern coast of Borneo (p. 617)

Bujumbura (booh-juhm-BOOHR-uh) (3°S, 29°E) the capital of Burundi (p. 393)

Bukhara (40°N, 64°E) an ancient city along the Silk Road in Central Asia; it has long been an important trade and cultural center in the region (p. 248)

Burkina Faso (boor-KEE-nuh FAH-soh) a landlocked country in West Africa (p. 371)

Burundi (buh-ROON-dee) a landlocked country in East Africa (p. 393)

C

Cairo (30°N, 31°E) the capital of Egypt (p. 349)

California a state on the west coast of the United States (p. 704)

Cambodia a country in Southeast Asia (p. 616)

Cameroon a country in Central Africa south of Nigeria (p. 413)

Canaan (KAY-nuhn) a region in what is now Israel near the Mediterranean coast; according to the Bible, Abraham settled in Canaan and his Hebrew descendants lived there for many years (p. 145)

Canada the country north of the United States that occupies most of northern North America (p. 710)

Canberra (35°S, 149°E) the capital of Australia (p. 642)

Cape of Good Hope a cape at the southern tip of Africa (p. 437)

Cape Town (34°S, 18°E) the legislative capital of South Africa (p. 435)

Cape Verde (VUHRD) an island country off the coast of West Africa (p. 370)

Caribbean Sea an arm of the Atlantic Ocean between North and South America (p. 710)

Carthage (KAHR-thij) (37°N, 10°E) a key trade center built by the Phoenicians on the northern coast of Africa (p. 137)

Caspian Sea an inland sea located between Europe and Asia; it is the largest inland body of water in the world (p. 245)

Central African Republic a landlocked country in Central Africa south of Chad (p. 413)

Chad a landlocked country in West Africa located east of Niger (p. 371)

Chang Jiang (Yangzi River) a major river in China (p. 567)

Chao Phraya (chow PRY-uh) a river in Thailand (p. 716)

China a large country in East Asia (p. 565)

Colombo (7°N, 80°E) the capital of Sri Lanka (p. 541)

Colorado a state in the western United States (p. 704)

Comoros (KAH-muh-rohz) an island country in the Indian Ocean off the coast of Africa (p. 435)

Conakry (KAH-nuh-kree) (10°N, 14°W) the capital of Guinea (p. 370)

Congo Basin a large flat area on the Congo River in Central Africa (p. 415)

Congo, Democratic Republic of the the largest and most populous country in Central Africa (p. 413)

Congo, Republic of the a country in Central Africa on the Congo River (p. 413)

Congo River the major river of Central Africa (p. 415)

Constantinople (41°N, 29°E) the capital of the Byzantine Empire, located between the Black Sea and Mediterranean Sea; the modern city of Istanbul (p. 161)

Córdoba (KAWR-doh-bah) (38°N, 5°W) a city in southern Spain; it was a center of Muslim rule in Spain (p. 182)

Côte d'Ivoire (KOHT dee-VWAHR) a country in West Africa between Liberia and Ghana (p. 371)

Cuba a Communist country and the largest island in the Caribbean Sea (p. 711)

D

Dakar (15°N, 17°W) the capital of Senegal (p. 370)

Damascus (34°N, 36°E) the capital of Syria (p. 195)

Dardanelles (dahrd-uhn-ELZ) a strait between the Aegean Sea and the Sea of Marmara; part of a waterway that connects the Black Sea and the Mediterranean Sea (p. 197)

Dar es Salaam (7°S, 39°E) the capital of Tanzania (p. 393)

Darfur a region in western Sudan; because of genocide, millions of people have fled from Darfur (p. 405)

Dead Sea the saltiest lake and lowest point on Earth; located on the border between Israel and Jordan and fed by the Jordan River (p. 197)

Deccan a large plateau in southern India (p. 543)

Delhi a city in northern India that was the capital of the Mughal Empire (p. 541)

GAZETTEER

GAZETTEER

Dhaka (DA-kuh) (24°N, 90°E) the capital of Bangladesh (p. 541)

Dili (8°N, 125°E) the capital of East Timor (p. 617)

Djibouti (ji-BOO-tee) a country in East Africa on the Horn of Africa (p. 393)

Djibouti (12°N, 43°E) the capital of Djibouti (p. 393)

Dodoma (6°S, 36°E) the capital of Tanzania (p. 393)

Doha (26°N, 51°E) the capital of Qatar (p. 219)

Drakensberg a mountain range in Southern Africa (p. 437)

Dushanbe (39°N, 69°E) the capital of Tajikistan (p. 243)

E

East Timor an island country in Southeast Asia (p. 617)

Eastern Ghats (gawts) a mountain range in India (p. 543)

Eastern Hemisphere the half of the globe between the prime meridian and 180° longitude that includes most of Africa and Europe as well as Asia, Australia, and the Indian Ocean (p. H3)

Egypt a country in North Africa between Libya and the Red Sea (p. 349)

Elburz Mountains a mountain range in northern Iran south of the Caspian Sea (p. 221)

equator the imaginary line of latitude that circles the globe halfway between the North and South Poles (p. H2)

Equatorial Guinea a country in Central Africa between Cameroon and Gabon (p. 413)

Eritrea (er-uh-TREE-uh) an East African country north of Ethiopia (p. 393)

Esfahan (es-fah-HAHN) (33°N, 52°E) ancient capital of the Safavid Empire; now a city in central Iran (p. 184)

Ethiopia an East African country located on the Horn of Africa (p. 393)

Euphrates River a river in Southwest Asia (p. 221)

Europe the continent between the Ural Mountains and the Atlantic Ocean (p. 714)

F

Fergana Valley a fertile plains region of Uzbekistan in Central Asia (p. 245)

Fertile Crescent a large arc of fertile lands between the Persian Gulf and the Mediterranean Sea; the world's earliest civilizations began in the region (p. 117)

Fiji an island country in the Pacific (p. 643)

Florida a state in the southern United States bordered mostly by the Atlantic Ocean and Gulf of Mexico (p. 705)

Freetown (9°N, 13°W) the capital of Sierra Leone (p. 370)

Fuji (FOO-jee) (35°N, 135°E) a volcano and Japan's highest peak (p. 593)

Funafuti (9°S, 179°E) the capital of Tuvalu (p. 656)

Gabon (gah-BOHN) a country in Central Africa between Cameroon and the Democratic Republic of the Congo (p. 413)

Gaborone (24°S, 26°E) the capital of Botswana (p. 435)

Gambia a country in West Africa surrounded on three sides by Senegal (p. 370)

Ganges River (GAN-jeez) a major river in northern India (p. 543)

Gangetic Plain a broad plain in northern India formed by the Ganges River (p. 543)

Gao (16°N, 0°) the capital of the Songhai Empire (p. 335)

Gaul an ancient region in Western Europe that included parts of modern France and Belgium (p. 158)

Gaza (32°N, 34°E) a city in southwestern Israel on the Mediterranean Sea (p. 207)

Ghana a country in West Africa between Côte d'Ivoire and Togo (p. 371)

Gobi (GOH-bee) a desert in China and Mongolia (p. 567)

Grand Canal a canal linking northern and southern China (p. 519)

Grand Canyon a large canyon in the southwestern United States (p. 702)

Great Barrier Reef a huge coral reef off the northeastern coast of Australia (p. 645)

Great Lakes the largest freshwater lake system in the world; located in North America (p. 703)

Great Plains a grassland region in the central United States (p. 703)

Great Rift Valley a series of valleys in East Africa caused by the stretching of Earth's crust (p. 395)

Great Zimbabwe an ancient walled town in Southern Africa (p. 441)

Guinea a country in West Africa north of Sierra Leone (p. 370)

Guinea-Bissau a country in West Africa north of Guinea (p. 370)

Gulf Stream a warm ocean current that flows north along the east coast of the United States (p. 52)

H

Hanoi (21°N, 106°E) the capital of Vietnam (p. 616)

Harappa (huh-RA-puh) an ancient city in the Indus Valley in modern Pakistan (p. 473)

Harare (hah-RAH-ray) (18°S, 31°E) the capital of Zimbabwe (p. 435)

Hawaii an island state of the United States located in the Pacific Ocean (p. 704)

Himalayas the highest mountains in the world; they separate the Indian Subcontinent from China (p. 543)

Hindu Kush a group of mountains that separates the Indian Subcontinent from Central Asia (p. 543)

Ho Chi Minh City a large city in southern Vietnam (p. 616)

Hokkaido (hoh-KY-doh) the northernmost of Japan's four major islands (p. 593)

Hong Kong (22°N, 115°E) a city in southern China (p. 565)

Honiara (9°S, 160°E) the capital of the Solomon Islands (p. 656)

Honshu (HAWN-shoo) the largest of Japan's four major islands (p. 593)

Huang He (Yellow River) a major river in northern China (p. 507)

India a large country in South Asia (p. 541)

Indian Ocean the world's third-largest ocean; it is located between Asia and Antarctica (pp. 706–707)

Indochina Peninsula a large peninsula in Southeast Asia (p. 619)

Indonesia the largest country in Southeast Asia (p. 617)

Indus River a major river in Pakistan (p. 543)

Indus Valley a river valley in Pakistan that was home to the ancient Harappan civilization (p. 543)

Iran a country in the Persian Gulf region; it includes the ancient region of Persia (p. 219)

Iraq a country in the Persian Gulf region; it includes the ancient region of Mesopotamia (p. 219)

Islamabad (34°N, 73°E) the capital of Pakistan (p. 541)

Israel a country between the Mediterranean Sea and Jordan; it was the homeland of the ancient Hebrews (p. 195)

Istanbul (41°N, 29°E) the largest city in Turkey; formerly known as Constantinople and was the capital of the Byzantine Empire and Ottoman Empire (p. 203)

Jakarta (6°S, 107°E) the capital of Indonesia (p. 616)

Japan an island country in East Asia (p. 591)

Java a large island in Indonesia (p. 619)

Jerusalem (32°N, 35°E) the capital of Israel; it contains holy sites of Christianity, Islam, and Judaism (p. 207)

Jordan River a river between Israel and Jordan that empties into the Dead Sea (p. 197)

Jordan a country east of Israel and the Jordan River (p. 195)

Kabul (35°N, 69°E) the capital of Afghanistan (p. 243)

Kampala (0°, 32°E) the capital of Uganda (p. 393)

Kara-Kum (kahr-uh-koom) a desert in Central Asia east of the Caspian Sea (p. 246)

Kashmir a disputed region between India and Pakistan (p. 541)

Kathmandu (kat-man-DOO) (28°N, 85°E) the capital of Nepal (p. 541)

Kazakhstan a country in Central Asia; it was part of the Soviet Union until 1991 (p. 246)

Kenya a country in East Africa south of Ethiopia (p. 393)

Khartoum (16°N, 33°E) the capital of Sudan (p. 393)

Kigali (2°S, 30°E) the capital of Rwanda (p. 393)

Kinshasa (4°S, 15°E) the capital of the Democratic Republic of the Congo (p. 413)

Kiribati an island country in the Pacific (p. 643)

Kolkata (Calcutta) a major city in eastern India (p. 541)

Kongo Kingdom a powerful kingdom that once ruled much of Central Africa (p. 420)

Kopet-Dag a group of mountains in northern Iran bordering Turkmenistan (p. 221)

Korean Peninsula a peninsula on the east coast of Asia (p. 591)

Koror (9°N, 138°E) the capital of Palau (p. 656)

Kuala Lumpur (3°N, 102°E) the capital of Malaysia (p. 616)

Kuwait a small country on the Persian Gulf (p. 219)

Kyoto (KYOH-toh) (35°N, 136°E) the ancient capital of Japan (p. 591)

Kyrgyzstan a country in Central Asia; it was part of the Soviet Union until 1991 (p. 246)

Kyushu (KYOO-shoo) the southernmost of Japan's four major islands (p. 593)

Kyzyl Kum (ki-ZIL KOOM) a vast desert region in Uzbekistan and Kazakhstan (p. 246)

Lagos (6°N, 3°E) a city in Nigeria; the most populous city in West Africa (p. 371)

Lake Victoria the largest lake in Africa (p. 395)

Laos (LOWS) a landlocked country in Southeast Asia (p. 616)

Lebanon a country on the Mediterranean Sea north of Israel (p. 195)

Lesotho (luh-SOH-toh) a country completely surrounded by South Africa (p. 435)

Liberia a country in West Africa between Sierra Leone and Côte d'Ivoire (p. 371)

GAZETTEER

Libreville (0°, 9°E) the capital of Gabon (p. 413)

Libya a country in North Africa between Egypt and Algeria (p. 349)

Lilongwe (li-LAWN-gway) (14°S, 34°E) the capital of Malawi (p. 413)

Lomé (6°N, 1°E) the capital of Togo (p. 371)

Lower Egypt the northern part of ancient Egypt, downriver from Upper Egypt (p. 279)

Luanda (9°S, 13°E) the capital of Angola (p. 413)

Lusaka (15°S, 28°E) the capital of Zambia (p. 413)

Lydia an ancient region in western Anatolia (p. 137)

M

Madagascar a large island country off the southeastern coast of Africa (p. 435)

Majuro (7°N, 171°E) the capital of the Marshall Islands (p. 656)

Malabo (mah-LAH-boh) (4°N, 9°E) the capital of Equatorial Guinea (p. 413)

Malawi a landlocked country in Central Africa located south of Tanzania (p. 413)

Malay Archipelago (muh-LAY) a large group of islands in Southeast Asia (p. 619)

Malay Peninsula (muh-LAY) a narrow peninsula in Southeast Asia (p. 619)

Malaysia a country in Southeast Asia (p. 616)

Maldives an island country south of India (p. 717)

Male (5°N, 72°E) the capital of the Maldives (p. 717)

Mali a country in West Africa on the Niger River (p. 371)

Manama (26°N, 51°E) capital of Bahrain (p. 219)

Manila (15°N, 121°E) the capital of the Philippines (p. 617)

Maputo (27°S, 33°E) the capital of Mozambique (p. 435)

Mariana Trench the world's deepest ocean trench; located in the Pacific Ocean (p. 37)

Marshall Islands an island country in the Pacific (p. 643)

Maseru (29°S, 27°E) the capital of Lesotho (p. 435)

Mauritania a country in West Africa located between Mali and the Atlantic Ocean (p. 371)

Mauritius (maw-RI-shuhs) an island country east of Madagascar (p. 719)

Mbabane (uhm-bah-BAH-nay) (26°S, 31°E) the capital of Swaziland (p. 435)

Mecca (21°N, 40°E) an ancient city in Arabia and the birthplace of Muhammad (p. 219)

Medina (muh-DEE-nuh) (24°N, 40°E) a city in western Saudi Arabia north of Mecca; people there were among the first to accept Islam (p. 173)

Mediterranean Sea a sea surrounded by Europe, Asia, and Africa (p. 351)

Mekong River a major river in Southeast Asia (p. 619)

Melanesia a huge group of Pacific islands that stretches from New Guinea to Fiji (p. 653)

Memphis (30°N, 31°E) the ancient capital of Egypt (p. 279)

Meroë (MER-oh-wee) (17°N, 34°E) an ancient capital of Kush (p. 316)

Mesopotamia (mes-uh-puh-TAY-mee-uh) the region in Southwest Asia between the Tigris and Euphrates rivers; it was the site of some of the world's earliest civilizations (p. 117)

Mexico a country in North America, south of the United States (p. 711)

Micronesia a large group of Pacific islands located east of the Philippines (p. 653)

Micronesia, Federated States of an island country in the western Pacific (p. 642)

Mid-Atlantic Ridge a mid-ocean ridge located in the Atlantic Ocean (p. 37)

Middle East the region around the eastern Mediterranean, northeastern Africa, and Southwest Asia that links the continents of Europe, Asia, and Africa (p. 220)

Mogadishu (2°N, 45°E) the capital of Somalia (p. 393)

Mohenjo Daro (mo-HEN-joh DAR-oh) (27°N, 68°E) an ancient city of the Harappan civilization in modern Pakistan (p. 473)

Mongolia a landlocked country in East Asia (p. 565)

Monrovia (6°N, 11°W) the capital of Liberia (p. 371)

Morocco a country in North Africa south of Spain (p. 348)

Moroni (12°S, 43°E) the capital of Comoros (p. 435)

Mount Everest the highest mountain in the world at 29,035 feet (8,850 km); it is located in India and Nepal (p. 543)

Mount Kilimanjaro (3°S, 37°E) the highest mountain in Africa at 19,341 feet (5,895 m); it is in Tanzania near the Kenya border (p. 395)

Mozambique a country in Southern Africa south of Tanzania (p. 435)

Mumbai (Bombay) a major city in western India (p. 541)

Muscat (24°N, 59°E) the capital of Oman (p. 219)

Myanmar (Burma) (MYAHN-mahr) a country in Southeast Asia (p. 616)

N

Nairobi (1°S, 37°E) the capital of Kenya (p. 393)

Namib Desert a desert in southwestern Africa (p. 437)

Namibia a country on the Atlantic coast of Southern Africa (p. 435)

Nauru an island country in the Pacific (p. 643)

N'Djamena (uhn-jah-MAY-nah) (12°N, 15°E) the capital of Chad (p. 371)

Negev (NE-gev) an arid region of southern Israel (p. 197)

Nepal a landlocked country in South Asia (p. 541)

New Delhi (29°N, 77°E) the capital of India (p. 541)

New Guinea the world's second-largest island; located in Southeast Asia (p. 619)

New York (41°N, 74°W) a city in the northeastern United States; the largest city in the United States (p. 705)

New Zealand an island country southeast of Australia (p. 643)

Niamey (14°N, 2°E) the capital of Niger (p. 371)

Niger (NY-juhr) a country in West Africa north of Nigeria (p. 371)

Nigeria a country on the Atlantic coast of West Africa (p. 371)

Niger River the major river of West Africa (p. 373)

Nile River the world's longest river; flows through East Africa, Egypt, and into the Mediterranean Sea (p. 351)

North China Plain a plains region of northeastern China (p. 567)

Northern Hemisphere the northern half of the globe, between the equator and the North Pole (p. H3)

North Korea a country in East Asia (p. 591)

North Pole (90°N) the northern point of Earth's axis (p. 721)

Nouakchott (nooh-AHK-shaht) (18°N, 16°W) the capital of Mauritania (p. 370)

Nubia a region in North Africa located on the Nile River south of Egypt (p. 311)

Nuku'alofa (21°N, 174°E) the capital of Tonga (p. 656)

O

Oman a country on the Arabian Peninsula (p. 219)

Osaka (oh-SAH-kuh) (35°N, 135°E) a city in Japan (p. 606)

Ouagadougou (wah-gah-DOO-goo) (12°N, 2°W) the capital of Burkina Faso (p. 371)

Outback the dry interior region of Australia (p. 645)

P

Pacific Ocean the world's largest ocean; located between Asia and the Americas (p. 720)

Pakistan a country in South Asia northwest of India (p. 541)

Palau an island country in the Pacific (p. 642)

Palestine a region between the Jordan River and the Mediterranean Sea in modern Israel (p. 205)

Palikir (6°N, 158°E) the capital of the Federated States of Micronesia (p. 656)

Pamirs a highland region in Central Asia, mainly in Tajikistan (p. 244)

Papua New Guinea a country on the island of New Guinea (p. 642)

Persian Gulf a body of water located between the Arabian Peninsula and the Zagros Mountains in Iran; it has enormous oil deposits along its shores (p. 219)

Perth (32°S, 116°E) a city in western Australia (p. 642)

Philippines an island country in Southeast Asia (p. 617)

Phnom Penh (puh-NAWM pen) (12°N, 105°E) the capital of Cambodia (p. 616)

Phoenicia (fi-NI-shuh) an ancient region on the shores of the Mediterranean Sea in the Fertile Crescent; modern Lebanon includes most of the area (p. 136)

Plateau of Tibet a high plateau in western China (p. 477)

Polynesia the largest group of islands in the Pacific Ocean (p. 653)

Port Louis (20°S, 58°E) the capital of Mauritius (p. 719)

Port Moresby (10°S, 147°E) the capital of Papua New Guinea (p. 642)

Porto-Novo (6°N, 3°E) the capital of Benin (p. 371)

Port-Vila (18°S, 169°E) the capital of Vanuatu (p. 656)

Praia (PRY-uh) (15°N, 24°W) the capital of Cape Verde (p. 370)

Pretoria (26°S, 28°E) the administrative capital of South Africa (p. 435)

prime meridian an imaginary line that runs through Greenwich, England, at 0° longitude (p. H2)

Pyongyang (pyuhng-YANG) (39°N, 126°E) the capital of North Korea (p. 591)

Q, R

Qatar (KUH-tahr) a country on the Arabian Peninsula (p. 219)

Rabat (34°N, 7°W) the capital of Morocco (p. 348)

Red Sea a sea between the Arabian Peninsula and Africa (p. 221)

Riyadh (25°N, 47°E) the capital of Saudi Arabia (p. 219)

Rome (42°N, 13°E) the capital of Italy; in ancient times it was the capital of the Roman Empire (p. 146)

Ross Ice Shelf the largest ice shelf in Antarctica (p. 643)

Rub' al-Khali (ROOB ahl-KAH-lee) a huge sandy desert on the Arabian Peninsula; its name means "empty quarter" (p. 222)

Russia a huge country that extends from Eastern Europe to the Pacific Ocean; it is the largest country in the world (p. 249)

Rwanda a country in East Africa between Tanzania and the Democratic Republic of the Congo (p. 393)

S

Sahara the world's largest desert; it dominates much of North Africa (p. 351)

Sahel a semiarid region between the Sahara and wetter areas to the south (p. 374)

GAZETTEER

Samarqand (40°N, 67°E) an ancient city on the Silk Road in modern Uzbekistan (p. 248)

Samoa an island country in the Pacific (p. 643)

São Tomé (sow too-MAY) (1°N, 6°E) the capital of São Tomé and Príncipe (p. 413)

São Tomé and Príncipe (PREEN-see-pee) an island country located off the Atlantic coast of Central Africa (p. 413)

Saudi Arabia a country occupying much of the Arabian Peninsula in Southwest Asia (p. 219)

Sea of Marmara (MAHR-muh-ruh) a small sea in Turkey; with the Bosporus and the Dardanelles, it forms a waterway that separates Europe and Asia and connects the Mediterranean Sea and Black Sea (p. 197)

Senegal a country in West Africa south of Mauritania (p. 370)

Seoul (38°N, 127°E) the capital of South Korea (p. 591)

Serengeti Plain a large plain in East Africa that is famous for its wildlife (p. 395)

Seychelles an island country located east of Africa in the Indian Ocean (p. 719)

Shanghai (31°N, 121°E) a major port city in eastern China (p. 572)

Shikoku (shee-koh-koo) the smallest of Japan's four major islands (p. 593)

Sierra Leone a West African country located south of Guinea (p. 370)

Silk Road an ancient trade route from China through Central Asia to the Mediterranean Sea (p. 248)

Singapore an island country at the tip of the Malay Peninsula in Southeast Asia (p. 616)

Solomon Islands an island country in the Pacific (p. 643)

Somalia an East African country located on the Horn of Africa (p. 393)

Songhai an ancient empire in West Africa (p. 335)

South Africa a country located at the southern tip of Africa (p. 435)

South America a continent in the Western and Southern hemispheres (p. 712)

Southern Hemisphere the southern half of the globe, between the equator and the South Pole (p. H3)

South Korea a country in East Asia (p. 591)

South Pole (90°S) the southern point of Earth's axis (p. 721)

Sri Lanka an island country located south of India (p. 541)

Sudan a country in East Africa; it is the largest country in Africa (p. 393)

Suez Canal a canal in Egypt that links the Mediterranean and Red seas (p. 351)

Sumatra a large island in Indonesia (p. 619)

Sumer (SOO-muhr) the region in southern Mesopotamia where the world's first civilization developed (p. 122)

Suva (19°S, 178°E) the capital of Fiji (p. 656)

Swaziland a country in Southern Africa almost completely surrounded by South Africa (p. 435)

Sydney (34°S, 151°E) the largest city in Australia (p. 642)

Syr Darya (sir duhr-YAH) the longest river in Central Asia; it flows through the Fergana Valley and Kazakhstan, Tajikistan, and Uzbekistan on its way to the Aral Sea (p. 245)

Syria a country on the eastern Mediterranean Sea (p. 195)

Syrian Desert a desert in Southwest Asia covering much of the Arabian Peninsula between the Mediterranean coast and the Euphrates River (p. 134)

T

Taipei (25°N, 122°E) the capital of Taiwan (p. 565)

Taiwan (ty-wahn) an island country southeast of China (p. 565)

Tajikistan a country in Central Asia; it was part of the Soviet Union until 1991 (p. 245)

Tanzania (tan-zuh-NEE-uh) an East African country south of Kenya (p. 393)

Tarawa the capital of Kiribati (p. 656)

Tashkent (41°N, 69°E) the capital of Uzbekistan; in ancient times it was an important trading city along the Silk Road in Central Asia (p. 243)

Tasmania a large island off the southern coast of Australia (p. 642)

Taurus Mountains a mountain range in southern Turkey along the Mediterranean Sea (p. 197)

Tehran (36°N, 51°E) the capital of Iran (p. 219)

Thailand (TY-land) a country in Southeast Asia (p. 616)

Thar Desert (TAHR) a desert in western India and eastern Pakistan (p. 543)

Thimphu (28°N, 90°E) the capital of Bhutan (p. 541)

Tigris River a river in Southwest Asia (p. 716)

Timbuktu (17°N, 3°W) a major cultural and trading city in the Mali and Songhai empires (p. 335)

Togo a country in West Africa between Ghana and Benin (p. 371)

Tokyo (36°N, 140°E) the capital of Japan (p. 591)

Tonga an island country in the Pacific (p. 643)

Tripoli (33°N, 13°E) the capital of Libya (p. 349)

Tropic of Cancer the parallel 23.5° north of the equator; parallel on the globe at which the sun's most direct rays strike Earth during the June solstice (p. 28)

Tropic of Capricorn the parallel at 23.5° south of the equator; parallel on the globe at which the sun's most direct rays strike Earth during the December solstice (p. 28)

Tunis (37°N, 10°E) the capital of Tunisia (p. 349)

Tunisia a country in North Africa on the Mediterranean Sea (p. 349)

Turkey a country on the eastern Mediterranean, it includes the regions of Anatolia and Asia Minor (p. 202)

Turkmenistan a country in Central Asia; it was part of the Soviet Union until 1991 (p. 246)

Tuvalu an island country in the Pacific (p. 643)

Uganda a country in East Africa located west of Kenya (p. 393)

Ulaanbaatar (oo-lahn-BAH-tawr) (48°N, 107°E) the capital of Mongolia (p. 565)

Uluru a huge natural rock formation in central Australia; also called Ayers Rock (p. 645)

United Arab Emirates a country on the Arabian Peninsula (p. 219)

Ur a city in ancient Sumer on the Euphrates River near the Persian Gulf; it was one of the largest cities of ancient Mesopotamia (p. 124)

Uzbekistan a country in Central Asia; it was part of the Soviet Union until 1991 (p. 246)

V, W

Vanuatu an island country in the Pacific (p. 643)

Victoria (1°S, 33°E) the capital of Seychelles (p. 719)

Vientiane (vyen-THAN) (18°N, 103°E) the capital of Laos (p. 616)

Vietnam (vee-ET-NAHM) a country in Southeast Asia (p. 616)

Washington a state in the northwestern United States (p. 704)

Washington, D.C. (39°N, 77°W) the capital of the United States (p. 705)

Wellington (41°S, 175°E) the capital of New Zealand (p. 643)

West Bank a disputed territory in eastern Israel (p. 207)

Western Ghats (GAWTS) a mountain range in India (p. 543)

Western Hemisphere the half of the globe between 180° and the prime meridian that includes North and South America and the Pacific and Atlantic oceans (p. H3)

Windhoek (VINT-hook) (22°S, 17°E) the capital of Namibia (p. 435)

Xi'an (34°N, 109°E) a city in western China (p. 519)

Yamoussoukro (yah-moo-SOO-kroh) (7°N, 5°W) the capital of Côte d'Ivoire (p. 371)

Yangon (Rangoon) (17°N, 96°E) the capital of Myanmar (Burma) (p. 616)

Yaoundé (yown-DAY) (4°N, 12°E) the capital of Cameroon (p. 413)

Yellow Sea a body of water between northeastern China and the Korean Peninsula (p. 567)

Yemen a country on the Arabian Peninsula bordering the Red Sea and the Gulf of Aden (p. 219)

Z

Zagros Mountains a mountain range in Iran; it forms the western boundary of the Plateau of Iran (p. 221)

Zambezi River a river in Central Africa that flows into the Indian Ocean (p. 415)

Zambia a country in Central Africa east of Angola (p. 413)

Zanzibar an island in Tanzania; once a major trading center (p. 395)

Zimbabwe a country in Southern Africa between Botswana and Mozambique (p. 435)

Biographical Dictionary

A

Ahmose the Great (ruled c. 1570–1546 BC) Egyptian pharaoh, he defeated the Hyksos. His reign marked the beginning of Egypt's New Kingdom. (p. 292)

Akbar (1542–1605) Mughal emperor, he conquered new lands and worked to make the Mughal government stronger. He also began a tolerant religious policy that helped unify the empire. (p. 548)

al-Gadhafi, Mu'ammar (1942–) Leader of Libya, he has ruled as a dictator and supported acts of violence against Israel and its neighbors. (p. 365)

Askia the Great (c. 1443–1538) Songhai ruler, he overthrew Sunni Baru. His reign was the high point of Songhai culture. (p. 337)

Asoka (uh-SOH-kuh) (ruled 270–232 BC) Ruler of the Mauryan Empire, he extended his control over most of India and promoted the spread of Buddhism. (p. 494)

Atatürk, Kemal (1881–1938) The first president of modern Turkey, he was given the name Atatürk, which means "father of the Turks." He worked to free Turkey from foreign control and began many programs to reform and modernize the country. (p. 201)

Avicenna (av-uh-SEN-uh) See Ibn-Sina.

B

Babur (BAH-boohr) (1483–1530) Indian emperor, he founded the Mughal Empire. (p. 548)

Buddha (BOO-duh) (c. 563–483 BC) Founder of Buddhism, he was an Indian prince originally named Siddhartha Gautama. He founded the Buddhist religion after a long spiritual journey through India. (p. 484)

Bush, George W. (1946–) President of the United States, he was president on September 11, 2001, when the country suffered its worst terrorist attack. This attack led to the country's War on Terror. (p. 231)

C

Candra Gupta II (kuhn-druh GOOP-tuh) (300s–400s) Gupta emperor, he ruled India during the height of Gupta power. (p. 492)

Candragupta Maurya (kuhn-druh-GOOP-tuh MOUR-yuh) (late 300s BC) Mauryan ruler, he founded the Mauryan Empire in northern India. (p. 490)

Chiang Kai-Shek (chang ky-SHEK) (1887–1975) Chinese general and leader of the Nationalists, he lost China's civil war and fled with his supporters to Taiwan. (p. 572)

Cleopatra (69–30 BC) Queen of Egypt, she tried to drive the Romans out of Egypt. (p. 355)

Confucius (551–479 BC) Chinese philosopher, he was the most influential teacher in Chinese history. His teachings, called Confucianism, focused on morality, family, society, and government. (p. 511)

Constantine (KAHN-stuhn-teen) (c. 280–337) Roman emperor, he was the first Roman emperor to become a Christian. Constantine moved the empire's capital from Rome to Constantinople and removed bans on Christianity. (p. 159)

E

Ebadi, Shirin (1947–) Iranian lawyer, writer, and teacher, she was the first Muslim woman to win the Nobel Prize for Peace in 2003. She has received many awards for her work to promote democracy and human rights, especially for women and children in Iran. (p. 237)

Enheduanna (c. 2350 BC) Daughter of Sargon, she is the first known female writer in history. Two of her hymns still exist and she may have helped to start a collection of songs dedicated to the temples of Babylonia. (p. 126)

Ezana (AY-zah-nah) (c. 300s) Aksumite ruler, he destroyed Meroë and took over the kingdom of Kush around AD 350. (p. 319)

F

Fay, Michael (1956–) American scientist, he walked 2,000 miles through the forests of Central Africa collecting data to make maps and determine land use patterns. (p. 418)

G

Gandhi, Mohandas (1869–1948) Indian nationalist and spiritual leader, he used nonviolence to protest British rule of India and helped the country achieve independence. (p. 548)

Genghis Khan (JENG-uhs KAHN) (c. 1162–1227) Ruler of the Mongols, he led his people in attacks against China and other parts of Asia. His name means "universal leader." (p. 528)

H

Hammurabi (ham-uh-RAHB-ee) (ruled c. 1792–1750 BC) Babylonian ruler, he was a brilliant military leader who brought all of Mesopotamia into the Babylonian Empire. Hammurabi is known for a unified code of 282 laws, the earliest known set of written laws, that was produced during his reign. (p. 133)

Hatshepsut (ruled c. 1503–1482 BC) Egyptian queen, she worked to increase trade with places outside of Egypt and ordered many impressive monuments and temples built during her reign. (p. 292)

Hirohito (1901–1989) Emperor of Japan, he led the country through World War II and the transition to a democratic government. (p. 598)

Hussein, Saddam (1937–) Iraqi dictator, he became president of Iraq in 1979 and led Iraq into two devastating wars. Known for his brutal suppression of opposition, Hussein was overthrown and captured by the United States in 2003 as part of the War on Terror. (p. 231)

Ibn-Sina (c. 980–1037) Muslim doctor, he wrote an influential book on medicine that was used throughout Europe until the 1600s. He is known in the West as Avicenna. (p. 187)

Jesus (c. AD 1–30) Founder of Christianity, he taught about kindness and love for God. His teachings eventually spread throughout the Roman Empire and the world. (p. 152)

Justinian (juh-STIN-ee-uhn) (c. 483–565) Byzantine emperor, he reunited the Roman Empire, simplified Roman laws with Justinian's Code, and ordered Hagia Sophia built in Constantinople. (p. 160)

Khufu (KOO-foo) (ruled 2500s BC) Egyptian pharaoh, he ruled during Egypt's Old Kingdom and is known for the many monuments built to honor him. (p. 284)

Kim Il Sung (1912–1994) Leader of North Korea, he established a Communist government there and attacked South Korea in 1950, launching the Korean War. (p. 610)

Kim Jong Il (1941–) Leader of North Korea, he took power after his father died in 1994. (p. 610)

Kublai Khan (KOO-bluh KAHN) (1215–1294) Mongol ruler, he completed the conquest of China and founded the Yuan dynasty. (p. 535)

Lalibela (ruled late 1100s–early 1200s) Christian emperor of Ethiopia, he is best known for the 11 rock churches he built during his reign. (p. 398)

Li Bo (701–762) One of China's greatest poets, he lived during the Tang dynasty. (p. 521)

Li Qingzhao (ching-ZHOW) (1081–1141) One of China's greatest poets, she lived during the Song dynasty. (p. 521)

Liu Bang (lee-OO bang) (256–195 BC) First emperor of the Han dynasty, he was born a peasant but led an army that gained control of China. As emperor, he lowered taxes and relied on educated officials to help him rule. (p. 510)

Mahfouz, Naguib (1911–) Egyptian writer, he became the first Arab writer to win the Nobel Prize for Literature. (p. 359)

Mandela, Nelson (1918–) South African president and Nobel Peace Prize winner, he worked to improve the living conditions of black South Africans. Before becoming president, he protested against apartheid and was imprisoned for 26 years. (p. 447)

Mansa Musa (MAHN-sah moo-SAH) (died c. 1332) Ruler of Mali, he was Mali's greatest and most famous ruler. Mansa Musa was a devout Muslim who made a pilgrimage to Mecca that helped spread Mali's fame. (pp. 335, 339)

Mao Zedong (MOW ZUH-DOOHNG) (1893–1976) Leader of China, he led the Communist takeover of China in 1949 and was head of the government until 1976. (p. 572)

Mehmed II (1432–1481) Ottoman sultan, he captured Constantinople in 1453, which brought an end to the Byzantine Empire. Later, he worked to restore Constantinople, which the Ottomans called Istanbul. (p. 183)

Menes (MEE-neez) (c. 3100 BC) Legendary Egyptian ruler, he unified the kingdoms of Upper and Lower Egypt and built a new capital city at Memphis. (p. 281)

Mobutu Sese Seko (1930–1997) African dictator, he became rich and used violence against his opponents while his country's economy collapsed. (p. 424)

Moses (c. 1200s BC) Biblical figure, according to the Bible he led the Hebrew people out of slavery in Egypt and back to Canaan in the Exodus. During this journey, Moses received the Ten Commandments from God. (p. 145)

Muhammad (c. 570–632) Founder of Islam, he spread Islam's teachings to the people of Arabia. Muhammad's teachings make up the Qur'an. (p. 172)

Nebuchadnezzar (neb-uh-kuhd-NEZ-uhr) (ruled c. 605–561 BC) Chaldean king, he rebuilt Babylon into a beautiful city, which featured the famed Hanging Gardens. (p. 135)

Nkrumah, Kwame (1909–1972) Leader of Ghana, he believed that Africa would be better off united instead of split into separate countries after independence from European colonial powers. (p. 377)

BIOGRAPHICAL DICTIONARY

Piankhi (PYANG-kee) (c. 751–716 BC) Ruler of Kush, he was one of Kush's most successful military leaders. His army captured all of Egypt. (p. 313)

Polo, Marco (1254–1324) Italian trader, he traveled to China and later wrote a book about his trip. During his time in China he served as a government official in Kublai Khan's court. (p. 530)

Ramses the Great (RAM-seez) (late 1300s and early 1200s BC) Egyptian pharaoh, he expanded the kingdom and built massive temples at Karnak, Luxor, and Abu Simbel. Ramses the Great is often considered one of Egypt's greatest rulers. (pp. 293, 297)

Sargon (c. 2300 BC) King of Akkad, a land north of Sumer, he built the world's first empire after defeating Sumer and northern Mesopotamia. (p. 123)

Shackelton, Ernest (1874–1922) Antarctic explorer, he led several expeditions to Antarctica including one in which he and his companions were stranded and had to make a daring journey to get help. (p. 660)

Shanakhdakheto (shah-nahk-dah-KEE-toh) (ruled 170–150 BC) Ruler of Kush, historians think she was the first woman to rule Kush. Her tomb is one of the largest pyramids at Meroë. (p. 317)

Shi Huangdi (SHEE hwahng-dee) (259–210 BC) Ruler of China, he united China for the first time, built roads and canals, began the Great Wall of China, and imposed a standard system of laws, money, weights, and writing. (p. 508)

Soyinka, Wole (1934–) Nigerian writer, he has written plays, novels, and poems about life in West Africa. He is a winner of the Nobel Prize for Literature. (p. 387)

Suleyman I (soo-lay-MAHN) (c. 1494–1566) Ottoman ruler, he governed the empire at its height. During his rule, the empire included much of the eastern Mediterranean and parts of Europe. (p. 182)

Sundiata (soohn-JAHT-ah) (died 1255) Founder of the Mali Empire, his reign is recorded in legends. (p. 334)

Sunni Ali (SOOH-nee ah-LEE) (died 1492) Emperor of Songhai, he conquered Mali and made Songhai into a powerful state. (p. 337)

Suu Kyi, Aung San (1945–) Human rights advocate in Myanmar, she protested against the country's military government and won the Nobel Peace Prize in 1991. (p. 630)

Tunka Manin (TOOHN-kah MAH-nin) (ruled c. 1068) Ruler of Ghana, his kingdom was visited by Muslim writers. (p. 330)

Tutankhamen (too-tang-KAHM-uhn) (c. 1300 BC) Egyptian pharaoh, he died while still a young king. The discovery of his tomb in 1922 has taught archaeologists much about Egyptian culture. (p. 303)

Wu (625–705) Empress of China during the Tang dynasty, she ruled ruthlessly and brought prosperity to China. (p. 519)

Wudi (WOO-dee) (156–87 BC) Emperor of China, he made Confucianism the official government philosophy. (p. 511)

Zheng He (juhng HUH) (c. 1371–c. 1433) Chinese admiral during the Ming dynasty, he led great voyages that spread China's fame throughout Asia. (p. 531)

Zhu Yuanzhang (JOO yoo-ahn-JAHNG) (1368–1398) Emperor of China and founder of the Ming dynasty, he led an army that overthrew the Mongols. (p. 530)

English and Spanish Glossary

Phonetic Respelling and Pronunciation Guide

Many of the key terms in this textbook have been respelled to help you pronounce them. The letter combinations used in the respelling throughout the narrative are explained in this phonetic respelling and pronunciation guide. The guide is adapted from *Merriam-Webster's Collegiate Dictionary, Eleventh Edition; Merriam-Webster's Geographical Dictionary;* and *Merriam-Webster's Biographical Dictionary.*

MARK	AS IN	RESPELLING	EXAMPLE
a	alphabet	a	*AL-fuh-bet
ā	Asia	ay	AY-zhuh
ä	cart, top	ah	KAHRT, TAHP
e	let, ten	e	LET, TEN
ē	even, leaf	ee	EE-vuhn, LEEF
i	it, tip, British	i	IT, TIP, BRIT-ish
ī	site, buy, Ohio	y	SYT, BY, oh-HY-oh
	iris	eye	EYE-ris
k	card	k	KAHRD
kw	quest	kw	KWEST
ō	over, rainbow	oh	OH-vuhr, RAYN-boh
ü	book, wood	ooh	BOOHK, WOOHD
ò	all, orchid	aw	AWL, AWR-kid
òi	foil, coin	oy	FOYL, KOYN
aü	out	ow	OWT
ə	cup, butter	uh	KUHP, BUHT-uhr
ü	rule, food	oo	ROOL, FOOD
yü	few	yoo	FYOO
zh	vision	zh	VIZH-uhn

*A syllable printed in small capital letters receives heavier emphasis than the other syllable(s) in a word.

A

Aborigines (a-buh-RIJ-uh-nees) the original inhabitants of Australia (p. 647)
  aborígenes habitantes originales de Australia (pág. 647)

absolute location a specific description of where a place is located; absolute location is often expressed using latitude and longitude (p. 12)
  ubicación absoluta descripción específica del lugar donde se ubica un punto; con frecuencia se define en términos de latitud y longitud (pág. 12)

acupuncture (AK-yoo-punk-cher) the Chinese practice of inserting fine needles through the skin at specific points to cure disease or relieve pain (p. 515)
  acupuntura práctica china de insertar agujas finas en la piel en puntos específicos para curar enfermedades o aliviar el dolor (pág. 515)

Afrikaners (a-fri-KAH-nuhrz) Dutch, French, and German settlers and their descendants in South Africa (p. 442)
  afrikaners colonizadores holandeses, franceses y alemanes y sus descendientes en Sudáfrica (pág. 442)

afterlife life after death (p. 286)
  la otra vida vida después de la muerte (pág. 286)

alloy a mixture of two or more metals (p. 498)
  aleación mezcla de dos o más metales (pág. 498)

alphabet a set of letters that can be combined to form words (p. 137)
  alfabeto conjunto de letras que pueden combinarse para formar palabras (pág. 137)

animism the belief that bodies of water, animals, trees, and other natural objects have spirits (p. 378)
  animismo creencia de que las masas de agua, los animales, los árboles y otros objetos de la naturaleza tienen espíritu (pág. 378)

apartheid South Africa's government policy of separation of races that was abandoned in the 1980s and 1990s; apartheid means "apartness" (p. 442)
  apartheid política gubernamental de Sudáfrica de separar las razas, abandonada en las décadas de 1980 y 1990; apartheid significa "separación" (pág. 442)

arable land that is suitable for growing crops (p. 255)
  cultivable tierra buena para el cultivo (pág. 255)

archipelago a large group of islands (p. 618)
  archipiélago gran grupo de islas (pág. 618)

architecture the science of building (p. 130)
  arquitectura ciencia de la construcción (pág. 130)

Association of Southeast Asian Nations (ASEAN) an organization that promotes economic development and social and cultural cooperation among the countries of Southeast Asia (p. 629)
Asociación de Naciones de Asia del Sudeste (ASEAN por sus siglas en inglés) organización que promueve el desarrollo económico y social y la cooperación cultural entre los países del sureste asiático (pág. 629)

astronomy the study of stars and planets (p. 499)
astronomía estudio de las estrellas y los planetas (pág. 499)

atoll a ring-shaped coral island that surrounds a lagoon (p. 654)
atolón isla de coral en forma de anillo que rodea una laguna (pág. 654)

B

basin a generally flat region surrounded by higher land such as mountains and plateaus (p. 414)
cuenca región generalmente llana rodeada de tierras más altas, como montañas y mesetas (pág. 414)

Bedouins Arabic-speaking nomads that live mostly in the deserts of Southwest Asia (p. 212)
beduinos nómadas que hablan árabe y viven principalmente en los desiertos del suroeste de Asia (pág. 212)

Berbers members of an ethnic group who are native to North Africa and speak Berber languages (p. 357)
bereberes miembros de un grupo étnico del norte de África que hablan lenguas bereberes (pág. 357)

Bible the holy book of Christianity (p. 152)
Biblia libro sagrado del cristianismo (pág. 152)

birthrate the annual number of births per 1,000 people (p. 88)
índice de natalidad número de nacimientos por cada 1,000 personas en un año (pág. 88)

Boers Afrikaner frontier farmers in South Africa (p. 442)
bóers agricultores afrikaners de la frontera en Sudáfrica (pág. 442)

Buddhism a religion based on the teachings of the Buddha that developed in India in the 500s BC (p. 550)
budismo religión basada en las enseñanzas de Buda, originada en la India en el siglo VI a. C. (pág. 550)

bureaucracy a body of unelected government officials (p. 526)
burocracia grupo de empleados no electos del gobierno (pág. 526)

Byzantine Empire (bi-zuhn-teen) the society that developed in the eastern Roman Empire after the west fell (p. 162)
Imperio bizantino sociedad que surgió en el Imperio romano de oriente tras la caída del Imperio romano de occidente (pág. 162)

C

caliph (KAY-luhf) a title that Muslims use for the highest leader of Islam (p. 180)
califa título que los musulmanes le dan al líder supremo del Islam (pág. 180)

calligraphy decorative writing (p. 189)
caligrafía escritura decorativa (pág. 189)

canal a human-made waterway (p. 118)
canal vía de agua hecha por el ser humano (pág. 118)

cartography the science of making maps (p. 19)
cartografía ciencia de crear mapas (pág. 19)

caste system the division of Indian society into groups based on birth or occupation (p. 550)
sistema de castas división de la sociedad india en grupos basados en el nacimiento o la profesión (pág. 550)

cataracts rapids along a river, such as those along the Nile in Egypt (p. 279)
rápidos fuertes corrientes de un río, como las del Nilo en Egipto (pág. 279)

chariot a wheeled, horse-drawn cart used in battle (p. 134)
carro de guerra carro tirado por caballos usado en las batallas (pág. 134)

Christianity a major world religion based on the teachings of Jesus (p. 152)
cristianismo una de las principales religiones del mundo, basada en las enseñanzas de Jesús (pág. 152)

city-state a political unit consisting of a city and its surrounding countryside (p. 122)
ciudad estado unidad política formada por una ciudad y los campos que la rodean (pág. 122)

civil service service as a government official (p. 526)
servicio público servicio como empleado del gobierno (pág. 526)

climate a region's average weather conditions over a long period of time (p. 50)
clima condiciones del tiempo promedio de una región durante un período largo de tiempo (pág. 50)

colony a territory inhabited and controlled by people from a foreign land (p. 548)
colonia territorio habitado y controlado por personas de otro país (pág. 548)

command economy an economic system in which the central government makes all economic decisions (pp. 94, 577)
economía autoritaria sistema económico en el que el gobierno central toma todas las decisiones económicas (págs. 94, 577)

communism a political system in which the government owns all property and dominates all aspects of life in a country (p. 92)
    comunismo sistema político en el que el gobierno es dueño de toda la propiedad y controla todos los aspectos de la vida de un país (pág. 92)

compass an instrument that uses Earth's magnetic field to indicate direction (p. 522)
    brújula instrumento que utiliza el campo magnético de la Tierra para indicar la dirección (pág. 522)

Confucianism a philosophy based on the ideas of Confucius that focuses on morality, family order, social harmony, and government (p. 574)
    confucianismo filosofía basada en las ideas de Confucio que se concentra en la moralidad, el orden familiar, la armonía social y el gobierno (pág. 574)

continent a large landmass that is part of Earth's crust; geographers identify seven continents (p. 36)
    continente gran masa de tierra que forma parte de la corteza terrestre; los geógrafos identifican siete continentes (pág. 36)

coral reef a chain of rocky material found in shallow tropical waters (p. 645)
    arrecife de coral cadena de material rocoso que se encuentra en aguas tropicales de poca profundidad (p. 645)

cultural diffusion the spread of culture traits from one region to another (p. 85)
    difusión cultural difusión de rasgos culturales de una región a otra (pág. 85)

cultural diversity having a variety of cultures in the same area (p. 83)
    diversidad cultural existencia de una variedad de culturas en la misma zona (pág. 83)

culture the set of beliefs, values, and practices that a group of people have in common (p. 80)
    cultura conjunto de creencias, valores y costumbres compartidas por un grupo de personas (pág. 80)

culture region an area in which people have many shared culture traits (p. 82)
    región cultural región en la que las personas comparten muchos rasgos culturales (pág. 82)

culture trait an activity or behavior in which people often take part (p. 81)
    rasgo cultural actividad o conducta frecuente de las personas (pág. 81)

cuneiform (kyoo-NEE-uh-fohrm) the world's first system of writing; it developed in Sumer (p. 127)
    cuneiforme primer sistema de escritura del mundo; desarrollado en Sumeria (pág. 127)

Daoism a philosophy that developed in China and stressed the belief that one should live in harmony with the Dao, the guiding force of all reality (p. 574)
    taoísmo filosofía que se desarrolló en China y que enfatizaba la creencia de que se debe vivir en armonía con el Tao, la fuerza que guía toda la realidad (pág. 574)

deforestation the clearing of trees (p. 69)
    deforestación tala de árboles (pág. 69)

delta a landform at the mouth of a river created by sediment deposits (pp. 279, 543)
    delta accidente geográfico que se forma en la desembocadura de un río, creado por depósitos de sedimento (págs. 279, 543)

demilitarized zone an empty buffer zone created to keep two countries from fighting (p. 609)
    zona desmilitarizada zona vacía que se crea como barrera entre dos países para evitar que luchen (pág. 609)

democracy a form of government in which the people elect leaders and rule by majority (p. 91)
    democracia sistema de gobierno en el que el pueblo elige a sus líderes y gobierna por mayoría (pág. 91)

desertification the spread of desert-like conditions (pp. 65, 374)
    desertización ampliación de las condiciones desérticas (págs. 65, 374)

developed countries countries with strong economies and a high quality of life (p. 95)
    países desarrollados países con economías sólidas y una alta calidad de vida (pág. 95)

developing countries countries with less productive economies and a lower quality of life (p. 95)
    países en vías de desarrollo países con economías menos productivas y una menor calidad de vida (pág. 95)

dialect a regional variety of a language (pp. 422, 573)
    dialecto variedad regional de una lengua (págs. 422, 573)

Diaspora the scattering of the Jewish population outside of Israel (p. 204)
    Diáspora dispersión de la población judía fuera de Israel (pág. 204)

dictator a ruler who has almost absolute power (p. 365)
    dictador gobernante que tiene poder casi absoluto (pág. 365)

Diet the name for Japan's elected legislature (p. 602)
    Dieta nombre de la asamblea legislativa electa de Japón (pág. 602)

disciples (di-SY-puhls) followers (p. 153)
    discípulos seguidores (pág. 153)

ENGLISH AND SPANISH GLOSSARY

division of labor an arrangement in which each worker specializes in a particular task or job (p. 118)
división del trabajo organización mediante la que cada trabajador se especializa en un trabajo o tarea en particular (pág. 118)

domino theory the idea that if one country fell to Communism, neighboring countries would follow like falling dominoes (p. 626)
teoría del efecto dominó idea de que si un país cae en manos del comunismo, los países vecinos lo seguirán como fichas de dominó que caen una tras otra (pág. 626)

droughts periods when little rain falls and crops are damaged (p. 396)
sequías períodos en los que los cultivos sufren daños por la falta de lluvia (pág. 396)

dryland farming farming that relies on rainfall instead of irrigation (p. 255)
cultivo de secano cultivo que depende de la lluvia en vez de la irrigación (pág. 255)

dynasty a series of rulers from the same family (pp. 281, 571)
dinastía serie de gobernantes pertenecientes a la misma familia (págs. 281, 571)

E

earthquake a sudden, violent movement of Earth's crust (p. 38)
terremoto movimiento repentino y violento de la corteza terrestre (pág. 38)

ebony a dark, heavy wood (p. 312)
ébano madera oscura y pesada (pág. 312)

ecosystem a group of plants and animals that depend on each other for survival, and the environment in which they live (p. 63)
ecosistema grupo de plantas y animales que dependen unos de otros para sobrevivir, y el ambiente en el que estos viven (pág. 63)

edicts laws (p. 491)
edictos leyes (pág. 491)

elite (AY-leet) people of wealth and power (p. 287)
élite personas ricas y poderosas (pág. 287)

embargo a limit on trade (p. 231)
embargo límite impuesto al comercio (pág. 231)

empire land with different territories and peoples under a single rule (p. 123)
imperio zona que reúne varios territorios y pueblos bajo un mismo gobierno (pág. 123)

enclave a small territory surrounded by foreign territory (p. 448)
enclave territorio pequeño rodeado de territorio extranjero (pág. 448)

engineering the application of scientific knowledge for practical purposes (p. 288)
ingeniería aplicación del conocimiento científico para fines prácticos (pág. 288)

environment the land, water, climate, plants, and animals of an area; surroundings (pp. 12, 62)
ambiente la tierra, el agua, el clima, las plantas y los animales de una zona; los alrededores (págs. 12, 62)

epics long poems that tell the stories of heroes (p. 128)
poemas épicos poemas largos que narran relatos de héroes (pág. 128)

erosion the movement of sediment from one location to another (p. 39)
erosión movimiento de sedimentos de un lugar a otro (pág. 39)

escarpment a steep face at the edge of a plateau or other raised area (p. 436)
acantilado cara empinada en el borde de una meseta o de otra área elevada (pág. 436)

ethnic group a group of people who share a common culture and ancestry (p. 83)
grupo étnico grupo de personas que comparten una cultura y una ascendencia (pág. 83)

Exodus the journey in which Moses led his people out of Egypt (p. 145)
Éxodo viaje en el que Moisés guió a su pueblo para salir de Egipto (pág. 145)

exports items sent to other regions for trade (p. 316)
exportaciones productos enviados a otras regiones para el intercambio comercial (pág. 316)

extended family a family group that includes the father, mother, children, and close relatives (p. 379)
familia extendida grupo familiar que incluye al padre, la madre, los hijos y los parientes cercanos (pág. 379)

extinct no longer here; a species that has died out has become extinct (p. 64)
extinto que ya no existe; una especie que ha desaparecido está extinta (pág. 64)

F

famine an extreme shortage of food (p. 385)
hambruna grave escasez de alimentos (pág. 385)

fasting going without food for a period of time (p. 485)
ayunar dejar de comer durante un período de tiempo (pág. 485)

Fertile Crescent an area of rich farmland in Southwest Asia where the first civilizations began (p. 117)
Creciente Fértil zona de ricas tierras de cultivo situada en el suroeste de Asia, en donde comenzaron las primeras civilizaciones (pág. 117)

fishery a place where lots of fish and other seafood can be caught (p. 595)

pesquería lugar donde suele haber muchos peces y mariscos para pescar (pág. 595)

Five Pillars of Islam five acts of worship required of all Muslims (p. 176)

los cinco pilares del Islam cinco prácticas religiosas que los musulmanes tienen que observar (pág. 176)

Forbidden City a huge palace complex built by China's Ming emperors that included hundreds of imperial residences, temples, and other government buildings (p. 532)

Ciudad Prohibida enorme complejo de palacios cons-truido por orden de los emperadores Ming de China que incluía cientos de residencias imperiales, templos y otros edificios del gobierno (pág. 532)

fossil fuels nonrenewable resources that formed from the remains of ancient plants and animals; coal, petroleum, and natural gas are all fossil fuels (p. 69)

combustibles fósiles recursos no renovables formados a partir de restos de plantas y animales antiguos; el carbón, el petróleo y el gas natural son combustibles fósiles (pág. 69)

fossil water water underground that is not being replaced by rainfall (p. 223)

aguas fósiles agua subterránea que no es reemplazada por el agua de lluvia (pág. 223)

free port a city in which almost no taxes are placed on goods (pp. 365, 636)

puerto libre ciudad donde hay muy pocos impuestos sobre los bienes (págs. 365, 636)

freshwater water that is not salty; it makes up only about 3 percent of our total water supply (p. 31)

agua dulce agua que no es salada; representa sólo alrededor del 3 por ciento de nuestro suministro total de agua (pág. 31)

front the place where two air masses of different temperatures or moisture content meet (p. 53)

frente lugar en el que se encuentran dos masas de aire con diferente temperatura o humedad (pág. 53)

genocide the intentional destruction of a people (p. 405)

genocidio destrucción intencional de un grupo de personas (pág. 405)

geography the study of the world, its people, and the landscapes they create (p. 4)

geografía estudio del mundo, de sus habitantes y de los paisajes creados por el ser humano (pág. 4)

geothermal energy energy produced from the heat of Earth's interior (p. 403)

energía geotérmica energía generada por el calor del interior de la Tierra (pág. 403)

ger a large, circular, felt tent used in Mongolia (p. 583)

ger gran tienda circular de fieltro usada en Mongolia (pág. 583)

glacier a large area of slow moving ice (p. 31)

glaciar gran bloque de hielo que avanza con lentitud (pág. 31)

globalization the process in which countries are increasingly linked to each other through culture and trade (p. 97)

globalización proceso por el cual los países se encuentran cada vez más interconectados a través de la cultura y el comercio (pág. 97)

globe a spherical, or ball-shaped, model of the entire planet (p. 8)

globo terráqueo modelo esférico, o en forma de bola, de todo el planeta (pág. 8)

Grand Canal a canal linking northern and southern China (p. 518)

Gran Canal canal que conecta el norte y el sur de China (pág. 518)

Great Wall a barrier built to protect China from invasion that stood near China's northern border (p. 509)

Gran Muralla barrera construida cerca de la frontera norte de China para proteger a China de las invasiones (pág. 509)

green revolution a program that encouraged farmers to adopt modern agricultural methods to produce more food (p. 555)

revolución verde programa que animó a los agricultores a adoptar métodos de agricultura modernos para producir más alimentos (pág. 555)

griot (GREE-oh) a West African storyteller (p. 340)

griot narrador de cuentos de África occidental (pág. 340)

gross domestic product (GDP) the value of all goods and services produced within a country in a single year (p. 95)

producto interior bruto (PIB) valor de todos los bienes y servicios producidos en un país durante un año (pág. 95)

groundwater water found below Earth's surface (p. 32)

agua subterránea agua que se encuentra debajo de la superficie de la Tierra (pág. 32)

gunpowder a mixture of powders used in guns and explosives (p. 522)

pólvora mezcla de polvos utilizada en las armas de fuego y los explosivos (pág. 522)

habitat the place where a plant or animal lives (p. 64)

hábitat lugar en el que vive una planta o animal (pág. 64)

Hammurabi's Code a set of 282 laws governing daily life in Babylon; the earliest known collection of written laws (p. 133)

Código de Hammurabi conjunto de 282 leyes que regían la vida cotidiana en Babilonia; la primera colección de leyes escritas conocida (pág. 133)

hieroglyphics (hy-ruh-GLIH-fiks) the ancient Egyptian writing system that used picture symbols (p. 298)

jeroglíficos sistema de escritura del antiguo Egipto, en el cual se usaban símbolos ilustrados (pág. 298)

Hindu-Arabic numerals the number system we use today; it was created by Indian scholars during the Gupta dynasty (p. 498)

numerales indoarábigos sistema numérico que usamos hoy en día; fue creado por estudiosos de la India durante la dinastía Gupta (pág. 498)

Hinduism the main religion of India; it teaches that everything is part of a universal spirit called Brahman (p. 550)

hinduismo religión principal de la India; sus enseñanzas dicen que todo forma parte de un espíritu universal llamado Brahma (pág. 550)

human geography the study of the world's people, communities, and landscapes (p. 18)

geografía humana estudio de los habitantes, las comunidades y los paisajes del mundo (pág. 18)

human rights rights that all people deserve, such as rights to equality and justice (p. 630)

derechos humanos derechos que toda la gente merece como derechos a la igualdad y la justicia (pág. 630)

humanitarian aid assistance to people in distress (p. 100)

ayuda humanitaria ayuda a personas en peligro (pág. 100)

humus (HYOO-muhs) decayed plant or animal matter; it helps soil support abundant plant life (p. 65)

humus materia animal o vegetal descompuesta; contribuye a que crezca una gran cantidad de plantas en el suelo (pág. 65)

hydroelectric power the production of electricity from waterpower, such as from running water (p. 70)

energía hidroeléctrica producción de electricidad generada por la energía del agua, como la del agua corriente (pág. 70)

iceberg a floating mass of ice that has broken off a glacier (p. 659)

iceberg masa de hielo flotante que se ha desprendido de un glaciar (pág. 659)

ice shelf a ledge of ice that extends over the water (p. 659)

banco de hielo saliente de hielo que se extiende sobre el agua (pág. 659)

imperialism an attempt to dominate a country's government, trade, and culture (p. 399)

imperialismo intento de dominar el gobierno, el comercio y la cultura de un país (pág. 399)

imports goods brought in from other regions (p. 316)

importaciones bienes que se traen a un país de otras regiones (pág. 316)

inflation the rise in prices that occurs when currency loses its buying power (p. 427)

inflación aumento de los precios que ocurre cuando la moneda de un país pierde poder adquisitivo (pág. 427)

inoculation (i-nah-kyuh-LAY-shuhn) injecting a person with a small dose of a virus to help build up defenses to a disease (p. 498)

inoculación inyectarle una pequeña dosis de un virus a una persona para ayudarla a crear defensas contra una enfermedad (pág. 498)

interdependence the reliance of one country on the resources, goods, or services of another country (p. 99)

interdependencia dependencia que tiene un país de los recursos, los bienes y los servicios de otro (pág. 99)

irrigation a way of supplying water to an area of land (p. 118)

irrigación método para suministrar agua a un terreno (pág. 118)

Islam a religion based on the messages that Muhammad is believed to have received from God (p. 172)

Islam religión basada en los mensajes que se cree que Mahoma recibió de Dios (pág. 172)

isolationism a policy of avoiding contact with other countries (p. 534)

aislacionismo política de evitar el contacto con otros países (pág. 534)

ivory a white material made from elephant tusks (p. 312)

marfil material blanco procedente de los colmillos de los elefantes (pág. 312)

Janissary an Ottoman slave soldier (p. 182)

jenízaro soldado esclavo otomano (pág. 182)

jihad (ji-HAHD) to make an effort, or to struggle; has also been interpreted to mean holy war (p. 175)

yihad esforzarse o luchar; se ha interpretado también con el significado de guerra santa (pág. 175)

Judaism (JOO-dee-i-zuhm) the religion of the Hebrews; it is the world's oldest monotheistic religion (p. 144)

judaísmo religión de los hebreos; es la religión monoteísta más antigua del mundo (pág. 144)

ENGLISH AND SPANISH GLOSSARY

kampong a traditional village in Indonesia; also the term for crowded slums around Indonesia's large cities (p. 635)
kampong aldea tradicional de Indonesia; término que también se usa para los barrios pobres y superpoblados que rodean las grandes ciudades de Indonesia (pág. 635)

karma in Buddhism and Hinduism, the effects that good or bad actions have on a person's soul (p. 482)
karma en el budismo y el hinduismo, efectos que las buenas o malas acciones producen en el alma de una persona (pág. 482)

kente (ken-TAY) a hand-woven, brightly colored West African fabric (p. 343)
kente tela de África occidental, tejida a mano y muy colorida (pág. 343)

kibbutz (kih-BOOHTS) in Israel, a large farm where people share everything in common (p. 206)
kibbutz en Israel, granja grande donde las personas comparten todo (pág. 206)

kimchi a traditional Korean food made from pickled cabbage and spices (p. 601)
kimchi comida tradicional coreana hecha con repollo en vinagre y especias (pág. 601)

kimono a traditional robe worn in Japan (p. 600)
kimono bata tradicional usada en Japón (pág. 600)

klong a canal in Bangkok (p. 630)
klong canal de Bangkok (pág. 630)

kosher a term used to refer to Jewish dietary laws; it means "acceptable" in Hebrew (p. 206)
kosher término utilizado para referirse a las leyes alimenticias judías; en hebreo significa aceptable (pág. 206)

L

landform a shape on the planet's surface, such as a mountain, valley, plain, island, or peninsula (p. 35)
accidente geográfico forma de la superficie terrestre, como una montaña, un valle, una llanura, una isla o una península (pág. 35)

landlocked completely surrounded by land with no direct access to the ocean (p. 244)
sin salida al mar que está rodeado completamente por tierra, sin acceso directo al océano (pág. 244)

landscape all the human and physical features that make a place unique (p. 4)
paisaje todas las características humanas y físicas que hacen que un lugar sea único (pág. 4)

latitude the distance north or south of Earth's equator (p. 27)
latitud distancia hacia el norte o el sur desde el ecuador (pág. 27)

lava magma that reaches Earth's surface (p. 37)
lava magma que llega a la superficie terrestre (pág. 37)

loess (LES) fertile, yellowish soil (p. 568)
loess suelo amarillento y fértil (pág. 568)

malaria a disease spread by mosquitoes that causes fever and pain (p. 428)
malaria enfermedad transmitida por los mosquitos que causa fiebre y dolor (pág. 428)

malnutrition a condition of not getting enough nutrients from food (p. 429)
desnutrición estado producido al no obtener suficientes nutrientes de los alimentos (pág. 429)

mandate of heaven the idea that heaven chose China's ruler and gave him or her power (p. 508)
mandato divino idea de que el cielo elegía al gobernante de China y le daba el poder (pág. 508)

Maori (MOWR-ee) the original inhabitants of New Zealand (p. 647)
maoríes primeros habitantes de Nueva Zelanda (pág. 647)

map a flat drawing that shows all or part of Earth's surface (p. 8)
mapa representación plana que muestra total o parcialmente la superficie de la Tierra (pág. 8)

market economy an economic system based on free trade and competition (p. 94)
economía de mercado sistema económico basado en el libre comercio y la competencia (pág. 94)

meditation deep continued thought that focuses the mind on spiritual ideas (p. 485)
meditación concentración profunda y continua que enfoca la mente en ideas espirituales (pág. 485)

mercenary a hired soldier (p. 490)
mercenario soldado a sueldo (pág. 490)

merchant a trader (p. 316)
mercader comerciante (pág. 316)

Messiah in Judaism, a new leader that would appear among the Jews and restore the greatness of ancient Israel (p. 152)
Mesías en el judaísmo, nuevo líder que aparecería entre los judíos y restablecería la grandeza del antiguo Israel (pág. 152)

metallurgy (MET-uhl-uhr-jee) the science of working with metals (p. 498)
metalurgia ciencia de trabajar los metales (pág. 498)

ENGLISH AND SPANISH GLOSSARY

meteorology the study of weather and what causes it (p. 20)

meteorología estudio de las condiciones del tiempo y sus causas (pág. 20)

Middle Kingdom the period of Egyptian history from about 2050 to 1750 BC marked by order and stability (p. 292)

Reino Medio período de la historia de Egipto desde aproximadamente 2050 a 1750 a. C., caracterizado por el orden y la estabilidad (pág. 292)

migration the movement of people from one place to live in another (p. 89)

migración movimiento de personas de un lugar para ir a vivir a otro lugar (pág. 89)

minaret a narrow tower from which Muslims are called to prayer (p. 189)

minarete torre fina desde la que se llama a la oración a los musulmanes (pág. 189)

missionary someone who works to spread religious beliefs (p. 488)

misionero alguien que trabaja para difundir creencias religiosas (pág. 488)

monotheism the belief in only one God (p. 147)

monoteísmo creencia en un solo Dios (p. 147)

monsoons seasonal winds that bring either dry or moist air to an area (pp. 58, 545)

monzones vientos estacionales que traen aire seco o húmedo a una región (págs. 58, 545)

mosaic a picture made with pieces of colored stone or glass (p. 163)

mosaico dibujo hecho con trozos de piedra o cristal de colores (pág. 163)

mosque (MAHSK) a building for Muslim prayer (pp. 173, 337)

mezquita edificio musulmán para la oración (págs. 173, 337)

mummy a specially treated body wrapped in cloth for preservation (p. 286)

momia cadáver especialmente tratado y envuelto en tela para su conservación (pág. 286)

Muslim a follower of Islam (p. 172)

musulmán seguidor del Islam (pág. 172)

N

natural resource any material in nature that people use and value (p. 68)

recurso natural todo material de la naturaleza que las personas utilizan y valoran (pág. 68)

New Kingdom the period from about 1550 to 1050 BC in Egyptian history when Egypt reached the height of its power and glory (p. 292)

Reino Nuevo período de la historia egipcia desde aproximadamente 1550 a 1050 a. C., en el que Egipto alcanzó el punto máximo de su poder y su gloria (pág. 292)

nirvana in Buddhism, a state of perfect peace (p. 486)

nirvana en el budismo, estado de paz perfecta (pág. 486)

noble a rich and powerful person (p. 284)

noble persona rica y poderosa (pág. 284)

nomads people who move often from place to place (p. 250)

nómadas personas que se trasladan frecuentemente de un lugar a otro (pág. 250)

nonrenewable resource a resource that cannot be replaced naturally; coal and petroleum are examples of nonrenewable resources (p. 69)

recurso no renovable recurso que no puede reemplazarse naturalmente; el carbón y el petróleo son ejemplos de recursos no renovables (pág. 69)

nonviolence the avoidance of violent actions (p. 483)

no violencia rechazo de las acciones violentas (pág. 483)

O

oasis a wet, fertile area in a desert where a spring or well provides water (pp. 222, 352)

oasis zona húmeda y fértil en el desierto con un manantial o pozo que proporciona agua (págs. 222, 352)

obelisk (AH-buh-lisk) a tall, pointed, four-sided pillar in ancient Egypt (p. 300)

obelisco pilar de cuatro caras, alto y puntiagudo en el antiguo Egipto (pág. 300)

ocean currents large streams of surface seawater; they move heat around Earth (p. 52)

corrientes oceánicas grandes corrientes de agua de mar que fluyen en la superficie del océano; transportan calor por toda la Tierra (pág. 52)

Old Kingdom the period from about 2700 to 2200 BC in Egyptian history that began shortly after Egypt was unified (p. 283)

Reino Antiguo período de la historia egipcia desde aproximadamente 2700 a 2200 a. C. que comenzó poco después de la unificación de Egipto (pág. 283)

OPEC an international organization whose members work to influence the price of oil on world markets by controlling the supply (p. 225)

OPEP organización internacional cuyos miembros trabajan para influenciar el precio del petróleo en los mercados mundiales controlando la oferta (pág. 225)

oral history a spoken record of past events (p. 340)

historia oral registro hablado de hechos ocurridos en el pasado (pág. 340)

ozone layer a layer of Earth's atmosphere that protects living things from the harmful effects of the sun's ultraviolet rays (p. 660)

capa de ozono capa de la atmósfera de la Tierra que protege a los seres vivos de los efectos dañinos de los rayos ultravioleta del sol (pág. 660)

pagoda a Buddhist temple based on Indian designs (p. 575)

pagoda templo budista basado en diseños de la India (pág. 575)

pans low, flat areas (p. 438)

depresiones áreas bajas y planas (pág. 438)

papyrus (puh-PY-ruhs) a long-lasting, paper-like material made from reeds that the ancient Egyptians used to write on (p. 298)

papiro material duradero hecho de juncos, similar al papel, que los antiguos egipcios utilizaban para escribir (pág. 298)

partition division (p. 549)

partición división (pág. 549)

periodic market an open-air trading market that is set up once or twice a week (p. 417)

mercado periódico mercado al aire libre que funciona una o dos veces a la semana (pág. 417)

permafrost permanently frozen layers of soil (p. 61)

permafrost capas de tierra congeladas permanentemente (pág. 61)

pharaoh (FEHR-oh) the title used by the rulers of Egypt (p. 281)

faraón título usado por los gobernantes de Egipto (pág. 281)

physical geography the study of the world's physical features—its landforms, bodies of water, climates, soils, and plants (p. 16)

geografía física estudio de las características físicas de la Tierra: sus accidentes geográficos, sus masas de agua, sus climas, sus suelos y sus plantas (pág. 16)

pictograph a picture symbol (p. 128)

pictograma símbolo con imágenes (pág. 128)

plate tectonics a theory suggesting that Earth's surface is divided into a dozen or so slow-moving plates, or pieces of Earth's crust (p. 36)

tectónica de placas teoría que sugiere que la superficie terrestre está dividida en unas doce placas, o fragmentos de corteza terrestre, que se mueven lentamente (pág. 36)

polar desert a high-latitude region that receives little precipitation (p. 659)

desierto polar región a una latitud alta que recibe pocas precipitaciones (pág. 659)

polytheism the worship of many gods (p. 124)

politeísmo culto a varios dioses (pág. 124)

popular culture culture traits that are well known and widely accepted (p. 98)

cultura popular rasgos culturales conocidos y de gran aceptación (pág. 98)

population the total number of people in a given area (p. 86)

población número total de personas en una zona determinada (pág. 86)

population density a measure of the number of people living in an area (p. 86)

densidad de población medida del número de personas que viven en una zona (pág. 86)

porcelain a thin, beautiful pottery invented in China (p. 521)

porcelana cerámica bella y fina creada en China (pág. 521)

precipitation water that falls to Earth's surface as rain, snow, sleet, or hail (p. 31)

precipitación agua que cae a la superficie de la Tierra en forma de lluvia, nieve, aguanieve o granizo (pág. 31)

prevailing winds winds that blow in the same direction over large areas of Earth (p. 51)

vientos preponderantes vientos que soplan en la misma dirección sobre grandes zonas de la Tierra (pág. 51)

priest a person who performs religious ceremonies (p. 125)

sacerdote persona que lleva a cabo ceremonias religiosas (pág. 125)

proverb a short saying of wisdom or truth (p. 341)

proverbio refrán breve que expresa sabiduría o una verdad (pág. 341)

pyramid a huge triangular tomb built by the Egyptians and other peoples (p. 288)

pirámide tumba triangular y enorme construida por los egipcios y otros pueblos (pág. 288)

Qur'an (kuh-RAN) the holy book of Islam (p. 172)

Corán libro sagrado del Islam (pág. 172)

rabbi a Jewish religious teacher (p. 149)

rabino maestro religioso judío (pág. 149)

reforestation planting trees to replace lost forestland (p. 69)

reforestación siembra de árboles para reemplazar los bosques que han desaparecido (pág. 69)

region a part of the world that has one or more common features that distinguish it from surrounding areas (p. 6)

región parte del mundo que tiene una o más características comunes que la distinguen de las áreas que la rodean (pág. 6)

ENGLISH AND SPANISH GLOSSARY

reincarnation a Hindu and Buddhist belief that souls are born and reborn many times, each time into a new body (p. 481)

reencarnación creencia hindú y budista de que las almas nacen y renacen muchas veces, cada vez en un cuerpo nuevo (pág. 481)

relative location a general description of where a place is located; a place's relative location is often expressed in relation to something else (p. 12)

ubicación relativa descripción general de la posición de un lugar; la ubicación relativa de un lugar suele expresarse en relación con otra cosa (pág. 12)

renewable resource a resource that Earth replaces naturally, such as water, soil, trees, plants, and animals (p. 69)

recurso renovable recurso que la Tierra reemplaza por procesos naturales, como el agua, el suelo, los árboles, las plantas y los animales (pág. 69)

Resurrection (re-suh-REK-shuhn) in Christianity, Jesus's rise from the dead (p. 153)

Resurrección en el cristianismo, la vuelta a la vida de Jesús (pág. 153)

revolution the 365 ¼ day trip Earth takes around the sun each year (p. 27)

revolución viaje de 365 ¼ días que la Tierra hace alrededor del Sol cada año (pág. 27)

revolution a drastic change in a country's government and way of life (p. 235)

revolución cambio drástico en el gobierno y la forma de vida de un país (pág. 235)

rift valleys places on Earth's surface where the crust stretches until it breaks (p. 394)

valles de fisura puntos de la superficie de la Tierra en los que la corteza se estira hasta romperse (pág. 394)

Rosetta Stone a huge stone slab inscribed with hieroglyphics, Greek, and a later form of Egyptian that allowed historians to understand Egyptian writing (p. 299)

piedra Roseta gran losa de piedra con inscripciones en jeroglíficos, en griego y en una forma tardía del idioma egipcio que permitió a los historiadores entender la escritura egipcia (pág. 299)

rotation one complete spin of Earth on its axis; each rotation takes about 24 hours (p. 26)

rotación giro completo de la Tierra sobre su propio eje; cada rotación toma 24 horas (pág. 26)

S

safari an overland journey to view African wildlife (p. 403)

safari excursión por tierra con el fin de ver animales salvajes en África (pág. 403)

Sahel (SAH-hel) a strip of land that divides the Sahara from wetter areas (p. 374)

Sahel franja de tierra que divide el Sahara de zonas más húmedas (pág. 374)

saint a person known and admired for his or her holiness (p. 156)

santo persona conocida y admirada por su santidad (pág. 156)

samurai (SA-muh-ry) a trained professional warrior in feudal Japan (p. 598)

samurai guerrero profesional entrenado del Japón feudal (pág. 598)

sanctions economic or political penalties imposed by one country on another to try to force a change in policy (p. 446)

sanciones penalizaciones económicas o políticas que un país impone a otro para obligarlo a cambiar su política (pág. 446)

Sanskrit the most important language of ancient India (p. 477)

sánscrito el idioma más importante de la antigua India (pág. 477)

savanna an area of tall grasses and scattered trees and shrubs (pp. 58, 374)

sabana zona de pastos altos con arbustos y árboles dispersos (págs. 58, 374)

scholar-official an educated member of China's government who passed a series of written examinations (p. 526)

funcionario erudito miembro culto del gobierno de China que aprobaba una serie de exámenes escritos (pág. 526)

scribe a writer (p. 128)

escriba escritor (pág. 128)

secede to break away from the main country (p. 382)

separarse dividirse del territorio principal del país (pág. 382)

secular the separation of religion and government; nonreligious (p. 203)

secular separación entre la religión y el gobierno; no religioso (pág. 203)

seismograph a device that measures the strength of an earthquake (p. 514)

sismógrafo aparato que mide la fuerza de un terremoto (pág. 514)

shah a Persian title that means "king" (p. 235)

sha título persa que significa "rey" (pág. 235)

Sherpas an ethnic group from the mountains of Nepal (p. 556)

sherpas grupo étnico de las montañas de Nepal (pág. 556)

Shia Muslims who believe that true interpretation of Islamic teaching can only come from certain religious and political leaders called imams; they make up one of the two main branches of Islam (p. 224)

chiítas musulmanes que creen que la interpretación correcta de las enseñanzas islámicas solo puede provenir de ciertos líderes religiosos y políticos llamados imanes; forman una de las dos ramas principales del Islam (pág. 224)

shogun a general who ruled Japan in the emperor's name (p. 598)

shogun general que gobernaba a Japón en nombre del emperador (pág. 598)

silent barter a process in which people exchange goods without contacting each other directly (p. 328)

trueque silencioso proceso mediante el cual las personas intercambian bienes sin entrar en contacto directo (pág. 328)

silt finely ground fertile soil that is good for growing crops (pp. 117, 350)

cieno tierra fértil de partículas finas que es buena para el crecimiento de los cultivos (págs. 117, 350)

social hierarchy the division of society by rank or class (p. 125)

jerarquía social división de la sociedad en clases o niveles sociales (pág. 125)

social science a field that focuses on people and the relationships among them (p. 5)

ciencias sociales campo de estudio que se enfoca en las personas y en las relaciones entre ellas (pág. 5)

solar energy energy from the sun (p. 26)

energía solar energía del Sol (pág. 26)

souk (SOOK) a marketplace or bazaar in the Islamic world (p. 365)

zoco mercado o bazar del mundo islámico (pág. 365)

sphinx (sfinks) an imaginary creature with a human head and the body of a lion that was often shown on Egyptian statues (p. 300)

esfinge criatura imaginaria con cabeza humana y cuerpo de león que se representaba a menudo en las estatuas egipcias (pág. 300)

steppe a semidry grassland or prairie; steppes often border deserts (p. 59)

estepa pradera semiárida; las estepas suelen encontrarse en el límite de los desiertos (pág. 59)

subcontinent a large landmass that is smaller than a continent (p. 542)

subcontinente gran masa de tierra, más pequeña que un continente (pág. 542)

Sufism (soo-fi-zuhm) a movement in Islam that taught people they can find God's love by having a personal relationship with God (p. 187)

sufismo movimiento perteneciente al Islam que enseñaba a las personas que pueden hallar el amor de Dios si establecen una relación personal con Él (pág. 187)

sultan the supreme ruler of a Muslim country (p. 637)

sultán gobernante supremo de un país musulmán (pág. 637)

sundial a device that uses the position of shadows cast by the sun to tell the time of day (p. 514)

reloj de sol aparato que utiliza la posición de las sombras que proyecta el sol para indicar las horas del día (pág. 514)

Sunnah (SOOH-nuh) a collection of writings about the way Muhammad lived that provides a model for Muslims to follow (p. 175)

Sunna conjunto de escritos sobre la vida de Mahoma que proporciona un modelo de comportamiento para los musulmanes (pág. 175)

Sunni Muslims who believe in the ability of the majority of the community to interpret Islamic teachings; they make up one of the two main branches of Islam (p. 224)

sunitas musulmanes que creen en la capacidad de la mayor parte de la comunidad de interpretar las enseñanzas islámicas; forman una de las dos ramas principales del Islam (pág. 224)

surface water water that is found in Earth's streams, rivers, and lakes (p. 31)

agua superficial agua que se encuentra en los arroyos, ríos y lagos de la Tierra (pág. 31)

surplus more of something than is needed (p. 118)

excedente más cantidad de algo de lo que se necesita (pág. 118)

Taliban a radical Muslim group that rose to power in Afghanistan in the mid-1990s (p. 253)

talibanes grupo radical musulmán que llegó al poder en Afganistán a mediados de la década de 1990 (pág. 253)

tariff a fee that a country charges on imports or exports (p. 603)

arancel tarifa que impone un país a las importaciones y exportaciones (pág. 603)

territory an area that is under the authority of another government (p. 655)

territorio zona que está bajo el control de otro gobierno (pág. 655)

theocracy a government ruled by religious leaders (p. 236)

teocracia gobierno dirigido por líderes religiosos (pág. 236)

tolerance acceptance (p. 182)

tolerancia aceptación (pág. 182)

Torah the most sacred text of Judaism (p. 148)

Torá el texto más sagrado del judaísmo (pág. 148)

townships crowded clusters of small homes in South Africa outside of cities where black South Africans live (p. 443)
distritos segregados grupos de pequeñas viviendas amontonadas ubicadas en las afueras de las ciudades de Sudáfrica, donde vivían los sudafricanos negros (pág. 443)

trade network a system of people in different lands who trade goods back and forth (p. 316)
red comercial sistema de personas en diferentes lugares que comercian productos entre sí (pág. 316)

trade route a path followed by traders (p. 293)
ruta comercial camino seguido por los comerciantes (pág. 293)

trade surplus when a country exports more goods than it imports (p. 603)
excedente comercial cuando un país exporta más bienes de los que importa (pág. 603)

tropics regions close to the equator (p. 29)
trópicos regiones cercanas al ecuador (pág. 29)

tsunami (sooh-NAH-mee) a destructive and fast-moving wave (p. 594)
tsunami ola rápida y destructiva (pág. 594)

United Nations an organization of countries that promotes peace and security around the world (p. 99)
Naciones Unidas organización de países que promueve la paz y la seguridad en todo el mundo (pág. 99)

urbanization the increase in the percentage of people who live in cities (p. 554)
urbanización aumento del porcentaje de personas que vive en las ciudades (pág. 554)

veld (VELT) open grassland areas in South Africa (p. 438)
veld praderas descampadas en Sudáfrica (pág. 438)

wadi a dry streambed (p. 223)
uadi cauce seco de un río o arroyo (pág. 223)

wat a Buddhist temple that also serves as a monastery (p. 626)
wat templo budista que sirve también como monasterio (pág. 626)

water cycle the movement of water from Earth's surface to the atmosphere and back (p. 33)
ciclo del agua circulación del agua desde la superficie de la Tierra hacia la atmósfera y de regreso a la Tierra (pág. 33)

water vapor water occurring in the air as an invisible gas (p. 32)
vapor de agua agua que se encuentra en el aire en estado gaseoso e invisible (pág. 32)

weather the short-term changes in the air for a given place and time (p. 50)
tiempo cambios a corto plazo en la atmósfera en un momento y lugar determinados (pág. 50)

weathering the process by which rock is broken down into smaller pieces (p. 39)
meteorización proceso de desintegración de las rocas en pedazos pequeños (pág. 39)

woodblock printing a form of printing in which an entire page is carved into a block of wood, covered with ink, and pressed to a piece of paper to create a printed page (p. 522)
xilografía forma de impresión en la que una página completa se talla en una plancha de madera, se cubre de tinta y se presiona sobre un papel para crear una página impresa (pág. 522)

work ethic a belief that work in itself is worthwhile (p. 603)
ética de trabajo creencia de que el trabajo tiene valor propio (pág. 603)

yurt a movable round house made of wool felt mats hung over a wood frame (p. 250)
yurt tienda redonda y portátil de fieltro de lana que se coloca sobre una armazón de madera (pág. 250)

ziggurat a pyramid-shaped temple in Sumer (p. 130)
zigurat templo sumerio en forma de pirámide (pág. 130)

Zionism a movement that began in the late 1800s and called for Jews to establish a country or community in Palestine (p. 205)
sionismo movimiento que comenzó a fines del siglo XIX y que llamaba a los judíos a establecer un país o comunidad judía en Palestina (pág. 205)

zonal organized by zone (p. 374)
zonal organizado por zonas (pág. 374)

Index

KEY TO INDEX

c = **chart**	*m* = **map**
f = **feature**	*p* = **photo**

A

Aborigines, 647, 649
Abraham, 144, 144p
absolute location, 12
Abu Bakr, 180
Abu Simbel, 300
Academic Words, H19
Active Reading, H14–H17
acupuncture, 515, 515p
Afghanistan, 112c, 253–54. *See also* Central Asia; alphabet, 251; as developing country, 94c, 95; standard of living in, 256c; Taliban, 253–54
Africa. *See also* Central Africa; East Africa; North Africa; Southern Africa; West Africa: Africa and the World, 272–273c; Africa's Growing Population, 275c; Africa's Largest Cities, 383c; climate zones of, 57m; energy production, 70c; Facts about Countries, 272–75; Geographical Extremes: Africa, 267c; Islam's spread to, 181; Malawi, 17p; physical map of, 718m; political map of, 268m, 719m; regional atlas of, 266–71m; resources map of, 269m; Size Comparison: The United States and Africa, 267c
African National Congress, 442, 447
African plate, 36p
African Union, 451
Afrikaans, 442
Afrikaners, 442
agriculture: ancient Egypt, 280, 280p, 295–96; Angola, 427; Australia, 646, 648; Cambodia, 633; Central Africa, 429; Chad, 385; China, 577p, 578; Chinese civilization beginnings and, 506; Côte d'Ivoire, 385; dry farming, 365p; dryland farming, 255; early civilization in Nubia, 311; early farming village in Turkey, 200–201f; Eastern Mediterranean, 198–99; Egypt, 362–63, 364; Ethiopia, 406; Farmland in Central Asia, 263m; Ghana empire, 326–27, 330, 331p; green revolution in India, 555; Harappan civi-

lization, 473; India, 555; Indian Subcontinent, 545; Indonesia, 621, 627; Israel, 206; Japan, 603, 606; Jordan, 213; Kenya, 403; Kush, 319; Kyrgyzstan, 255; Laos, 633; Malawi, 427; Malaysia, 621; Mali, 334, 386, 386p; Mauritania, 385; Mesopotamia, 117–19, 118–19p; New Zealand, 646, 648; Niger, 385; North Africa, 362–63, 364m, 365, 365p; North Korea, 611; oasis agriculture, 365p; Pacific Islands, 654, 657; periodic markets in Central Africa, 417; Philippines, 637p; River Valley Civilizations, 120–21f; Southeast Asia, 621; South Korea, 609; Sudan, 405; Tanzania, 403; Togo and Benin, 385; Turkmenistan, 255; Uganda, 405; Vietnam, 633; West Africa, 375; Zambia, 427
Ahmadinejad, Mahmoud, 237
AIDS, 429
Ajanta, 495
Akbar, 184–85, 547c, 548
Akkadian empire: rise of, 123–24
Aksum: rise of, 319
al-Assad, Hafiz, 210
Alawites, 211
Alexander the Great, 230, 355
Alexandria, 355, 357, 364
al-Gadhafi, Mu'ammar, 365
algebra, 225
Algeria, 272c. *See also* North Africa; agriculture, 364m, 365, 365p; Berbers, 358; cities, 365; European influence and independence, 357; features and landscape of, 5p; government, 364; natural resources of, 353; oil, 365; physical geography of, 350–53, 351m; today, 366
Algiers, 365, 366p
Almoravids, 330
alphabet: Afghanistan, 251; Cyrillic, 251; Latin, 251; Phoenician, 137
Amazon River: rain forest, 58
Amman, 213
Amon-Re, 285, 285p
Amu Darya, 245, 247
Amundsen, Roald, 660p
Anatolia, 182, 196
Andes, 37
Angkor Wat, 624, 625p
Angola, 272c. *See also* Central Africa; agriculture, 427; Angola, 2000, 430c; economy of, 427; independence, 421; language in, 422; natural resources of, 427; physi-

cal geography of, 414–15, 415m; today, 427
animism, 378, 401
Ankara, 202
An Nafud, 222
Antarctica: climate, 659; early explorers, 660, 660p; environmental concerns, 661; glaciers, 39; ice shelves, 659, 661p; latitude of, 27; natural resources, 659; physical geography, 658–59; scientific research, 660–61
Antarctic Circle: midnight sun, 29, 29p
Antarctic plate, 36p
Anubis, 285
apartheid, 442–43; end of, 446–47
Apostles, 156
Arabia, 170, 180; Islam spreads in, 173; life in, 171, 171p; physical features and climate of, 170
Arabian Peninsula, 170; Bahrain and Qatar, 226; climate of, 222, 222m; desert in, 220–22, 222m; Kuwait, 226; natural resources of, 223; Oman, 227; physical geography of, 220–21, 221m; political map of, 218–19m; Saudi Arabia, 224–26; United Arab Emirates, 227; vegetation of, 222; Yemen, 227
Arabian Sea, 220, 221m
Arabic, 337, 340
Arabs, 235; creation of Israel and, 205; in history of North Africa, 355, 357; influence in Central Asia, 249c; in Iraq, 232; Palestine and, 204
Aral Sea, 245, 247, 256, 258–59f
architecture: ancient Egypt, 294; Buddhist temples in Japan, 599, 599p; Christian churches in Lalibela, 398, 399p; Egyptian pyramids, 288–90; Egyptian temples, 300, 301p; Hagia Sophia, Constantinople, 163p, 164; houses in West Africa, 379, 379p; Islamic achievements in, 189; Ndebele Village, 444–45p; temples of Ancient India, 495–96, 496p; Village Architecture, 426–27f
Arctic Circle: midnight sun, 29, 29p
Armenians, 211
art: ancient China, 521; ancient Egypt, 294–95, 302–3; Byzantine, 163; Central Africa, 422–23; Fertile Crescent, 115p; Gupta Empire, 493p; Han dynasty, 514, 515p; Harappan civilization, 476, 476p; Islamic achievements in,

189; Japan, 600; Music from Mali to Memphis, 342f; North Africa, 358–59; religious art of ancient India, 495–96, 496p; Shang dynasty, 507; Southern Africa, 445; Sumerian society, 130–31, 130–31p; West Africa, 342–43; West African masks, 378f

art connection: Masks, 378; Music from Mali to Memphis, 342

Aryans, 476–77; Brahmanism, 480; caste system, 479; government and society, 476–77; influence of, 546c, 547; language of, 477, 547; rajas, 476–77; sudras, 479; *varnas,* 478, 479p

Asia: climate zones of, 57m; energy production, 70c; monsoons, 58; physical map of, 716m; political map of, 717m

Askia the Great, 337–38

Asoka, 494; as Mauryan ruler, 491, 546c, 547; spread of Buddhism and, 488, 491, 494, 546c, 547

Assyrian Empire, 134–35, 134m

Assyrians: end of Kushite rule of Egypt, 314, 314p

astronomy: ancient Indian achievements, 499, 499p; Chaldeans, 135; Islamic achievements in, 186p, 187

Aswan High Dam, 351

Atatürk, Kemal, 201, 202

Atlantic Ocean: hurricanes, 53; Panama Canal, 40

Atlas: Africa Regional Atlas, 266–71m; Pacific Regional Atlas, 464–65m; South and East Asia Regional Atlas, 458–63m; Southwest and Central Asia Regional Atlas, 106–11m; world, 702–21m

Atlas Mountains, 351m, 352

atoll, 654, 655f

Augrabies Falls, 437

Australia, 466c; Aborigines, 647, 649; Adelaide, 88c; agriculture, 646, 648; Australian Sports, 648f; as British colony, 647; Ayers Rock, 644; cities, 649; climate, 646, 665m; as developed country, 94c; early settlers of, 647; economy of, 648; ethnic groups, 648–49, 649c; government, 648; Great Barrier Reef, 645; history, 647; industry, 648; Outback, 648; physical geography of, 644–45, 645m; population, 648–49; Settling the Pacific, 650–51f; wildlife and natural resources of, 646

ayatollahs, 236

B

Babur, 184, 547c, 547p, 548

Babylon: Chaldeans, 135; Hammurabi's rule, 132–33; Hittite, 134

Babylonian Empire, 134m; Hammurabi's Code, 133; rise of, 132

Baghdad, 232p; as capital of Islamic Empire, 182; rebuilding, 233

Bahrain, 112c, 226

Bali, 637

Bamako, 373p

Bangkok, 630, 631f

Bangladesh, 466c; Average Monthly Precipitation, Dhaka, Bangladesh, 551c; climate and resources of, 545; floods, 33, 558; independence, 549; people of, 558; physical geography of, 542–44, 543m; rate of natural increase, 90; religion, 556, 557m

Bantu language, 422, 423f, 440, 444

bar graph: analyzing, 74

baseball: cultural diffusion of, 84m, 85

bauxite, 384; West Africa, 375

Bedouins, 357

Beijing, 505p, 530, 580, 581p; Forbidden City, 532–33f; National Day, 574–75f; Tiananmen Square, 578

Belgium: colonization of Central Africa, 421

Benin, 272c, 385. *See also* West Africa

Berbers, 181, 326, 337, 357, 358f

Bethlehem, 152

Bhutan, 466c; challenges of, 558–59; climate and resources of, 545; economy of, 558; government of, 558; physical geography of, 542–44, 543m; religion, 556, 557m

Bible: Christian, 152; Hebrew, 148, 149p

Biography: Askia the Great, 337f; Asoka, 494f; Atatürk, Kemal, 201f, 202; Cleopatra, 355f; Ebadi, Shirin, 237f; Eratosthenes, 18f; Gandhi, Mohandas, 548f; Hatshepsut, Queen, 292f, 293; Hirohito, 598f; Kublai Khan, 535f; Maathai, Wangari, 69f; Mandela, Nelson, 447f; Mansa Musa, 339f; Mehmed II, 183f; Piankhi, 313f; Ramses the Great, 297f; Sargon, 123f; Shackleton, Sir Ernest, 660f; Shanakhdakheto, Queen, 317f; Shi Huangdi, 508f, 508p; Suu Kyi, Aung San, 630f; Tunka Manin, 330f; Wegener, Alfred, 37f

birthrate: population changes and, 88–89; in Russia, 88

Bishkek, 260m

blizzards, 53

Blue Mosque, 188f

Blue Nile, 350, 396

Boers, 442

Bollywood, 555f

Borneo: physical geography of, 619

Bosporus, 196, 198p

Botswana, 272c. *See also* Southern Africa; art, 445; climate, 438; economy of, 449; ethnic groups, 444; language, 444; natural resources, 439; tourism, 437, 449, 450c

Brahmanism, 480–81

Brahmaputra River, 544

Brahmins, 478, 479p; Buddha's challenge to, 487

Brazzaville, 427

Brunei, 466c. *See also* Southeast Asia; independence, 634; natural resources, 637; per capita GDP, 638c; religion in, 627

Buddhism, 471p; Asoka and spread of, 488, 491, 494f, 546c, 547; Buddha finds enlightenment, 485; caste system and, 487; in China today, 574; The Eightfold Path, 487c; First Sermon, 485; Four Noble Truths, 486; Great Departure, 485p; Gupta Empire, 492; India, 550; Indian Subcontinent, 556, 557m; Japan, 597, 599, 599p; Koreas, 597, 600; Mahayana, 489; origins of, 484–89; Siddhartha's quest for answers, 484–85; Southeast Asia, 626–27; split in, 489; spread of, 488–89, 488m; teachings of, 486–87, 487c; temples, 495–96; Thailand, 627f; Theravada, 489; Tree of Wisdom, 485

Bukhara, 248

bullet trains, 606

Burkina Faso, 272c. *See also* West Africa; economy of, 386

Burundi, 270c; ethnic conflict, 405; population, 405; population density, 404c

Byblos, 136, 137m

Byzantine Empire, 143p, 160–64; Christianity, 163–64; government of, 162–63; new society of, 162–63

Byzantium, 160, 162, 201

Cabinda, 427
Cairo, 357, 363, 383c
calendar: Shang dynasty, 507
caliph, 180, 183
calligraphy, 189
Cambodia, 466c, 633. *See also* Southeast Asia; agriculture, 633; cities, 633; communism, 626; independence of, 625; modern history of, 625; physical geography of, 618–19, 619m
Cameroon, 103c, 272c. *See also* Central Africa; economy of, 426; government, 426; physical geography of, 414–15, 415m; today, 426
Canaan, 144, 145
Canada: ethnic group conflict, 83; government in, 91
canals: in Fertile Crescent, 118, 118p
Candra Gupta I, 492
Candra Gupta II, 492
Candragupta Maurya, 490
Cape of Good Hope, 441–42
Cape Town, 448–49p
Cape Verde, 272c, 384. *See also* West Africa
capitalism, 94
Caribbean South America: political map of, H8
Carter, Howard, 302p
Carthage, 136
cartography, 19; computer mapping, 19, 19m
Casablanca, 365
Casbah, 365
Case Study: Mapping Central Africa's Forests, 418–19f; Oil in Saudi Arabia, 228–29f; Ring of Fire, 42–43f; Tsunami, 622–23f
Caspian Sea, 220, 221m, 245
caste system: in ancient India, 479; Buddha's teachings on, 487; Gupta Empire, 492–93; Hinduism and, 481–82; in India, 550
Catal Hüyük, Turkey, 200–201f
cataracts, 279, 311, 311p
Catholics: in Central Africa, 422; in Philippines, 627
celadon, 521
Central Africa: agriculture, 429; Angola, 427; art, 422–23; Central Africa's National Parks, 416m; civil war in, 428; climate, vegetation and animals, 416–17; colonization of, by European countries, 421; culture, 422–23; Democratic Republic of the Congo, 424–25,

425p; disease and, 428–29; environmental issues, 429; ethnic groups, 421, 422; Gabon, 427; history of, 420–21; independence, 421; issues and challenges of, 428–29; ivory trade, 420–21, 421p; Kinshasa's Growing Population, 425c; language, 422; Malaria in Central Africa, 428m; malnutrition, 429; Mapping Central Africa's Forests, 418–19f; natural resources of, 417; physical geography, 414–17, 415m; political map of, 412–13m; religion, 422; Republic of the Congo, 427; slave trade, 421; tropical forests of, 416–17; Village Architecture, 426–27f; Zambia, 427
Central African Republic, 272c. *See also* Central Africa; economy of, 425; natural resources of, 425; physical geography of, 414–15, 415m; today, 425
Central Asia: Afghanistan, 253–54; climate of, 111m, 246; culture of, 250–52; ethnic groups of, 251, 252p; farmland in, 263m; history of, 248–49; invasions, 249; issues and challenges for, 256–57; Kazakhstan, 254; Kyrgyzstan, 255; land use and resources, 246m, 247; language of, 251, 251m; Largest Oil Reserves by Country, 113c; natural resources of, 109m, 247; physical geography of, 106m, 244–45, 245m; political instability of, 257; political map of, 108m, 242–43m; population of, 110m; religion, 251–52; Russian and Soviet Rule, 249; Size Comparison: The United States and Southwest and Central Asia, 107m; standard of living in, 256c; Tajikistan, 255; today, 253–57; trade, 248; traditional nomadic lives, 250; Turkmenistan, 255; Uzbekistan, 256; vegetation of, 246; water, 247; yurt, 250, 250p
Chad, 272c. *See also* West Africa; agriculture, 385; petroleum, 385
Chaldeans, 135
Chang'an, 520
Chang Jiang, 506, 568, 581
Chao Phraya, 630
Chart and Graph Skills, 74, 209, 238, 360, 430, 551
charts and graphs: Africa and the World, 275c; Africa's Growing Population, 275c; Africa's Largest Cities, 383c; Angola, 2000, 430c;

Average Annual Precipitation by Climate Region, 74c; Average Monthly Precipitation, Dhaka, Bangladesh, 551c; Basic Jewish Beliefs, 147c; China's Early Dynasties, 570–71c; China's Projected Urban Population, 579g; Chinese Inventions, 522c; Climate Graph for Nice, France, 59; Per Capita GDP in Island Southeast Asia and United States, 638c; Developed and Developing Countries, 103c; Developed and Less Developed Countries, 700c; Development of Writing, 128c; Earth Facts, 698; Economic Activity, 93c; Economic Powers, 469c; Egypt's Population, 2003, 362c; The Eightfold Path, 487c; Essential Elements and Geography Standards, 13c; Ethnic Groups in Australia and New Zealand, 649c; Ethnic Groups in China, 573c; Ethnic Groups in Indonesia, 2005, 641c; Facts about Countries: Africa, 272–75; Facts about Countries: South and East Asia and the Pacific, 466–68c; Field Notes, 418c; Geographical Extremes: Africa, 267c; Geographical Extremes: The Pacific World, 465c; Geographical Extremes: South and East Asia, 459c; Geographical Extremes: Southwest and Central Asia, 107c; India's History, 2300 BC–1947, 546–47c; Irish Migration to the United States, 1845–1855, 89c; Kinshasa's Growing Population, 425c; Largest Oil Reserves by Country, 113c; Life in Iran and the United States, 236c; Literacy Rates in Southwest Asia, 238c; Major Beliefs of Hinduism, 480c; Major Eruptions in the Ring of Fire, 42c; Major Oil Producers, 241c; Major Religions of Central Africa, 422c; Making Decisions, 344c; Origin of Israel's Jewish Population, 206c; Percentage of Students on High School Soccer Teams by Region, 9c; Population Density in East Africa, 404c; Population Giants, 469c; Population Growth in Japan, 606c; Population Growth in Indian Subcontinent, 563c; Reforms in Afghanistan, 254c; Religions of the Indian Subcontinent, 557c; Saudi Arabia's Exports, 229c; Saudi Arabia's Oil Production, 228c; Sources of Islamic Beliefs, 177c; Southwest and Central Asia: Facts

INDEX

INDEX

about Countries, 112–13c; Standard of Living in Central Asia, 256c; Tensions between China and Taiwan, 584c; Time Line: Beginnings of Islam, 172; Time Line: Major Events in the Fertile Crescent, 138c; Top Five Aluminum Producers, 2000, 77c; Tourism in Southern Africa, 450c; The Western Roman and Byzantine Empires, 164c; World Climate Regions, 56c; World Energy Production Today, 70c; World Languages, 701c; World Oil Reserves, 113c; World Population, 700c; World Population Growth, 90c; World Religions, 701c

Chiang Kai-shek, 572, 584

China, 466c; agriculture, 577p, 578; Beijing's National Day, 574–75f; Buddhism, 574; China's Early Dynasties, 570–71c; climate of, 568–69, 568m; communist China since Mao, 573; communist China under Mao, 573; culture, 575–76; Daoism, 574; economy of, 577–78; environmental challenges, 580m, 581; ethnic groups in, 573–74, 573c; flooding in, 568, 569p; food, 81; government, 92, 578–79; Great Wall, 571; human rights, 579; importance of family, 574; industry, 578; influence on Japan and Koreas, 597; language, 573–74; martial arts, 576; natural resources of, 568–69; physical geography of, 566–68, 567m; political map of, 564–65m; population of, 572m, 573; precipitation, 568–69, 568m, 589m; religion, 574; results of economic growth, 578; rural China, 579; Taiwan and, 572, 584; Tensions between China and Taiwan, 584c; urban China, 579–80, 579g, 579p

ancient China, 1600 BC–AD 1450: artists and poets, 521; Chinese Dynasties, 589–1279, 519m; city life, 520; civil service, 526, 534; Confucianism's influence, 525–26; Early Dynasties of China, 507m; Empress Wu, 519; Five Dynasties and Ten Kingdoms, 519; Forbidden City, 532–33f; Grand Canal, 518, 520–21p, 520m; Great Wall, 509, 533, 571c; gunpowder, 522, 522c; Han dynasty, 510–15; inventions, 522–23, 522f; isolationism, 534, 571; map of, 504–5m; Ming

dynasty, 531–34; Mongol Empire, 528–30, 529m; porcelain, 521, 522c; Qin dynasty, 509; Shang dynasty, 507; Shi Huangdi, 508f, 509; silk, 521; The Silk Road, 516–17f; Song dynasty, 519; Sui dynasty, 518; Tang dynasty, 519; trade, 520–21; woodblock printing, 522, 522f; writing system, 507; Yuan dynasty, 529–30; Zhou dynasty, 508

Christianity, 143p; acceptance of, 159; Aksum, 319; Apostles, 156; Byzantine Empire, 163–64; Central Africa, 422, 422c; China, 574; Christmas, 154f; Constantine, 159; defined, 152; denomination, 155; East Africa, 398, 401; Easter, 154f; Eastern Orthodox Church, 164; Egypt, 357; Ethiopia, 406; Gospels, 156; growth of the early church, 158–59; Jesus of Nazareth, 152–53, 152p; Jesus's acts and teachings, 154–55; The Jewish and Christian Worlds, 2000 BC–AD 1453, 142–43m; Judea, 153m; Koreas, 600; origins of, 152–59; Pacific Islands, 656; parables, 154–55; Paul of Tarsus, 156–57, 157m; persecution, 158; pope, 159; Resurrection, 153; Sermon on the Mount, 155f; Southeast Asia, 627; Southern Africa, 444; spread of, 158–59, 158m; West Africa, 378

cities: Africa's Largest Cities, 383c; Alexandria, 364; Algiers, 366p; ancient China, 520; Babylon, 132; Baghdad, 233; Bangkok, 630, 631f; Beijing, 580, 581p; Benghazi, 365; Cairo, 363; Cape Town, 448, 448–49p; Casablanca, 365; Chang'an, 520; China, 579–80; city-state, 122–23; Constantinople, 162–63f; Dar es Salaam, 404; Delhi, 554m, 554p; Dodoma, 404; Dubai, 226p; The Forbidden City, 532–33f; Gao, 386; Great Zimbabwe, 441, 441p; growth of in Muslim world, 182; Hanoi, 633; Harappan civilization, 475; Ho Chi Minh City, 633; Hong Kong, 580; Istanbul, 202–3p; Jakarta, 637; Jerusalem, 205p; Kaifeng, 520; Khartoum, 405; Kinshasa, 425, 425p; Kuala Lumpur, 635; Malaysia, 635; Manila, 638–39; Mesopotamia, 119; Mohenjo Daro, 474–75f; Myanmar, 630; Nairobi, 404; Osaka, 606; Ouagadougou, 386;

Philippines, 637; Phnom Penh, 633; Pyongyang, 611, 611p; Seoul, 609; Shanghai, 580; Singapore, 636, 636p; Taipei, 584, 585p; Taiwan, 584–85, 585p; Tangier, 365; Timbuktu, 386; Tokyo, 604–5f; Tunis, 365; Ulaanbaatar, 583

civilization: development of, in ancient Egypt, 280–81; early civilization in Nubia, 311; rise of, in Fertile Crescent, 117; rivers support growth of, 116–17; River Valley Civilizations, 120–21f; Sumer, 122–31

Cleopatra, 355

climate, 48p, 75p; Antarctica, 659; of Arabia, 170; of Arabian Peninsula, Iraq, and Iran, 222, 222m; Australia, 646; Average Annual Precipitation by Climate Region, 74c; Central Africa, 416–17; Central Asia, 246; China, Mongolia, and Taiwan, 568–69, 568m; defined, 50; dry, 59; East Africa, 396–97; Eastern Mediterranean, 198, 199m; highland, 60p, 61; Indian Subcontinent, 545; Japan and Koreas, 594; large bodies of water, 53; limits on life and, 62; Mediterranean Climate, 59c; mountains and, 54, 54p; New Zealand, 646; North Africa, 353; ocean currents, 52, 52m; overview of major climate zones, 55, 56–57c, 56–57m; Pacific Islands, 654; polar, 61; rain shadow, 54, 54p; Southeast Asia, 620–21, 620m; Southern Africa, 438; sun and location, 51; temperate, 59–60; tropical, 58; vs. weather, 50; West Africa, 374, 374m; wind and, 51–52; zonal, 374

climate maps: Arabian Peninsula, Iraq, and Iran, 222m; Australia and New Zealand, 665m; Eastern Mediterranean, 199m; Madagascar, 455m; South and East Asia, 462m; Southeast Asia, 620m; Southwest and Central Asia, 111m; West Africa, H9, 374m; World Climate Regions, 56–57m

colonization: The Atlantic Slave Trade, 380–81f; of Central Africa, 421; of East Africa, 399–400; India as British colony, 548; of Southeast Asia, 625, 626m; of West Africa, 377

colony, 548

Colorado: mining industry, 40

Colorado River: erosion, 40, 40p

INDEX

command economy, 94, 577; in North Korea, 610

communism, 92; Cambodia, 626; China, 572–73; Laos, 633; Mongolia, 583; North Korea, 598, 601, 610; Vietnam, 626, 633

Comoros, 272c, 450. *See also* Southern Africa

compass, 522

compass rose, H7

condensation, 32–33f, 33

Confucianism: Buddhism's popularity and, 525; in China today, 574; Confucian ideas, 525; Han dynasty, 511; importance of family, 513; influence of, 525; Koreas, 600; Neo-Confucianism during Song dynasty, 526; Period of Disunion, 525, 524–25p; social class and, 512

Congo Basin, 415, 415m

Congo, Democratic Republic of the, 272c, 424–25, 425p. *See also* Central Africa; cities, 425, 425p; copper belt, 417; independence, 425; language in, 422; Mobutu's rule, 424–25; natural resources of, 425; physical geography of, 414–15, 415m

Congo, Republic of the, 272c

Congo River, 413p, 415, 415m

conic projection, H5

coniferous trees, 60

Constantine, 159

Constantinople, 161m, 201. *See also* Istanbul; captured by Mehmed, 182; emperors rule from, 160–61; empire after Justinian, 161; glory of, 162–63f; Justinian, 160–61

constitutional monarchy: Japan as, 602

continental drift, 36–38

continental plates, 36, 36p

continents, H3, 36; facts about, 699; movement of, 36–37

Cook, James, 647, 655

copper belt, 417

coral reef, 645; formation of an atoll, 655f

Córdoba, Spain, 181p, 182

Côte d'Ivoire, 272c, 385. *See also* West Africa

Critical Thinking Skills: Analyzing Primary and Secondary Sources, 304; Analyzing Visuals, 628; Identifying Bias, 320; Making Decisions, 344; Making Economic Choices, 536; Sequencing and Using Time Lines, 138

crucifixion, 153

Crusades, 204

Cuba: economy of, 94; government of, 92

cultural diffusion, 84m, 85

cultural diversity, 83

cultural traits, 81

culture, 78p, 80–85; Australian Sports, 648f; The Berbers, 358f; Central Africa, 422–23; Central Asia, 246, 250–52; China, 575–76; Chinese Martial Arts, 576; Christian holidays, 154; cultural diffusion, 84m, 85; cultural diversity, 83; cultural traits, 81; culture groups, 82–83; culture regions, 82; defined, 80; development of, 81–82; East Africa, 400–401; globalization and popular culture, 97–98; how culture changes, 84; India, 550; innovation and, 84; Iran, 235–36; Iraq, 232–33; Islamic achievements in, 186–89; Japan, 599–600; Koreas, 600–601; landforms and, 40; Maori Culture, 646–47f; midnight sun, 29, 29p; Mongolia, 583; Music of South Africa, 443; North Africa, 357–59; Pacific Islands, 656; popular culture, 98; The Sacred Ganges, 482; Saudi Arabia, 224–25; Southeast Asia, 626–27; Southern Africa, 443–45; The Swahili, 400f; Taiwan, 584; Thai Teenage Buddhist Monks, 627; Tuareg of Sahara, 58; Turkey, 203; Turkmen Carpets, 255p; as a way of life, 80–82; West Africa, 378–79

culture groups, 82–83

culture region, 82; Arab Culture Region, 82, 82m, 82p; Japan, 82; Mexico, 82

cuneiform, 115p, 127–28, 127p, 128p

customs. *See also* traditions: Japan, 600; Koreas, 601; North Africa, 358; Southeast Asia, 627; Southern Africa, 443–45

cylindrical projection, H4

Cyprus, 112c

Cyrillic alphabet, 251

daily life: India, 552–53; Japan, 604–6, 605f; North Korea, 611; South Korea, 609

Dakar, 371p

Dalai Lama, 579

Dalits, 550

Damascus, 181, 210

Daoism: in China today, 574

Dardanelles, 196

Dar es Salaam, 404

Darfur, 405

Dead Sea, 197, 197p

death rate: population changes and, 88–89; Russia, 88

Deccan Plateau, 543m, 544

deforestation, 69; China, 581

Delhi, 186p, 548, 554m, 554p

delta, 40, 279; inland delta, 373; of Nile River, 279, 351

demilitarized zone, 609, 609m

democracy, 91–92; South Korea, 598, 608, 610

Democratic Republic of the Congo. *See* Congo, Democratic Republic of the

demography. *See* population

Deng Xiaoping, 573

desert climate, 59; average annual precipitation, 74c; East Africa, 397; North Africa, 353; overview of, 56c; West Africa, 374

desertification, 65, 374

deserts: Arabian Peninsula, Iraq, Iran, 220–22, 222m; Gobi, 567, 567m; Kalahari, 438; Namib, 438, 438p; Sahara, 332–33f, 349p, 350, 352–53f, 374, 376p, 377, 386; Southern Africa, 438, 438f

developed countries, 94c, 95, 103c, 700c

Dhaka, 558

dharma, 482

dialect, 422

Diaspora, 146, 204

dictatorship, 92

Diwali, 553, 553p

Djenné, 338

Djibouti, 272c; ethnic groups, 407; French control of, 407; language, 407; Muslims, 407

domino theory, 626

Drakensberg, 436, 437p

Druze Muslims, 211

dry farming, 365p

dryland farming, 255

Du Fu, 521

Dutch East Indies, 625

dynasty: defined, 281, 571; Egyptian, 281–83; Han, 510–15, 571c; Kushite, 313–14; Ming, 531–34; Qin, 509, 571, 571c; Qing, 571; Shang, 507, 570c; Song, 519; Sui, 518; Tang, 519; Third Dynasty, 283; Yuan, 529–30; Zhou, 508, 570c

Earth: axis, 26–27; facts about, 698; mapping, H2–H3; movement of, 26–27; revolution of, 27; rotation of, 26–27; solar energy and movement of, 26–27; structure of, 698; tilt and latitude, 27; water supply of, 30–32

earthquakes: defined, 38; Japan and the Koreas, 594, 594m; plate tectonics and, 38, 38p; Ring of Fire, 38, 42–43f; tsunami, 622–23f

Earth's plates, 36–38, 36m

Earth's surface: erosion, 39–40; forces below, 36–38; forces on, 39–40; landforms, 35; plate tectonics, 36–38, 36p; weathering, 39

East Africa: climate and vegetation, 396–97; colonization, 399–400; culture, 400–401; Djibouti, 407; Eritrea, 406; Ethiopia, 406; European influence and conflict, 399–400; history, 398–400; Kenya, 403–4; language, 400, 401m; physical geography, 394–96, 395m; political map of, 392–93m; population, 404, 404c, 404m; religion, 401; Rwanda and Burundi, 405; Somalia, 407; The Swahili, 400f; Tanzania, 403–4

Easter Island, 643p

Eastern Ghats, 543, 543m

Eastern Hemisphere: defined, H3; map of, H3

Eastern Mediterranean: agriculture, 198–99; climate and vegetation, 198, 199m; Israel, 204–8; Jordan, 212–13; Lebanon, 211–12; natural resources of, 198–99; physical geography, 196–98, 197m; political map of, 194–95m; Syria, 210–11; Turkey, 200–203

Eastern Orthodox Church, 164

East India Company, 547c, 548

East Timor, 466c, 637. *See also* Southeast Asia

Ebadi, Shirin, 237

economic activity, 93–94, 93p

economic geography, 19

economic indicators, 95

economics, 93–95; command economy, 94, 577; China 572; developed and developing countries, 94c, 95; economic activity, 93–94, 93p; economic indicators, 95; Economic Powers, 469c; Economics Handbook, 693–97; Egypt, 362–63; global economy, 98–99f; India, 554–55; Iran, 236–

37; Iraq, 233; market economy, 94; natural resources and wealth, 72; Saudi Arabia, 225–26; silent barter, 328; South Africa, 447–48; South Korea, 608, 609; systems of, 94; traditional economy, 94

economics connection: Bollywood, 555f; The Paper Trail, 523f; Tourism in Southern Africa, 450f

ecosystems, 63, 63f

education: Mali empire, 335, 337; Songhai empire, 338; Sumerian society, 124–26

Egypt, 273c. *See also* North Africa; agriculture, 362–63, 364, 364m; art and literature, 358–59; cities, 363–64; economy of, 362–63; ethnic groups, 357; European influence and independence, 357; Exodus, 145; government, 361–62; history of, 354–57; language, 357; natural resources of, 353, 362–63; physical geography of, 350–53, 351m; population, 361, 362c, 362m; poverty, 362; role of Islam in, 361–62; society, 361–62; Suez Canal, 363; today, 361–64; tourism, 362

ancient Egypt: achievements of, 298–303; art of, 302–3; burial practices, 286–87, 287p; civilization develops in, 280–81; Egyptian Trade, c. 1400 BC, 293m; Egyptian writing, 298–99, 299p; emphasis on afterlife, 286–87; family life in, 296; farming in, 280, 280p; floods of Nile, 279; gods of, 285, 285p; invasions of, 293; kings unify Egypt, 281–82; King Tutankhamen, 356f; King Tut's tomb, 302–3p, 303; Kush's rule of, 313–14; Lower Egypt, 278, 280–81; map of, 276–77m, 279m; Menes, 281–82; Middle Kingdom, 291–92; New Kingdom, 292–96; Old Kingdom, 283–90; Periods of Egyptian History Time Line, 291; pharaohs of, 283–84, 290, 292; physical geography of, 278–79; pyramids, 288–90; religion, 285–87; social order in, 284, 294–96; temples of, 300, 301p; trade and, 284, 292–93, 293m; Upper Egypt, 278, 280–81

Eightfold Path, 486, 487c

Elburz Mountains, 221, 221p

electricity: from coal, 70; hydroelectric energy, 70

elephants, 419p; ivory trade, 420–21, 421p

embargo, 231

empires: Akkadian, 123–24; defined, 123; Fertile Crescent, 114p; Ghana, 326–31, 377; Gupta, 492–93, 546c, 547; Khmer, 624; Mali, 334–37, 377; Mauryan, 490–91, 546c, 547; Mongol, 528–30, 529m, 582; Mughal, 184–85, 185m, 547c, 548; Muslim, 180–85; in New Kingdom of ancient Egypt, 292–93; Ottoman, 182, 183m, 201–2; Persian, 234; Safavid, 183–84, 184m; Songhai, 337–38, 377

energy: from fossil fuels, 69–70; geothermal, 403; nonrenewable energy resources, 69–70; nuclear, 71; renewable energy resources, 70–71; solar, 27, 71; from water, 34, 70; from wind, 70; World Energy Production Today, 70c

engineering: Harappan civilization, 475; pyramids, 288; roads and canals in Qin dynasty, 509

environmental issues: Antarctica, 661; Central Africa, 429; China and challenges of, 580m, 581; environmental issues of Central Asia, 256–57; Pacific Islands, 657; Southern Africa, 451

Environment and Society, 13–14, 13c

environments, 12, 49p, 75p; changes to, 64; defined, 62; development of culture and, 81–82; Earth's changing environments, 66–67f; ecosystems, 63, 63f; limits on life, 62; soil and, 64–65

equator: climate at, 58; defined, H2; prevailing wind, 51m, 52, 52m

Equatorial Guinea, 273c, 426

Eratosthenes, 18

Eritrea, 273c, 406

erosion, 39–40, 39p, 40p; glaciers, 39; water, 40, 40p; wind, 39, 39p

Esfahan, 184

Esma'il, 183–84

essential elements of geography, H12–H13, 13–14

Ethiopia, 273c; agriculture, 406; Christianity and, 398, 401, 406; economy of, 406; food/eating habits, 81; Lalibela, 398, 399p;

Ethiopian Highlands, 395, 395p, 396

ethnic groups, 83; Australia, 648–49, 649c; Burundi, 405; Central Africa, 421, 422; Central Asia, 251, 252p; China, 573–74, 573c; Djibouti, 407; Indian Subcontinent, 556; Iran, 235; Iraq, 232; New Zealand, 648–49, 649c; Nigeria, 382; North Africa, 357;

Pacific Islands, 656; Rwanda, 405; Somalia, 407; Southeast Asia, 626; Southern Africa, 443–44; Sri Lanka, 559; Sudan, 405; Syria, 211; Turkey, 202; West Africa, 378

Euphrates River, 197, 220, 221p; Fertile Crescent and, 117; transportation, 120p; urban civilization and, 40

Eurasian plate, 36p, 37

Europe: climate zones of, 57m; energy production, 70c; food/eating habits, 81; physical map of, 714m; political map of, 715m

Exodus, 145, 149; Possible Routes of the Exodus, 151m

exotic rivers, 220

Ezana, King, 319

family: Confucianism and, 513; family life in ancient Egypt, 296; importance in China, 574; importance in Han dynasty, 513; West Africa and extended family, 379

famine, 385

farming. *See* agriculture

Fergana Valley, 245

Fertile Crescent, 117; appearance of cities, 119; Assyrian Empire, 134–35, 134m; Babylonian Empire, 132–33; Chaldeans, 135; city-state of Ur, 124–25f; The Fertile Crescent, 7000–500 BC, 115m; food surpluses, 118; geography of, 116–19, 117m; Hittites, 134; irrigation and controlling water, 118; Kassites, 134; as land between two rivers, 117; map of, 115m, 117m; Phoenicians, 136–37, 136–37m; rise of civilization, 117; rivers support growth of civilization, 116–17; Sargon's Empire, c. 2330 BC, 123m; Sumer, 122–31; Time Line: Major Events in the Fertile Crescent, 138c

Fès, 357, 359p

fireworks, 505p, 522, 522c

fishing: Japan, 595; Sri Lanka, 559; Namibia, 449; Pacific Islands, 657; Southeast Asia, 621; West Africa, 385, 385p

Five Pillars of Islam, 176, 176p

five themes of geography, H12–H13, 10–12, 11f

fjords: New Zealand, 645

flat-plane projection, H5

floods/flooding, 33, 53p; Bangladesh, 558; China, 568, 569f; floodplain, 40; Huang He, 568; Nile River, 279, 350–51; North Africa, 350–51, 353; Southeast Asia, 620

Forbidden City, 532–33f

forests: deforestation, 69; ecosystem, 63, 63f; Mapping Central Africa's Forests, 418–19f; reforestation, 69; Southeast Asia, 620–21; tropical forests of Central Africa, 416–17

Formosa, 584

fossil fuels, 69–70

fossil water, 223

Four Noble Truths, 486

France: colonial rule in Southeast Asia, 625, 626m; colonization in West Africa, 377; colonization of Central Africa, 421; colonization of Pacific Islands, 655; control of Djibouti, 407; control of Lebanon, 211; control of Syria, 210; Mediterranean climate, 59f, 59c; rate of natural increase, 90

free port, 636

freshwater, 31, 31p

Fuji, 591p, 593

Fulani, 378

Gabon, 273c. *See also* Central Africa; economy of, 427; physical geography of, 414–15, 415m

Gambia, 273c, 383–84. *See also* West Africa

Gandhi, Mohandas, 482, 548f, 549

Ganges River, 471p, 543–44, 543m; Aryans, 476; The Sacred Ganges, 482f

Gangetic Plain, 543m, 544; climate, 545

Gao, 337, 386

Gaya, 485, 486p

Gaza, 207, 207m

Genghis Khan, 528–29, 582

Geographic Dictionary, H10–H11

geography: defined, 4; economic, 19; essential elements, 13–14, 13f; five themes of, 10–12, 11f; global level, 7, 7p; human, 3, 17p, 18, 21; human-environment interaction theme, 10–12, 11f; Islamic achievements in, 187; landscape, 4; local level, 6, 6p; location theme, 10–12, 11f; movement theme, 10–12, 11f;

physical, 3, 16–17, 16p, 21; place theme, 10–12, 11f; regional level, 6–7, 7p; region theme, 10–12, 11f; as science, 5; as social science, 5; studying, 4–9; urban, 19

Geography and History: The Aral Sea, 258–59f; The Atlantic Slave Trade, 380–81f; Crossing the Sahara, 332–33f; Earth's Changing Environments, 66–67f; The Hajj, 178–79f; River Valley Civilizations, 120–21f; Settling the Pacific, 650–51f; The Silk Road, 516–17f

Geography and Map Skills Handbook: Geographic Dictionary, H10–H11; Geography Themes and Elements, H12–H13; Map Essentials, H6–H7; Mapmaking, H4–H5; Mapping the Earth, H2–H3; Working with Maps, H8–H9

Geography Skills: Analyzing a Cartogram, 209; Analyzing a Precipitation Map, 388; Analyzing Satellite Images, 15; Comparing Maps, 500; Doing Fieldwork and Using Questionnaires, 408; Interpreting a Population Pyramid, 430; Interpreting a Route Map, 151; Using a Physical Map, 44; Using a Topographic Map, 596; Using Scale, 260

geothermal energy: Kenya, 403

Germany: colonization in West Africa, 377; colonization of Central Africa, 421; as developed country, 95

gers, 583, 583p

Ghana, 273c, 385. *See also* West Africa; agriculture, 375

Ghana Empire, 326–31; agriculture, 326–27; beginnings of, 326–27; Crossing the Sahara, 332–33f; *Dausi,* 341; decline of, 330–31; expansion of, 330; Ghana Empire, c. 1050, 327m; gold and salt trade, 327–28, 328p, 329; growth of trade, 328–29; in history of West Africa, 377; invasion of, 330; iron tools and weapons, 327; military, 328, 329, 330; overgrazing, 330, 331p; taxes and, 329

Gilgamesh, 123, 128

Giza, 277p, 288–89p, 354

glaciers, 31; erosion by, 39

globalization, 79p, 97–100; defined, 97; global economy, 98–99f; global trade, 99; interdependence and, 99; popular culture and, 97–98; world community, 99–100

global warming: Pacific Islands and, 657

global wind system, 51–52, 51m

globe, H2; as geographer's tool, 8–9

Gobi, 567, 567m

gods: of ancient Egypt, 285, 285p; Kush, 317; polytheism, 285

gold: Ghana empire and trade, 327–28, 328p, 329, 332f

Gospels, 156

government: China, 578–79; city-state, 122–23; command economy, 577; communism, 92; democracy, 91–92; dictatorship, 92; imperialism, 399–400; India, 554–55; Iran, 236–37; Iraq, 233; Israel, 205–6; Japan, 602; monarchy, 92; North Korea, 598, 610; Saudi Arabia, 225–26; secular state, 203; South Africa, 447–48; South Korea, 598, 608; theocracy, 236; types of, 91–92; world governments, 92m;

Grand Canal, 518, 520–21p, 520m

Grand Canyon: erosion and, 40

Grand Mosque, 178f

graphs. *See* charts and graphs

grasslands, 60

Great Barrier Reef, 645

Great Britain: colonial rule in Southeast Asia, 625, 626m; colonization in West Africa, 377; colonization of Australia, 647; colonization of East Africa, 399; colonization of New Zealand, 647; colonization of Pacific Islands, 655; colony in Cape of Good Hope, 442; control of India, 547c, 548–49; control of Iraq, 230; control of Jordan, 212

Great Indian Desert, 544

Great Lakes, 31; temperature in Michigan, 53

Great Plains, 60

Great Pyramid of Khufu, 288–89p, 354–55

Great Rift Valley, 394, 396f

Great Salt Lake, Utah, 31

Great Sphinx, 354p

Great Wall, 565p, 571; in Ming dynasty, 533; Qin dynasty, 509

Great Zimbabwe, 441, 441p

Greeks: in history of North Africa, 355, 357

Greenland: glaciers, 39

griots, 325p, 340–41, 384

gross domestic product (GDP), 95

Guam, 655

Guinea, 273c, 384. *See also* West Africa

Guinea-Bissau, 273c, 384. *See also* West Africa

Gulf of Guinea, 372, 373m

Gulf of Oman, 220, 221m

Gulf Stream, 52, 52m

gunpowder, 522, 522c

Gupta Empire, 546c, 547; art, 493p; caste system, 492–93; Gupta Empire, c. 400, 492m, 500m; paintings and sculpture, 496; women, 492–93

Hagia Sophia, Constantinople, 143p, 163p, 164, 182

hajj, 176, 178–79f

Hammurabi, 132–33

Hammurabi's Code, 133f

Han dynasty, 571c; art and literature, 514, 515p; Confucianism, 511; government, 511; Han Dynasty, c. 206 BC–AD 220, 511m; inventions and advances, 514–15; Liu Bang, 510–11; lives of rich and poor, 512; mandate of heaven, 510; rise of, 510–11; social classes, 512; taxes, 511; time line of, 510; Wudi, 511

Hanging Gardens, 135

Hanoi, 628p, 633

Hanukkah, 149

Harappa, 473, 473m, 474–76

Harappan civilization, 472–76; agriculture, 473; artistic achievements of, 476, 476p; cities of, 475; contact with other cultures, 473; life in Mohenjo Daro, 474–75f; map of, 473m; trade, 473; writing system, 474

Hatshepsut, Queen, 292f, 293

Hausa, 378

Hawaiian Islands, 653. *See also* Pacific Islands; as high island, 654, 654p; latitude of, 27

Hebrew Bible, 148, 149p

Hebrews, 204. *See also* Judaism; in Canaan and Egypt, 144–45

Hebron, 207, 207m

hegira, 173

Herodotus, 278

hieroglyphics: Egyptian, 298–99, 299p, 355

High Holy Days, 150

highland climate, 61; overview of, 56–57m, 57c

highlands: East Africa, 395; Ethiopian Highlands, 395, 395p

Himalayas, 543, 543m, 543p, 566, 567m, 567p; climate, 545; plate tectonics and, 37, 38p

Hinduism, 471p; Brahma, 480p, 481; Brahman, 481; caste system and, 481–82; in Central Africa, 422; in Gupta Empire, 492; in India, 550; Indian Subcontinent, 556, 557m; karma, 482; Major Beliefs of Hinduism, 480c, 481; Mughal Empire and, 185; origins of, 478–83; partition of India, 549; reincarnation, 481, 482; The Sacred Ganges, 482f; Siva, 481, 481p; Southeast Asia, 626; temples, 495, 496p; Vishnu, 481, 481p; women and, 482

Hindu Kush, 244, 245m, 542, 543m

Hirohito, 598f

Hiroshima, 598

history: The Aral Sea, 258–59f; The Atlantic Slave Trade, 380–81f; Australia, 647; Bantu Languages, 423; Central Africa, 420–21; Central Asia, 248–49; China, 570–73; Crossing the Sahara, 332–33f; development of culture and, 81; early history of Judaism, 144–46; Earth's Changing Environments, 66–67f; East Africa, 398–400; hajj, 178–79f; India, 546–48; Iran, 234–35; Iraq, 230–31; of Israel, 204–5; Japan, 597–98; Jordan, 212; Koreas, 597–98; Lebanon, 211; Mongolia, 582; New Zealand, 647; North Africa, 355–57; Pacific Islands, 655; preserving history of West Africa, 340–42; River Valley Civilizations, 120–21f; Settling the Pacific, 650–51f; The Silk Road, 516–17f; Southeast Asia, 624–26; Southern Africa, 440–43; Syria, 210–11; Taiwan, 584; Turkey, 200–202; West Africa, 376–77

history connection: Bantu Languages, 423

Hittites, 134, 293

HIV/AIDS, 429, 451

Ho Chi Minh City, 633

Hokkaido, 592, 593m

holy days: Christian, 154f; of Judaism, 149–50

Holy Land, 204

Hong Kong, 580

Hong River, 633

Honshu, 592, 593m

Huang He Valley, 506; Shang dynasty, 507

Huang He (Yellow River), 568

INDEX

human-environment interaction theme, 10–12, 11f
human geography, 16–17, 18, 21; defined, 3, 18; facts about, 700–701; uses of, 18
humanitarian aid, 100
Human Systems, 13–14, 13c
humid continental climate, 56c, 60
humid subtropical climate, 56c, 60
humid tropical climate, 56c, 58
Huns, 493
hurricanes, 53–54
Hussein, King, 212
Hutu, 405
hydroelectric power, 70; Koreas, 595
hydrology, 20
Hyksos, 292

Ibn Battutah, 187, 193, 342
ice: erosion by glaciers, 39; as weathering, 39
icebergs, 659
ice cap climate, 61; overview of, 57c
Iceland: plate tectonics and, 37, 38p
ice shelf, 659, 661p
Igbo, 378, 382
immigrants: new cultural traits, 81
imperialism, 399–400
Inanna, 124
India, 466c; agriculture, 555; ancient civilizations and early empires, 546c, 546, 547; Bollywood, 555f; British control of, 547c, 548–49; Buddhism, 550; caste system, 550; cities, 552, 554, 554p; climate and resources of, 545; culture, 550; daily life, 552–53; economy of, 554–55; festivals, 553, 553p; foreign control of, 548; Gandhi, Mohandas, 548f, 549; government, 554–55; green revolution, 555; Hinduism, 550; independence, 549; industry, 555; monsoons, 29; Mughal Empire, 184–85, 185m, 547c, 548; partition of, 549, 549p; physical map of, H9, 500m; population of, 554, 554m; relations with Pakistan, 557; religion, 550, 556, 557m; rivers and plains of, 543–44; rural life in, 552; social structure, 550; today, 552–55; urban life in, 552
ancient India, 2300 BC–AD 500: Aryans, 476–80, 477m, 546c, 547; Brahmanism, 480; caste

system, 479, 487; Early Spread of Buddhism, 488m; The Eightfold Path, 487c; Gupta Empire, 492–93, 492m, 500m, 546c, 547; Harappan civilization, 472–76, 473m, 546c, 547; Jainism, 483; karma, 482; language, 547; map of, 470–71m; Mauryan Empire, 490–91, 491m, 546c, 547; origins of Buddhism, 484–89; origins of Hinduism, 478–83; reincarnation, 481, 482; The Sacred Ganges, 482f; Sanskrit literature, 497; scientific advances in, 498–99, 498–99p; sudras, 479; *varnas,* 478, 479p; Vedas, 476, 477, 480; women and Hinduism, 482
Indian plate, 36p, 37
Indian Subcontinent: agriculture, 545; Bangladesh, 558; Bhutan, 558–59; climate, 545; ethnic groups, 556; history and culture of India, 546–50; Sri Lanka, 559; natural resources, 545; Nepal, 558; Pakistan, 557; physical geography of, 542–45, 543m; political map of, 540–41m; Population Growth in Indian Subcontinent, 563c; precipitation map of, 544m; religion, 556, 557c, 557m; Sri Lanka, 559
Indonesia, 466c. *See also* Southeast Asia; agriculture, 621; cities, 637; communism, 626; Ethnic Groups in Indonesia, 2005, 641c; Krakatau, 42c; Muslims, 627; natural resources, 637; per capita GDP, 638c; physical geography of, 618–19, 619m; plants, 620; population, 637; religion in, 627; tourism, 637
Indus River, 472, 543m, 544
Indus River Valley, 121p, 543m, 544; Aryans, 476; Harappan civilization, 472–73
industry: Australia, 648; China, 578; India, 555; Indonesia, 637; Japan, 603; Mongolia, 583; South Korea, 608, 609; Taiwan, 585
infrared images, 15, 15p
Internet Activities, 22, 46, 76, 102, 140, 166, 192, 216, 240, 262, 306, 322, 346, 368, 390, 410, 432, 454, 502, 538, 562, 588, 614, 640, 664
Inyanga Mountains, 436
Iran, 112c, 234–37; climate of, 222, 222m; culture, 235–36; desert in, 220–22, 222m; economy of, 236–37; government of, 236–37; history, 234–35; invasion by Iraq,

231; Life in Iran and the United States, 236c; literacy rate in, 238c; natural resources of, 223; oil in, 113c, 236, 241c; people of, 235; Persian Empire, 234; political map of, 218–19m; Shah and Islamic Revolution, 235; today, 235–37; vegetation of, 222
Iran-Iraq War, 231
Iraq, 112c, 230–33; climate of, 222, 222m; culture, 232–33; desert in, 220–22, 222m; early history, 230–31; economy of, 233; ethnic groups, 232; government, 92, 233; Great Britain control of, 230; invasion of Iran, 231; invasion of Kuwait, 231; literacy rate in, 238c; Mesopotamia and Sumer, 231m; natural resources of, 223; oil in, 113c, 233, 241c; Persian Gulf War, 231; political map of, 218–19m; rebuilding Baghdad, 233; religion, 232–33; Saddam and regional conflicts, 231; today, 233; vegetation of, 222; war with United States, 231
iron: Ghana empire, 327; in Kush, 316
irrigation, 120f; in ancient Egypt, 280; Central Asia, 247; defined, 118; in Fertile Crescent, 118
Isis, 285, 285p
Islam, 168p. *See also* Muslims; beliefs of, 174; caliph, 180, 183; Central Africa, 422, 422c; China, 574; conflict of Shia and Sunni Muslims, 183; cultural achievements of, 186–89; defined, 172; East Africa, 398–99, 401; Egypt, 361–62; Five Pillars of Islam, 176, 176p; growth of cities, 182; hadith, 175; hajj, 176, 178–79f; hegira, 173; influence in Saudi Arabia, 224; Islamic law, 177; The Islamic World, AD 550–1650, 168–69m; jihad, 175; life and Muhammad's teachings, 172–73; literature and art achievements, 188–89; Mogadishu, 399; Mombasa, 399; mosque, 173; Mughal Empire, 184–85, 185m; Muslim armies conquer lands, 180–81; North Africa, 357; origins of Islam, 170–73; Ottoman Empire, 182, 183m; Qur'an, 172, 174–75, 175p; Ramadan, 176; Safavid Empire, 183–84, 184m; science and philosophy achievements, 186–87, 186–87p; Shariah, 177; Sources of Islamic Beliefs, 177c; Southeast Asia, 627; Southern

Africa, 444–45; spreads in Arabia, 173; Sunnah, 175–77; trade helps Islam spread, 181–82; West Africa, 378

Islamic Revolution, 235

islands. *See also* Pacific Islands: atoll, 654, 655f; defined, 35; high, 654, 654p; Island Southeast Asia, 619; Japan, 592, 593m; low, 654, 654p

Island Southeast Asia, 619; Brunei, 637; cities, 634–35; Current Per Capita GDP in Island Southeast Asia and United States, 638c; East Timor, 637; economy of, 634–35; Indonesia, 637; Malaysia, 635–38; Philippines, 638

isolationism: in ancient China, 571; in Ming dynasty, 534

Israel, 112c, 145–46, 204–8; agriculture of, 206; climate and vegetation of, 198, 199m; creation of, 205; Diaspora, 204; economy of, 205–6; Gaza, 207, 207m; government, 205–6; history of, 204–5; Holy Land, 204; Israeli Jewish culture, 206; language, 206; natural resources of, 198–99; Origin of Israel's Jewish Population, 206c; Palestinian territories, 207–8, 207m; physical geography of, 196–98, 197m; today, 205–6; West Bank, 207, 207m

Israelites, 145

Issa, 407

Istanbul, 182, 198f, 202–3p, 203. *See also* Constantinople

ivory, 312, 420–21, 421p, 442

Jainism, 483; in Gupta Empire, 492

Jakarta, 637

janissaries, 182

Japan, 466c, 615m; agriculture, 603, 606; art, 600; Buddhism, 489, 597, 599, 599p; cities, 604–6; climate, 594; culture, 599–600; culture region, 82; customs and traditions, 600; daily life in, 604–6, 605f; economy of, 603, 607; government, 602; history, 597–98; industry, 603; issues and challenges, 607; language, 599; natural disasters in, 594, 594p; natural resources, 595, 603; physical features of, 592–93, 593m; political map of, 590–91m; population, 604–6,

606m; technology in, 603f, 607p; today, 602–7; Tokyo, 604–5f; World War II, 598

Jerusalem, 145–46, 204, 205p; control of, 207

Jesus of Nazareth, 152–53, 153p; acts and teaching of, 154–55; birth of, 152–53; crucifixion of, 153; disciples of, 153; followers of, 156–57; message of, 155; miracles, 154; parables, 154–55; resurrection of, 153

jihad, 175

Johannesburg, 448

Jordan, 112c, 212–13, 212m; agriculture, 213; climate and vegetation of, 198, 199m; economy of, 212–13; history and government of, 212; natural resources of, 198–99; people of, 212–13, 213p; physical geography of, 196–98, 197m

Jordan River, 194p, 197

Judah, 146

Judaism, 142p; Abraham, 145, 144p; beginnings in Canaan and Egypt, 145; commentaries, 149, 149p; defined, 144; Diaspora, 146, 204; early history of, 144–46; Exodus, 145; Hanukkah, 149; Hebrew Bible, 148, 149p; Israeli Jewish culture, 206; The Jewish and Christian Worlds, 2000 BC–AD 1453, 142–43m; Jewish beliefs, 147, 147c; Jewish Migration after AD 70, 146m; Jewish texts, 148–49; Mosaic law, 147; Moses, 144p, 145; origins of, 144–50; Possible Routes of the Exodus, 151m; series of invasions, 145–46; Talmud, 149; Ten Commandments, 145, 147; Torah, 148, 148p; traditions and holy days, 149–50; Zionism, 205

K2, 541p, 543, 543m

Kaaba, 176

kabuki, 600

Kabul, 253

Kalahari, 438

kami, 599

kanji, 599

Kao-hsiung, 584

Kara-Kum, 246

karma, 482

Kashmir, 557

Kashta, 313

Kassites, 134

Kathmandu, 558

Kazakh, 251

Kazakhstan, 112c, 254. *See also* Central Asia; standard of living in, 256c

Kente cloth, 343, 371p

Kenya, 273c; agriculture, 403; cities, 404; climate of, 396; eating habits, 81p; economy of, 403; European settlers in, 400; Nairobi, 88c; population, 404, 404m; population density, 404c; reforestation, 69f; terrorism, 404; tourism, 403

Kerma, 311, 312, 313

Khafre: pyramid for, 288–89p

Khartoum, 383c, 396, 405

Khmer Empire, 624

Khoisan, 440, 444

kibbutz, 206

kimchi, 601

Kim Il Sung, 591p, 610

Kim Jong Il, 610

kimonos, 600, 601p

King Tut. *See* Tutankhamen, King

Kinshasa, 383c, 425, 425p

Kiribati, 466c

Kish, 123, 123m

klongs, 630

Knesset, 205

Kobe, 594p

Kolkata (Calcutta), 552

Komodo dragon, 620

Kongo Kingdom, 420

Kopet-Dag, 221

kora, 342

Korean Peninsula, 593

Korean War, 598

Koreas, 615m. *See also* North Korea; South Korea; Buddhism, 489, 597, 600; Christianity, 600; climate, 594; clothing, 600p; Confucianism, 600; culture, 600–601; customs and traditions, 601; The Demilitarized Zone, 609m; food, 601; history, 597–98; language, 600; natural disasters in, 594, 594p; natural resources, 595; physical features of, 593, 593m; political map of, 590–91m; religion, 600; reunification, 612; volcanoes and earthquakes in, 594, 594m

kosher, 206

Koumbi Saleh, 329, 330

Kouyaté, Soriba, 342

Krakatau, Indonesia, 42c

Kshatriyas, 478, 479p
Kuala Lumpur, 635
Kublai Khan, 529–30, 535f
kung fu, 576f
Kurds, 82, 202, 211, 235; in Iraq, 231, 232
Kush: agriculture, 319; ancient Kush, 308–9m, 311m; culture, 317; decline and defeat of, 319; early civilization in Nubia, 311; economy of, 315–16; Egypt's control of, 292, 312; iron industry, 316; Kush's Trade Network, 316f; land of Nubia, 310; language, 317; loss of resources, 319; metalwork, 315p, 316; physical geography of, 310–11; pyramids, 308p, 314, 317, 318p; rise of Aksum, 319; rulers of, 317, 318f; ruling Egypt, 313–14; trade, 316, 316m, 319; women in, 317
Kuwait, 112c, 226; invasion by Iraq, 231; oil in, 113c, 241c
Kyoto, 606
Kyoto Protocol, 607
Kyrgyz, 251, 252p
Kyrgyzstan, 112c, 255, 260m. *See also* Central Asia; standard of living in, 256c
Kyushu, 592, 593m
Kyzyl Kum, 246

Lagos, 383, 383p
Lake Balkhash, 245
Lake Nyasa, 414, 415m
Lake Tanganyika, 414, 415m
Lalibela, 398, 399p
landforms, 45p; Central Africa, 415; defined, 35; erosion of, 39–40; examples of, H10–H11; influence on life, 40–41; types of, 35; weathering of, 39
land use and resources maps: South and East Asia, 463m; Southeast Asia, 632m; West Africa, 384m
language: Afrikaans, 442; Aryans, 477, 547; Bantu, 422, 423; Central Africa, 422; Central Asia, 251, 251m; China, 573–74; as cultural trait, 81; dialect, 422; Djibouti, 407; East Africa, 400, 401m; English as global language, 98; Israel, 206; Japanese, 599; Koreas, 600; Kush, 317; and landforms, 40; North Africa, 357; Official Languages of East Africa, 401m; Sanskrit, 547; Senegal and Gambia,

384; Southeast Asia, 626; Southern Africa, 444; West Africa, 378; of world, 701c
Laos, 466c, 633. *See also* Southeast Asia; agriculture, 633; economics, 633; independence of, 625; modern history of, 625; physical geography of, 618–19, 619m
latitude: defined, H2, 27; solar energy and, 27
laws, 81; Islamic, 177; Jewish beliefs and, 147; Justinian, 160
Lebanon, 112c, 211–12, 212m; civil war, 212; climate and vegetation of, 198, 199m; economy of, 212; history, 211; natural resources of, 198–99; people of, 211, 213p; physical geography of, 196–98, 197m; religion, 211; today, 212
Lesotho, 273c, 448–49. *See also* Southern Africa
Liberia, 273c, 384. *See also* West Africa
Li Bo, 521
Libya, 273c. *See also* North Africa; agriculture, 364m, 365; cities, 365; European influence and independence, 357; natural resources of, 353; oil, 365; physical geography of, 350–53, 351m; today, 365
Lima, 88c
Limpopo River, 437
Li Qingzhao, 521
literature: *Aké: The Years of Childhood* (Soyinka), 387; *Antarctic Journal: Four Months at the Bottom of the World* (Dewey), 662; Han dynasty, 514; Islamic achievements in, 189; North Africa, 358–59; *River, The* (Gary Paulsen), 73; Sanskrit literature of ancient India, 497; *Shabanu: Daughter of the Wind* (Staples), 560; Sumerian society, 128; Vedas, 476, 477
Liu Bang, 510–11
location: absolute, 12; relative, 12
location theme, 10–12, 11f; defined, 12
locator map, H7
loess, 568
London, 6p, 7p
longitude: defined, H2
Luxor, 300

Maathai, Wangari, 69
Macau, 580

Madagascar, 273c. *See also* Southern Africa; climate, 438, 455m; deforestation, 451; economy of, 450; ethnic groups, 444; government, 450; natural resources, 438
Maghreb, 364
magma, 37
Mahayana Buddhists, 489
Mahfouz, Naguib, 359
Mainland Southeast Asia, 618, 629–33; Bangkok Canal, 631f; Cambodia, 633; cities in, 630; Laos, 633; Myanmar, 630; Thailand, 630, 632; Vietnam, 633
malaria, 428, 428m
Malawi, 17p, 273c. *See also* Central Africa; agriculture, 427; economy of, 427; language in, 422; physical geography of, 414–15, 415m; today, 427
Malay Peninsula, 618
Malaysia, 466c. *See also* Southeast Asia; agriculture, 621, 635; cities, 635; independence, 635; natural resources, 635; per capita GDP, 638c; religion in, 627; rubber tree plantations, 635p
Maldives, 466c; climate and resources of, 545; independence, 549; physical geography of, 542–44, 543m
Mali, 274c. *See also* West Africa; agriculture, 386, 386p; economy of, 386; rate of natural increase, 88–89
Mali Empire: agriculture, 334; education, 335, 337; in history of West Africa, 377; Mansa Musa, 335–37, 339; map of, 335m; Muslim influence, 335–37; Sundiata, 334–35; Timbuktu, 335, 336f, 337
mandate of heaven, 508, 510
Mandela, Nelson, 447f
Manila, 638
Mansa Musa, 335–37, 339f, 377
Maori, 646–47f, 647, 649
Mao Zedong, 572
map projections, H4–H5; conic projection, H5; cylindrical projection, H4; flat-plane projection, H5; Mercator projection, H4
maps. For a complete list of maps, see pp. xxv–xxvii. *See also* climate maps, land use and resources maps, Map Skills, physical maps, political maps, population maps.
Map Skills: Analyzing Satellite Images, 15; Comparing Maps, 500; How to Read a Map, H6–H7; Interpreting a Route Map, 151; interpreting maps, 52m, 57m, 87, 106m, 108m, 109m, 110m,

111m, 115m, 117m, 121m, 123m, 134m, 137m, 143m, 146m, 157m, 158m, 161m, 169m, 179m, 183m, 184m, 185m, 195m, 197m, 199m, 202m, 207m, 218m, 221m, 222m, 231m, 242m, 245m, 251m, 268m, 269m, 276m, 279m, 293m, 309m, 311m, 325m, 327m, 335m, 349m, 351m, 362m, 364m, 371m, 373m, 384m, 393m, 413m, 415m, 416m, 428m, 435m, 437m, 439m, 460m, 461m, 462m, 463m, 464m, 471m, 473m, 477m, 489m, 491m, 505m, 511m, 517m, 519m, 529m, 540m, 554m, 557m, 565m, 567m, 590m, 593m, 594m, 606m, 617m, 619m, 632m, 643m, 645m, 656m; Map Activity, 322; Analyzing a Cartogram, 209; Analyzing a Precipitation Map, 388; Understanding Map Projections, H4–H5; Using a Physical Map, 44; Using a Topographic Map, 596; Using Different Kinds of Maps, H8–H9; Using Latitude and Longitude, H2–H3; Using Scale, 260

marine west coast climate, 56c, 60

market economy, 94

Maronites, 211

Marshall Islands, 467c

math connection: Calculating Population Density, 88; Muslim contributions to math, 225

mathematics: ancient Indian achievements, 498, 499p; Hindu-Arabic numerals, 498; Islamic achievements in, 187; Muslim contributions to, 225f; Sumerian society, 129

Mauritania, 274c, 385. *See also* West Africa

Mauritius, 64, 274c

Mauryan Empire, 490–91, 546c, 547; Asoka, 491; Mauryan Empire, c. 320–185 BC, 491m; military, 490–91

Mawsynram, India, 58

Mecca, 171p, 172, 218p; hajj, 176, 178–79f

medicine: acupuncture, 515, 515p; ancient Indian achievements, 498–99; inoculation, 498; Islamic achievements in, 187; Sumerian society, 129

Medina, 173

meditation, 485

Mediterranean climate, 59, 59p; Average Annual Precipitation, 74c; overview of, 56c

Mediterranean Sea: Nile River, 350–51; Suez Canal, 351, 351m

Mehmed II, 182, 183f

Mekong River, 619, 619p

Melanesia, 653, 655. *See also* Pacific Islands

Memphis (Egypt), 120p, 281

Menes, 281, 282p

menorah, 149

Mercator projection, H4

merchants, 316; in ancient Egypt, 295

meridians: defined, H2

Meroë, 316, 317

Mesopotamia, 117m, 141m, 220; agriculture in, 117–19; Babylonian Empire, 132–33; city-state of Ur, 124–25f; 118p; division of labor, 118; as early history of Iraq, 230; farming and cities, 118–19, 118–19p; food surpluses, 118; invasions of, 134–35; Mesopotamia and Sumer, 231m; northern, 117; rise of civilization, 117; Sargon's Empire, c. 2330 BC, 123m; Sumer, 122–31

Messiah, 152

Mexico: culture region, 82; government in, 91

Micronesia, 467c, 653, 655. *See also* Pacific Islands

Mid-Atlantic Ridge, 37

Middle East. *See also* Eastern Mediterranean: energy production, 70c; term of, 196

Middle Kingdom, 291–92

migration, 89; Irish, 89, 89c, 89p

mihrab, 188p

military: in ancient Egypt, 295; Assyrian, 135, 135p; city-state, 122–23; Ghana empire, 327, 328, 329, 330; Hittites, 134; Israel, 205; janissaries, 182; Muslim armies conquer many lands, 180–81; Ottoman Empire, 182

minarets, 188p, 189

Mobutu Sese Seko, 424–25

Mogadishu, 399, 407

Mohenjo Daro, 473, 473m, 475; life in, 474–75f

moksha, 482

Mombasa, 399

monarchy, 92; constitutional, 602

money: paper money in Song dynasty, 521, 522c, 523f; Qin dynasty, 509

Mongol Ascendancy, 529

Mongol Empire, 528–30, 529m, 582; rule in China, 529–30

Mongolia, 467c; climate of, 568–69, 568m; culture, 583; government, 583; history of, 582; natural resources of, 569; nomadic life in, 583, 583p; physical geography of, 566–68, 567m; political map of, 564–65m; precipitation, 568–69, 568m, 589m; Soviet Union's influence over, 583

Mongols, 249

monotheism, 147

monsoons, 58; defined, 545; India, 29; Indian Subcontinent, 544p, 545; Southeast Asia, 620

Moors, 181

Morocco, 274c, 338. *See also* North Africa; agriculture, 364, 365, 365p; Berbers, 358f; cities in, 359p; European influence and independence, 357; music, 359; natural resources of, 353; physical geography of, 350–53, 351m; tourism, 365

Mosaic law, 147

mosaics, 163

Moses, 143p, 144p, 145, 147, 148

mosque, 173; Blue Mosque, Istanbul, 188f

mountains: Arabian Peninsula, Iraq and Iran, 221; of Central Asia, 244; China, Mongolia, and Taiwan, 566, 567m; climate, 54, 54p; defined, 35; East Africa, 395; in Eastern Mediterranean, 197–98; glaciers and, 39; Indian Subcontinent, 542–43, 543m; Japan, 593; Koreas, 593; North Africa, 352; plate tectonics and, 37, 38p; rain shadow, 54, 54p; Southeast Asia, 618; Southern Africa, 436, 437p

Mount Everest, 543, 543m, 566

Mount Kilimanjaro, 393p, 395, 403

Mount Pinatubo, 42c, 43

movement: of Earth and solar energy, 26–27; migration, 89; plate tectonics, 36–38

movement theme of geography, 10–12, 11f

Mozambique, 274c. *See also* Southern Africa; economy of, 450; independence, 443; language, 444; Tourism in Southern Africa, 450c

Mughal Empire, 184–85, 185m, 547c, 548

Muhammad: life and, 172; teachings of, 172–73

Muhammad Ture, 338

Mumbai (Bombay), 545, 552, 555f

mummies, 286–87, 287p
music: Central Africa, 423; Music from Mali to Memphis, 342f; Music of South Africa, 443f; North Africa, 358–59; Sumerian, 131; West Africa, 343
Muslim empires, 180–85; beginnings of, 181; growth of cities, 182; growth of empire, 181; literature and art achievements, 188–89; mix of cultures, 182; Mughal Empire, 184–85, 185m; Ottoman Empire, 182, 183m, 201–2; Safavid Empire, 183–84, 184m; science and philosophy achievements, 186–87, 186–87p
Muslims, 169p. *See also* Islam; contributions to math, 225f; defined, 172; Indian Subcontinent, 556, 557m; Indonesia, 627; invasion of Ghana empire, 330; in Israel, 205; Mali empire, 335–37; Mughal empire in India, 548; partition of India, 549; Shia, 183; Songhai empire, 337, 338; Sunni, 183
Myanmar, 467c. *See also* Southeast Asia; cities, 630; government, 630; natural resources, 630; physical geography of, 618–19, 619m

N

Nablus, 207, 207m
Nagasaki, 598
Nairobi, 88c, 404, 404p
Namib Desert, 438, 438p
Namibia, 274c. *See also* Southern Africa; art, 445; economics, 449; environmental protection, 451; government, 449; independence, 443; language, 444; natural resources, 439, 449; religion, 444; Tourism in Southern Africa, 450c
Nanna, 124
Napata, 313
natural environments. *See* environments
natural gas: as nonrenewable energy resources, 69–70; North Africa, 353; Pakistan, 545
natural resources, 49p, 68–72, 75p; Antarctica, 659; of Arabian Peninsula, Iraq, and Iran, 223; Australia, 646; Brunei, 637; Central Africa, 417; Central African Republic, 425–26; Central Asia, 246m, 247; China, Mongolia, and

Taiwan, 568–69; Côte d'Ivoire, 385; defined, 68; Democratic Republic of the Congo, 425; economic activity and, 93, 93p; Egypt, 362–63; energy resources, 69–71; Ghana, 385; Guinea, 384; Indian Subcontinent, 545; Indonesia, 637; Japan, 595, 603; Koreas, 595; Malaysia, 638; managing, 69; map: Africa, 269m; mineral resources, 71; Myanmar, 630; Namibia, 449; New Zealand, 646; Nigeria, 383; nonrenewable, 69–70; North Africa, 353, 362–63, 365; North Korea, 610; nuclear energy, 71; Pacific Islands, 654; people using in daily life, 71; Philippines, 638; renewable, 69, 70–71; Sierra Leone and Liberia, 384; Southeast Asia, 621; Southern Africa, 438–39; Southwest and Central Asia, 109m; Swaziland, 449; Tanzania, 403; Thailand, 630, 632; Top Five Aluminum Producers, 2000, 77c; types of, 69; wealth and, 72; West Africa, 375
Nauru, 653
Nazareth, 153
Nazca plate, 36, 36p, 37
Nebuchadnezzar, 135
Nefertiti, Queen, 294p
Negev, 198
Nepal, 3p, 467c, 558p; climate and resources of, 545; environmental challenges, 558; physical geography of, 542–44, 543m; religion, 556, 557m
Netherlands: colonial rule in Southeast Asia, 625, 626m; in Southern Africa, 441–42
New Caledonia, 654
New Guinea, 653. *See also* Pacific Islands; Southeast Asia; as high island, 654; language and landforms, 40; physical geography of, 618–19, 619m
New Kingdom (Egypt): building an empire, 292; growth and effects of trade, 292–93; invasions of, 292; work and daily life in, 294–96
New Zealand, 467c, 647; agriculture, 646, 648; as British colony, 647; cities, 649; climate, 646, 665m; early settlers of, 647; economy of, 648; ethnic groups, 648–49, 649c; government, 648; history, 647; Maori Culture, 646–47f, 647, 649; physical geography of, 644–45, 645m; population, 648–49; Settling the Pacific, 650–51f; wildlife and natural resources of, 646

Niger, 274c. *See also* West Africa; agriculture, 385; famine, 385
Nigeria, 178f, 274c. *See also* West Africa; as developing country, 95; economy of, 383; ethnic groups, 378, 382; government, 382; petroleum, 375, 383; rate of natural increase, 90
Niger River, 373, 373m, 373p; Ghana Empire, 326, 327; Mali Empire, 335
Nile River, 276p, 350–51, 351m, 351p, 396; cataracts, 279, 311p; delta of, 279; Egypt's agriculture and, 363, 363f; floods of, 279; gift of, 278–79; land of Nubia, 310; location and physical features of, 278–79; satellite image of, 363p
nirvana, 486
Noh, 600
nomads, 171, 250; Mongolia, 583, 583p
nonrenewable energy resources, 69–70
North Africa, 369m; agriculture, 362–63, 364m, 365, 365p; Arab culture region, 82, 82m, 83p; arts and literature, 358–59; cities of, 363–64, 365; climate and resources, 353; culture, 357–59; Egypt, 361–64; ethnic groups, 357; food, 357–58; government and economy, 361–62, 364–65; history of, 355–57; holidays and customs, 358; language, 357; natural resources, 362–63, 365; Nile River, 350–51; physical geography of, 350–53, 351m; political map of, 348–49m; religion, 357
North America: climate zones of, 56m; energy production, 70c; physical map of, 710m; political map of, 711m
North American plate, 36p, 37
North Atlantic Drift, 52, 52m
North China Plain, 88, 568, 573
Northern Hemisphere: defined, H3; map of, H3; seasons in, 28–29
North Korea, 467c. *See also* Koreas; agriculture, 611; cities, 611, 611p; communism, 598, 610; daily life in, 611; The Demilitarized Zone, 609, 609m; economy of, 94, 610–11; formation of, 598; government, 598, 610; issues and challenges of, 611; lack of technology, 611; nuclear weapons, 611; population, 611; Pyongyang, 611, 611p; religion, 611; resources of, 610; reunification, 612

INDEX

North Pole, 721m; climate, 61; prevailing winds, 52

note-taking skills, 4, 10, 16, 26, 30, 35, 50, 55, 62, 68, 86, 91, 97, 116, 122, 127, 132, 144, 152, 160, 170, 174, 180, 186, 196, 200, 210, 220, 224, 230, 234, 244, 248, 253, 278, 283, 291, 298, 310, 315, 326, 334, 340, 350, 354, 361, 372, 376, 382, 394, 398, 402, 414, 420, 424, 436, 440, 446, 472, 478, 484, 490, 495, 506, 510, 518, 524, 528, 542, 546, 552, 556, 566, 570, 577, 582, 592, 597, 602, 608, 618, 624, 629, 634, 644, 653, 658

Nubia, 284, 398; early civilization in, 311; land of, 310

nuclear energy, 71

nuclear weapons: North Korea, 611; tensions between Pakistan and India, 557

oasis, 170, 222, 352; A Sahara Oasis, 352–53f

ocean currents, 52, 52m; world's major, 52m

ocean plates, 36, 36p

oceans: erosion, 40; salt water, 31

ocean trenches: plate tectonics and, 37

oil. *See* petroleum

Okavango River, 434p, 437

Old Kingdom (Egypt), 283–90; burial practices, 286–87, 287p; early pharaohs of, 283–84; Egyptian society, 284, 284p; emphasis on afterlife, 286–87; First and Second Dynasties of, 281–83; gods of, 285, 285p; mummies, 286–87, 287p; pyramids, 288–90; religion, 285–87; social order, 284, 284p; Third Dynasty, 283

Oman, 112c, 227; literacy rate in, 238c

Omar Khayyám, 189

oral history, 340, 376; of West Africa, 340–42

Orange River, 437

Organization of Petroleum Exporting Countries (OPEC), 225, 229

Osaka, 606

Osiris, 285, 285p

Ottoman Empire, 182, 183m, 201–2, 210

Ottoman Turks, 161

Ouagadougou, 386

Outback, 648

overgrazing: in Ghana empire, 330, 331p

ozone layer, 660–61

Pacific Islands: agriculture, 654, 657; climate of, 654; colonization of, 655; culture, 656; economy of, 657; environmental concerns, 657; ethnic groups, 656; fishing, 657; high and low islands, 653–54, 654p, 655f; history, 655; island regions of, 653; natural resources, 654; physical geography, 653–54, 654p; political map of, 656m; population, 656; religion, 656; Settling the Pacific, 650–51f; today, 657; traditions, 656

Pacific Ocean: Panama Canal, 40; Ring of Fire, 42–43f; tsunami, 622–23f, 623m; typhoons, 53

Pacific plate, 36p, 37

Pacific Ring of Fire, 619

Pacific World: Antarctica, 658–62; Australia, 644–49; Australia and New Zealand: Climate, 665m; Australia and New Zealand: Physical, 645m; Ethnic Groups in Australia and New Zealand, 649c; facts about, 466–68c; Geographical Extremes: The Pacific World, 465c; Maori Culture, 646–47f, 647; New Zealand, 644–49; Pacific Islands, 653–57; The Pacific Islands: Political, 656m; physical map of, 645m; political map of, 642–43m, 720; regional atlas of, 464–65m; Settling the Pacific, 650–51f, 651m

pagodas, 575

Pakistan, 467c; as ally of U.S., 557; challenges and issues of, 557; climate and resources of, 545; partition of India and, 549, 549p; physical geography of, 542–44, 543m; relations with India, 557; religion, 556, 557m

Palau, 467c

Palestine: creation of Israel and, 205; Gaza, 207, 207m; Roman Empire and, 204; West Bank, 207, 207m

Palestinian Authority, 207

Pamirs, 244

Panama Canal, 40

Pangaea, 66f, 66m

paper: invention of, 514, 522, 522c; paper money, 521, 522c, 523; woodblock printing, 522, 522c

Papua New Guinea, 467c, 654, 657

papyrus, 298

parables, 154–55

parallels: defined, H2

parliament: Australia and New Zealand, 648

partition, 549

Pashto, 251

Passover, 149–50, 206

Paul of Tarsus, 156–57; journeys of, 157m

Pearl Harbor, 598

Peking, 580

peninsula, 35; Indochina Peninsula, 618; Korean, 593; Malay Peninsula, 618

per capita GDP, 95

periodic markets, 417

Period of Disunion, 518; Confucianism and, 525, 525p

permafrost, 61

Persepolis, 219p, 234

Persian Empire: as early history of Iran, 234

Persian Gulf, 220, 221m

Persian Gulf War, 231

petroleum: Angola, 427; Arabian Peninsula, 223, 225–27; Bahrain, 226; Brunei, 637; Central Asia, 247; Chad, 385; Equatorial Guinea, São Tomé and Príncipe, 426; Gabon, 427; Iran, 223, 236; Iraq, 223, 233; Kuwait, 226; Libya, 365; Major Oil Producers, 241c; Nigeria, 375, 383; as nonrenewable energy resources, 69–70; North Africa, 353; Oman and Yemen, 227; Pakistan, 545; in Qatar, 226; Republic of the Congo, 427; Saudi Arabia, 225, 228–29f; World Oil Reserves, 113c

pharaoh, 281; early pharaohs of Old Kingdom, 283–84; Middle Kingdom, 292; of New Kingdom, 292–93; pyramids and, 290

Philippines, 467c. *See also* Southeast Asia; agriculture, 638; cities, 638; communism, 626; economy of, 638; independence, 625; Mount Pinatubo, 42c, 43; natural resources, 638; per capita GDP, 638c; population, 638; religion in, 628; U.S. control of, 625, 626m

Phnom Penh, 633

Phoenicia, 136–37, 136–37m; alphabet, 137; geography of, 136

physical features, H10–H11; and climate, 54; in geography, 16–17; on land, 35–41

physical geography, 16–17, 16p, 21; ancient Egypt, 278–79; Antarctica, 658–59; Arabia, 170; Arabian Peninsula, 220–21, 221m; Australia, 644–45, 645m; Central Africa, 414–15, 415m; Central Asia, 244–45, 245m; China, 566–68, 567m; defined, 3, 16; East Africa, 394–96, 395m; Eastern Mediterranean, 196–98, 197m; Indian Subcontinent, 542–44; Iran, 220–21, 221m; Iraq, 220–24, 221m; Japan and the Koreas, 592–94, 593m; Kush, 310–11; Mongolia, 566–68, 567m; New Zealand, 644–45, 645m; North Africa, 350–52, 351m; Pacific Islands, 653–54, 654p; Southeast Asia, 618–19, 619m; Southern Africa, 436–37, 437m; Taiwan, 566–68, 567m; uses of, 17; West Africa, 372–73, 373m; of the world, 698–99

physical maps, 46m; Africa, 718m; Arabian Peninsula, Iraq, and Iran, 221m; Asia, 716m; Australia, 645m; Central Africa, 415m; Central Asia, 245m; China, Mongolia, and Taiwan, 567m; defined, H9; East Africa, 395m; Eastern Mediterranean, 197m; Europe, 714m; India, 44, 500m; Indian Subcontinent, H9; Japan and the Koreas, 593m; New Zealand, 645m; North Africa, 351m; North America, 710m; The North Pole, 721m; The Pacific, 720m; The Pacific World, 464–65m; South America, 712m; Southeast Asia, 619m; Southern Africa, 437m; South Pole, 721m; Southwest and Central Asia, 106m; Turkey, 217m; United States, 702–3m; using, 44; West Africa, 373m; World, 706–7m

Physical Systems, 13–14, 13c

Piankhi, King, 313, 317

pictographs, 128

Places and Regions, 13–14, 13c

place theme, 10–12, 11f

plains, 35; Central Asia, 245; East Africa, 395; in Eastern Mediterranean, 197–98; Indian Subcontinent, 543–44; Koreas, 593; North China Plain, 568; Serengeti Plain, 395; Southern Africa, 437; West Africa, 372

plants: in ecosystem, 63, 63f; as renewable resource, 69; soil and, 64–65

plateau: Arabian Peninsula, Iraq and Iran, 221; Central Asia, 245; Plateau of Tibet, 567, 567m; Southern Africa, 436

Plateau of Tibet, 567, 567m, 569

plate tectonics, 36–38, 36p; continental drift, 36–37; erosion, 39–40, 39p, 40p; mountains, 37; ocean trenches, 37; plates collide, 37; plates separate, 37, 38p; plates slide, 38, 38p

poetry: in ancient China, 521; Han dynasty, 514; Islamic achievements in, 189

polar climate, 57p, 61; overview of, 56–57, 57c

Polar easterlies, 51m, 53

political maps: Africa, 268m, 719m; The Arabian Peninsula, Iraq, and Iran, 218–19m; Asia, 717m; Caribbean South America, H8; Central Africa, 412–13m; Central Asia, 242–43m; China, Mongolia, and Taiwan, 564–65m; defined, H8; East Africa, 392–93m; Eastern Mediterranean, 194–95m; Europe, 715m; Indian Subcontinent, H9, 540–41m, 543m; Japan and the Koreas, 590–91m; North Africa, 348–49m; North America, 711m; The Pacific Islands, 656m; South America, 713m; South and East Asia, 460m; Southeast Asia, 616–17m; Southern Africa, 434–35m; Southwest and Central Asia, 108m, 209m; United States, 704–5m; West Africa, 370–71m; World, 708–9m

pollution: China, 581; Japan, 607; South Korea, 609; water, 33

Polo, Marco, 530

Polynesia, 653, 655. *See also* Pacific Islands

polytheism, 124, 285

Pontic Mountains, 197, 197m

popular culture: defined, 98; globalization and, 98

population, 79p, 86–90; Africa's Growing Population, 275c; Angola, 2000, 430c; birthrate, 88; death rate, 88–89; defined, 86; Interpreting a Population Pyramid, 430c; Kinshasa's Growing Population, 425c; migration, 89; Origin of Israel's Jewish Population, 206c; population density, 86–87; Population Density in East Africa, 404c; Population Growth in Indian Subcontinent, 563c; Population Growth in Japan, 606c; rate of natural increase, 88; tracking population changes, 88–89; of world, 700c; world population growth, 89–90, 90c; world trends in, 89–90

population density, 86–87; of Australia, 86; calculating, 88f; defined, 86; of Japan, 87; where people live and, 87–88; world population density, 87m

population maps: China, 572m; East Africa, 404m; Egypt, 362m; India, 554m; Japan, 606m; South and East Asia, 461m; Southwest and Central Asia, 110m; Turkey, 202m; West Africa, 391m; World Population Density, 87m

porcelain, 521, 522c

Portugal: colonial rule in Southeast Asia, 625, 626m; colonization in West Africa, 377; colonization of Central Africa, 421; slave trade and, 399; in Southern Africa, 441

poverty: Democratic Republic of the Congo, 425; Egypt, 362, 363, 364; Indian, 555; Lesotho, 449; Mozambique, 450; Nigeria, 383; São Tomé and Príncipe, 426; Southern Africa, 451

precipitation: Average Annual Precipitation by Climate Region, 74c; defined, 31; East Africa, 396–97; mountains and, 54; Southern Africa, 438; water cycle and, 32–33f, 33

precipitation maps: China, Mongolia, and Taiwan, 568m, 589m; Indian Subcontinent, 544m; West Africa, 388m

prevailing winds, 51–52, 51m

Primary Sources: *Adoration of Inanna of Ur,* 126; Analyzing Primary and Secondary Sources, 304; *Bhagavad Gita,* 503; *The Book of Routes and Kingdoms* (al-Bakri), 330; *Charter of the United Nations, The,* 100; *A Chinese City* (Polo), 530; *Crossing Antarctica* (Steger), 659; *The Dead Sea Scrolls,* 205; *Geography for Life,* 14; *Geography* (Strabo), 317, 320, 323; *Hammurabi's Code,* 133; *History Begins at Sumer,* 128; *The History of Nations: India,* 486; *A History of West Africa* (Davidson), 347; *I Speak of Freedom* (Nkrumah), 377; Kidnapped and Taken to a Slave Ship (Baquaqua), 381f; *The Koran,* 172; *Medieval Russia's Epics, Chronicles, and Tales,* 529; *Panchatantra,* 497; "Quiet Night Thoughts" (Li Bo), 521; *Roots of the Western Tradition* (Hollister), 304; *The Sermon on the Mount,* 155; *Shabanu: Daughter*

of the Wind (Staples), 560; "The Snows of Kilimanjaro" (Hemingway), 411; *Time Enough for Love*, 50; *The Travels*, 193; *Wings of the Falcon: Life and Thought of Ancient Egypt* (Kaster), 304
prime meridian, H2
printing: movable type, 522–23, 522c; woodblock, 522, 522c
projections. *See* map projections
prophets: Jewish, 148; Muhammad, 172
proverbs, 148; West Africa, 341
Pyongyang, 611, 611p
pyramids, 355; An Egyptian Pyramid, 360p; Egyptian, 288–89p, 289–90; Kush, 308p, 314, 317, 318p

Q

Qatar, 112c, 226; literacy rate in, 238c
Quick Facts, 21, 45, 75, 101, 139, 165, 191, 215, 239, 261, 305, 321, 345, 367, 389, 409, 431, 453, 501, 537, 561, 587, 613, 639, 663
Qur'an, 172, 174–75, 175p, 224

R

rainfall: limits on life and, 62; seasons and, 29; tropics, 29
rain forest, 58; clearing, 64; Madagascar, 438; Southeast Asia, 620–21; West Africa, 374, 375p
rain shadow, 54, 54p
Ramadan, 176, 358
Ramallah, 207, 207m
Ramses the Great, 293, 297f, 312; Abu Simbel Temple, 300
rate of natural increase, 88; Bangladesh, 90; France, 90; Japan, 88; Mali, 88–89; Nigeria, 90
Re, 285, 285p, 290
Reading Skills: Active Reading, H14–H17; Asking Questions, 308, 679; Categorizing, 276, 678; Drawing Conclusions, 642, 692; Identifying Implied Main Ideas, 564, 689; Identifying Supporting Details, 392, 683; Making Generalizations, 434, 685; Paraphrasing, 114, 672; Re-Reading, 218, 676; Sequencing, 168, 470, 674, 686; Setting a Purpose,

194, 675; Summarizing, 348, 681; Understanding Cause and Effect, 48, 324, 670, 680; Understanding Chronological Order, 504, 687; Understanding Comparison-Contrast, 370, 682; Understanding Fact and Opinion, 590, 690; Understanding Implied Main Ideas, 142, 689; Understanding Main Ideas, 78, 671; Using Context Clues, 242, 677; Using Context Clues-Definitions, 616, 691; Using Prior Knowledge, 2, 668; Using Word Parts, 24, 412, 669, 684; Visualizing, 540, 688
Red Sea, 220, 221m; Suez Canal, 351, 351m
reforestation, 69
region, 6–7
region theme, 10–12, 11f
reincarnation, 481, 482; Buddha's teachings on, 487
relative location, 12
religion. *See also* Buddhism; Christianity; Hinduism; Islam; Judaism; Muslims: ancient Egypt, 285–87; animism, 378, 401; Brahmanism, 480; Central Africa, 422; Central Asia, 251, 253; China, 574; development of culture and, 81; East Africa, 401; India, 550; Indian Subcontinent, 556, 557m; Iraq, 232–33; Israel, 204–06; Jainism, 483; Japan, 599; Koreas, 600; Lebanon, 211; Major Religions of Central Africa, 422c; monotheism, 147; North Africa, 357; North Korea, 611; Pacific Islands, 656; polytheism, 124; Religions of the Indian Subcontinent, 557c; Shinto, 599; Southeast Asia, 626–27; Southern Africa, 444–45; Sufism, 187; Sumerian, 124–25; Syria, 211; West Africa, 378; of world, 701c
renewable energy resources, 70–71
Republic of the Congo. *See also* Central Africa: economy of, 427; petroleum, 427; physical geography of, 414–15, 415m; urbanization, 427
rift valley: East Africa, 394–95; Great Rift Valley, 394, 396f
Ring of Fire, 38, 42–43f
rivers, 31; Central Africa, 415, 415m; of Central Asia, 245; China, Mongolia, and Taiwan, 568; East Africa, 396; erosion, 40, 40p; exotic, 220; Indian Subcontinent, 543–44; Koreas, 593; as means of transportation, 120p; North

Africa, 350–51; River Valley Civilizations, 120–21f; Southern Africa, 437; support growth of civilization, 116–17
Roman Empire: Byzantine Empire, 160–64, 201; Palestine and, 204; Silk Road, 516–17f
Rosetta Stone, 299
Rosh Hashanah, 150
rotation of Earth, 26–27; prevailing winds, 52
Rub' al-Khali, 222, 222p
Russia: birthrate, 88; death rate, 88; influence in Kazakhstan, 254; rule of Central Asia, 249
Russian Orthodox Church, 251
Russian Revolution, 249
Rwanda, 274c; ethnic conflict, 83, 405; population, 405; population density, 404c

S

Saddam Hussein, 92, 231
Safavid Empire, 183–84, 184m
Sahara, 349p, 350, 352–53f, 376p; climate and vegetation of, 374; Crossing the Sahara, 332–33f; Mali, 386; A Sahara Oasis, 352–53f; trade and history of West Africa, 377
Sahel: climate and vegetation of, 374, 374p; countries of, 385–86
salt: Ghana empire and trade, 327–28, 328p, 332f
Samarqand, Uzbekistan, 243p, 248, 256
Samoa, 467c, 653. *See also* Pacific Islands
samurai: in Japan, 598
San Andreas Fault, 38
Sanskrit: Aryans, 477, 547
São Tomé and Príncipe, 274c, 426
sarcophagus, 287, 287p
Sargon (Akkadian emperor): empire of, 123–24, 123m, 123f
satellite images: Africa, 265; analyzing, 15; Antarctica's Ice Shelves, 661p; Flooding in China, 569p; as geographer's tool, 9; Great Rift Valley, 396p; infrared images, 15, 15p; Namib Desert, 438p; The Nile River, 363p; South and East Asia and the Pacific, 457p; true color, 15, 15p
Saudi Arabia, 112c, 224–26; economy of, 225–26; government of, 92, 225–26; literacy rate in, 238c;

oil in, 113c, 225, 228–29f, 241c; people and customs, 224–25

savanna, 58; East Africa, 396; Southeast Asia, 620; Southern Africa, 438; West Africa, 374, 375p

scale, map, H7; using, 260

Scavenger Hunt, xxxii

science and technology: advanced technology in Japan, 603, 603f, 607p; ancient India and, 498–99, 498–99p; in ancient Mesopotamia, 118–19; compass, 522, 522c; computer mapping, 19, 19f; electronics in Japan, 603f; gunpowder, 522, 522c; Han dynasty and, 514–15; Hindu-Arabic numerals, 498; inoculation, 498; inventions in ancient China, 522–23; metalworking in ancient India, 498; movable type, 522–23, 522c; Muslim Contributions to Math, 225f; in Muslim world, 186–87; North Korea, 611; Oil in Saudi Arabia, 228–29f; Pacific Tsunami Warning Center, 622; paper, 514, 522, 522c; porcelain, 521, 522c; scientific research at Antarctica, 660; soil factory, 64f; Sumerian society, 129; wheel, 129f; woodblock printing, 522, 522c

science connection: The Formation of an Atoll, 655f

Sea of Marmara, 196, 198p

seasons, 28–29

secondary industry, 93, 93p

secondary sources: Analyzing Primary and Secondary Sources, 304

Second Temple of Jerusalem, 143p

seismograph, 514–15

Seljuk Turks, 201

Senegal, 274c, 383–84. See also West Africa

Senegal River, 326

Seoul, 609, 610p

Serengeti National Park, 402–3f

Serengeti Plain, 395

Sermon on the Mount, 155

Seychelles, 274c

Shabaka, 314

Shackleton, Sir Ernest, 660f

Shah Jahan, 185

Shah of Iran, 235

Shanakhdakheto, Queen, 317f

Shang dynasty, 507, 570c; writing system, 507

Shanghai, 580

Shariah, 177

Sherpa, 556

Shia Muslims, 183, 211; in Iran, 235; in Saudi Arabia, 224

Shi Huangdi, 508, 508p, 509, 571

Shikoku, 592, 593m

Shinto, 599

shoguns: in Japan, 598

Shona, 440–41

Siam, 625

Siddhartha Gautama, 484–85, 550. See also Buddhism

Sidon, 136, 137m

Sierra Leone, 274c, 384. See also West Africa

silent barter, 328

Silk Road, 248, 333f; Han dynasty, 516–17f; spread of Buddhism and, 489

silt, 117, 120f, 350

Sima Qian, 514

Sinai Peninsula, 351, 351m

Singapore, 103c, 467c, 636, 636p. See also Southeast Asia; cities, 636; economy of, 636; government, 636; independence, 634; per capita GDP, 638c

Sinhalese, 559

Siva, 481, 481p

slaves: Hebrews in Egypt, 145; Muslims, 175; Sumerian society, 125

slave trade: The Atlantic Slave Trade, 380–81f; Central Africa, 421; East Africa, 399; West Africa, 377, 380–81f; Zanzibar, 399

social structure. See also caste system: Aryans, 476–77; caste system in ancient India, 478–79; Han dynasty, 512; India, 550; in Sumer, 125–26; Zhou dynasty, 508

Social Studies Skills

  Chart and Graph Skills: Analyzing a Bar Graph, 74; Analyzing a Cartogram, 209; Analyzing a Diagram, 360; Analyzing a Line Graph, 551; Analyzing Tables and Statistics, 238; Interpreting a Population Pyramid, 430

  Critical Thinking Skills: Analyzing Points of View, 586; Analyzing Primary and Secondary Sources, 304; Analyzing Visuals, 628; Identifying Bias, 320; Making Decisions, 344; Making Economic Choices, 536; Sequencing and Using Time Lines, 138

  Geography Skills: Analyzing a Cartogram, 209; Analyzing a Precipitation Map, 388; Analyzing Satellite Images, 15; Comparing Maps, 500; Doing Fieldwork and Using Questionnaires, 408; Interpreting a Population Pyramid, 430; Interpreting a Route Map, 151; Using a Physical Map, 44;

Using a Topographic Map, 596; Using Scale, 260

  Study Skills: Analyzing Primary and Secondary Sources, 304; Evaluating a Web Site, 452; Locating Information, 652; Organizing Information, 96; Outlining, 190

Social Studies Words, H18

soil: environments, 64–65; humus, 64, 65; layers of, 65p; limits on life and, 62; losing fertility, 65; as renewable resource, 69; soil factory, 64f

solar energy, 45p; defined, 26; in ecosystem, 63, 63f; movement of Earth and, 27; as renewable energy, 71; seasons and, 28–29; water cycle and, 32–33f, 33; weather and climate, 51

Somalia, 274c; climate of, 396; ethnic groups, 407; government, 407; Muslims, 399, 401

Somali language, 400

Song dynasty, 505p, 519; art and artists, 521; cities in, 521; Confucianism and government, 524–27; inventions during, 522–23; Neo-Confucianism, 526; scholar-officials, 526–27, 527p; trade, 521

Songhai Empire: Askia the Great, 337–38; education, 338; fall of, 338; in history of West Africa, 377; map of, 335m; Muslim influence in, 337, 338; Timbuktu, 338

souks, 365

South Africa, 275c. See also Southern Africa; apartheid, 442–43, 446–47; Cape Town, 448–49p; economy of, 447–48; government, 447–48; natural resources, 439; religion, 444; Tourism in Southern Africa, 450c

South America: Andes, 37; climate zones of, 56m; energy production, 70c; physical map of, 712m; political map of, 713m

South American plate, 36p, 37

South and East Asia: climate map of, 462m; facts about, 466–68c; land use and resources map, 463m; political map of, 460m; Population Giants, 469c; population map of, 461m; regional atlas of, 458–62m; Size Comparison: The United States and South and East Asia, 459c

Southeast Asia: agriculture, 621; animals of, 620–21; Bangkok Canal, 631f; Brunei, 637f; Buddhism,

colonial rule of, 625, 626m; culture, 626–27; Per Capita GDP in Island Southeast Asia and United States, 638c; early history of, 624–25; East Timor, 637; ethnic groups, 626; independence, 625; Indian Ocean Tsunami, 623m; Indonesia, 637; Land Use and Resources, 632m; language, 626; Laos, 633; 629–33; Malaysia, 635; modern history of, 625–26; Myanmar, 630; natural resources, 621; Philippines, 638; physical geography of, 618–19, 619m; plants of, 620–21; political map of, 616–17m; rain forest, 620–21; religion, 626–27; Singapore, 636, 636p; terrace farming, 40–41; Thailand, 630–31; Thai Teenage Buddhist Monks, 627; traditions and customs, 627f; Tsunami, 622–23f; Vietnam, 633

Southern Africa: apartheid, 442–43; Botswana, 449; Cape of Good Hope, 441–42; climate and vegetation, 438, 439m; Comoros, 450; culture, 443–45; customs and art, 445; disease, 451; early history of, 440–41; environmental destruction, 451; ethnic groups, 444; Europeans in, 441–42; Great Zimbabwe, 441, 441p; history, 440–43; independence, 443; issues and challenges of, 451; ivory trade, 442; language, 444; Lesotho, 448–49; Madagascar, 450, 455m; Mozambique, 450; Music of South Africa, 443f; Namibia, 449; natural resources, 438–39; Ndebele Village, 444–45p; physical geography, 436–37, 437m; plains and rivers, 437; political map of, 434–35m; poverty, 451; religion, 444–45; South Africa, 446–48; Swaziland, 448–49; Tourism in Southern Africa, 450f; Zimbabwe, 449–50

Southern Hemisphere: defined, H3; map of, H3; seasons in, 28–29

South Korea, 468c. *See also* Koreas; agriculture, 609; cities, 609, 610p; The Demilitarized Zone, 609, 609p; economy of, 608, 609; formation of, 598; government, 598, 608; industry, 608, 609; pollution, 609; population, 609; reunification, 612; Seoul, 609, 610p

South Pole, 721m; climate, 61; prevailing winds, 51

Southwest and Central Asia: regional atlas of, 106–11m

Southwest Asia: Arab culture region, 82, 82m, 82p; climate of, 111m; Geographical Extremes: Southwest and Central Asia, 107c; Largest Oil Reserves by Country, 113c; natural resources of, 109m; physical map of, 106m; political map of, 108m; population of, 110m; Size Comparison: The United States and Southwest and Central Asia, 107m

Soviet Union: Central Africa and Cold War, 421; influence over Mongolia, 583; North Korea and, 598; rule of Central Asia, 249; war with Afghanistan, 253

Spain: colonial rule in Southeast Asia, 625; colonization of Central Africa, 421; colonization of Pacific Islands, 655; government of, 92; Islam's spread to, 181; Muslim conquest of, 181

Sphinx, 277p, 288p, 300

sports: Australian Sports, 648f; in China, 576, 576f

Sri Lanka, 468c, 559p; Buddhism in, 488, 489p; challenges of, 559; climate and resources of, 545; ethnic conflict, 559; independence, 549; physical geography of, 542–44, 543m; religion, 556, 557m

Standardized Test Practice, 23, 47, 77, 103, 141, 167, 193, 217, 241, 263, 307, 323, 347, 369, 391, 411, 433, 455, 503, 539, 563, 589, 615, 641, 665

steppe climate, 59; average annual precipitation, 74c; overview of, 56c

storm surges, 54

Strait of Gibraltar, 136

Study Skills: Analyzing Primary and Secondary Sources, 304; Evaluating a Web Site, 452; Identifying Bias, 320; Locating Information, 652; Organizing Information, 96; Outlining, 190

stupas, 496

subarctic climate, 57c, 61

Sudan, 275c; agriculture, 405; climate of, 396; Darfur, 405; ethnic conflict, 405; Muslims, 401, 405; population density, 404c

Sudras, 478, 479p

Suez Canal, 351, 351m, 357; Egypt's economy and, 363

Sufism, 187, 189

Sui dynasty, 518

Suleyman I, 182

sultan, 637

Sumer, 122–31; architecture of, 130, 131p; arts, 130–31, 130–31p; city-states of, 122–23; cuneiform, 127–28, 127p, 128p; cylinder seals, 130–31, 130p; invention of wheel, 129, 129f; invention of writing, 127–28; math and science, 129; Mesopotamia and Sumer, 231m; religion of, 124–25; rise of Akkadian empire, 123–24; social order, 125; technical advances of, 129; trade, 125; ziggurat, 115p, 130, 131p

sundial, 514

Sundiata, 334–35, 341–42

Sunnah, 175–77

Sunni Ali, 337

Sunni Baru, 337–38

Sunni Muslims, 183, 211; in Iran, 235; in Saudi Arabia, 224

sutras, 479

Suu Kyi, Aung San, 630

Swahili, 400, 440f, 441

Swaziland, 275c, 449. *See also* Southern Africa; natural resources, 449

Syr Darya, 245, 247

Syria, 112c, 210–11, 211p, 212m; climate and vegetation of, 198, 199m; Egypt's conquest of, 292; history and government of, 210–11; natural resources of, 198–99; people of, 211, 212p; physical geography of, 196–98, 197m

Tahiti: as high island, 654

Taipei, 584, 585p

Taiwan, 468c; Chinese Nationalists and, 572, 584; cities, 584–85, 585p; climate of, 568–69, 568m; culture, 584; history of, 584; industry and agriculture, 585; natural resources of, 569; physical geography of, 566–68, 567m; political map of, 564–65m; population, 584; precipitation, 568–69, 568m, 589m; Tensions between China and Taiwan, 584c; today, 584–85

Tajikistan, 113c, 255. *See also* Central Asia; standard of living in, 256c

Taj Mahal, 185, 541p, 548

Taklimakan Desert, 567

Taliban, 253–54

Talmud, 149

Tamil, 556, 559

Tanach, 148

Tang dynasty, 505p, 519; art and artists, 521; cities in, 520; inventions during, 522–23; trade, 520–21

Tangier, 365

Tanzania, 275c; agriculture, 403; cities, 404; economy of, 403; ethnic groups of, 83; natural resources, 403; population, 404, 404m; population density, 404c; Serengeti National Park, 402–3f; terrorism, 404; tourism, 403

Tarai, 544

Taurus Mountains, 197, 197m

taxes: in ancient Egypt, 295; Ghana empire, 329; Han dynasty, 511; Mongol empire, 530

technology. *See* science and technology

technology connection: Building Small, 603f; the wheel, 129f

tectonic plates, 698

Tehran, 235

Tel Aviv, 207

temperate climate, 59–60; overview of, 56–57m, 56c

temperature: large bodies of water, 53; limits on life and, 62; mountains and, 54

Temple of Karnak, 300–301f

temples: of ancient India, 495–96, 496p; Buddhist temples in Japan, 599, 599p

Ten Commandments, 145, 147

Test-Taking Skills, H20–H24

Thailand, 468c. *See also* Southeast Asia; Buddhism, 626–27, 627f; cities, 630; economy of, 632; natural resources, 632; physical geography of, 618–19, 619m; tourism, 632

Thar Desert, 543m, 544

Thebes, 293

theocracy, 236

Theodora, 161

Theravada Buddhists, 489

Third Dynasty, 283

Thoth, 285

Three Gorges Dam, 581

thunderstorms, 48–49p, 53

Thutmose I, 312

Tiananmen Square, 578; Beijing's National Day, 574–75f

Tibesti Mountains, 372, 373m

Tibet, 579

Tigris River, 220; Fertile Crescent and, 117, 117m; urban civilization and, 40

Timbuktu, 335, 336p, 337, 338, 377, 386

Time Line: Beginnings of Islam, 172; The Han Dynasty, 510; Major Events in the Fertile Crescent, 138; Periods of Egyptian History, 291; Sequencing and Using Time Lines, 138

Togo, 275c, 385. *See also* West Africa

Tokyo, 595p, 602; daily life in, 604, 604–5f

Tonga, 468c, 653. *See also* Pacific Islands

topographic maps: Awaji Island: Topographic Map, 596m; Using a Topographic Map, 596

Torah, 148, 148p

tornadoes, 53, 53p

tourism: Botswana, 437, 449; Cape Verde, 384; Egypt, 363; Indonesia, 637; Kenya, 403; Morocco, 365; Pacific Islands, 657; Senegal and Gambia, 383; Tanzania, 403; Thailand, 632; Tourism in Southern Africa, 450c; Tunisia, 365

townships, 443

trade: ancient China, 520–21; ancient Egypt, 284, 292–93, 293m; Arabia, 171; The Atlantic Slave Trade, 380–81f; Cape of Good Hope, 441–42; Central Asia, 248; Constantinople, 162; Egyptian Trade, c. 1400 BC, 293m; Fertile Crescent, 115p; Ghana empire, 327–29; globalization and, 99; great kingdoms of West Africa and, 377; Great Zimbabwe, 441; Harappan civilization, 473; helps Islam spread, 181–82; ivory trade, 420–21, 421p, 442; Japan, 603; Kongo Kingdom, 420; Kush, 316, 316m, 319; Mongol empire, 530; Phoenicians, 136–37, 136–37m; silent barter, 328; Singapore, 636; slave trade, 377, 380–81f, 399, 421; Sumer, 125; Swahili, 441

trade routes, 293; Egyptian Trade, c. 1400 BC, 293m; Ghana empire, 327m, 327–28

trade surplus, 603

trade winds, 51m

traditional economy, 94

traditions. *See also* customs: Japan, 600, 601p; Pacific Islands, 656; Southeast Asia, 627

trains: bullet trains in Japan, 606; in Tokyo, 604, 604p

transportation: between cities in Japan, 606; Kenya, 404; klong, 630; Niger River, 373; in North Korea, 611; on rivers, 120p; roads in Qin dynasty, 509

tributary, 31

Tripoli, 365

tropical climate, 56p, 58; overview of, 56–57m, 56c

tropical forests: Central Africa, 416–17; Mapping Central Africa's Forests, 418–19f

tropical humid climate, 58; average annual precipitation, 74c

tropical rain forest: Madagascar, 438; Southeast Asia, 620–21; West Africa, 374, 375p

tropical savanna: Central Africa, 417; East Africa, 396

tropical savanna climate, 58; average annual precipitation, 74c; overview of, 56c

Tropic of Cancer, 58

Tropic of Capricorn, 58

tropics: defined, 29; rainfall and, 29; Southeast Asia, 620

true color images, 15, 15p

tsunami, 619, 622–23f, 623m; Japan, 594

Tuareg, 337

tundra climate, 61; average annual precipitation, 74c; overview of, 57c

Tunis, 365

Tunisia, 275c. *See also* North Africa; agriculture, 364m, 365; cities, 365; European influence and independence, 357; physical geography of, 350–52, 351m; today, 366; tourism, 365

Tunka Manin, 330

Turkey, 113c, 200–203; Anatolia, 196; climate and vegetation of, 198, 199m; culture of, 202; early farming village, 200–201f; economy of, 203; government of, 203; history of, 200–202; invasions of, 201; modernization of, 202; natural resources of, 198–99; Ottoman Empire, 201–2; people of, 202; physical geography of, 196–98, 197m, 217m; population map, 202m; today, 203

Turkic, 251

Turkmen, 251, 252p

Turkmenistan, 113c, 255. *See also* Central Asia; carpets of, 255f; standard of living in, 256c

Turks, 235

Turpan Depression, 567

Tutankhamen, King, 303, 349p, 356f; tomb, 302–3p, 303

Tutsi, 405

Tuvalu, 468c

typhoons, 53; Japan and Koreas, 594

Tyre, 136, 137m

Uganda, 275c; agriculture, 405; climate of, 396–97; economy of, 405; government, 405; population density, 404c
Ukraine, 103c
Ulaanbaatar, 583
Uluru, 642p, 644
Umayyads, 181
Unas, 290
United Arab Emirates, 113c, 227; Largest Oil Reserves by Country, 113c; oil in, 241c
United Kingdom: colonization of Central Africa, 421
United Nations (UN), 99; *Charter of the United Nations, The,* 100
United Nations Children's Fund (UNICEF), 100
United States, 23m, 112c, 113c, 272c, 466c; Central Africa and Cold War, 421; control of Philippines, 625, 626m; Current Per Capita GDP in Island Southeast Asia and United States, 638c; as developed country, 95; economic system of, 94; ethnic groups in, 83; global popular culture and, 98; government in, 91; High School Soccer Participation, 8m; Life in Iran and the United States, 236c; Pakistan as ally, 557; Persian Gulf War, 231; physical map of, 702–3m; political map of, 704–5m; population density, 404c; South Korea and, 598; standard of living in, 256c; war with Iraq, 231
untouchables, 479, 482
Upanishads, 480
Upper Egypt, 278, 279m, 280–81
Ur, 123, 123m, 124, 132; city-state of, 124–25f
urban geography, 19
urbanization: China, 579–80; India, 552; Mainland Southeast Asia, 630; Republic of the Congo, 427
urban life: Japan, 604–6; North Korea, 611; South Korea, 609
Uruguay, 103c
Uruk, 123, 123m
Uses of Geography, 13–14, 13c
Uzbek, 184, 251, 252p
Uzbekistan, 113c, 243p, 256. *See also* Central Asia; standard of living in, 256c

Vaisyas, 478, 479p
Vanuatu, 468c, 657p
Vedas, 476, 477, 480
Vedic texts, 480
vegetation: of Arabian Peninsula, Iraq, and Iran, 222; Central Africa, 416–17; Central Asia, 246; East Africa, 396–97; Eastern Mediterranean, 198, 199m; Southern Africa, 438, 439m
veld, 438
Victoria Falls, 17p, 415, 415p
Vietnam, 468c, 633. *See also* Southeast Asia; agriculture, 633; cities, 633; communism, 626; economy of, 633; independence of, 626; modern history of, 626; physical geography of, 618–19, 619m
Viewing Skills: Presenting and Viewing a Weather Report, 48; Presenting and Viewing a Travelogue, 540; Viewing a TV News Report, 434
Vishnu, 481, 481p, 496p
volcanoes: formation of high islands, 654; Island Southeast Asia, 619; Japan and the Koreas, 594, 594m; Mount Fuji, 593; Pacific Ring of Fire, 619; Ring of Fire, 38, 42–43f; stratovolcanoes, 43; tsunami, 619, 622–23

W

wadis, 223
Wake Island, 654
warfare/weapons. *See also* military: arquebus, 338; Aryans, 476–77; civil war in Central Africa, 428; Ghana empire, 327; gunpowder, 522; Moroccan, 338
Warring States period, 508
water, 30–34, 45p; Arabian Peninsula, Iraq and Iran, 223; benefits of, 33–34; canals, 118, 118p; Central Asia, 247; controlling, in Fertile Crescent, 118–19, 118p; drought, 33; erosion, 40, 40p; food and, 33–34; fossil water, 223; freshwater, 31–32, 31p; groundwater, 32; hydroelectric energy, 70; importance of, 30; irrigation, 118, 118p; limits on life and, 62; pivot-irrigated fields, 223f; pollution, 33; as renewable resource, 69; salt, 31,

31p; surface, 31; in water cycle, 32–33, 32f; weathering, 39
water cycle, 32–33, 32–33f, 47p
water vapor, 32–33, 32–33f
wats, 626
wealth: in Arabian Peninsula, from oil, 225–27; natural resources and, 72
weather: blizzards, 53; vs. climate, 50; defined, 50; front, 53; hurricanes, 53–54; ocean currents, 52, 52m; storm surges, 54; sun and location, 51; thunderstorms, 53; tornadoes, 53, 53p; wind and, 51–52
weathering, 39
Wegener, Alfred, 37f
West Africa: Africa's Largest Cities, 383c; agriculture, 375; art, 378f; The Atlantic Slave Trade, 380–81f; climate and vegetation, H9, 374, 374m; clothing, 371p, 378; colonial era and independence, 377; Côte d'Ivoire, 385; culture, 378–79; fishing, 385, 385p; Gambia, 383–84; Ghana, 385; great kingdoms of, 377; history, 376–77; homes, 379, 379p; land use and resources, 384m; language, 378; natural resources, 375; Nigeria, 382–83; Niger River, 373, 373m; physical geography, 372–73, 373m; political map of, 370–71m; population map of, 391m; precipitation map of, 388m; religion, 378; Senegal, 383–84; slave trade, 377, 380–81f; today, 382–86
   History of West Africa, 500 BC–AD 1650: art, music and dance of, 342–43; Crossing the Sahara, 332–33f; epics, 341–42; Ghana empire, 326–31; griots, 325p, 340–41; Mali empire, 334–37, 335m; map of, 324–25m; Muslims in, 325p; oral history of, 340–42; salt, 324p; Songhai empire, 335m, 337–38
West Bank, 207, 207m
Westerlies, 51m, 52, 52m, 53
Western Ghats, 543, 543m
Western Hemisphere: defined, H3; map of, H3
Western Rift Valley, 414, 415m
White Nile, 350, 396
wildlife: Australia and New Zealand, 646; Central Africa, 416–17; in ecosystem, 63, 63f; New Zealand, 646; as renewable resource, 69; Serengeti National Park, 402–3f; Southeast Asia, 620–21; Southern Africa, 437

wind: climate and, 51–52; erosion, 39, 39p; monsoons, 58; ocean currents, 52, 52m; prevailing, 51–52, 51m; as renewable energy, 70

Windhoek, 449

women: ancient Egypt, 296; in communist China under Mao, 572; Gupta Empire, 492–93; Hinduism and, 482; Kush, 317; Saudi Arabia, 225; Sumerian society, 126; Taliban and, 253; Turkey, 202

world facts, 698–701

World in Spatial Terms, 13–14, 13c

World War I: Ottoman Empire, 201–2

World War II: Japan in, 598

writing: cuneiform, 127–28, 127p, 128p; invention of, 127–28, 127p, 128p; Phoenicians' alphabet, 137; pictographs, 128; Sumerian society and, 127–28

Writing Skills: Compare and Contrast Places, 264; Composing a Five-Line Poem, 590; Creating a Brochure, 642; Creating a Geographer's Log, 218; Creating a Poster, 78, 114, 470; Designing a Web Site, 168; Explaining a Process, 104; Explaining Cause or Effect, 456; Giving an Oral Description, 370; A Journal Entry, 324; Presenting an Interview, 616; A Travel Presentation, 242; Writing a Description, 194; Writing a Fictional Narrative, 308;

Writing a Haiku, 24; Writing a Job Description, 2; Writing a Legend, 564; Writing a Letter, 142; Writing a Letter Home, 392; Writing a Magazine Article, 504; Writing a Myth, 348; Writing an Acrostic, 412; Writing a Riddle, 276

writing system: Harappan civilization, 474; Japanese, 599; Korean, 600; Qin dynasty, 509; Sanskrit and Aryans, 477; Shang dynasty, 507

writing systems: Egyptian hieroglyphics, 298–99, 299p, 355

Wu Daozi, 521

Wu, Empress, 519

Wudi, Emperor, 511

Xi'an, 509, 520

Yang Jian, 518

Yangon (Rangoon), 617p, 630

Yangzi River, 568

Yellow River, 568

Yemen, 113c, 227, 227p

Yom Kippur, 150, 206

Yoruba, 378

Yuan dynasty, 505p, 529–30

Yunnan, China, 577p

yurt, 250, 250p

Zagros Mountains, 221

Zaire, 424–25

Zambezi, 415, 415p

Zambia, 275c, 427p. *See also* Central Africa; agriculture, 427; copper belt, 417; economy of, 427; language in, 422; Muslims in, 422; physical geography of, 414–15, 415m; today, 427

Zanzibar, 399

Zheng He, 531–32, 531p, 534

Zhou dynasty, 508, 570c; mandate of heaven, 508; social structure, 508

Zhuang Chinese, 573c

Zhu Yuanzhang, 530

ziggurat, 115p, 130, 131p

Zimbabwe, 275c. *See also* Southern Africa; art, 445; economy of, 449; government, 449–50; independence, 443; land reform, 449; language, 444; Tourism in Southern Africa, 450c

Zionism, 205

Zulu, 442, 444

INDEX

Credits and Acknowledgments

Images; 56 (tl), Age FotoStock/SuperStock; 57 (tr), AlaskaStock; 57 (cr), Royalty-free/Corbis; 58 (cl), Martin Harvey/Corbis; 59 (br), Ingram/PictureQuest; 60 (t), Sharna Balfour/Gallo Images/Corbis; 64 (b), Carl and Ann Purcell/Corbis; 69 (tr), William Campbell/Peter Arnold, Inc.; 69 (c), Adrian Arbib/Corbis; 71 (tr), James L. Amos/Corbis; 71 (cr), Creatas/PictureQuest; 72 (tl), Sarah Leen/National Geographic Image Collection; 73 (tr), James Randklev/Getty Images; 75 (tl), Royalty-free/Corbis; 75 (tr), James L. Amos/Corbis.

Chapter 4: 78 (b), Liu Liqun/Corbis; 78–79 (t), Getty Images; 79 (br), Shahn Rowe/Stone/Getty Images; 79 (cl), Richard I'Anson/Lonely Planet Images; 81 (tl), Tom Wagner/Corbis; 81 (tr), Knut Mueller/Peter Arnold, Inc.; 83 (bl), Peter Armenia; 83 (br), Sebastian Bolesch/Peter Arnold, Inc.; 84 (cl), Reuters/Corbis; 84 (cr), Timothy A. Clary/AFP/Getty Images; 84–85 (t), Courtesy Library of Congress; 89 (t), The Granger Collection; 90 (cr), Marcel & Eva Malherbe/The Image Works; 90 (t), Peter Beck/Corbis; 93 (t), Owaki-Kulla/Corbis; 93 (tc), Rosenfeld Images, Ltd./Photo Researchers, Inc.; 93 (bc), Kevin Fleming/Corbis; 93 (b), Michelle Garrett/Corbis; 94 (bl), Glen Allison/Stone/Getty Images; 94 (br), Carl & Ann Purcell/Corbis; 101 (tl), Sebastian Bolesch/Peter Arnold, Inc.; 101 (tc), Rosenfeld Images Ltd./Photo Researchers, Inc.; 101 (tr), Reuters/Corbis.

Unit 2: A, Robert Frerck/Odyssey Productions, Inc.; B (tl), Ustinenko Anatoly/ITAR-TASS/Corbis; B (bl), Hans Christian Heap/Taxi/Getty Images; 401, Gavin Hellier/Robert Harding World Imagery/Getty Images; 403, Steve Vidler/SuperStock.

Chapter 5: 114 (br), AKG London; 115 (bl), Musée du Louvre/Dagli Orti/The Art Archive (The Picture Desk); 115 (br), Private Collection, Beirut/Dagli Orti/The Art Archive (The Picture Desk); 115 (c), Gianni Dagli Orti/Corbis; 116–117 (b), Reuters/Corbis; 120 (tr), Nik Wheeler/Corbis; 120 (bl), Carmen Redondo/Corbis; 121 (tl), Corbis; 121 (tr), Lowell Georgia/Corbis; 126 (t), The Trustees of The British Museum; 128 (tl), Gianni Dagli Orti/Corbis; 129 (bl), Scala/Art Resource, NY; 129 (br), Bob Krist/Corbis; 130 (tc), Musée du Louvre, Paris/Album/Joseph; 130 (tr), The British Museum/Topham-HIP/The Image Works; 131 (tc), Ancient Art & Architecture Collection, Ltd.; 131 (tr), Ancient Art & Architecture Collection, Ltd.; 134–135 (b), Gianni Dagli Orti/Corbis; 139 (tc), Gianni Dagli Orti/Corbis; 139 (tr), Gianni Dagli Orti/Corbis.

Chapter 6: 142 (br), David Sanger/DanitaDelimont.com; 143 (bl), Scala/Art Resource, NY; 143 (br), Archivo Iconografico/Corbis; 144 (br), Chiesa della Certosa di San Martino, Naples/SuperStock; 144 (bl), National Gallery Budapest / Dagli Orti (A)/Art Archive; 145 (cr), Erich Lessing/Art Resource, NY; 148 (t), ASAP Ltd./Index Stock Imagery; 149 (tl), Ronald Sheridan/Ancient Art & Architecture Collection, Ltd.; 150, Ellen B. Senisi; 153, Alinari/Art Resource, NY; 154, Kathy McLaughlin/The Image Works; 155, Erich Lessing/Art Resource, NY; 156, Alinari/Art Resource, NY; 165 (l), ASAP Ltd./Index Stock Imagery; 165 (c), Alinari/Art Resource, NY.

Chapter 7: 168 (br), Kamal Kishore/Reuters/Corbis; 169 (bl), Bibliotheque Nationale de Cartes/Bridgeman Art Library; 169 (br), Age Fotostock/SuperStock; 173 (tr), Paul Almasy/Corbis; 175 (t), J A Giordano/Corbis; 175 (tr), Art Directors & TRIP Photo Library; 178 (tr), EPA/Mike Nelson/AP/Wide World Photos; 178 (br), Kamran Jebreili/AP/Wide World Photos; 179 (tr), Vahed Salemi/AP/Wide World Photos; 179 (cl), EPA/Mike Nelson/AP/Wide World Photos; 181 (br), Vanni Archive/Corbis; 181 (bl), Ian Dagnall/Alamy Images; 183 (cl), The Granger Collection, New York; 186 (br), R&S Michaud/Woodfin

Camp & Associates; 188 (tr), Art Directors & TRIP Photo Library; 188 (tl), Helene Rogers/Art Directors & TRIP Photo Library; 188 (cl), Robert Frerck/Odyssey/Chicago; 191 (tl), Art Directors & TRIP Photo Library; 191 (tc), Ian Dagnall/Alamy Images; 191 (tr), R&S Michaud/Woodfin Camp & Associates.

Chapter 8: 194, Jon Arnold Images/Photolibrary. com; 195 (bl), Wolfgang Kaehler Photography; 195 (br), Robert Frerck/Stone/Getty Images; 197 (b), Hanan Isachar/Corbis; 197 (cr), Woodfin Camp & Associates; 198, Worldsat; 201 (tr), The Granger Collection, New York; 202-203, Danny Lehman/Corbis; 205 (b), The Image Bank/Getty Images; 205 (tr), West Semitic Research/Dead Sea Scrolls Foundation/Corbis; 206, Russell Mountford/Lonely Planet Images; 208, Seedsofpeace.org; 211. Alison Wright/Corbis; 212, Siegfried Tauqueur/eStock Photo; 213 (tl), Ayman Trawi, *Beirut's Memory*; 213 (tr), Anthony Ham/Lonely Planet Images; 215 (l), Hanan Isachar/Corbis; 215 (c), The Image Bank/Getty Images; 215 (r) Ayman Trawi, *Beirut's Memory*.

Chapter 9: 218, Nader/Sygma/Corbis; 219 (br), K.M. Westermann/Corbis; 219 (bl) Corbis; 221 (cr) Corbis; 221 (b), Nik Wheeler/Corbis; 222, Chris Mellor/Lonely Planet Images; 223, Worldsat; 225, The Granger Collection, New York; 226 (bl), Ludovic Maisant/Corbis; 226(br), Frank Perkins/Index Stock Imagery/PictureQuest; 227, Abbie Enock; Travel Ink/Corbis; 229, Topham/The Image Works; 232 (cr), Ivan Sekretarev/AP/Wide World Photos; 232 (tr), Faleh Kheiber/Reuters; 232 (cl), Andrew Parsons/AP/Wide World Photos; 235, Michael Yamashita/IPN; 236 (bl), Kaveh Kazemi/ Corbis; 236 (br), Bob Daemmrich/Stock Boston; 237, Scanpix/Tor Richardsen/Reuters/ Corbis; 239 (l), Chris Mellor/Lonely Planet Images; 239 (r), Michael Yamashita/IPN; 239 (c), Andrew Parsons/AP/Wide World Photos.

Chapter 10: 242, Martin Moos/Lonely Planet Images; 243 (bl), Wolfgang Kaehler Photography; 243 (br), SuperStock; 245, Francoise de Mulder/Corbis; 246, Zylberman Laurent/Corbis Sygma; 249 (tr), Gérard Degeorge/ Corbis; 249 (br), Christine Osborne/Lonely Planet Images; 252 (tl), Martin Moos/Lonely Planet Images; 252 (tc), Michele Molinari/DanitaDelimont.com; 252 (tr), Nevada Wier/ Corbis; 254, Reuters/Corbis; 255, Martin Moos/Lonely Planet Images; 256 (bl), David Samuel Robbins/Corbis; 257, David Mdzinarishvili/Corbis; 258 (cl), Reuters; 258 (b) Howell Paul/Corbis; 259 (r), Worldsat; 261 (l), Zylberman Laurent/Corbis Sygma; 261 (r), Howell Paul/Corbis.

Unit 3: A, Celia Mannings/Alamy; B (tl), Digital Vision/Getty Images; B (b), Photodisc Red/Getty Images; 265, Joseph Van Os/The Image Bank/Getty images; 266–267, Sharna Balfour/Corbis.

Chapter 11: 276, Anders Blomqvist/Lonely Planet Images; 277 (br), Erich Lessing/Art Resource, NY; 277 (bl), SIME s.a.s/eStock Photo; 280 (b), Erich Lessing/Art Resource, NY; 281 (c), Josef Polleross/The Image Works, Inc.; 285 (bl), Scala/Art Resource, NY; 285 (bcr), Erich Lessing/Art Resource, NY; 285 (bcl), Araldo de Luca/Corbis; 285 (br), Réunion des Musées Nationaux/Art Resource, NY; 286 (t), Musee du Louvre, Paris/SuperStock; 287 (tr), HIP/The Image Works, Inc.; 287 (tl), Archivo Iconografico, S.A./Corbis; 287 (br), British Museum, London, UK/Bridgeman Art Library; 294 (bl), Bildarchiv Preussischer Kulturbesitz/Art Resource, NY; 294 (br), Gianni Dagli Orti/Corbis; 295 (bl), Gianni Dagli Orti/Corbis; 297 (br), HIP/Scala/Art Resource, NY; 299 (tr), Robert Harding Picture Library; 302 (tr), Scala/Art Resource, NY; 302 (tl), Time Life Pictures/Getty Images; 303 (tr), Egyptian National Museum, Cairo, Egypt/SuperStock; 305 (l), Josef Polleross/The Image Works, Inc.;

305 (c), Egyptian National Museum, Cairo, Egypt/SuperStock; 305 (r), Robert Harding Picture Library.

Chapter 12: 308 (bl), Robert Caputo/Aurora Photos; 309 (bl), Bildarchiv Preussischer Kulturbesitz/Art Resource, NY; 309 (br), Sandro Vannini/CORBIS; 312 (bl), The British Museum/Topham-HIP/The Image Works, Inc.; 313 (br), Erich Lessing/Art Resource, NY; 314, Erich Lessing/Art Resource, NY; 315 (bc), Nubian, Meroitic Period, 100 B.C.-A.D. 300. Object Place: Sudan, Nubia. Iron. Length: 9.7 cm (3 13/16 in.) Museum of Fine Arts, Boston. MFA- University of Pennsylvania Exchange. 1991.1119; 315 (br), Egyptian, Ram's Head Amulet, ca. 770-657 B.C.E; Dynasty 25; late Dynastic period, gold; 1 5/8 x 1 3/8 in. (4.2 x 3.6 cm): The Metropolitan Museum of Art, Gift of Norbert Schimmel Trust, 1989 (1989.281.98) Photograph © 1992 The Metropolitan Museum of Art; 315 (bl), Nubian, Meroitic Period, 100 B.C.-A.D. 300. Object Place: Sudan, Nubia. Iron. Width x length: 1.5 x 7.6 cm (9/16 X 3 in.) Museum of Fine Arts, Boston. MFA- University of Pennsylvania Exchange. 1991.1116; 321 (l), The British Museum/Topham-HIP/The Image Works, Inc.; 321 (c), Erich Lessing/Art Resource, NY; 321 (l), 1989.281.98 Egyptian, Ram's Head Amulet, ca. 770-657 B.C.E; Dynasty 25; late Dynastic period, gold; 1 5/8 x 1 3/8 in. (4.2 x 3.6 cm): The Metropolitan Museum of Art, Gift of Norbert Schimmel Trust, 1989 (1989.281.98) Photograph © 1992 The Metropolitan Museum of Art.

Chapter 13: 324, Nik Wheeler/Corbis; 325 (bl), Christy Gavitt/Woodfin Camp & Associates; 325 (br), PhotoDisc; 328, John Elk III Photography; 329 (cl), Carol Beckwith&Angela Fisher/HAGA/The Image Works, Inc.; 331, Steve McCurry/Magnum Photos; 332 (cl), Dagli Orti (A)/The Art Archive; 332 (br), Dagli Orti (A)/The Art Archive; 332 (bl), Aldo Tutino/Art Resource, NY; 332 (tr), Nik Wheeler/CORBIS; 333 (cr), Reza; Webistan/ Corbis; 333 (tr), HIP/Scala/Art Resource, NY; 335 (tr), Private Collection, Credit: Heini Schneebeli/Bridgeman Art Library; 339 (t), The Granger Collection, New York; 341, Pascal Meunier/Cosmos/Aurora Photos; 342 (br), AFP/Getty Images; 342 (bc), Reuters/Corbis; 345 (l), Carol Beckwith&Angela Fisher/HAGA/The Image Works, Inc.; 345 (c), The Granger Collection, New York; 345 (r), AFP/Getty Images.

Chapter 14: 348, Jon Arnold/DanitaDelimont. com; 349 (tr), Claudia Adams/DanitaDelimont. com; 349 (bl), Age Fotostock/SuperStock; 351, Steve Vidler/eStock Photo; 354, Age Fotostock/SuperStock; 355 (t), Sandro Vannini/Corbis; 356 (br), Kenneth Garrett/National Geographic Image Collection; 356 (bl), Art by Elizabeth Daynes/National Geographic Image Collection; 356 (t), Kenneth Garrett/National Geographic Image Collection; 358, Patrick Ward/Corbis; 359 (r), Abbas/Magnum Photos; 359 (l), Karim Selmaoui/EPA/Landov; 362, Age Fotostock/ SuperStock; 363, NASA/Science Photo Library; 365 (tl), Moritz Steiger/The Image Bank/Getty Images; 365 (cl), Frans Lemmens/Peter Arnold, Inc.; 366, Digital Vision/Getty Images; 367 (c), Age Fotostock/SuperStock; 367 (r), Digital Vision/Getty Images.

Chapter 15: 370, David Else/Lonely Planet Images; 371 (bl), Owen Franken/CORBIS; 371 (br), David Else/Lonely Planet Images; 373, Bruno Morandi/Robrt Harding World Imagery/Getty Images; 374, M. ou Me. Desjeux/Corbis; 375 (tl), Jane Sweeney/Lonely Planet Images; 375 (tr), Kings College/Art Directors & TRIP Photo Library; 376, Margaret Courtney-Clarke/Corbis; 377, Bettmann/ Corbis; 377 (tl), Royalty-Free/Corbis; 378 (bl), Imagestate/PictureQuest; 378 (bc), Face mask with plank, Bwa People, Burkina Faso (wood & pigment), African, (20th century), Indianapolis Museum of Art, USA, Gift of Mr and Mrs Harrison Eiteljorg; / Bridgeman Art